The Modern Elements
Volume I: Advance Field Mathematics

authorHOUSE®

Dedications

In silent, contemplative, memory of my son Daniel, since his thoughts where cut so tragically short he was not able to have thoughts of his own maybe the ideas I have laid down here might have been his. Certainly they are his know.

Also to Frank J. Janza a very special teacher who had the patience to teach me tensors in his E.M. courses, Radio Frequency Engineer, Professor, at California State of Sacramento without whom this book would never have gotten started. This dedication was from him to me in his book "Manual of Remote Sensing" Vol. I, published by *American Society of Photogrammetrym*, 1975:

To a special student who has the right approach to analysis. May this R.S. Manual give him the challenges to extend his talents in analysis.

   Sincerely,

   Fank J Janza

A Thought, a Reminder

"What clever monkeys we are", said Mr. Robert to Mr. Edward.

Introduction to *The Modern Elements*

About twenty-three hundred years ago Euclid brought together all of the mathematics of his time into thirteen books called *The Elements*, so too I have brought together all of the primary mathematics and science of modern time into two volumes called *The Modern Elements*.

I have spent some years in trying to write this dissertation and would create bits and pieces of it only to go a little ways with a particular proof and then realize that I could not go any farther, because the foundation had not been laid down. I would more often than not find that I had developed a proof placing me on a precipice with nowhere to go, the proofs where spotty and not very well written. As an example; in developing hypersheets of parallel surfaces for tensors was impossible, because it had not really been developed with a proper foundation for Gaussian Curvature using Riemannian Tensors. There was also the problem that from book-to-book, author-to-author, there was never any continuity if I wished to call out a particular theorem or definition to reference for an idea I was developing, don't even bother. I also found that the majority of the authors, for expedience, had done spotty work with gaps in a proof that you would have to fill in. Often they would miss crucial ideas and pretty much if one person hit on a correct idea everybody copied and carried it along, without question, gaps and all. Since no one talked to one another, having lived and died within current history, spanning a hundred to hundred and fifty years, well just forget it, there was no way those people could be dug up for their opinion. I was frustrated that there was no centralized source, which coordinated everybody's efforts or brought to bear all of the tools that are needed to develop good sound mathematical theory. I would have to hunt through a dozen books and papers just to find a simple or maybe not so simple ungrounded proof.

Eventually I came to realize what I needed was a reference book that started from basics and carried out a sound, systematic, logical development in the algebras and physical science. Just as Newton, Maxwell and Einstein built their theories from first principles, there needed to be a book to allow others to do the same. As this idea began to develop I became aware that there where dozens of reference books on mathematical equations, tables of integrals and differentials, products, series and so on. However the mathematician and engineer had no book that encompassed all of the formulas and theorems needed to develop theories. There was no reference book that laid out theorems by the number for the development of mathematical and scientific proof writing.

The last great work of mathematics and geometry that did that was Euclid's book *The Elements* 300 BCE, which summarized all of the known mathematics of his time. As late as the 1920's one could still find people that could quote Euclid's definitions, axioms and propositions by the numbers. But from about 1680 to 1682 strange new mathematics such as Newton's and Leibniz's Calculus, new kinds of numbers called complex, as explored by John Wallis (1685), Euler (1750), Caspar Wessel (1799), and Argand (1806). Euler (1736) with his Königsberg bridge introduces graph theory and thereby topology, Galois (1830) with Group Theory, finally the great liberation of algebra with W.R. Hamilton breaking the Commutative Axiom of Algebra, things started to drift from their classical anchorage.

Tensor Calculus & Physics: A General Treatise

Then with the advent of Carl Friedrich (Johann) Gauss of Germany (1777-1855), Janos Bolyai (1802-1860) of Hungary and Nicolai Ivanovitch Lobachevsky (1793-1856) of Russia with his version of the parallel postulate. When this premise replaced Euclid's it created new, independent and consistent geometries that where not Euclidean. These new geometries would lead Riemann to his calculus of non-Euclidean spaces, thereby unifying all differential geometry. These major discoveries, gradually occurring over three hundred years has left mathematics without its original rudder *The Elements*. This lack of continuity in the progress of mathematics has left behind holes in its development from canyon size to potholes. As an example the distribution of the dot product operator of vector analysis over addition was always taken for granted and has never been truly proven, until the projective geometric proof of irregular polygons in this paper. Only with a consistent development of a numerically ordered mathematical system can these glaring discrepancies, become apparent and hopefully filled in.

There was also an educational gap, a lack of sense of history, which is critical for backwards accountability and intern leads the proof writer to be able to work forwards towards validating a proof. Without this gut feeling of where you have been there is no claxton to be sounded when you're headed for the rocky shoals of a paradox, or a bad conclusion that might not be apparent. So as far as I am concerned a solid foundation in math history and a solid base to build upon are essential for good proof writing.

Some may criticize me for going to far back and starting with the root algebras as over kill, but in going all the way to the axioms and theorems of the basic algebras it gave me away of self-testing my own work. As I went deeper into these more sophisticated proofs it turned out it was exactly what I needed, since about 20 to as much as 90% of these proofs rely heavily on the elementary axioms and propositions in order to manipulate variables. These processes taught me that there are patterns stemming all away back to logical systems in the shape of Groups, which intern I used to check my work. They made sure that the axioms and theorems I was developing constituted as near a complete structured algebraic system as possible.

As I have said completeness, or as close as one can truly ever approach, was developed by use of Groups based on the postulates of algebra. Without these notions the thoroughness that such an enterprise requires would not have been possible. Also the variety of theorems would not be so large and diverse, in that case the user of this reference document might not have been able to find the theorem they may require, which in the end is the one of the goals of this work.

Rigor is another problem, how can one maintain rigor and accuracy topic-to-topic and theorem-to-theorem? The only people in history that might have been up to it, having the intellectual presence and discipline of mind, might have been Newton, Bertrand Russell or David Hilbert. Since they cannot be here to help out I have had to rely on Euclid's fundamental principles of step-by-step logical deduction and concepts of modern time, notation, standardization, patterns and symmetry. Also all theorems, unless otherwise stipulated, use a standard frame; see T3.2.11 "The Framework of a Theorem".

This treatise was not meant to be placed on a bookshelf and collect dust, but to be a living document. As new theorems and topics are discovered, if they follow standard development, as I have defined it, they can be added through the appendices associated with the appropriate chapter. In this way the knowledge base can continue to expand and still be centralized. Also the main branch of this thesis is of course tensor calculus, but any branch of mathematics can be added as a separate book using these volumes as the root and trunk of the tree. These books will come under the title of *The Modern Elements* and each branch, as this one, will start with the <u>Branch Title: A General Treatise In ...</u> .

Tensor Calculus & Physics: A General Treatise

What I am **not** looking for is a project like Bertrand Russell's great work *Principia Mathematica*, which after all the incredible work he put into it, in order to obtain certainty and perfection turned out to be only some prelude to a great overture. When in 1931, Kurt Gödel with his published book *"Über formal unentscheidbare Sätze der Principia Mathematica und verwandter Systeme"* (On Formally Undecidable Propositions in *Principia Mathematica* and Related Systems), which contained his so-called theorems of undecidability. This work devastated Russell, because it meant he would never be able to find certainty in mathematics, to do that would require an infinite set of axioms, an impossibility. This is also why I am so liberal with non-classical axioms for the various disciplines throughout this paper my, concern is not finding the fewest set of atomized axioms, but provide a starting platform upon, which to stand and start the journey. Note when talking about rigor, certainty and consistency are treated in the **wide sense**, not in the **narrow or strict sense**, which could lead to uncertainty.

With Gödel's theorems showing, that certainty could never be attained, than there could never be a guarantee that a choice of axioms would not lead to paradoxes. This was the thing that broke Russell, because no matter how carefully he chose his set of axioms there would be no way he could ever prove consistency for his axiomatic system, hence there would always be an uncertainty in his mathematical development. He had built his theory of mathematics on a foundation of sand.

This document is not meant to prove certainty, but be the embodiment of what man has learned, to teach, and be a reference for future development, not be an icon to the rigor of certainty. After all certainty implies the game is over and there's nothing else to be learned and where would the fun in that be?

Randall H. Shaw

Tensor Calculus & Physics: A General Treatise

Table of Content

Tensor Calculus & Physics: A General Treatise

AuthorHouse™
1663 Liberty Drive
Bloomington, IN 47403
www.authorhouse.com
Phone: 1 (800) 839-8640

Published by AuthorHouse: 08/27/2015

The Modern Elements   Second Edition

ISBN: 978-1-4969-2211-3 (sc)
ISBN: 978-1-4969-2210-6 (e)

Print information available on the last page.

This book is printed on acid-free paper.

Tensor Calculus & Physics: A General Treatise

Table of Contents

Tensor Calculus & Physics: A General Treatise

List of Tables

List of Figures

Tensor Calculus & Physics: A General Treatise

Chapter 1 Symbols, Notation and Origin Stories

Section 1.1 The Treatise's Foundation of Symbols and Their Origin Stories

The following symbols define the mathematical constructions that are used throughout this treatise they come from set theory [LIP64] [BEY88], symbolic logic [COP68], and calculus.

Table 1.1.1: Immediate Mathematical Symbols

Symbol	Name and Description
( )	**parentheses**, delineates quantities closed within are to be taken together.
{ }	**braces**, delineates quantities closed within are to be taken together.
[ ]	**square brackets**, delineates quantities closed within are to be taken together.
⟨ ⟩	**angle brackets**, indicating quantities closed within are an ordered set of objects.
$\sum q$	**summation**, over a set of discrete quantities-q.
$\int S dq$	**integral summation**, over an infinite set S of infinitesimal quantities-dq.
$\prod q$	**product**, multiply a set of discrete quantities-q.
Δq	**difference**, between two quantities-q.
$\sum_{(i<j)} a_i^j$	$\sum_{i=1}^{n-1} \sum_{j=i+1}^{n} a_i^{\ j}$
$\prod_{(i<j)} a_i^j$	$\prod_{i=1}^{n-1} \prod_{j=i+1}^{n} a_i^{\ j}$
iff	**if, and only if,**
→ or ←	**associated with**, or **belonging to**, or **relationship with**
∴	**Hence, Therefore**
∵	**Since**
∋	**Such that**
∃	**There exists**
~∃	**There does not exist**
∃|	**There exists uniquely**
∀x	**For every x**
∈	**(is) a member (of)**, or **(does) belong to**
∉	**(is) not a member (of)**, or **(does) not belong to**
~∨∈	**(is or does) not necessarily [a member (of), or belong to]**[1.1.1] a contraction of $\{\in \vee \notin\}$
∪	**Union.**
∩	**Intersection.**
⊂	Contains (or containing) as a proper **subclass, subset.** A ⊂ B; [A] is a subset of [B]
⊃	Contains (or containing) as a proper **superclass, superset.** B ⊃ A, [B] contains [A]
⊆	Contains (or containing) as an **equivalent or subclass, subset**
⊇	Contains (or containing) as an **equivalent or superclass, superset**
⊄	**Does not contain** (or contains) as a proper **subclass, subset**

Symbol	Description and Name
$\varnothing$	**Null** or **empty class or set**
$\rightarrow$	Set [A] **maps to** set [B]
$\leftarrow\rightarrow$	quantity [A] **corresponds to** quantity [B]. The existence of a relationship between quantities.
$\propto$	quantity [A] is **proportional to** quantity [B]. Relationship of parts to each other or to the whole. Having a relationship by ratio of parts to one another. Relationship by harmonious symmetry between two quantities.
$=$	quantity [A] **equal to** quantity [B]. Quantities are the same and interchangeable.
$\equiv$	quantity [A] **identical to** quantity [B]. Equality by definition or by identity.
$\cong$	quantity [A] **congruent to** quantity [B]. Unique quantities are the same (shape, field type, etc.), but not interchangeable.
$>$	quantity [A] **greater than** quantity [B]. Quantities are the same and interchangeable.
$\geq$	quantity [A] **greater than or equal to** quantity [B].
$<$	quantity [A] **less than** quantity [B]. Quantities are the same and interchangeable.
$\leq$	quantity [A] **less than or equal to** quantity [B].
$\perp$	quantity [A] **is perpendicular to** or **to all** quantity(ies) [B]
$\sim>$	quantity [A] **not greater than** quantity [B].
$\sim\geq$	quantity [A] **not greater than or equal to** quantity [B].
$\sim<$	quantity [A] **not less than** quantity [B].
$\sim\leq$	quantity [A] **not less than or equal to** quantity [B].
$\sim=$ or $\neq$	quantity [A] **not equal to** quantity [B].
$\sim\equiv$ or $\not\equiv$	quantity [A] **not identical to** quantity [B].
$\sim\cong$ or $\not\cong$	quantity [A] **not congruent to** quantity [B].
$\sim\vee=$	quantity [A] **not necessarily equal to** quantity [B] [1.1.1] a contraction of $(=\vee\neq)$
$\sim\vee\cong$	quantity [A] **not necessarily congruent to** quantity [B] [1.1.1] a contraction of $(\cong\vee\not\cong)$
$\sim\vee\perp$	quantity [A] **not necessarily perpendicular to** quantity [B] [1.1.1] a contraction of $(\perp\vee\not\perp)$
$\approx$	quantity [A] **approximately equal to** quantity [B]. Quantities are interchangeable.
$\wedge$	Logical **AND** (name; over a letter *circumflex* separator *caret*) *<binary set builder operator>*
$\vee$	Logical **OR** *<binary set builder operator>*
$+$	Logical **Inclusive-OR** (in context)
$\oplus$	Logical **Exclusive-OR**
$*$	Logical **AND** (in context) (Name, *asterisks*)
$\sim$	**Complement** of a set, all elements *not* in the set, external to it is a logically **NOT** the negated of validity. For geometry means similar. (Name, *tilde*)
$\ldots$	Ellipsis, "and so on", or "and so forth", or "and so forth up to".
∇	Covariant Differential Operator, DEL [9.6.1]
$\square$	Covariant Quaternion Partial Operator, D'Alembert and Poincare [9.6.9.1A]

Symbol	Name and Description
Q.E.D.	Quod Erat Demonstradum is *that which was to be proved*.
G.C.M.	Greatest Common Multiple
L.C.M.	Least Common Multiple
L.C.D.	Least Common Denominator
E.S.N.	Einsteinian Summation Notation does apply.
~E.S.N.	Einsteinian Summation Notation does not apply.

[1.1.1]Notice: Negation of the logical OR operation implies an uncertainty, hence such operations are known as ambiguous operators and are exploited in Fuzzy Set Theory where an element x ~∨∈ A, a set.

Section 1.2 *The Origin of the Vector and Related Ideas*

The notion of a directed quantity may have been known in prior civilizations, possibly prehistory, "as necessity is the mother of invention" it would have been discovered only to be lost and then rediscovered again. It has been known that since the days of the Greeks the idea of a quantity having magnitude and direction was understood. While the modern mathematical form was unknown, the vector in action is plainly seen in the ancient's geometry and practical applications in war and civil engineering.

If we take four slats of wood, any two pieces of the same length, and bolt them together at their ends a free moving parallelogram at its vertices is formed. That is, if we grab any of the two opposing sides than shift them opposite and parallel to one another the internal angles can be altered to almost any degree, yet the sum of the angles always add to a total of 360°degrees. When a structural shape can be altered, like the parallelogram, it is said to be ***deformable***.

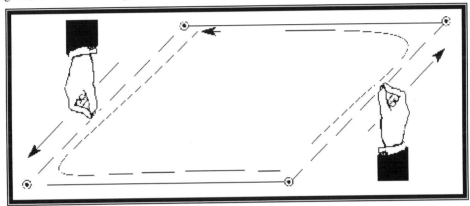

Figure 1.2.1: Deformable Parallelogram Under Compression

Now bolt a wood slat of the appropriate length across the diagonal of the parallelogram. We now have a geometric structure that is immovable, since the diagonal slate opposes any possible movement. The geometric parallelogram with its diagonal slat is a physical thing, hence its imperfect with lose fitting bolts, compressible wood, that flexes and so on. So, even though its mostly rigid we can feel the opposition of the diagonal brace as we stress it by trying to move the opposing sides of the diagonal in opposite directions.

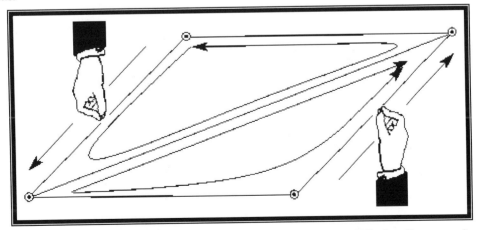

Figure 1.2.2: Rigid Parallelogram with Diagonal Brace and Under Compression

Since the diagonal is opposing motion, preventing the collapse of the parallelogram, the parallelogram is said to be **rigid**. This rigid geometrical shape can be thought of as being comprised of two triangles the one with the strut being compressed opposing collapse as under **compression** and the other triangle with its opposing side being pulled on as being under **tension**. The Greeks recognized that it was the triangles composing the parallelogram and their common side, the diagonal, which provided rigidity. In fact if one were to perform a simple experiment and remove any of the two sides the triangle that is left would still have two sides in tension and one under compression. As with the parallelogram, try to deform it, it remains perfectly ridged, because under any given external stress on any given side it is either under compression, or tension, opposing movement. In summary any of the three sides can act, as an opposing diagonal brace under compression towards the combination of the other two sides in tension, which leaves no possible degree of freedom for motion. This gives rise to the notion that there exists some kind of directed quantity in opposition to the sum of two other directed quantities. This is stated in what we now call the **Parallelogram Law** and is formalized in Newton's famous <u>Corollary I</u> in his Principia and restated later in this treatise.

So, the Greeks thought of the triangle as a fundamental geometric subcomponent of the parallelogram and as such the strongest of all geometric constructions. Clear evidence of its practical use has been recorded in the Roman Engineering (400 BCE to 500 AD) projects such as in the fabricated bridges they built to ford rivers, as seen below in Julius Caesars' bridge over the river Rhine.

Museo della Civilta in Rome

The rivers Rhine and Danube are huge streams, which for most of the time formed the northern border of the empire. Though at times the Romans did set across to conquer some of the lands beyond these great rivers. The Romans still built enormous bridges across rivers in order to easily re-supply their troops. Hence, the barbarians soon learned to see the bridges themselves as another Roman weapon. Julius Caesar did one of these famous feats of Roman engineering when he had such a bridge built across the Rhine in only 10 days. Also, the Trojans built a huge bridge across the Danube into Dacia. These were fetes of incredible engineering skill, performed with only the most basic tools. No other civilization, but the Romans could have achieved this at the time.

Figure 1.2.3: Julius Caesars' Bridge over the river Rhine

Also we see it arising in their siege machines such as the Onager:

A model of an *onager* catapult at the Museo della Civilta in Rome. This machine would be the heavy artillery to the ancient world. The handles to the left (rear of the catapult) are in fact levers by which the soldiers would wind the throwing arm back. On the right, you see a cushion at the front of the catapult. No doubt, it was there to soften the blow of the throwing arm and so help to prevent the machine from tearing itself apart. Also the triangular back would have provided the structural strength to with stand such blows from the throwing arm.

Museo della Civilta in Rome

Figure 1.2.4: The Onager

And to this day, the triangle provides the functional element in the truss a mechanical structure used for building bridges and buildings [MER71].

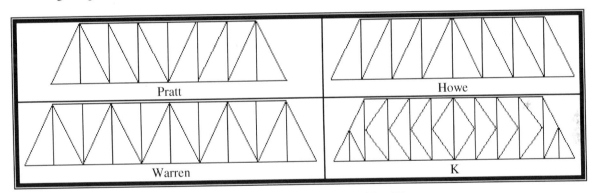

Figure 1.2.5: Commonly Used Bridge Trusses

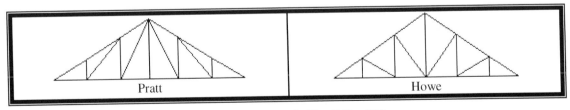

Figure 1.2.6: Commonly Used Roof Trusses

Another property imposed by the physical world on geometry, not so popular, or well understood, but still important, is the notion of a sense of direction in a plane. From arithmetic, the idea of a sense of direction on a one-dimensional Euclidian line is developed in the following way. If A and B are two points on a straight line, we can specify two opposite senses of directions on the line, which is equivalent to a directed line. One sense is from A to B, and the other sense is from B to A, as indicated by the directing arrow.

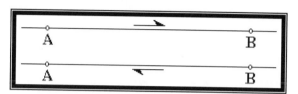

Figure 1.2.7: Direction on a Euclidean Line

In a two-dimensional Euclidean Plane a sense of direction can also be established by barrowing our physical notation of rigidity for a triangle ΔABC.

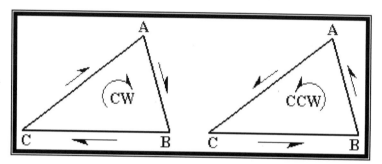

Figure 1.2.8: Direction about two ridged triangles

When two rigid triangles ΔABC, not necessarily congruent to ΔABD, but flipped relative to one another, share a common side; a strange result happens in determining their respective directions.

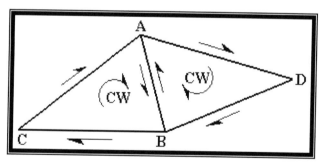

Figure 1.2.9: Two Triangles of Common Sides have Opposite Direction

The common side acts as a brace setting up an opposing direction in order to maintain rigidity for both triangles. This intern sets the sense of that triangle counter to the original triangle. However, the Quadrilateral ABCD around its perimeter has the same sense as the original triangles so internally the sense of direction cancels out leaving an outer common orientation of direction. This idea can be extended to an entire Euclidian plane made up of a network of irregular triangles known as a ***Triangular Net*** so that the entire plane has a web with a sense of direction. Since the triangles are ridged than, the plane must also be ridge and is called a ***ridged plane***, hence also has a net sense of direction. Such a plane having a surface made from a triangular net is said to be ***oriented*** acquiring a sense of direction from, a seed, or original, triangle. A Euclidean plane having an orient-able surface is said to be an ***Oriented*** or a ***Ridged Euclidean plane***.

Another notion about sense of direction for two plane triangles comes about from the ***symmetry of transformation*** by reflection. Let's take our two triangles with a common side, but let $\triangle ABC \cong \triangle ABD$.

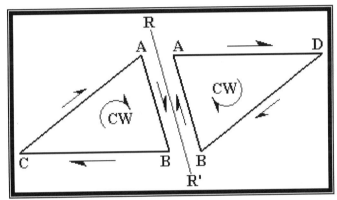

Figure 1.2.10: Parallelogram Triangles have Mirror Image Sense of Direction

Splitting them apart and constructing a line down the center RR' $\triangle ABC$ has a sense of direction and $\triangle ABD$ has a relative ***anti-sense of direction***. Now imagine RR' being the edge of a flat plane mirror sticking out perpendicular to the plane of the drawing, either triangle is the mirror image of the other. Or it can be imagined that the clockwise triangle is a transformation rotation about RR' point-for-point onto the anti-directed triangle being equivalent to the original directed sense only rotated through three-space 180°.

When an image is reflected about a line, such as RR', that line is called the ***line of symmetry*** if symmetrical and ***line of asymmetry***, if otherwise, then the resulting reflection is called the ***anti- or mirror image*** of the original object. If such an axis can be found through an object, that object is said to be ***symmetrical*** about the line of symmetry. If the axis of symmetry splits the object into two apparent independent objects, each object is said to be ***asymmetrical*** by themselves, which can be seen in the following figure of counter spiral seashells.

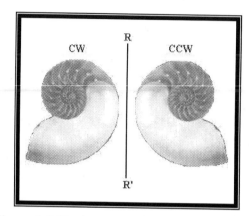

Figure 1.2.11: Asymmetrical Spiral Seashells

Clearly compared to themselves, there is a sense of symmetry, but alone that symmetry is lost, hence they are said to be asymmetrical. For three-dimensional objects the only way, that symmetry can be realized is by rotating them through a four-space 180° and finding their anti-image. This idea adds a new notion of the existence of dimensional spaces higher than our world of three-dimensions, which is exploited in n-dimensional geometries.

All of these ideas on sense of direction and the anti-sense of direction play an important role in the development of advance geometry, which is why I've spent some time and thought by presenting these concepts so the reader will be exposed to them.

Section 1.3 *Modern Origins of The Vector*[1.3.1]

Vector (ve • ktar)

[a. L. *vector*, agent-noun **f**, *vehĕre* to carry. So (in sense 1) Sp. and Pg. *vecter*, F. *vecteur*.]

†**1**. *Astr.* (See quotation 1704.) Also *vector radius*, = vector.radius Radius ɪe, *obj*.

1704 J. Harris *Lex. Techn.* I. s.v., A Line supposed to be drawn from any Planet moving around a center, or the Focus of an Ellipsis, to that Center or Focus, is by some Writers of the New Astronomy, called vector; because 'tis that Line by which the Planet seems to be carried around its Center. **1796** Morse *Amer. Geog.* 1. 28 If a right line called by some the vector radius, be drawn from the sun through any planet, and supposed to revolve around the sun with the planet [etc.].

[OXF71, Vol. II pg 3597][1,*.1]

Clearly, the word vector predates Newton's time and surprisingly has further meaning analogous to the mythology of the Greek Titan, Atlas, as a directed radius to carry upon its shoulders a planet about the sun.

Closer to the time of Newton is Stevin (Stevinus); Simon (1548–1620) in his paper on "Statics and Hydrostatics" (1586) enunciated the important theorem of the triangle of forces. Gave a new impetus to the study of statics (the study of rigid bodies or non-compressible fluids under pressure), which had previously been developed early in the theory of the lever by Archimedes (287–212 BCE) [BRI60, Vol. 21 pg 404].

While it is not clear weather Newton was influenced by Stevinus, in his use of the vector-force and the Parallelogram Law is unknown, but certainly Stevinus predates him and certainly Newton used vector-force parallelogram diagrams to solve his problems in terrestrial and celestial mechanics. It seems Newton acquired most of his contemporary information from his immediate friends in the Royal Society to know what was going on in the scientific world around him so maybe he would have known of Stevinus indirectly.

If one looks at Newton's "Principia Mathematica", we see example after example of geometric proofs of force diagrams, which are definite Euclidean constructions. His notion of creating a logical deductive system for his physics reflects his rigorous training in Euclidean geometry. It's true that the formalized notation of vector arithmetic was not known to Newton, but he relied on the pictorial geometric representation of the parallelogram law to provide the arithmetic for vectors to work through his problems and represent the quantity of force. In fact, in ***Corollary I*** from the Principia, it states exactly that:

A body, acted on by two forces simultaneously, will describe the diagonal of a parallelogram in the same time as it would describe the sides by those forces separately. [CAJ71, Vol. I pg 14]

So, through the Greeks and the Romans, Stevinus and Newton all acquire the notion of a geometric quantity that posses both direction of action and the magnitude of a scalar number, with a well-defined arithmetic by the geometry of ridged parallelograms. The name of this geometric quantity of course is called a ***vector*** in modern times.

[1.3.1]Note: Modern in this sense does not mean a relative time to my time in the 21st century, but relative to Newton's time in the 17th century.

The symbolism and notation of vectors was not to occur till sometime later certainly from Newton's work he does not refer to force as an arrow showing direction or the notation of a mathematical quantity that symbolizes a directed line segment.

Section 1.4 *Modern Origins of Vector Symbolism*

In the geometric constructions of Newton none of the symbols show directed line symbols such as the flying arrow pointing the direction of the field of force $A\xrightarrow{\text{Vector}}B$, or the arrow crowning the variable for a directed segment of line as shown by $\overrightarrow{AB}$. In fact, not even a straight line crowning a variable as a segment of line is shown by Newton $\overline{AB}$. The modern angle prefix symbol for a Euclidian angle construction is not shown either [$\underline{\angle AOB}$ or $\angle AOB$]. Newton, simply required the reader of his work to know what it is by the context in which the group of construction points [**A, B, C, ...**] are found.

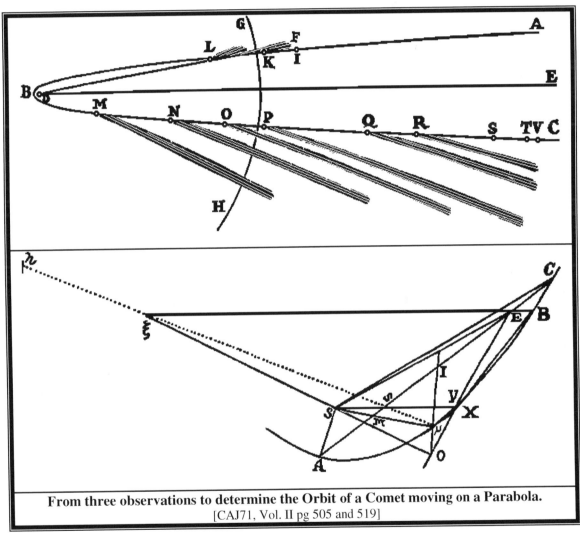

From three observations to determine the Orbit of a Comet moving on a Parabola.
[CAJ71, Vol. II pg 505 and 519]

Figure 1.4.1: Newtonian-Euclidian Symbolism

Hence the line segment **AI**, the angle of **AIO**, the triangle **AIO**, parallelogram **ASCO** and so on are simply stated and are only known in the wording and context of the problem they are established in.

This is one of the reasons that reading Newton's "Principia Mathematica" is so difficult you have to be aware at all times what is going on in the geometry with Newton's use of the variables. A major step in the evolution of the symbolism is not seen until James Clark Maxwell's famous treatises on Electromagnets 1873 that we start to see the modern symbolism of vectors in Clark's proofs and woodcuts.

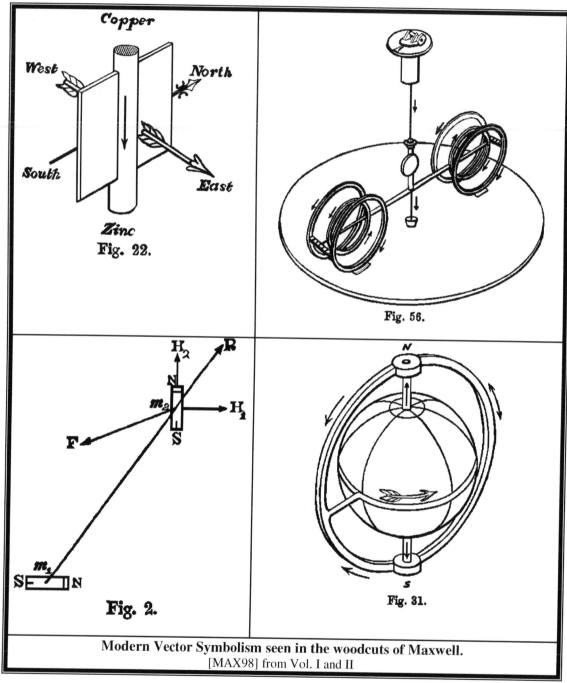

Modern Vector Symbolism seen in the woodcuts of Maxwell.
[MAX98] from Vol. I and II

Figure 1.4.2: Maxwellian Vector Symbolism

The magnet may be the reason for the adoption of the new symbolism, it sort of jumps out right at you. Faraday and Maxwell where studying electromagnets with their North and South poles and both men being from a Sea Faring Nation, England, where quick to adapt, maybe even subconsciously, the symbolism drawn on the mariner's loadstone compass and navigational charts as the direction to proceed, as demonstrated by the Cartographer's 16th century map.

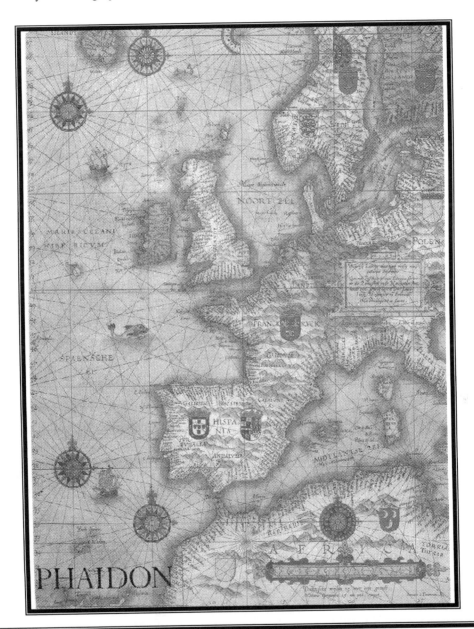

Chart of Europe
Leyden, 1586. From Claatt the first engraved sea atlas, the *Spiegel* der *Zeevaerdt*. Originally published in 1584 in Dutch there were many editions of the atlas in different languages until 1615.

Figure 1.4.3: Possible Origins of Directed Line Segment Symbolism

Clearly we see the fleur-de-lis pointing to north on the compass of the above map and the directed lines of the compass pointing the way for the mariner to navigate by compass, or as Maxwell used it in a spearhead pointing south.

Now the following woodcut comes from Sturgeon's time 1825 his invention of the electromagnet clearly the vector symbolism is missing, though he does label a North and South Pole.

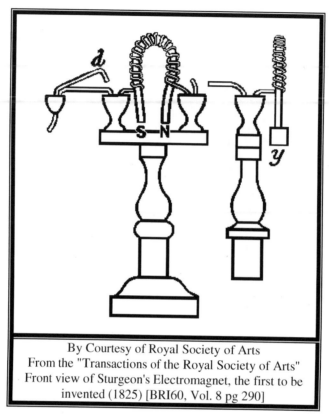

By Courtesy of Royal Society of Arts
From the "Transactions of the Royal Society of Arts"
Front view of Sturgeon's Electromagnet, the first to be
invented (1825) [BRI60, Vol. 8 pg 290]

Figure 1.4.4: Setting a Bottom Limit to the Origins of Vector Symbolism

From the woodcuts of Benjamin Franklin "Experiments and observations on electricity, made at Philadelphia in America", 1752, and Joseph Priestley "The History and Present State of Electricity", 1769, no arrowed flow of electricity is shown, it would follow than that sometime between 1825 and 1873 these symbols where adopted. Most likely it was Maxwell, possible Faraday, but the pathway was most unlikely not through the mathematicians such as William Rowan Hamilton (1788-1856) published works in 1843 and Hermann Günther Grassmann (1809-1877) published works in 1844[1.3.1] the founders of modern vector algebra and calculus. Without the resources to investigate the original papers and drawings of Faraday, Ampère, Weber, Green (1793-1841), Stokes (1819-1903) true contemporaries of Maxwell at Cambridge School of Mathematical Physics, and others of that time period its impossible to tell the only thing that is certain is its use in Maxwell's treatises.

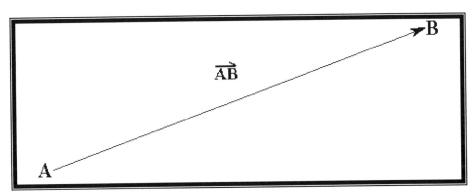

Figure 1.4.5: Maxwell-Newtonian Vector Construction

Another thing that Maxwell did was to use Gothic or Germanic capital lettering to represent vectors this solves the problem of trying to print text with arrows over them. In Maxwell's day this would have added cost to the publication for such none standard print. Today computer applications using equation builders, which may-or-may not support multiple lettering with vectors crowning them. In ether, case Germanic text was standardized in Maxwell's time and is available with computer font sets.

618.] In this treatise we have endeavoured to avoid any process demanding from the reader knowledge of the Calculus of Quaternions. At the same time we have not scrupled to introduce the idea of a vector when it was necessary to do so. When we have had occasion to denote a vector by a symbol, we have used a German letter, the number of different vectors being so great that Hamilton's favorite symbols would have been exhausted at once. Whenever therefore a German letter is used it denotes a Hamiltonian vector, and indicates not only its magnitude but its direction. The constituents of a vector are denoted by Roman or Greek letters.

The principal vectors which we have to consider are

	Symbol of Vector.	Constituents.		
The radius vector of a point	ρ	x	y	z
The electromagnetic momentum at a point	$\mathfrak{A}$	F	G	H
The magnetic induction	$\mathfrak{B}$	a	b	c
The (total) electric current	$\mathfrak{C}$	u	v	w
The electric displacement	$\mathfrak{D}$	f	g	h
The electromotive intensity	$\mathfrak{E}$	P	Q	R
The mechanical force	$\mathfrak{F}$	X	Y	Z
The velocity at a point	$\mathfrak{G}$ or $\dot{\rho}$	$\dot{x}$	$\dot{y}$	$\dot{z}$
The magnetic force	$\mathfrak{H}$	α	β	γ
The intensity of magnetization	$\mathfrak{J}$	A	B	C
The current of conduction	$\mathfrak{K}$	p	q	r

Quaternion Expressions for the Electromagnetic Equations [MAX98] Vol II pg 257

Figure 1.4.6: Symbols as Maxwell used them

Tensor Calculus & Physics: A General Treatise

In papers spanning sets, matrices, vectors and tensors everybody uses bold face or capital Roman lettering, or some combination thereof, very confusing. So, in this paper I establish symbol use as follows:

Table 1.4.1: Standard for Symbols used in this Treatise

Quantity	Case	Font	Constituents	Ornaments	How to read
Sets	Upper	Roman	A, B, C …	None	NA
Spaces and Fields	Upper	Roman	A, B, C …	E^n, C^n	Field of n-products
Specialty Sets	NA	Hebrew	א, ב, ג …	$\aleph_0$, ל$_0$ [1.4.2]	Set aleph-subzero
Euclidian Points	Upper	Roman	A, B, C …	None	NA
Euclidian Line Segment	Upper	Roman	A, B, C …	$\overline{AB}$	NA
Euclidian Vector	Upper	Roman	A, B, C …	$\overrightarrow{AB}$	·Directed A to B
Euclidian Angle	Upper	Roman	A, B, C …	$\angle AOB$	Keystone vertex center letter
Euclidian Triangle	Upper	Roman	A, B, C …	ΔABC	Keystone vertex first letter
Euclidian Parallelogram	Upper	Roman	A, B, C …	SPQR	Keystone vertex first letter
Fields	Upper	Roman	A, B, C …	$F_m{}^n$	Field of m quantities by n products
Manifolds	Upper	Germanic	𝕬, 𝕭, 𝕮 …	𝔐^n [1.4.3]	Type-t Manifold in n-Space
Vector, unit	Lower	Any	a, b, c …	$\hat{a}$	Unit vector
Vector	Either	Any	Any	$\overrightarrow{A}$, $\overrightarrow{a}$, **A**, **a**	Vector A, base a
Vector and Tensor Coefficients	Lower	Roman	a, b, c …	None	NA[1.4.5]
Tensors	Upper	Roman	A, B, C …	$\overset{*}{A}{}^m$ [1.4.6]	Tensor A, rank-m
Variables and set elements	Lower	Roman	a, b, c …	None	UOS[1.4.4]
Variables and subsets	Lower	Greek	α, β, γ, …	$\angle\alpha$, $\kappa[i, k]$	Angles
Constants	Either	Any	Any	Specialty	As Defined

[1.4.1]Note: It looks like a classical case of inventive parallelism Hamilton publishing in 1843 and Grassmann in 1844 neither one knowing of the other, assuming Grassmann might have read Hamilton's paper, and have put something together as sophisticated as his work, in less than a year, would seem unlikely. Grassmann's work did not establish its self at first due to poor publication and distribution so Hamilton took the lead, but Grassmann had the better product and notation, so in the long run won out. Though Hamilton's work still is important as a historical stepping-stone in the development of modern vectors, it seems to have been forgotten. Surprisingly it reemerges in Quantum Mechanics in a very practical way, see Section 9.8.

[1.4.2] Note: Hebrew Font set by Insert→Symbol→Lucida Sans Unicode or "Alchemy related true fonts"

[1.4.3] Note: Germanic Font by The Walden Font Co. "The Gutenberg Press" or by German American Corner German, Fraktur. Install fonts to Start→Settings→Control Panel→Fonts→File→Install New Font …

[1.4.4] Note: UOS is Unless Otherwise Specified.

[1.4.5] Note: NA is Not Applicable.

[1.4.6] Note: Crowning asterisks denotes tensor; symbolically representing multidirectional vectors comprising a tensor quiver. Analogous to the crowning directed vector over a quantity denoting a vector, see vector symbol in the above table.

Tensor Calculus & Physics: A General Treatise

Section 1.5 *Modern Matrix Notation*

The notation of matrix algebra in this document will draw heavily from Tensor Calculus, because as it turns out matrix algebra is a subset of tensors. It deals with special tensors of rank-2 called metrics and their additive and manipulative operations between rows and column matrix elements.

Metrics are true tensors that contain the quaint essential attributes that describe the geometrical properties for any given coordinate system where distance is measurable. These unique tensors can be broken apart into three distinct categories, matrix type rows called ***covariant*** and matrix type columns called ***contravariant***. The third category is the intersection between covariant and contravariant a geometric nether world the orthogonal center of the system. This deals with the mixing of the covariant and contravariant as typified by the classical ***mix-metric*** the Kronecker Delta. Tensor Calculus is the study of tensors and their unique ability to transform from one coordinate system to another leaving the tensor unaltered by the law of invariance.

Matrix algebra deals with the special center world of tensors that are mix tensors of rank-2. This distinction is either not well understood, or simply not at all. For this reason the Kronecker Delta is loosely bantered about. This part of the treatise is a prelude to tensor calculus; therefore, it behooves to start as soon as possible with the tensor calculus notation and the correct use of the Kronecker Delta. The place to start than is with the tensor subscripted rows and superscripted columns.

- Matrix will be denoted by upper case letters A, B, C, D, E ... .
- Scalar number elements within the matrix will be represented by lower case letters a, b, c, d, e ... .
- A one column or row matrix is called a vector and written as such

$$A = (a, b, c, d) \qquad \text{for Row (4 x 1)} \qquad \text{EQ 1.4.1}$$

or

$$B = \begin{pmatrix} e \\ f \\ g \\ h \end{pmatrix} \qquad \text{for Column (1 x 4)} \qquad \text{EQ 1.4.2}$$

In general tensor-vector subscripted and superscripted notation is represented as follows:

$$A = (a_i) \qquad \text{for } \textbf{\textit{Covariant}} \qquad \text{Row} \quad \text{(n x 1)} \qquad \text{EQ 1.4.3}$$

or

$$B = (b^j) \qquad \text{for } \textbf{\textit{Contravariant}} \qquad \text{Column (1 x m)} \qquad \text{EQ 1.4.4}$$

where it is understood that [i and j] are indexed by element from i = 1, 2, 3, ..., n and j = 1, 2, 3, ..., m.

Or as a covariant-contravariant ***mixed*** tensor of rank-2:

$$A = (a_i^j) \qquad \text{for Row by Column (n x m)} \qquad \text{EQ 1.4.5}$$

The orthogonal center of the system the Identity Matrix comes about because of the existence of the invariability of a square matrix allowing for the inverse operation of the multiplication of matrices:

$$I \equiv AA^{-1} \text{ where} \qquad \text{EQ 1.4.6}$$

$$A^{-1} \equiv adj(A) / |A| \equiv (A^{ij}) / a, \text{ such that} \qquad \text{EQ 1.4.7}$$

- Adj(A) is known as the **classical adjoint** or simply **adjoint** of A.
- A^{ij} is the **cofactor element** of the adjoint, or the determinate of A with the i-row and j-column missing times the sign coefficient $(-1)^{i+j}$ all of which can only be true iff $a \neq 0$, see T6.9.18 "Inverse Square Matrix for Multiplication", also see [LIP68, pg 176].

$$I \equiv (\delta_i{}^j) \text{ for an (n x n) matrix, such that} \qquad \text{EQ 1.4.8}$$

$$\delta_i{}^j \equiv \begin{cases} 1 & i = j \\ 0 & i \neq j \end{cases} \text{ for all i and j.} \qquad \text{EQ 1.4.9}$$

Known as the Kronecker Delta, see D6.1.6 "Identity Matrix".

Since the basis vectors spanning the Riemannian n-space are non-collinear, linearly independent, they form a matrix algebraic system in their own right, a set of elements that correspond to an (n x n) square matrix G, called the **Riemannian Covariant Spatial Metric**, such that

$$G \equiv (g_{ij}) \text{ and} \qquad \text{EQ 1.4.10}$$

$$G^{-1} \equiv (g^{ij}), \text{ where} \qquad \text{EQ 1.4.11}$$

$$g^{ij} \equiv G^{ij} / g \text{ the Contravariant Metric, } G^{ij} \text{ cofactor and} \qquad \text{EQ 1.4.12}$$

$$g \equiv |G|, \text{ the determinate of the metric measuring the degree from dependence} \qquad \text{EQ 1.4.13}$$

$$g \neq 0, \qquad \text{EQ 1.4.14}$$

hence it is invertible and the covariant and contravariant metrics converge onto their orthogonal identity.

$$\delta_i{}^j \equiv g_{ik} g^{kj} \text{ summation implied over k,} \qquad \text{EQ 1.4.15}$$

altogether,

$$I \equiv G G^{-1}. \qquad \text{EQ 1.4.16}$$

From EQ 1.4.15 the Kronecker Delta is the end contraction of the contra and covariant Riemannian metric tensor as a result is in itself the Riemannian mix-metric tensor of rank-2 as such the center of Riemann's world of metric tensors. In most text on tensors, this point is rarely brought forward, yet it plays a critical role in the manipulation, expansion and contraction the tensors.

Tensor Calculus & Physics: A General Treatise

Section 1.6 *Modern Origins of the Tensor*

Tensor (te'nsōr, -ər). [a. mod. L. *tensor*, agent-n. from *tendĕre* to stretch.]

1. *Anat.* (also see *tensor muscle*): A muscle that stretches or tightens some part. Opp. to *Laxator*. In mod. use, distinguished from an extensor by not altering the direction of the part.

1704 J. Harris *Lex, techn.* 1, *Tensors* or *Extensors*, are those common Muscles that that serve to extend the Toes, and have their tendons inserted into all the lesser Toes. **1799** Home in *Phil. Trans.* XC. to The combined action of the tensor and laxator muscles varying the degree of its [the membranes tympani] tension. **1808** Barclay Muscular motions 384 The biceps .. being a flexor and supinator of the fore-arm, and at the same time a tensor of its fascia. **1879** *St. George's Hosp. Rcp.* XI. 591 The functions of the adductors and tensors are more delicate.

2. *Math.* In Quaternions, a quantity expressing the ratio in which the length of a vector is increased.

1853 Hamilton *Elem. Quaternions u.* i. (1866) 108 The former elements of the complex relation .. because .. two lines or vectors [*vis.* their relative length], is .. represented by a simple ratio .. , or by number expressing that ratio. *Note,* This number, which we shall .. call the tensor of the quotient, .. may always be equated .. to a positive scalar. **1886** W.S. Alvis *Solid Geom.* xiv. (ed. 4) 235 Since the operation denoted by a quaternion consists of two parts, one of rotating OA into the position OB and the other of extending OA into the length of OB, a quaternion may be .. represented as the product of two factors, .. the versor .. and the tensor of the quaternion.

b. *Comb.*, *tensor – twist*, in Clifford's biquaternions, a twist multiplied by a tensor.

[OXF71, Vol. II pg 3261][1.*.1]

The name tensor has had sort of a circuitous evolution, starting out very clearly in anatomy to mean, an ex<u>tensor</u> muscle pulling or ***stretching*** the laxator muscle so it's <u>tense</u>. Hamilton's use of it in Quaternions is a simple scalar ratio between the magnitudes of two vectors representing the ***stretched or extended*** ratio of their length.

With Hamilton, it rested until the German mathematician Georg Friedrich Bernhard Riemann (1826-1866) saw something. Riemann was born in 1826 in a small village in Hanover, the son of a Lutheran pastor. In manner, he was always shy; in health, he was always frail. In spite of the very modest circumstances of his father, Riemann managed to secure a good education, first at the University of Berlin and then at the University of Göttingen. He took his doctoral degree at the latter institution with a brilliant thesis in the field of complex-function theory. In this thesis, one finds the so-called Cauchy-Riemann differential equations (known, though, before Riemann's time) that guarantee the analyticity of a function of a complex variable and the highly fruitful concept of a Riemann surface, which introduced topological considerations into analysis. Riemann clarified the concept of integrability by the definition of what we now know as the Riemann integral, which led, in the twentieth century, to the more general Lebesgue integral, and thence to further generalizations of the integral.

Tensor Calculus & Physics: A General Treatise

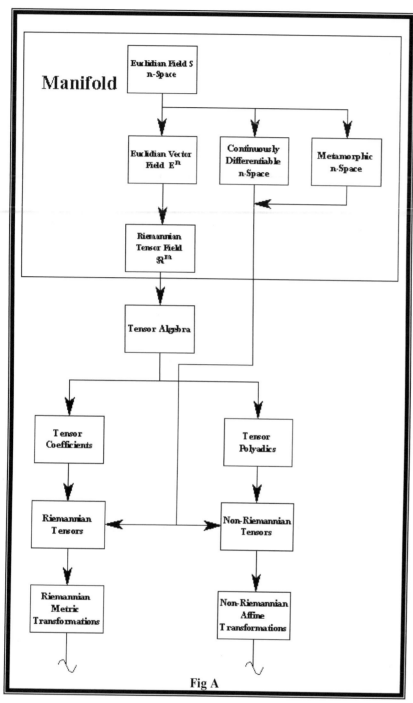

Manifold

- Euclidian Field S n-Space
- Euclidian Vector Field E^n
- Continuously Differentiable n-Space
- Metamorphic n-Space
- Riemannian Tensor Field $\Re^m$
- Tensor Algebra
- Tensor Coefficients
- Tensor Polyadics
- Riemannian Tensors
- Non-Riemannian Tensors
- Riemannian Metric Transformations
- Non-Riemannian Affine Transformations

Fig A

In 1854 Riemann became Privatdocent (officially an unpaid lecturer) at Göttingen, and for this privilege presented his famous probationary lecture on the hypotheses, which lie at the foundations of geometry. This has been considered the richest paper of comparable size ever presented in the history of mathematics; in it appears a broad generalization of space and geometry. Riemann's point of departure was the formula for distance between two infinitesimally close points, the embarkation point for his theory of tensors. In 1857 Riemann was appointed assistant professor and then, in 1859, full professor, succeeding Dirichlet in the chair once occupied by Gauss. Riemann died of tuberculosis in 1866 in northern Italy, where he had gone to seek an improvement in his health. [EVE76, pg 428]

In the above schematic Riemann initiated Tensors by exploring n-space transformations applied to Tensor Coefficients. Elwin Christoffel in 1869 while looking at differential changes in vector positions discovered quantities that looked and behaved liked tensors, but where not. In 1917 Levi-Civita put forward the concept of parallel displacement of a vector with respect to a curve (Differential Geometric Affine Transformations).

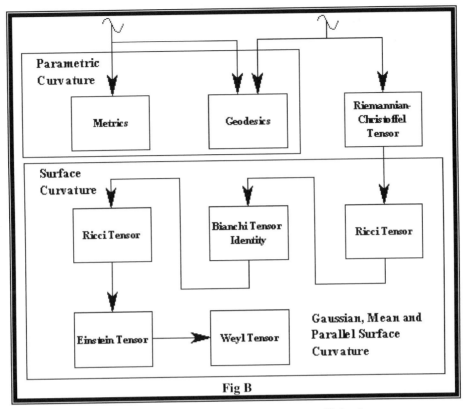

Figure 1.6.1: Development of Tensor Calculus

In 1918 saw the beginning of epoch-making generalizations of the Riemannian tensor analysis. It was first observed that many of the results of the Riemannian tensor calculus were independent of the metric structure of the Christoffel symbols $\{\iota, \kappa, \lambda\}$ in terms of the metrical tensor $g_{\iota\kappa}$, but were merely dependent on the law of transformations of the Christoffel symbols under a transformation of coordinates. There thus arose the fruitful conceptions of non-Riemannian geometries and tensor calculi in which the fundamental geometric object is not a metrical tensor $g_{\iota\kappa}$, but a linear connection with components $\Gamma_{\iota\kappa}{}^{\lambda}$, which under goes non-tensorial transformation.

The development of non-Riemannian tensor analysis is associated with the names of L. Berwald, E. Bortolotti, E. Cartan, A.S. Eddington, A. Einstein, L.P. Eisenhart, V. Hlavatry, A.D. Michal, J.A. Schouten, E. Schrödinger, T.Y. Thomas, D. Van Dantzig, O. Veblen, H. Weyl, J.H.C. Whitehead and many others. [BRI60, Vol.21 pg 942]

As can be seen from the drawings below tensors can be used to represent a verity of physical phenomena.

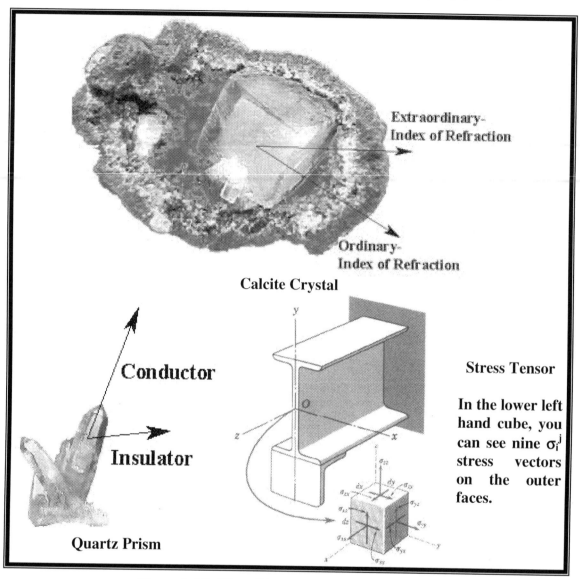

Extraordinary-Index of Refraction

Ordinary-Index of Refraction

Calcite Crystal

Conductor

Insulator

Quartz Prism

Stress Tensor

In the lower left hand cube, you can see nine σ_i^j stress vectors on the outer faces.

Figure 1.6.2: Three Physical examples of Tensors

Tensors appear to have a verity of applications, optics, thermodynamics, mechanics, etc. Also, from the above examples, tensors have something to do with sets of vectors having different attributes and how they're bound together by their geometry and a common theme; therein lies what a tensor is.

Section 1.7 *Tensor Notation*

Definition 1.7.1 Covariant and Contravariant Tensors

From Einstein's original paper "THE FOUNDATION OF THE GENERAL THERORY OF RELATIVITY" [PER52II pg 122] he defines a tensor in terms of a dual system of orthogonality as m-compound vectors having four components, in order to denote a four-dimensional world of space and time, hence [4^m] components all together, however for an n-dimensional space there will be [n^m] components,

$$\prod_{j=1}^{m} \left(\sum_{j=1}^{n} {}_y A_j \, \mathbf{b}^j \right) \equiv \prod_{j=1}^{m} \left(\sum_{i=1}^{n} x_{i,j} \, {}_x A_i \, \mathbf{b}^i \right) \qquad \textit{covariant} \qquad \text{EQ A}$$

and conversely

$$\prod_{i=1}^{m} \left(\sum_{i=1}^{n} {}_x B^i \, \mathbf{b}_i \right) \equiv \prod_{i=1}^{m} \left(\sum_{j=1}^{n} y_{i,j} \, {}_y B^j \, \mathbf{b}_j \right) \qquad \textit{contravariant}, \qquad \text{EQ B}$$

To denote that duality he adopts an index notation by following Ricci and Levi-Civita for covariant by placing the index below in the vector component, see equation A and the contravariant above see equation B. The difference in index subscript and superscript lies in the transformation between the two dualities.

Definition 1.7.2 Tensors of Rank-m

A tensor comprised of m-vectors is said to be of rank-m.

Definition 1.7.3 Set Summation Notation over Long Indices

From A10.2.1 a Riemannian Tensor Field a tensor is analytically represented as the product of a vector set or its quiver bound together by a common theme of properties or attributes associated with each vector and handled as a single mathematical quantity as follows:

$$\overset{+}{\mathbf{A}} \equiv \prod_i^m \vec{A}^i \equiv \prod_i^m \mathbf{A}^i \qquad \text{EQ A}$$

for $\vec{A}^i$ a vector. Applying D8.1.3 and A3.1.3:

$$\overset{+}{\mathbf{A}} = \prod_i^m \left(\sum_j^n A_{ij} \mathbf{b}^{ij} \right) \qquad \text{EQ B}$$

Following through just before expansion by distribution over all the products by A3.1.16:

$$\overset{+}{\mathbf{A}} = {}_1\sum{}^n {}_2\sum{}^n {}_3\sum{}^n \ldots {}_m\sum{}^n \prod_i^m A_{ij[i]}\mathbf{b}^{ij} = \bullet {}_m\sum_j^n \bullet \prod_i^m A_{ij[i]}\mathbf{b}^{ij} \qquad \text{EQ C}$$

The above summation is clumsy, however if it is expanded it can be seen one term at a time in a single summation over the ordered **Universal or Long Index Set-**L_r:[1.6.1]

$$\overset{\rightharpoonup}{\mathbf{A}} = \prod_i^m A_{i\lambda[i,\,1]}\mathbf{b}^{i\lambda[i,\,1]} + \prod_i^m A_{i\lambda[i,\,2]}\mathbf{b}^{i\lambda[i,\,2]} + \prod_i^m A_{i\lambda[i,\,3]}\mathbf{b}^{i\lambda[i,\,3]} + \prod_i^m A_{i\lambda[i,\,4]}\mathbf{b}^{i\lambda[i,\,4]} + \dots$$
$$+ \prod_i^m A_{i\lambda[i,\,k]}\mathbf{b}^{i\lambda[i,\,k]} + \dots + \prod_i^m A_{i\lambda[i,\,r]}\mathbf{b}^{i\lambda[i,\,r]} \qquad \text{for } r = n^m \text{ and where} \qquad \text{EQ D}$$

$L_r =$
{

$$
\begin{array}{llllllll}
\lambda[1,\,1] & \lambda[2,\,1] & \lambda[3,\,1] & \lambda[4,\,1] & \dots & \lambda[i,\,1] & \dots & \lambda[m,\,1] \\
\lambda[1,\,2] & \lambda[2,\,2] & \lambda[3,\,2] & \lambda[4,\,2] & \dots & \lambda[i,\,2] & \dots & \lambda[m,\,2] \\
\lambda[1,\,3] & \lambda[2,\,3] & \lambda[3,\,3] & \lambda[4,\,3] & \dots & \lambda[i,\,3] & \dots & \lambda[m,\,3] \\
\lambda[1,\,4] & \lambda[2,\,4] & \lambda[3,\,4] & \lambda[4,\,4] & \dots & \lambda[i,\,4] & \dots & \lambda[m,\,4] \\
\vdots \\
\lambda[1,\,k] & \lambda[2,\,k] & \lambda[3,\,k] & \lambda[4,\,k] & \dots & \lambda[i,\,k] & \dots & \lambda[m,\,k] \\
\vdots \\
\lambda[1,\,r] & \lambda[2,\,r] & \lambda[3,\,r] & \lambda[4,\,r] & \dots & \lambda[i,\,r] & \dots & \lambda[m,\,r]
\end{array}
$$

}

where $\lambda[i,\,k]$ is called a **long or box index**.

Generalizing the indices, the set can be reduced in size
$L_r =$
{   for $k = 1, 2, 3, \dots, n^m$

$$\lambda[1,\,k]\,\lambda[2,\,k]\,\lambda[3,\,k]\,\lambda[4,\,k]\,\dots\,\lambda[i,\,k]\,\dots\,\lambda[m,\,k] \qquad\qquad \text{EQ E}$$

or

$$k_1 \qquad k_2 \qquad k_3 \qquad k_4 \qquad \dots k_i \qquad \dots k_m \qquad\qquad \text{EQ F}$$

or

$$\rho(k,\,m) = \{\lambda[1,\,k]\,\lambda[2,\,k]\,\lambda[3,\,k]\,\lambda[4,\,k]\,\dots\,\lambda[i,\,k]\,\dots\,\lambda[m,\,k]\} \qquad \text{EQ G}$$
}

such that
$$\rho(k,\,m) \subset L_r \qquad\qquad\qquad\qquad \text{EQ H}$$
or
$$\rho(k,\,m) \in L_r \qquad\qquad\qquad\qquad \text{EQ I}$$
now in its compact index form
$$\overset{\rightharpoonup}{\mathbf{A}} = \sum_k^r \prod_i^m A_{i\lambda[i,\,k]}\mathbf{b}^{\lambda[i,\,k]} \qquad\qquad\qquad \text{EQ J}$$

[1.6.1]Note: Long could stand for the **longest possible ordered set of indices**, but in fact I've always had a thing for Viking Longboats since my childhood. When my parents would walk my sister and I to the ocean at the end of Golden Gate Park, in San Francisco California, where the Swedish government had presented the park with an actual longboat. Unfortunately, even though it was somewhat sheltered from the weather, under a canopy of an open-air pavilion, it started to rot, as I remember with a very bad smell. The ambassador from Sweden noticed this as well and asked for it back, since the San Franciscans didn't seem to know how to take care of this important historical artifact. So sadly, it is not there today. I guess I've always wanted to name something after that Viking Longboat.

However if we realize this is actually the set of ρ-ordered indices being summed over all indices [k]. The count [m] is defined in the product notation, yet the same for all products, hence redundant, so it can be dropped in the summation over the set L_r. Now substituted into the summation:

$$\overset{+}{\mathbf{A}} = \Sigma_{\rho(k) \in Lr} \textstyle\prod_i^m A_{i\lambda[i,\,k]} \mathbf{b}^{i\lambda[i,\,k]} \qquad\qquad \text{EQ K.}$$

Finally from the original definition:

$$\textstyle\prod_j^m \mathbf{A}^i = \Sigma_{\rho(k) \in Lr} \textstyle\prod_i^m A_{i\lambda[i,\,k]} \mathbf{b}^{i\lambda[i,\,k]} \qquad\qquad \text{EQ L.}$$

Definition 1.7.4 Set Summation Notation over Combinatorial Indices

Notice by altering the set other types of summations can be considered. Reflect on the elimination of repetitive indices from the product and no preferred orientation than the set L_r contracts to the **Combinatorial Set-C_r** from n^m to $_nT_m$ terms, hence $r = {}_nT_m$, see Appendix A Section 1, where $\kappa(s, t)$ are the exponents of the number of repeated indices. The exponents form a new index set $\zeta(s, t)$ over which to sum, hence

$$\Sigma_{\rho(k) \in Lr} \textstyle\prod_i^m A_{i\lambda[i,\,k]} \equiv \Sigma_{s=1}^n \Sigma_{t=1}^{Q(m,\,s)} {}_mK_{\kappa(s,\,t)} \Sigma_{u=1}^{S(n,\,s)} \textstyle\prod_{v=1}^s A_{\gamma[s,\,t,\,u,\,v]}^{\eta[s,\,t,\,v]}.$$

From AA.6.3

Definition 1.7.5 Set Summation Notation over Permutation Indices

Again reflect on the elimination of repetitive indices from the product, but this time the indices have a preferred orientation than L_r contracts to the **Permutation Set-S_r** from n^m to $n! / (n - m)!$ terms.

$$\Sigma_{\rho(k) \in Lr} \textstyle\prod_i^m A_{i\lambda[i,\,k]} \equiv \Sigma_{\sigma(k) \in Sr} \textstyle\prod^m A_{k\gamma[i,k]} \qquad\qquad \text{EQ A.}$$
$$\text{where } r \equiv n! / (n - m)! \qquad\qquad\qquad \text{EQ B}$$

For a specific treatises on permutated summations see chapter 6 section 5 for determinates where m = n.

$$\text{In summary } L_r \supset (C_r \wedge S_r)^{\,[1.6.2]}. \qquad\qquad \text{EQ C}$$

Definition 1.7.6 Aggregate Index Notation

When Riemann developed tensors he noticed that the Tensor coefficient, an agreaget of indices for vector components, could be treated as a composite, a single quantity dependent on the indices only. This alowed him to focus only on the tensor quantity and their properties while simplifying the manipulation. This can be written in a number of different ways.

or

$$A_{\lambda[1, k]\, \lambda[2, k]\, ...\lambda[i, k]\, ...\, \lambda[m, k]} \quad \equiv \prod{}^{m} A_{i\lambda[i, k]} \qquad\qquad \text{EQ A}$$

or

$$A_{\bullet\lambda[i, k, m]\bullet} \quad \equiv \prod{}^{m} A_{i\lambda[i, k]} \qquad\qquad \text{EQ B}$$

or

$$A_{\bullet\lambda[i, k]\bullet} \quad \equiv \prod{}^{m} A_{i\lambda[i, k]} \qquad \text{of rank-m} \qquad \text{EQ C}$$

or

$$A_{\rho(k, m)} \quad \equiv \prod{}^{m} A_{i\lambda[i, k]} \qquad\qquad \text{EQ D}$$

or

$$A_{\rho(k)} \quad \equiv \prod{}^{m} A_{i\lambda[i, k]} \qquad \text{of rank-m} \qquad \text{EQ E}$$

or

$$A_{i_1 i_2 ... i_k ... i_m} \quad \equiv \prod{}^{m} A_{i\, i_k} \qquad\qquad \text{EQ F}$$

or

$$A_{\bullet i_k \bullet} \quad \equiv \prod{}^{m} A_{i\, i_k} \qquad\qquad \text{EQ G}$$

or

$$A_{\bullet i_{[k,m]}\bullet} \quad \equiv \prod{}^{m} A_{i\, i_k} \qquad \text{of rank-m} \qquad \text{EQ H}$$

Definition 1.7.7 Dot or Repeated Index Notation

As seen in definitions D2.16.13 "Ordered n–Tuple as a Point", Table 10.1.1 "Tensor Polyadic Types" and D10.1.1 "Tensor", the definition for n-space, the black dots bounding an index or listed quantities means it is understood there are m-elements as specified by the rank. Where the subscripted index can take on values for k = 1, 2, 3, … m or in n-space i = 1, 2, 3, … n. This dot notation is short hand symbolism for repeated index quantities and has found effective use with tensors, manifolds, special coordinates, etc. This is a new notation having no bases in what Levi-Civita originally devised for contra and covariant notation. This notation came about because it was tiring to write out $(x_1, x_2, x_3, \, ... , x_n)$ when it could simply be written $(\bullet\, x_i\, \bullet)$, thereby eliminating the list and ellipsis notation.

Also notice that the coordinate variables in the n-tuple are denoted with an index superscript not to be confused with a contravariant superscript, since they are not directly related to the contravariant base vectors of the manifold. Confusing, yes, but there are only so many ways of indexing a variable so context is the only way of making a distinction.

To avoid some of the confusing decreasing or increasing sub-sub and super-super scripting, **box dote notation** is used $i_j = \lambda[j, k]$ where $\lambda[...]$ is a one-dimensional array, [j] is array element and [k] is the permutation index, as an example see D1.6.6E "Aggregate Index Notation".

[1.6.2]Note: For more examples and a complete development of Long, Combinatorial and Permutated indices see Appendix A section A.5.

Definition 1.7.8 **Einstein's Summation Notion**

Einstein noticed that for tensors such as Riemann metrics measuring arc length, summation always occurred with double or repeated indices, such as

$$ds^2 = \sum_n \sum_n g_{ij}\, dx_i\, dx_j \qquad \text{for n-space} \qquad\qquad \text{EQ A}$$

Hence it is redundant to support, where the indices are understood to vary from 1, 2, ... , n spatially. So he reasoned lets simply drop the summation symbols:

$$ds^2 = g_{ij}\, dx_i\, dx_j \qquad \text{for n-space} \qquad\qquad \text{EQ B}$$

and if there should be an exception to this rule it could simply be specified as such.

Section 1.8 *Tensor Differential Notation*

As can be seen from all of these definitions on tensors, tensors are a study in notational abbreviation. The height of this notational art can be found in how differential calculus is expressed.[1..7.1]

Definition 1.8.1 Differential Notation: Zero Parametric

$$\phi^0(t) \equiv \phi(t)$$

Definition 1.8.2 Differential Notation: Multiple Parametric

$$\phi^m(t) \equiv \frac{d^m \varphi(t)}{dt^m} \qquad \text{for } 0 < m$$

Definition 1.8.3 Differential Notation: Partial

$$f_{,i} \equiv f(._\bullet x^i._\bullet),i \equiv \frac{\partial f(._\bullet x^i._\bullet)}{\partial x^i}$$

To the right of the equality the partial differential is a cumbersome notation when all that is needed is the index referring to the variable being differentiated and some type of delineator expressing differentiation by the partial, for a tensor and here the comma [,] is used.

Definition 1.8.4 Differential Notation: Multiple Partials

$$h_{,i,j,\dots} = h_\bullet,i_k._\bullet = h_\bullet,\lambda[i, k]._\bullet = h_{,\rho(k)} = h_{,\rho_k}$$

with multiple partials defined by $\rho(k) \in L_r$ and $[k]$ a free index of the long index set L_r. This way any combination $[\zeta(k) \in C_r]$ or even permutation $[\sigma(k) \in P_r]$ ordering of the partials can be specified.

Definition 1.8.5 Christoffel 3-Index symbol of the First Kind

$$[ij, k] \equiv \tfrac{1}{2}(g_{ik,j} + g_{jk,i} - g_{ij,k}) \qquad \text{for } g_{ij} \text{ the metric-tensors of a Riemannian space.}$$

Definition 1.8.6 Christoffel 3-Index Symbols of the Second Kind

$$\begin{Bmatrix} k \\ ij \end{Bmatrix} \equiv [ij, \alpha]\, g^{i\alpha}.$$

In component form

$$\Gamma^k_{ij} \equiv \Gamma_{ij,\alpha}\, g^{\alpha k}$$

summation notation applies over α.

Definition 1.8.7 Christoffel 3-Index Symbols of the Third Kind

$$\Gamma^{ki}_j \equiv \Gamma^k_{\alpha j},\, g^{\alpha i}$$

Definition 1.8.8 Differential Notation: Partial of a Tensor for Rank-1

$\mathbf{A}^{\cdot j} \quad \equiv A_{i;j}\,\mathbf{a}^i$ partial differentiation, covariant   vector

$\mathbf{A}_{\cdot j} \quad \equiv A^{i;j}\,\mathbf{a}_i$ partial differentiation, contravariant vector

summation notation applies over i, where

$A_{i;j} \quad \equiv A_{i,j} - \Gamma^k_{\ ij}\,A_k$ partial differentiation, covariant   tensor component[1.7.2]

$A^{i;j} \quad \equiv A^{i,j} + \Gamma^i_{\ kj}\,A^k$ partial differentiation, contravariant tensor component[1.7.2]

summation notation applies over k.

[1.7.1]Note: For strict definitions on differential notation see Appendix K definition DxK.3.2 "Definitions on Differential Notation"

[1.7.2]Note: The New Differential delineator the semicolon [;] for differential of a tensor component. Also increases the rank of the tensor by one.

Section 1.9 *Modern Origins of the Manifold*

Manifold (mæ'nifŏuld), *α.*, *adv.*, and *sb.* Now *literary*. Forms: a. 1 maniō-, moniō-, mæniō-, meniōf(e)ald, 1-2 mænifeald, 2 manifald, 2-3 –feald, 2-6 monifald, (4 monyfaulde), 4-5 many', monyfald (e, (6 many-, mony-, monie-fauld). *β.* 1 meni(ō)fæld, -feld, 3-4 manifeld. *γ.* (2 monifold, 3 maniuold, maniōefold), 3-7 , 9 manyfold, (4 manye-), 4-6 manyfolde (5 maiefoold, mony-, manye-, 6 manniefolde), (7 manyfould), 3- manifold. [Common Teut: OE. *manigfeald* = OFris *manichfald*, OS. *managfeld* (MLG. *mannichvolt*, MDu. *menichvout*), OHG. *manacfalt* (MHG. *manecvalt*, mod. G. *mannigfalt*), ON. margfaldr (OSw. *marghfalder, mangfalder*, Sw. *mångfalt*), Goth. *managfalps*: see Many *α.* and –FOLD. A form with adj. suffix (= -Y) occurs as MLG. *mannichvoldech*, MDu. *menichvoudich* (Du. *menigvuldig*), G. *manunichfaltig*, Sw. *mångfaldig*, Da. *mangfoldich*.]

1c. *adj.* In technical and commercial use.

1851 *Offic. Catal. Gt. Exhib.* II. 597 A manifold bell-pull constructed on an entirely new plan, by which one pull is made to ring bells in any number of rooms. **1857** Tregelles tr. *Gesnius' Heb. Lex.* s.v. רכוגב, *Anbubaja* (i.e. *tibicina* Hor.).. a double or manifold pipe, and instrument composed of many pipes. **1879** Stainer *Music of bible* 95 Two classes of 'manifold-pipes can exist, the one .. a collection of *flauti traversi* the other .. of *flûtes à bec*. **1900** *Westni. Gaz.* 25 May 4/2 A model military balloon of the regulation-varnished manifold goldbeater s-skin variety.

3. *Math.* = Multiple. Manifold to = a *multiple* of. *Obs.*

1557 Recorde *Whetst.* Biv b, There is one kinde of proportion, that is named *multiplex*, or manyfolde. **1660** Borrow Euclid VII Post. 1 That numbers equal or manifold to any number may be taken at pleasure.

2. That which is manifold.

a. *spec.* In the Kantian philosophy, the sum of the particulars furnished by sense before they have been unified by the synthesis of the understanding.
This renderes G. mannigfaltiges, mannigfaltigkeit. Some earlier English translations of Kant's works have Multifarious, Multiple, Multiplex. **1855** Meiklejohn tr. *Kant's Crit. Pure Reason* 63 By means of the synthetical unity of the manifold in intuition. **1877** E. Caird Philos. Kant it. i. 199 The activity of the mind must bring with it certain principles of relation, under which the manifold of sense must be brought.

3. Math. = Manifodness 2.

1890 in Century Dict. **1902** R.A. W. Russell in Encycl. Brit. XXVIII. 666/1 Riemann's work contains two fundamental conceptions, that of a manifold, and that of the measure of curvature of a continuous manifold possessed of what he calls flatness in the smallest parts .. Conceptions of magnitude, he explains, are only possible where we have a general conception capable of determination in various ways. The manifold consists of all these various determinations, each of which is an element of the manifold. **1902** G.B. Mathews *ibid*. XXXI. 281/2 A manifold may consist of a single element.

5. Mech. (see quot.)

1891 Patterson Nant. Dict. 332 Manifold, a pipe or chamber to which are connected several branch suction pipes with their valves and one or more main suctions to pump.

[OXF71, Vol. I pg 1716][1.*.1]

Tensor Calculus & Physics: A General Treatise

So, what is a manifold? From the above definition, sub-definitions and their context it has something to do with an object that is made up of many parts such as a pipe organ. Or in the Riemannian sense some sort of topological n-space that is a composite of analytic points and the collection of associated properties, such as curvature, continuously differentiability and other attributes. Actually, a manifold is all of these and much, much more. The best way to think of a manifold is, as a mathematical toolbox containing all of the tools a tensor needs to carry out its task of describing and manipulating quantities for either a mathematical or physical system. So if a tensor is working with a manifold that has genus as a propriety then that becomes part of the attributes the tensor can draw upon, or a metamorphic manifold having a eccentricity for a propriety, and so on.

In tensor calculus, manifolds are treated as a collection of ordered points forming an n-space or subspace (hypersurface) of (n–1)-dimensions. These special spaces have Euclidian coordinates mapped point-to-point, into or onto, the space and are continuous at those points by virtue of being continuously differentiable within the their neighborhood. Variations of these themes come from the characterizing properties of the manifold, which vary from the basic, to the classical, to whatever the imagination of the mathematician can dream up. Typically geometric and topologic manifold characteristics are:

Table 1.9.1: Typical Manifold Properties

Typical Manifold Properties	Found in Topic of Study
Continuous unbroken surface,	Topological
Continuously differentiability in the neighborhood of a manifold point,	Analytic Geometry
Open hypersurfaces,	Geometrical, Sheets and Planes
Closed hypersurfaces,	Topologically Deformable
Segmental surface Euclidian Solids containing a collection of vertices, edges and faces.	Geometrically Combinatorial [BRI60, vol.14, pg 804]
Hypersurfaces and Trajectory Curvature	Topological Riemannian and Geometric
Hypersurfaces Enclosing a Volume	Geometrically Closed Surfaces
Differentiable,	Topologically Transformable
Metric defined (distance),	Geometric
Single sided surfaces with only one edge, such as a Möbius Strip, or no edges at all such as a Klein Bottle.	Topologically Orientation [ADL66, pg 97]
Genus (Number of holes in the manifold, a doughnut or torrid has a genus of 1)	Topological Deformable and Transformable
Eccentricity of Metamorphosis,	Geometrical Transformable
Symmetry	Geometric Dimensional Contraction

So, in one way a manifold can be thought of as comprised of many parts or attributes analogous to the pipe organ.

Tensor Calculus & Physics: A General Treatise

Not all of these attributes are of concern when it comes to this treatise. Tensor manifolds will be based on the basic notations of topological geometry as follows:

Table 1.9.2: Tensor Manifold Properties

Tensor Manifold Properties	Found in Topic of Study
Fields Comprising a Manifold	
Euclidian Fields,	Geometrical Transformable
Manifold Fields, Euclidian Field mapped to a Manifold.	Geometrical Transformable
Euclidian Vector Fields,	Vector Algebra
Riemannian Tensor Fields,	Tensor Algebra
Typical Attributes of a Manifold	
Continuous unbroken surfaces,	Topological
Open hypersurfaces,	Geometrical, Sheets and Planes
Closed hypersurfaces,	Topologically Deformable
Differentiable,	Topologically Transformable
Metric space defined as a Pythagorean distance,	Geometric
Magnitude invariant under Transformation,	Tensor Calculus
Affine Geometry (Direction invariant under Transformation),	Tensor Calculus
Hypersurfaces and Trajectory Curvatures,	Topological Riemannian and Geometric
Hypersurfaces Enclosing a Volume,	Geometrically Closed Surfaces
Eccentricity of Metamorphosis,	Geometrical Transformable
Symmetry	Geometric Dimensional Contraction

So, a manifold can take on any architecture depending on the Fields they are constructed from. Here fields are analogous to the tools in the manifold toolbox and the attribute the function of the tools.

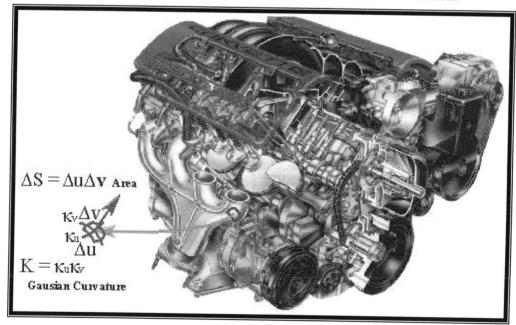

$$\Delta S = \Delta u \Delta v \text{ Area}$$

$$K = K_u K_v$$

Gausian Curvature

Figure 1.9.1: Exhaust Manifold Multi-Attribute Surface Element [*Corvette Engine*]

Section 1.10 *Rules of Annotation for this Treatise*

Annotation 1) Footnotes: Unlike most books footnotes, for flexibility, appending and removing sections, are local to within that section and sequentially indexed in this treatise, hence referenced by chapter, section and index delineated by a period [Chapter.Section.Index]. If a note is repeatedly used in multiple sections throughout a single chapter the section number is substituted with an asterisks [Chapter.*.Index], signifying that the note is multiply, used in different sections and can be found at the bottom of the chapter. This also gives a way to cross reference notes between chapters economizing their use.

There is an exception to this rule: if there should be multiple topics within in a given section in which case a prefix an alphabetically sequence letter will be used to differentiate footnotes applying to the various topics [Chapter.Topic (lower case letter) Section.Index].

Annotation 2) Reference notes: Follows modern annotated notation by boxing the reference with square brackets, listing the first three letters of the first (if more than one) authors last name and last two digits of the year of publication. This creates a unique reference code that can be used anywhere within this text, independent of chapter and alphabetized in sequence by authors name in references. This rule only varies if the author publishes two or more books within the same year than an indexing number in the order of publication replaces the author's last initial. As an example [LI164] and [LI264] verses a single publication [LIP64]. If however there should be a conflict between two authors having the same first three letters in their last name the fourth letter will be used next or the next after that, or if identical names start with the first letter of the first name and so on.

With the advent of the new millennia marking the year lets say for [04] for 2004 is in conflict with a paper that was written in 1904 [04] such as Lorentz's 1904 paper, so the Greek lower case letter will be used for the prefix numbers, hence $\alpha \equiv 19$ and $\beta \equiv 20$. For 1800, 1700, 1600, etc. the full four-digit year must be written out.

Sometimes this notation can be augmented with volume and/or page number such as [BRI60, Vol. 21 pg 942] to help the reader find specific references.

Since there are two books at this time if a reference is to be found in that particular book's list of references it will be marked by the Roman numeral of the book just after the date of publication, such as [LAW62I pg 19] [I] being book-I.

[1.*.1]Note: Oxford English Dictionary definitions are out of context consequently internal references that are used within the definitions are out of context as well. Also only, those parts of the definitions that are relevant to this treatise are considered.

Tensor Calculus & Physics: A General Treatise

Table of Contents

Tensor Calculus & Physics: A General Treatise

Tensor Calculus & Physics: A General Treatise

List of Tables

List of Figures

List of Observations

Chapter 2 **Principles of Set Theory**[2.*.*]

This review of set theory is not meant to be complete, but follow the main thread of the discipline that directly applies to this treatise. So a full development of functions is not present nor Cantor's great work on infinite sets and their comparability. Also, it is tailored toward tensors laying down the foundation for Manifolds (see chapter 11) and Multiplicative Sets to be developed. These aspects of set theory are not considered in most dissertations on sets.

[2.*.*]Note: [LIP64, pg 1-5, 17-20]

Section 2.1 *Definition of a Set*

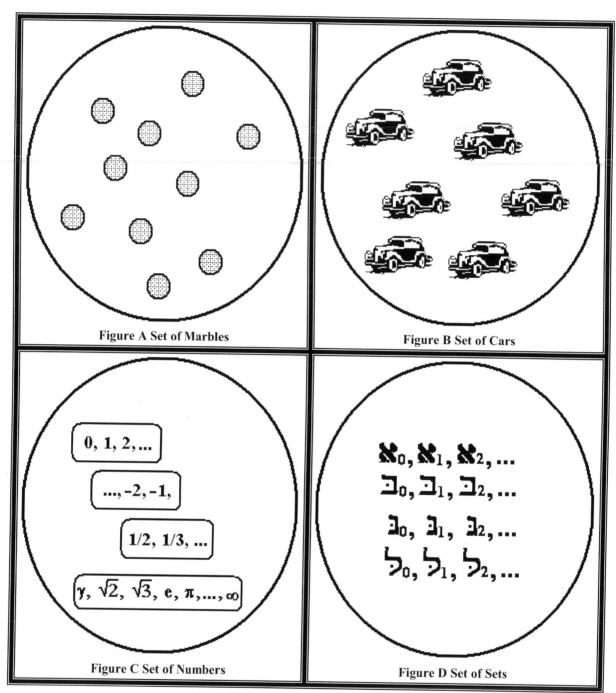

Figure A Set of Marbles

Figure B Set of Cars

Figure C Set of Numbers

Figure D Set of Sets

Figure 2.1.1 A Set is a Collection of Objects or Elements

Definition 2.1.1 Set

A set is a collection of objects or elements.

Tensor Calculus & Physics: A General Treatise

Section 2.2 *Basic Set Definitions*

Sets or classes of objects will be denoted by upper case letters A, B, C, D, E ... .
Elements, members, or objects will be represented by lower case letters a, b, c, d, e ... .

Definition 2.2.1 Tabular Form of a Set

A set written as a list of elements within braces is called a tabular form, such as
$$A = \{a, b, c, d, e\}$$

Definition 2.2.2 Set-Builder Form of a Set

A set written as a list of its attributes or properties is called a set-builder form, such as

$$A = \{\text{a statement of the properties in question}\} \hspace{2cm} \text{Equation A}$$

It can also be represented with a generic element and a list of properties representing that element, such as

$$A = \{x \mid (\text{a collection of } \textbf{\textit{property(ies)}} \text{ of x}) \; P_1(x), P_2(x), ... \} \hspace{1cm} \text{Equation B}$$

Read as, *A is the set of all elements x and y such that it contains $P_1(x)$, and $P_2(x)$, and ... .*

$$A = \{x, y \mid (\text{a collection of } \textbf{\textit{property(ies)}} \text{ for x and y})$$
$$P_1(x), P_2(x), ... , Q_1(y), Q_2(y), ... R_1(x, y), R_2(x, y), ... \} \hspace{1cm} \text{Equation C}$$

Where P and Q are independent properties and R an intersection of common properties.

Definition 2.2.3 Element Belong to a Set

A generic element [x] belonging to a set [A] can be written in the following form $x \in A$

Definition 2.2.4 Element Not Belong to a Set

A generic element [x] not belonging to a set [A] can be written in the following form $x \notin A$

Definition 2.2.5 Finite Set

A finite set has a limited number of elements and is represented as follows
$$A = \{a_1, a_2, ..., a_N\}$$

Definition 2.2.6 Infinite Set

An infinite set has an unlimited number of elements and is represented as follows
$$A = \{a_1, a_2, ... \}$$

Definition 2.2.7 Equal Sets

Two sets, for $(. \, a_i \, .) \in A$ and $(. \, b_i \, .) \in B$, are equal iff they contain the same elements or objects, in the same number, with the same equivalent properties and represented as follows
$$A = B \hspace{2cm} \text{Equation A}$$
and
$$P_1(a_1) \equiv P_1(b_1), P_2(a_2) \equiv P_2(b_2), ..., P_N(a_N) \equiv P_N(b_N) \hspace{1cm} \text{Equation B}$$
such that for all pairs of elements
$$a_1 = b_1, a_2 = b_2, ..., a_N = b_N \hspace{2cm} \text{Equation C}$$

Definition 2.2.8 Non-Equal Sets

Sets A, B are $A \neq B$ if just one pair of their elements do not match $a_i \neq b_i$ or $P_i(a_i) \not\equiv P_i(b_i)$

Definition 2.2.9 Congruent Sets

Two sets, for $(\bullet\, a_i\, \bullet) \in A$ and $(\bullet\, b_i\, \bullet) \in B$, are congruent but independent of one another iff they have a one-to-one correspondence of elements with the same congruent properties being represented as follows

$$A \cong B \qquad\qquad\qquad\text{Equation A}$$

for

and

$$a_1 \longleftrightarrow b_1,\ a_2 \longleftrightarrow b_2,\ \ldots,\ a_N \longleftrightarrow b_N \qquad\qquad \text{Equation B}$$

$$P_1(a_1) \cong P_1(b_1),\ P_2(a_2) \cong P_2(b_2),\ \ldots,\ P_N(a_N) \cong P_N(b_N) \qquad \text{Equation C}$$

but for all pairs of elements

$$a_1 \neq b_1,\ a_2 \neq b_2,\ \ldots,\ a_N \neq b_N \qquad\qquad\qquad \text{Equation D}$$

Definition 2.2.10 Null Set

A Null set is the simplest set having no elements or objects within and represented as follows
$$A \equiv \varnothing \qquad \text{iff A has no elements within.}$$

Definition 2.2.11 Unitary Set

A unitary set is the second simplest set being comprised of only one element and represented as follows
$$A = \{a\}$$

Definition 2.2.12 Set of Sets or Superset

A set of sets or superset is when the elements themselves are sets and represented as follows
$$\mathcal{A} = \{B_1,\ B_2,\ \ldots,\ B_N\} \text{ [2.2.1]}$$

Definition 2.2.13 Universal Set

The universal set [U] is a special superset that encompasses a particular group of sets under investigation.

[B.2.1]Note: Microsoft Monotype Corsiva True Font upper case used for superset notation.

Definition 2.2.14 Complement of a Set

The complement of A are those set of elements that do not belong to A, that is,

$$\sim A = U - A \text{ Fig B}$$

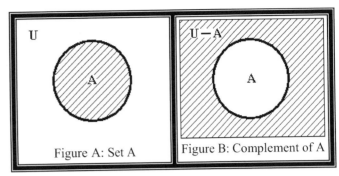

Figure A: Set A   Figure B: Complement of A

Figure 2.2.1 Complement of a Set

Definition 2.2.15 Exponential Notation for Set Operators

Exponential notation is the symbolic representation of the products of n-quantities.

$$A^n \equiv \Pi \text{ o } A, \text{ for any set operation [o]}$$

where [A] is any set $\in$ U and [n] is natural number called the **exponent**.

Definition 2.2.16 The Dual or Duality of a Statement

If $\cap$ and $\cup$ are interchanged and U and $\varnothing$ are interchanged as well in any statement about sets, then the new statement is called the dual of the original one.

Definition 2.2.17 Group

A group is a set S together with a binary operator [o] along with the existence of an inverse element $[a^{-1}]$ and a unique identity element [e], having the following properties acting on the elements within the set, such that:

Propositions of Quantity: Expanded Galois Group[4.1.4]
a) Existence: then (a, b, c, a^{-1}, e) $\in$ S
b) Equality: then a $=$ b $\wedge$ b $=$ a
c) Substitution: for a $=$ b $\wedge$ b $=$ c then a $=$ c
Propositions of Operation: Formal Galois Group[4.1.4]
c) Closure: then (a o b $\wedge$ b o a) $\in$ S
d) Associative: then a o (b o c) $=$ (a o b) o c
e) Identity: then e o a $=$ a o e $=$ a
f) Inverse: then a o a^{-1} $=$ a^{-1} o a $=$ e

Definition 2.2.18 Commutative or Abelian Group

Commutative: then a o b $=$ b o a

Definition 2.2.19 Ring

A ring is the interaction of two groups with two distinct binary operators [o] and [◊], such that

- a)   At least one of the binary operators [◊] is an abelian group, hence **abelian operator**, and
- b)   Distributive: then a ◊ (b o c) = (a ◊ b) o (a ◊ c) since the operators are complementary to one another the operator distributed across [o] is called the **complementary operator** of [◊].

As an example, real numbers are a ring under the operation of addition and multiplication. Likewise, vectors are a ring under the operation of addition and the inner or dot product.

Definition 2.2.20 Finite Field

A finite field is a set made up of a finite number of quantities that follow the rules of operation for a group. The subscript [m] symbolically denotes the maximum number of elements found in the set. As an element in a product of fields it is marked by an indexed superscript [k], F_m^k.

Definition 2.2.21 Infinite Field

An infinite field is a field made up of an infinite number of quantities and denoted by an infinite number [∞] as a subscript, F_∞^k. This is a cumbersome to manipulate, so the infinity sign is simply dropped, F^k.

Definition 2.2.22 Primary Field

A primary field is a set of quantities, such that it is an ordered product of F^1. This is a cumbersome to manipulate, so the superscripted is simply dropped, F.

Definition 2.2.23 Ordered Field

An ordered field is a finite or infinite field having all quantities ordered, such that for any selected pairs of independent numbers (a, b), that one might want to chose, [a] precedes [b], or the size of [a] is said to be **less than** [b], or b − a is a positive number.

$$(a, b) \in F \qquad \text{for } a < b.$$

Definition 2.2.24 Field of Ordered Products-N

It is a set of primary fields of ordered products of [n] denoted as a superscript, such that

$$P^n \equiv \Pi^{\,n} \times F^k.$$

Definition 2.2.25 Integer Field

An integer field is defined as follows:

$(0, n, -m) \in IG$ where [n] and [m] represent natural numbers. Since any number in the integer field is ordered it follows that the integer field is ordered.

Definition 2.2.26 Real Field

A real field is defined as follows:

(0, n, c/d, τ, −b) ∈ **R** where [n] represent natural numbers, [c/d] is a rational number, [τ] irrational numbers and [−b] all negative numbers where b > 0. Since any number in the real field is ordered it follows that the real field is ordered. In the real field numbers also have the property that under the operators of addition, subtraction, multiplication and division they satisfy closure within the real number system;

$$\textbf{(a + b, a − b, ab, a/b)} \in \textbf{R}.$$

Definition 2.2.27 Imaginary Field

An Imaginary field is defined as follows:

(b ∈**R, i** ≡ **√-1, ib)** ∈ **IM** where [b] represent any real number and [i] is called a ***pure imaginary number*** and [ib] simply ***imaginary***. Since **b** ∈**R** and R is an ordered field than the imaginary field is an ordered field.

$$i^2 = -1 \qquad\qquad \text{Equation A}$$
$$i = \sqrt{-1} \qquad\qquad \text{Equation B}$$
$$j = \pm i \qquad\qquad \text{Equation C}$$

Section 2.3 *Axioms on Sets in Common*

Table 2.3.1: Axioms for Existence of Sets[2.*.1]

Axiom	Axioms of Sets	Description
Axiom 2.3.1 Existence	$\exists\mid$ a set A, B and P $\ni$ B = {a | a $\in$ A, P(a) is true}	Let P(x) be any statement and let A be any set.
Axiom 2.3.2 Equality	A = B Equation A B = A Equation B	Equality of sets
Axiom 2.3.3 Congruency	A $\cong$ B Equation A B $\cong$ A Equation B	Congruency of sets
Axiom 2.3.4 Substitution	A = B and B = C $\therefore$ A = C	Equal sets can be substituted one for the other all being one and the same.
Axiom 2.3.5 Transitivity	A $\cong$ B and B $\cong$ C $\therefore$ A $\cong$ C	Congruent sets can be substituted one for the other.
Axioms by Group Quantities		Description
$\exists\mid$ a field G $\ni$, quantities where (e, a, a^{-1}) $\in$ G		Existence of a Group Field where [e] is called the identity element and has the unique identity property of leaving quantities being left unaltered under the operation of the group system.
a = b Equation A b = a Equation B		Equality of quantities
b = a and c = b $\therefore$ c = a		Substitution of quantities

Table 2.3.2: Axioms for Complementation of Sets

Axiom	Axioms of Sets	Description
Axiom 2.3.6 Closure Comp	$(\sim A) \in U$	Closure with respect to complement
Axiom 2.3.7 Identity Comp	$\sim\sim A = A$	Identity by complement
Axiom 2.3.8 Inverse Comp	$\sim\varnothing = U$ Equation A $\sim U = \varnothing$ Equation B	Inverse by complement

Table 2.3.3: Axiom of Correspondence

Axiom 2.3.9 Correspondence	Axioms balancing sets across equalities hold true for congruency of sets as well.	Iff congruency holds than the expressions on either side hold true by any required conditions. (See A2.5.4, A2.5.5, A2.8.5, A2.11.9, A2.11.10)
Axiom 2.3.10 Principle of Duality	If certain axioms imply their own duals, then the dual of any theorem that is a consequence of the axioms is also a consequence of the axioms. For, given any theorem and its proof, the dual of the theorem can be proven in the same way by using the dual of each step in the original proof.	

[2.*.1]Note:
In order for these axioms to be comprehensive, they follow the format of groups whose origins can be found in Axioms 4.2. Most sources axioms are only partial, hence neither systematic nor complete. Also, congruency is added here to be consistent or bridge across to axiomatic, similar, sets found in geometry, sets that are independent, but contain the same number and types of elements.

Section 2.4 *Definitions on Intersection of Sets*

Definition 2.4.1 Intersection of Sets

Sets share common elements represented as follows

$$B = A_1 \cap A_2 \dots \cap A_N \qquad \text{Equation A}$$

read as "the intersecting set B is the intersection of A_1, intersecting A_2, … intersecting A_N", hence

(b) $\in$ B where   Equation B
(b) $\in A_i$ for all i   Equation C

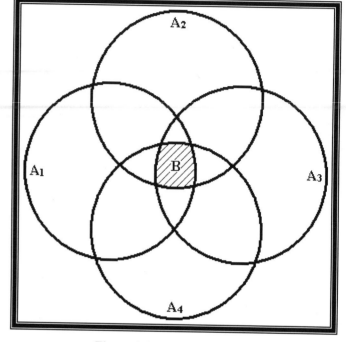

Figure 2.4.1 Intersection of Sets

Section 2.5 *Axioms of Intersection Sets*

Venn diagrams are used in these sections on axioms to demonstrate the validity of an axiomatic statement, but since they lie outside the deductive system, they in their own right, are not formal proofs. They simply lend support-giving confidence in the axiom's validity.

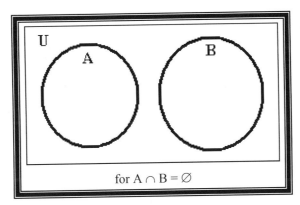

for $A \cap B = \varnothing$

Figure 2.5.1 Disjoined Sets

Table 2.5.1: Axioms for Intersection of Sets [2.*.1][2.5.1]

Axiom	Axioms of Sets	Description
Axiom 2.5.1 Closure Inter	$(A \cap B \wedge B \cap A) \in U$	Closure with respect to intersection
Axiom 2.5.2 Commutative Inter	$A \cap B = B \cap A$	Commutative by intersection
Axiom 2.5.3 Associative Inter	$A \cap (B \cap C) = (A \cap B) \cap C$	Associative by intersection
Axiom 2.5.4 Identity Inter	$A \cap U = A$ Equation A $A \cap A = A$ Equation B	Identity by intersection
Axiom 2.5.5 Inverse Inter	$A \cap \varnothing = \varnothing$ Equation A $A \cap \sim A = \varnothing$ Equation B	Inverse by intersection sets see Fig. 2.2.1

Axioms by Group Operator	Description
$(a \circ b \wedge b \circ a) \in G$	Closure with respect to Group Operator
$a \circ b = b \circ c$	Commutative by Group Operator
$a \circ (b \circ c) = (a \circ b) \circ c$	Associative by Group Operator
$a \circ e = a$	Identity by Group Operator
$a \circ a^{-1} = e$	Inverse Group Operator by reciprocal Quantity

Table 2.5.2: Axioms Distribution by Intersection and Complement across Sets[2.*.1]

Axiom	Axioms by Complement	Description
Axiom 2.5.6 Distribution Inter	$A \cap (B \cup C) = (A \cap B) \cup (A \cap C)$	Distribution by intersection across union
Axiom 2.5.7 De Morgan Inter	$\sim(A \cap B) = \sim A \cup \sim B$ Equation A	Distribution by complement across intersection,

Axioms by Complement	Description
$a \lozenge (b \circ c) = (a \lozenge b) \circ (a \lozenge c)$	Distribution by complementary operator across group operator

[2.5.1]Note: [JAIβ04, pg 6]

Section 2.6 *Theorems on Intersection of Sets*

Theorem 2.6.1 Complementation of Sets: Uniqueness

1g 2g	Given	$(a, b) \in U$ $A = B$
Steps	Hypothesis	$\sim A = \sim B$
1	From 1g and A2.3.6 Closure Union	$\sim A = F$
2	From 2g and A2.3.4 Substitution	$\sim B = F$
$\therefore$	From 1, 2, A2.3.2B Equality and A2.3.4 Substitution	$\sim A = \sim B$

Theorem 2.6.2 Intersection of Sets: Uniqueness

1g 2g 3g	Given	$(a, b, c, d. f) \in U$ $A = B$ $C = D$
Steps	Hypothesis	$A \cap C = B \cap D$
1	From 1g and A2.5.1 Closure Union	$A \cap C = F$
2	From 2g, 3g and A2.3.4 Substitution	$B \cap D = F$
$\therefore$	From 1, 2, A2.3.2B Equality and A2.3.4 Substitution	$A \cap C = B \cap D$

Theorem 2.6.3 Intersection of Sets: Right Distributive Law

Steps	Hypothesis	$(B \cup C) \cap A = (B \cap A) \cup (C \cap A)$
1	From A2.3.2 Equality	$(B \cup C) \cap A = (B \cup C) \cap A$
2	From 1 and A2.5.2 Commutative Inter	$(B \cup C) \cap A = A \cap (B \cup C)$
3	From 2 and A2.5.6 Distribution Inter	$(B \cup C) \cap A = (A \cap B) \cup (A \cap C)$
$\therefore$	From 3 A2.5.2 Commutative Inter	$(B \cup C) \cap A = (B \cap A) \cup (C \cap A)$

Theorem 2.6.4 Intersection of Sets: Identity by Complement

Steps	Hypothesis	$(A \cup B) \cap (A \cup \sim B) = A$
1	From A2.3.2 Equality	$(A \cup B) \cap (A \cup \sim B) = (A \cup B) \cap (A \cup \sim B)$
2	From 1 and A2.8.6 Distribution Union	$(A \cup B) \cap (A \cup \sim B) = A \cup (B \cap \sim B)$
3	From 2 and A2.5.4B Inverse Inter	$(A \cup B) \cap (A \cup \sim B) = A \cup \varnothing$
$\therefore$	From 3 A2.8.4A Identity Union	$(A \cup B) \cap (A \cup \sim B) = A$

Section 2.7 *Definitions on Union of Sets*

Definition 2.7.1 Union of Sets

Combining subsets into a single superset is represented as follows

$$B = A_1 \cup A_2 \ldots \cup A_N \qquad \text{Equation A}$$

read as "the unifying set B is the union of A_1, union A_2, ..., union A_N" such that

$$\text{Equation B}$$
$$A_1 = B \cap A_1 \text{ and } A_1 \subset B$$
$$A_2 = B \cap A_2 \text{ and } A_2 \subset B$$
$$\vdots$$
$$A_N = B \cap A_N \text{ and } A_N \subset B$$

for

$$(a) \in A_i \text{ then} \qquad \text{Equation C}$$
$$(a) \in B \qquad \text{Equation D}$$

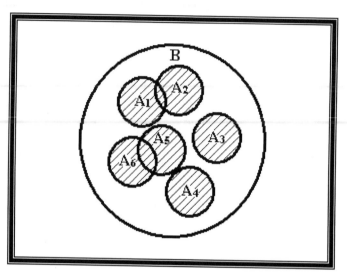

Figure 2.7.1 Union of Sets

Section 2.8 *Axioms on Union of Sets*

Table 2.8.1: Axioms for Union of Sets[2.*.1]

Axiom	Axioms of Sets	Description
Axiom 2.8.1 Closure Union	$(A \cup B \wedge B \cup A) \in U$	Closure with respect to union
Axiom 2.8.2 Commutative Union	$A \cup B = B \cup A$	Commutative by union
Axiom 2.8.3 Associative Union	$A \cup (B \cup C) = (A \cup B) \cup C$	Associative by union
Axiom 2.8.4 Identity Union	$A \cup \varnothing = A$ Equation A $A \cup A = A$ Equation B	Identity by union
Axiom 2.8.5 Inverse Union	$A \cup U = U$ Equation A $A \cup {\sim}A = U$ Equation B $A \cup B = U$ Equation C	Inverse by union and with conditions, see Fig. 2.5.1B

Axioms by Group Operator	Description
$(a \lozenge b \wedge b \lozenge a) \in G$	Closure with respect to Group Operator
$a \lozenge b = b \lozenge c$	Commutative by Group Operator
$a \lozenge (b \lozenge c) = (a \lozenge b) \, o \, c$	Associative by Group Operator
$a \lozenge e = a$	Identity by Group Operator
$a \lozenge a^{-1} = e$	Inverse Group Operator by reciprocal Quantity

Table 2.8.2: Axioms Union and Complement using Distribution of Sets

Axiom	Axioms by Complement	Description
Axiom 2.8.6 Distribution Union	$A \cup (B \cap C) = (A \cup B) \cap (A \cup C)$	Distribution by union across intersection
Axiom 2.8.7 De Morgan Union	${\sim}(A \cup B) = {\sim}A \cap {\sim}B$	Distribution by complement across union

Axioms by Complement	Description
$a \lozenge (b \, o \, c) = (a \lozenge b) \, o \, (a \lozenge c)$	Distribution by complementary operator across group operator

Section 2.9 *Theorems on Union of Sets*

Theorem 2.9.1 Union of Sets: Uniqueness

1g 2g 3g	Given	$(a, b, d, f) \in U$ $A = B$ $C = D$
Steps	Hypothesis	$A \cup C = B \cup D$
1	From 1g and A2.8.1 Closure Union	$A \cup C = F$
2	From 2g, 3g and A2.3.4 Substitution	$B \cup D = F$
∴	From 1, 2, A2.3.2B Equality and A2.3.4 Substitution	$A \cup C = B \cup D$

Theorem 2.9.2 Union of Sets: Right Distributive Law

Steps	Hypothesis	$(B \cap C) \cup A = (B \cap A) \cup (C \cap A)$
1	From A2.6.2 Equality	$(B \cap C) \cup A = (B \cap C) \cup A$
2	From 1 and A2.8.2 Commutative Union	$(B \cap C) \cup A = A \cup (B \cap C)$
3	From 2 and A2.8.6 Distribution Union	$(B \cap C) \cup A = (A \cup B) \cap (A \cup C)$
∴	From 3 A2.8.2 Commutative Union	$(B \cap C) \cup A = (B \cup A) \cap (C \cup A)$

Theorem 2.9.3 Union of Sets: Identity by Complement

Steps	Hypothesis	$(A \cap B) \cup (A \cap \sim B) = A$
1	From A2.3.2 Equality	$(A \cap B) \cup (A \cap \sim B) = (A \cap B) \cup (A \cap \sim B)$
2	From 1 and A2.5.6 Distribution Inter	$(A \cap B) \cup (A \cap \sim B) = A \cap (B \cup \sim B)$
3	From 2 and A2.8.5B Inverse Union	$(A \cap B) \cup (A \cap \sim B) = A \cap U$
∴	From 3 A2.8.4A Identity Union	$(A \cap B) \cup (A \cap \sim B) = A$

Section 2.10 *Definitions on Differences of Sets*

Definition 2.10.1 Difference between Sets

The difference of sets A and B is the set of elements which belong to A, but which do not belong to B and represented as follows

    C = A – B iff
    (a) ∈ A but (a) ∉ B Equation A

and

    D = B – A iff
    (b) ∈ B but (b) ∉ A Equation B

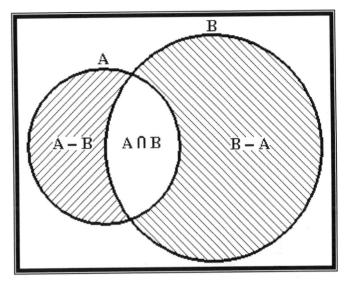

Figure 2.10.1 Difference between Sets

Section 2.11 *Axioms of Differences of Sets*

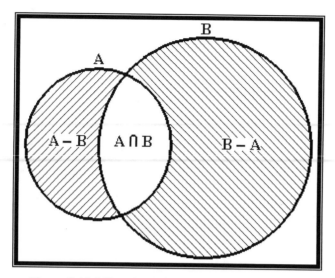

Figure 2.11.1 Non-Commutative Difference of Sets

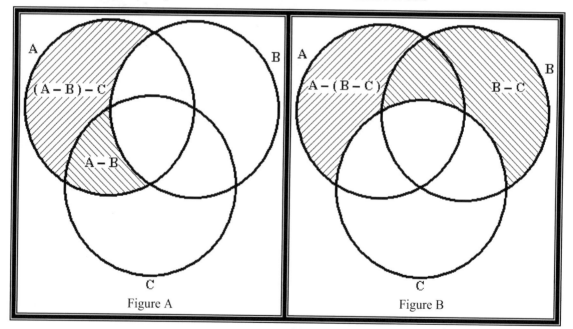

Figure A   Figure B

Figure 2.11.2 Associative for Difference of Sets

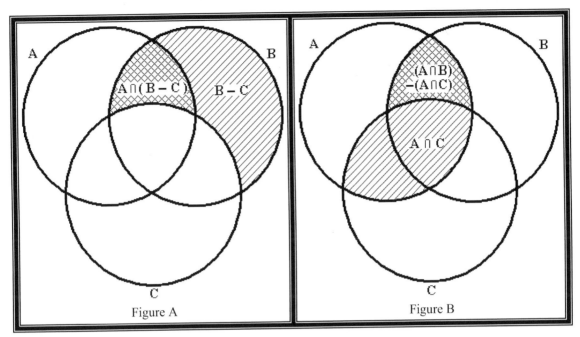

Figure 2.11.3 Distribution of Intersection Across Difference

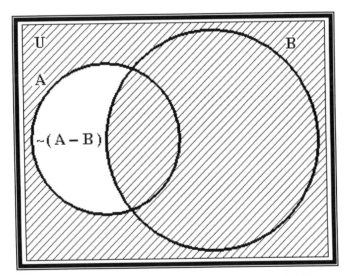

Figure 2.11.4 Distribution of Complement Across Difference

Table 2.11.1: Axioms for Difference of Sets[2.*.1] [2.11.1]

Axiom	Axioms of Sets	Description
Axiom 2.11.1 Closure Diff	$(A - B \wedge B - A) \in U$	Closure with respect to difference
Axiom 2.11.2 Commutative Diff	$A - B \neq B - A$ iff $A \neq B$ EQ A $A - B = B - A$ iff $A = B$ EQ B	Commutative by difference Fig. 2.11.1
Axiom 2.11.3 Associative Diff	$A - (B - C) = (A - B) - C$	Associative by difference Fig. 2.11.2
Axiom 2.11.4 Identity Diff	$A - \varnothing = A$ EQ A $(A - B) \cup (A \cap B) = A$ EQ B	Identity by difference
Axiom 2.11.5 Inverse Diff	$A - A = \varnothing$	Inverse by difference
Axioms by Group Operator	Description	
$(a \, o \, b \wedge b \, o \, a) \in G$	Closure with respect to Group Operator	
$a \, o \, b = b \, o \, c$	Commutative by Group Operator	
$a \, o \, (b \, o \, c) = (a \, o \, b) \, o \, c$	Associative by Group Operator	
$a \, o \, e = a$	Identity by Group Operator	
$a \, o \, a^{-1} = e$	Inverse Group Operator by reciprocal Quantity	

Table 2.11.2: Axioms with Complementary Operations[2.*.1] [2.11.1]

Axiom	Axioms by Complement	Description
Axiom 2.11.6 Distribution	$A \cap (B - C) = (A \cap B) - (A \cap C)$ EQ A $\sim(A - B) = U - (A - B)$ EQ B	Distribution of intersection Fig. 2.11.3 and complement across difference Fig. 2.11.4
Axiom 2.11.7 De Morgan	$A - (B \cup C) = (A - B) \cap (A - C)$ EQ C $A - (B \cap C) = (A - B) \cup (A - C)$ EQ D	Distribution by difference across union and intersection
Axioms by Complement	Description	
$a \lozenge (b \, o \, c) = (a \lozenge b) \, o \, (a \lozenge c)$	Distribution by complementary operator across group operator	

Table 2.11.3: Axioms with Conditional Differences[2.11.1]

Axiom	Axioms by Complement	Description
Axiom 2.11.8 Material Diff	$A - B = A \cap \sim B$ Equation A $A - B = \sim B - \sim A$ Equation B	Material intersection by complement for difference
Axiom 2.11.9 Identity Cond	iff $A \cap B = \varnothing$ $A - B = A$ Equation A	Identity by condition for difference

[2.11.1]Note: [JAIβ04, pg 6]

Section 2.12 *Theorems on Differences of Sets*

Theorem 2.12.1 Difference of Sets: Uniqueness

1g	Given	$(a, b, c, d, f) \in U$
2g		$A = B$
3g		$C = D$
Steps	Hypothesis	$A - C = B - D$
1	From 1g and A2.11.1 Closure Diff	$A - C = F$
2	From 2g, 3g, 1 and A2.3.4 Substitution	$B - D = F$
∴	From 1, 2, A2.3.2B and A2.3.4 Substitution	$A - C = B - D$

Theorem 2.12.2 Difference, Inclusion of an Element

1g	Given	$(a) \in A$	
2g		$A - B$	
3g		$A \cap B$	
Steps	Hypothesis	$(a) \in A - B$ given $(a) \in A$ and $(a) \notin B$	
1	From 1g, 2g and D2.10.1A Difference between Sets	$(a) \in A$	$(a) \notin B$
2	From 1g, 3g and D2.4.1C Intersection of Sets	$(a) \in A$	$(a) \in B$
3	From 1 given 2g is a contradiction hence	$(a) \notin A \cap B$	
∴	From 3 the only other part of the set A, [a] can be found in would have to be	$(a) \in A - B$ given $(a) \in A$ and $(a) \notin B$	

Theorem 2.12.3 Distribution of Complement Across Difference

1g	Given	$(a) \in A$
2g		$(b) \in B$
Steps	Hypothesis	$\sim(A - B) = \sim A \cup B$
1	From A2.3.2A Equality	$\sim(A - B) = \sim(A - B)$
2	From 1, A2.11.8A Material Diff, and A2.3.4 Substitution	$\sim(A - B) = \sim(A \cap \sim B)$
3	From 2 and A2.5.7A De Morgan Inter	$\sim(A - B) = \sim A \cup (\sim \sim B)$
∴	From 3 and A2.3.7 Identity Comp	$\sim(A - B) = \sim A \cup B$

Section 2.13 *Definitions for Subsets*

Definition 2.13.1 Subset of Sets

One set within another or all elements of the lesser set are within the superset and represented as follows

$$A \subset B \qquad \text{Equation A}$$

for

$$n_a < n_b \qquad \text{Equation B}$$

and

$$\text{if } (x) \in A \text{ then } (x) \in B \qquad \text{Equation C}$$

the number of elements in each set respectively.

read as "A is a subset of B"

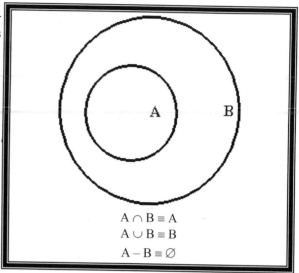

$$A \cap B \equiv A$$
$$A \cup B \equiv B$$
$$A - B \equiv \varnothing$$

Figure 2.13.1 Subset

Section 2.14 *Axioms for Subsets*

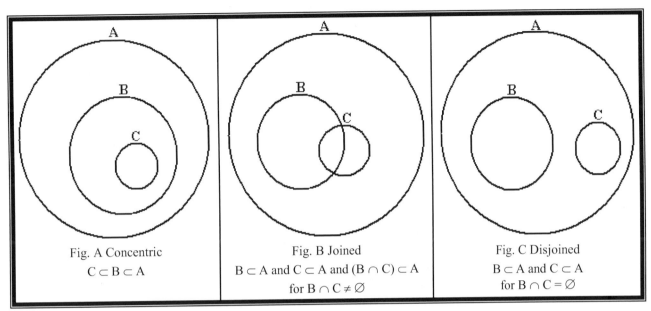

Fig. A Concentric	Fig. B Joined	Fig. C Disjoined
$C \subset B \subset A$	$B \subset A$ and $C \subset A$ and $(B \cap C) \subset A$ for $B \cap C \neq \varnothing$	$B \subset A$ and $C \subset A$ for $B \cap C = \varnothing$

Figure 2.14.1 Possible Ternary Subsets

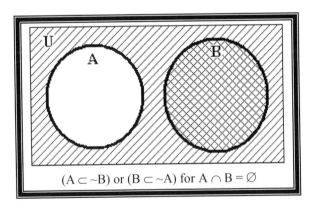

$(A \subset \sim B)$ or $(B \subset \sim A)$ for $A \cap B = \varnothing$

Figure 2.14.2 Complement of Disjoined Sets as Subsets

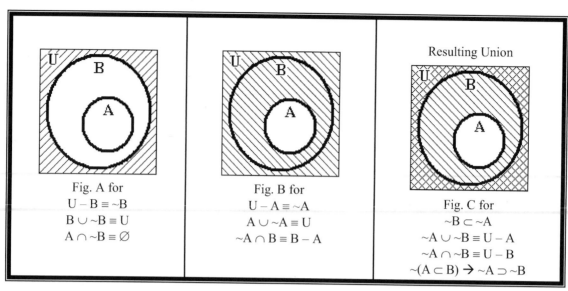

Figure 2.14.3 Distribution of Complement Across a Subset[2.14.1]

Table 2.14.1: Axioms for Conditional Subsets

Axiom	Axioms of Sets	Description
Axiom 2.14.1 Concentric Subsets	iff $C \subset B \subset A \vee$ $A \subset B \wedge$ $B \subset C$ then $A \subset C$ Equation A	Concentric subsets can be substituted one for the other see Fig. 2.14.1A
Axiom 2.14.2 Joined Subsets	iff $A \cap B \neq \varnothing \wedge$ $B \subset A \wedge$ $C \subset A$ then $(B \cap C) \subset A$ Equation A	Joined subsets fork independently, hence unlike concentric subsets they cannot be substituted one for the other see Fig. 2.14.1B
Axiom 2.14.3 Disjoined Subsets	iff $A \cap B = \varnothing \wedge$ $B \subset A \wedge$ $C \subset A$ then $(B \cap C) \subset A \wedge$ Equation A iff $(A \subset {\sim}B) \vee$ $(B \subset {\sim}A)$ then $A \subset (A \cup B) \vee$ Equation B $B \subset (A \cup B)$ Equation C	see Fig. 2.14.1C see Fig 2.14.2
Axiom 2.14.4 Non-Commutative Difference in Subset	iff $A \subset B$ then $A - B \equiv \varnothing$ Equation A $B - A \equiv {\sim}A \cap B$ Equation B	see Fig. 2.13.1 and Fig. 2.14.3B
Axiom 2.14.5 Identity Subset	iff $A \subset B$ then $A \cap B = A$ Equation A $A \cup B = B$ Equation B iff $A \cap B \neq \varnothing \wedge$ $A \subset B$ then ${\sim}A \cup {\sim}B \equiv {\sim}A$ Equation C ${\sim}A \cap {\sim}B \equiv {\sim}B$ Equation D and $U - A \equiv {\sim}A$ Equation E $U - B \equiv {\sim}B$ Equation F	see Fig. 2.13.1 Identity by complement of subset see Fig 2.14.3C and see Fig 2.14.3B,A
Axiom 2.14.6 Inverse Subset	iff $A \subset B$ then $A \cap {\sim}B \equiv \varnothing$	Inversed by condition for difference see Fig 2.14.3A

Table 2.14.2: Axioms Distribution of Complement across Subsets[2.*.1]

Axiom	Axioms by Complement	Description
Axiom 2.14.7 De Morgan Subset	iff $A \subset B$ then $\sim(A \subset B) \rightarrow \sim A \supset \sim B$	Distribution by complement across subset, see Fig. 2.14.3C for proof[2.14.2]

[2.14.1]Note:

That the complement of the set taking up the balance of U now encompasses more elements than the set it was a subset to. This in turn swaps the sets around were the original parent set is now the subset. This is analogous to the algebraic inequality sign being flipped when multiplying a cross by a negative number.

[2.14.2]Note:

Equality cannot be established because subset is a nonlinear convolution of union and intersection of those sets the best that can be said the negation of a subset can only imply the distribution of the negation operation.

Section 2.15 *Theorems for Subsets*

Theorem 2.15.1 Difference as a Subset

1g 2g 3g 4g	Given	Venn diagram Fig. 2.2.4 Difference between Sets n_a elements $\in A$ n_d elements $\in \{A - B\}$ n_i elements $\in \{A \cap B\}$	
Steps	Hypothesis	$(A - B) \subset A$ $\quad$ for $n_d < n_a$	
1	From D2.10.1A Difference between Sets	(a) $\in A$	(a) $\notin B$
2	From 1 and T2.12.2 Difference, Inclusion of an Element	(a) $\in A - B$	
3	From 1g, 2g, 3g, 4g and A2.11.4A Identity Diff	$n_a = n_d + n_i$	
4	From 3 and T4.3.3B Equalities: Right Cancellation by Addition	$n_i = n_a - n_d$	
5	From 4 and D4.1.1 Positive Integers [+n]: Natural or Countable numbers	$0 < n_i$ a counted number	
6	From 4, 5 and A4.2.3 Substitution	$0 < n_a - n_d$	
7	From 6 and D4.1.7 Greater Than Inequality	$n_d < n_a$	
∴	From 1, 2, 7 and D2.13.1A,B Subset of Sets	$(A - B) \subset A$ $\quad$ for $n_d < n_a$	

Theorem 2.15.2 De Morgan Theorem for Union and Intersection of a Subset

1g 2g	Given	$D = (A \cup B)$ $E = (A \cap B)$
Steps	Hypothesis	$\sim[(A \cap B) \subset (A \cup B)] \rightarrow [(\sim A \cap \sim B) \subset (\sim A \cup \sim B)]$
1	From F2.14.3C	$\sim B \subset \sim A$
2	From 1, A2.14.1C,D Identity Subset and A2.3.4 Substitution	$(\sim A \cap \sim B) \subset (\sim A \cup \sim B)$
3	From 2, A2.8.7 De Morgan Union, and A2.5.7 De Morgan Inter	$\sim(A \cup B) \subset \sim(A \cap B)$
4	From 1g, 2g, 3 and A2.3.4 Substitution	$\sim D \subset \sim E$
5	From 4 and A2.14.3C De Morgan Subset	$\sim(E \subset D) \rightarrow (\sim D \subset \sim E)$
∴	From 1g, 2g, 5 and A2.3.4 Substitution	$\sim[(A \cap B) \subset (A \cup B)] \rightarrow [(\sim A \cap \sim B) \subset (\sim A \cup \sim B)]$

Section 2.16 *Definitions for Product Sets*

The following definitions are the naming nomenclature of numbered elements in a multiple set or n–tuple.

Table 2.16.1: Naming Polytuples Formal and Informal

Name[2.16.1]	Definition
Definition 2.16.1 Dituple or pair tuple or dual tuple	2-element set such that $(\bullet \, a_i \, \bullet) \in A$ for $i = 1, 2$
Definition 2.16.2 Trituple or triplet	3-element set such that $(\bullet \, a_i \, \bullet) \in A$ for $i = 1, 2, 3$
Definition 2.16.3 Quadrituple	4-element set such that $(\bullet \, a_i \, \bullet) \in A$ for $i = 1, 2, 3, 4$
Definition 2.16.4 Penatuple or quintuple	5-element set such that $(\bullet \, a_i \, \bullet) \in A$ for $i = 1, 2, 3, 4, 5$
Definition 2.16.5 Hexatuple or sextuple	6-element set such that $(\bullet \, a_i \, \bullet) \in A$ for $i = 1, 2, 3, 4, 5, 6$
Definition 2.16.6 Heptatuple	7-element set such that $(\bullet \, a_i \, \bullet) \in A$ for $i = 1, 2, 3, 4, 5, 6, 7$
Definition 2.16.7 Octatuple	8-element set such that $(\bullet \, a_i \, \bullet) \in A$ for $i = 1, 2, 3, 4, 5, 6, 7, 8$
Definition 2.16.8 Nonatuple	9-element set such that $(\bullet \, a_i \, \bullet) \in A$ for $i = 1, 2, 3, 4, 5, 6, 7, 8, 9$
Definition 2.16.9 Decatuple	10-element set such that $(\bullet \, a_i \, \bullet) \in A$ for $i = 1, 2, 3, 4, 5, 6, 7, 8, 9, 10$
Definition 2.16.10 Hendecatuple or undecatuple	11-element set such that $(\bullet \, a_i \, \bullet) \in A$ for $i = 1, 2, 3, 4, 5, 6, 7, 8, 9, 10, 11$
Definition 2.16.11 Dodecatuple or duodecituple	12-element set such that $(\bullet \, a_i \, \bullet) \in A$ for $i = 1, 2, 3, 4, 5, 6, 7, 8, 9, 10, 11, 12$
Definition 2.16.12 n–tuple	n-element set such that $(\bullet \, a_i \, \bullet) \in A$ for $i = 1, 2, \ldots, n$

[2.16.1]Note: The origins of tuple names lead with a Greek numerical prefix name.

Definition 2.16.13 Ordered n–Tuple as a Point

An *order n-tuple* is a set called a *point* of n-elements iff it has a biased order, numbering from left to right

$P^n \equiv (\bullet \ a_i \ \bullet)$ or if [n] is known in the system that it belongs to, it can be dropped and

$P \equiv (\bullet \ a_i \ \bullet)$.

The concept of a product set can be extended to more than two sets in a natural way as investigated by Descartes. Let's take an axial field set A_1 with n_1 elements such that $e_{1i} \in A_1$ for *order index* i = 1, 2, ..., n_1, another A_2 with n_2 elements such that $e_{2i} \in A_2$ for order index i = 1, 2, ..., n_2, another and so on to, A_m with n_m elements such that $e_{mi} \in A_m$ for order index i = 1, 2, ..., n_m , in general this can be written as a super set $\{\bullet \ A_k \ \bullet\}$ for *axis index* k = 1, 2, ..., m. What Descartes realized is that if he took the product of these sets in the following way he could generate new sets of n-tuples $\{e_{1\lambda[1, 1]}, e_{2\lambda[2, 1]}, ..., e_{n\lambda[n, 1]}\}$, $\{e_{1\lambda[1, 2]}, e_{2\lambda[2, 2]}, ..., e_{n\lambda[n, 2]}\}$, ..., $\{e_{1\lambda[1, r]}, e_{2\lambda[2, r]}, ..., e_{n\lambda[n, r]}\}$ or generally written $\{\bullet \ e_{i\lambda[i, j]} \ \bullet\}$ for i = 1, 2, ..., n and j = 1, 2, ..., r. Where $\lambda[i, j]$ is a unitary index set made up of one integer number, such that over the index [i] it constitutes a combination of a long set of integers from the arrangement of the positional place of the axial field element. For [j], denoting the list of all possible combinations from the product of the axial sets terminating with a numbering to [r]. The terminating index [r] is the count of all possible n-tuples given by $r = \prod_{k=1}^m n_k$.

Definition 2.16.14 Positional Unitary Index Set

The $\lambda[i, j]$ dual index is a unitary index set made up of one positional integer number, such that over the index [i] it constitutes a combinatorial set of integers, with order and replacement of indices, forming the combination of the positional place of an element in the axial field. The unitary index can also be represented by an indexed of a sub-index, like so

$$\lambda[i, j] \equiv i_j.$$

Other books use this notation, but do to the difficulty of printing sub-indices it is not commonly used in this treatise except for special tensor relations for backwards compatibility to the classical presentations.

Definition 2.16.15 Set of Coordinate Points; ρ-ordered indices

The collection of unitary index set for a given n-tuple defines the combination of <u>ρ(j)-ordered positional axial indexes</u>, called *jth coordinate point*:

$$\rho(j) = \{\lambda[1, j], \lambda[2, j], ... \lambda[i, j], ..., \lambda[n, j]\}$$

$$\rho(j) = \{\bullet \ \lambda[i, j] \ \bullet\} \qquad \text{for i = 1, 2, ..., n and j = 1, 2, ..., r.}$$

Definition 2.16.16 Ordered Universal or Long Index Set-L$_r$

The collection of coordinate points, ρ-ordered indices, is called the ordered Universal or Long Index Set-L$_r$.

$$\rho(j) \in L_r \qquad\qquad \text{for } j = 1, 2, \ldots, r \text{ and } r = \prod_{k=1}^{m} n_k.$$

for the origins of the name *long index set* see footnote [1.7.1].

It is from here that the set notation used for tracking indices springs from. It is used in summation notation over sets of indices; see section 1.7 and Appendix A on "The Mathematics of Multinomial Summation Notation".

Definition 2.16.17 Cartesian Product

The Cartesian[2.16.2] product yields a set of ordered n-tuples,

$$P_j \equiv (\bullet, e_{i\lambda[i,\,j]}\,\bullet) \text{ called } \textbf{\textit{coordinates points}}$$
$$\text{for } i = 1, 2, \ldots, n \text{ and } j = 1, 2, \ldots, r \qquad \text{Equation A}$$

from the ordered superset

$$\mathcal{P} \equiv \{\, \prod_{k=1}^{m} \times A_k \,\} \equiv \{\bullet, P_j, \bullet\} \qquad\qquad \text{Equation B}$$

where A_k is denoted as the kth–*axial field set*. If each axial set contains [n_k] elements than the number of coordinates or number of points are

$$r = \prod_{k=1}^{m} n_k. \qquad\qquad \text{Equation C}$$

Clearly, if any one axial set is infinite than the superset or coordinate field is infinite as well.[2.16.3]

Definition 2.16.18 Power Set

For any set [A] than $n^m \equiv \{\prod^m \times A\}$ is called the power set including [A, $\varnothing$]. This new set has n^m non-ordered subsets starting with n-elements and decreasing to 1-element and than none the empty set.

[2.16.2]Note:

The word *Cartesiānus* is the latinized, French *Cartesius*, name of Descartes [OXF71, Vol. I, pg 345] from the French philosopher and mathematician, Rene Descartes, (1596–1650) as applied to his studies of ordered paired product sets. He published his ideas in 1637 in a treatise called *La géométrie* (*Geometry*).

[2.16.3]Note:

Descartes only dealt with ordered pairs of points in his field coordinates, since this paper deals in the generality with n-dimensions the Cartesian Product is not an ordered pair of points, but generalized to an ordered n–tuple of point.

Tensor Calculus & Physics: A General Treatise

Section 2.17 *Axioms on Products of Sets*

Building product ordered sets by use of a tree diagram:
Given: Axial sets A = {a, b}, B = {1, 2, 3} and C = {x, y} then by a tree diagram;

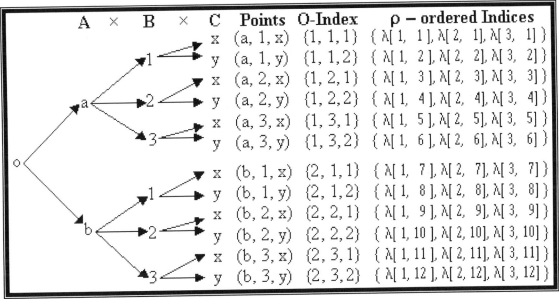

Figure 2.17.1 Building Product Sets by Tree Diagram for Asymmetrical Unequal Sets

Table 2.17.1: Axioms for Products of Sets[2.*.1]

Axiom	Axioms of Sets	Description
Axiom 2.17.1 Closure Product	$(A \times B \land B \times A) \in U$	Closure with respect to products
Axiom 2.17.2 Commutative Product	$A \times B = B \times A$ Equation A iff symmetrical product operation $A \times B \neq B \times A$ Equation B iff asymmetrical product operation	Commutative by products
Axiom 2.17.3 Associative Product	$A \times (B \times C) = (A \times B) \times C$	Associative by products
Axiom 2.17.4 Identity Product	$A \times \varnothing = A$	Identity by products
Axiom 2.17.5 Inverse Product	$A \times U = U$ Equation A $A \times \sim A = U$ Equation B	Inverse by products

Axioms by Group Operator	Description
$(a \lozenge b \land b \lozenge a) \in G$	Closure with respect to Group Operator
$a \lozenge b = b \lozenge c$	Commutative by Group Operator
$a \lozenge (b \lozenge c) = (a \lozenge b) \, o \, c$	Associative by Group Operator
$a \lozenge e = a$	Identity by Group Operator
$a \lozenge a^{-1} = e$	Inverse Group Operator by reciprocal Quantity

Section 2.18 *Theorems on Products of Sets*

Theorem 2.18.1 Product of Sets: Uniqueness

1g	Given	$(A, B, F) \in U$
2g		$A = B$
3g		$C = D$
Steps	Hypothesis	$A \times C = B \times D$
1	From 1g and A2.17.1 Closure Product	$A \times C = F$
2	From 2g, 3g, 1 and A2.3.4 Substitution	$B \times D = F$
∴	From 1, 2 and A2.3.4 Substitution	$A \times C = B \times D$

Theorem 2.18.2 Product as a Superset

1g	Given	$(a) \in A$
2g	and	$(b) \in B$
Steps	Hypothesis	$(a, b) \in \{ A \times B \}$
∴	From 1g, 2g and D2.16.17(A, B) Cartesian Product	$(a, b) \in \{ A \times B \}$

Theorem 2.18.3 Product Distribution Across Intersection Operator

1g	Given	$(x) \in A$	
2g	and	$(y) \in B \cap C$	
Steps	Hypothesis	$A \times (B \cap C) = (A \times B) \cap (A \times C)$ for all points	
1	From 1g, 2g and T2.18.2 Product as a Superset	$(x, y) \in \{ A \times (B \cap C) \}$ for all points	
2	From 2g and D2.4.1(B, C) Intersection of Sets	$(y) \in B$	$(y) \in C$
3	From 1g, 2 and T2.18.2 Product as a Superset	$(x, y) \in \{A \times B\}$	$(x, y) \in \{A \times C\}$
4	From 3 and D2.4.1B Intersection of Sets	$(x, y) \in \{(A \times B) \cap (A \times C)\}$ for all points	
∴	From 1, 4 and D2.2.7 Equal Sets	$A \times (B \cap C) = (A \times B) \cap (A \times C)$ for all points	

Theorem 2.18.4 Product Distribution Across Union Operator, Case I

1g	Given	$(x) \in A$	
2g		$(y) \in B$	
3g		$(y) \notin C$ for $B \neq C$, case I	
Steps	Hypothesis	$A \times (B \cup C) = (A \times B) \cup (A \times C)$ for all points for $B \neq C$	
1	From 2g and D2.7.1(C, D) Union of Sets	$(y) \in B \cup C$	
2	From 1g, 1 and T2.18.2 Product as a Superset	$(x, y) \in \{ A \times (B \cup C) \}$ for all points	
3	From 2g, 3g and T2.18.2 Product as a Superset	$(x, y) \in \{A \times B\}$	$(x, y) \notin \{A \times C\}$
4	From 3 and D2.7.1D Union of Sets	$(x, y) \in \{(A \times B) \cup (A \times C)\}$ for all points for $B \neq C$	
∴	From 2, 4 and D2.2.7 Equal Sets	$A \times (B \cup C) = (A \times B) \cup (A \times C)$ for all points for $B \neq C$	

Theorem 2.18.5 Product Distribution Across Union Operator, Case II

1g	Given	$(x) \in A$
2g		$(y) \in B$
3g		$(y) \in C$ for $B = C$, case II
Steps	Hypothesis	$A \times (B \cup C) = (A \times B) \cup (A \times C)$ for all points for $B = C$
1	From 2g and D2.7.1(C, D) Union of Sets	$(y) \in B \cup C$
2	From 1g, 1 and T2.18.2 Product as a Superset	$(x, y) \in \{ A \times (B \cup C) \}$ for all points
3	From 2g, 3g and T2.18.2 Product as a Superset	$(x, y) \in \{A \times B\}$ $\qquad$ $\big|$ $\qquad$ $(x, y) \in \{A \times C\}$
4	From 3 and D2.7.1D Union of Sets	$(x, y) \in \{(A \times B) \cup (A \times C)\}$ for all points for $B = C$
$\therefore$	From 2, 4 and D2.2.7 Equal Sets	$A \times (B \cup C) = (A \times B) \cup (A \times C)$ for all points for $B = C$

Theorem 2.18.6 Product Distribution Across Union Operator

1g	Given	$(x) \in A$
2g		$(y) \in B$
Steps	Hypothesis	$A \times (B \cup C) = (A \times B) \cup (A \times C)$ for all points
$\therefore$	From 1g, 2g, T2.18.4 Product Distribution Across Union Operator, Case I and T2.18.5 Product Distribution Across Union Operator, Case II	$A \times (B \cup C) = A \times B \cup A \times C$ for all points

Theorem 2.18.7 Product Distribution Across Difference Operator

1g	Given	$(x) \in A$
2g		$(y) \in B$
Steps	Hypothesis	$A \times (B - C) = (A \times B) - (A \times C)$ for all points in difference
1	From 2g and T2.12.2 Difference, Inclusion of an Element	$(y) \in \{B - C\}$ iff $(y) \notin C$
2	From 1g, 1 and T2.18.2 Product as a Superset	$(x, y) \in \{ A \times (B - C) \}$ for all points in difference
3	From 1g, 2g, 1 and T2.18.2 Product as a Superset	$(x, y) \in \{A \times B\}$ $\qquad$ $\big|$ $\qquad$ $(x, y) \notin \{A \times C\}$
4	From 3 and T2.12.2 Difference, Inclusion of an Element	$(x, y) \in \{(A \times B) - (A \times C)\}$ for all points in difference
$\therefore$	From 2, 4 and D2.2.7 Equal Sets	$A \times (B - C) = (A \times B) - (A \times C)$ for all points in difference

Theorem 2.18.8 Product, Uniqueness of Subsets

1g	Given	$(x) \in A$
2g		$(y) \in C$
3g		$A \subset B$
4g		$C \subset D$
Steps	Hypothesis	$(A \times C) \subset (B \times D)$
1	From 1g, 2g, T2.18.2 Product as a Superset	$(x, y) \in \{A \times C\}$
2	From 1g, 3g and D2.13.1C Subset of Sets	$(x) \in B$
3	From 2g, 4g and D2.13.1C Subset of Sets	$(y) \in D$
4	From 2, 3, T2.18.2 Product as a Superset	$(x, y) \in \{B \times D\}$
∴	From 1, 4 and D2.13.1A,C Subset of Sets	$(A \times C) \subset (B \times D)$

Theorem 2.18.9 Product, Distribution of Complement Across a Product

1g	Given	$(x) \in A$
2g		$(y) \in C$
3g		$A \subset B$
4g		$C \subset D$
Steps	Hypothesis	$\sim(A \times C) = \sim A \times \sim B$????
1		
2		
3		
4		
∴		

When the axes are common:

Theorem 2.18.10 Product of the Power Set

1g	Given	$(a) \in A$ for n-elements
Steps	Hypothesis	$A^m \equiv \Pi \times A$ for number of coordinate points n^m
1	From 1g and D2.2.15 Exponential Notation for Set Operators	$A^m \equiv \Pi \times A$ for m products
2	From 1g, 1 and D2.16.17C Cartesian Product	$N = \prod_{k=1}^{m} n_k$ for $n = n_k$ for all k, number of coordinate points
3	From 2 and A4.2.3 Substitution	$N = \prod_{k=1}^{m} n$
4	From 3 and D4.1.17 Exponential Notation	$N = n^m$
∴	From 1 and 4	$A^m \equiv \Pi \times A$ for number of coordinate points n^m

Theorem 2.18.11 Product of the Power Set Defined

Steps	Hypothesis	$n^m = \{A^m\}$ for n elements
$\therefore$	From T2.27.1 Product of the Power Set, D2.16.18 Power Set and A2.3.4 Substitution	$n^m = \{A^m\}$ for n elements

When the axes are uncommon:

Section 2.19 *Definitions on Multiplicative Sets*

In this axiomatic system, sets are treated as numbers under multiplication like operators with numerical certainty. The interest of Multiplicative Sets is that you can take something of a known quantity and combine them by a multiplication like operator in a product creating a totally new quantity. Also, it allows a general way of exploring the properties of products, which are found throughout mathematics, such as multinomial products, polynomials, real number exponents, logical products, finite and infinite product formulas, tensors and so on.

Definition 2.19.1 Multiplicative Set

A *Multiplicative Set* $[A^m]$ is a product of sets $[A_i]$ under some kind of multiplication operator $[\lozenge]$.

Definition 2.19.2 Rank of a Multiplicative Set

The *rank [m] of a multiplicative set* is the number of sets in the product. Since rank is a measure of the count of sets within the product, it is a measurable quantity, hence countable.

Definition 2.19.3 Multiplicative Set: Rank-0

Multiplication Product of rank-0 is a scalar so $[A^0 \equiv s]$, such that $[s \in (R \vee C^2)]$.

Definition 2.19.4 Multiplicative Set Superset

The set of a multiplication product can be thought of as its own set hence $\mathcal{A}^m = \{ A^m \}$ are called the *Multiplicative Set Superset*.

Section 2.20 *Axioms on Multiplicative Sets*

In this axiomatic system, sets are treated as numbers under multiplication like operators with numerical certainty. The interest of Multiplicative Sets is that you can take something of a known quantity and combine them by a multiplication like operator in a product creating a totally new quantity. Also, it allows a general way of exploring the properties of products, which are found throughout mathematics, such as multinomial products, polynomials, real number exponents, logical AND products, finite and infinite product formulas, tensors and so on.

Table 2.20.1: Axioms on Existence for Multiplicative Sets

Axiom	Axioms on Multiplicative Sets	Description
Axiom 2.20.1 Existence of Product and Operator	$\exists \mid$ a field M called the universal product set $\ni$, the quantities of products $A^m \equiv \prod^m \lozenge A_i$ EQ A $(A^m) \in M$ EQ B since rank is countable it is positive only. $0 < m$ EQ C.	Existence of the product set under some operator $[\lozenge]$.
Axiom 2.20.2 Equality	$A^m = B^m$ Equation A $B^m = A^m$ Equation B iff $A_i = B_i$ for all i	Equality of product of sets
Axiom 2.20.3 Congruency	$A^m \cong B^m$ Equation A $B^m \cong A^m$ Equation B iff $A_i \cong B_i$ for all i	Congruency of product of sets

Table 2.20.2: Axioms on Multiplicative of Sets

Axiom	Axioms on Multiplicative Sets	Description
Axiom 2.20.4 Closure for Multiplicative Sets	$(A^p \lozenge B^q \wedge B^q \lozenge A^p) \in M$ the universal product set	Closure with respect to multiplication of any product is also a product set of countable rank.
Axiom 2.20.5 Commutative	$A_i \lozenge B_i = B_i \lozenge A_i$ for all i EQ A iff symmetrical product operation $A_i \lozenge B_i \neq B_i \lozenge A_i$ for all i EQ B iff asymmetrical product operation	Commutative by Multiplicative Operators
Axiom 2.20.6 Set as a Number Reciprocal[2.23.1]	$A^m / B^p \equiv A^m \lozenge (1 / B^p)$ Equation A $\equiv A^m \lozenge B^{-p}$ Equation B iff $p < m$ and Equation C $\mathcal{B}^p \subseteq \mathcal{A}^m$ Equation D	Reciprocal by Multiplicative Operators
Axiom 2.20.7 Identity	$1 \lozenge A_i = A_i$ for any i	Identity by Multiplicative Operators
Axiom 2.20.8 Inverse	$A_i / A_i = 1$ iff $A_i = A_i$ or $A_i \cong A_i$ for all i	Inverse by Multiplication and congruency for Multiplicative Operators.

[2.23.1]Note: Only valid if the operator also supports a reciprocal operation.

Table 2.20.3: Axioms on Addition for Multiplicative Sets

Axiom	Axioms on Multiplicative Sets	Description
Axiom 2.20.9 Closure for Multiplicative Sets	$(A^m + B^m \wedge B^m + A^m) \in M$ the universal product set	Closure with respect to addition for multiplication products is also a product set of countable rank.
Axiom 2.20.10 Commutative	$A^m + B^m = B^m + A^m$	Commutative by Multiplication of products
Axiom 2.20.11 Associative	$A^m + (B^m + C^m) = (A^m + B^m) + C^m$	Associative by Multiplication of products
Axiom 2.20.12 Identity	$A^m + 0 = A^m$	Identity by Multiplication of products
Axiom 2.20.13 Inverse	$A^m - A^m = 0$	Inverse by Multiplication of products.

Section 2.21 *Theorems on Multiplicative Sets*

Theorem 2.21.1 Multiplicative Sets: Substitution for Rank-m

1g 2g	Given	$A^m = B^m$ $B^m = C^m$
Steps	Hypothesis	$A^m = C^m$
1	From 2g and A2.20.2A Equality	$B^i = C^i$ for all i
2	From 1g, 1 and A2.20.1A Existence of Product and Operator	$A^m = \Pi_{i=1}{}^m \lozenge B_i$
3	From 1, 2 and A2.3.4 Substitution	$A^m = \Pi_{i=1}{}^m \lozenge C_i$
∴	From 3 and A2.20.1A Existence of Product and Operator	$A^m = C^m$

Theorem 2.21.2 Multiplicative Sets: Transitivity for Rank-m

1g 2g	Given	$A^m \cong B^m$ $B^m \cong C^m$
Steps	Hypothesis	$A^m \cong C^m$
1	From 2g and A2.20.3A Congruency	$B^i \cong C^i$ for all i
2	From 1g, 1 and A2.20.1A Existence of Product and Operator	$A^m = \Pi_{i=1}{}^m \lozenge B_i$
3	From 1, 2 and A2.3.4 Substitution	$A^m = \Pi_{i=1}{}^m \lozenge C_i$
∴	From 3 and A2.20.1A Existence of Product and Operator	$A^m \cong C^m$

Theorem 2.21.3 Multiplicative Sets: Equality

1g	Given	$F^m = G^m$
Steps	Hypothesis	$A^p \lozenge B^q = C^p \lozenge D^q$
1	From A2.20.1 Existence of Product and Operator	$F^m = A^p \lozenge B^q$
2	From A2.20.1 Existence of Product and Operator	$G^m = C^p \lozenge D^q$
∴	From 1g, 1, 2 and T2.21.1 Multiplicative Sets: Substitution for Rank-m	$A^p \lozenge B^q = C^p \lozenge D^q$

Theorem 2.21.4 Multiplicative Sets: Parsing of Rank-m

Steps	Hypothesis	$A^m = A^p \lozenge A^{m-p}$
1	From A2.20.2A Equality	$A^m = A^m$
2	From 2 and A2.20.1A Existence of Product and Operator	$A^m = \Pi_{i=1}{}^m \lozenge A_i$
3	From 2 and splitting product into parts of p and m − p products	$A^m = \Pi_{j=1}{}^p \lozenge A_j \lozenge \Pi_{k=p+1}{}^m \lozenge A_k$
∴	From 3 and A2.20.1A Existence of Product and Operator	$A^m = A^p \lozenge A^{m-p}$

Theorem 2.21.5 Multiplicative Sets: Addition of Rank-p+q

1g	Given	$A_k = C_{k-p}$
2g		$A_j = B_j$
3g		$h = k - p$
4g		$m - p = q$
Steps	**Hypothesis**	$A^m = B^p \lozenge C^q \qquad \text{for } m = p + q$
1	From T2.21.4 Multiplicative Sets: Parsing of Rank-m	$A^m = A^p \lozenge A^{m-p}$
2	From 2 and A2.20.1A Existence of Product and Operator	$A^m = \Pi_{j=1}{}^p \lozenge A_j \lozenge \Pi_{k=p+1}{}^m \lozenge A_k$
3	From 1g, 2g, 2 and T2.21.1 Multiplicative Sets: Substitution for Rank-m	$A^m = \Pi_{j=1}{}^p \lozenge B_j \lozenge \Pi_{k=p+1}{}^m \lozenge C_{k-p}$
4	From 3 and re-indexing the count from k to h	$A^m = \Pi_{j=1}{}^p \lozenge B_j \lozenge \Pi_{h=1}{}^{m-p} \lozenge C_h$
5	From 4 and A2.20.1A Existence of Product and Operator	$A^m = B^p \lozenge C^{m-p}$
∴	From 4g, 5, A4.2.3 Substitution and T4.3.6A Equalities: Reversal of Left Cancellation by Addition	$A^m = B^p \lozenge C^q \qquad \text{for } m = p + q$

Theorem 2.21.6 Multiplicative Sets: Union of Supersets

Steps	Hypothesis	$\mathcal{A}^m = \mathcal{B}^p \cup \mathcal{C}^q \qquad \text{for } m = p + q$
1	From T2.21.5 Multiplicative Sets: Addition of Rank-p+q	$A^m = B^p \lozenge C^q \qquad \text{for } m = p + q$
2	From 1, D2.19.4 Multiplicative Set Superset, D2.7.1A Union of Sets and A2.8.1 Closure Union	$\{ A^m \} = \{ B^p \} \cup \{ C^q \}$
∴	From 1, 2 and D2.19.4 Multiplicative Set Superset	$\mathcal{A}^m = \mathcal{B}^p \cup \mathcal{C}^q \qquad \text{for } m = p + q$

Theorem 2.21.7 Multiplicative Sets: Parsed as Subsets

Steps	Hypothesis	$\mathcal{B}^p \subseteq \mathcal{A}^m$	$\mathcal{C}^q \subseteq \mathcal{A}^m$
1	From T2.21.6 Multiplicative Sets: Union of Supersets	$\mathcal{A}^m = \mathcal{B}^p \cup \mathcal{C}^q$ for $m = p + q$	
2	From 1 and T4.3.3 Equalities: Right Cancellation by Addition	$q = m - p$	$p = m - q$
3	From 2 and A2.20.1C Existence of Product and Operator, is a Countable Number	$0 < q$	$0 < p$
4	From 2, 3 and A4.2.3 Substitution	$0 < m - p$	$0 < m - q$
5	From 3 and D4.1.7 Greater Than Inequality [>]	$p < m$	$q < m$
6	From 1	$\mathcal{B}^p \in \mathcal{A}^m$	$\mathcal{C}^q \in \mathcal{A}^m$
$\therefore$	From 4, 5 and D2.13.1B, C Subset of Sets	$\mathcal{B}^p \subseteq \mathcal{A}^m$	$\mathcal{C}^q \subseteq \mathcal{A}^m$

Theorem 2.21.8 Multiplicative Sets: Uniqueness for Sets

1g	Given	$(A^p, B^p, C^q, D^q, F^q) \in M$
2g		$A^p = C^q$
3g		$B^p = D^q$

Steps	Hypothesis	$A^p \lozenge C^q = B^p \lozenge D^q$
1	From 1g and A2.20.4 Closure for Multiplicative Sets	$A^p \lozenge C^q = F^p$
2	From 2g, 3g and T2.21.1 Multiplicative Sets: Substitution for Rank-m	$B^p \lozenge D^q = F^p$
$\therefore$	From 1, 2 and T2.21.1 Multiplicative Sets: Substitution for Rank-m	$A^p \lozenge C^q = B^p \lozenge D^q$

Theorem 2.21.9 Multiplicative Sets: Commutative for Sets

Steps	Hypothesis	$A^p \lozenge B^q = B^q \lozenge A^p$
1	From A2.21.3 Multiplicative Sets: Equality	$A^p \lozenge B^q = A^p \lozenge B^q$
2	From 1 and A2.20.1A Existence of Product and Operator	$A^p \lozenge B^q = \Pi_{i=1}^p \lozenge A_i \lozenge \Pi_{i=1}^q \lozenge B_i$
3	From 2, A2.20.5 Commutative by commuting by pairs $p(p+q)$ times	$A^p \lozenge B^q = \Pi_{i=1}^p \lozenge B_i \lozenge \Pi_{i=1}^q \lozenge A_i$
$\therefore$	From 3 and A2.20.1A Existence of Product and Operator	$A^p \lozenge B^q = B^q \lozenge A^p$

Theorem 2.21.10 Multiplicative Sets: Binary Distribution of Rank

Steps	Hypothesis	$(A \Diamond B)^m = A^m \Diamond B^m$
1	From A2.20.2A Equality	$(A \Diamond B)^m = (A \Diamond B)^m$
2	From 1 and A2.20.1A Existence of Product and Operator	$(A \Diamond B)^m = \prod^m \Diamond (A_i \Diamond B_i)$
3	From 2, A2.20.5 Commutative and commuting sets two at a time for [m] factors requires $_mC_2$ combinations separate sets.	$(A \Diamond B)^m = \prod_{i=1}^m \Diamond A_i \Diamond \prod_{i=1}^m \Diamond B_i$
∴	From 3 and A2.20.1A Existence of Product and Operator	$(A \Diamond B)^m = A^m \Diamond B^m$

The two above theorems can be extended to multinomial of n-quantities the number of commuting factors becomes m^n and $_mC_n$ respectively. These theorems of generality are not developed in this treatise because they are simply not used.

Theorem 2.21.11 Multiplicative Sets: Association for Sets and Rank

1g 2g	Given	$D^u = B^q \Diamond C^r \qquad u = q + r$ $E^v = A^p \Diamond B^q \qquad v = p + q$
Steps	Hypothesis	$A^p \Diamond (B^q \Diamond C^r) = (A^p \Diamond B^q) \Diamond C^r$ for $p + (q + r) = (p + q) + r$
1	From A2.20.2A Equality	$F^m = F^m$
2	From 1 and T2.21.5 Multiplicative Sets: Addition of Rank-p+q	$A^p \Diamond D^u = E^v \Diamond C^r$ for $m = p + u$ and $m = v + r$
3	From 1g, 2g, 2, T2.21.1 Multiplicative Sets: Substitution for Rank-m and A4.2.3 Substitution	$A^p \Diamond (B^q \Diamond C^r) = (A^p \Diamond B^q) \Diamond C^r$ for $m = p + (q + r)$ and $m = (p + q) + r$
∴	From 3 and A4.2.2 Equality	$A^p \Diamond (B^q \Diamond C^r) = (A^p \Diamond B^q) \Diamond C^r$ for $p + (q + r) = (p + q) + r$

Theorem 2.21.12 Multiplicative Sets: Reduction by Subtraction of Rank-m–q

Steps	Hypothesis	$A^m / C^q = B^p$ for $m - q = p$ and $C^q \subseteq \mathcal{A}^m$
1	From T2.21.5 Multiplicative Sets: Addition of Rank-p+q and T2.21.6 Multiplicative Sets: Union of Supersets	$A^m = B^p \lozenge C^q$ for $m = p + q$ and $\mathcal{A}^m = \mathcal{B}^p \cup C^q$
2	From 1, A2.20.6 Set as a Number Reciprocal, T2.21.8 Multiplicative Sets: Uniqueness for Sets	$A^m / C^q = (B^p \lozenge C^q) / C^q$
3	From 2 and T2.21.11 Multiplicative Sets: Association for Sets and Rank	$A^m / C^q = B^p \lozenge (C^q / C^q)$
4	From 3, A2.20.1A Existence of Product and Operator, A2.20.8 Inverse and A4.2.12 Identity Multp	$A^m / C^q = B^p \lozenge 1$
5	From 4, A2.20.1A Existence of Product and Operator, and A2.20.7 Identity	$A^m / C^q = B^p$
6	From 1, A4.2.6 Associative Add and T4.3.3A Equalities: Right Cancellation by Addition	$m - q = p$
7	From 1, A2.20.6C Set as a Number Reciprocal and D4.1.8 Less Then Inequality [$<$]	$0 < m - q$
8	From 6, 7 and A4.2.3 Substitution	$0 < p$
$\therefore$	From 5, 7, 8 and A2.20.6D Set as a Number Reciprocal	$A^m / C^q = B^p$ for $m - q = p > 0$ and $C^q \subseteq \mathcal{A}^m$

Section 2.22 *Definitions for Transformation of Sets*

The concept of transformational mappings can be generalized into the following definitions, axioms and theorems.

Definition 2.22.1 Transformation between sets

In general, given a set of elements A and B, not necessarily of the same number of elements in either set, a way can be found to map one set of elements onto another set this operation of converting one set to another is called a transformation <T> of the set A to B. It is denoted by

$$<T>: A \rightarrow B \qquad\qquad\qquad \text{EQA}$$

The above expression is read as the transformation of A to B.

How the transformation takes place, which is the operation of mapping point for point between sets is called the function of mapping or simply function [f]. It is denoted by

$$f<T>: A \rightarrow B \qquad\qquad\qquad \text{EQB}$$

The above expression is read as; function transforms set A to set B.

Definition 2.22.2 Domain of a Function

Given f<T>: A → B, A is said to be the domain of the function.

Definition 2.22.3 Codomain of a Function

Given f<T>: A → B, B is said to be the codomain of the function.

Definition 2.22.4 Image of a Domain Element

Iff (a) ∈ A, the domain, than the element in the codomain (b) ∈ B is assigned to [a] under transformation of f<T>A → B this is called the image of [a] and denoted by

$$f(a) = b$$

The above expression is read as; the function [f] of [a] is equal to [b] or [b] is the image of [a].

Definition 2.22.5 Range of a Function

Not every element in A may be mapped into every element in B forming an incomplete set of images. The range of [f] consists precisely of those elements mapped from A into B and is denoted by

range: elements in set A={a_1, a_2, ..., a_n}, f(A) = B, image: elements in set B = {b_1, b_2, ..., b_m}

The above expression is read as the function [f] of [A] having a range of elements spanning [B].

Definition 2.22.6 One to One Functions

The function [f] is called a one to one (one-one or 1–1) function if different elements in B are assigned to different elements in A, that is, if there is a one-one correspondence between elements in A having the same image in B, as an example.

If f(x) = x^2 then f(2) = 4 and f(–2) = 4, hence is ~1–1, because not all the range is found in the image.
If f(x) = x^3 then f(2) = 8 and f(–2) = –8, hence is 1–1, because all range is to be found in the image.

Definition 2.22.7 Onto Functions

The function [f] is called onto function if the domain is a subset of the range or at least every element in some way are images of some or all elements in B.

Given f(A) $\supseteq$ B implying f^{-1}(B) $\supseteq$ A

test

if f^{-1}(f(A)) $\equiv$ A TRUE then onto
if f^{-1}(f(A)) $\not\equiv$ A FALSE then ~onto

The above expression is read as; the function [f] of A is a subset of B, or equal set of B likewise [f^{-1}] of B is a subset of A, or equal set of A, as an example.

If f(x) = x^3 then f(2) = 8 and f(–2) = –8, all elements in the image are found in the range.
If f(x) = x^2 then f(2) = 4 and f(–2) = 4, not all elements in the image are found in the range.

Definition 2.22.8 Constant Functions

The function [f] is called a constant function if the domain is an image of one element in the range.

f<T>:A $\rightarrow$ (only one element [b]) $\in$B

The above expression is read as the function [f] of A maps all elements in A into a singular or constant element in B.

Definition 2.22.9 Invariance Under Transformation

If properties are associated with sets under transformation C = P $\wedge$ A and D = P $\wedge$ B for P = {• p_i •}, such that f<T>: C $\rightarrow$ D and f<T>: A $\rightarrow$ B holds valid than P is left unaltered under transformation and set P is said to be invariant.

An example of invariant properties, the scalar quantity of magnitude remaining the same, or a directed quantity (vector or tensor) maintains the same direction between any two given points between manifolds.

Section 2.23 *Axioms on Transformation of Sets*

In this presentation on transformation of sets, a new tack is taken to develop this theory. In classical developments, it is always assumed that transformations meet the format of groups from which an orderly axiomatic system is developed. Here the axiomatic system is developed from first principles starting with set theory's notation of functional mapping, but extended to include the method (operator) of transformation. Acknowledging the type operation on transformation makes the notation more general and hopefully more universal with all the different possible operators that one can imagine. With this new notation, we start outside the notion of a group axiomatic system and those axioms now become theorems in this approach.

Axiom 2.23.1 Existence of Transformation

$\exists|$ a universal set of transformations U_T, such that all quantities of transformation $(\bullet f_i, <T_i>\bullet) \in U_T$ for all possible transformations.

Iff such quantities exist than set A maps to set B by some method, or process, or function [f] of correspondence to the operation of [T] transforming one set to another.

$$f<T>: A \rightarrow B$$

It is read as; function [f] of the transform $<T>$ set mapped from set A to set 2.

Axiom 2.23.2 Equivalency of Transformation

If $[f<T_f>]$ and $[g<T_g>]$ are functions under transformations T_f and T_g defined on the same domain-A and if $f(a) = g(a)$ for every $(a) \in A$, then the functions f and g are equal, also their transformations T_f and T_g equal one another and onto the same codomain-B, denote by

$$f<T_f>: A \rightarrow B \text{ and } g<T_g>: A \rightarrow B \qquad \text{Equation A}$$
$$f \equiv g \qquad \text{Equation B}$$
$$T_f \equiv T_g \qquad \text{Equation C}$$

It is read as; the function [f] is equal to [g] and the transformation $<T_f>$ set is equal to the transformation $<T_g>$ set.

Axiom 2.23.3 Substitution of Transformation with Closure

If $[f<T_f>]$ and $[g<T_g>]$ are functions under the transformations T_f and T_g defined on alternate domains-A and B, then the functions f and g can map alternately from A to B and B to C and their respective codomain,

$$f<T_f>:A \rightarrow B \text{ and} \qquad \text{Equation A}$$
$$g<T_g>:B \rightarrow C \text{ therefore} \qquad \text{Equation B}$$
$$g<T_g>:(f<T_f>:A) \rightarrow C \text{ where} \qquad \text{Equation C}$$
$$(f \bullet g)<T_f \bullet T_g>:A \equiv g<T_g>:(f<T_f>:A) \qquad \text{Equation D}$$

Creating a new function of transformation under the substitution operator [$\bullet$] can be rewritten for simplification

$$h<T_h>:A \rightarrow C \text{ where} \qquad \text{Equation D}$$
$$h \equiv f \bullet g \qquad \text{Equation E}$$
$$T_h \equiv T_f \bullet T_g \qquad \text{Equation F}$$
$$\text{for } (h, <T_h>) \in U_T.$$

Axiom 2.23.4 Composition of Transformation with Closure

$f<T_f>:A \rightarrow B$ and   Equation A

$g<T_g>:B \rightarrow C$ by the product   Equation B

$(f \circ g)<T_f \circ T_g>:A \rightarrow C$ Equation C

Creating a new function of transformation under the composition operator [$\circ$] can be rewritten for simplification

$h<T_h>:A \rightarrow C$ where   Equation D

$h \equiv f \circ g$ Equation E

$T_h \equiv T_f \circ T_g$ Equation F

for $(h, <T_h>) \in U_T$.

It is read as; the product of functions [f] and [g] of the product transformations f and g mapped from the set A to the set C.

Axiom 2.23.5 Identity Transformation

Set A maps to itself A by some identity method, or process, or function [e] of correspondence the net operation [E] transforms it back.

$e<E>: A \rightarrow A$

It is read as; identity function [e] of the identity transform <E> set mapped from set A back to set A.

Axiom 2.23.6 Inverse Transformation

Given a set A such that it maps onto a set B by some process or function [f] under transformation <T> then $\exists|$ an inverse transformation iff set B can be mapped back to the original set A by some inverse function denoted by [f^{-1}] and having a corresponding transformation <T^{-1}>.

$f^{-1}<T^{-1}>: B \rightarrow A$

It is read as; inverse function [f^{-1}] of the transform inverse < T^{-1}> set mapped from set B back to set A.

Section 2.24 *Theorems on Transformations of Sets*

Theorem 2.24.1 Uniqueness of Transformation

1g 2g 3g	Given	$(f{<}T_f{>}, g{<}T_g{>}, h{<}T_h{>}, r{<}T_r{>}) \in U_T$ $f{<}T_f{>}: A \rightarrow g{<}T_g{>}: A \rightarrow B$ $h{<}T_h{>}: B \rightarrow r{<}T_r{>}: B \rightarrow C$	$f \equiv g$ $h \equiv r$	$T_f \equiv T_g$ $T_h \equiv T_r$
Steps	Hypothesis	$f \circ h \equiv g \circ r$	$T_f \circ T_h \equiv T_g \circ T_r$	
1	From 1g and A2.23.4C Composition of Transformation with Closure	$(f \circ h){<}T_f \circ T_h{>}:A \rightarrow F$		
2	From 1g and A2.23.4C Composition of Transformation with Closure	$(g \circ r){<}T_g \circ T_r{>}:A \rightarrow F$		
3	From 1, 2 and A2.3.2A Equality	$(f \circ h){<}T_f \circ T_h{>}:A \rightarrow (g \circ r){<}T_g \circ T_r{>}:A$		
∴	From 3 and A2.23.2(B,C) Equivalency of Transformation	$f \circ h \equiv g \circ r$	$T_f \circ T_h \equiv T_g \circ T_r$	

Theorem 2.24.2 Chaining Images, Functions within Functions

1g 2g 3g	Given	$(a) \in A, (b) \in B$ and $(c) \in C$ $f{<}T_f{>}: A \rightarrow B$ $g{<}T_g{>}: B \rightarrow C$
Steps	Hypothesis	$g(f(a)) = c$
1	From 1g, 2g and A2.22.4 Image of a Domain Element	$f(a) = b$
2	From 1g, 3g and A2.22.4 Image of a Domain Element	$g(b) = c$
∴	From 1, 2, D2.2.11 Unitary Set and A2.3.4 Substitution	$g(f(a)) = c$

Theorem 2.24.3 Equivalency of Composition and Chained Images

1g 2g	Given	$(a) \in A, (b) \in B$ and $(c) \in C$ $f{<}T_f{>}: A \rightarrow B$ and $g{<}T_g{>}: B \rightarrow C$
Steps	Hypothesis	$(f \circ g)(a) = g(f(a))$
1	From T2.24.2 Chaining Images, Functions within Functions	$g(f(a)) = c$
2	From 1, D2.2.11 Unitary Set and A2.3.2B Equality	$c = g(f(a))$
3	From 2g and A2.23.3C Composition of Transformation with Closure	$(f \circ g){<}T_f \circ T_g{>}:A \rightarrow C$
4	From 1g, 3 and A2.23.3E Composition of Transformation with Closure	$(f \circ g)(a) = c$
∴	From 2, 4, D2.2.11 Unitary Set and A2.3.4 Substitution	$(f \circ g)(a) = g(f(a))$

Theorem 2.24.4 Inverse with Respect to Composition

1g 2g	Given	(a) $\in$ A and (b) $\in$ B f<T>: A $\rightarrow$ B and f^{-1}<T^{-1}>: B $\rightarrow$ A	
Steps	Hypothesis	e<E>: A $\rightarrow$ A e = f $\circ$ f^{-1}	E = T $\circ$ T^{-1}
1	From 1g, 2g and A2.23.3C Composition of Transformation with Closure	(f $\circ$ f^{-1})<T $\circ$ T^{-1}>:A $\rightarrow$ A	
2	From A2.23.5 Identity Transformation	e<E>: A $\rightarrow$ A	
∴	From 1, 2 and A2.23.2A Equivalency of Transformation From 1, 2 and A2.23.2B Equivalency of Transformation	e<E>: A $\rightarrow$ A e = f $\circ$ f^{-1}	E = T $\circ$ T^{-1}

Theorem 2.24.5 Right Handed Identity with Respect to Composition

1g 2g	Given	(a) $\in$ A and (b) $\in$ B f<T$_f$>: A $\rightarrow$ B	
Steps	Hypothesis	f $\circ$ e = f	T$_f$ $\circ$ E = T$_f$
1	From 1g and A2.23.5 Identity Transformation	e<E>: B $\rightarrow$ B	
2	From 2g, 1 and A2.23.3C Composition of Transformation with Closure	(f $\circ$ e)<T$_f$ $\circ$ E>:A $\rightarrow$ B	
3	From 2g, 2 and A2.23.3E Composition of Transformation with Closure	f $\circ$ e = f	
4	From 2g, 2 and A2.23.3F Composition of Transformation with Closure	T$_f$ $\circ$ E = T$_f$	
∴	From 3 From 4	f $\circ$ e = f	T$_f$ $\circ$ E = T$_f$

Theorem 2.24.6 Left Handed Identity with Respect to Composition

1g 2g	Given	(a) $\in$ A and (b) $\in$ B f<T$_f$>: A $\rightarrow$ B	
Steps	Hypothesis	e $\circ$ f = f	E $\circ$ T$_f$ = T$_f$
1	From 1g and A2.23.5 Identity Transformation	e<E>: A $\rightarrow$ A	
2	From 1, 2g and A2.23.3C Composition of Transformation with Closure	(e $\circ$ f)<E $\circ$ T$_f$ >:A $\rightarrow$ B	
∴	From 2 and A2.23.3E Composition of Transformation with Closure From 2 and A2.23.3F Composition of Transformation with Closure	e $\circ$ f = f	E $\circ$ T$_f$ = T$_f$

Theorem 2.24.7 Identity with Respect to Composition[2.21.1]

Steps	Hypothesis	$f \circ e = e \circ f = f$	$T_f \circ E = E \circ T_f = T_f$
∴	From T2.24.5 Right Handed Identity with Respect to Composition and T2.24.6 Left Handed Identity with Respect to Composition	$f \circ e = e \circ f = f$	$T_f \circ E = E \circ T_f = T_f$

Theorem 2.24.8 Composition Chains to the Right

1g 2g 3g 4g	Given	(a) $\in$ A, (b) $\in$ B and (c) $\in$ C f<T_f>: A $\to$ B g<T_g>: B $\to$ C Φ general representation for any set under transformation
Steps	Hypothesis	g<T_g>: (f<T_f>: Φ) $\equiv$ (f $\circ$ g)<$T_f \circ T_g$>: Φ Composition chains to the right
1	From 2g, 3g and A2.3.4 Substitution	g<T_g>: (f<T_f>: A) $\to$ C
2	From 2g, 3g and A2.23.3C Composition of Transformation with Closure	(f $\circ$ g)<$T_f \circ T_g$>:A $\to$ C
∴	From 4g, 2, 3 and A2.3.2A Equality	g<T_g>: (f<T_f>: Φ) $\equiv$ (f $\circ$ g)<$T_f \circ T_g$>: Φ Composition chains to the right

Theorem 2.24.9 Composition Does Not Chain to the Left

1g 2g 3g 4g	Given	(a) $\in$ A, (b) $\in$ B and (c) $\in$ C g<T_g>: B $\to$ C f<T_f>: A $\to$ B A $\neq$ C	
Steps	Hypothesis	f<T_f>: (g<T_g>: Φ) $\neq$ (g $\circ$ f)<$T_g \circ T_f$>:Φ for A $\neq$ C ~∃ a g $\circ$ f to map A onto C	~∃ a $T_g \circ T_f$ either
1	From 2g, 3g and A2.23.3C Composition of Transformation with Closure	(g $\circ$ f)<$T_g \circ T_f$>:B $\nrightarrow$ B not defined by any axiom.	
2	From 2g, 3g, 4g and A2.3.4 Substitution	f<T_f>: (g<T_g>: B) $\nrightarrow$ B for A $\neq$ C substitution cannot be carried out, hence a contradiction.	
3	From 1, 2 and A2.3.2A Equality	f<T_f>: (g<T_g>: Φ) $\neq$ (g $\circ$ f)<$T_g \circ T_f$>:Φ for A $\neq$ C equality does not hold, so composition to the left likewise does not hold.	
4	From 4g, 3 and A2.23.3(E, F) Composition of Transformation with Closure	g $\circ$ f, hence no function to map from A onto C exists	$T_g \circ T_f$, and likewise no transformation
∴	From 3 and 4	f<T_f>: (g<T_g>: Φ) $\neq$ (g $\circ$ f)<$T_g \circ T_f$>:Φ for A $\neq$ C ~∃ a g $\circ$ f to map A onto C	~∃ a $T_g \circ T_f = \varnothing$

Theorem 2.24.10 Composition is Non-Commutative[2.18.1]

1g 2g 3g 4g	Given	(a) $\in$ A, (b) $\in$ B and (c) $\in$ C $g<T_g>$: B $\rightarrow$ C $f<T_f>$: A $\rightarrow$ B A $\neq$ C
Steps	Hypothesis	$(f \circ g)<T_f \circ T_g>$: $\Phi \neq (g \circ f)<T_g \circ T_f>$:$\Phi$ for A $\neq$ C hence, non-commutative $f \circ g \neq g \circ f$ $\quad\mid\quad T_f \circ T_g \neq T_g \circ T_f$
1	From T2.24.8 Composition Chains to the Right	$g<T_g>$: $(f<T_f>$: $\Phi) \equiv (f \circ g)<T_f \circ T_g>$: Φ
2	From 4g and T2.24.9 Composition Does Not Chain to the Left	$f<T_f>$: $(g<T_g>$: $\Phi) \neq (g \circ f)<T_g \circ T_f>$:$\Phi$ for A $\neq$ C
$\therefore$	From 1, 2, A2.3.2 Equality and A2.23.3(E, F) Composition of Transformation with Closure	$(f \circ g)<T_f \circ T_g>$: $\Phi \neq (g \circ f)<T_g \circ T_f>$:$\Phi$ for A $\neq$ C hence, non-commutative $f \circ g \neq g \circ f$ $\quad\mid\quad T_f \circ T_g \neq T_g \circ T_f$

Theorem 2.24.11 Composition that Does Chain to the Left

1g 2g 3g 4g	Given	(a) $\in$ A, (b) $\in$ B and (c) $\in$ C $g<T_g>$: B $\rightarrow$ C $f<T_f>$: A $\rightarrow$ B A = C
Steps	Hypothesis	$f<T_f>$: $(g<T_g>$: $\Phi) = (g \circ f)<T_g \circ T_f>$:$\Phi$ for A = C $\exists$ a $g \circ f$ to map A onto C $\quad\mid\quad \exists$ a $g \circ f$ to map A onto C
1	From 2g, 3g, 4g, A2.3.4 Substitution and A2.23.3C Composition of Transformation with Closure	$(g \circ f)<T_g \circ T_f>$:B $\rightarrow$ B
2	From 2g, 3g, 4g and A2.3.4 Substitution	$f<T_f>$: $(g<T_g>$: B) $\rightarrow$ B for A = C
3	From 1, 2 and A2.3.2A Equality	$f<T_f>$: $(g<T_g>$: $\Phi) = (g \circ f)<T_g \circ T_f>$:$\Phi$ for A = C
4	From 4g, 3 and A2.23.3(E, F) Composition of Transformation with Closure	$g \circ f$ there exists a function to map from A onto C $\quad\mid\quad T_g \circ T_f$, and likewise there exists a transformation to map from A onto C
$\therefore$	From 3 and 4	$f<T_f>$: $(g<T_g>$: $\Phi) = (g \circ f)<T_g \circ T_f>$:$\Phi$ for A = C $\exists$ a $g \circ f$ to map A onto C $\quad\mid\quad \exists$ a $T_g \circ T_f \neq \emptyset$

Theorem 2.24.12 Composition is Commutative[2.18.1]

1g	Given	(a) ∈ A, (b) ∈ B and (c) ∈ C	
2g		f<T_f>: A → B	
3g		g<T_g>: B → C	
4g		A = C	
Steps	Hypothesis	(f ° g)<T_f ° T_g>: Φ = (g ° f)<T_g ° T_f>:Φ for A = C	f ° g = g ° f = e
		f ° g = g ° f = e	
1	From 2g, 3g, T2.24.8 Composition Chains to the Right	(f ° g)<T_f ° T_g>:A → C	
2	From 3g, 2g, 4g and T2.24.11 Composition that Does Chain to the Left	(g ° f)<T_g ° T_f>:B → B for A = C	
3	From 1, A2.23.3(E, F) Composition of Transformation with Closure, A2.23.5 Identity Transformation and A2.6.2A Equality	f ° g = e	T_f ° T_g = E
4	From 2, A2.23.3(E, F) Composition of Transformation with Closure, A2.23.5 Identity Transformation and A2.3.2A Equality	g ° f = e	T_g ° T_f = E
∴	From 3, 4 and A2.23.3C Composition of Transformation with Closure	(f ° g)<T_f ° T_g>: Φ = (g ° f)<T_g ° T_f>:Φ for A = C / f ° g = g ° f = e	T_f ° T_g = T_g ° T_f = E

Theorem 2.24.13 Associative with Respect to Composition[2.21.1]

1g	Given	(a) ∈ A, (b) ∈ B, (c) ∈ C and (d) ∈ D	
2g		f<T_f>: A → B	
3g		g<T_g>: B → C	
4g		h<T_h>: C → D	
Steps	Hypothesis	(f ° g) ° h = f ° (g ° h)	(T_f ° T_g) ° T_h = T_f ° (T_g ° T_h)
1	From 2g, 3g and A2.23.3C Composition of Transformation with Closure	(f ° g)<T_f ° T_g>:A → C	
2	From 1, 4g and A2.23.3C Composition of Transformation with Closure	((f ° g) ° h)< (T_f ° T_g) ° T_h >:A → D	
3	From 3g, 4g and A2.23.3C Composition of Transformation with Closure	(g ° h)<T_g ° T_h>:B → D	
4	From 2g, 3 and A2.23.3C Composition of Transformation with Closure	(f ° (g ° h))< T_f ° (T_g ° T_h) >:A → D	
∴	From 2, 4 and A2.23.2B Equivalency of Transformation / From 2, 4 and A2.23.2C Equivalency of Transformation	(f ° g) ° h = f ° (g ° h)	(T_f ° T_g) ° T_h = T_f ° (T_g ° T_h)

Observation 2.24.1 Operating on The Properties of a Set under Transformation

While a set is not a proposition the properties of a set are, hence logical functions can operate on them. Likewise mapping ($\rightarrow$) one set to another is an implied implication ($\rightarrow$). These two symbols should not be confused. Logical operations can be implemented over sets as long as it is understood that they are acting on the properties of the set.

P op (f<T>: A $\rightarrow$ B) $\rightarrow$ (f<T>: P op A $\rightarrow$ P op B)

Theorem 2.24.14 Distributing Property by Logical AND Under Transformation

1g	Given	(a) $\in$ A, (b) $\in$ B, (c) $\in$ C and (d) $\in$ D
2g		f<T>: A $\rightarrow$ B
3g		f<T>: C $\rightarrow$ D
4g	for	C = P $\wedge$ A
5g	for	D = P $\wedge$ B
6g		P property of sets A and B under transformation
Steps	Hypothesis	{P $\wedge$ (f<T>: A $\rightarrow$ B)} $\rightarrow$ {f<T>: (P $\wedge$ A) $\rightarrow$ (P $\wedge$ B) }
1	From 2g and 6g acting on the transformation	f<T>: P $\wedge$ (A $\rightarrow$ B)
2	From 3g, 4g, 5g and A2.6.4 Substitution	f<T>: (P $\wedge$ A) $\rightarrow$ (P $\wedge$ B)
3	From 1, 2 and Lx3.7.5.1 DAI	$\dfrac{P \wedge (A \rightarrow B)}{\therefore (P \wedge A) \rightarrow (P \wedge B)}$
$\therefore$	From 1, 2, 3, A3.7.9 Therefore as an Implication and O2.24.1 Operating on The Properties of a Set under Transformation	{P $\wedge$ (f<T>: A $\rightarrow$ B)} $\rightarrow$ {f<T>: (P $\wedge$ A) $\rightarrow$ (P $\wedge$ B) }

Theorem 2.24.15 Distributing Property by Logical OR Under Transformation

1g	Given	(a) $\in$ A, (b) $\in$ B, (c) $\in$ C and (d) $\in$ D
2g		f<T>: A $\rightarrow$ B
3g		f<T>: C $\rightarrow$ D
4g	for	C = P $\vee$ A
5g	for	D = P $\vee$ B
6g		P property of sets A and B under transformation
Steps	Hypothesis	{P $\vee$ (f<T>: A $\rightarrow$ B)} $\rightarrow$ {f<T>: (P $\vee$ A) $\rightarrow$ (P $\vee$ B) }
1	From 2g and 6g acting on the transformation	f<T>: P $\vee$ (A $\rightarrow$ B)
2	From 3g, 4g, 5g and A2.6.4 Substitution	f<T>: (P $\vee$ A) $\rightarrow$ (P $\vee$ B)
3	From 1, 2 and Lx3.7.35 DOI	$\dfrac{P \vee (A \rightarrow B)}{\therefore (P \vee A) \rightarrow (P \vee B)}$
$\therefore$	From 1, 2, 3, A3.7.9 Therefore as an Implication and O2.24.1 Operating on The Properties of a Set under Transformation	{P $\vee$ (f<T>: A $\rightarrow$ B)} $\rightarrow$ {f<T>: (P $\vee$ A) $\rightarrow$ (P $\vee$ B) }

Theorem 2.24.16 Distributing Property by Logical XOR Under Transformation

1g	Given	$(a) \in A, (b) \in B, (c) \in C$ and $(d) \in D$
2g		$f<T>: A \rightarrow B$
3g		$f<T>: C \rightarrow D$
4g	for	$C = P \oplus A$
5g	for	$D = P \oplus B$
6g		P property of sets A and B under transformation
Steps	Hypothesis	$\{P \oplus (f<T>: A \rightarrow B)\} \rightarrow \{f<T>: (P \oplus A) \rightarrow (P \oplus B)\}$
1	From 2g and 6g acting on the transformation	$f<T>: P \oplus (A \rightarrow B)$
2	From 3g, 4g, 5g and A2.6.4 Substitution	$f<T>: (P \oplus A) \rightarrow (P \vee B)$
3	From 1, 2 and Lx3.7.36 DXOI	$\dfrac{P \oplus (A \rightarrow B)}{\therefore (P \oplus A) \rightarrow (P \oplus B)}$
$\therefore$	From 1, 2, 3, A3.7.9 Therefore as an Implication and O2.24.1 Operating on The Properties of a Set under Transformation	$\{P \oplus (f<T>: A \rightarrow B)\} \rightarrow \{f<T>: (P \oplus A) \rightarrow (P \oplus B)\}$

[2.21.1]Note: In spite of what most books say about transformation axioms, as seen in the above postulates, they are not based on Galois Groups. Since these axioms exist outside such postulates, it is possible to prove what would have been considered a Galois Axiom as theorem. The following theorems, **Inverse** T2.24.4 "Inverse with Respect to Composition", **Identity** T2.24.7 "Identity with Respect to Composition", **Non-Commutative** T2.24.10 "Composition is Non-Commutative" and **Associative** T2.24.11 "Composition that Does Chain to the Left" cannot be atomized, hence are not axioms.

Tensor Calculus & Physics: A General Treatise

Table of Contents

Tensor Calculus & Physics: A General Treatise

List of Tables

List of Figures

Chapter 3 Principles of Mathematical Proof Writing

Holmes: "You will not apply my precept," he said, shaking his head. "How often have I said to you that when you have eliminated the impossible, whatever remains, *how-ever improbable*, must be the truth? We know that he did not come through the door, the window, or the chimney. We also know that he could not have been concealed in the room, as there is no concealment possible. When, then, did he come?"

Watson: "He came through the hole in the roof!" I cried.

From Sherlock Holmes *The Sign of Four* [DOY30I pg111]

Section 3.1 Definitions of Logic

Definition 3.1.1 Atom or Atom of Logic

An **atom** being an indivisible proposition that cannot be derived by any other proposition or set of propositions **within** the system of logic[3.1.1].

Definition 3.1.2 Axiom

It is a self-evident proposition from a set of implied statements that form the bases for a deductive system. These propositions are to be assumed but not proven to prevent a logical digression. However, outside the deductive system they can have a valid foundation based on experience. Note in physics axioms are called Laws. [ADL66][SPI63]

A simple test to see if you have an axiom is by expressing it as a **property** of an object and an atom of the logical system.

Definition 3.1.3 Hypothesis (Conjecture)

A conjecture also called a **hypothesis**, may be a very deep, very meaningful and an important statement but without a proof logically validating it, it is not a proven theorem [ACZ96].

Definition 3.1.4 Theorem

A theorem is a proposition (such as a mathematical statement) whose proof is derived by deductive or inductive process of reason. The proof of a theorem is a rigorous justification of the veracity of it in such a way that it cannot be disputed by anyone who follows the rules of logic, and who accepts a set of axioms put forth as the basis for the logic system. [ACZ96]

Definition 3.1.5 Lema

A lema is a preliminary proven statement, which then leads to a more profound theorem [ACZ96].

Definition 3.1.6 Corollary

Is a proven result that follows a theorem [ACZ96].

[3.1.1]Note: Adopted as an adjective from physical science, to denote the indivisibility, or fundamentalness, for a proposition in a logical system.

Definition 3.1.7 Undefined and Defined Terms

Undefined terms are words that take their meaning from the context of the axiom they are found. They are not to be defined inside the axiomatic system to avoid a possible infinite regression of circular definitions; such as the word A defines B and B defines A. However all other terms in the axiomatic system get their meaning from these elementary terms [ADL66].

A test for when a definition should be defined as such is that a definition is a ***description*** *of an object* helping to provide clarity in a logical proof. Also, if a definition is being used in a proof an inordinate number of times as a property, it probably should be rewritten as an axiom.

Definition 3.1.8 Proof by Deductive Reasoning

Is a collection of elementary propositions and/or axioms in which an attempt is made to validate each new proposition by logically deducing it from previous axioms and theorems, furthermore to the extent that it is possible, defining the terms that are used [ADL66]. It should also be said that each affirmed premises in an unbroken logical sequence should lead also to an irrevocable valid conclusion in a deductive proof.

Definition 3.1.9 A Theory

A theory is a set of conjectures that lead to a conclusion or prediction. If the conclusion is proven incorrect than it follows that the theory is incorrect or at best some part is erroneous. A theory is a form of indirect Reasoning.

Definition 3.1.10 Truth Table Tautology

In a proof of a truth table the arguments end up with the finial column being valid, that is a column of all TRUE values. This method of proof for proving that the parts of an argument come together correctly and result in an entire column as TRUE is called an ***Tautology*** [COP68] and represented symbolically as [**T**].

Definition 3.1.11 Truth Table Inverse Tautology

In a proof of a truth table when the arguments end up with the finial column being invalid that is a column of all FALSE values, this type of proof is called an ***Inverse Tautology*** and represented symbolically as [**F**].

The Tautology and its inverse have their algebraic correspondence in the ***inverse laws*** of algebra [JOH69], for multiplication $A / A = 1$ and addition $A - A = 0$, terminating with a singular value.

Axiom 3.1.1 Simple Logical Premise

A *simple logical premise* is a permutation of the validity (T or F) of one or two variables (A, B) combined together, from a set of fundamental logical operations, such that they result in some logical conclusion f(A) or f(A, B).

A	f(A)
T	TF
F	TF

A	B	f(A, B)
T	T	TF
T	F	TF
F	T	TF
F	F	TF

Axiom 3.1.2 Logical Premise

A *truth table* is generalized, simple, logical premise made from a set of n-variables that concludes in a single *logical premise*.

A_1	...	A_i	...	A_n	$f(_\bullet A_{j\bullet})$
$[TF]_1$	...	$[TF]_i$	...	$[TF]_n$	$[TF]$

Axiom 3.1.3 Deductive Theroem

A *deductive theorem* is a sequence of logical premises $[A_j]$ bound together by an [AND] operator in a multiple product resulting in a tautology [T] that is a generalized logical premise $g(_\bullet A_{j\bullet})$.

$_\bullet f_i(_\bullet A_{j\bullet})_\bullet$	$\prod_i \wedge f_i(_\bullet A_{j\bullet}) \rightarrow g(_\bullet A_{j\bullet})$
$[TF]_i$	$[T]$

Here it is *deduced* from one-premise to another, step-by-step, till a conclusion is reached, which must also be valid, hence a tautology.

In this development the truth table is a fundamental building block in logic, however large number of multiple variables can become so cumbersome that the truth table may not be able to be manipulated. So other methods must be adopted and considered. Here a logical premise is a single column from a truth table whose veracity comes from a formula operation on the permutated variables. In classical logic a premise is just a statement having some type of provable validated associated with it then, as each premise is proven true in an unbroken chain, the implication is made that the conclusion must also be valid.

Axiom 3.1.4 Deductive System

A deductive system is a set $_L D$ containing all logically, chained, deductive theorems $^L[_\bullet G \rightarrow H_\bullet]$ forming a *complete* unbroken path called a *proof*. Deductive systems also go by the name of *deductive reasoning*.

$$^L[_\bullet G \rightarrow H_\bullet] \in {_L D}$$

Now a proof of the validity for theorem G implies a proof of validity for theorem H. This type of proof is characterized as stratified logic known as order one; $\backsigma_1{}^{[3.1.2]}$.

[3.1.2]Note: see D3.10.2 "Stratification logic order one [$\backsigma_1$]"

Gathering information (data) and looking for patterns and relations (observation) a working (tentative) hypothesis H can be formulated and when all of its premises have been validated a finial hypothesis G can be concluded.

Axiom 3.1.5 Proof by Indirect Reasoning

Indirect reasoning is where two theorems can be shown to share the same truth table $^L[_\bullet G \Leftrightarrow H_\bullet]$. First by validating verification of all of its premises in truth table H. Now the truth table of H is the truth table of the finial hypothesis G so H becomes *indirectly* a proof for the validity of G. Since this type of proof is also comprised of a path of theorems, it is also a member of the stratified logic known as order one; $\mathbf{b}_1$.

Note that this type of proof sometimes goes by the name *inductive reasoning*, but this name is not a very accurate description of the reasoning process, since inductive reasoning may include other methods of reasoning not necessarily an indirect proof, see A3.1.16 "Proof by Contradiction of Indirect Reasoning".

At the beginning of this chapter there is a famous quote from Sir Arthur Conan Doyle's story "The Sign of Four". Here the main character Sherlock Holmes is seen reasoning his way through a problem in his investigation to solve a crime. The method he uses is indirect reasoning by which he has four possible conclusions C_1, C_2, C_3 and C_4 that might satisfy the working hypothesis-H on how the criminal entered the room. One-by-one he eliminates them by testing against investigative clues that he has collected, till in his crucible of logic all falsies have been burned away, and Watson is forced to exclaim the correct conclusion, thereby establishing the finial and valid hypothesis-G. This *indirect* or *inductive* reasoning process while found in a fictional story is a very real and valid progression of thought, which makes a powerful tool in the logic of investigative science, see Section 3.14 "Newtonian Physical Deductive Systems of Logic" *the critical experiment*.

Axiom 3.1.6 Proof by Contradiction of Indirect Reasoning

Contradiction by indirect reasoning is where two theorems or premises, one the contradiction of the other, can be shown not to share the same truth table [*assume* G ≡ ~H :: *prove* G ≠ ~H :: ∴ G ≡ H *by contradiction (by default)*]. Now a proof of ~H being invalid indirectly deduces the validity of G. Since this type of proof is also comprised of a path of theorems, it is also a member of the stratified logic known as order one; $\mathbf{b}_1$. This is kind of like a proof by identity except the starting assumption is a contradiction.

Axiom 3.1.7 Proof by Counter Example

A counter example is the one case that invalidates the conclusion to an inductive proof of reasoning.

Axiom 3.1.8 Proof by Exhaustion

Proof by exhaustion is for a reasonable sized set of conjectures that can be proved as valid, case-by-case, until all conjectures are exhausted, thereby conclusively proving the conclusion. This clearly eliminates the possibility that no counter example exists.

Axiom 3.1.9 Inductive System

A inductive system is a set $_LI$ containing all logically, chained, indirect theorems $^L[_\bullet G \Leftrightarrow H_\bullet]$ forming a verity of **complete** unbroken paths called a ***proof***.

$$^L[_\bullet G \Leftrightarrow H_\bullet] \in {}_LI$$

Now a proof of the validity for theorem H implies a proof of validity for theorem G. This type of proof is characterized as stratified logic known as order one; $\mathbf{b}_1{}^{[3.1.2]}$.

Axiom 3.1.10 Universal System

A universal system $[_LU]$ is the total collection of all appropriate deductive and inductive systems constituting a complete proof:

$$_LU \equiv {_LD} \cup {_LI}$$

Tensor Calculus & Physics: A General Treatise

Section 3.2 Theorem: the Framework of a Proof

For every mathematician there is a unique way to write a theorem, for this thesis I've established my own version. This structure has followed me through school since my first exposure to it in High School Geometry and has always held me in good stead. When tackling very complex word problems, that at first glance baffle me, this is the crowbar that I use to start the process of analysis that breaks a very big problem into a set of smaller problems that are identifiable and than can be solved.

All most, all theorems in this paper are broken into four major parts,

- the given,
- the objective of the proof or hypothesis, which is also a conjecture of what might be true,
- the steps to move to the final solution or proof of the hypothesis, and
- the results or conclusion confirming the hypothesis and converting it from a conjecture to a valid proposition within the logical system.

This is all built into the following frame:

Theorem 3.2.1 The Framework of a Theorem

1g 2g 3g 4g	Given	Assumptions
Steps	Hypothesis	Objective
1 2 3 4 5	Justification List (List of justifiable arguments)	The object of justification.
$\therefore$		Objective confirmed

Theorem 3.2.2 An example from Chapter 4: Equalities: Uniqueness of Addition

1g 2g 3g	Given	$(a, b, c, d, f) \in R$ $a = b$ $c = d$
Steps	Hypothesis	$a + c = b + d \qquad$ for $a = b$ and $c = d$
1	From 1g and A4.2.4 Closure Add	$a + c = f$
2	From 2g, 3g and A4.2.3 Substitution	$b + d = f$
$\therefore$	From 1, 2 and A4.2.3 Substitution	$a + c = b + d \qquad$ for $a = b$ and $c = d$

Observe this is like a little game you are not allowed to move to the next step until you have conclusively proven that step and so on. Also, if these theorems take more than a page you are doing something wrong. Go back and break it down into smaller theorems this action will probably reveal what you did wrong. Giant monolithic theorems have the potential to hid problems; they make things difficult to follow and all in all are just bad proof writing.

Of course, this is not the only frame a theorem might be placed in. Another example would be a truth table that would provide the necessary structure. There are many such theorem frames and one should use whatever can be used to get the job done in a systematic way. So, the above frame is not an absolute, but here I rule.

Section 3.3 Diagramming Two Categorical Propositions

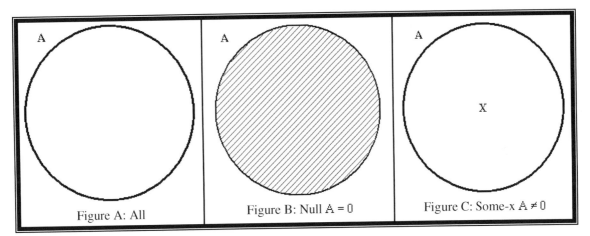

Figure 3.3.1 Venn Diagram Logical Building Blocks

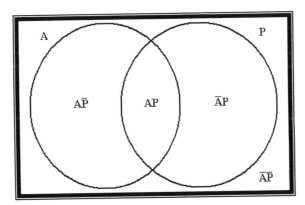

Figure 3.3.2 Product Diagram

Theorem 3.3.1 Truth Table: Categorical Syllogisms by Twos

A	P	AP	$A\bar{P}$	$\bar{A}P$	$\bar{A}\bar{P}$
T	T	T	F	F	F
T	F	F	T	F	F
F	T	F	F	T	F
F	F	F	F	F	T

From the above truth table, the following identities can be established.

Table 3.3.1 Identities of Categorical Syllogisms by Twos

Num	Identities		
L1	AP	$\equiv$	$\sim(A \rightarrow \sim P)$
L2	$A\bar{P}$	$\equiv$	$\sim(A \rightarrow P)$
L3	$\bar{A}P$	$\equiv$	$\sim(P \rightarrow A)$
L4	$\bar{A}\bar{P}$	$\equiv$	$\sim(\sim A \rightarrow P)$

The above table of identities clearly shows that the schematic products of sets are representative of the implied statement by implication from the syllogism. This allows the syllogism to be diagrammatically represented by Venn Diagrams as follows:

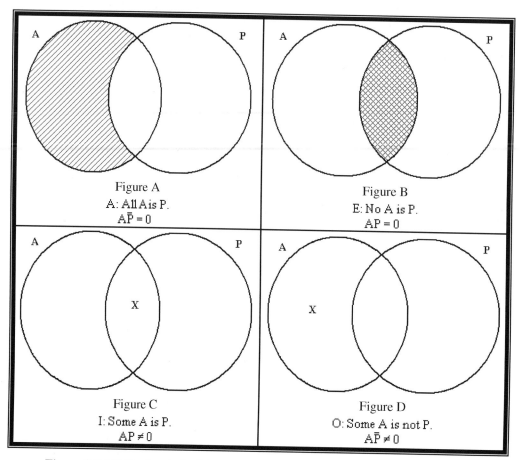

Figure A
A: All A is P.
$A\bar{P} = 0$

Figure B
E: No A is P.
$AP = 0$

Figure C
I: Some A is P.
$AP \neq 0$

Figure D
O: Some A is not P.
$A\bar{P} \neq 0$

Figure 3.3.3 Syllogism to be Diagrammatically represented by Venn Diagrams

The [x] for *some* represents the idea that there is at least one (some) element from the set that it is placed in some diagrammatic region belonging to those two sets. The grayed out area represents the absence of any element from the opposing set.

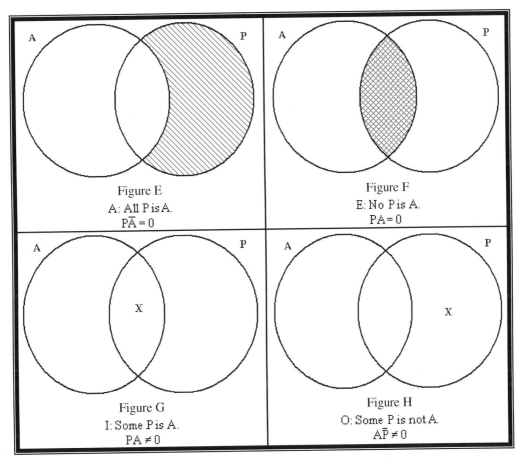

Figure 3.3.4 Diagrammatic Logical Representations

From the four standard-forms of categorical propositions can be combined to form a variety of syllogisms, however not all combinations result in a correct conclusion, but contradict one another. The following Boolean Square of Opposition represents these contradictions and should be avoided.

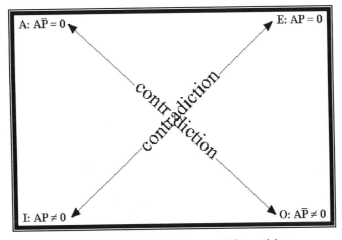

Figure 3.3.5 Boolean Square of Opposition

Section 3.4 Closed Number Systems

[3.4.1]We are all used to an open numbering system. That is a system of countable integers without bound. So we can count to 10 (ten) and beyond.

0	1	2	3	4	5	6	7	8	9	10	11	12	...

With base ten when I wish to add 9 + 1 we have a representative number to move beyond 9, 10 and 10 + 1, 11 and so on.

However there are number systems that cannot count beyond their base, as an example base three. The number system base three is made up of only three numbers 0, 1 and 2. Hence, 2 + 1 is **not defined** because no other numbers exist beyond these three integers. The only thing we can do to make sense of all of this is let 2 + 1 = 2, not 3 which is not defined. Multiplying 2 * 2 does not equal 4 but just 2. In this world of base-numbers, the number systems are closed with respect to the largest number defined.

The most fundamental numbering system is base two this is the number system that computers use because the only thing a computer can truly understand is on or off, one or zero, true or false, etc.

In the number system base, two it is only made up of 0 and 1. We can construct a table of addition and multiplication for this world of the binary as follows:

Binary Addition Table

A	B	A + B
1	1	1
1	0	1
0	1	1
0	0	0

Binary Multiplication Table

A	B	A * B
1	1	1
1	0	0
0	1	0
0	0	0

[3.4.1]Note: This section comes from a lecture series given at Evergreen College and UC Santa Cruz 2000 spring semester (quarter) respectively.

Tensor Calculus & Physics: A General Treatise

Multiplication and addition are pretty, standard operators, but in this world of closed number systems, other operators can be defined, such as the toggle operation of negation.

Binary Negation Table

A	~A
1	0
0	1

A bar can represent sometimes the negation operator by being placed over the variable, $\overline{A}$. Though this is not common, due to the lack of fonts with the capability of writing letters with a bar over the symbol. Notice that the concept of a negative number is not defined here, however the negation operator and others take the place of finding the inverse of addition and multiplication.

This is all that one can really do with binary numbers as far as operating on them. There are other operations but they can always be broken down into these three basic forms, addition, multiplication and negation.

Section 3.5 Where is the logic in all of this?

The English mathematician George Boole (1815–1864) applied the binary closed number system to solving problems in logic. In his pamphlet called Mathematical Analysis of Logic (1847) he develops algebra for logic, which to this day bears his name called Boolean Algebra. His objective was to clarify difficult Aristotelian logic. This system is used widely today playing a major role in computer logic and programming. It was and still is used as a tool to help develop sound reasoning in logical arguments. Boole's basic idea was that if the simple propositions of logic could be represented by precise symbols, the relations between two propositions could be read as precisely as an algebraic equation.

Section 3.6 What do we mean by logic?

The Greeks devised a system of organized thinking called logic. As an example:

> All men are mortal.
> Socrates is a man.
> Therefore, Socrates is mortal.

This argument is called a syllogism. It has the following logical form:

Let   [A] represent the proposition,   ***All men are mortal.***
Also let [B] represent the proposition,   ***Socrates is a man.***
With   [C] representing the concluding statement, ***Socrates is mortal.***

IF A is TRUE **AND** B is TRUE **THEN** we must concluded C is TRUE.

With this clear razor sharp logic all the fuzzy thinking that preceded this in history was laid waste. If Boole could get rid of the words and just have left the pure form of the argument he would have a powerful tool to handle very sophisticated and complex logic.

Here is what Boole saw in all of this, notice that the above **IF THEN** statement and the internal **AND** are sentence connectives and the only things of consequence tying together the propositions. Also, note the use of TRUE providing the state of validation for the proposition. With TRUE there is FALSE, which could be represented by the binary numbers 1 and 0. Consider all the possible permutations of TRUE and FALSE for the connective **AND** with its propositions A and B, becomes it is the binary multiplication operator we can construct a binary TRUTH table as follows:

Table 3.6.1 Logical AND

A	B	A **AND** B
TRUE	TRUE	TRUE
TRUE	FALSE	FALSE
FALSE	TRUE	FALSE
FALSE	FALSE	FALSE

Or we can simplify by just taking the first letters of TRUE and FALSE leaving:

A	B	A **AND** B
T	T	T
T	F	F
F	T	F
F	F	F

For the connective **OR** all possible permutations of TRUE and FALSE for the two propositions are found to be equivalent to the addition operator.

Table 3.6.2 Logical Inclusive OR

A	B	A **OR** B	
T	T	T	(Included)
T	F	T	
F	T	T	
F	F	F	

Negation of a proposition in Logic becomes known as a **NOT** operation.

Table 3.6.3 Logical NOT

A	NOT A
T	F
F	T

Looking at these truth tables gives no real intuitive understanding of **AND,** and **OR** operation. In order to have a better understand of these ideas, let's consider the **Adventures of Lassie**.

Parable 1)
Originally Lassie shared a duplex with two families so family A **AND** family B carried for her jointly. So, the possibilities for Lassie are limited.

Lassie is owned by family A **AND** Lassie is owned by family B.

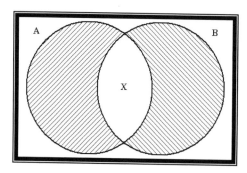

Figure 3.6.1Exclusive Intersection of Element-X between Sets

A special set diagram called a Venn Diagram can visually represent the above prepositional statement. Here the two families have a common responsibility for the **element-x** called Lassie.

Parable 2)
While riding with family A she was accidentally left behind at a rest stop when she wondered off. Don't worry a kind farmer; let's call him A took her in. Lassie was happy she got steak every two nights. Soon she found another farmer let's call B. Lassie is a very happy dog having steak every night, alternately at farmer A's house than at farmer B's house. Farmer A **AND** B find out about Lassie going over to each other's house, and being good friends agree to jointly take care of Lassie. Again, what are the possibilities for Lassie?

Lassie can be owned by farmer A, **OR**
Lassie can be owned by farmer B, **OR**
Lassie can be owned by farmer A **AND** B.

Here there are three possibilities for the element-x, Lassie, can be associated with:

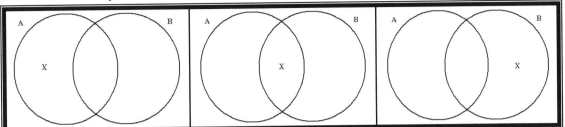

Figure 3.6.2 Inclusion of Element-X for Intersection between Sets

Parable 3)

Let's alter our story to spice things up. Farmer A hates farmer B's guts. Farmer A finds out about Lassie going over to farmer B's house and confronts farmer B with his shotgun, click-clack goes the shotgun. Farmer B not to take this laying down gets his shotgun out click-clack goes the shotgun. Farmer A and Farmer B have a Mexican standoff. Lassie is not going to get steak every night. Now what are the possibilities?

Lassie can be owned by farmer A, **OR**
Lassie can be owned by farmer B, **OR**
Lassie **cannot** be owned by farmer A **AND** B.

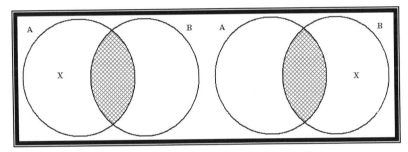

Figure 3.6.3 Exclusion of Element-X for Intersection between Sets

This last story refers to a special kind of **OR**, since this **OR** excludes any possibility of ownership of Lassie by both Farmer A **AND** B; it is called an **Exclusive-OR**, sometimes abbreviated as **XOR**. It has a special type of truth table:

Table 3.6.4 Logical XOR

A	B	A **XOR** B	
T	T	F	(Excluded)
T	F	T	
F	T	T	
F	F	F	

Here the possibility of A = B = TRUE is excluded.

Finally the self-evident truth table of equivalent propositions:

Table 3.6.5 Logical Equivalency

A	B	$A \equiv B$
T	T	T
T	F	F
F	T	F
F	F	T

Now let's consider the implication from the original framework of the syllogism. Implication actually relates to the initial proposition, called the **hypothesis**, as being a subset of the object, called the **conclusion**, of the implication. Case in point:

IF [The animal in question is a prairie dog] **THEN** [The animal is a rodent]

Here prairie dog is a species of rodent. In Venn-Diagram form:

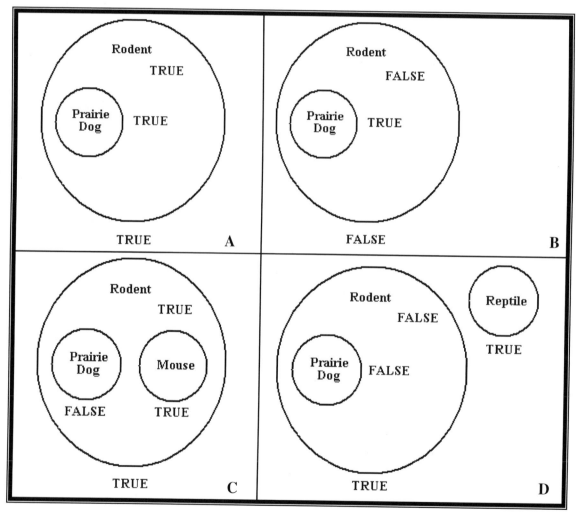

Figure 3.6.4 Implication, Demonstrated by Subsets

Let's take the proposition [The animal in question is a prairie dog] as TRUE and [The animal is a rodent] as TRUE than the implication that a prairie dog is a rodent remains valid.

However if [The animal in question is a prairie dog] is TRUE, but the proposition [The animal is a rodent] is FALSE, this is a contradiction because a prairie dog is still a rodent, then implication is in error, hence invalid.

However if [The animal in question is a prairie dog] is FALSE, but the proposition [The animal is a rodent] is TRUE, the animal in question is a mouse, the conclusion is still valid, hence the implication remains TRUE.

However if [The animal in question is a prairie dog] is FALSE, but the proposition [The animal is a rodent] is also FALSE, the animal really in question was a reptile, reptiles not being a member of the set of animals called rodent, the implication still holds.

So, implication can be represented by the following TRUTH Table.

Table 3.6.6 Logical IMPLICATION

A	B	IF A THEN B
T	T	T
T	F	F
F	T	T
F	F	T

We now have enough of an understanding to go back to the original syllogism and actually prove its credibility.

IF A THEN B
IF B THEN C
Therefore **IF A THEN** C
All of which implies:
IF [x ≡ (**IF A THEN** B)] **AND** [y ≡ (**IF B THEN** C)] **THEN** [z ≡ (**IF A THEN** C)]

A	B	C	IF A THEN B	IF B THEN C	IF A THEN C	IF x AND y THEN z
T	T	T	T	T	T	**T**
T	T	F	T	F	F	**T**
T	F	T	F	T	T	**T**
T	F	F	F	T	F	**T**
F	T	T	T	T	T	**T**
F	T	F	T	F	T	**T**
F	F	T	T	T	T	**T**
F	F	F	T	T	T	**T**

These Logical Formulas can be broken down into simpler, more fundamental logical operations called identities.

A **XOR** B	≡	[(**NOT** A) **AND** B] **OR** [A **AND** (**NOT** B)]
IF A **THEN** B	≡	(**NOT** A) **OR** B

Section 3.7 Truth Tables as Logical Building Blocks of Theorems

A theorem leads with a conditional statement and a proof built up around it with a variety of identities and other supporting theorems. This section starts by developing the fundamental ideas required to start the process for a conditional statement, identity and logical frame.

Symbolic Logic [COP68]:
The symbolic logic relies heavily upon set theory symbols, however in Copi [COP68] he uses the contains superset symbol [⊃] to denote implication. Implication is a grammatical representation of the logical statement:

If (Given) A than B is implied, or simply A implies B.

Since *implies* means something totally different than *contains* then in order to avoid grammatical confusion the arrow symbol [→] for implementation will be used. [LIP64]

$$A \to B$$

This is a **conditional statement** of logic and can be grammatically written in different ways:

Definition 3.7.1 Grammatical Conditional Statement   **only-if**
         [hypothesis], only if [conclusion] [NOL98, pg 48]

Definition 3.7.2 Grammatical Conditional Statement   **if-then**
         If [hypothesis] then [conclusion]

Definition 3.7.3 Grammatical Conditional Statement   **iff** (read as: if and only if)
         also known as a biconditional statement
         [hypothesis] iff [conclusion] [NOL98, pg 48]

A	(read as: B on the condition of A and their validity)	A iff B
T	T	T
T	F	F
F	T	F
F	F	T

Certain propositions and related statements in mathematics have the base [if-than] form. They have specific meaning and continuously reoccur; therefore, special names are given to these variations. Let [A] be the [hypothesis] and [B] the [conclusion] than:

Definition 3.7.4 Logical Conditional Statement   ($\exists$ **propositions** $A \wedge B$) ∋ ($A \to B$).

Definition 3.7.5 Logical Conditional Converse   ($\exists$ **propositions** $A \wedge B$) ∋ ($B \to A$).

Definition 3.7.6 Logical Conditional Inverse   ($\exists$ **propositions** $A \wedge B$) ∋ ($\sim A \to \sim B$).

Definition 3.7.7 Logical Conditional Contrapositive   ($\exists$ **propositions** $A \wedge B$) ∋ ($\sim B \to \sim A$).

Tensor Calculus & Physics: A General Treatise

These ideas on the fundamental components on Boolean Logic can be summarized in the following Axiomatic Table:

A	B	Axiom 3.7.1 A AND B	Axiom 3.7.2 A OR B	Axiom 3.7.3 A XOR B	Axiom 3.7.4 A ≡ B	Axiom 3.7.5 A iff B	Axiom 3.7.6 A → B	Axiom 3.7.7 A only if B
T	T	T	T	F	T	T	T	T
T	F	F	T	T	F	F	F	F
F	T	F	T	T	F	F	T	T
F	F	F	F	F	T	T	T	T

The most elementary argument that was used by the Greeks to comprise a *deductive* theorem is the syllogism. Given propositions A, B and C where:

$$A \rightarrow B$$
$$\underline{B \rightarrow C}$$
$$\therefore A \rightarrow C.$$

Most theorems follow this classical form, however there are other ways of proving an idea using other methods. Reviews of these patterns of reasoning are summarized in the following tables:

Axiom 3.7.8 Valid Argument for a Logical Theorem

An argument is valid if, and only if; the conjunction of (all) the premises implies a valid conclusion as a Tautology.

Axiom 3.7.9 Therefore as an Implication in General Form

"Therefore" [∴] the conclusion is the result for implication of the products of the leading arguments, hence

$\underline{\prod_i \textbf{AND } A_i}$ $\qquad$ converts to: $\prod_i \textbf{AND } A_i \rightarrow B$ $\qquad$ a tautology [T]

$\therefore B$ a tautology [T]

Table 3.7.1 Rules of Inference [COP68]

Ref	Name	Acronym	Method of Proof
Lx3.7.1.1	Modus Ponens	MP	$A \rightarrow B$ $\underline{A}$ $\therefore B$
Lx3.7.1.2	Modus Tollens	MT	$A \rightarrow B$ $\underline{\sim B}$ $\therefore \sim A$
Lx3.7.1.3	Hypothetical Syllogism	HS	$A \rightarrow B$ $\underline{B \rightarrow C}$ $\therefore A \rightarrow C$
Lx3.7.1.4	Disjunctive Syllogism	DS	$A \vee B$ $\underline{\sim A}$ $\therefore B$
Lx3.7.1.5	Constructive Dilemma	CD	$(A \rightarrow B) \wedge (C \rightarrow D)$ $\underline{A \vee C}$ $\therefore B \vee D$
Lx3.7.1.6	Absorption	AB	$\underline{A \rightarrow B}$ $\therefore A \rightarrow (A \wedge B)$
Lx3.7.1.7	Simplification	SP	$\underline{A \wedge B}$ $\therefore A$
Lx3.7.1.8	Conjunction	CJ	A $\underline{B}$ $\therefore A \wedge B$
Lx3.7.1.9	Addition	AD	$\underline{A}$ $\therefore A \vee B$

As per axiom A3.7.8 "Valid Argument for a Logical Theorem" these patterns of reasoning are used in constructing proofs of mathematical theorems, as an example let's build a Truth Table for Modus Ponens:

A	B	$A \rightarrow B$	$(A \rightarrow B) \wedge A$	B	$[(A \rightarrow B) \wedge A] \rightarrow B$
T	T	T	T	T	**T**
T	F	F	F	F	**T**
F	T	T	F	T	**T**
F	F	T	F	F	**T**

From the above example, ANDing all premises for the logical pattern and implying the conclusion yields a TRUE column [**T**] tautology.

In support of proving logical proofs other than using truth tables, a set of the most common relationships have been assembled. Some are proven by truth table others by identity, which is where the title comes from.

Table 3.7.2 Identities of Inference [COP68]

Ref	Name	Acronym	Identities		
Lx3.7.2.1a	De Morgan's Theorems	DM	$\sim (A \vee B)$	$\equiv$	$\sim A \wedge \sim B$
Lx3.7.2.1b			$\sim (A \wedge B)$	$\equiv$	$\sim A \vee \sim B$
Lx3.7.2.2a	Commutation	CM	$A \vee B$	$\equiv$	$B \vee A$
Lx3.7.2.2b			$A \wedge B$	$\equiv$	$B \wedge A$
Lx3.7.2.2c			$A \oplus B$	$\equiv$	$B \oplus A$
Lx3.7.2.3a	Association	AC	$A \vee (B \vee C)$	$\equiv$	$(A \vee B) \vee C$
Lx3.7.2.3b			$A \wedge (B \wedge C)$	$\equiv$	$(A \wedge B) \wedge C$
Lx3.7.2.3c			$A \oplus (B \oplus C)$	$\equiv$	$(A \oplus B) \oplus C$
Lx3.7.2.4a	Distribution	DS	$A \wedge (B \vee C)$	$\equiv$	$(A \wedge B) \vee (A \wedge C)$
Lx3.7.2.4b			$A \vee (B \wedge C)$	$\equiv$	$(A \vee B) \wedge (A \vee C)$
Lx3.7.2.5	Double Negation	DN	A	$\equiv$	$\sim\sim A$
Lx3.7.2.6a	Transposition	TR	$A \rightarrow B$	$\equiv$	$\sim B \rightarrow \sim A$
Lx3.7.2.6b			$A \rightarrow B$	$\equiv$	$[T] \rightarrow (A \rightarrow B)$
Lx3.7.2.6c			$A \rightarrow B$	$\equiv$	$\sim(A \rightarrow B) \rightarrow (A \rightarrow B)$
Lx3.7.2.7	Material Implication	MI	$A \rightarrow B$	$\equiv$	$\sim A \vee B$
Lx3.7.2.8a	Material Equivalence	ME	$(A \equiv B)$	$\equiv$	$(A \rightarrow B) \wedge (B \rightarrow A)$
Lx3.7.2.8b			$(A \equiv B)$	$\equiv$	$(\sim A \vee B) \wedge (A \vee \sim B)$
Lx3.7.2.8c			$(A \equiv B)$	$\equiv$	$\sim((A \wedge \sim B) \vee (\sim A \wedge B))$
Lx3.7.2.8d			$(A \equiv B)$	$\equiv$	$\sim(A \oplus B)$
Lx3.7.2.8e			$(A \equiv B)$	$\equiv$	$(\sim A \wedge \sim B) \vee (A \wedge B)$
Lx3.7.2.9	Exportation	EX	$(A \wedge B) \rightarrow C$	$\equiv$	$A \rightarrow (B \rightarrow C)$
Lx3.7.2.10a	Tautology	TA	A	$\equiv$	$A \vee A$
Lx3.7.2.10b			A	$\equiv$	$A \wedge A$
Lx3.7.2.10c			A	$\equiv$	$(A \equiv [T])$
Lx3.7.2.10d			$\sim A$	$\equiv$	$(A \equiv [F])$
Lx3.7.2.10e			A	$\equiv$	$A \rightarrow [T]$
Lx3.7.2.10f			$\sim A$	$\equiv$	$A \rightarrow [F]$
Lx3.7.2.10g			A	$\equiv$	$A \vee [F]$
Lx3.7.2.10h			A	$\equiv$	$A \wedge [T]$
Lx3.7.2.10i			$\sim A$	$\equiv$	$A \rightarrow \sim A$
Lx3.7.2.10j			A	$\equiv$	$\sim A \rightarrow A$

Table 3.7.2 Identities of Inference [COP68] (Continued)

Ref	Name	Acronym	Identities		
Lx3.7.2.11a	Inverse Tautology	IT	[F]	$\equiv$	$A \wedge \sim A$
Lx3.7.2.11b			[T]	$\equiv$	$A \vee \sim A$
Lx3.7.2.11c			[T]	$\equiv$	$A \equiv A$
Lx3.7.2.11d			[T]	$\equiv$	$A \rightarrow A$
Lx3.7.2.11e			[T]	$\equiv$	$\sim A \rightarrow \sim A$
Lx3.7.2.11f			[T]	$\equiv$	$[T] \rightarrow A$
Lx3.7.2.11g			[T]	$\equiv$	$[F] \rightarrow A$
Lx3.7.2.11h			[T]	$\equiv$	$[F] \rightarrow (A \rightarrow B)$
Lx3.7.2.11i			[F]	$\equiv$	$[F] \wedge A$
Lx3.7.2.11j			[T]	$\equiv$	$[T] \vee A$
Lx3.7.2.12a	Material XOR	MX	$A \oplus B$	$\equiv$	$(A \wedge \sim B) \vee (\sim A \wedge B)$
Lx3.7.2.12b			$A \oplus B$	$\equiv$	$\sim ((\sim A \vee B) \wedge (A \vee \sim B))$

To have a true arithmetic of logical operators the uniqueness of ORing, ANDing, Implementation and Exclusive ORing is established. That is the balance of quantities on either side of the equality must be maintained to keep the logic consistent. What one does to one side of the equality then the something must be done to the other side to maintain balance.

Table 3.7.3 Uniqueness of Logical Operations

Ref	Name	Acronym	Method of Proof
Lx3.7.3.1	Substitution	SES	$A \equiv B$ $\underline{B \equiv C}$ $\therefore A \equiv C$
Lx3.7.3.2	ORed (Additive)	OES	$A \equiv B$ $\underline{C \equiv C}$ $\therefore A \vee C \equiv B \vee C$
Lx3.7.3.3	AND (Multiplicative)	AES	$A \equiv B$ $\underline{C \equiv C}$ $\therefore A \wedge C \equiv B \wedge C$
Lx3.7.3.4a Lx3.7.3.4b	Implication	IES	$A \equiv B$ $\underline{C \equiv C}$ $\therefore A \rightarrow C \equiv B \rightarrow C$ or $\therefore C \rightarrow A \equiv C \rightarrow B$
Lx3.7.3.5	Exclusive ORed	XES	$A \equiv B$ $\underline{C \equiv C}$ $\therefore A \oplus C \equiv B \oplus C$

Theorem 3.7.1 Truth Table for Biconditional IFF

1g	Given			(A iff B) ≡ (A ←→ B) ≡ ((A → B) ∧ (B → A)) biconditional [NOL98 pg 48]			
A	B	A → B	B → A	(A → B) ∧ (B → A)	A ≡ B	[a] (A iff B) ≡ (A ≡ B)	[b] (A iff B) ←→ (A ≡ B)
T	T	T	T	T	T	**T**	**T**
T	F	F	T	F	F	**T**	**T**
F	T	T	F	F	F	**T**	**T**
F	F	T	T	T	T	**T**	**T**

Theorem 3.7.2 Truth Table for Equivalency of ONLY IF

1g	Given			(A only if B) ≡ (A → B) [NOL98 pg 48]	
A	B	A → B	A only if B	(A only if B) ≡ (A → B)	(A only if B) ←→ (A → B)
T	T	T	T	**T**	**T**
T	F	F	F	**T**	**T**
F	T	T	T	**T**	**T**
F	F	T	T	**T**	**T**

Theorem 3.7.3 Truth Table for Substitution

A	B	C	A ≡ B	B ≡ C	(A ≡ B) ∧ (B ≡ C)	A ≡ C	(A ≡ B) ∧ (B ≡ C) → (A ≡ C)
T	T	T	T	T	T	T	**T**
T	T	F	T	F	F	F	**T**
T	F	T	F	F	F	T	**T**
T	F	F	F	T	F	F	**T**
F	T	T	F	T	F	F	**T**
F	T	F	F	F	F	T	**T**
F	F	T	T	F	F	F	**T**
F	F	F	T	T	T	T	**T**

Lema 3.7.3.1 Truth Table for ORed

A	B	C	A ≡ B	C ≡ C	A ∨ C	B ∨ C	(A ≡ B) ∧ (C ≡ C)	A ∨ C ≡ B ∨ C
T	T	T	T	T	T	T	T	T
T	T	F	T	T	T	T	T	T
T	F	T	F	T	T	T	F	T
T	F	F	F	T	T	F	F	F
F	T	T	F	T	T	T	F	T
F	T	F	F	T	F	T	F	F
F	F	T	T	T	T	T	T	T
F	F	F	T	T	F	F	T	T

A	B	C	(A ≡ B) ∧ (C ≡ C) → (A ∨ C ≡ B ∨ C)
T	T	T	**T**
T	T	F	**T**
T	F	T	**T**
T	F	F	**T**
F	T	T	**T**
F	T	F	**T**
F	F	T	**T**
F	F	F	**T**

Lema 3.7.3.2 Truth Table for AND

A	B	C	$A \equiv B$	$C \equiv C$	$A \wedge C$	$B \wedge C$	$(A \equiv B) \wedge (C \equiv C)$	$A \wedge C \equiv B \wedge C$
T	T	T	T	T	T	T	T	T
T	T	F	T	T	F	F	T	T
T	F	T	F	T	T	F	F	F
T	F	F	F	T	F	F	F	T
F	T	T	F	T	F	T	F	F
F	T	F	F	T	F	F	F	T
F	F	T	T	T	F	F	T	T
F	F	F	T	T	F	F	T	T

A	B	C	$(A \equiv B) \wedge (C \equiv C) \rightarrow (A \wedge C \equiv B \wedge C)$
T	T	T	**T**
T	T	F	**T**
T	F	T	**T**
T	F	F	**T**
F	T	T	**T**
F	T	F	**T**
F	F	T	**T**
F	F	F	**T**

Since Implication has an asymmetrical truth table, we have to verify that it is communicative.

Lema 3.7.3.3 Truth Table for Right Handed Implication

A	B	C	$A \equiv B$	$C \equiv C$	$A \rightarrow C$	$B \rightarrow C$	$(A \equiv B) \wedge (C \equiv C)$	$A \rightarrow C \equiv B \rightarrow C$
T	T	T	T	T	T	T	T	T
T	T	F	T	T	F	F	T	T
T	F	T	F	T	T	T	F	T
T	F	F	F	T	F	T	F	F
F	T	T	F	T	T	T	F	T
F	T	F	F	T	T	F	F	F
F	F	T	T	T	T	T	T	T
F	F	F	T	T	T	T	T	T

A	B	C	$(A \equiv B) \wedge (C \equiv C) \rightarrow (A \rightarrow C \equiv B \rightarrow C)$
T	T	T	**T**
T	T	F	**T**
T	F	T	**T**
T	F	F	**T**
F	T	T	**T**
F	T	F	**T**
F	F	T	**T**
F	F	F	**T**

Lema 3.7.3.4 Truth Table for Left Handed Implication

A	B	C	$A \equiv B$	$C \equiv C$	$C \to A$	$C \to B$	$(A \equiv B) \wedge (C \equiv C)$	$C \to A \equiv C \to B$
T	T	T	T	T	T	T	T	T
T	T	F	T	T	T	T	T	T
T	F	T	F	T	T	F	F	F
T	F	F	F	T	T	T	F	T
F	T	T	F	T	F	T	F	F
F	T	F	F	T	T	T	F	T
F	F	T	T	T	F	F	T	T
F	F	F	T	T	T	T	T	T

A	B	C	$(A \equiv B) \wedge (C \equiv C) \to (C \to A \equiv C \to B)$
T	T	T	**T**
T	T	F	**T**
T	F	T	**T**
T	F	F	**T**
F	T	T	**T**
F	T	F	**T**
F	F	T	**T**
F	F	F	**T**

Lema 3.7.3.5 Truth Table for Exclusive ORed

A	B	C	$A \equiv B$	$C \equiv C$	$A \oplus C$	$B \oplus C$	$(A \equiv B) \wedge (C \equiv C)$	$A \oplus C \equiv B \oplus C$
T	T	T	T	T	F	F	T	T
T	T	F	T	T	T	T	T	T
T	F	T	F	T	F	T	F	F
T	F	F	F	T	T	F	F	F
F	T	T	F	T	T	F	F	F
F	T	F	F	T	F	T	F	F
F	F	T	T	T	T	T	T	T
F	F	F	T	T	F	F	T	T

A	B	C	$(A \equiv B) \wedge (C \equiv C) \to (A \oplus C \equiv B \oplus C)$
T	T	T	**T**
T	T	F	**T**
T	F	T	**T**
T	F	F	**T**
F	T	T	**T**
F	T	F	**T**
F	F	T	**T**
F	F	F	**T**

Table 3.7.4 Reversibility of a Logical Argument

Ref	Name	Acronym	Method of Proof
Lx3.7.4.1	Reversibility of a Logical Argument	RLA	A $\underline{A \rightarrow B}$ $\therefore B$ $B \rightarrow A$

Theorem 3.7.4 Reversibility of a Logical Argument

A	B	$A \rightarrow B$	$A \wedge (A \rightarrow B)$	$B \rightarrow A$	$B \wedge (B \rightarrow A)$	$A \wedge (A \rightarrow B) \rightarrow B \wedge (B \rightarrow A)$
T	T	T	T	T	T	T
T	F	F	F	T	F	T
F	T	T	F	F	F	T
F	F	T	F	T	F	T

Under the operators of addition and multiplication, this is only true if the order of manipulation remains the same. This is constructed from axiom A2.3.9 "Principle of Duality" of sets, which allows theorems to be bi-directional under certain operations that in themselves have bi-directional properties.

Table 3.7.5 Distribution of AND Over Implication

Ref	Acronym	Name	Method of Proof
Lx3.7.5.1	DAI	Distribution of AND Over Implication	$C \wedge (A \rightarrow B)$ $\therefore (C \wedge A) \rightarrow (C \wedge B)$
Lx3.7.5.2	DOI	Distribution of OR Over Implication	$C \vee (A \rightarrow B)$ $\therefore (C \vee A) \rightarrow (C \vee B)$
Lx3.7.5.3	DXOI	Distribution of XOR Over Implication	$C \oplus (A \rightarrow B)$ $\therefore (C \oplus A) \rightarrow (C \oplus B)$

Theorem 3.7.5 Truth Table for Distribution of AND Over Implication

A	B	C	$A \to B$	$C \wedge (A \to B)$	$C \wedge A$	$C \wedge B$	$(C \wedge A) \to (C \wedge B)$
T	T	T	T	T	T	T	T
T	T	F	T	F	F	F	T
T	F	T	F	F	T	F	F
T	F	F	F	F	F	F	T
F	T	T	T	T	F	T	T
F	T	F	T	F	F	F	T
F	F	T	T	T	F	F	T
F	F	F	T	F	F	F	T

A	B	C	$C \wedge (A \to B) \to [(C \wedge A) \to (C \wedge B)]$
T	T	T	T
T	T	F	T
T	F	T	T
T	F	F	T
F	T	T	T
F	T	F	T
F	F	T	T
F	F	F	T

Theorem 3.7.6 Truth Table for Distribution of OR Over Implication

A	B	C	$A \to B$	$C \vee (A \to B)$	$C \vee A$	$C \vee B$	$(C \vee A) \to (C \vee B)$
T	T	T	T	T	T	T	T
T	T	F	T	T	T	T	T
T	F	T	F	T	T	T	T
T	F	F	F	F	T	F	F
F	T	T	T	T	T	T	T
F	T	F	T	T	F	T	T
F	F	T	T	T	T	T	T
F	F	F	T	T	F	F	T

A	B	C	$C \vee (A \to B) \to [(C \vee A) \to (C \vee B)]$
T	T	T	T
T	T	F	T
T	F	T	T
T	F	F	T
F	T	T	T
F	T	F	T
F	F	T	T
F	F	F	T

Theorem 3.7.7 Truth Table for Distribution of XOR Over Implication

A	B	C	$A \to B$	$C \oplus (A \to B)$	$C \oplus A$	$C \oplus B$	$(C \oplus A) \to (C \oplus B)$
T	T	T	T	F	F	F	T
T	T	F	T	T	T	T	T
T	F	T	F	T	F	T	T
T	F	F	F	F	T	F	F
F	T	T	T	F	T	F	F
F	T	F	T	T	F	T	T
F	F	T	T	T	T	T	T
F	F	F	T	F	F	F	T

A	B	C	$C \oplus (A \to B) \to [(C \oplus A) \to (C \oplus B)]$
T	T	T	**T**
T	T	F	**T**
T	F	T	**T**
T	F	F	**T**
F	T	T	**T**
F	T	F	**T**
F	F	T	**T**
F	F	F	**T**

Tensor Calculus & Physics: A General Treatise

Section 3.8 Proof by Identity using an Indirect Proof by Contradiction

Any proof that falls into the following form is called an identity. Identity proofs are the workhorse of mathematics, because they are one of the most commonly used proofs. Its simple form follows the more classical indirect proof by contradiction. Case in point the following trigonometry identity:

Theorem 3.8.1 Proof by Identity: Sine of Ninety Degrees and Some Angle

1g	Given	$\cos(x) = \sin(90^\circ + x)$ *by assumption* or simply *assume*
Steps	Hypothesis	$\cos(x) = \sin(90^\circ + x)$
1	From 1g	$\cos(x) = \sin(90^\circ + x)$
2	From 1 and LxE.3.1.22 sine sum of angles	$\cos(x) = \sin(90^\circ)\cos(x) + \cos(90^\circ)\sin(x)$
3	From 2, DxE.1.6.7 Sine of Ninety Degrees, DxE.1.6.7 Cosine of Ninety Degrees and A4.2.3 Substitution	$\cos(x) = 1 * \cos(x) + 0 * \sin(x)$
4	From3, A4.2.12 Identity Multp, T4.4.1 Equalities: Any Quantity Multiplied by Zero is Zero and A4.2.7 Identity Add	$\cos(x) = \cos(x)$
∴	From 1g, 4 *and by identity*	$\cos(x) = \sin(90^\circ + x)$

Here an assumption is made, step-1g than either one side or the other of the equality or both is proven as in step-4. Since step-4 holds the conclusion must be that the original assumption is valid as well.

So the form always starts with a given assumption, (*by assumption* or simply *assume*) and ends with the conclusion (*and by identity*).

Section 3.9 Proof by Indirection

From the Rules of Inference deferent kinds of proofs can be created to prove theorems. One such proof is called **indirect reasoning** [JOH69]. The following two proofs are classic examples of indirect reasoning; however the question arises is there a common theme that runs between them?

Theorem 3.9.1 Indirect Reasoning by Contrapositive:

1g	Given	Propositions A and B
2g		$A \to B \equiv \sim B \to \sim A$ assume
Step	Hypothesis	$A \to B \equiv \sim B \to \sim A$ hence has the same truth table.
1	From 1g and 2g	$A \to B \equiv \sim B \to \sim A$
2	From 1, Lx3.7.2.6a Transposition and T3.7.3 Truth Table for Substitution	$A \to B \equiv A \to B$
∴	From 3g, 2 and by identity	$A \to B \equiv \sim B \to \sim A$

Another type of proof is the **reductio ad absurdum**. [JOH69]

As a consideration, consider a proof by contradiction from geometry:

A	= {two independent lines.}
B	= {two lines intersect at point P only.}
~B	= {two lines intersect at points P and Q.}
C	= {for every two different points, there is exactly one line that contains both points.}
~C	= {for every two different points, there are two lines that contain both points.}

where

$$\begin{array}{l} A \\ \underline{\sim B} \\ \therefore C \wedge \sim C \end{array}$$

This argument is absurd because it implies a conclusion that is in contradiction to its self. However, the implication of A to B is logical given the above definition of the propositions. It would seem that in some form of opposite logic they are related and if the above argument could be proven the other would follow suite.

Theorem 3.9.2 Reductio Ad Absurdum:

1g	Given	Propositions A, B and C
2g		$A \to B \equiv (A \wedge \sim B) \to (C \wedge \sim C)$ assume
Step	Hypothesis	$A \to B \equiv (A \wedge \sim B) \to (C \wedge \sim C)$
		hence has the same truth table.
1	From 1g and 2g	$A \to B \equiv (A \wedge \sim B) \to (C \wedge \sim C)$
2	From 1, Lx3.7.2.11a Inverse Tautology and T3.7.3 Truth Table for Substitution	$A \to B \equiv (A \wedge \sim B) \to [F]$
3	From 2, Lx3.7.2.10f and T3.7.3 Truth Table for Substitution	$A \to B \equiv \sim(A \wedge \sim B)$
4	From 3, Lx3.7.2.1b De Morgan's Theorems and Lx3.7.2.5 Double Negation	$A \to B \equiv \sim A \vee B$
5	From 4, Lx3.7.2.7 Material Implication and T3.7.3 Truth Table for Substitution	$A \to B \equiv A \to B$
∴	From 2g, 5 and by identity	$A \to B \equiv (A \wedge \sim B) \to (C \wedge \sim C)$

The two indirect proofs do seem to have a common thread; for prepositional argument $G \equiv \sim H$ having equivalent truth tables than a proof for H being invalid, becomes indirectly a proof for the theorem G as valid, this than is the general underlying principle for indirect logic.

Section 3.10 Proof Writing Using Stratification of Logical Systems

This chapter started out looking at a finite group of specific ways one might prove theorems only to find that there is an unlimited set of logical systems, which can be used to solve them. These methods used to solve problems are plans of attack on these conundrums and how the proof writer might provide a structured logical path to solve them. These logical methods, called **stratification of logic**, can be categorized and used as building blocks to develop very complex schemes or paths of logic to solve these quandaries. The balance of this chapter deals with identifying these logic blocks and how to use them.

The direct proof illustrates the first levels of stratification of logic for proving theorems by abstracting the relationship between truth tables. As an example instead of equating truth tables lets have one imply the other, such as $G \to H$, which for a set of sequential arguments is what a deductive logical system is. This can be written in a more general way as **the theorem G has a logical relationship with the theorem H by some logical operator** or symbolically $^L[.G \Leftrightarrow H.]$, where the broad arrow and superscripted [L] together form a logical operator or relationship to be specified as a property of the system. And the logical set $[_LT]$, or as it will be called from here on a **logical system**, contains all $^L[.G \Leftrightarrow H.]$ forming a logical path of proof;

$$^L[.G \Leftrightarrow H.] \in {}_LT \text{ and constitutes a path for a proof.}$$

To differentiate the different stratified levels of logical structures (paths) will let the Hebrew letter ל (Lamed) subscripted with zero [0] represent the first logical level using propositions $ל_0$ and $ל_1$ that use theorems and are defined as follows:

Definition 3.10.1 Stratification logic order zero [$ל_0$]
Unbroken sequential logical structures (paths) based on sets of propositions.

Definition 3.10.2 Stratification logic order one [$ל_1$]
Unbroken sequential logical structures (paths) based on sets of theorems.

Theorem 3.10.1 Order of Stratification of Logic

1g	Given	$^L(.G \Leftrightarrow H.) \in {}_LP$ where $[_LP]$ is a set of logically related propositions comprising a theorem.
2g		$^L[.G \Leftrightarrow H.] \in {}_LT$ which constitutes a path for proof.
Steps	Hypothesis	$ל_0 \subset ל_1$
1	From 1g and D3.10.1 the logical relationship for theorems a similar symbolism for propositions can be extrapolated, such that	$_LP \in ל_0$
2	From 2g and D3.10.2	$_LT \in ל_1$
3	From 1 and 2	$_LP \in {}_LT$ comprise paths of proofs
4	From 3 as a set;	$_LP \subset {}_LT$
∴	From 4 by levels of stratified logic.	$ל_0 \subset ל_1$

Section 3.11 Theorems Based on Properties of Countable Numbers

Definition 3.11.1 Stratification logic order two [$\flat_2$]

Logical structures based on induction property of counting integers.

Logic patterns provide methods for constructing theorems, however there are other inductive proofs that can be found in algebra that are based on the properties of numbers. One such proof is based on the unique property found in integers. The technique is called the **Induction Property of Integers**. [JOH69] This property of integers has no parallel in rational, real, or complex number systems, because it is based on the notion of counting, a unique fundamental property of integers only. In most textbooks on algebra this property of integer numbers is stated as a theorem, this is a misnomer, because this method is used to derive theorems, which parallel it, it is never derived. The property is an atomized logical structure, hence must be an axiom forming its own unique and distinct logical strata of order two, $\flat_2$. So, in this treatise it will always be referenced as a logical axiom.

Axiom 3.11.1 Induction Property of Integers

1g	Given	S is an infinite set of elements [$i \leftrightarrow e_i \in$ S] and [e_i] countable by positive integers [i] starting with 1 (one) or some other positive number.
2g		Assume $e_k \in$ S.
Step	Hypothesis	S contains all countable positive integers and hence must be true for all countable elements e_i as well.
1	From 1g establish a finite set of elements i = 1, 2, 3, … are valid for S.	$(e_1, e_2, e_3) \in$ S.
2	From 1 and 2g	Extrapolate to the kth element e_k.
3	From 2	By some ($^L[K_0 \Leftrightarrow S] \in {}_LS$) $\rightarrow (e_k \in$ S)? Iff Yes, proof continues. No, poof fails.
4	From 3 does $(k+1)^{th}$ element follow?	By ($^L[K_1 \Leftrightarrow S] \in {}_LS$) $\rightarrow (e_{k+1} \in$ S)? Iff Yes, conclude hypothesis. No, poof fails.
∴	From 4, Now there is a one-to-one correspondence between countable positive integers [i] and elements [e_i]. Since every positive integer can be reached k to k+1 than it follows e_k to e_{k+1} can also be reached.	It is concluded then S contains all countable positive integers and hence must be true for all countable elements e_i as well.

In the development of n-dimensional objects or manifolds in Tensor Calculus, this theorem is one of the most powerful and important tools in exploring that realm.

Most axioms are simply a statement of the property or algorithmic steps that comprise it. Here those logical steps have been placed in a logic framework for a theorem. This does not imply the axiom is a theorem, but that any theorem based on this axiom will parallel these steps in this format so a specific theorem can be constructed from it. Not only is this not a theorem, but also the axiom specifically states that there is the possibility that the logical strata might not be possible. At every step, the set S must be satisfied in order for it to work. Also some logical path of proof $^L[K \Leftrightarrow S] \in {}_LS$ must be devised from steps 1 to 3 to reaffirm the sequence of the count. It is **not** enough to simply construct a count $(e_1, e_2, e_3) \in$ S and extrapolate a kth element from this set and say that's how the sequence behaves. Lets say 1, 2, 3 are done, [k] is derived and from that [k+1], but later k = 4 is calculated and diverges from the prediction of k+1. A brake from the inductive count of positive integers invalidates the argument. In this type of proof this happens quite often, hence at this point a formal logical proof is required to validate $e_{k+1} \in S$, which may or may not be possible, hence making the axiom difficult to use.

The question does come up, how many finite elements are sufficient? Certainly e_1 is necessary, but is it sufficient? This is determined by two things, how many are necessary to validate the conjecture for the kth element, and can a logical proof showing $e_k \in S$ be derived, thereby ending the need for another count and showing that the number is also sufficient.

In an account of Andrew Wiles' proof of the Taniyama-Shimura Conjecture for "Fermat's Last Theorem" [SIN97, pg 210], he masterfully uses this axiom, but not quite in a recognizable form as it is classically used. For the path he took, to establish step 3 and prove step 4 involved advanced concepts in number theory so it's somewhat obscured, but it is still there.

Tensor Calculus & Physics: A General Treatise

Section 3.12 Theorems Based on Properties of Geometric Constructed Symmetries

There is another unique and different path taken in logical development that it deserves its own order of logical stratification:

Definition 3.12.1 Stratification logic order three [ℶ₃]

Logical structures based on sets of Geometric Constructed Symmetries.

Just as there is an inductive proof based on the properties of numbers in algebra so to a counterpart can be found in geometry, thereby completing the symmetry between the two disciplines and can be shown to have a one-to-one correspondence. The idea is quite simple yet profound and finds use in geometric proofs in countable n-spaces.

Axiom 3.12.1 Equivalency Property of Constructible Geometric Components

$\exists$ a set of G_{cc} (Geometric Constructible Components) made up of n-[$g_{\ddot{a}}$] components, such that

$$g_{ci} \in G_{cc} \qquad \text{for i = 1, 2, 3, ... ,n.}$$

or

$$G_{cc} \equiv \{ \, (_\bullet \, g_{ci} \, _\bullet) \text{ for n-base objects} \}$$

where

$$g_{ci} \equiv \{(_\bullet \, p_{ij} \, _\bullet) \text{ for m-properties of construction } \}$$

and if the construction of an element [g_i] is identical to each element [g_{cj}], for i ≠ j, or it is said

$$(_\bullet = g_{ci} \, _\bullet),$$

than a proof for one [g_j] is a proof for all elements whether it be done by construction or by property. Also, G_{cc} is a *circular closed set*, iff

$$g_{c1} = g_{c(n+1)}.$$

As an example, take a starflower pattern, or the martial arts Shuriken throwing star, made up of identical rhombuses:

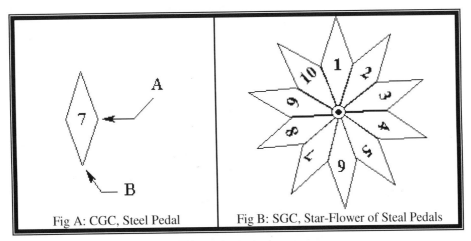

| Fig A: CGC, Steel Pedal | Fig B: SGC, Star-Flower of Steal Pedals |

Figure 3.12.1 Star Pattern

Tensor Calculus & Physics: A General Treatise

It is self evident that the star being made up of identical steel petals that each petal has the same geometric properties independent of the star system and circular since pedal-1 is congruent to pedal-10, so:

Theorem 3.12.1 Geometric Property of a Steel Pedal

Step	Hypothesis	$180 = A + B$ for all pedals
1	From F3.12.1A Parallelogram T5.20.6 Irregular Polygons: Accumulated Internal Angle-n=4 and T5.12.1 Opposing Angels of a Parallelogram are Congruent	$360 = 2\,A + 2\,B$
2	From 1, D4.1.19 Primitive Definition for Rational Arithmetic and A4.2.14 Distribution	$180\cdot 2 = (A + B)\,2$
$\therefore$	From 2, T4.4.3 and A3.12.1 Equivalency Property of Constructible Geometric Components	$180 = A + B$ for all pedals

Theorem 3.12.2 Acute Vertex Angle of a Star-Flower System

1g	Given	an n-steel pedal star system
Step	Hypothesis	$B = 360 / n$
1	From 1g, F3.12.1B Star Pattern and adding all vertex angles	$360 = \sum_n B$
2	From 1 and A2.2.12 Identity Multp	$360 = \sum B \cdot 1$
3	From 2 and A4.2.14 Distribution	$360 = B \sum 1$
4	From 3 and LxE.3.1.37 Sums of Powers of 0 for the First n-Integers	$360 = B\,n$
$\therefore$	From 4 and T4.4.4B Equalities: Right Cancellation by Multiplication	$B = 360 / n$

Theorem 3.12.3 Oblique Interior angle of a Star-Flower System

Step	Hypothesis	$A = 180\,(1 - 2 / n)$
1	From T3.12.2 Acute Vertex Angle of a Star-Flower System	$180 = A + B$
2	From 1 and T4.3.4B Equalities: Right Cancellation by Addition	$A = 180 - B$
3	From 2, T3.12.2 Acute Vertex Angle of a Star-Flower System and A4.2.3 Substitution	$A = 180 - 360 / n$
4	From 3 and D4.1.19 Primitive Definition for Rational Arithmetic	$A = 180 - 2\cdot180 / n$
5	From 4 and A4.2.12 Identity Multp	$A = 1\cdot180 - 2\cdot180 / n$
6	From 5, D4.1.20B Negative Coefficient, A4.2.10 Commutative Multp and A4.2.11 Associative Multp	$A = 1\cdot180 - (2 / n)\,180$
$\therefore$	From 6 and A2.2.14 Distribution	$A = 180\,(1 - 2 / n)$

Now this is all very fine you might say, intuitively obvious, but the crucial point that makes T3.12.1 "Geometric Property of a Steel Pedal" so powerful is unobtrusively overlooked in step 1 of T3.12.3 "Oblique Interior angle of a Star-Flower System". If the ***inductive property of integers*** where required the property for pedal-1 would have had to be proven, then for pedal-2 and so on to have a clear and valid count established in countable integer set. The next step would have been to establish the sum up to some value-k than add an element to generate k+1 and watch $\sum_{k+1} B$ as we add around the star reaching 360°. Notice that T3.12.1 "Geometric Property of a Steel Pedal" and the net geometry of the system establish all this, thereby eliminating all of these laborious steps. Working the ***inductive property of integers*** proof for n-space may not be easily used, or even possible, but if a geometric subunit can be established than the ***Equivalency Property of Geometric Constructible Components*** becomes a very powerful tool in n-dimensional proof writing.

Also, notice that this type of proof has exploited the sense of symmetry. Not the classical meaning of symmetry, but the symmetry of the component system. Such as the growth into the resulting symmetry of the starflower, axial mirror symmetries, thereby matching each element, starting with element-1, element-2 and so on around the set until element-1 is reached and matched completing the symmetry of a circle. Various kinds of set symmetries can be considered based on circular element matching, thereby grouping elements into subsets. From this various logical proofs might be constructed to solve intricate problems in otherwise what might appear to be impossible solvable geometries. This meaning of symmetry is why I call this section ***Theorems Based on Properties of Geometric Constructed Symmetries***.

Section 3.13 Fermat's Last Theorem and Langlands' Hypothesis

Case in point, within resent times one of the most famous problems in mathematics **Fermat's Last Theorem** ($z^n = x^n + y^n$ for n > 2 does not have whole number triplet solutions as n = 2 has) was proven by an inductive logical method developed by Andrew Wiles of Cambridge England and others. He did this by proving Taniyama-Shimura conjecture.

In 1922, the English mathematician Louis J. Mordell discovered what he thought was a very strange connection between the solutions of algebraic equations and topology, since he wasn't able to prove it, it became known as Mordell's conjecture. If the surface of solutions, using complex numbers for greatest generality, has two or more holes (geniuses) in a continuous topological surface, then the equation has a finite set of **whole-number solutions**.

In 1983, a twenty-seven year old German mathematician, Gered Faltings, at that time at the University of Wuppertal, was able to prove Mordell's conjecture, which bears his name Falting's Proof. Toward the solution of Fermat's equations it is known that they have a genus of 2 or more for n > 3. This was reassuring because this meant that if whole number triplet solutions did exist than the number of sets is finite.

Soon after Falting's find lead Granville and Heath-Brown, to use his results and showed that if triplet solutions existed, that they decreased as the exponent [n] increased and for [n] large they didn't exist at all. So if Fermat's Last Theorem was TRUE than solutions where far and few between.

These findings tantalizingly alluded to the possible solution of Fermat's Last Theorem.

The piece of the jigsaw puzzle that would make it possible came from the **Taniyama-Shimura Conjecture** "every elliptic equation of rational coefficients has a corresponding modular form in the complex plane" to prove this Wiles had to carry out a series of complicated transformations. These metamorphoses went from a projection of elliptic equations of Diophantus, onto the modular surface of a topological torrid, genus one, of the complex plane, which intern maps into the modular Poincaré functions of automorphic form.

If the conjecture where TRUE then the surface of the torrid will hold all solutions to elliptic equations over the rational numbers, these in turn arising from the equations of Diophantus. The connection to Fermat's equations is that if solutions did exist, they would have to lie on the complex surface of this special torrid.

A fuzzy logical path of thought came from Gerhard Frey, University of Heidelberg, to try to firmly tie the Taniyama-Shimura Conjecture to Fermat's Last Theorem. Ken Ribet clarified it into a theorem, which established the following implication:

$$\text{Taniyama-Shimura Conjecture} \rightarrow \text{Fermat's Last Theorem}$$

What Frey had hypothesized was this:

1) Suppose Fermat's Last Theorem is FALSE than there is a set of integer triplets (a, b, c) that will satisfy Fermat's equation for some n > 2.
2) These triplets (a, b, c) would than give rise to a very specific elliptic curve having a general form that Frey specified and is called the Frey's curve. The important thing about this curve is that it is not modular.
3) So, if the Taniyama-Shimura Conjecture where proven TRUE then all elliptic curves must be modular, however Frey's curve is not.
4) Therefore since Frey's curve is the solution, but not modular than it does not exist, if it does not exist there cannot be any solutions to Fermat's equations.
5) QED Fermat's Last Theorem is proven.

What Ken Ribet did was to formally derive Frey's curve so what started out as a hypothesis was now established as a theorem. [ACZ96]

Now let,

- FLT   be Fermat's Last Theorem
- TSC   be the Taniyama-Shimura Conjecture
- FRC   be Frey's curve
- ELL   be elliptic equations of Diophantus
- MOD   be modular Poincaré functions of automorphic form
- ABC   be a set of ordered triplets (a, b, c) that satisfies Fermat's equation of n > 2.

Symbolically the argument goes as follows:

1) $\sim$FLT[F] $\to$ iff $\exists$ ABC
2) ABC $\to$ FRC
3) As a property of; MOD $\notin$ FRC
4) TSC [T] $\to$ for all (MOD $\equiv$ ELL) it follows than
5) MOD $\in$ TSC
6) $\therefore$FRC $\notin$ TSC [T]
7) Hence $\sim$ABC
8) QED FLT [T].

Clearly, this proof falls into the category of A3.1.6 "Proof by Contradiction of Indirect Reasoning" with the use of Frey- Ribet's Theorem. This logic is still messy, so let's clean out all the conditionals on the propositions and simplifying the variable names further, abstracting it more so:

1) $\sim$A $\to$ B
2) B $\to$ C
3) D $\notin$ C
4) D $\in$ E
5) $\underline{C \notin E}$
6) $\therefore$E $\to$ A Frey's original objective.

Now if we realize this is a proof within a proof and we take out steps 3, 4 and 5 and let them stand on their own. This leads to the Taniyama-Shimura Conjecture being proven TRUE implying Frey's Curve being impossible to construct in this space, hence FALSE.

7) $D \notin C$
8) $D \in E$
9) $\underline{C \notin E}$
10) $\therefore E \rightarrow \sim C$
This result can be substituted back into the original path of the proof.
11) $\sim A \rightarrow B$
12) $B \rightarrow C$
13) $\underline{E \rightarrow \sim C}$
14) $\therefore E \rightarrow \sim(\sim A)$
QED:
15) $\underline{E \rightarrow \sim(\sim A)}$
16) $\therefore E \rightarrow A$ by Lx3.7.2.5 Double Negation

Thanks to Ken Ribet, Gerhard Frey's logic was proven actually as being crystal clear and all that remained was Andrew Wiles to prove the validity of the Taniyama-Shimura Conjecture, which he did.

Here is a good example demonstrating a system of inductive logic involving conjectures incomplete deductive logic and topological-geometric theorems providing a circuitous path of proof to the solution of "Fermat's Last Theorem".

But more to the point, it demonstrates another method for problem-solver's. Notice that for **Taniyama-Shimura Conjecture** "every elliptic equation of rational coefficients has a corresponding modular form in the complex plane" forming a bridge from one side of a math system to another. What appeared to be two separate systems of math were actually different sides of the same coin. So, a proof that may not be provable on one side of the coin is clearly provable on the other. It follows that problems in one domain could be solved by analogy with problems in the parallel domain.

What Wiles had done was to take the first step toward Robert Langlands' grander scheme of unification --- the **Langlands' Program**. During the 1960's Robert Langlands, at the Institute for Advanced Study, Princeton was struck by the potency of the Taniyama-Shimura conjecture. Even though the conjecture had not been proved, Langlands believed that it was just one element of a much grander scheme of unification. He went about finding and collecting other conjectures that might lead to bridges that would eventually unify all of mathematics. Proving these conjectures became known as Langlands' program [SIN97, pg 192]. If Langlands dream could be achieved than, the reward would be enormous. Any insoluble problem in one area of mathematics could be transformed into an analogous problem in another area, where a whole new arsenal of techniques could be brought to bear on it. If a solution was still elusive, the problem could be transformed and transported to yet another area of mathematics, and so on, until it was solved. One day, according to the Langlands' program, mathematicians would be able to solve their most esoteric and intractable problems by shuffling them around the mathematical landscape.

There were also important implications for the applied sciences and engineering. Whether it is modeling the interactions between colliding quarks or discovering the most efficient way to organize a telecommunications network, often the key to the problem is performing a mathematical calculation. In some areas of science and technology the complexity of the calculations is so immense that they cannot be completed, and consequently progress in those subjects has been severely hindered. So, if Langlands' program could be completed there would be new ways to solve real-world problems, as well as abstract ones. However, any provable conjecture would satisfy Langlands' hypothesis as seen in the following table on theorems from other fields of mathematics.

Definition 3.13.1 Stratification logic order four [ℶ₄]

Langlands's hypothesis is any property in the domain that can be associated within the codomain.

Axiom 3.13.1 Langlands' Hypothesis: Associated Properties between Domain and Codomain

1g	Given	C	Conjecture relationship bridging Domain and Codomain
2g		P_{dom}	Property of Domain
3g		S	Domain set of all properties of the system
4g		S'	Codomain set of all properties of the system
5g		$f< C >: S \rightarrow S'$	
6g		$P_{dom} \in S$	
Steps	Hypothesis	$P_{dom} \in S'$	for $f< C >: S \rightarrow S'$ valid
1	From 5g iff conjecture is proven valid [True], such that	$f< C >: S \rightarrow S'$	
∴	From 6g and 1	$P_{dom} \in S'$	

Table 3.13.1 Examples of ℶ₄ Solutions

Codomain S'	Domain S	Conjecture C	Domain Property P_{dom}
Elliptic equations of rational coefficients	Modular Forms in the complex plane	Taniyama-Shimura	$z^n = x^n + y^n$ for $n > 2$ does not have whole number triplet solutions [Algebra]
Irrational Polygons	Average Regular Polygons	$(\exists S \ni) \subset S'$	Zero Property applies to all odd and even n-sided polygons [Geometry]
Riemannian Manifold space	Geodesic Manifold space	$(\exists S \ni) \subset S'$	D14.3.1A Geodesic Space $\mathfrak{G}^n$ and T14.3.1 Bianchi Identity Covariant; Bianci relation [Tensors]

Section 3.14 Newtonian Physical Deductive Systems of Logic

While Newton emulated Euclid by creating a logical deductive system, **The Science of Physics**, his system of logic being based on the physical world, required other kinds of logical proof. Any logical system stands on the foundation of its axioms, which may or may not be valid, but are accepted to initiate the system of thought. Confidence in the axioms is another whole affair requiring proof external to the logical system if one wishes to have some degree of confidence in them. In the Newtonian logic system, this is done by experimentation. As an example there is Newton's Law (Axiom) of motion, LAW I: [CAJ71, Vol. I pg 13]

Everybody continues in its state of rest, or of uniform motion in a right line, unless it is compelled to change that state by forces impressed upon it.

In an interview with Dr. Richard P. Feynman [SYK93, interview at his home], Doctor Feynman relates, when he was a child, how his father Melville Feynman understood and explained Newton's Law of Inertia to him. So based on what his father said Feynman proceeds to prove the Law by going outside and placing a small ball in his toy wagon, sees that it does not roll when stationary. He then carefully sits alongside it, pulls the wagon handle accelerating the wagon. All the time he watches the ball, and notices the ball, relative to the sidewalk, for a brief moment, stays its position and then as friction kicks in starts to roll. Of course, eventually the ball hits the back of the wagon causing the ball to move with the wagon, but it did remain at rest for a little while resisting the motion of the wagon. He then relates how he reversed the experiment by setting the ball in the wagon so it was moving with it and then stopping the wagon and watching the ball continue to move till it hit the front of the wagon stopping the ball.

Clearly a simple child's experiment proves the law independent of Newton's logical system and it can be done at any time, any place, in any century, hence our confidence in the Law is established. This is one of the awesome beauties of Newton's system of logical proof that the simplicity of experimentation works hand in hand to <u>almost</u>[3.14.1] make it perfectly self-evident.

Newton in his famous set of experiments on the refraction of light 1666, declares at one moment that he creates **the critical experiment** [BRO73, pg 227]. That is once the light had been diffracted and filtered by slits into a single pure color no matter how he tried with lens, interferometers and other prisms he could no longer change the color or convert it back to the original aggregate of white light, except of course by recombining the original components of light.

For Einstein the critical experiment occurred not to provide confidence in his great work on Relativity and General Relativity at the beginning, but all most after the fact. This experiment was done on the 29th of May 1919, three years after the publication of General Relativity, when the Royal Society sent to Brazil and the west coast of Africa expeditions to test the theory by observing star light being bent around the sun do to its gravitational field.

[3.14.1]Note: I say <u>almost</u> because this experiment over simplifies the physics somewhat. The tangential velocity imparted to the ball as it rolls is the same and opposite as is imparted to the velocity of the wagon, which could only be true if the inertia or mass of the ball where indeed resisting the motion of the wagon. Of course, the ball being perfectly round has ideal symmetry; hence, the mass acts as an ideal point at its center of gravity. Also since gravitational weight and friction are perpendicular forces to the linear motion and are nullified by other opposing forces creating independent factors and as such do not need to be considered. Also you could consider the ball as non-elastic or a hardball; one could say Feynman was playing hardball. On the other hand, what does a young boy know about all of this any way?

Fellows of the Royal Society rushed the news to one another, Eddington by telegram to the mathematician Littlewood, and Littlewood in a hasty note to Bertrand Russell, [BRO73]

Dear Russell:

Einstein's theory is completely confirmed. The predicted displacement was 1"• 72 and the observed 1"• 75 ± • 06.

Yours,
J.E.L.

Experimentation was also critical to Einstein's work, but as seen above not quite in the way it was for Newton.

Echoing Newton, *if I can see farther than others is because I have stood on the shoulders of giants*. Einstein himself spoke repeatedly in later life of his debt to others that preceded him – "the four men who laid the foundations of physics on which I have been able to construct my theory are Galileo, Newton, Maxwell, and Lorentz" he said during his visit to the United States in 1921. Michelson and Morley are strangely absent from this list. Their great experiment has been linked to Einstein's work since it had something to do with light, guilt by association. The experiment involved establishing the existence of a media through which light had to travel in, the aether. Serendipitously they proved that the yard sticks they used (actually a bi-directional interferometer) shrank in the direction of motion, hence light in any frame of reference was always measured having the velocity of [c = 299,792,510 meters per second or 186,348.3 miles per second] [ITT75, pg 3-11]. Einstein states the impact of this experiment on his work in a letter he wrote from the Institute for Advanced Study at Princeton on March 17, 1942, to one of Michelson's biographers. "It is no doubt that Michelson's experiment was of considerable influence upon my work insofar as it strengthened my conviction concerning the validity of the principle of the Special Theory of Relativity," he said. "On the other side I was pretty much convinced of the validity of the principle before I did know this experiment and its result. In any case, Michelson's experiment removed practically any doubt about the validity of the principle in optics and showed that a profound change of the basic concepts of physics was inevitable." [3.14.2] [CLA71 pg 95]. While Einstein would say he was not aware of the Michelson and Morley experiments he really did know, in [PER52, pg 1] the translated reprint of "Michelson's Interference Experiment" by H.A. Lorentz 1895 and Lorentz's other paper, in 1904, "Electromagnetic Phenomena in a System Moving with any Velocity less than that of Light", which on page one references Michelson and his experiment, one of Einstein's favorite pre-relativity authors, here was a paper that he did read, but why his indifference? First his Relativity papers were derived from first principles the basic premises explaining what the results of such an experiment would have to be. Second the experiment was based on the notion of the existence of an all permeating either to propagate light something Einstein had immediately thrown out. In his mind the experiment was simply not relevant to his work.

Here the experiment was actually the critical experiment for Maxwell's solution of his free space wave equations that the velocity of light is always constant and will be measured the same in any frame of reference. It took a powerful mind like Einstein's to break his head on it to explain the obvious, too what Maxwell had failed see in his own work, and expound upon.

Strange that many years later when Einstein was visiting Mount Wilson California and the Palomar Observatory in 1931 he actual met a very old and ailing Michelson, sadly no one seemed to notice or record their conversation.

Tensor Calculus & Physics: A General Treatise

This is what Einstein established in his famous [3.14.3]LAW 2.II: [PER52 pg 41]

Any ray of light moves in the "stationary" system of co-ordinates with the determined velocity c, whether the ray be emitted by a stationary or by a moving body. Hence

Velocity = light path / time interval

Where time interval is to be taken in the sense of the definition in § LAW I.

So, Einstein's work unlike Newton's was not based upon experimentation, but his faith in the Maxwell's Treatises of (1873). Which was intern soundly based on the works of many other experimentalists, Maxwell (1831–1879) himself, J. Henry (1797–1878), J. Neumann (1798–1895), W. Thomson (1853–1937), W. E. Weber (1804–1891), predecessors Gauss (1777–1855), Ampère (1775–1836) and Faraday (1791–1867) all working with electrical and magnetic phenomena. Einstein's Theory simply picks up where Maxwell left off, so Einstein's critical experiment came for General Relativity and indirectly for Special Relativity with the 1919 Royal Society expedition. Needless to say Einstein was very relieved when he got the news of the expedition's success, *the critical experiment* proved the theories correct.

So for physical systems when is an experiment a law or just a theorem?

- Given a physical system S with a set of atomistic laws L and a set of theorems T such that (L, T) ∈ S.
- If the experiment, A[T], so experiment [A] is valid [T], than A ∈ L hence A is a law. If and only if no other physical or atomistic logical development of experiment-B or theorem can be found, such that [B] is not valid and thereby contradicting it.
- If it is unsure that a critical experiment-A[T] is valid, but experiment-B[T] is found demonstrating mathematical logical development, B ∈ L, and if B[T] → A[T], then the conclusion is forced that A ∈ L, this can only happen if the initial premises are complete and valid.

So the logic of physics requires something a little more than what is needed for proof writing in a mathematical system. Physics requires that confidence must be established for the axioms by experimentation and/or some critical experiment indirectly establishing the validity of the logical system in general. In Volume II on physics, physical proof writing will be based on an a system of law, while non-axiomatic experiments will be considered as theorems along with regular theorems, since an experiment is a type of proof. While for those experiments that are atomized, they will be the axioms of the system and known as *laws*.

In summary what Newton and Einstein have shown is that physics can be rendered into a deductive system, if and only if, first principles can be identified and validated, than all else must follow.

Unfortunately most physicists think that if they shotgun experiments that somehow their conjectures will be validated. A lot of billion dollar atom smashers have been funded based on that idea, but appropriate experimentation, in the right order and logic must go hand-in-hand.

[3.14.2]Note: In Lorentz's 1904 paper "Electromagnetic phenomena in a system moving with any velocity less than that of light" [PER52 pg 11], a reprint of an English translation. Lorentz starts out at the very beginning of his paper that the old physics of his day is in conflict with Michelson's well-known interference experiment, which leads Fitzgerald and him to investigate the issue. It is interesting that Einstein claims not to have known about this paper at the time he wrote his paper on Relativity [PER52 pg 38 footnote].

Still another account supports this claim in a letter, by R.S. Shankland, who visited Einstein on February 4, 1950, from the Case Institute of Technology, while preparing a historical account of the experiment. "When I asked him how he had learned of the Michelson-Morley experiment he told me he had become aware of it through the writings of H.A. Lorentz, but only after 1905 had it come to his attention." "Otherwise," he said, "I would have mentioned it in my paper." [CLA71 pg 96] His fountainhead had arisen as he quoted so often from James Clark Maxwell's work, which was the foundation for Einstein's paper on relativity and including key ideas for his own mathematics. Still there is the onus that he had to have known about the experiment if he had read Lorentz's work the two papers parallel each other very strongly, apparently after 45 years, pride and time had taken their toll. It also occurs to me that inventive parallelism might have been at work here; when something is ready to be invented, everybody invents it simultaneously.

[3.14.3]Note: In Einstein's original paper *On the Electrodynamics of Moving Bodies*, 1905 it is not delineated as "LAW" but simply numbered. I titled it as "LAW" to be consistent with Newton's use, thereby clarifying it as an axiom for Einstein's Physical System of Logic. Also, it is index with a prefix by section in his paper.

Section 3.15 Thoughts on Developing Axiomatic Systems

Axiomatic systems are good and bad. Good because they give us a place to start inquires into the topic at hand without creating circular logical arguments. Bad, because it creates a premise of isolation, they do not need to be questioned prior to the investigation, making them appear even further distant or uncoupled from everything else. However, this insulation against prior investigation can work against the developer, because it puts blinders on preventing a view of the grander panorama of the vast landscape that is mathematics. It also creates what appear to be islands of logic, apparently independent of anything else. Yet, if Robert Langlands is correct everything is connected, however in spite of this mathematics has, to some existent, been successfully developed over the years, in an irony do to its insular nature. What appear to be islands of logic can be bridged under universal topics or provable conjectures for example; the discipline of algebra is independent of geometry except of course under the umbrella of analytical geometry. However, if one carefully looks at those systems the sets of axioms that they are built on, and not surprisingly, because of connectivity, certain consistencies seem to arise forming common patterns. These patterns can be summarized by the use of a group or groups characterizing the system. Also, if appropriate an increase in complexity can be developed by using of the ideas of the Abelian group and/or binding them with the notion of a Ring.

In algebra, the axioms start with an "Existence Axiom" describing the properties of the system being studied. The second axiom type is an equality or reflexive axiom between quantities that defines what constitutes the "Equality Axiom". Finally followed by the third the "Substitution Axiom" based upon the notion of equating one quantity to another. The "Transition Axiom" replaces substitution, where separate quantities are congruent, but not equivalent, such as two dodecahedrons of the same radius.

All known systems appear to initiate with some form of these three types of axioms and of course radically diverge from that point on do to unique and different properties of the logic. So, in setting up, any new system to be studied a safe and complete systematic approach would be to find, if appropriate, representation of these three axioms; -- Existence, Equality and Substitution.

This also gives rise to the notation that if there, are an infinite number of rules in any given system, and so far, no one has found a limit to the collection of all systems, it would take an infinite number of axioms to completely describe all of mathematics. So, the idea that a small finite set of universal axioms can represent mathematics in creating the ideal logical system would simply not be possible [SIN97, pg 139] [EVE76, pg 483].

While theorems of course vary, being dependent on the specifics of the system, there seems to be basic theorems that are universal such as the "Uniqueness Theorem" under the equality of the system's group operator. Mistakenly, since the Uniqueness Theorem is so fundamental and found at the beginning of the axiomatic system, it is sometimes been erroneously used for a basic propriety or axiom. So far, in the study of these various systems in math it has always been derivable, hence not an axiom. In the end that is the crucial test of whether a proposition is an axiom or a theorem, if it is derivable within the confines of the system.

When you're outside of the Galois Group axiomatic structure, theorems can be constructed proving such propositions. This happens at times, because the group axioms may not be the most fundamental way of stating the properties for the logical system. Examples can be found in the axioms on transformations (see Note[3.18.1]) in chapter 2 and again for tensors in chapter 10, but good proof writing should always attempt to pattern theorems corresponding to groups if there not axioms, because this provides a systematic way of making sure you have developed a thorough set of theorems representing your system.

Section 3.16 Landmines in Logic: When Logic isn't Logical.

This chapter expounds on the successes of logic and the mathematical system, but there are times when the logic fails us. This usually comes about for one or two reasons. One, logical contradictions, or not enough information is present to properly understand the problem. Two, our knowledge about the foundation, for an axiom set, which the system is built, is incorrect [False] or simply incomplete.

Not enough information leading to a wrong conclusion is pretty much, self-evident, so let's consider the first issue logical contradictions, giving rise to conflicts.[3.16.1] Consider the following syllogistic argument:

>All A is (are) P.
>All B is (are) A.
>∴All B is (are) P.

which by the mechanics of our logic is perfectly valid. Now let

>A ≡ Greeks
>P ≡ men
>B ≡ Athenians

used in the sense of the plural verb [are].

>All Greeks are men.
>All Athenians are Greeks.
>∴All Athenians are men.

and it logically holds that is makes sense to the reader. Now let

>A ≡ sodium salts
>P ≡ water-soluble substances
>B ≡ soaps

>All sodium salts are water-soluble substances.
>All soaps are sodium salts.
>∴All soaps are water-soluble substances.

[3.16.1]Note: Contradictions are a broad topic in logical philosophy and only certain aspects are considered here. [COP68, pg 146, 244, 271]

So, by substitution with other propositions it is found that the structure of the argument still holds. It would appear that a valid syllogism is a formally valid argument, valid by virtue of its form alone. ***This implies that if a given syllogism is valid, another syllogism of the same form will all so be valid.*** And it would follow that if a syllogism were invalid; any other syllogism of the same form would be invalid. Let's try again, now let

> All A is (are) P.
> Some B is (are) P.
> ∴ Some B is (are) A.

and

> A ≡ communists
> P ≡ proponents of socialized medicine
> B ≡ members of the administration

> All communists are proponents of socialized medicine.
> Some members of the administration are proponents of socialized medicine.
> ∴ Some members of the administration are communists.

The validity of this argument is not exactly clear and it may-or-may not follow that some administrators are communists by guilt of association only. The best way of exposing its fallacious character would be to construct another argument having exactly the same form, but whose invalidity was immediately apparent. So, we might reply by saying, "You might as well argue that" by letting:

> A ≡ rabbits
> P ≡ very fast runners
> B ≡ horses

> All rabbits are very fast runners.
> Some horses are very fast runners.
> ∴ Some horses are rabbits.

Which cannot be seriously defended, hence is fallacious. It can be immediately seen that it is false, because our real world experience, which comes from outside the logical system, very quickly reviles that horses and rabbits are clearly very different species, hence cannot be the same. So here the form is valid, but the premises lead to a false conclusion. Also this kind of substitution is not foolproof you might not be able to find a set of premises to test that would lead to a false conclusion, yet that doesn't mean there is not such a set. Since the syllogism is based on three premises, a technique(s) is needed to test three at a time.

Let's expand the diagrammatic two categorical propositions of Section 3.3 to a three-premise system looking categorically at all possible configurations at once. The following Venn Diagram fits this requirement nicely.

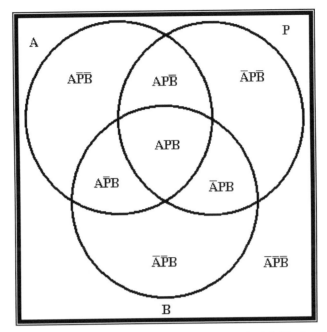

Figure 3.16.1 Venn Diagram for a Categorical Syllogism

We have three basic regions in the Venn Diagram representing the three propositions ($A\overline{P}\overline{B}$, $A\overline{P}B$ and $\overline{A}PB$) of the syllogism. Regions that do not include a proposition are negated with a bar across the top of the proposition label.

Theorem 3.16.1 Truth Table: Categorical Syllogisms by Threes

A	P	B	$AP\overline{B}$	$A\overline{P}B$	$\overline{A}PB$	$A\overline{P}\overline{B}$	$\overline{A}\overline{P}B$	$\overline{A}P\overline{B}$	$\overline{A}\overline{P}\overline{B}$	APB
T_g	T_g	T_g	F	F	F	F	F	F	F	T_g
T_y	T_y	F_y	T_y	F	F	F	F	F	F	F
T_r	F_r	T	F	T_r	F	F	F	F	F	F
T_y	F_y	F	F	F	F	T_y	F	F	F	F
F_r	T_r	T	F	F	T_r	F	F	F	F	F
F	T_y	F_y	F	F	F	F	F	T_y	F	F
F	F_r	T_r	F	F	F	F	T_r	F	F	F
F_b	F_b	F_b	F	F	F	F	F	F	T_b	F

From the above truth table, the following identities can be established.

Table 3.16.1 Identities of Categorical Syllogisms by Threes

Ref	Identities		
Lx3.16.1.1	$A P \bar{B}$	$\equiv$	$\sim(P \rightarrow B)$ A
Lx3.16.1.2	$A \bar{P} B$	$\equiv$	$\sim(A \rightarrow P)$ B
Lx3.16.1.3	$\bar{A} P B$	$\equiv$	$\sim(P \rightarrow A)$ B
Lx3.16.1.4	$A \bar{P} \bar{B}$	$\equiv$	$\sim(A \rightarrow P)$ ~B
Lx3.16.1.5	$\bar{A} \bar{P} B$	$\equiv$	$\sim(B \rightarrow P)$ ~A
Lx3.16.1.6	$\bar{A} P \bar{B}$	$\equiv$	$\sim(P \rightarrow B)$ ~A
Lx3.16.1.7	$\bar{A} \bar{P} \bar{B}$	$\equiv$	$A \rightarrow$ $(P \rightarrow$ ~B)
Lx3.16.1.8	$A P B$	$\equiv$	$\sim(A \rightarrow \sim(P \rightarrow$ ~B))

As with Section 3.7 the identities establish a syllogism to be diagrammatically represented by Venn Diagrams.

Let:

All B is (are) P.	$\bar{A} P B$
All A is (are) B.	$A \bar{P} B$
∴ All A is (are) P.	$A P \bar{B}$

Now the syllogism is valid if and only if the two premises imply or entail the conclusion. That is, if together they assert what is asserted by the conclusion. Consequently, diagramming the premises of a valid argument should suffice to show what its conclusion asserts also, with no further marking of the circles needed. To diagram the conclusion "All A is P" is to shade out both the portion labeled $A \bar{P} \bar{B}$ and the portion labeled, $A \bar{P} B$. Inspecting the diagram, which represents the two premises, we see that it does diagram the conclusion also. From this we can conclude the syllogism is valid.

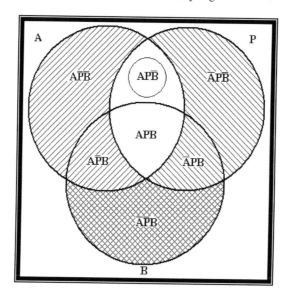

Figure 3.16.2 Diagramming a Valid Syllogism

All B is (are) P. $\bar{A} P B$ the shaded area excludes A: $\bar{A} P = 0$ as demonstrated in F3.3.3A
All A is (are) B. $A \bar{P} B$ the shaded area excludes P: $A \bar{P} = 0$ as demonstrated in F3.3.3A

So

All A is (are) P. $A P \bar{B}$

implying the circled conclusion in the above Venn Diagram is correct because there is no shaded conflict.

Lets look at an invalid diagram:

All dogs are mammals. $D \bar{C} \bar{M}$
All cats are mammals. $\bar{D} C \bar{M}$
∴ All cats are dogs. $D C \bar{M}$

Diagramming the syllogism:

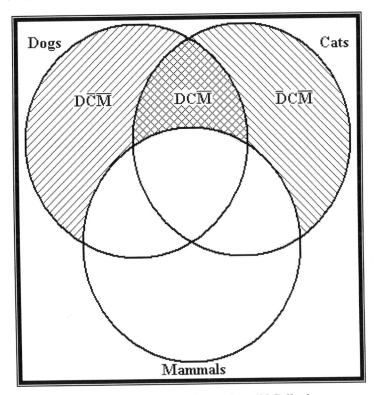

Figure 3.16.3 Diagramming an Invalid Syllogism

Unlike the diagrammatic conclusion of the valid syllogism, where there was no over lapping shaded region here there is a visual conflict in the conclusion of the logical area $D C \bar{M}$.

In summary if a Venn Diagram shows no shaded conflict the logic is correct.

Section 3.17 The Case of the Incomplete Set of Axioms

Now let's look at what happens when our knowledge about the foundation, the set of axioms, upon which the system is built, is incorrect [False] or incomplete. Kurt Gödel who proved that trying to create a complete system of axioms to some depth was an impossible task. His ideas could be encapsulated in two statements. [SIN97, pg 139] [EVE76, pg 483]

Axiom 3.17.1 Gödel's First Proposition of Undecidability

If axiomatic set theory is consistent, there exist theorems that can neither be proved nor disproved.

Axiom 3.17.2 Gödel's Second Proposition of Undecidability

There is no constructive procedure that will prove axiomatic theory to be consistent.

Essentially Gödel's first statement said that no matter what set of axioms were being used there would be questions that the discipline of mathematics could not answer, completeness could never be achieved. Worse still, the second statement said that mathematicians could never even be sure that their choice of axioms would not lead to a contradiction, consistency could never be proved.

This would be analogous to Heisenberg's uncertainty principle in physics, the closer one tries to measure the position of a nuclear particle the greater ones uncertainty to its true location, because the photons used to measure the position interfere with the measurement itself, thereby degrading it. Here because we can never prove consistency in our axiomatic set we can never be, absolutely, certain about the validity of our math.

Section 3.18 Fallacies of Arguments

The Greeks knew that logic was riddled with illogical arguments in public speaking so they hit upon the idea of classifying or organizing those arguments so they could be more easily identified proving the arguments of the speaker as false. Common to all arguments, which commit fallacies of relevance, except for the fallacy of Petitio Principii, or begging the question is the circumstance that their premises are logically irrelevant to, and therefore incapable of establishing the truth of, their conclusions. [COP68 pg 60]

Definition 3.18.1 Argumentum ad Baculum (appeal to force)

The *argumentum ad baculum* is the fallacy committed when one appeals to force or the threat of force to cause acceptance of a conclusion.

Definition 3.18.2 Argumentum ad Hominem (abusive)

The phrase *argumentum ad hominem* translates literally as "argument directed to the man." It is susceptible to two interpretations, whose interrelationship will be explained after the two are discussed separately. We may designate this fallacy on the first interpretation as the "abusive" variety. It is committed when, instead of trying to disprove the truth of what is asserted, one attacks the man who made the assertion. Thus it may be argued that Bacons's philosophy is untrustworthy because he was removed from this chancellorship for dishonesty. This argument is fallacious, because the personal character of a man is logically irrelevant to the truth or false hood of what he says or the correctness or incorrectness of his argument.

The classic example of this fallacy has to do with British law procedure. There the practice of law is divided between solicitor, who prepare the cases for trial, and barristers, who argue or "plead" the cases in court. Ordinarily their cooperation is admirable, but sometimes it leaves much to be desired. On one such latter occasion, the barrister ignored the case completely until the day it was to be presented at court, depending upon the solicitor to investigate the defendant's case and prepare the brief. He arrived at court just a moment before the trial was to begin and was handed his brief by the solicitor. Surprised at its thinness, he glanced inside to find written: "No case; abuse the plaintiff's attorney!"

Definition 3.18.3 Argumentum ad Hominem (circumstantial)

Argumentum ad hominem is an argument made from a belief to the circumstances. An argument is made from a person's beliefs to their circumstances. If one's opponent is, a Republican, one may argue, not that a certain proposition is true, but that he ought to assent to it because it is implied by the tenets of his party.

Definition 3.18.4 Argumentum ad Ignorantiam (argument from ignorance)

The argumentum ad ignorantiam something is true because it cannot be proven. There must be extraterrestrials because nobody has ever been able prove there aren't any.

Definition 3.18.5 Argumentum ad Misericordiam (appeal to pity)

The *argumentum ad misericordiam* is the fallacy committed when pity is appealed to for the sake of getting a conclusion accepted.

Definition 3.18.6 Argumentum ad Populum (appeal to the mass)

The ***argumentum ad populum*** fallacy is an attempt to win popular assent to a conclusion by arousing the feelings and enthusiasms of the multitude. This is a favorite device with propagandist, the demagogue, and the advertiser.

Definition 3.18.7 Argumentum ad Verecundiam (appeal to authority)

The argumentum ad verecundiam is the appeal to authority – that is, to the feeling of respect people have for the famous – to win assent to a conclusion.

Definition 3.18.8 Accident

The fallacy of ***accident*** consists in applying a general rule to a particular case whose "accidental" circumstances render the rule inapplicable. pg 67

Definition 3.18.9 Converse Accident (hasty generalization)

Out of a set of cases one considers lets say only exceptional cases and hastily generalizes to a rule that fits them alone the fallacy committed is that of converse accident.

Definition 3.18.10 False Cause

First type of ***false cause*** is ***non causa pro causa*** meaning to mistake what is not the cause of a given effect for its real cause.

Second type of false cause is ***post hoc ergo propter hoc*** to inference that one event is the cause of another from the bare fact that the first occurs earlier than the second, so any argument that incorrectly attempts to establish a causal connection as an instance of the fallacy of false cause.

Definition 3.18.11 Petitio Principii (begging the question)

Assumes the premise of the argument as the conclusion, the committed fallacy is ***petitio principii***. This circular argument assumes the conclusion as the premise that one is trying to prove.

Definition 3.18.12 Complex Question

A ***complex question*** cannot have a simple answer of yes or no, because in saying so implies your involvement either way. As an example "Have you stopped beating your wife?" if the answer is yes then you were beating your wife. If no then you are still beating your wife.

Definition 3.18.13 Ignoratio Elenchi (irrelevant conclusion)

The fallacy of ***ignoratio elenchi*** is committed when an argument purporting to establish a particular conclusion is directed to proving a different conclusion. This misdirected argument assumes the validity of a conclusion and then uses it to prove another conclusion.

Tensor Calculus & Physics: A General Treatise

Table of Contents

Tensor Calculus & Physics: A General Treatise

Tensor Calculus & Physics: A General Treatise

List of Figures

List of Tables

List of Observations

List of Conjectures

Chapter 4 Foundation of Mathematics: The Algebraic Axiomatic System

This chapter deals with the axiomatic system upon which the arithmetic of numbers and their abstraction algebra rests. Other disciplines are discussed in other chapters, but if they are not in the direct path of this treatise, than they are set-aside in the appropriate chapter appendix. As an example in the next two chapters, Geometry and Linear Algebra are discussed while Trigonometry and Series are simple lookup tables in Appendix-E.

In Appendix-D, this is associated with this chapter, deals with the algorithms of operations in arithmetic. Addition, Subtraction with carry from one place holder to another, Multiplication and Division, which require quick ways of implementing those operations have lead to development of algorithms over the centuries. These common techniques are found in our everyday lives and taken for granted. When the following definitions refer to addition, subtraction, multiplication and division, their implementation is implied by these methods. Also included are the algorithms for finding the nth root as found in A4.2.25 "Reciprocal Exp".

Tensor Calculus & Physics: A General Treatise

Section 4.1 The Algebraic Primitive Definitions

Definitions of the Real Number System:

In our modern system of being able to write any possible number, positional numbers have been developed, which can represent rational and irrational numbers. Yet, these aggregate numbers are made up of zero, countable numbers and non-rational numbers. In order to talk about definitions D4.1.1 "Positive Integers [+n]: Natural or Countable numbers" to 4.1.28 "Least Common Denominator (LCD)" they have to be defined first in simple specifics of positional numbers in order to understand the concepts involved. So the situation of a circular definition creeps into these designations, for those situations it is understood that the positional number is left as an undefined term, see D3.1.7, in a sense, till definition D4.1.28 "Least Common Denominator (LCD)" formally defines decimal number.

The number system as it is known today is a result of a gradual development over a number of millennia the results of which are synthesized into the following definitions. [SPI64]

Definition 4.1.1 Positive Integers [+n]: Natural or Countable numbers

Countable numbers such as 1, 2, 3, 4, ..., are called *positive integers* sometime also called *whole or natural numbers*, and were first used in counting. The symbols varied with the culture and the times, e.g. the Romans used I, II, III, IV, ... Algebraically it permits a solution to the equation $0 \leq x$.

The positive integers also are comprised of subsets of special positive numbers having various properties. These numbers can be categorized into two major types, the subset of *prime numbers*, numbers that can only be divided by one and themselves, the balance are *factorable numbers*. Euclid proved that the set of prime numbers is infinite in book IX proposition 20. [EVE76, pg 119] Euclid also showed, by the fundamental theorem of arithmetic, that all other integers are comprised of prime numbers, the building blocks, which the factorable numbers are multiplicatively made. [EVE76, pg 119, 434]. By definition, since a prime number is only divisible by itself it would follow that they are not divisible by two, except 2 which is a prime itself, hence are an odd number. The primes 1, 2 and 3 are the only primes that are contiguous; as primes become larger, they are spaced farther and farther apart. So prime numbers greater than 2 are bound on either side by two even numbers, since odd and even number alternate. These bounding numbers, not being prime numbers, are factorable numbers as odd and even numbers alternate. From these notions of what a prime is a general form for bounding a prime number can be established. Given any prime number-p than it is bounded as follows:

$$2^m f_h > p > 2^n f_w \text{ for } p > 2 \qquad \text{Equation A}$$

Where (f_w, f_h) are factorable numbers, while (n, m) are exponents creating two even leading coefficients. Since these bounding numbers can only differ from the prime number by one than the two numbers can represent a single prime number:

$$p \equiv 2^n f_w + 1 \text{ for } p > 2 \qquad \text{Equation B}$$
$$p \equiv 2^m f_h - 1 \text{ for } p > 2 \qquad \text{Equation C}$$

So far no one has proven the form or shown an equation for calculating a prime number in general, it follows that equations B and C are as close a representation as one can come. However there are special cases, where there are unique forms. Consider the subset of primes:

$$p_w = 2^n + 1 \text{ for } f_w \equiv 1 \text{ in addition to} \qquad \text{Equation D}$$
$$p_h = 2^m - 1 \text{ for } f_h \equiv 1 \qquad \text{Equation E}$$

So, even for the prime numbers, there are subsets of subsets with special characteristics.

Notice that p is not allowed to be 2. So, the above formula can be written, since all primes are odd except for the prime number 2, which is also even the exception to the rule of what prime is. Specifically 2 is bounded by the odd numbers 1 and 3, this little corner property of prime numbers is often forgotten thus far is the simplest subset of the primes in its own right.

Let's consider patterns found in factorable numbers using an integer operation, such as the Totient function $\phi(n)$. The Totient operation is for [n] relatively primed divisors, which do not exceed [n]. So if n = 10 the divisible prime numbers, between 1 and 10, into 10 are (2, 5) and the total number of divisions are $[1]_1$, $[2]_2$, $[5]_3$ and $[10]_4$, hence $\phi(10) = 4$. These relatively primed numbers in themselves form a special subset.

How many such sets are there? Since the positive integers are infinite, and so far, no objection or limitation has been placed on them, a plausible conjecture would be that, there are probably an infinite number of sets and corresponding subsets with unique features. The intricate interlacing of subsets for positive integers creates a richly woven tapestry with its infinite rainbow of colors, which sets the foundation of the advanced algebras. These disciplines, relating to the ability to count, such as combinatorial math, play a major role in many branches of learning, such as information and coding theory.

The positive integers have within them the notion for a shortcut method of addition called **multiplication**. It follows than if [a] and [b] are natural numbers than the process of multiplication has operators that can be written as

$c \equiv a \times b,$	Multiplication Operator Definition A
$c \equiv a * b,$	Multiplication Operator Definition B
$c \equiv (a)(b)$ or	Multiplication Operator Definition C
$c \equiv ab$	Multiplication Operator Definition D

where the resultant of the multiplication [c] is called the **product** of [a] and [b], [a] called the **multiplicand** and [b] the **multiplier**. Classically multiplication is represented in its vertical form as follows:

$$\begin{array}{ll} a & \text{multiplicand} \\ \underline{x\,b} & \underline{\text{multiplier}} \\ c & \text{product} \end{array}$$

Multiplication Operator Definition E

and the multiplier is the number of times the multiplicand is added to itself, hence represents a shortcut algorithm for handling repeated additions.

Definition 4.1.2 The Number Zero

1. Denoted by [0] [4.1.1] allows a person counting to know when there is nothing left to count.
2. The absence of a number, empty.
3. Algebraically it permits a solution to the equation x = 0.
4. Zero also performs another purpose in a positional numbering system, it acts as a place holder, hence as an example the number 52 can be distinguished from 502 by the position held by zero in the 10th place and forcing 5 to shift upward to the 100th place. In general for any positional number

$$N = \bullet + a_{n-1} \times b^{n-1} + 0 \times b^n + a_{n+1} \times b^{n+1} + \bullet$$

zero holds a place in the exponential position [n] for any base [b].

Definition 4.1.3 Negative Integers [–n]

Denoted by –1, –2, –3, –4, …, allows for a reversal of counting beyond zero. Algebraically it permits a solution to the equation x + b = a, for [a] and [b] any natural numbers. Negative Integers allow a return to the start of the count by the existence of the operation of *subtraction* or *inverse addition*.

Definition 4.1.4 Rational Numbers

Also called *fractions* or ***whole fractions*** and denoted by ½, 1/3, 2/3, ¼, ¾, 1/8, 3/8, 5/8, 7/8, …, allows for a partial count out of a set of countable objects, hence a count of 21 out of 50, or 21/50, or 21:50[4.1.2]. As long as a partial count out of nothing, or zero, is not tried because that would be meaningless, hence a count of 31 out of 0 or 31/0 has no meaning. There can also be a count in excess of all the countable objects so negative rational numbers must also be considered –1/3, –5/51, … . So far any real or negative numbers, such as [a] and [b], will always have the form a/b where b ≠ 0. Algebraically it permits a solution to bx = a for b ≠ 0. The ***simplest ratio*** is 1/1 that is counting one object from only one object. Rational Numbers allow a return to the simplest ratio by the existence of the operation of *division* or *inverse multiplication*.

Observe that Natural Numbers are a subset of the Rational Numbers if b = 1 for the ratio a/b or using Cantor's technique placing fractions in a one-to-one correspondence to the countable numbers so they can be enumerated.

Rational numbers imply the notion of the inverse operation of multiplication, division. It follows than if [a] and [b] are rational numbers than the process of division has operators that can be written as

$$c \equiv a / b \qquad \text{Division Operator Definition A}$$
$$c \equiv a (1/b) \qquad \text{Division Operator Definition B}^{[4.1.3]}$$
$$c \equiv \frac{a}{b} \qquad \text{Division Operator Definition C}$$
$$c \equiv a\left(\frac{1}{b}\right) \qquad \text{Division Operator Definition D}^{[4.1.3]}$$
$$c \equiv a \div b \qquad \text{Division Operator Definition E}$$

where the resultant of the division [c] is called the ***quotient*** of [a], [a] called the ***dividend*** and [b] the ***divisor*** or sometimes called a ***factor***. Where the parts of the fraction, [a] is called the ***numerator***, and [b] the ***denominator***. Classically division is represented as follows:

$$b \text{ (divisor) } \overline{)a \text{ (dividend)}}^{\;c\;\text{(quotient)}} \text{ Division Operator Definition F}$$

and the quotient is the number of times the divisor is subtracted from the dividend.

A fraction of the following form is called a ***proper fraction*** a/b for [a] and [b] integers, such that a < b. While for a > b the fraction is called an ***improper fraction***.

There are also other rules, hidden rules for rational numbers that deal with the positive coefficient. These rules are a mathematician's short hand notation that always implies a positive one is the leading coefficient weather there or not.

$$a \equiv +a \qquad \text{addition operator definition F}$$
$$+a \equiv (+1)a \qquad \text{addition operator definition G}$$

Another positive coefficient is used for another type of number called a ***mixed number***. The positive coefficient lies between the integer and proper fraction and always implies a positive one exists between the two numbers.

$$n\ a/b \equiv n + a/b \qquad \text{mixed operator definition H}$$
$$n + a/b \equiv n + (+1)\ a/b \qquad \text{mixed operator definition I}$$

[4.1.1]Note: Zero was devised in a number of cultures independent of one another such as the Maya people who lived in central America, the Chinese in the Tang and Sung dynasties including the modern symbol-O, even the Greeks had it, it would take root for a while than fade away. It came at an early stage for Al-Khwarizmi who wrote Al'Khwarizmi in the "Hindu Art of Reckoning", which describes the Indian place-value system of numerals based on 1, 2, 3, 4, 5, 6, 7, 8, 9, and 0. The lost mathematical works of the ancients, the Indian and Greeks mathematicians, were translated by the Islamic and Arabic mathematicians into their works and finally passed into Europe as it came out of the dark ages. It is seen in about 1200 AD Fibonacci introduces zero to Europe in his paper "Liber Abaci", but it wasn't really established till about the 1600's.

[4.1.2]Note: The colon [:] used in this way as a delineator between the numerator and denominator is read as ***proportional to*** and found most commonly in classical writings on fractions, but in modern times has gone out of favor for the use of the forward slash [/] or horizontal bar [—]. It is from this nomenclature that the inquiry into ratios acquires its name the study of proportions.

[4.1.3]Note: Definitions B and D are properties of fractions allowing them to be disassociated into a product of two quantities, thereby relating the fraction back to multiplication as an inverse operation.

Definition 4.1.5 Irrational Numbers

Are comprised of what is left over in the real number system for those numbers that can only be made out of rational numbers, if they have an infinitely large ratio of a numerator and denominator. In a positional numbering system it satisfies the condition that there is no periodic group of positional numbers that occur repetitiously before the set terminates at infinity. As an example 7.15609246597196238 15609246597196238 15609246597196238 … repeats infinitely or to denote periodic repetition 7.15609 24659 71962 38, the part that repeats is underlined. The number 7.1560 92465 97196 238 is not irrational because it repeats before reaching infinity and can be rendered into a rational number with finite ratio of a numerator and denominator, however numbers, such as

$$\sqrt{2} = 1.41421\ 35623\ 73095\ 04880\ \ldots$$
$$\pi = 3.14159\ 26535\ 89793\ 23646\ \ldots$$
$$e = 2.71828\ 18284\ 59045\ 23536\ \ldots$$

are never periodic these are irrational numbers and sometimes are called ***transcendental numbers*** (discovered by the Pythagorean society (School of Pythagoras) [EVE76, pg 63] [ADL66, pg 29]. These numbers were formally identified by Joseph Liouville, 1809-1882 [MAR64]), because they are unending it seemed to him the property was irrational.

From the above definition of an irrational number, the following conjecture is made in an attempt to reconcile the rational with the irrational. One can visualize a rational number that is periodically growing within well defined bounded set of conditions, and that periodicity becoming larger and larger until it only repeats at infinity, hence becoming irrational. If such a situation could exist, the number would have to be comprised as a ratio of two finite polynomials, which for finite parametric numbers form rational numbers. As these polynomials become progressively larger with the increase of the their parametric parameters the periodicity of the rational number would become larger and larger until the periodicity reaches infinity and converges onto the desired irrational number[4.1.4]].

Conjecture 4.1.1: ∃ a Finite Polynomial Ratio that Converges onto an Irrational Number

Let τ be the transcendental number of convergence than there exists two finite rational polynomials $N(\eta)$ and $D(\delta)$ for (η, δ), **not necessarily equal**, where

$$\frac{\lim_{\eta \to \infty} N(\eta)}{\lim_{\delta \to \infty} D(\delta)} = \tau$$

, such that (η, δ) are related by the existence of $\eta(\delta)$ and $\delta(\eta)$.

The idea is that starting out with the limit first guarantees the existence of the two rational polynomials, such that they generate an infinitely large numerator and denominator in just the right way as η and δ grow to infinity satisfying the limit τ. Now η and δ being invertible, with respect to one other, are dependent, thereby forcing convergence at an uneven rate and an uneven growth between the two limits.

[4.1.4]]Note: The operation of multiplication and division apply equally as well to the irrational numbers as the rational, hence all notation follows as well: D4.1.1A, B, C, D and E and D4.1.4A, B, C, D, E and F.

Definitions of Algebra:

Definition 4.1.6 Identity element

The quantity [e] has the property in a group to allow the existence of the identity and inverse axioms.

Definition 4.1.7 Greater Than Inequality [>]

The quantities $b > a$ is true iff, $0 < b - a$

Definition 4.1.8 Less Then Inequality [<]

The quantities $a < b$ is true iff, $0 < b - a$

Definition 4.1.9 Greater Than and Equal to [≥]

The quantities $a \geq b$ is true iff, the two cases $a > b$ is true or the possibility $a = b$ is required to be considered.

Definition 4.1.10 Less Than and Equal to [≤]

The quantities $a \leq b$ is true iff, the two cases $a < b$ is true or the possibility $a = b$ is required to be considered.

Definition 4.1.11 Absolute Value | a |

The absolute value of the quantity [a] is taken, such that there is always a non-negative value associated with the absolute value | a |.

$$|a| = \begin{cases} +a & a \geq 0 \\ -a & a < 0 \end{cases}$$
  Equation A

or

$$0 \leq |a|$$
  Equation B

Definition 4.1.12 ANDing Greater Than Inequalities

For quantities $a > x$ AND $x > b$ inclusive they can be rewritten in a short hand form as

$$a > x > b$$

This notation can be extended to any combination of greater than and equal to inequality, [≥].

Definition 4.1.13 ANDing Less Than Inequalities

For quantities $a < x$ AND $x < c$ inclusive they can be rewritten a short hand form as

$$a < x < b$$

This notation can be extended to any combination of less than and equal to inequality, [≤].

Definition 4.1.14 ORing Greater Than Inequalities

For quantities x > a OR b > y are exclusive such that a ≥ b they are said to be

$$x > a \text{ OR } b > y.$$

This notation can be extended to any combination of greater than and equal to inequality, [≥].

Definition 4.1.15 ORing Less Than Inequalities

For quantities x < a OR b < y are exclusive such that a ≤ b they are said to be

$$x < a \text{ OR } b < y.$$

This notation can be extended to any combination of less than and equal to inequality, [≤].

In summary:

Number Line Geometry	Description	Inequality Formula
	**Open interval. Does not include the point a or b.**	**a < b** Can be written in terms of the end points as (a, b).
	**Closed interval. Does include the point a and b.**	**a ≤ b** Can be written in terms of the end points as [a, b].
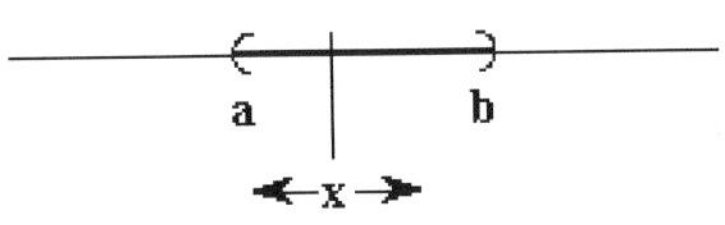	**Pick a number [x] in the following open interval:**	**a < x AND x < b** or short hand **a < x < b**
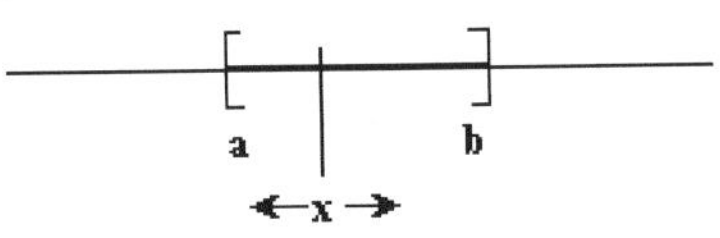	**Pick a number [x] in the following closed interval:**	**a ≤ x AND x ≤ b** or short hand **a ≤ x ≤ b**
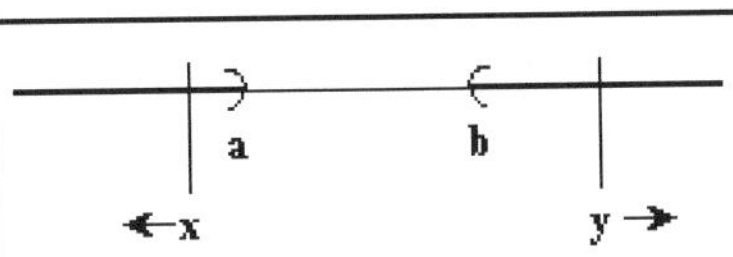	**Pick an [x] and [y] for the following open and unbounded interval:**	**x < a OR b < y** Let y = x. **x < a OR b < x**
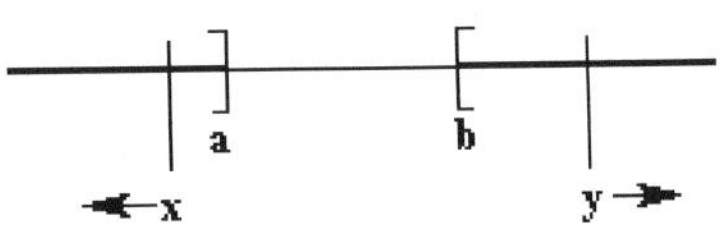	**Pick an [x] and [y] for the following closed and unbounded interval:**	**x ≤ a OR b ≤ y** Let y = x. **x ≤ a OR b ≤ x**

Figure 4.1.1: Inequalities on a Number Line

Definition 4.1.16 Distance or Magnitude on a Number Line

Distance or magnitude on a number line is within a closed interval between two points and always positive.

Distance Formula over $[P_1(x_1)P_0(x_0)] = |P_1(x_1)P_0(x_0)|$ Equation A

       $|P_1(x_1)P_0(x_0)| = |x_1 - x_0|$ Equation B

       $|P_1(x_1)P_0(x_0)| = x_1 - x_0$ for $x_0 < x_1$ Equation C

Definition 4.1.17 Exponential Notation

Exponential notation is the symbolic representation of the products of n-quantities.

$$\mathbf{a}^n \equiv \Pi^n \, o \, \mathbf{a}, \text{ for any operation [o]}$$

where [a] is any quantity $\in$ R and [n] is natural number called the *exponent*.

Definition 4.1.18 Negative Exponential

Allowing for the situation a negative exponent notation it is represented as follows:

$$a^{-n} \equiv 1 / a^n \equiv \frac{1}{a^n} \qquad \text{for a} \in \text{R but a} \neq 0$$

Definition 4.1.19 Primitive Definition for Rational Arithmetic

+	-2	-1	+0	+1	+2
-2	-4	-3	-2	-1	+0
-1	-3	-2	-1	+0	+1
+0	-2	-1	+0	+1	+2
+1	-1	+0	+1	+2	+3
+2	+0	+1	+2	+3	+4

−	-2	-1	+0	+1	+2
-2	+0	-1	-2	-3	-4
-1	+1	+0	+1	-2	-3
+0	+2	+1	+0	-1	-2
+1	+3	+2	-1	+0	-1
+2	+4	+3	+2	+1	+0

×	-2	-1	+0	+1	+2
-2	+4	+2	+0	-2	-4
-1	+2	+1	+0	-1	-2
+0	+0	+0	+0	+0	+0
+1	-2	-1	+0	+1	+2
+2	-4	-2	+0	+2	+4

÷	-2	-1	+0	+1	+2
-2	+1	+2	U	-2	-1
-1	+½	+1	U	-1	-½
+0	+0	+0	U	+0	+0
+1	-½	-1	U	+1	+½
+2	-1	-2	U	+2	+1

U is undefined for 0/0, ±A/0

The problem with the case of division by zero is that for a field of rational numbers they are not continuous between values, so no limit can be defined, see D4.1.4 "Rational Numbers", unlike calculus for real number fields, which if properly defined would allow the division to be evaluated. So, the only thing that can be done is to do nothing and leave them undefined.

Definition 4.1.20 Negative Coefficient

The negative coefficient is a short hand notation that always implies a negative one is the leading coefficient.

$$-a \equiv (-1)a \text{ or}$$
$$a - b \equiv a + (-b)$$

subtraction operator definition A

subtraction operator definition B

Definition 4.1.21 Positional Number System

Any written, aggregate, number N made up of an infinite series of coefficients and exponential bases, marking position, constitutes a system for writing any real number and having the following form:

$$N = \sum_{n=-\infty}^{\infty} a_n \times b^n$$

where $[a_n]$ is a positive integer coefficient counted from the sequential set 0, 1, 2, ..., b − 1 holding a position determined by the exponent [n] for any base [b].

Definition 4.1.22 Decimal Point

A positional number has one other property characterizing the written number a delineating symbol called the decimal point, which separates the integer part of the number from its fractional part. It is placed just after the first placeholder for the exponent n = 0, since the base [b] raised to a negative exponent constitutes a fraction. In the American standard system that symbol is represented by a period placed on the written line [.] while in the English or some European system by a period raised to the center of the written line [·]. The word decimal in Greek means ten and does not necessarily imply that the number is base-ten, this can be clarified by a modifying prefix, as an example hexadecimal for a base-sixteen number. So for a base-ten number the correct designation is decadecimal, which is cumbersome so most people simply know a base-ten number as a decimal number. This confusing conflict in nomenclature for base-ten can only be resolved by the context in, which the number is used. This is unfortunate, but nobody in the literature, past or present, has proposed a better solution. In general, a decimal point is represented as follows:

$$N = \sum_{n=1} a_{+n} \times b^{+n} + a_{0 \, [\bullet]} + \sum_{n=-1} a_n \times b^n$$

As an example the number pi:

$$\pi = 3.14159 \ 26535 \ 89793 \ 23646 \ ...$$

where 3 is the characterizing integer part prior to the decimal point and 14159 26535 89793 23646 ... the characterizing fractional part post decimal point.

For a formal history of the decimal point, see Appendix A.3 *The Positional Numeral and the Decimal System*.

Definition 4.1.23 Ratio of Quantities

Two quantities [A and B] that correspond to one another (A $\longleftrightarrow$ B) are said to be ***proportional quantities*** having a real ratio that can be represented as follows:

A $\longleftrightarrow$ B ***correspond to*** one another than	Equation A
A $\propto$ B, read as [A] ***proportional to*** [B],	Equation B
iff $\exists$ a ratio, $\ni$ the quantities k $\in$ R, while (A, B) $\{\in \vee \notin \}$ R, than	
A / B $\equiv$ A : B $\equiv$ k the ***ratio of Quantities***	Equation C

Definition 4.1.24 Constant of Proportional Quantities

Two or more pairs of quantities that have real ratios and are equal to one another are said to be ***ratios of constant proportionality*** that can be represented as follows:

A $\longleftrightarrow$ B and C $\longleftrightarrow$ D than	Equation A
A $\propto$ B and C $\propto$ D,	Equation B
iff $\exists$ a set of ratios, $\ni$ the quantities (k, k') $\in$ R, while (A, B, C, D) $\sim \vee \in$ R, than	
A / B $\equiv$ k and C/D $\equiv$ k' such that	Equation C
k $\equiv$ k' are ratios of constant proportionality.	Equation D

Definition 4.1.25 Proportional Statement

Proportional Statement is two or more ratios of constant proportionality.

$$(. = A_i / B_i .) = k \qquad \text{for all i and k} \in R$$

Definition 4.1.26 Greatest Common Multiple (GCM)

The Greatest Common Multiple for $(. a_i, p_h, e_h .) \in$ IG where p_h are distinct primes, which divide all a_i's and e_h's, the smallest exponent to which the prime is raised, then

$$GCM(. a_i .) = \prod p_h^{e(h)} \text{, such that}$$
$$p_h \in (. a_i .) \text{ for all i and h. [BRI60 Vol. 2, pg 356]}$$

Definition 4.1.27 Least Common Multiple (LCM)

Least Common Multiple for $(. a_i, p_t, n_t .) \in$ IG where p_h are distinct primes, which divide all a_i's and n_t's, the largest exponent to which the prime is raised, then

$$LCM(. a_i .) = \prod p_t^{n(t)} \text{, such that}$$
$$p_t \in (. a_i .) \text{ for all i and t. [BRI60 Vol. 2, pg 356]}$$

Definition 4.1.28 Least Common Denominator (LCD)

Least Common Denominator for $(m_k, M_k) \in IG$ and where m_k is not necessarily prime, but comprised of such numbers for any given k^{th} number.

Now given a series of a rational fractions

$$\sum \frac{1}{M_k}$$

There is a set of lowest common integer numbers m_k's factored from M_k's, where

$$\sum \frac{m_k}{m_k M_k}$$

such that

$$\sum \frac{m_k}{LCD} = \frac{1}{LCD} \sum m_k$$

for

$$LCD = (\bullet = M_k \, m_k \, \bullet) \text{ for all } k \qquad\qquad \text{Equation A}$$

where $(\bullet \, m_k \, \bullet)$ are made up the lowest distinct multiple primes, such that

$$LCM(\bullet \, M_k \, \bullet) \equiv \prod_k m_k \qquad\qquad \text{Equation B}$$

Table 4.2.1: Axioms for Quantities

Axiom	Axioms of Quantities	Description
Axiom 4.2.1 Existence [4.2.1]	∃ a field R ∋, quantities $(0, n, -m, a/b, \tau) \in R$	Existence of an arithmetic field
Axiom 4.2.2 Equality [4.2.2]	$a = b$ Equation A $b = a$ Equation B	Equality of quantities
Axiom 4.2.3 Substitution [4.2.3]	$b = a$ and $c = b$ $\therefore c = a$	Substitution of quantities

Axioms by Group Quantities [4.2.4]	Description
∃ a field G ∋, quantities where $(e, a, a^{-1}) \in G$	Existence of a Group Field where [e] is called the identity element and has the unique identity property of leaving quantities unaltered under the operation of the group system.
$a = b$ Equation A $b = a$ Equation B	Equality of quantities
$b = a$ and $c = b$ $\therefore c = a$	Substitution of quantities

Table 4.2.2: Axioms for Addition

Axiom	Axioms by Addition	Description
Axiom 4.2.4 Closure Add	$[(a + b \wedge b + a) \wedge$ $(a - b \wedge b - a)] \in R$	Closure with respect to Addition and Subtraction
Axiom 4.2.5 Commutative Add [4.2.5]	$a + b = b + a$	Commutative by Addition
Axiom 4.2.6 Associative Add	$a + (b + c) = (a + b) + c$	Associative by Addition
Axiom 4.2.7 Identity Add	$a + 0 = a$	Identity by Addition
Axiom 4.2.8 Inverse Add [4.2.6]	$a - a = 0$	Inverse by Addition

Axioms by Group Operator	Description
$(a \circ b \wedge b \circ a) \in G$	Closure with respect to Group Operator
$a \circ b = b \circ c$	Commutative by Group Operator
$a \circ (b \circ c) = (a \circ b) \circ c$	Associative by Group Operator
$a \circ e = a$	Identity by Group Operator
$a \circ a^{-1} = e$	Inverse Group Operator by reciprocal Quantity

Table 4.2.3: Axioms for Multiplication

Axiom	Axioms by Multiplication	Description
Axiom 4.2.9 Closure Multp	$[\,(ab \wedge ba)\, \wedge$ $(a \div b \wedge b \div a)] \in R$	Closure with respect to Multiplication and Division
Axiom 4.2.10 Commutative Multp[4.2.5]	$ab = bc$	Commutative by Multiplication
Axiom 4.2.11 Associative Multp	$a(bc) = (ab)\,c$	Associative by Multiplication
Axiom 4.2.12 Identity Multp	$a\,1 = a$	Identity by Multiplication
Axiom 4.2.13 Inverse Multp[4.2.6]	$a\,/\,a = 1$ for $a \neq 0$	Inverse by Multiplication

Axioms by Group Operator	Description
$(a \lozenge b \wedge b \lozenge a) \in G$	Closure with respect to Group Operator
$a \lozenge b = b \lozenge c$	Commutative by Group Operator
$a \lozenge (b \lozenge c) = (a \lozenge b)\, o\, c$	Associative by Group Operator
$a \lozenge e = a$	Identity by Group Operator
$a \lozenge a^{-1} = e$	Inverse Group Operator by reciprocal Quantity

Table 4.2.4: Axiom with Complementary Operation

Axiom	Axioms by Complement	Description
Axiom 4.2.14 Distribution[4.2.7]	$a\,(b + c) = ab + ac$	Distribution by multiplication across addition

Axioms by Complement	Description
$a \lozenge (b\, o\, c) = (a \lozenge b)\, o\, (a \lozenge c)$	Distribution by complementary operator across group operator

[4.2.1]Note: The existence axiom makes sure our toolbox has everything in it we need to do the job.

[4.2.2]Note: This axiom is always taken for granted, and almost, always forgotten in proof writing, yet without it nothing in a proof could start. This axiom sets the stage for algebraic equality and identity, it is the balance bar of the fulcrum and lever and without it, the jugglers, on either side, have no place to stand to perform their act.

[4.2.3]Note: Substitution is actually a classic logical proof from the Greeks called a Syllogism. Like A4.2.2 it is usually little understood and sometimes glossed over, yet without it the speed of proof writing would be slow if not impossible to do in some cases.

[4.2.4]Note: The reference to Expanded Galois Group verses Formal Galois Group refers to the classical definition of group as stated in the proper sense, while the expanded form simply acknowledges the more encompassing notion of the axiomatic system with the extended properties of existence, equality and substitution. These things Galois might not have considered, since he had more pressing matters on his mind like going out and being shoot to death in a pointless dual, by his political enemies.

[4.2.5]Note: This is only true iff, the group operator supports this relationship.

[4.2.6]Note: This is only true iff, the field of the group contains a quantity $[a^{-1} \in G^n]$ with the property to permit an inverse operation. A4.2.1 guarantees this unless other wised specified.

[4.2.7]Note: This is only true iff, two groups constitute a ring having complementary operators $[\lozenge]$ and $[o]$.

Table 4.2.5 Axioms for Inequalities

Axiom	Axioms by Complement	Description
Axiom 4.2.15 Inequality	$\exists\,(a, b) \in R$ such that $a < b$	Inequality of quantities
Axiom 4.2.16 The Trichotomy Law of Ordered Numbers	$\exists\,(a, b, x) \in R$ such that a number line has the following form: $-b < -a < 0 < a < b$ or specifically stated in three parts $\forall x\ \exists\mid$ one and only one true statement such that: $0 < x$ Equation A or $x = 0$ Equation B or $x < 0$ Equation C	Order of quantities on a number line.
Axiom 4.2.17 Correspondence of Equality and Inequality	A4.2.1 to 14 hold true for inequalities as long as the order of the inequality is left unchanged.	Correspondence of equality axioms with inequalities.

Table 4.2.6: Axioms for Exponentials

Axiom	Axioms by Exponents	Description
Axiom 4.2.18 Summation Exp	$a^{m+n} = a^m a^n$	Summation of exponents.
Axiom 4.2.19 Difference Exp	$a^{m-n} = \dfrac{a^m}{a^n}$	Difference between exponents.
Axiom 4.2.20 Commutative Exp	$(a^m)^n = (a^n)^m$	Commutative Property of exponents.
Axiom 4.2.21 Distribution Exp	$(ab)^n = a^n b^n$	Distribution of an exponent across a product.
Axiom 4.2.22 Product Exp	$a^{mn} = (a^m)^n$	Product of exponents
Axiom 4.2.23 Identity Exp	$a^1 = a$	Identity by exponent
Axiom 4.2.24 Inverse Exp	$a^0 = 1$ and $(-1)^0 = 1$	Inverse by exponent
Axiom 4.2.25 Reciprocal Exp	Iff, $b^n = a$ and $0 < n$ than $\exists \mid b = a^{1/n} = \sqrt[n]{a}$ the inverse operation of an exponent.	Reciprocal of an exponent The inverse operation of an exponent is called the *n^{th} root* of [a] and [$\sqrt[n]{}$] is called the *radical of n* for [a]. For simplicity when n = 2 the two is dropped from the radical and it is implied that it is there.
Axiom 4.2.26 Logarithms	Iff, $b^x = y$ for $(b, x, y) \in R$ than $\exists \mid x = \log_b y$ for $0 < b$ the inverse operation of an exponent	The logarithm of an exponent.
Axiom 4.2.27 Correspondence Exp	A4.2.18 to 24 hold true for real exponents in A4.2.26.	Correspondence between integer and real exponent axioms.

Section 4.3 Theorems on Equalities: Addition

Theorem 4.3.1 Equalities: Uniqueness of Addition[4.3.1]

1g	Given	$(a, b, c, d, f) \in R$
2g		$a = b$
3g		$c = d$
Steps	Hypothesis	$a + c = b + d$ for $a = b$ and $c = d$
1	From 1g and A4.2.4 Closure Add	$a + c = f$
2	From 2g, 3g, 1 and A4.2.3 Substitution	$b + d = f$
∴	From 1, 2, A4.2.2B Equality and A4.2.3 Substitution	$a + c = b + d$ for $a = b$ and $c = d$

Theorem 4.3.2 Equalities: Cancellation by Addition

1g	Given	$a + c = b + c$
Steps	Hypothesis	$a = b$
1	From A4.2.2A Equality	$-c = -c$
2	From 1, 1g and T4.3.1 Equalities: Uniqueness of Addition	$(a + c) -c = (b + c) -c$
3	From 2 and A4.2.6 Associative Add	$a + (c -c) = b + (c -c)$
4	From 3 and A4.2.8 Inverse Add	$a + 0 = b + 0$
∴	From 4 and A4.2.7 Identity Add	$a = b$

Theorem 4.3.3 Equalities: Right Cancellation by Addition

1g	Given	$a = b + c$			
Steps	Hypothesis	$a - c = b$ Equation A		$b = a - c$ Equation B	
1	From A4.2.2A Equality	$-c = -c$			
2	From 1, 1g and T4.3.1 Equalities: Uniqueness of Addition	$a - c = (b + c) -c$			
3	From 2 and A4.2.6 Associative Add	$a - c = b + (c -c)$			
4	From 3 and A4.2.8 Inverse Add	$a - c = b + 0$			
∴	From 4 and A4.2.7 Identity Add	$a - c = b$ by A4.2.2A		$b = a - c$ by A4.2.2B	

Theorem 4.3.4 Equalities: Reversal of Right Cancellation by Addition

1g	Given	$b = a - c$			
Steps	Hypothesis	$b + c = a$	Equation A	$a = b + c$	Equation B
∴	From T4.3.3B Equalities: Right Cancellation by Addition and T3.7.2 Reversibility of a Logical Argument	$b + c = a$	by A4.2.2A	$a = b + c$	by A4.2.2B

Theorem 4.3.5 Equalities: Left Cancellation by Addition

1g	Given	$b + c = a$			
Steps	Hypothesis	$a - c = b$	Equation A	$b = a - c$	Equation B
1	From 1g and A4.2.2B Equality	$a = b + c$			
∴	From 1 and A4.2.4 Closure Add	$a - c = b$	by A4.2.2A	$b = a - c$	by A4.2.2B

Theorem 4.3.6 Equalities: Reversal of Left Cancellation by Addition

1g	Given	$a - c = b$			
Steps	Hypothesis	$a = b + c$	Equation A	$b + c = a$	Equation B
∴	From T4.3.5A Equalities: Left Cancellation by Addition and T3.7.2 Reversibility of a Logical Argument	$a = b + c$	by A4.2.2A	$b + c = a$	by A4.2.2B

Theorem 4.3.7 Equalities: Uniqueness of Subtraction[4.3.1]

1g	Given	$a = b$	
2g		$c = d$	
Steps	Hypothesis	$a - c = b - d$	for $a = b$ and $c = d$
1	From A4.2.4 Closure Add	$a - c = f$	
2	From 1g, 2g, 1 and A4.2.3 Substitution	$b - d = f$	
∴	From 1, 2 and A4.2.2A Equality	$a - c = b - d$	for $a = b$ and $c = d$

Theorem 4.3.8 Equalities: Nullification of Subtraction

1g	Given	$c = a + b$
Steps	Hypothesis	$c = a - (-b) = a + b$
1	From 1g and A4.2.12 Identity Multp	$c = a + (+1)b$
2	From 2 and D4.1.19 Primitive Definition for Rational Arithmetic	$c = a + (-1)(-1)b$
3	From 3 and D4.1.20A Negative Coefficient	$c = a + (-1(-b))$
∴	From 1g, 4 and D4.1.20B Negative Coefficient	$c = a - (-b) = a + b$

Theorem 4.3.9 Equalities: Summation of Repeated Terms

1g	Given	$a_i = a$ for all i
2g		$n\,a = \sum_{i=1}^{n} a_i$ assume
Steps	Hypothesis	$n\,a = \sum_{i=1}^{n} a_i$ for $a_i = a$ for all n-terms
1	From 2g	$n\,a = \sum_{i=1}^{n} a_i$
2	From 1g, 1 and A4.2.3 Substitution	$n\,a = \sum_{i=1}^{n} a$
3	From 2 and A4.2.12 Identity Multp	$n\,a = \sum_{i=1}^{n} (a\,1)$
4	From 3 and A4.2.14 Distribution	$n\,a = a \sum_{i=1}^{n} (1)$
5	From 4, LxE.3.29 and A4.2.10 Commutative Multp	$n\,a = n\,a$
∴	From 1, 5 and by identity	$n\,a = \sum_{i=1}^{n} a_i$ for $a_i = a$ for all n-terms

Theorem 4.3.10 Equalities: Summation of Repeated Terms by 2

1g	Given	$n = 2$
Steps	Hypothesis	$2a = a + a$ for $n = 2$
1	From A4.2.2A Equality	$a + a = a + a$
∴	From 1g and T4.3.9 Equalities: Summation of Repeated Terms	$2a = a + a$ for $n = 2$

Theorem 4.3.11 Equalities: Summation of Repeated Terms by 3

1g	Given	$n = 3$
Steps	Hypothesis	$3a = a + a + a$ for $n = 3$
1	From A4.2.2A Equality	$a + a + a = a + a + a$
∴	From 1g and T4.3.9 Equalities: Summation of Repeated Terms	$3a = a + a + a$ for $n = 3$

Theorem 4.3.12 Equalities: Summation of Repeated Terms by 4

1g	Given	$n = 4$
Steps	Hypothesis	$4a = a + a + a + a$ for $n = 4$
1	From A4.2.2A Equality	$a + a + a + a = a + a + a + a$
∴	From 1g and T4.3.9 Equalities: Summation of Repeated Terms	$4a = a + a + a + a$ for $n = 4$

[4.3.1]Note: The uniqueness axioms complement the equality axiom this is where the real work of proof writing takes place and the famous expression, ***"what is done to one side must be done to the other side"***. Without it, the jugglers could not maintain their balance.

Theorem 4.3.13 Equalities: Equality by Difference

1g	Given	$0 = b - a$
Steps	Hypothesis	$a = b$
1	From 1g and T4.3.4 Equalities: Reversal of Right Cancellation by Addition	$0 + a = b$
∴	From 1 and A4.2.7 Identity Add	$a = b$

Theorem 4.3.14 Equalities: Equality by Difference Reversed

1g	Given	$a = b$
Steps	Hypothesis	$0 = b - a$
1	From 1g, A4.2.2A Equality and T4.3.7 Equalities: Uniqueness of Subtraction	$a - a = b - a$
∴	From 1 and A4.2.8 Inverse Add	$0 = b - a$

Theorem 4.3.15 Equalities: Reversal of Identity and Inverse of Addition

1g	Given	$b = b + (a - a)$ Equation A	$b = (a - a) + b$ Equation B
Steps	Hypothesis	$b = b$	
1	From 1g A and B	$b = b + (a - a)$	$b = (a - a) + b$
2	From 1 and A4.2.8 Inverse Add	$b = b + 0$	$b = 0 + b$
∴	From 2 and A4.2.7 Identity Add	$b = b$	

Theorem 4.3.16 Equalities: Addition of Zero

1g	Given	$a = b$
Steps	Hypothesis	$a = b + (c - c)$
1	From 1g	$a = b$
2	From 1 and A4.2.7 Identity Add	$a = b + 0$
∴	From 2 and A4.2.8 Inverse Add	$a = b + (c - c)$

Section 4.4 Theorems on Equalities: Multiplication

Theorem 4.4.1 Equalities: Any Quantity Multiplied by Zero is Zero

Steps	Hypothesis	$a\,0 = 0$
1	From A4.2.8 Inverse Add	$a - a = 0$
2	From 1 and A4.2.14 Distribution	$a\,(1 - 1) = 0$
∴	From 2 and D4.1.19 Primitive Definition for Rational Arithmetic	$a\,0 = 0$

Theorem 4.4.2 Equalities: Uniqueness of Multiplication[4.3.1]

1g	Given	$(a, b, c, d, f) \in R$
2g		$a = b$
3g		$c = d$
Steps	Hypothesis	$ac = bd$ for $a = b$ and $c = d$
1	From 1g and A4.2.9 Closure Multp	$ac = f$
2	From 2g, 3g, 1 and A4.2.3 Substitution	$bd = f$
∴	From 1, 2, A4.2.2B Equality and A4.2.3 Substitution	$ac = bd$ for $a = b$ and $c = d$

Theorem 4.4.3 Equalities: Cancellation by Multiplication

1g	Given	$ac = bc$
Steps	Hypothesis	$a = b$
1	From A4.2.2A Equality	$(1/c) = (1/c)$
2	From 1, 1g and T4.4.2 Equalities: Uniqueness of Multiplication	$(ac)\,(1/c) = (bc)\,(1/c)$
3	From 2, D4.1.4(B, A) Rational Numbers and A4.2.11 Associative Multp	$a\,(c/c) = b\,(c/c)$
4	From 3 and A4.2.13 Inverse Multp	$a\,1 = b\,1$
∴	From 4 and A4.2.12 Identity Multp	$a = b$

Theorem 4.4.4 Equalities: Right Cancellation by Multiplication

1g	Given	$a = bc$			
Steps	Hypothesis	$a / c = b$ Equation A		$b = a / c$	Equation B
1	From A4.2.2A Equality	$(1/c) = (1/c)$			
2	From 1, 1g and T4.4.2 Equalities: Uniqueness of Multiplication	$a(1/c) = (bc)(1/c)$			
3	From 2, D4.1.4(B, A) Rational Numbers and A4.2.11 Associative Multp	$a / c = b(c/c)$			
4	From 3 and A4.2.13 Inverse Multp	$a / c = b \, 1$			
∴	From 4 and A4.2.12 Identity Multp	$a / c = b$ by A4.2.2A		$b = a / c$	by A4.2.2B

Theorem 4.4.5 Equalities: Reversal of Right Cancellation by Multiplication

1g	Given	$b = a / c$			
Steps	Hypothesis	$bc = a$	Equation A	$a = bc$	Equation B
∴	From T4.4.4B Equalities: Right Cancellation by Multiplication and T3.7.2 Reversibility of a Logical Argument	$bc = a$	by A4.2.2A	$a = bc$	by A4.2.2B

Theorem 4.4.6 Equalities: Left Cancellation by Multiplication

1g	Given	$bc = a$			
Steps	Hypothesis	$a / c = b$ Equation A		$b = a / c$	Equation B
1	From 1g and A4.2.2B Equality	$a = bc$			
∴	From 1, T4.4.5B Equalities: Reversal of Right Cancellation by Multiplication and T3.7.2 Reversibility of a Logical Argument	$a / c = b$ by A4.2.2A		$b = a / c$	by A4.2.2B

Theorem 4.4.7 Equalities: Reversal of Left Cancellation by Multiplication

1g	Given	$a / c = b$			
Steps	Hypothesis	$bc = a$ Equation A		$a = bc$	Equation B
∴	From T4.4.6A Equalities: Left Cancellation by Multiplication and T3.7.2 Reversibility of a Logical Argument	$bc = a$ by A4.2.2A		$a = bc$ by A4.2.2B	

Theorem 4.4.8 Equalities: Cross Product of Proportions

1g	Given	$a / d = c / b$
Steps	Hypothesis	$a\, b = c\, d$
1	From A4.2.2A Equality	$(d\, b) = (d\, b)$
2	From 1, 1g and T4.4.2 Equalities: Uniqueness of Multiplication	$(a / d)\,(d\, b) = (c / b)\,(d\, b)$
3	From 2, D4.1.4B Rational Numbers and A4.2.10 Commutative Multp	$a\,(1/d)\,(d\, b) = c\,(1/b)\,(b\, d)$
4	From 3 and A4.2.11 Associative Multp	$a\,((1/d)\, d)\, b = c\,((1/b)\, b)\, d$
5	From 4 and D4.1.4A Rational Numbers	$a\,(d / d)\, b = c\,(b/b)\, d$
6	From 5 and A4.2.13 Inverse Multp	$a\, 1\, b = c\, 1\, d$
∴	From 6 and A4.2.12 Identity Multp	$a\, b = c\, d$

Theorem 4.4.9 Equalities: Any Quantity Divided by One is that Quantity

Steps	Hypothesis	$a / 1 = a$
1	From A4.2.2A Equality	$a / 1 = a / 1$
2	From 1, A4.2.12 Identity Multp, A4.2.11 Associative Multp and D4.1.4(B, A) Rational Numbers	$a / 1 = a\,(1 / 1)$
3	From 2 and A4.2.13 Inverse Multp	$a / 1 = a\, 1$
∴	From 3 and A4.2.12 Identity Multp	$a / 1 = a$

Theorem 4.4.10 Equalities: Zero Divided by a Non-Zero Number

1g	Given	$a \neq 0$	
Steps	Hypothesis	$0 / a = 0$	for $a \neq 0$
1	From A4.2.2A Equality	$0 / a = 0 / a$	
2	From 1g, 1 and D4.1.4(A,B) Rational Numbers	$0 / a = (1/a) \, 0$	for $a \neq 0$
$\therefore$	From 2 and T4.4.1 Equalities: Any Quantity Multiplied by Zero is Zero	$0 / a = 0$	for $a \neq 0$

Theorem 4.4.11 Equalities: Uniqueness of Division[4.3.1]

1g 2g	Given	$a = b$ $c = d$	
Steps	Hypothesis	$a / c = b / d$	for $a = b$ and $c = d$
1	From 1g, 1, A4.2.2B Equality and T4.4.2 Equalities: Uniqueness of Multiplication	$a \, d = b \, c$	
2	From A4.2.2 Equality	$(1/c)(1/d) = (1/c)(1/d)$	
3	From 2, 3 and T4.4.2 Equalities: Uniqueness of Multiplication	$(1/c)(1/d) \, a \, d = b \, c \, (1/c)(1/d)$	
4	From 4, A4.2.10 Commutative Multp, D4.1.4(B, A) Rational Numbers and A4.2.11 Associative Multp	$(a/c)(d/d) = (c/c)(b/d)$	
$\therefore$	From 1g, 2g, 5, A4.2.13 Inverse Multp and A4.2.12 Identity Multp	$a / c = b / d$	for $a = b$ and $c = d$

Theorem 4.4.12 Equalities: Reversal of Cross Product of Proportions

1g	Given	$a \, b = c \, d$
Steps	Hypothesis	$a / d = c / b$
$\therefore$	From T4.4.2 Equalities: Uniqueness of Multiplication and T3.7.2 Reversibility of a Logical Argument	$a / d = c / b$

Theorem 4.4.13 Equalities: Cancellation by Division

1g	Given	$a / c = b / c$
Steps	Hypothesis	$a = b$
1	From A4.2.2A Equality	$c = c$
2	From 1g, 1 and T4.4.2 Equalities: Uniqueness of Multiplication	$(a / c) c = (b / c) c$
3	From 2 and D4.1.4B Rational Numbers	$a (1 / c) c = b (1 / c) c$
4	From 3 and A4.2.10 Commutative Multp	$a c (1 / c) = b c (1 / c)$
5	From 4 and D4.1.4A Rational Numbers	$a (c / c) = b (c / c)$
6	From 5 and A4.2.13 Inverse Multp	$a 1 = b 1$
∴	From 6 and A4.2.12 Identity Multp	$a = b$

Theorem 4.4.14 Equalities: Formal Cross Product

1g	Given	$a / c = b / d$
Steps	Hypothesis	$d / b = c / a$
1	From 1g and T4.4.8 Equalities: Cross Product of Proportions	$a d = b c$
∴	From 1 and T4.4.12 Equalities: Reversal of Cross Product of Proportions	$d / b = c / a$

Theorem 4.4.15 Equalities: Product by Division

Steps	Hypothesis	$a (1 / b) = a / b$ Equation A	$(1 / b) a = a / b$ Equation B
1	From A4.2.2A Equality	$a (1 / b) = a (1 / b)$	
∴	From 1, D4.1.4(D, C) and A4.2.10 Commutative Multp	$a (1 / b) = a / b$ Equation A	$(1 / b) a = a / b$ Equation B

Theorem 4.4.16 Equalities: Product by Division-Common Factor

Steps	Hypothesis	$a (1 / a) = 1$ Equation A	$(1 / a) a = 1$ Equation B
1	From A4.2.2A Equality	$a (1 / a) = a (1 / a)$	
2	From 1 and D4.1.4(D, C)	$a (1 / a) = a / a$	
∴	From 2, A4.2.13 Inverse Multp and A4.2.10 Commutative Multp	$a (1 / a) = 1$ Equation A	$(1 / a) a = 1$ Equation B

Theorem 4.4.17 Equalities: Product by Division-Factorization

Steps	Hypothesis	b = a (b/a)	Equation A	b = (1/a)(a b)	Equation B
1	From A4.2.2A Equality and A4.2.12 Identity Multp	b = 1 b			
2	From 1 and A4.2.13 Inverse Multp	b = (a/a) b			
3	From 2 and D4.1.4(C, B) Rational Numbers	b = a (1/a) b			
4	From A4.2.11 Associative Multp / A4.2.10 Commutative Multp and A4.2.11 Associative Multp	b = a ((1/a) b)		b = (1/a) (a b)	
∴	From 4 and D4.1.4(D, C) Rational Numbers	b = a (b/a)	Equation A	b = (1/a) (a b)	Equation B

Theorem 4.4.18 Equalities: Product by Division-Factorization by 2

1g	Given	a = 2			
Steps	Hypothesis	b = 2 (b/2)	Equation A	b = (1/2)(2 b)	Equation B
1	From 1g, T4.4.17 Equalities: Product by Division-Factorization and A4.2.3 Substitution	b = 2 (b/2)	Equation A	b = (1/2)(2 b)	Equation B
∴	From 1 and D4.1.19 Primitive Definition for Rational Arithmetic	b = 2 (b/2)	Equation A	b = (1/2)(2 b)	Equation B

Theorem 4.4.19 Equalities: Product by Division-Factorization by 3

1g	Given	a = 3			
Steps	Hypothesis	b = 3 (b/3)	Equation A	b = (1/3)(3 b)	Equation B
1	From 1g, T4.4.17 Equalities: Product by Division-Factorization and A4.2.3 Substitution	b = 3 (b/3)	Equation A	b = (1/3)(3 b)	Equation B
∴	From 1 and D4.1.19 Primitive Definition for Rational Arithmetic	b = 3 (b/3)	Equation A	b = (1/3)(3 b)	Equation B

Theorem 4.4.20 Equalities: Product by Division-Factorization by 5

1g	Given	$a = 5$			
Steps	Hypothesis	$b = 5 \, (b/5)$	Equation A	$b = (1/5)(5 \, b)$	Equation B
1	From 1g, T4.4.17 Equalities: Product by Division-Factorization and A4.2.3 Substitution	$b = 5 \, (b/5)$	Equation A	$b = (1/5)(5 \, b)$	Equation B
∴	From 1 and D4.1.19 Primitive Definition for Rational Arithmetic	$b = 5 \, (b/5)$	Equation A	$b = (1/5)(5 \, b)$	Equation B

Theorem 4.4.21 Equalities: Product by Division-Factorization by 7

1g	Given	$a = 7$			
Steps	Hypothesis	$b = 7 \, (b/7)$	Equation A	$b = (1/7)(7 \, b)$	Equation B
1	From 1g, T4.4.17 Equalities: Product by Division-Factorization and A4.2.3 Substitution	$b = 7 \, (b/7)$	Equation A	$b = (1/7)(7 \, b)$	Equation B
∴	From 1 and D4.1.19 Primitive Definition for Rational Arithmetic	$b = 7 \, (b/7)$	Equation A	$b = (1/7)(7 \, b)$	Equation B

Theorem 4.4.22 Equalities: Reversal of Identity and Inverse of Multiplication

1g	Given	$b = b \, (a / a)$	Equation A	$b = (a / a) \, b$	Equation B
Steps	Hypothesis	$b = b$			
1	From 1g A and B	$b = b \, (a / a)$		$b = (a / a) \, b$	
2	From 1 and A4.2.13 Inverse Multp	$b = b \, 1$		$b = 1 \, b$	
∴	From 2 and A4.2.12 Identity Multp	$b = b$			

Theorem 4.4.23 Equalities: Multiplication of a Constant

1g	Given	$a = b$	
Steps	Hypothesis	$ac = bc$	for [c] a constant
1	From A4.2.2A Equality	$c = c$	for [c] a constant
∴	From 1g, 1 and T4.4.2 Equalities: Uniqueness of Multiplication	$ac = bc$	for [c] a constant

Theorem 4.4.24 Equalities: Division by a Constant

1g	Given	$a = b$	
Steps	Hypothesis	$a/c = b/c$	for [c] a constant and $c \neq 0$
1	From A4.2.2A Equality	$c = c$	for [c] a constant and $c \neq 0$
∴	From 1g, 1 and T4.4.10 Equalities: Uniqueness of Division	$a/c = b/c$	for [c] a constant and $c \neq 0$

Section 4.5 Theorems on Equalities: Reciprocal Products

Theorem 4.5.1 Equalities: Reciprocal Products

1g 2g	Given	$(a, b, e) \in R$ for $(a, b) \neq 0$ $e \equiv (1 / a) (1 / b)$
Steps	Hypothesis	$1 / a b = (1 / a) (1 / b)$
1	From 2g and T4.4.5 Equalities: Reversal of Right Cancellation by Multiplication	$a e = (1 / b)$
2	From 2, T4.4.5 Equalities: Reversal of Right Cancellation by Multiplication and A4.2.11 Associative Multp	$(a b) e = 1$
3	From 2 and T4.4.6B Equalities: Left Cancellation by Multiplication	$e = 1 / a b$
∴	From 2g, 3 and A4.2.3 Substitution	$1 / a b = (1 / a) (1 / b)$

Theorem 4.5.2 Equalities: Inverse of Reciprocal Products

Steps	Hypothesis	$a / b = (a / h) / (b / h)$
1	From A4.2.2A Equality	$a / b = a / b$
2	From 1, D4.1.4B Rational Numbers and A4.2.12 Identity Multp	$a / b = a (1 / b) 1$
3	From 2 and A4.2.13 Inverse Multp	$a / b = a (1 / b) [(1 / h) / (1 / h)]$
4	From 3 and A4.2.10 Commutative Multp	$a / b = a (1 / h) (1 / b) [1 / (1 / h)]$
5	From 4 and A4.2.11 Associative Multp	$a / b = [a (1 / h)] ((1 / b) [1 / (1 / h)])$
6	From 5 and D4.1.4A Rational Numbers	$a / b = (a / h) [(1 / b) (1 / h)]$
7	From 6 and T4.5.1 Equalities: Reciprocal Products	$a / b = (a / h) [1 / (b h)]$
∴	From 7 and D4.1.4A Rational Numbers	$a / b = (a / h) / (b / h)$

Theorem 4.5.3 Equalities: Compound Reciprocal Products

Steps	Hypothesis	$(a / c) / (b / d) = (a\,d) / (b\,c)$
1	From A4.2.2A Equality	$(a / c) / (b / d) = (a / c) / (b / d)$
2	From 1 and D4.1.4B Rational Numbers	$(a / c) / (b / d) = [a\,(1 / c)]\,(1 / [b\,(1 / d)])$
3	From 2 and A4.2.12 Identity Multp	$(a / c) / (b / d) = [a\,(1 / c)]\,(1 / [b\,(1 / d)])\,1$
4	From 3 and A4.2.13 Inverse Multp	$(a / c) / (b / d) = [a\,(1 / c)]\,(1 / [b\,(1 / d)])\,(d/d)$
5	From 4, D4.1.4B Rational Numbers and A4.2.10 Commutative Multp	$(a / c) / (b / d) = [a\,(1 / c)]\,(\,(((d)\,1) / [b\,(1 / d)\,(d)])$
6	From 5, A4.2.12 Identity Multp, A4.2.11 Associative Multp and D4.1.4A Rational Numbers	$(a / c) / (b / d) = [a\,(1 / c)]\,(\,d / [b\,(d / d)])$
7	From 6 and A4.2.12 Identity Multp	$(a / c) / (b / d) = [a\,(1 / c)]\,(\,d / [b\,1])$
8	From 7 and A4.2.12 Identity Multp	$(a / c) / (b / d) = [a\,(1 / c)]\,(\,d / b)$
9	From 8 and D4.1.4B Rational Numbers	$(a / c) / (b / d) = [a\,(1 / c)]\,[\,d\,(1 / b)]$
10	From 9 and A4.2.11 Associative Multp	$(a / c) / (b / d) = a\,d\,(1 / b)\,(1 / c)$
11	From 10 and T4.5.1 Equalities: Reciprocal Products	$(a / c) / (b / d) = a\,d\,(1 / b\,c)$
∴	From 11 and D4.1.4A Rational Numbers	$(a / c) / (b / d) = (a\,d) / (b\,c)$

Theorem 4.5.4 Equalities: Product of Two Rational Fractions

Steps	Hypothesis	$(a/c)\,(b / d) = (a\,b) / (c\,d)$
1	From Hypothesis	$(a / c)\,(b / d) = (a / c)\,(b / d)$
2	From 1 and D4.1.4B Rational Numbers	$(a / c)\,(b / d) = [a\,(1 / c)]\,[b\,(1 / d)]$
3	From 2, A4.2.10 Commutative Multp and A4.2.11 Associative Multp	$(a / c)\,(b / d) = (a\,b)\,(1 / c)\,(1 / d)$
4	From 3 and T4.5.1 Equalities: Reciprocal Products	$(a / c)\,(b / d) = (a\,b)\,(1 / c\,d)$
∴	From 4 and D4.1.4A Rational Numbers	$(a / c)\,(b / d) = (a\,b) / (c\,d)$

Theorem 4.5.5 Equalities: Addition of Two Rational Fractions

Steps	Hypothesis	$(a/c) + (b/d) = (a\,d + b\,c)/cd$
1	From A4.2.2A Equality	$(a/c) + (b/d) = (a/c) + (b/d)$
2	From 1 and A4.2.12 Identity Multp	$(a/c) + (b/d) = 1\,(a/c) + (b/d)\,1$
3	From 2 and A4.2.13 Inverse Multp	$(a/c) + (b/d) = (d/d)\,(a/c) + (b/d)\,(c/c)$
4	From 3 and T4.5.4 Equalities: Product of Two Rational Fractions	$(a/c) + (b/d) = (ad/cd) + (bc/cd)$
5	Form 4 and D4.1.4B Rational Numbers	$(a/c) + (b/d) = (ad)\,(1/cd) + (bc)\,(1/cd)$
6	From 5 and A4.2.14 Distribution	$(a/c) + (b/d) = (ad + bc)\,(1/cd)$
$\therefore$	From 6 and D4.1.4A Rational Numbers	$(a/c) + (b/d) = (ad + bc)/cd$

Theorem 4.5.6 Equalities: Contraction of a Rational Chain

1g 2g	Given	$(a, b, c_k) \in R$ for $(b, c_k) \neq 0$ for all k $a/b = \prod_{k=1}^{n}(c_k/c_{k+1})$ for $c_k = c_{k+1}$ for all k, for n factors assume $c_1 \equiv a$, and $c_{n+1} \equiv b$

Steps	Hypothesis	$a/b = \prod_{k=1}^{n}(c_k/c_{k+1})$ for $c_k \equiv c_{k+1}$ for all k
1	From 2g	$a/b = a\,[\,\prod_{k=1}^{n}(c_k/c_{k+1})\,]\,(1/b)$
2	From 2g, A4.2.2 Equality, A4.2.3 Substitution, A4.2.13 Inverse Multp and A4.2.12 Identity Multp	$c_k/c_{k+1} = 1$
3	From 1, 2 and A4.2.3 Substitution	$a/b = a\,[\,\prod_{k=1}^{n}(1)\,]\,(1/b)$
4	From 3, D4.1.19 Primitive Definition for Rational Arithmetic	$a/b = a\,(1)\,(1/b)$
5	From 3, A4.2.12 Identity Multp and D4.1.4A Rational Numbers	$a/b = a/b$
$\therefore$	From 2g, 5 and by identity	$a/b = \prod_{k=1}^{n}(c_k/c_{k+1})$

Theorem 4.5.7 Equalities: Product and Reciprocal of a Product

Steps	Hypothesis	$a\,b = b\,/\,(1/a)$
1	From A4.2.2A Equality	$a\,b = a\,b$
2	From 1 and A4.2.10 Commutative Multp	$a\,b = b\,a$
3	From 2, A4.2.12 Identity Multp and A4.2.13 Inverse Multp	$a\,b = b\,a\,(1/a)\,/\,(1/a)$
4	From 3, A4.2.11 Associative Multp and A4.2.13 Inverse Multp	$a\,b = b\,1\,/\,(1/a)$
∴	From 4 and A4.2.12 Identity Multp	$a\,b = b\,/\,(1/a)$

The above theorems are sometimes called short cut theorems because if used properly they can eliminate a lot of steps in a proof. Also, any divisor in these theorems cannot be zero by D4.1.19.

Theorem 4.5.8 Equalities: Multiplication of Unity

1g	Given	$a = b$	
Steps	Hypothesis	$a = b\,(c/c)$	for [c] a constant
1		$c = c$	for [c] a constant and not zero
∴		$a = b\,(c/c)$	for [c] a constant and not zero

Theorem 4.5.9 Equalities: Ratio and Reciprocal of a Ratio

1g	Given	$a\,/\,b = 1\,/\,(b\,/\,a)$	assume
Steps	Hypothesis	$a\,/\,b = 1\,/\,(b\,/\,a)$	
1	From 1g	$a\,/\,b = 1\,/\,(b\,/\,a)$	
2	From 2, A4.2.12 Identity Multp and A4.2.13 Inverse Multp	$a\,/\,b = [\,1\,/\,(b\,/\,a)\,]\,(a\,/\,b)\,/\,(a\,/\,b)$	
3	From 2, A4.2.12 Identity Multp and A4.2.10 Commutative Multp	$a\,/\,b = (a\,/\,b)\,[\,1\,/\,(b\,/\,a)\,]\,[\,1\,/\,(a\,/\,b)\,]$	
4	From 3 and T4.5.1 Equalities: Reciprocal Products	$a\,/\,b = (a\,/\,b)\,[\,1\,/\,(b\,/\,a)\,(a\,/\,b)\,]$	
5	From 4 and T4.5.4 Equalities: Product of Two Rational Fractions	$a\,/\,b = (a\,/\,b)\,[\,1\,/\,(a\,b\,/\,b\,a)\,]$	
6	From 5 and A4.2.10 Commutative Multp	$a\,/\,b = (a\,/\,b)\,[\,1\,/\,(a\,b\,/\,a\,b)\,]$	
7	From 6 and A4.2.13 Inverse Multp	$a\,/\,b = (a\,/\,b)\,[\,1\,/\,1\,]$	
8	From 7, A4.2.13 Inverse Multp and A4.2.12 Identity Multp	$a\,/\,b = a\,/\,b$	
∴	From 1g, 8 and by identity	$a\,/\,b = 1\,/\,(b\,/\,a)$	

Section 4.6 Theorems on Equalities: Absolute Quantities

Theorem 4.6.1 Equalities: Uniqueness of Absolute Value

1g 2g	Given	$a = b$ for $a \geq 0$ and $b \geq 0$ $a = b$ for $a < 0$ and $b < 0$
Steps	Hypothesis	$|a| = |b|$
$\therefore$	From 1g, 2g, A4.2.2 Equality and D4.1.11 Absolute Value $|a|$	$|a| = |b|$

Theorem 4.6.2 Equalities: Square of an Absolute Value

Steps	Hypothesis	$|a|^2 = a^2$
1	From T4.6.1 Equalities: Uniqueness of Absolute Value and T4.8.6 Integer Exponents: Uniqueness of Exponents	$|a|^2 = |a|^2$
2	From 1 and D4.1.11 Absolute Value $|a|$ for $a \geq 0$	$|a|^2 = (+a)^2$
3	From 2, D4.1.17 Exponential Notation and D4.1.19 Primitive Definition for Rational Arithmetic	$|a|^2 = a^2$
4	From 1 and D4.1.11 Absolute Value $|a|$ for $a < 0$	$|a|^2 = (-a)^2$
5	From 4, D4.1.17 Exponential Notation and D4.1.19 Primitive Definition for Rational Arithmetic	$|a|^2 = a^2$
$\therefore$	From 3 and 4	$|a|^2 = a^2$

Theorem 4.6.3 Equalities: Absolute Product is the Absolute of the Products

Given		$|a|\,|b| > 0$	$|ab| > 0$	$|a|\,|b| = |ab|$	Hypothesis
$a > 0$	$b > 0$	$(+a)\,(+b) > 0$	$+(ab) > 0$	$(+a)\,(+b) = +(ab)$	From D4.1.11, D4.1.19 and A4.2.16
$a > 0$	$b < 0$	$(+a)\,(-b) > 0$	$-(ab) > 0$	$(+a)\,(-b) = -(ab)$	From D4.1.11, D4.1.19 and A4.2.16
$a < 0$	$b > 0$	$(-a)\,(+b) > 0$	$-(ab) > 0$	$(-a)\,(+b) = -(ab)$	From D4.1.11, D4.1.19 and A4.2.16
$a < 0$	$b < 0$	$(-a)\,(-b) > 0$	$+(ab) > 0$	$(-a)\,(-b) = +(ab)$	From D4.1.11, D4.1.19 and A4.2.16
	$\therefore$			$|a|\,|y| = |ay|$	From D2.1.14 Proof by Exhaustion

Theorem 4.6.4 Equalities: Absolute Reciprocal is the Reciprocal of the Absolute Value

Steps	Hypothesis	$|1/a| = 1/|a|$
1	From T4.6.1 Equalities: Uniquness of Absolute Value	$|1/a| = |1/a|$
2	From 1 and D4.1.11 Absolute Value $|a|$	$|1/a| = 1/+a$ for $a > 0$
3	From 1 and D4.1.11 Absolute Value $|a|$	$|1/a| = 1/-a$ for $a < 0$
∴	From 2, 3 and D4.1.11 Absolute Value $|a|$	$|1/a| = 1/|a|$

Theorem 4.6.5 Equalities: Absolute Quotient is the Absolute Value of [a], [b] of the Quotient

1g	Given	$c = 1/b$
Steps	Hypothesis	$|a|/|b| = |a/b|$
1	From T4.6.3 Equalities: Absolute Product is the Absolute of the Products	$|a||c| = |ac|$
2	From 1, 1g and A4.2.3 Substitution	$|a||1/b| = |a(1/b)|$
3	From 2 and T4.6.4 Equalities: Absolute Reciprocal is the Reciprocal of the Absolute Value	$|a|(1/|b|) = |a(1/b)|$
∴	From 3 and the product	$|a|/|b| = |a/b|$

Section 4.7 Theorems on Inequalities

Theorem 4.7.1 Inequalities: Transitive Law

1g 2g	Given	$a < b$ $b < c$
Steps	Hypothesis	$a < b$ and $b < c$ hence $a < c$
$\therefore$	From 1g, 2g and A4.2.15 Inequality	$a < c$

Theorem 4.7.2 Inequalities: Uniqueness of Addition by a Positive Number

1g 2g	Given	$a < b$ $0 < c$
Steps	Hypothesis	$a + c < b + c$ shifts inequality up equally leaving order unaltered
1	From 1g and D4.1.7 Greater Than Inequality [>]	$0 < b - a$
2	From 2g, 1, A4.2.7 Identity Add and A4.2.8 Inverse Add	$0 < b + (c - c) - a$
3	From 2, A4.2.6 Associative Add, D4.1.20 Negative Coefficient and A4.2.14 Distribution	$0 < (b + c) - (c + a)$
$\therefore$	From 3, A4.2.5 Commutative Add and D4.1.7 Greater Than Inequality [>]	$a + c < b + c$ shifts inequality up equally leaving order unaltered

Tensor Calculus & Physics: A General Treatise

Theorem 4.7.3 Inequalities: Uniqueness of Subtraction by a Positive Number, Low

1g	Given	$0 < c < a < b$	
Steps	Hypothesis	$a - c < b - c$ for $0 < c < a < b$	leaving order unaltered
1	From 1g and D4.1.8 Less Then Inequality [<]	$0 < a - c$	
2	From 1g and D4.1.8 Less Then Inequality [<]	$0 < b - a$	
3	From 1g and D4.1.8 Less Then Inequality [<]	$0 < b - c$	
4	From 2, A4.2.7 Identity Add and A4.2.8 Inverse Add	$0 < b + (c - c) - a$	
5	From 4 and A4.2.5 Commutative Add	$0 < b - c + c - a$	
6	From 5, A4.2.6 Associative Add, D4.1.20 Negative Coefficient, A4.2.12 Identity Multp, T4.8.2 Integer Exponents: Negative One Squared and A4.2.14 Distribution	$0 < (b - c) - (a - c)$	
∴	From 6 and A4.2.15 Inequality	$a - c < b - c$ for $0 < c < a < b$	leaving order unaltered

Theorem 4.7.4 Inequalities: Uniqueness of Subtraction by a Negative Number

1g	Given	$c < 0 < a < b$	
Steps	Hypothesis	$a - c < b - c$ for $c < 0 < a < b$	leaving order unaltered
1	From 1g and D4.1.8 Less Then Inequality [<]	$0 < a - c$	
2	From 1g and D4.1.8 Less Then Inequality [<]	$0 < b - c$	
3	From 1g and D4.1.8 Less Then Inequality [<]	$0 < b - a$	
4	From 3, A4.2.7 Identity Add and A4.2.8 Inverse Add	$0 < b + (c - c) - a$	
5	From 4 and A4.2.5 Commutative Add	$0 < b - c + c - a$	
6	From 5, A4.2.6 Associative Add, D4.1.20 Negative Coefficient, A4.2.12 Identity Multp, T4.8.2 Integer Exponents: Negative One Squared and A4.2.14 Distribution	$0 < (b - c) - (a - c)$	
∴	From 6 and A4.2.15 Inequality	$a - c < b - c$ for $c < 0 < a < b$	leaving order unaltered

Tensor Calculus & Physics: A General Treatise

Theorem 4.7.5 Inequalities: Uniqueness of Subtraction by a Positive Number, Interim

1g	Given	$0 < a < c < b$	
Steps	Hypothesis	$-(c - a) < b - c$ for $0 < a < c < b$	leaving order unaltered
1	From 1g and D4.1.8 Less Then Inequality [<]	$0 < b - a$	
2	From 1g and D4.1.8 Less Then Inequality [<]	$0 < b - c$	
3	From 1g and D4.1.8 Less Then Inequality [<]	$0 < c - a$	
4	From 1, A4.2.7 Identity Add and A4.2.8 Inverse Add	$0 < b + (c - c) - a$	
5	From 4 and A4.2.5 Commutative Add	$0 < b - c + c - a$	
6	From 5 and A4.2.6 Associative Add	$0 < (b - c) + (c - a)$	
7	From 6, T4.8.2 Integer Exponents: Negative One Squared and D4.1.20 Negative Coefficient	$0 < (b - c) - [-(c - a)]$	
∴	From 7 and D4.1.8 Less Then Inequality [<]	$-(c - a) < b - c$ for $0 < a < c < b$	leaving order unaltered

Theorem 4.7.6 Inequalities: Uniqueness of Subtraction by a Positive Number, High

1g	Given	$0 < a < b < c$	
Steps	Hypothesis	$c - b < c - a$ for $0 < a < c < b$	
1	From 1g and D4.1.8 Less Then Inequality [<]	$0 < b - a$	
2	From 1g and D4.1.8 Less Then Inequality [<]	$0 < c - b$	
3	From 1g and D4.1.8 Less Then Inequality [<]	$0 < c - a$	
4	From 1, A4.2.7 Identity Add and A4.2.8 Inverse Add	$0 < b + (c - c) - a$	
5	From 4 and A4.2.5 Commutative Add	$0 < b - c + c - a$	
6	From 5 and A4.2.6 Associative Add	$0 < (b - c) + (c - a)$	
7	From 6, T4.8.2 Integer Exponents: Negative One Squared, D4.1.20 Negative Coefficient and A4.2.14 Distribution	$0 < -(c - b) + (c - a)$	
8	From 7 and A4.2.5 Commutative Add	$0 < (c - a) - (c - b)$	
∴	From 8 and D4.1.8 Less Then Inequality [<]	$c - b < c - a$ for $0 < a < c < b$	

Theorem 4.7.7 Inequalities: Multiplication by Positive Number

1g 2g	Given	$a < b$ $0 < c$
Steps	Hypothesis	$ca < cb$ leaves order unaltered
1	From 1g	$0 < a$
2	From 1g, and D4.1.7 Greater Than Inequality [>]	$0 < b - a$
3	From 2g, 1 and A4.2.9 Closure Multp	$0 < c(b - a)$
4	From 3 and A4.2.14 Distribution	$0 < cb - ca$
∴	From 4 and, D4.1.7 Greater Than Inequality [>] or A4.2.17 Correspondence of Equality and Inequality, Equality and T4.3.13 Equalities: Equality by Difference	$ca < cb$ leaves order unaltered

Theorem 4.7.8 Inequalities: Multiplication by Negative Number Reverses Order

1g 2g	Given	$a < b$ $c < 0$
Steps	Hypothesis	$ca > cb$ order is altered
1	From 1g and D4.1.7 Greater Than Inequality [>] or 1g, A4.2.17 Correspondence of Equality and Inequality, and T4.3.13 Equalities: Equality by Difference	$0 < b - a$
2	From 2g and D4.1.7 Greater Than Inequality [>]	$0 < 0 - c$
3	From 2 and A4.2.7 Identity Add	$0 < -c$
4	From 1, 3 and A4.2.9 Closure Multp	$0 < -c(b - a)$
5	From 2 and A4.2.14 Distribution	$0 < -cb - (-c)a$
6	From 5, D4.1.20A&B Negative Coefficient and D4.1.19 Primitive Definition for Rational Arithmetic	$0 < -cb + ca$
7	From 6 and A4.2.5 Commutative Add	$0 < ca - cb$
∴	From 3 and D4.1.7 Greater Than Inequality [>]	$ca > cb$ order is altered

Theorem 4.7.9 Inequalities: Addition of (a < b) + (c < d)

1g	Given	$(a, b, c, d) \in R$
2g		$a < b$
3g		$c < d$
Steps	Hypothesis	$a + c < b + d$ shifts inequality up leaving order unaltered
1	From 2g and A4.2.15 Inequality	$a < b$
2	From 1 and D4.1.7 Greater Than Inequality [>]	$0 < b - a$
3	From 3g and A4.2.15 Inequality	$c < d$
4	From 3 and D4.1.7 Greater Than Inequality [>]	$0 < d - c$
5	From 2, 4, A4.2.17 Correspondence of Equality and Inequality, and T4.3.1 Equalities: Uniqueness of Addition	$0 + 0 < (b - a) + (d - c)$ for either side
6	From 5, A4.2.7 Identity Add, A4.2.5 Commutative Add, A4.2.6 Associative Add, D4.1.20 Negative Coefficient and A4.2.14 Distribution	$0 < (b + d) - (a + c)$
∴	From 6 and D4.1.7 Greater Than Inequality [>]	$a + c < b + d$ shifts inequality up leaving order unaltered

Theorem 4.7.10 Inequalities: Subtraction of (c < d) – (a < b) for (a < c) and (b < d)

1g	Given	$(a, b, c, d) \in R$
2g		$a < b$ any inequality and
3g		$c < d$ any other inequality
4g		$a < c$ condition
5g		$b < d$ condition
Steps	Hypothesis	$c - a > -(d - b)$ order is altered for $a < c$ and $b < d$
1	From 4g and D4.1.7 Greater Than Inequality [>]	$0 < c - a$
2	From 5g and D4.1.7 Greater Than Inequality [>]	$0 < d - b$
3	From 1, 2, A4.2.17 Correspondence of Equality and Inequality, and T4.3.1 Equalities: Uniqueness of Addition	$0 + 0 < (c - a) + (d - b)$ for either side
4	From 3, A4.2.7 Identity Add, D4.1.19 Primitive Definition for Rational Arithmetic and D4.1.20 Negative Coefficient	$0 < (c - a) + (-1)((-1)(d - b))$
5	From 4 and A4.2.14 Distribution	
6	From 5 and D4.1.20 Negative Coefficient	$0 < (c - a) - (-(d - b))$
7	From 6 and A4.2.5 Commutative Add	$0 < (c - a) - (-(d - b))$
8	From 7 and D4.1.7 Greater Than Inequality [>]	$-(d - b) < c - a$
∴	From 8 and A4.2.5 Commutative Add	$c - a > -(d - b)$ order is altered for $a < c$ and $b < d$

Theorem 4.7.11 Inequalities: Subtraction of (c < d) – (a < b) for (a < c) and (b > d)

1g 2g 3g 4g 5g	Given	$(a, b, c, d) \in R$ $a < b$ any inequality and $c < d$ any other inequality $a < c$ condition $b > d$ condition
Steps	Hypothesis	$c - a > d - b$ order is altered for $a < c$ and $b > d$
1	From 4g and D4.1.7 Greater Than Inequality [>]	$0 < c - a$
2	From 5g and D4.1.7 Greater Than Inequality [>]	$0 < b - d$
3	From 1, 2, A4.2.17 Correspondence of Equality and Inequality, and T4.3.1 Equalities: Uniqueness of Addition	$0 + 0 < (c - a) + (b - d)$ for either side
4	From 3, A4.2.7 Identity Add, D4.1.19 Primitive Definition for Rational Arithmetic and D4.1.20 Negative Coefficient	$0 < (c - a) + ((-1)(-1)b + (-1)d)$
5	From 4 and A4.2.14 Distribution	$0 < (c - a) + (-1)((-1)b + d)$
6	From 5 and D4.1.20 Negative Coefficient	$0 < (c - a) - (-b + d)$
7	From 6 and A4.2.5 Commutative Add	$0 < (c - a) - (d - b)$
8	From 7 and D4.1.7 Greater Than Inequality [>]	$c - a > d - b$
∴	From 8 and A4.2.5 Commutative Add	$c - a > d - b$ order is altered for $a < c$ and $b > d$

Theorem 4.7.12 Inequalities: Subtraction of (c < d) – (a < b) for (a > c) and (b < d)

1g	Given	$(a, b, c, d) \in R$
2g		$a < b$ any inequality and
3g		$c < d$ any other inequality
4g		$a > c$ condition
5g		$b < d$ condition
Steps	Hypothesis	$c - a < d - b$ order is unaltered for $a > c$ and $b < d$
1	From 4g and D4.1.7 Greater Than Inequality [>]	$0 < a - c$
2	From 5g and D4.1.7 Greater Than Inequality [>]	$0 < d - b$
3	From 1, 2, A4.2.17 Correspondence of Equality and Inequality, and T4.3.1 Equalities: Uniqueness of Addition	$0 + 0 < (a - c) + (d - b)$ for either side
4	From 3, A4.2.7 Identity Add, D4.1.19 Primitive Definition for Rational Arithmetic and D4.1.20 Negative Coefficient	$0 < ((-1)(-1)a + (-1)c) + (d - b)$
5	From 4 and A4.2.14 Distribution	$0 < (-1)((-1)a + c) + (d - b)$
6	From 5 and D4.1.20 Negative Coefficient	$0 < -(-a + c) + (d - b)$
7	From 6 and A4.2.5 Commutative Add	$0 < (d - b) - (c - a)$
8	From 7 and D4.1.7 Greater Than Inequality [>]	$c - a < d - b$
∴	From 8 and A4.2.5 Commutative Add	$c - a < d - b$ order is unaltered for $a > c$ and $b < d$

Theorem 4.7.13 Inequalities: Subtraction of (c < d) – (a < b) for (a > c) and (b > d)

	Given	$(a, b, c, d) \in R$	
1g		$a < b$	any inequality and
2g		$c < d$	any other inequality
3g		$a > c$	condition
4g		$b > d$	condition
5g			
Steps	Hypothesis	$c - a < -(d - b)$ order is unaltered for $a > c$ and $b > d$	
1	From 4g and D4.1.7 Greater Than Inequality [>]	$0 < a - c$	
2	From 5g and D4.1.7 Greater Than Inequality [>]	$0 < b - d$	
3	From 1, 2, A4.2.17 Correspondence of Equality and Inequality, and T4.3.1 Equalities: Uniqueness of Addition	$0 + 0 < (a - c) + (b - d)$ for either side	
4	From 3, A4.2.7 Identity Add, D4.1.19 Primitive Definition for Rational Arithmetic and D4.1.20 Negative Coefficient	$0 < ((-1)(-1)a + (-1)c) + ((-1)(-1)b + (-1)d)$	
5	From 4 and A4.2.14 Distribution	$0 < (-1)((-1)a + c) + (-1)((-1)b + d)$	
6	From 5 and D4.1.20 Negative Coefficient	$0 < -(-a + c) + (-(-b + d))$	
7	From 6 and A4.2.5 Commutative Add	$0 < (-(d - b)) - (c - a)$	
8	From 7 and D4.1.7 Greater Than Inequality [>]	$c - a < -(d - b)$	
∴	From 8 and A4.2.5 Commutative Add	$c - a < -(d - b)$ order is unaltered for $a > c$ and $b > d$	

Theorem 4.7.14 Inequalities: Subtraction of (c < d) – (a < b) for (a < c) and (b = d)

1g	Given	$(a, b, c, d) \in R$
2g		$a < b$ any inequality and
3g		$c < d$ any other inequality
4g		$a < c$ condition
5g		$b = d$ condition
Steps	Hypothesis	$c - a > 0$ order is altered for $a < c$ and $b = d$
1	From T4.7.10 Inequalities: Subtraction of (c < d) – (a < b) for (a < c) and (b < d)	$c - a > b - d$ order is altered for $a < c$ and $b < d$
∴	From 5g, 1, A4.2.3 Substitution and A4.2.8 Inverse Add	$c - a > 0$ order is altered for $a < c$ and $b = d$

Theorem 4.7.15 Inequalities: Subtraction of (c < d) – (a < b) for (a > c) and (b = d)

1g	Given	$(a, b, c, d) \in R$
2g		$a < b$ any inequality and
3g		$c < d$ any other inequality
4g		$a > c$ condition
5g		$b = d$ condition
Steps	Hypothesis	$c - a < 0$ order is unaltered for $a > c$ and $b = d$
1	From T4.7.10 Inequalities: Subtraction of (c < d) – (a < b) for (a > c) and (b < d)	$c - a < d - b$ order is unaltered for $a > c$ and $b < d$
∴	From 5g, 1, A4.2.3 Substitution and A4.2.8 Inverse Add	$c - a < 0$ order is unaltered for $a > c$ and $b = d$

Theorem 4.7.16 Inequalities: Subtraction of (c < d) – (a < b) for (a = c) and (b < d)

1g	Given	$(a, b, c, d) \in R$
2g		$a < b$ any inequality and
3g		$c < d$ any other inequality
4g		$a = c$ condition
5g		$b < d$ condition
Steps	Hypothesis	$0 < d - b$ order is unaltered for $a = c$ and $b < d$
1	From T4.7.10 Inequalities: Subtraction of (c < d) – (a < b) for (a < c) and (b < d)	$c - a > -(d - b)$ order is altered for $a < c$ and $b < d$
∴	From 4g, 1, A4.2.3 Substitution, A4.2.8 Inverse Add and T4.7.6 Inequalities: Multiplication by Negative Number Reverses Order	$0 < d - b$ order is unaltered for $a = c$ and $b < d$

Theorem 4.7.17 Inequalities: Subtraction of (c < d) – (a < b) for (a = c) and (b > d)

1g	Given	$(a, b, c, d) \in R$	
2g		$a < b$	any inequality and
3g		$c < d$	any other inequality
4g		$a = c$	condition
5g		$b < d$	condition
Steps	Hypothesis	$0 > d - b$ order is altered for $a = c$ and $b > d$	
1	From T4.7.9 Inequalities: Subtraction of (c < d) – (a < b) for (a < c) and (b > d)	$c - a > d - b$ order is altered for $a < c$ and $b > d$	
∴	From 4g, 1, A4.2.3 Substitution, A4.2.8 Inverse Add and T4.7.6 Inequalities: Multiplication by Negative Number Reverses Order	$0 > d - b$ order is altered for $a = c$ and $b > d$	

Theorem 4.7.18 Inequalities: Cancellation by Addition

1g	Given	$(a, b, c) \in R$
2g		$a + c < b + c$
Steps	Hypothesis	$a < b$
1	From 2g and T4.7.4 Inequalities: Uniqueness of Subtraction by a Negative Number	$(a + c) - c < (b + c) - c$
2	From 1, A4.2.6 Associative Add and A4.2.8 Inverse Add	$a + 0 < b + 0$
∴	From 2 and A4.2.7 Identity Add	$a < b$

Theorem 4.7.19 Inequalities: Cancellation by Multiplication for a Positive Number

1g	Given	$(a, b, c) \in R$
2g		$ac < bc$ for $0 < c$
Steps	Hypothesis	$a < b$
1	From A4.2.2A Equality	$1/c = 1/c$
2	From 2g, 1 and T4.7.7 Inequalities: Multiplication by Positive Number	$ac(1/c) < bc(1/c)$
3	From 2 and A4.2.11 Associative Multp	$a(c(1/c)) < b(c(1/c))$
4	From 3, D4.1.4(A,B) Rational Numbers and A4.2.13 Inverse Multp	$a\,1 < b\,1$
∴	From 4 and A4.2.12 Identity Multp	$a < b$

Theorem 4.7.20 Inequalities: Cancellation by Multiplication for a Negative Number Reverses Order

1g	Given	$(a, b, c) \in R$
2g		$ac < bc$ for $0 > c$
Steps	Hypothesis	$a > b$
1	From A4.2.2A Equality	$1/c = 1/c$
2	From 2g, 1 and T4.7.6 Inequalities: Multiplication by Negative Number Reverses Order	$ac(1/c) > bc(1/c)$
3	From 2 and A4.2.11 Associative Multp	$a(c(1/c)) > b(c(1/c))$
4	From 3, D4.1.4(A,B) Rational Numbers and A4.2.13 Inverse Multp	$a\,1 > b\,1$
∴	From 4 and A4.2.12 Identity Multp	$a > b$

Theorem 4.7.21 Inequalities: Multiplication for Positive Numbers

1g	Given	$(a, b, c, d) \in R$
2g		$0 < a < b$
3g		$0 < c < d$
Steps	Hypothesis	$0 < ac < bd$
1	From 2g, 3g, A4.2.9 Closure Multp and A4.2.16A The Trichotomy Law of Ordered Numbers	$0 * 0 < ac < bd$
∴	From 1 and T4.4.1 Equalities: Any Quantity Multiplied by Zero is Zero	$0 < ac < bd$

Theorem 4.7.22 Inequalities: Multiplication for Negative Numbers

1g	Given	$(a, b, c, d) \in R$
2g		$a > b > 0$
3g		$c > d > 0$
Steps	Hypothesis	$0 < ac < bd$
1	From 2g, T4.4.1 Equalities: Any Quantity Multiplied by Zero is Zero and T4.7.6 Inequalities: Multiplication by Negative Number Reverses Order	$0 < ac < bc$
2	From 3g, T4.4.1 Equalities: Any Quantity Multiplied by Zero is Zero and T4.7.6 Inequalities: Multiplication by Negative Number Reverses Order	$0 < bc < bd$
∴	From 1, 2 and T4.7.1 Inequalities: Transitive Law	$0 < ac < bd$

Theorem 4.7.23 Inequalities: Multiplication for Negative and Positive Numbers

1g	Given	$(a, b, c, d) \in R$
2g		$b > a > 0$
3g		$d < c < 0$
Steps	Hypothesis	$bd < ac < 0$
1	From 2g, T4.4.1 Equalities: Any Quantity Multiplied by Zero is Zero and T4.7.6 Inequalities: Multiplication by Negative Number Reverses Order	$db < da < 0$
2	From 3g, T4.4.1 Equalities: Any Quantity Multiplied by Zero is Zero and T4.7.7 Inequalities: Multiplication by Positive Number	$ad < ac < 0$
∴	From 1, 2, A4.2.10 Commutative Multp and T4.7.1 Inequalities: Transitive Law	$bd < ac < 0$

Theorem 4.7.24 Inequalities: Multiplication for Mixed and Positive Numbers

1g	Given	$(a, b, c, d) \in R$
2g		$b < 0 < a$
3g		$0 < c < d$
Steps	Hypothesis	$bd < 0 < ac$
1	From 2g, T4.4.1 Equalities: Any Quantity Multiplied by Zero is Zero and T4.7.7 Inequalities: Multiplication by Positive Number	$bc < 0 < ac$
2	From 3g, T4.4.1 Equalities: Any Quantity Multiplied by Zero is Zero and T4.7.6 Inequalities: Multiplication by Negative Number Reverses Order	$bd < bc < 0$
∴	From 1, 2 and T4.7.1 Inequalities: Transitive Law	$bd < 0 < ac$

Theorem 4.7.25 Inequalities: Multiplication for Mixed and Negative Numbers

1g 2g 3g	Given	$(a, b, c, d) \in R$ $b < 0 < a$ $d < c < 0$
Steps	Hypothesis	$bc > 0 > ad$
1	From 2g, T4.4.1 Equalities: Any Quantity Multiplied by Zero is Zero and T4.7.6 Inequalities: Multiplication by Negative Number Reverses Order	$bc > 0 > ac$
2	From 3g, T4.4.1 Equalities: Any Quantity Multiplied by Zero is Zero and T4.7.7 Inequalities: Multiplication by Positive Number	$ad < ac < 0$
∴	From 1, 2 and T4.7.1 Inequalities: Transitive Law	$bc > 0 > ad$

Theorem 4.7.26 Inequalities: Multiplication for Mixed and Mixed Numbers

1g 2g 3g	Given	$(a, b, c, d) \in R$ $b < 0 < a$ $d < 0 < c$	
Steps	Hypothesis	$0 < ac < bd$ iff $|a| < |b|$ and $|c| < |d|$ or $|a||c| < |b||d|$	$0 < bd < ac$ iff $|b| < |a|$ and $|d| < |c|$ or $|b||d| < |a||c|$
1	From 2g, T4.4.1 Equalities: Any Quantity Multiplied by Zero is Zero and T4.7.7 Inequalities: Multiplication by Positive Number	$ad < 0 < ac$	
2	From 3g, T4.4.1 Equalities: Any Quantity Multiplied by Zero is Zero and T4.7.6 Inequalities: Multiplication by Negative Number Reverses Order	$ad < 0 < bd$	
∴	From 1, 2 and T4.7.1 Inequalities: Transitive Law	$0 < ac < bd$ iff $|a| < |b|$ and $|c| < |d|$ or $|a||c| < |b||d|$	$0 < bd < ac$ iff $|b| < |a|$ and $|d| < |c|$ or $|b||d| < |a||c|$

Theorem 4.7.27 Inequalities: Non-Existence of Equal Values About a Greater Quantity

1g	Given	$(a, b) \in R$
2g		$-a > b > a$
3g		for $a > 0$
Steps	Hypothesis	$-a > b > a$ for $a > 0$ does not exist
1	From 2g and D4.1.12 ANDing Greater Than Inequalities	$+b > a$
2	From 2g and D4.1.12 ANDing Greater Than Inequalities	$-a > b$
3	From 3g, 2 and T4.7.2 Inequalities: Uniqueness of Addition by a Positive Number	$a - a > a + b$
4	From 4 and A4.2.8 Inverse Add	$0 > a + b$
5	From 3g, 1 and T4.7.2 Inequalities: Uniqueness of Addition by a Positive Number	$a + b > a + a$
6	From 5 and T4.3.10 Equalities: Summation of Repeated Terms by 2	$a + b > 2a$
7	From 4, 6 and D4.1.12 ANDing Greater Than Inequalities	$0 > a + b > 2a$
8	From 7	$0 > 2a$
9	From 8 and T4.4.1 Equalities: Any Quantity Multiplied by Zero is Zero	$(2)(0) > 2a$
10	From 9 and T4.4.3 Equalities: Cancellation by Multiplication	$0 > a$
∴	From 12 but by 3g is a contradiction	$-a > b > a$ for $0 > a$ is invalid, hence this relation cannot exist

Theorem 4.7.28 Inequalities: Existence of Equal Values About a Lesser Quantity

1g	Given	$(a, b) \in R$
2g		$-a < b < a$
3g		for $a > 0$
Steps	Hypothesis	$-a < b < a$ for $a > 0$ exist
1	From 2g and D4.1.12 ANDing Greater Than Inequalities	$+b < a$
2	From 2g and D4.1.12 ANDing Greater Than Inequalities	$-a < b$
3	From 3g, 2 and T4.7.2 Inequalities: Uniqueness of Addition by a Positive Number	$a - a < a + b$
4	From 4 and A4.2.8 Inverse Add	$0 < a + b$
5	From 3g, 1 and T4.7.2 Inequalities: Uniqueness of Addition by a Positive Number	$a + b < a + a$
6	From 5 and T4.3.10 Equalities: Summation of Repeated Terms by 2	$a + b < 2a$
7	From 4, 6 and D4.1.12 ANDing Greater Than Inequalities	$0 < a + b < 2a$
8	From 7	$0 < 2a$
9	From 8 and T4.4.1 Equalities: Any Quantity Multiplied by Zero is Zero	$(2)(0) < 2a$
10	From 9 and T4.4.3 Equalities: Cancellation by Multiplication	$0 < a$
∴	From 10 and 3g creates no contradiction	$-a < b < a$ for $0 < a$ remains valid, hence relation exist

Corollary 4.7.28.1 Inequalities: Equal Absolute Values About a Lesser Quantity

1g	Given	$(a, b) \in R$
Steps	Hypothesis	$-|a| < b < |a|$
1	From T4.7.28 Inequalities: Existence of Equal Values About a Lesser Quantity	$-a < b < a$ for $a > 0$
∴	From 1 and D4.1.11 Absolute Value $|a|$	$-|a| < b < |a|$

Corollary 4.7.28.2 Inequalities: Equal Absolute Values Bounding Itself

1g	Given	$b = a$
Steps	Hypothesis	$-|a| < a \leq |a|$
1	From C4.7.26.1 Inequalities: Equal Absolute Values About a Lesser Quantity	$-|a| < b < |a|$
$\therefore$	From 1, 2g and A4.2.3 Substitution	$-|a| < a \leq |a|$

Theorem 4.7.29 Inequalities: Equal Values About a Lesser Absolute Quantity

1g	Given	$(a, b) \in R$
Steps	Hypothesis	$|b| < a$ for $-a < b < a$
1	From T4.7.26 Inequalities: Existence of Equal Values About a Lesser Quantity	$-a < b < a$ for $a > 0$
2	From 1 and D4.1.13 ANDing Less Than Inequalities	$+b < a$
3	From 1 and D4.1.13 ANDing Less Than Inequalities	$-a < b$
	From 2 and T4.7.6 Inequalities: Multiplication by Negative Number Reverses Order	$-b < a$
$\therefore$	From 1, 3 and D4.1.11 Absolute Value $|a|$	$|b| < a$

Theorem 4.7.30 Inequalities: The Absolute Sum is Less than the Sum of Absolute Values

Steps	Hypothesis	$|a + b| \leq |a| + |b|$
1	From C4.7.26.2 Inequalities: Equal Absolute Values Bounding Itself	$-|a| < a \leq |a|$
2	From C4.7.26.2 Inequalities: Equal Absolute Values Bounding Itself	$-|b| < b \leq |b|$
3	From 1, 2 and T4.7.7 Inequalities: Addition of $(a < b) + (c < d)$	$-|a| - |b| < a + b \leq |a| + |b|$
4	From 3, A4.2.12 Identity Multp, A4.2.14 Distribution and D4.1.20 Negative Coefficient	$-(|a| + |b|) < a + b \leq |a| + |b|$
$\therefore$	From 4 and T4.7.29 Inequalities: Equal Values About a Lesser Absolute Quantity	$|a + b| \leq |a| + |b|$

Theorem 4.7.31 Inequalities: Absolute Summation is Less than the Sum of all Absolute Values

1g	Given	$(a_i) \in R$ for i = 1, 2, ..., n
Steps	Hypothesis	$\lvert \sum^n a_i \rvert \leq \sum^n \lvert a_i \rvert$
1	From C4.4.19.2 Inequalities: Equal Absolute Values Bounding Itself	$-\lvert a_i \rvert < a_i \leq \lvert a_i \rvert$ for all i
3	From 1 and T4.7.7 Inequalities: Addition of (a < b) + (c < d) for all i	$\sum^n (-\lvert a \rvert_i) < \sum^n a_i \leq \sum^n \lvert a_i \rvert$
4	From 3, A4.2.12 Identity Multp, A4.2.14 Distribution and D4.1.20 Negative Coefficient	$-(\sum^n \lvert a \rvert_i) < \sum^n a_i \leq \sum^n \lvert a_i \rvert$
∴	From 4 and T4.7.29 Inequalities: Equal Values About a Lesser Absolute Quantity	$\lvert \sum^n a_i \rvert \leq \sum^n \lvert a_i \rvert$

Theorem 4.7.32 Inequalities: Absolute Difference is Less than the Sum of Absolute Values

1g	Given	$c = -b$
Steps	Hypothesis	$\lvert a - b \rvert \leq \lvert a \rvert + \lvert b \rvert$
1	From T4.7.29 Inequalities: Absolute Summation is Less than the Sum of all Absolute Values	$\lvert a + c \rvert \leq \lvert a \rvert + \lvert c \rvert$
2	From 1, 1g and A4.2.3 Substitution	$\lvert a - b \rvert \leq \lvert a \rvert + \lvert -b \rvert$
∴	From 2 and D4.1.11 Absolute Value $\lvert a \rvert$	$\lvert a - b \rvert \leq \lvert a \rvert + \lvert b \rvert$

Theorem 4.7.33 Inequalities: Cross Multiplication with Positive Reciprocal Product

1g 2g	Given	$a < b$ $0 < a$	
Steps	Hypothesis	$1/b < 1/a$	for $0 < a$ and $a < b$
1	From 1g, 2g and T4.7.1 Inequalities: Transitive Law	$0 < b$	
2	From 2g, 1 and T4.7.21 Inequalities: Multiplication for Positive Numbers	$0 < ab$	
3	From 2, A4.2.16 The Trichotomy Law of Ordered Numbers and T4.7.7 Inequalities: Multiplication by Positive Number	$0 < (1/ab)$	
4	From 3 and T4.7.7 Inequalities: Multiplication by Positive Number	$(1/ab)a < (1/ab)b$	
5	From 4 and D4.1.4(A, D) Rational Numbers	$(1/a)(1/b)a < (1/a)(1/b)b$	
6	From 5 and A4.2.10 Commutative Multp	$(1/b)(1/a)a < (1/a)(1/b)b$	
7	From 6 and D4.1.4(D, A) Rational Numbers	$(1/b)(a/a) < (1/a)(b/b)$	
8	From 7 and A4.2.13 Inverse Multp	$(1/b)1 < (1/a)1$	
∴	From 1g, 2g, 8 and A4.2.12 Identity Multp	$1/b < 1/a$	for $0 < a$ and $a < b$

Theorem 4.7.34 Inequalities: Cross Multiplication with Negative Product

1g 2g	Given	$a < b$ $b < 0$
Steps	Hypothesis	$1/b < 1/a$ for $a < b$ and $b < 0$
1	From 1g, 2g and T4.7.1 Inequalities: Transitive Law	$a < 0$
2	From 2g, 1 and T4.7.22 Inequalities: Multiplication for Negative Numbers	$0 < ab$
3	From 2, A4.2.16 The Trichotomy Law of Ordered Numbers and T4.7.7 Inequalities: Multiplication by Positive Number	$0 < (1/ab)$
4	From 3 and T4.7.6 Inequalities: Multiplication by Negative Number Reverses Order	$(1/ab)a < (1/ab)b$
5	From 4 and D4.1.4(A, D) Rational Numbers	$(1/a)(1/b)a < (1/a)(1/b)b$
6	From 5 and A4.2.10 Commutative Multp	$(1/b)(1/a)a < (1/a)(1/b)b$
7	From 6 and D4.1.4(D, A) Rational Numbers	$(1/b)(a/a) < (1/a)(b/b)$
8	From 7 and A4.2.13 Inverse Multp	$(1/b)1 < (1/a)1$
∴	From 1g, 2g, 8 and A4.2.12 Identity Multp	$1/b < 1/a$ for $a < b$ and $b < 0$

Theorem 4.7.35 Inequalities: Ternary Cross Multiplication with Positive Products

1g 2g	Given	$a < b < c$ $0 < a$
Steps	Hypothesis	$(1/c) < (1/b) < (1/a)$ for $0 < a$ and $a < b < c$
1	From 1g	$a < b$
2	From 2g, 1 and T4.7.33 Inequalities: Cross Multiplication with Positive Reciprocal Product	$(1/b) < (1/a)$
3	From 1g	$b < c$
4	From 2g, 2 and T4.7.33 Inequalities: Cross Multiplication with Positive Reciprocal Product	$(1/c) < (1/b)$
∴	From 1g, 2g, 2, 4 and T4.7.1 Inequalities: Transitive Law	$(1/c) < (1/b) < (1/a)$ for $0 < a$ and $a < b < c$

Theorem 4.7.36 Inequalities: Ternary Cross Multiplication with Negative Products

1g 2g	Given	$a < b < c$ $c < 0$	
Steps	Hypothesis	$(1/c) < (1/b) < (1/a)$	for $a < b < c$ and $c < 0$
1	From 1g	$a < b$	
2	From 2g, 1 and T4.7.33 Inequalities: Cross Multiplication with Positive Reciprocal Product	$(1/b) < (1/a)$	
3	From 1g	$b < c$	
4	From 2g, 2 and T4.7.33 Inequalities: Cross Multiplication with Positive Reciprocal Product	$(1/c) < (1/b)$	
∴	From 1g, 2g, 2, 4 and T4.7.1 Inequalities: Transitive Law	$(1/c) < (1/b) < (1/a)$	for $a < b < c$ and $c < 0$

Since Theorem 4.7.35 "Inequalities: Ternary Cross Multiplication with Positive Products" and Theorem 4.7.35 "Inequalities: Ternary Cross Multiplication with Negative Products" depend on their development by the reciprocals of (a, b, c) and dividing by zero is not permitted so they stand-alone. It follows that no theorem could be devised to cross zero on the number line because of the singularity.

Theorem 4.7.37 Inequalities: Subtraction Across Ternary Inequality

1g	Given	$d + a < b < d + c$
2g		$0 < d < b$ Case I
3g		$0 < b < d$ Case II
Steps	Hypothesis	$a < \pm(b - d) < c$ for case I ($d < b$) and case II ($b < d$)
1	From 1g and D4.1.8 Less Then Inequality [<]	$0 < b - (d + a)$
2	From 1g and D4.1.8 Less Then Inequality [<]	$0 < (d + c) - b$
3	From 1, A4.2.14 Distribution and A4.2.5 Commutative Add	$0 < (b - d) + a$
4	From 3, T4.8.2 Integer Exponents: Negative One Squared, D4.1.20 Negative Coefficient	$0 < -[-(b - d)] + a$
5	From 4 and A4.2.5 Commutative Add	$0 < a - [-(b - d)]$ true iff case II
6	From 2 and A4.2.5 Commutative Add	$0 < c - b + d$
7	From 2g, 6, T4.8.2 Integer Exponents: Negative One Squared, D4.1.20 Negative Coefficient and A4.2.14 Distribution	$0 < c - (b - d)$ true iff case I
8	From 3g, 5 and D4.1.8 Less Then Inequality [<]	$a < -(b - d)$ true iff case II
9	From 7 and D4.1.8 Less Then Inequality [<]	$+(b - d) < c$ true iff case I
∴	From 8 and 9	$a < \pm(b - d) < c$ for case I ($d < b$) and case II ($b < d$)

Section 4.8 Theorems on Integer Exponents

Theorem 4.8.1 Integer Exponents: Unity Raised to any Integer Value

1g 2g	Given	$(m, n) \in IG$ $n = -m$ for $n < 0$ and $0 < m$
Steps	Hypothesis	$(+1)^n = \prod^n (+1) = +1$ for any integer n
1	From A4.2.2A Equality	$(+1)^n = (+1)^n$ for $0 < n$
2	From 1 and D4.1.17 Exponential Notation	$(+1)^n = \prod^n (+1)$ for $0 < n$
3	From 2 and multiplying out	$(+1)^n = +1$ for $0 < n$
4	From A4.2.24 Inverse Exp	$(+1)^0 = +1$ for $0 = n$
5	From A4.2.2A Equality	$(+1)^n = (+1)^n$ for $n < 0$
6	From 2g, 5 and A4.2.3 Substitution	$(+1)^n = (+1)^{-m}$ for $n < 0$
7	From 6 and D4.1.18 Negative Exponential	$(+1)^n = 1 / (+1)^m$ for $n < 0$
8	From 7 and 3	$(+1)^n = 1 / (+1)$ for $n < 0$
9	From 8 and A4.2.13 Inverse Multp	$(+1)^n = 1$ for $n < 0$
∴	From 2, 3,4 and 9	$(+1)^n = \prod^n (+1) = +1$ for any integer n

Theorem 4.8.2 Integer Exponents: Negative One Squared

Steps	Hypothesis	$(-1)^2 = +1$ EQ A	$(-1)(-1) = +1$ EQ B
1	From A4.2.2A Equality	$(-1)^2 = (-1)^2$	
2	From D4.1.17 Exponential Notation	$(-1)^2 = (-1)(-1)$	
∴	From 2 and D4.1.19 Primitive Definition for Rational Arithmetic	$(-1)^2 = +1$ EQ A	$(-1)(-1) = +1$ EQ B

Theorem 4.8.3 Integer Exponents: Negative One Raised to an Even Number

1g 2g	Given	$(m, n) \in IG$ $n = 2m$ for n even and $0 < m$
Steps	Hypothesis	$(-1)^n = +1$ for integer n even
1	From A4.2.2A Equality	$(-1)^n = (-1)^n$
2	From 2g, 1 and A4.2.3 Substitution	$(-1)^n = (-1)^{2m}$
3	From 2 and A4.2.20 Commutative Exp	$(-1)^n = ((-1)^2)^m$
4	From 3 and T4.8.2 Integer Exponents: Negative One Squared	$(-1)^n = (+1)^m$
∴	From 4 and T4.8.1 Integer Exponents: Unity Raised to any Integer Value	$(-1)^n = +1$

Theorem 4.8.4 Integer Exponents: Negative One Raised to an Odd Number

1g	Given	$(m, n) \in IG$
2g		$n = 2m + 1$ for n odd and $0 \le m$
Steps	Hypothesis	$(-1)^n = -1$ for integer n odd
1	From A4.2.2A Equality	$(-1)^n = (-1)^n$
2	From 2g, 1 and A4.2.3 Substitution	$(-1)^n = (-1)^{2m+1}$
3	From 2 and A4.2.18 Summation Exp	$(-1)^n = (-1)^{2m} (-1)^1$
4	From 3 and T4.8.3 Integer Exponents: Negative One Raised to an Even Number	$(-1)^n = (+1) (-1)^1$
5	From 4 and A4.2.23 Identity Exp	$(-1)^n = (+1) (-1)$
∴	From 5 and D4.1.19 Primitive Definition for Rational Arithmetic	$(-1)^n = -1$

Theorem 4.8.5 Integer Exponents: Zero Raised to the Positive Power-n

1g	Given	$n \in IG$ for $0 < n$			
Steps	Hypothesis	$(0)^n = 0$	EQA	$(0)^n = \prod^n 0$	EQB
1	From 1g and A4.2.2A Equality	$(0)^n = (0)^n$			
2	From 1 and D4.1.17 Exponential Notation	$(0)^n = \prod^n 0$			
∴	From 2 and D4.1.19 Primitive Definition for Rational Arithmetic	$(0)^n = 0$	EQA	$(0)^n = \prod^n 0$	EQB

Theorem 4.8.6 Integer Exponents: Uniqueness of Exponents[4.3.1]

1g	Given	$(a, b) \in R$
2g		$n \in IG$
Steps	Hypothesis	$a^n = b^n$ for $a = b$
1	From 1g and A4.2.2A Equality	$a = b$
2	From 1g, 2g, 1 and T4.4.2 Equalities: Uniqueness of Multiplication	$\prod^n a = \prod^n b$ repeating n-products of [a] and [b]
∴	From 1 and D4.1.17 Exponential Notation	$a^n = b^n$ for $a = b$

Theorem 4.8.7 Integer Exponents: Distribution Across a Rational Number

1g	Given	rational a/b for $b \neq 0$
2g		$n \in IG$
Steps	Hypothesis	$(a/b)^n = a^n/b^n$ for $b \neq 0$
1	From 1g and A4.2.2A Equality	$a/b = a/b$
2	From 1 and T4.8.6 Integer Exponents: Uniqueness of Exponents	$(a/b)^n = (a/b)^n$
3	From 2 and A4.2.11 Associative Multp	$(a/b)^n = (a\,(1/b)\,)^n$
4	From 3 and A4.2.21 Distribution Exp	$(a/b)^n = a^n\,(1/b)^n$
5	From 4 and D4.1.18 Negative Exponential	$(a/b)^n = a^n\,(b^{-1})^n$
6	From 5 and A4.2.20 Commutative Exp	$(a/b)^n = a^n\,(b^{(-1)n})$
7	From 6 multiplication	$(a/b)^n = a^n\,(b^{-n})$
8	From 7 and D4.1.18 Negative Exponential	$(a/b)^n = a^n\,(1/b^n)$
∴	From 8 multiplication	$(a/b)^n = a^n/b^n$

Theorem 4.8.8 Integer Exponents: Uniqueness of Unequal Exponents Raised to a Positive One

1g	Given	$n \neq m$ but (n, m) are even/odd numbers
2g		
Steps	Hypothesis	$(+1)^n = (+1)^m$
1	From 1g and T4.8.1 Integer Exponents: Unity Raised to any Integer Value	$(+1)^n = +1$
2	From 1g and T4.8.1 Integer Exponents: Unity Raised to any Integer Value	$(+1)^m = +1$
∴	From 1, 2 and A4.2.3 Substitution	$(+1)^n = (+1)^m$

Theorem 4.8.9 Integer Exponents: Uniqueness of Unequal Even Exponents Raised to a Negative One

1g	Given	$n \neq m$ but (n, m) are even numbers
2g		$a - b$
Steps	Hypothesis	$(-1)^n = (-1)^m$
1	From 1g and T4.8.3 Integer Exponents: Negative One Raised to an Even Number	$(-1)^n = +1$
2	From 1g and T4.8.3 Integer Exponents: Negative One Raised to an Even Number	$(-1)^m = +1$
∴	From 1, 2 and A4.2.3 Substitution	$(-1)^n = (-1)^m$

Theorem 4.8.10 Integer Exponents: Uniqueness of Unequal Odd Exponents Raised to a Negative One

1g	Given	$n \neq m$ but (n, m) are odd numbers
2g		$a - b$
Steps	Hypothesis	$(-1)^n = (-1)^m$
1	From 1g and T4.8.4 Integer Exponents: Negative One Raised to an Odd Number	$(-1)^n = -1$
2	From 1g and T4.8.4 Integer Exponents: Negative One Raised to an Odd Number	$(-1)^m = -1$
∴	From 1, 2 and A4.2.3 Substitution	$(-1)^n = (-1)^m$

Section 4.9 Theorems on Rational Exponents

Theorem 4.9.1 Rational Exponent: Integer of a Positive One

1g 2g	Given	$n \in IG$ $n = -m$ for $0 < m$
Steps	Hypothesis	$(+1)^{1/n} = +1$ for $n \neq 0$
1	From T4.8.1 Integer Exponents: Unity Raised to any Integer Value for $0 < n$	$(+1)^{n} = +1$ for $0 < n$
2	From 1 and A4.2.25 Reciprocal Exp	$+1 = (+1)^{1/n}$ for $0 < n$
3	From T4.8.1 Integer Exponents: Unity Raised to any Integer Value for $0 > n$	$(+1)^{n} = +1$ for $0 > n$
4	From 3, 2g and A4.2.3 Substitution	$(+1)^{-m} = +1$ for $0 > n$
5	From 4 and A4.2.25 Reciprocal Exp	$+1 = (+1)^{-1/m}$ for $0 > n$
6	From T4.8.1 Integer Exponents: Unity Raised to any Integer Value for $0 = n$	$(+1)^{n} = +1$ for $0 = n$
7	From 6 and A4.2.25 Reciprocal Exp	$+1 = (+1)^{1/n}$ for $0 = n$
8	From 7 and A4.2.3 Substitution	$+1 = (+1)^{1/0}$ for $0 = n$
9	From 8 and D4.1.19 Primitive Definition for Rational Arithmetic	$+1$ cannot be defined $= (+1)^{\text{Undefined}}$ for $0 = n$
∴	From 2, 5 and 9	$(+1)^{1/n} = +1$ for $0 \neq n$

Theorem 4.9.2 Rational Exponent: Even Integer of a Positive One

1g	Given	$n \in IG$
Steps	Hypothesis	$(+1)^{1/n} = \pm 1$ because origins are unknown, for n even.
1	From T4.8.3 Integer Exponents: Negative One Raised to an Even Number for n even	$(-1)^{n} = +1$ for n even
2	From 1 and A4.2.25 Reciprocal Exp	$-1 = (+1)^{1/n}$ for n even
∴	From 2 and T4.9.1 Rational Exponent: Integer of a Positive One	$(+1)^{1/n} = \pm 1$ because no information remains after multiplication on weather its origin was a $[-1]$ or not, for n even. So ± 1 are both considered accounting for either case.

Theorem 4.9.3 Rational Exponent: Odd Integer of a Negative One

1g	Given	$n \in IG$ for [n] odd
Steps	Hypothesis	$-1 = (-1)^{1/n}$ for [n] odd
1	From T4.8.4 Integer Exponents: Negative One Raised to an Odd Number	$(-1)^n = -1$
∴	From 1, 1g and A4.2.25 Reciprocal Exp	$-1 = (-1)^{1/n}$ for [n] odd

Theorem 4.9.4 Rational Exponent: Square Root of a Positive One

1g	Given	$j^2 \in IG$
2g		$j^2 = +1$
Steps	Hypothesis	$(+1)^{1/2} = \pm 1$ for $j^2 = +1$
1	From 2g and A4.2.25 Reciprocal Exp	$j = (+1)^{1/2}$
2	From 1, T4.8.1 Integer Exponents: Unity Raised to any Integer Value and T4.8.2 Integer Exponents: Negative One Squared	$j = \pm 1$ two solutions found $(+1)(+1)$ or $(-1)(-1)$
∴	From 2g, 1 and 2	$(+1)^{1/2} = \pm 1$ for $j^2 = +1$

Theorem 4.9.5 Rational Exponent: Square Root of a Negative One

1g	Given	$j^2 \in IG$	
2g		$j^2 = -1$	
Steps	Hypothesis	$(-1)^{1/2} = \pm i$ and $i = \sqrt{-1}$ EQ A	$j^2 = -1$ EQ B
1	From 2g and A4.2.25 Reciprocal Exp	$j = (-1)^{1/2}$	
2	From 1, D4.1.20 Negative Coefficient and A4.2.25 Reciprocal Exp, but nothing can be found to represent twin pair numbers for $\sqrt{-1}$.	$j = \pm i$. No, twin pair numbers (?)(?) can be found to represent the number $\sqrt{-1}$ in the IG number system so it is left as a unique, independent quantity requiring its own distinctive field IM to be defined ∋, $(b \in R, i \equiv \sqrt{-1}, ib) \in IM$ see D2.2.27 "Imaginary Field". Now let $i \equiv \sqrt{-1}$.	
∴	From 2g and 2	$(-1)^{1/2} = \pm i$ and $i = \sqrt{-1}$ EQ A	$j^2 = -1$ EQ B

Theorem 4.9.6 Rational Exponent: Uniqueness of Exponents[4.3.1]

1g	Given	$n > 0$
2g		$c^n = a$
Steps	Hypothesis	$a^{1/n} = a^{1/n}$ for $a = a$
1	From A4.2.2A Equality	$a = a$
2	From 2g and A4.2.25 Reciprocal Exp	$c = a^{1/n}$
∴	From 2 and A4.2.3 Substitution	$a^{1/n} = a^{1/n}$ for $a = a$

Theorem 4.9.7 Rational Exponent: Distribution Across a Product

1g	Given	$n > 0$
2g		$c^n = ab$
3g		$d^n = a$
4g		$e^n = b$
Steps	Hypothesis	$(ab)^{1/n} = a^{1/n} b^{1/n}$
1	From 2g and A4.2.25 Reciprocal Exp	$c = (ab)^{1/n}$
2	From 3g and A4.2.25 Reciprocal Exp	$d = a^{1/n}$
3	From 4g and A4.2.25 Reciprocal Exp	$e = b^{1/n}$
4	From 2g, 3g, 4g	$c^n = d^n e^n$
5	From 4 and A4.2.21 Distribution Exp	$(\,c\,)^n = (de)^n$
6	From 5, T4.9.6 Rational Exponent: Uniqueness of Exponents	$(\,(\,c\,)^n\,)^{1/n} = (\,(de)^n\,)^{1/n}$
7	From 6, A4.2.22 Product Exp, D4.1.4(B, A) Rational Numbers, A4.2.13 Inverse Multp and A4.2.23 Identity Exp	$c = de$
∴	From 7, 2, 3 and A4.2.3 Substitution	$(ab)^{1/n} = a^{1/n} b^{1/n}$

Theorem 4.9.8 Rational Exponent: Commutative Product

1g	Given	$(a, b, c) \in R$
2g		$(m, n) \in IG$
3g		$0 < n$
4g		$c^n = a$
Steps	Hypothesis	$(a^{1/n})^m = (a^m)^{1/n}$ for $0 < n$
1	From 4g and T4.8.6 Integer Exponents: Uniqueness of Exponents	$c^{nm} = a^m$
2	From 1 and A4.2.25 Reciprocal Exp	$c^m = (a^m)^{1/n}$
3	From 2 and A4.2.12 Identity Multp	$c^{1\,m} = (a^m)^{1/n}$
4	From 3 and A4.2.22 Product Exp	$(\,c^1\,)^m = (a^m)^{1/n}$
5	From 4 and A4.2.23 Identity Exp	$(\,c\,)^m = (a^m)^{1/n}$
6	From 4g and A4.2.25 Reciprocal Exp	$c = a^{1/n}$
∴	From 6 into 5	$(a^{1/n})^m = (a^m)^{1/n}$

Theorem 4.9.9 Rational Exponent: Distribution Across a Rational Number

	Given	
1g	Given	$(a, b, c) \in R$
2g		$(m, n) \in IG$
3g		$0 < n$
4g		$c = 1/b$
Steps	Hypothesis	$(a/b)^{1/n} = a^{1/n} / b^{1/n}$
1	From 1g and A4.2.2A Equality	$a = a$
2	From 1g and A4.2.2A Equality	$c = c$
3	From 2 and T4.3.9 Closure Multp	$ac = ac$
4	From 3 and T4.9.6 Rational Exponent: Uniqueness of Exponents	$(ac)^{1/n} = (ac)^{1/n}$
5	From 4 and T4.9.7 Rational Exponent: Distribution Across a Product	$(ac)^{1/n} = a^{1/n} c^{1/n}$
6	From 5 and 4g	$(a/b)^{1/n} = a^{1/n} (1/b)^{1/n}$
7	From 6 and D4.1.18 Negative Exponential	$(a/b)^{1/n} = a^{1/n} (b^{-1})^{1/n}$
8	From 7 and T4.9.8 Rational Exponent: Commutative Product	$(a/b)^{1/n} = a^{1/n} (b^{1/n})^{-1}$
∴	From 8 and D4.1.18 Negative Exponential	$(a/b)^{1/n} = a^{1/n} / b^{1/n}$

Theorem 4.9.10 Rational Exponent: Left Inverse Rational Product

	Given	
1g	Given	$0 < n$ for $n \in IG$
2g		$b^n = a$ for $(a, b) \in R$
Steps	Hypothesis	$a^{(1/n) n} = (a^{1/n})^n$ for $0 < n$
1	From 2g and A4.2.25 Reciprocal Exp	$b = a^{1/n}$
2	From 1, 1g and T4.8.6 Integer Exponents: Uniqueness of Exponents	$b^n = (a^{1/n})^n$
3	From 2, 2g and A4.2.23 Identity Exp	$a^1 = (a^{1/n})^n$
4	From 3 and A4.2.13 Inverse Multp	$1 = (1/n) n$
∴	From 3, 4 and A4.2.3 Substitution	$a^{(1/n) n} = (a^{1/n})^n$

Theorem 4.9.11 Rational Exponent: Right Inverse Rational Product

1g	Given	$0 < n$ for $n \in IG$
2g		$b^{1/n} = a$ for $(a, b) \in R$
Steps	Hypothesis	$a^{n(1/n)} = (a^n)^{1/n}$ for $0 < n$
1	From 2g and A4.2.25 Reciprocal Exp	$b = a^n$
2	From 1, 1g and T4.8.6 Integer Exponents: Uniqueness of Exponents	$b^{1/n} = (a^n)^{1/n}$
3	From 2, 2g and A4.2.23 Identity Exp	$a^1 = (a^n)^{1/n}$
4	From 3 and A4.2.13 Inverse Multp	$1 = n\,(1/n)$
∴	From 3, 4 and A4.2.3 Substitution	$a^{n(1/n)} = (a^n)^{1/n}$ for $0 < n$

Theorem 4.9.12 Rational Exponent: Right Product of Natural Numbers

1g	Given	$(a, b) \in R$
2g		$(m, n) \in IG$
3g		$0 < m$
4g		$b^m = a$
Steps	Hypothesis	$a^{n(1/m)} = (a^n)^{1/m}$ for $0 < m$
1	From 2g, 4g and T4.8.6 Integer Exponents: Uniqueness of Exponents	$(b^m)^n = a^n$
2	From 1 and A4.2.22 Product Exp	$b^{mn} = a^n$
3	From 2 and A4.2.10 Commutative Multp	$b^{nm} = a^n$
4	From 3 and A4.2.22 Product Exp	$(b^n)^m = a^n$
5	From 4 and A4.2.25 Reciprocal Exp	$b^n = (a^n)^{1/m}$
6	From 5 and A4.2.25 Reciprocal Exp	$b = ((a^n)^{1/m})^{1/n}$
7	From 4g, A4.2.25 Reciprocal Exp and 6	$a^{1/m} = ((a^n)^{1/m})^{1/n}$
8	From 7 and A4.2.12 Identity Multp	$a^{(1/m)\,1} = ((a^n)^{1/m})^{1/n}$
9	From 8 and A4.2.13 Inverse Multp	$a^{(1/m)\,(n/n)} = ((a^n)^{1/m})^{1/n}$
10	From 9 and A4.2.11 Associative Multp	$a^{((1/m)(n))(1/n)} = ((a^n)^{1/m})^{1/n}$
11	From 10 multiplication	$a^{n(1/m)\,(1/n)} = ((a^n)^{1/m})^{1/n}$
12	From 11, A4.2.27 Correspondence Exp and A4.2.22 Product Exp	$(a^{n(1/m)})^{(1/n)} = ((a^n)^{1/m})^{1/n}$
∴	From 12 and T4.9.6 Rational Exponent: Uniqueness of Exponents	$a^{n(1/m)} = (a^n)^{1/m}$

Theorem 4.9.13 Rational Exponent: Left Product of Natural Number

Steps	Hypothesis	$a^{n(1/m)} = (a^{1/m})^n$ for $0 < m$
1	From T4.9.12 Rational Exponent: Right product of Natural Numbers	$a^{n(1/m)} = (a^n)^{1/m}$
$\therefore$	From 1 and T4.9.8 Rational Exponent: Commutative Product	$a^{n(1/m)} = (a^{1/m})^n$

Theorem 4.9.14 Rational Exponent: Addition of Compound Rational Number

	Given	
1g		$(a, b) \in R$
2g		$(m, n) \in IG$
3g		$0 < m$
4g		$b^m = a^{nm+1}$ is true for [c] a mixed transcendental number.
Steps	Hypothesis	$a^{n + 1/m} = a^n a^{1/m}$
1	From 4g and A4.2.18 Summation Exp	$b^m = a^{nm} a^1$
2	From 1 and A4.2.25 Reciprocal Exp	$b = (a^{nm} a^1)^{1/m}$
3	From 2 and T4.9.7 Rational Exponent: Distribution Across a Product	$b = (a^{nm})^{1/m} (a^1)^{1/m}$
4	From 4g and A4.2.25 Reciprocal Exp	$b = (a^{nm+1})^{1/m}$
5	From 3 and 4	$(a^{nm+1})^{1/m} = (a^{nm})^{1/m} (a^1)^{1/m}$
6	From 5 and T4.9.12 Rational Exponent: Right Product of Natural Numbers	$a^{(nm+1)(1/m)} = a^{(nm)(1/m)} a^{1(1/m)}$
7	From 6 and A4.2.14 Distribution	$a^{(nm)(1/m)+1(1/n)} = a^{(nm)(1/m)} a^{1(1/m)}$
8	From 7 and A4.2.11 Associative Multp	$a^{n(m(1/m))+1(1/m)} = a^{n(m(1/m))} a^{1(1/m)}$
9	From 8 and A4.2.13 Inverse Multp	$a^{n1+1(1/m)} = a^{n1} a^{1(1/m)}$
$\therefore$	From 9 and A4.2.12 Identity Multp	$a^{n+1/m} = a^n a^{1/m}$

Theorem 4.9.15 Rational Exponent: Subtraction of Compound Rational Number

	Given	
1g	Given	$(a, b) \in R$
2g		$(m, n) \in IG$
3g		$0 < m$
4g		$b^m = a^{nm-1}$ is true for [c] a mixed transcendental number.
Steps	Hypothesis	$a^{n - 1/m} = a^n a^{-1/m}$
1	From 4g and A4.2.18 Summation Exp	$b^m = a^{nm} a^{-1}$
2	From 1 and A4.2.25 Reciprocal Exp	$b = (a^{nm} a^{-1})^{1/m}$
3	From 2 and T4.9.7 Rational Exponent: Distribution Across a Product	$b = (a^{nm})^{1/m} (a^{-1})^{1/m}$
4	From 4g and A4.2.25 Reciprocal Exp	$b = (a^{nm-1})^{1/m}$
5	From 3 and 4	$(a^{nm-1})^{1/m} = (a^{nm})^{1/m} (a^{-1})^{1/m}$
6	From 5 and T4.9.12 Rational Exponent: Right Product of Natural Numbers	$a^{(nm-1)(1/m)} = a^{(nm)(1/m)} a^{(-1)(1/m)}$
7	From 6 and A4.2.14 Distribution	$a^{(nm)(1/m) - 1(1/m)} = a^{(nm)(1/n)} a^{(-1)(1/m)}$
8	From 7 and A4.2.11 Associative Multp	$a^{n((1/m)m) - 1(1/m)} = a^{n((1/m)m)} a^{(-1)(1/m)}$
9	From 8 and A4.2.13 Inverse Multp	$a^{1 n - 1(1/m)} = a^{1 n} a^{(-1)(1/m)}$
$\therefore$	From 9 and A4.2.12 Identity Multp $a - b$	$a^{n - 1/m} = a^n a^{-1/m}$

Theorem 4.9.16 Rational Exponent: Addition of Rational Numbers

1g	Given	$(a, b) \in R$
2g		$(m, n) \in IG$
3g		$0 < m$ and $0 < n$
4g		$b^{mn} = a^{m+n}$ is true for [c] a mixed transcendental number.
Steps	Hypothesis	$a^{1/n + 1/m} = a^{1/n} a^{1/m}$
1	From 4g and A4.2.18 Summation Exp	$b^{mn} = a^m a^n$
2	From 1 and A4.2.25 Reciprocal Exp	$b = (a^m a^n)^{1/mn}$
3	From 2 and T4.9.7 Rational Exponent: Distribution Across a Product	$b = (a^m)^{1/mn} (a^n)^{1/mn}$
4	From 4g and A4.2.25 Reciprocal Exp	$b = (a^{m+n})^{1/mn}$
5	From 3 and 4	$(a^{m+n})^{1/mn} = (a^m)^{1/mn} (a^n)^{1/mn}$
6	From 5 T4.9.12 Rational Exponent: Right product of Natural Numbers	$a^{(m+n)(1/mn)} = a^{(m)(1/mn)} a^{n(1/mn)}$
7	From 6 A4.2.14 Distribution	$a^{m(1/mn)+n(1/mn)} = a^{m(1/mn)} a^{n(1/mn)}$
8	From 7 factor and group	$a^{(m(1/m)(1/n))+(n(1/m)(1/n))} = a^{(m(1/m)(1/n))} a^{(n(1/m)(1/n))}$
9	From 8 A4.2.10 Commutative Multp	$a^{(m(1/m)(1/n))+(n(1/n)(1/m))} = a^{(m(1/m)(1/n))} a^{(n(1/n)(1/m))}$
10	From 9 A4.2.11 Associative Multp	$a^{(m(1/m))(1/n)+(n(1/n))(1/m)} = a^{(m(1/m))(1/n)} a^{(n(1/n))(1/m)}$
11	From 10 and A4.2.13 Inverse Multp	$a^{1(1/n)+1(1/m)} = a^{1(1/n)} a^{1(1/m)}$
∴	From 11 and A4.2.12 Identity Multp	$a^{1/n + 1/m} = a^{1/n} a^{1/m}$

Theorem 4.9.17 Rational Exponent: Subtraction of Rational Numbers

1g	Given	$(a, b) \in R$
2g		$(m, n) \in IG$
3g		$0 < m$ and $0 < n$
4g		$b^{mn} = a^{m-n}$ is true for [c] a mixed transcendental number.
Steps	Hypothesis	$a^{1/n - 1/m} = a^{1/n} a^{-1/m}$
1	From 4g and A4.2.18 Summation Exp	$b^{mn} = a^m a^n$
2	From 1 and A4.2.25 Reciprocal Exp	$b = (a^m a^{-n})^{1/mn}$
3	From 2 and T4.9.7 Rational Exponent: Distribution Across a Product	$b = (a^m)^{1/mn} (a^{-n})^{1/mn}$
4	From 4g and A4.2.25 Reciprocal Exp	$b = (a^{m-n})^{1/mn}$
5	From 3 and 4	$(a^{m-n})^{1/mn} = (a^m)^{1/mn} (a^{-n})^{1/mn}$
6	From 5 T4.9.12 Rational Exponent: Right Product of Natural Numbers	$a^{(m-n)(1/mn)} = a^{(m)(1/mn)} a^{-n(1/mn)}$
7	From 6 and A4.2.14 Distribution	$a^{m(1/mn) - n(1/mn)} = a^{m(1/mn)} a^{-n(1/mn)}$
8	From 7 A4.2.11 Associative Multp (group)	$a^{(m(1/m)(1/n)) - (n(1/m)(1/n))} = a^{(m(1/m)(1/n))} a^{-(n(1/m)(1/n))}$
9	From 8 A4.2.10 Commutative Multp	$a^{(m(1/m)(1/n)) - (n(1/n)(1/m))} = a^{(m(1/m)(1/n))} a^{-(n(1/n)(1/m))}$
10	From 9 A4.2.11 Associative Multp	$a^{(m(1/m))(1/n) - (n(1/n))(1/m)} = a^{(m(1/m))(1/n)} a^{-(n(1/n))(1/m)}$
11	From 10 and A4.2.13 Inverse Multp	$a^{1(1/n) - 1(1/m)} = a^{1(1/n)} a^{-1(1/m)}$
∴	From 11 and A4.2.12 Identity Multp	$a^{1/n - 1/m} = a^{1/n} a^{-1/m}$

Theorem 4.9.18 Rational Exponent: Negative Exponent as a Reciprocal Quantity

1g	Given	$a \in R$
2g		$n \in IG$
3g		$0 < n$
4g		$a^{-1/n} = a^{-1/n}$
Steps	Hypothesis	$a^{-1/n} = 1 / a^{1/n}$
1	From 4g	$a^{-1/n} = a^{-1/n}$
2	From 1 and group	$a^{-1/n} = a^{(-1)(1/n)}$
3	From 2 and A4.2.10 Commutative Multp	$a^{-1/n} = a^{(1/n)(-1)}$
4	From 3 and T4.9.13 Rational Exponent: Left Product of Natural Number	$a^{-1/n} = (a^{1/n})^{-1}$
∴	From 4 and D4.1.18 Negative Exponential	$a^{-1/n} = 1 / a^{1/n}$

Theorem 4.9.19 Rational Exponent: Addition by Product and Rational Number

1g 2g	Given	$0 \le a$ for $(a) \in R$ $(m, n) \in IG$	
Steps	Hypothesis	$a^n \, a^{1/m} = [a^{(mn+1)}]^{1/m}$ EQ A	$a^n \, a^{1/m} = [a^{1/m}]^{(mn+1)}$ EQ B
1	From A4.2.2A Equality	$a^n \, a^{1/m} = a^n \, a^{1/m}$	
2	From 1 and A4.2.18 Summation Exp	$a^n \, a^{1/m} = a^{n + 1/m}$	
3	From 2, A4.2.12 Identity Multp and A4.2.13 Inverse Multp	$a^n \, a^{1/m} = a^{mn/m + 1/m}$	
4	From 3, A4.2.12 Identity Multp, D4.1.1(A, B) Rational Numbers and A4.2.14 Distribution	$a^n \, a^{1/m} = a^{(mn+1)/m}$	
∴	From 4, D4.1.1(A, B) Rational Numbers, A4.2.22 Product Exp and T4.9.8 Rational Exponent: Commutative Product	$a^n \, a^{1/m} = [a^{(mn+1)}]^{1/m}$ EQ A	$a^n \, a^{1/m} = [a^{1/m}]^{(mn+1)}$ EQ B

Theorem 4.9.20 Rational Exponent: Subtraction by Radical Number as Divisor

1g 2g	Given	$0 < a$ for $(a) \in R$ $(m, n) \in IG$	
Steps	Hypothesis	$a^n / a^{1/m} = [a^{(mn-1)}]^{1/m}$ EQ A	$a^n / a^{1/m} = [a^{1/m}]^{(mn-1)}$ EQ B
1	From A4.2.2A Equality	$a^n / a^{1/m} = a^n / a^{1/m}$	
2	From 1, D4.1.18 Negative Exponential	$a^n / a^{1/m} = a^n \, a^{-1/m}$	
3	From 2 and A4.2.18 Summation Exp	$a^n / a^{1/m} = a^{n - 1/m}$	
4	From 3, A4.2.12 Identity Multp and A4.2.13 Inverse Multp	$a^n / a^{1/m} = a^{mn/m - 1/m}$	
5	From 4, A4.2.12 Identity Multp, D4.1.1(A, B) Rational Numbers and A4.2.14 Distribution	$a^n / a^{1/m} = a^{(mn-1)/m}$	
∴	From 5, D4.1.1(A, B) Rational Numbers, A4.2.22 Product Exp and T4.9.8 Rational Exponent: Commutative Product	$a^n / a^{1/m} = [a^{(mn-1)}]^{1/m}$ EQ A	$a^n / a^{1/m} = [a^{1/m}]^{(mn-1)}$ EQ B

Theorem 4.9.21 Rational Exponent: Subtraction by Rational and Divisor

1g 2g	Given	$0 < a$ for $(a) \in R$ $(m, n) \in IG$	
Steps	Hypothesis	$a^{1/m} / a^n = [1 / a^{mn-1}]^{1/m}$ EQ A	$a^{1/m} / a^n = 1 / (a^{1/m})^{mn-1}$ EQ B
1	From A4.2.2A Equality	$a^{1/m} / a^n = a^{1/m} / a^n$	
2	From 1 and D4.1.18 Negative Exponential	$a^{1/m} / a^n = a^{1/m} \, a^{-n}$	
3	From 2 and A4.2.18 Summation Exp	$a^{1/m} / a^n = a^{1/m - n}$	
4	From 3, A4.2.12 Identity Multp and A4.2.13 Inverse Multp	$a^{1/m} / a^n = a^{1/m - mn/m}$	
5	From 4, A4.2.12 Identity Multp, D4.1.1(A, B) Rational Numbers and A4.2.14 Distribution	$a^{1/m} / a^n = a^{(1 - mn)/m}$	
6	From 5, D4.1.20 Negative Coefficient, T4.8.2B Integer Exponents: Negative One Squared, A4.2.14 Distribution D4.1.1(A, B) Rational Numbers, A4.2.22 Product Exp and T4.9.8 Rational Exponent: Commutative Product	$a^{1/m} / a^n = [a^{-(mn-1)}]^{1/m}$	$a^{1/m} / a^n = [a^{1/m}]^{-(mn-1)}$
∴	From 6 and D4.1.18 Negative Exponential	$a^{1/m} / a^n = [1 / a^{mn-1}]^{1/m}$ EQ A	$a^{1/m} / a^n = 1 / (a^{1/m})^{mn-1}$ EQ B

Section 4.10 Theorems on Radicals

Theorem 4.10.1 Radicals: Uniqueness of Radicals[4.3.1]

Steps	Hypothesis	$\sqrt[n]{a} = \sqrt[n]{a}$ for a = a
1	From T4.9.6 Rational Exponent: Uniqueness of Exponents	$a^{1/n} = a^{1/n}$ for a = a
∴	From 1 and A4.2.25 Reciprocal Exp	$\sqrt[n]{a} = \sqrt[n]{a}$ for a = a

Theorem 4.10.2 Radicals: Commutative Product

Steps	Hypothesis	
1g	Given	$a \in R$
2g		$n \in IG$
3g		$0 < n$
Steps	Hypothesis	$(\sqrt[n]{a})^m = \sqrt[n]{a^m}$ for $0 < n$
1	From T4.9.8 Rational Exponent: Commutative Product	$(a^{1/n})^m = (a^m)^{1/n}$
∴	From 1 and A4.2.25 Reciprocal Exp	$(\sqrt[n]{a})^m = \sqrt[n]{a^m}$ for $0 < n$

Theorem 4.10.3 Radicals: Identity Power Raised to a Radical

Steps	Hypothesis	$a = \sqrt[n]{a^n}$
1	From A4.2.2A Equality	$a = a$
2	From 1 and A4.2.13 Inverse Multp	$a = a^{n/n}$
3	From 2 and A4.2.11 Associative Multp	$a = a^{n(1/n)}$
4	From 2 and T4.9.11 Rational Exponent: Right Inverse Rational Product	$a = (a^n)^{(1/n)}$
∴	From 4 and A4.2.25 Reciprocal Exp	$a = \sqrt[n]{a^n}$

Theorem 4.10.4 Radicals: Identity Radical Raised to a Power

Steps	Hypothesis	$a = (\sqrt[n]{a})^n$
1	From T4.10.3 Radicals: Identity Power Raised to a Radical	$a = \sqrt[n]{a^n}$
∴	From 2 and T4.10.2 Radicals: Commutative Product	$a = (\sqrt[n]{a})^n$

Tensor Calculus & Physics: A General Treatise

Theorem 4.10.5 Radicals: Distribution Across a Product

Steps	Hypothesis	$\sqrt[n]{ab} = \sqrt[n]{a}\,\sqrt[n]{b}$
1	From T4.9.7 Rational Exponent: Distribution Across a Product	$(ab)^{1/n} = a^{1/n}\,b^{1/n}$
$\therefore$	From 1 and A4.2.25 Reciprocal Exp	$\sqrt[n]{ab} = \sqrt[n]{a}\,\sqrt[n]{b}$

Theorem 4.10.6 Radicals: Distribution Across a Rational Number

Steps	Hypothesis	$\sqrt[n]{a/b} = (\sqrt[n]{a}) / (\sqrt[n]{b})$
1	From T4.9.9 Rational Exponent: Distribution Across a Rational Number	$(a/b)^{1/n} = a^{1/n} / b^{1/n}$
$\therefore$	From 1 and A4.2.25 Reciprocal Exp	$\sqrt[n]{a/b} = (\sqrt[n]{a}) / (\sqrt[n]{b})$

Theorem 4.10.7 Radicals: Reciprocal Exponent by Positive Square Root

Steps	Given	
1g 2g	Given	$(a, b) \in R$ $b^2 = a$
Steps	Hypothesis	$b = \pm\sqrt{a}$ for $b^2 = a$
1	From 1g, 2g	$b^2 = a$
2	From 1 and A4.2.12 Identity Multp	$b^2 = 1\,a$
3	From 2 and A4.2.25 Reciprocal Exp	$b = \sqrt{(1\,a)}$
4	From 3 and T4.10.5 Radicals: Distribution Across a Product	$b = \sqrt{1}\,\sqrt{a}$
5	From 4 and A4.2.25 Reciprocal Exp	$b = (1)^{1/2}\,\sqrt{a}$
$\therefore$	From 1, 5 and T4.9.4 Rational Exponent: Square Root of a Positive One	$b = \pm\sqrt{a}$ for $b^2 = a$

Theorem 4.10.8 Radicals: Reciprocal Exponent by Negative Square Root

1g	Given	$(a, b) \in R$
2g		$b^2 = -a$
Steps	Hypothesis	$b = \pm i\sqrt{a}$ for $b^2 = -a$
1	From 1g, 2g	$b^2 = -a$
2	From 1 and D4.1.20 Negative Coefficient	$b^2 = -1\,a$
3	From 2 and A4.2.25 Reciprocal Exp	$b = \sqrt{(-1\,a)}$
4	From 3 and T4.10.5 Radicals: Distribution Across a Product	$b = \sqrt{-1}\,\sqrt{a}$
5	From 4 and A4.2.25 Reciprocal Exp	$b = (-1)^{½}\,\sqrt{a}$
∴	From 1, 5 and T4.9.5 Rational Exponent: Square Root of a Negative One	$b = \pm i\sqrt{a}$ for $b^2 = -a$

Theorem 4.10.9 Radicals: Squaring the Inverse by Positive Square Root

1g	Given	$(a, b) \in R$
2g		$b^2 = a$
Steps	Hypothesis	$b = \pm\sqrt{a}$ for $b^2 = a$
1	From 1g, 2g	$b^2 = a$
2	From 1, T4.10.4 Radicals: Identity Radical Raised to a Power and A4.2.3 Substitution	$b^2 = (\sqrt{a})^2$
3	From 2 and A4.2.12 Identity Multp	$b^2 = (\sqrt{(+1)}\,a)^2$
4	From 3, T4.10.5 Radicals: Distribution Across a Product and A4.2.25 Reciprocal Exp	$b^2 = ((+1)^{½}\,\sqrt{a})^2$
5	From 4 and T4.9.4 Rational Exponent: Square Root of a Positive One	$b^2 = (\pm\sqrt{a})^2$
∴	From 5 and A4.2.25 Reciprocal Exp	$b = \pm\sqrt{a}$ for $b^2 = a$

Theorem 4.10.10 Radicals: Squaring the Inverse by Negative Square Root

1g 2g	Given	$(a, b) \in R$ $b^2 = -a$
Steps	Hypothesis	$b = \pm i\sqrt{a}$ for $b^2 = -a$
1	From 1g, 2g	$b^2 = -a$
2	From 1, T4.10.4 Radicals: Identity Radical Raised to a Power and A4.2.3 Substitution	$b^2 = (\sqrt{-a})^2$
3	From 2 and D4.1.20 Negative Coefficient	$b^2 = (\sqrt{(-1)\,a})^2$
4	From 3, T4.10.5 Radicals: Distribution Across a Product and A4.2.25 Reciprocal Exp	$b^2 = ((-1)^{1/2}\,\sqrt{a})^2$
5	From 4 and T4.9.5 Rational Exponent: Square Root of a Negative One	$b^2 = (\pm i\sqrt{a})^2$
∴	From 5 and A4.2.25 Reciprocal Exp	$b = \pm i\sqrt{a}$ for $b^2 = -a$

General proofs for taking the nth root of a real number, positive one and negative one can be found in Chapter 7 theorems T7.4.32 "Complex Planes: De Moivre's Root of a Complex Number", T7.4.33 "Complex Planes: De Moivre's Root of One" and T7.4.34 "Complex Planes: De Moivre's Root of a Negative One"

Theorem 4.10.11 Radicals: Addition by Product and Radical Number

1g 2g	Given	$0 \le a$ for $(a) \in R$ $(m, n) \in IG$		
Steps	Hypothesis	$a^n\,{}^m\!\sqrt{a} = {}^m\!\sqrt{a^{nm+1}}$ EQ A		$a^n\,{}^m\!\sqrt{a} = ({}^m\!\sqrt{a})^{nm+1}$ EQ B
1	From 5 and T4.9.19 Rational Exponent: Addition by Product and Rational Number	$a^n\,a^{1/m} = [a^{(mn+1)}]^{1/m}$		$a^n\,a^{1/m} = [a^{1/m}]^{(mn+1)}$
∴	From 1 and A4.2.25 Reciprocal Exp	$a^n\,{}^m\!\sqrt{a} = {}^m\!\sqrt{a^{nm+1}}$ EQ A		$a^n\,{}^m\!\sqrt{a} = ({}^m\!\sqrt{a})^{nm+1}$ EQ B

Theorem 4.10.12 Radicals: Subtraction by Radical Number as Divisor

1g 2g	Given	$0 < a$ for $(a) \in R$ $(m, n) \in IG$		
Steps	Hypothesis	$a^n / {}^m\!\sqrt{a} = {}^m\!\sqrt{a^{nm-1}}$ EQ A		$a^n / {}^m\!\sqrt{a} = ({}^m\!\sqrt{a})^{nm-1}$ EQ B
1	From T4.9.20(A, B) Rational Exponent: Subtraction by Radical Number as Divisor	$a^n / a^{1/m} = [a^{(mn-1)}]^{1/m}$		$a^n / {}^m\!\sqrt{a} = [a^{1/m}]^{(mn-1)}$
∴	From 1 and A4.2.25 Reciprocal Exp	$a^n / {}^m\!\sqrt{a} = {}^m\!\sqrt{a^{nm-1}}$ EQ A		$a^n / {}^m\!\sqrt{a} = ({}^m\!\sqrt{a})^{nm-1}$ EQ B

Theorem 4.10.13 Radicals: Subtraction by Radical and Divisor

1g 2g	Given	$0 < a$ for $(a) \in R$ $(m, n) \in IG$	
Steps	Hypothesis	$\sqrt[m]{a} / a^n = 1 / \sqrt[m]{a^{mn-1}}$ EQ A	$\sqrt[m]{a} / a^n = 1 / (\sqrt[m]{a})^{mn-1}$ EQ B
1	From T4.9.21(A, B) Rational Exponent: Subtraction by Rational and Divisor	$a^{1/m} / a^n = [1 / a^{mn-1}]^{1/m}$	$a^{1/m} / a^n = 1 / (a^{1/m})^{mn-1}$
∴	From 1, D4.1.18 Negative Exponential and A4.2.25 Reciprocal Exp	$\sqrt[m]{a} / a^n = 1 / \sqrt[m]{a^{mn-1}}$ EQ A	$\sqrt[m]{a} / a^n = 1 / (\sqrt[m]{a})^{mn-1}$ EQ B

Section 4.11 Theorems on Real Logarithmic Exponents:

With integer exponents they can be plotted on a graph as b^n and than for rational $b^{(a/c)}$ its only a matter of extrapolating the graph to plot it for all in between numbers all real numbers are accounted for b^x as follows:

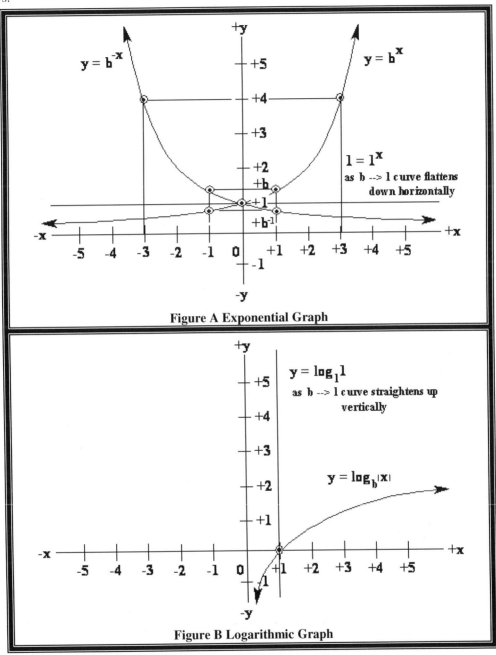

Figure 4.11.1: Exponent and Logarithmic Graphs for Real Numbers.

Likewise, for the inverse function it can be plotted as well as seen in the above graph, but only for absolute values for the argument otherwise it is driven into the complex plane. Since the exponent axioms hold for integer and rationales it can be conjectured that, they hold for all real numbers as well. If so the following theorems will hold.

Theorem 4.11.1 Real Exponents: Uniqueness of Logarithms[4.3.1]

1g	Given	$(x, y, a, b) \in R$	
2g		$x = b^a$	
3g		$y = b^a$	
Steps	Hypothesis	$\log_b x = \log_b y \quad$ for $x = y$	
1	From 2g, 3g and A4.2.3 Substitution	$x = y = b^a$	
2	From 1	$x = b^a$	$y = b^a$
3	From 2 and A4.2.26 Logarithms	$a = \log_b x$	$a = \log_b y$
∴	From 3 and A4.2.3 Substitution	$\log_b x = \log_b y \quad$ for $x = y$	

Theorem 4.11.2 Real Exponents: The Logarithm of Unity

1g	Given	Fig4.11.1B Logarithmic Graph
2g		$y = 1$
Steps	Hypothesis	if $y = 1$ then $0 = \log_b 1 = x$
∴	From 1g, 2g and A4.2.26 Logarithms	$0 = \log_b 1 = x$

Theorem 4.11.3 Real Exponents: The Logarithm of the Base

1g	Given	Fig4.11.1B Logarithmic Graph
2g		$y = b$
Steps	Hypothesis	if $y = b$ then $1 = \log_b b = x$
∴	From 1g, 2g and A4.2.26 Logarithms	$1 = \log_b b = x$

Theorem 4.11.4 Real Exponents: Exponent of Logarithm

1g	Given	$(x, y, z, a, b) \in R$
2g		$z = b^{ay}$
3g		$x = b^a$
Steps	Hypothesis	$y \log_b x = \log_b(x^y)$
1	From 3g and A4.2.26 Logarithms	$a = \log_b x$
2	From 2g, A4.2.27 Correspondence Exp and A4.2.22 Product Exp	$z = (b^a)^y$
3	From 2 and 3g	$z = x^y$
4	From 2g and A4.2.26 Logarithms	$ay = \log_b z$
5	From 4 and A4.2.10 Commutative Multp	$ya = \log_b z$
∴	From 1, 3 and 5	$y \log_b x = \log_b(x^y)$

Theorem 4.11.5 Real Exponents: Summation of Logarithms

1g	Given	$(u, v, w, x, y, z, b) \in R$
2g		$x = b^u$
3g		$y = b^v$
4g		$z = b^{u+v}$
Steps	Hypothesis	$\log_b xy = \log_b x + \log_b y$
1	From A4.2.27 Correspondence Exp and A4.2.18 Summation Exp	$b^{u+v} = b^u \, b^v$
2	From 1, 2g, 3g and 4g	$z = xy$
3	From 2g and A4.2.26 Logarithms	$\log_b x = u$
4	From 3g and A4.2.26 Logarithms	$\log_b y = v$
5	From 4g and A4.2.26 Logarithms	$\log_b z = u + v$
6	From 3, 4 and 5	$\log_b z = \log_b x + \log_b y$
∴	From 6 and 2	$\log_b xy = \log_b x + \log_b y$

Theorem 4.11.6 Real Exponents: Difference of Logarithms

1g	Given	$(x, y, z, b) \in R$
2g		$z = y^{-1}$
Steps	Hypothesis	$\log_b x/y = \log_b x - \log_b y$
1	From T4.11.5 Real Exponents: Summation of Logarithms	$\log_b xz = \log_b x + \log_b z$
2	From 1, 2g	$\log_b x \, y^{-1} = \log_b x + \log_b y^{-1}$
3	From 2 and T4.11.4 Real Exponents: Exponent of Logarithm	$\log_b x \, y^{-1} = \log_b x - \log_b y$
∴	From 3 and D4.1.18 Negative Exponential	$\log_b x/y = \log_b x - \log_b y$

Theorem 4.11.7 Real Exponents: Logarithm of a Reciprocal Base

1g	Given	$(x, y, b) \in R$
2g		$x = (1/b)^y = b^{-y}$
3g		$-1 = -1$
Steps	Hypothesis	$\log_{1/b} x = -\log_b x$
1	From 2g and A4.2.26 Logarithms	$y = \log_{1/b} x$
2	From 2g and A4.2.26 Logarithms	$-y = \log_b x$
3	From 2, 3g, T4.3.9, D4.1.19 Primitive Definition for Rational Arithmetic and D4.1.20 Negative Coefficient	$y = -\log_b x$
∴	From 1 and 3 and A4.2.3 Substitution	$\log_{1/b} x = -\log_b x$

Theorem 4.11.8 Real Exponents: Logarithm raised to its Base

1g	Given	$(x, b) \in R$
2g		$x = b^{\log_b x}$ assume
Steps	Hypothesis	$x = b^{\log_b x}$
1	From 2g	$x = b^{\log_b x}$
2	From 1g and T4.11.1 Real Exponents: Uniqueness of Logarithms	$\log_b x = \log_b b^{\log_b x}$
3	From 2 and T4.11.3 Real Exponents: The Logarithm of the Base	$\log_b x = (\log_b x)\log_b b$
4	From 3 and T4.11.2 Real Exponents: The Logarithm of Unity	$\log_b x = (\log_b x)1$
5	From 4 and A4.2.12 Identity Multp	$\log_b x = \log_b x$
∴	From 1, 5 and by identity	$x = b^{\log_b x}$

The users of logarithms should be aware of different bases.

Table 4.11.1: Definition of Logarithm Types

Name	Description of use	Base	Exponent	Log	Alternate
Binary log	Used in computer systems	2	2^x	$\log_2$	
Half-Life	Used in physics and biology	½	$(½)^x$	$\log_{1/2}$	
Common log	Used by alien creatures with 10 fingers and toes.	10	10^x	$\log_{10}$	log
Natural log	The nexus between differential and integral calculus.	e	e^x	$\log_e$	ln

Where e = 2.71828 18284 59045 23536 02874 71352 66249 77572 47093 69996 [BEY88, pg 5]

Section 4.12 Theorems on Polynomials:

Theorem 4.12.1 Polynomial Quadratic: The Perfect Square

Steps	Hypothesis	$(a + b)^2 = a^2 + 2ab + b^2$
1	From A4.2.2A Equality	$(a + b)^2 = (a + b)^2$
2	From 1 and D4.1.17 Exponential Notation	$(a + b)^2 = (a + b)(a + b)$
3	From 2 and A4.2.14 Distribution	$(a + b)^2 = a(a + b) + b(a + b)$
4	From 3 and A4.2.14 Distribution	$(a + b)^2 = aa + ab + ba + bb$
5	From 4 and D4.1.17 Exponential Notation	$(a + b)^2 = a^2 + ab + ba + b^2$
6	From 5 and A4.2.10 Commutative Multp	$(a + b)^2 = a^2 + ab + ab + b^2$
$\therefore$	From 6 and T4.3.10 Equalities: Summation of Repeated Terms by 2	$(a + b)^2 = a^2 + 2ab + b^2$

Theorem 4.12.2 Polynomial Quadratic: The Perfect Square by Difference

Steps	Hypothesis	$(a - b)^2 = a^2 - 2ab + b^2$
1	From A4.2.2A Equality	$(a - b)^2 = (a - b)^2$
2	From 1 and D4.1.17 Exponential Notation	$(a - b)^2 = (a - b)(a - b)$
3	From 2 and A4.2.14 Distribution	$(a - b)^2 = a(a - b) + b(a - b)$
4	From 3 and A4.2.14 Distribution	$(a - b)^2 = aa - ab - ba + bb$
5	From 4 and D4.1.17 Exponential Notation	$(a - b)^2 = a^2 - ab - ba + b^2$
6	From 5 and A4.2.10 Commutative Multp	$(a - b)^2 = a^2 - ab - ab + b^2$
$\therefore$	From 6, A4.2.14 Distribution, and T4.3.10 Equalities: Summation of Repeated Terms by 2	$(a - b)^2 = a^2 - 2ab + b^2$

Theorem 4.12.3 Polynomial Quadratic: Difference of Two Squares

Steps	Hypothesis	$(a + b)(a - b) = a^2 - b^2$
1	From A4.2.2A Equality	$(a + b)(a - b) = (a + b)(a - b)$
2	From 1 and A4.2.14 Distribution	$(a + b)(a - b) = a(a + b) - b(a + b)$
3	From 2 and A4.2.14 Distribution	$(a + b)(a - b) = aa + ab - ba - bb$
4	From 3 and D4.1.17 Exponential Notation	$(a + b)(a - b) = a^2 + ab - ba - b^2$
5	From 4 and A4.2.10 Commutative Multp	$(a + b)(a - b) = a^2 + ab - ab - b^2$
6	From 5, A4.2.8 Inverse Add and T4.4.1 Equalities: Any Quantity Multiplied by Zero is Zero	$(a + b)(a - b) = a^2 + 0 - b^2$
∴	From 6 and A4.2.7 Identity Add	$(a + b)(a - b) = a^2 - b^2$

Theorem 4.12.4 Polynomial Quadratic: Product of N-Quantities

Steps	Hypothesis	$(\sum^n a_i)(\sum^n b_i) = \sum^n a_i b_i + \sum_{i<j}(a_i b_j + a_j b_i)$
1	From A4.2.2A Equality	$(\sum^n a_i)(\sum^n b_i) = (\sum^n a_i)(\sum^n b_i)$
2	From 2 and A4.2.14 Distribution	$(\sum^n a_i)(\sum^n b_i) = \sum^n \sum^n a_i a_j$
3	From 3 and A4.2.6 Associative Add	$(\sum^n a_i)(\sum^n b_i) = \sum_{(i=j)} a_i b_j + \sum_{(i<j)} a_i b_j + \sum_{(i<j)} a_j b_i$
4	From 4 and let i = j	$(\sum^n a_i)(\sum^n b_i) = \sum^n a_i b_i + \sum_{(i<j)} a_i b_j + \sum_{(i<j)} a_j b_i$
∴	From 1g, 6 and A4.2.6 Associative Add	$(\sum^n a_i)(\sum^n b_i) = \sum^n a_i b_i + \sum_{i<j}(a_i b_j + a_j b_i)$

Theorem 4.12.5 Polynomial Quadratic: Perfect Square of N-Quantities

1g	Given	$a_i = b_i$ for all i
Steps	**Hypothesis**	$(\sum^n a_i)^2 = \sum^n a_i^2 + 2\sum_{i<j} a_i a_j$
1	From 1g, T4.12.4 and A4.2.3 Substitution	$(\sum^n a_i)(\sum^n a_i) = \sum_n a_i a_i + \sum_{i<j}(a_i a_j + a_j a_i)$
2	From 1 and D4.1.17 Exponential Notation	$(\sum^n a_i)^2 = \sum^n a_i^2 + \sum_{i<j}(a_i a_j + a_i a_j)$
3	From 3 and T4.3.10 Equalities: Summation of Repeated Terms by 2	$(\sum^n a_i)^2 = \sum^n a_i^2 + \sum_{i<j} 2\, a_i a_j$
∴	From 3 and A4.2.14 Distribution	$(\sum^n a_i)^2 = \sum^n a_i^2 + 2\sum_{i<j} a_i a_j$

Theorem 4.12.6 Polynomial Quadratic: Quadratic Formula (Factoring a Binomial)

1g	Given	$(b / 2a)^2 = (b / 2a)^2$
Steps	Hypothesis	$x = (-b \pm \sqrt{ b^2 - 4ac }) / 2a$
1	From quadratic equation solving for optimal roots	$ax^2 + bx + c = 0$
2	From 1, T4.4.24 Equalities: Division by a Constant, A4.2.14 Distribution, D4.1.4(A, B) Rational Numbers and T4.4.1 Equalities: Any Quantity Multiplied by Zero is Zero	$x^2 + (b / a)x + (c / a) = 0$
3	From 2, T4.3.5B Equalities: Left Cancellation by Addition and A4.2.7 Identity Add	$x^2 + (b / a)x = -(c / a)$
4	From 1g, 3, T4.3.1 Equalities: Uniqueness of Addition and A4.2.5 Commutative Add	$x^2 + 2 (b / 2a)x + (b / 2a)^2 = (b / 2a)^2 - (c / a)$
5	From 4, T4.8.7 Integer Exponents: Distribution Across a Rational Number, A4.2.21 Distribution Exp, A4.2.12 Identity Multp, A4.2.13 Inverse Multp and D4.1.17 Exponential Notation	$x^2 + 2 (b / 2a)x + (b / 2a)^2 = b^2 / 2^2 a^2 - (2^2 ac / 2^2 a^2)$
6	From 5, D4.1.17 Exponential Notation, D4.1.19 Primitive Definition for Rational Arithmetic and T4.12.1 Polynomial Quadratic: The Perfect Square	$(x + b / 2a)^2 = b^2 / 2^2 a^2 - (4ac / 2^2 a^2)$
7	From 6, A4.2.12 Identity Multp, D4.1.4(A, B) Rational Numbers, A4.2.10 Commutative Multp and A4.2.14 Distribution	$(x + b / 2a)^2 = (1/2^2 a^2) (b^2 - 4ac)$
8	From 7, T4.10.1 Radicals: Uniqueness of Radicals, T4.10.3 Radicals: Identity Power Raised to a Radical, T4.10.7 Radicals: Reciprocal Exponent by Positive Square Root, T4.8.1 Integer Exponents: Unity Raised to any Integer Value, T4.10.5 Radicals: Distribution Across a Product	$x + (b / 2a) = \pm\sqrt{(1^2/2^2 a^2)} \sqrt{ b^2 - 4ac }$

9	From 10, A4.2.21 Distribution Exp and T4.8.7 Integer Exponents: Distribution Across a Rational Number	$x + (b / 2a) = \pm\sqrt{(1/2a)^2} \sqrt{b^2 - 4ac}$
10	From 8 and T4.10.3 Radicals: Identity Power Raised to a Radical	$x + (b / 2a) = \pm (1/2a) \sqrt{b^2 - 4ac}$
∴	From 10, T4.3.5B Equalities: Left Cancellation by Addition, T4.2.5 Commutative Add, D4.1.4(A, B) Rational Numbers and A4.2.14 Distribution	$x = (-b \pm \sqrt{b^2 - 4ac}) / 2a$

Theorem 4.12.7 Polynomial Identity: Binomial Count of Terms-2^m

1g 2g 3g	Given	$(x + y)^m$ $x = y = 1$ $0 \le m$ and $m \in IG$
Steps	Hypothesis	$2^m = \sum_{k=0}^{m} \binom{m}{k}$
1	From 1g and LxE.3.5	$(y + x)^m = \sum_{k=0}^{m} \binom{m}{k} y^{m-k} x^k$
2	From 1, 2g and A4.2.3 Substitution	$(1 + 1)^m = \sum_{k=0}^{m} \binom{m}{k} (1)^{m-k} (1)^k$
3	From 2, 3g, T4.3.10 Equalities: Summation of Repeated Terms by 2 and T4.8.1 Integer Exponents: Unity Raised to any Integer Value	$2^m = \sum_{k=0}^{m} \binom{m}{k} 1$
∴	From 3 and A4.2.12 Identity Multp	$2^m = \sum_{k=0}^{m} \binom{m}{k}$

Theorem 4.12.8 Polynomial Identity: Right Multinomial Count of Terms n^m

1g	Given	$(y + x)^m$
2g		$y = 1$
3g		$x = n - 1$
4g		$0 \le (m, n)$ and $(m, n) \in IG$
Steps	Hypothesis	$n^m = \sum_{k=0}^{m} \binom{m}{k} (n - 1)^k$
1	From 1g and LxE.3.5	$(y + x)^m = \sum_{k=0}^{m} \binom{m}{k} y^{m-k} x^k$
2	From 1, 2g, 3g and A4.2.3 Substitution	$(1 + n - 1)^m = \sum_{k=0}^{m} \binom{m}{k} (1)^{m-k} (n - 1)^k$
3	From 2, A4.2.5 Commutative Add, A4.2.8 Inverse Add, A4.2.7 Identity Add, 4g and T4.8.1 Integer Exponents: Unity Raised to any Integer Value	$n^m = \sum_{k=0}^{m} \binom{m}{k} 1 (n - 1)^k$
$\therefore$	From 3 and A4.2.12 Identity Multp	$n^m = \sum_{k=0}^{m} \binom{m}{k} (n - 1)^k$

Theorem 4.12.9 Polynomial Identity: Left Multinomial Count of Terms n^m

1g	Given	$(y + x)^m$
2g		$y = n - 1$
3g		$x = 1$
4g		$0 \le (m, n)$ and $(m, n) \in IG$
Steps	Hypothesis	$n^m = \sum_{k=0}^{m} \binom{m}{k} (n - 1)^{m-k}$
1	From 1g and LxE.3.5	$(y + x)^m = \sum_{k=0}^{m} \binom{m}{k} y^{m-k} x^k$
2	From 1, 2g, 3g and A4.2.3 Substitution	$(1 + n - 1)^m = \sum_{k=0}^{m} \binom{m}{k} (n - 1)^{m-k} (1)^k$
3	From 2, A4.2.5 Commutative Add, A4.2.8 Inverse Add, A4.2.7 Identity Add, 4g and T4.8.1 Integer Exponents: Unity Raised to any Integer Value	$n^m = \sum_{k=0}^{m} \binom{m}{k} 1 (n - 1)^{m-k}$
$\therefore$	From 3 and A4.2.12 Identity Multp	$n^m = \sum_{k=0}^{m} \binom{m}{k} (n - 1)^{m-k}$

Theorem 4.12.10 Polynomial Identity: Centered Multinomial Count of Terms n^m

1g	Given	$(y + x)^m$
2g		$y = n - m$
3g		$x = m$
4g		$0 \leq (m, n)$ and $(m, n) \in IG$
Steps	Hypothesis	$n^m = \sum_{k=0}^{m} \binom{m}{k} (n - m)^{m-k} (m)^k$
1	From 1g and LxE.3.5	$(y + x)^m = \sum_{k=0}^{m} \binom{m}{k} y^{m-k} x^k$
2	From 1, 2g, 3g and A4.2.3 Substitution	$(n - m + m)^m = \sum_{k=0}^{m} \binom{m}{k} (n - m)^{m-k} (m)^k$
$\therefore$	From 2, A4.2.8 Inverse Add and A4.2.7 Identity Add	$n^m = \sum_{k=0}^{m} \binom{m}{k} (n - m)^{m-k} m^k$

Section 4.13 Landmines

In the real number system with its deductive logic, there are still little things the math-practitioner needs to keep an eye out for.

Subtraction of Inequalities:

The order inversion of inequalities is handled with close attention to what is happening on the number line however while,

$$4 < 9$$
$$\underline{1 < 2 \text{ subtract}}$$
$$3 < 7 \quad \text{holds under subtraction}$$

$$4 < 9$$
$$\underline{1 < 7 \text{ subtract}}$$
$$3 < 2 \quad \text{does not hold under subtraction!}$$

So, in order for the last example to hold the inequality would have to reverse its direction, which may or may not be permissible depending on the conditions of the problem it is being used.

So what happened? The two examples at a quick glance look like they both should be valid? In order to see the mechanism at work that caused the order reversal lets crank out some numbers on the number line [x]. First case will anchor one end of the bottom inequality at the constant [k] and reverse the process in the other case at [c]:

Case I: Of the lower bound anchor.

$a < b$

$p < x$ subtract

$k < b - x$ for $k = a - p$

			a				b									
U	U	U	G	G	G	3	R	R	R	R	R	R	R	R	R	R
y	U	U	G	G	G	3	R	R	R	R	R	R	R	R	R	R
y	y	U	G	G	G	3	R	R	R	R	R	R	R	R	R	R
y	y	y	G	G	G	3	R	R	R	R	R	R	R	R	R	R
y	y	y	y	G	G	3	R	R	R	R	R	R	R	R	R	R
y	y	y	y	y	G	3	R	R	R	R	R	R	R	R	R	R
y	y	y	y	y	y	3	R	R	R	R	R	R	R	R	R	R
y	y	y	y	y	y	3	R	R	R	R	R	R	R	R	R	R
y	y	y	y	y	y	3	y	R	R	R	R	R	R	R	R	R

Result: $k < b - x$

0	1	2	3	4	5	6	7	8	9	10	11	12	13	14	15	16	x
9	8	7	6	5	4	3	2	1	0	-1	-2	-3	-4	-5	-6	-7	$3 < 9 - x$
<	<	<	<	<	<	=	>	>	>	>	>	>	>	>	>	>	

Case II: Of the upper bound anchor.

$a < b$

$x < q$ subtract

$a - x < c$ for $c = b - q$

			a				b									
y	y	2	G	G	G	U	U	U	U	U	U	U	U	U	U	U
y	y	2	G	G	G	U	U	U	U	U	U	U	U	U	U	U
R	y	2	G	G	G	U	U	U	U	U	U	U	U	U	U	U
R	R	2	G	G	G	U	U	U	U	U	U	U	U	U	U	U
R	R	2	y	G	G	U	U	U	U	U	U	U	U	U	U	U
R	R	2	y	y	G	U	U	U	U	U	U	U	U	U	U	U
R	R	2	y	y	y	U	U	U	U	U	U	U	U	U	U	U
R	R	2	y	y	y	y	U	U	U	U	U	U	U	U	U	U
R	R	2	y	y	y	y	y	U	U	U	U	U	U	U	U	U

Result: $a - x < c$

0	1	2	3	4	5	6	7	8	9	10	11	12	13	14	15	16	x
4	3	2	1	0	-1	-2	-3	-4	-5	-6	-7	-8	-9	-10	-11	-12	$4 - x < 2$
>	>	=	<	<	<	<	<	<	<	<	<	<	<	<	<	<	

What observation can be gleaned from the above tables?

Axiom 4.13.1 Absolute Order of the Right Anchor Constant

Since there is no constraints on how the upper and lower inequality can be stacked a [k] can always be found, such that

$$0 < k,$$

and

[c] may or may not be positive.

Observation 4.13.1: Existence of a Guard Band

There exists a region between the upper and lower anchor constants that will always maintain the inequality leaving it un-reversed [U] called the Guard Band [G].

Observation 4.13.2: Lower Bound before Resulting Inequality Reveres

From Cass I [k = a − p] is the demarcation at which the inequalities are equal and below reversed [R].

Observation 4.13.3: Upper Bound before Resulting Inequality Reveres

From Cass II [c = b − q] is the demarcation at which the inequalities are equal and above reversed [R].

Conjecture 4.13.1: Guard Band Bounds

$$c < x < c \bullet k$$

If the observations are valid than the point at where a reversal might occur can be calculated allowing for inequality subtraction without reversal. Likewise, if Conjecture 4.13.1 is valid it can guarantee were subtraction holds.

Let's look at random inequalities below an established anchoring constant:

k = −7		k = −27	
4 < 9 11 < 15 subtract -7 < -6	holds under subtraction	4 < 9 31 < 17 subtract -27 < -8	hold under subtraction
4 < 9 11 < 35 subtract -7 < -26	does not holds under subtraction	4 < 9 31 < 37 subtract -27 < -28	does not hold under subtraction

So, the problem of subtracting inequalities is far more complicated than we probably wanted to deal with in the first place. Somehow an upper or lower anchor constant has to be established as in the above examples by at least two exploratory samples. So iff an anchor constant can be established a prediction can be made to as to whether two pairs of inequalities can be subtracted without reversal and what might follow.

What all of this means is that subtraction of inequalities is a constant problem that the developer has to be continually aware of. As for the observations and conjecture there needs to be some challenge for future math-practitioners. There may be a general solution or theorem that might make such a prediction, but at this time it is beyond the scope of this treatise.

The above examples blindly subtract inequalities. Theorems 4.7.10 to 17 are special cases and might help in guiding the mathematician, by looking at specific conditions.

Table 4.13.1: Inequalities: Subtraction of (a < b) – (c < d) for specific conditions

Theorem	for	and	Subtraction
T4.7.10	a < c	b < d	c – a > –(d – b)
T4.7.11	a < c	b > d	c – a > +(d – b)
T4.7.12	a > c	b < d	c – a < +(d – b)
T4.7.13	a > c	b > d	c – a < –(d – b)
T4.7.14	a < c	b = d	c – a > 0
T4.7.15	a > c	b = d	c – a < 0
T4.7.16	a = c	b < d	d – b > 0
T4.7.17	a = c	b > d	d – b < 0

Section 4.14 Falsified Axioms in Algebra

Axioms 4.1.18 to 24 (see Table 4.2.6) are not true atomized axioms, because they can be proven starting from the Exponential Notation Definition 4.1.16. Instead of developing them as theorems, it occurred to me that it would cause a major incompatible for textbooks written 600 years ago [EVE76, pg 214] see "Summa de arithmetica, geometrica, proportioni et proportionalita" by Luca Pacioli (1445-1509) at about, 1494. Of course, this statement is only true if the exponents are only integers.

Tensor Calculus & Physics: A General Treatise

Table of Contents

Tensor Calculus & Physics: A General Treatise

Tensor Calculus & Physics: A General Treatise

List of Tables

List of Figures

Tensor Calculus & Physics: A General Treatise

List of Observations

List of Givens

Chapter 5 Foundation of Mathematics: Euclidian and Hilbert Geometry

Section 5.1 The Geometric System of Definitions

Definition 5.1.1 Point

(Euclidian Geometry an Undefined Term)

1. Dimensional: A *point* is a geometrical object having Zero-Dimensions, no length, width or depth.
2. Modern Positional: A *point* indicates position. [LEF97]
3. Symbolic: Represented by a dot and an upper case letter [A].

Definition 5.1.2 Line

(Euclidian Geometry an Undefined Term)

1. Dimensional: A *line* is a geometrical object having One-Dimension.
2. Modern Parse: A *line* is a set of continuous points that extend indefinitely in either direction parsing a plane into two parts. [LEF97]
3. Modern Positional: A *straight line* is positional being defined in a plane or space by two points.
4. Symbolic: Represented by a geometric straight line with arrows on each end pointing outwardly, denoting extension of the line indefinitely and a lower case letter. Two points (A, B) define a line, symbolically [$\overleftrightarrow{AB}$][5.1.1].

Definition 5.1.3 Ray

A ray is a one-sided line consisting of a given point on the non-extended side, called an ***endpoint***. [LEF97] Two points (A, B) define a ray, end point-A and head point-B, symbolically [$\overrightarrow{AB}$][5.1.1].

Definition 5.1.4 Line Segment

1. Euclidian: Euclidian Geometry an Undefined Term used interchangeable with a line.
2. Geometric Object: A line segment is a part of a straight line consisting of two endpoints on each end labeled A and B. [LEF97]
3. Geometric Section: The section between two points on a straight line labeled A and B.
4. Symbolic: Two points (A, B) define a segment between them, symbolically [$\overline{AB}$][5.1.1].

Definition 5.1.5 Collinear and Non-collinear Points

Collinear points are points that lie on the same line and when they do not than they are called ***non-collinear***.

[5.1.1]Note, ***Symbolic Context Rule for Lines, Rays and Segments***:
In order to be backwards compatible with other authors and for ease of use in setting text, if these symbols $\overleftrightarrow{AB}$, $\overrightarrow{AB}$ and $\overline{AB}$ are defined by the context they are in than they can be simplified to just [AB]. The confusion can be considerable recommend use in relation with segments only.

Definition 5.1.6 Angle

1.   An ***angle*** is a measured arc that is a positive magnitude constructed between two rays having a common endpoint.

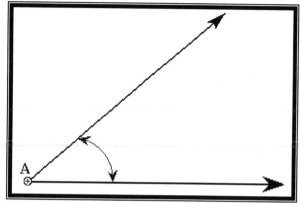

Figure 5.1.1 Angular Component

2.   An ***angle*** is a fractional measure of part of the circumference of a circle.

Definition 5.1.7 Right Angle

1.   A right angle a constructed perpendicular angle having a measure of exactly 90° or ½π. Sometimes called an ***orthogonal or perpendicular angle***, or square of the compass one of the quadratures of the equivalent four points, North-West-South-East.

2.   Any construct coming together at a right angle is said to be ***perpendicular***, such as two rays having a ***square angle***, and angle congruent with one of the vertexes of a square. Notice in classical geometry it is not a measured angle, but a perpendicular construct and a system constant.

Definition 5.1.8 Flat Angle

A flat angle is constructed from two adjacent perpendicular right angles having a total measure of exactly 180° or π, or can be represented by a straight line. Note in classical geometry it is not a measured angle but the sum of two perpendicular right angles.

Definition 5.1.9 Acute Angle

An acute angle has a measure between 0° and 90°.

Definition 5.1.10 Obtuse Angle

An obtuse angle has a measure between 90° and 180°.

Definition 5.1.11 Complementary Angles

Two angles are complementary if the net construction is congruent to a perpendicular having a measured angle of 90° or ½π.

Definition 5.1.12 Supplementary Angles

Two angles are supplementary if the net construction is congruent to two adjacent perpendicular angles having a measured angle of 180° or π.

Definition 5.1.13 Parallelogram

A parallelogram has opposite sides parallel.

Definition 5.1.14 Triangle

Three non-collinear points determine a triangle consisting of the three segments whose endpoints are the given points with positively measured magnitudes for interior angles.

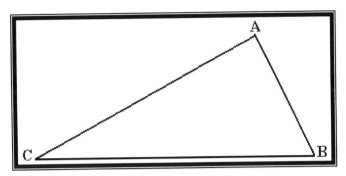

Figure 5.1.2 Triangle

Definition 5.1.15 Triangle; Equilateral

All sides are congruent.

Definition 5.1.16 Triangle; Equiangular

All angles are congruent.

Definition 5.1.17 Triangle; Isosceles

At least two sides are congruent.

Definition 5.1.18 Triangle; Scalene

No two sides are congruent.

Definition 5.1.19 Triangle; Acute

All angles measure less than 90°.

Definition 5.1.20 Triangle; Right

One-angle measures 90°.

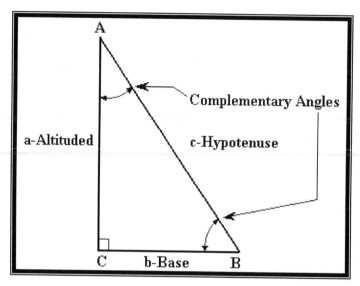

Figure 5.1.3 Right Triangle

Definition 5.1.21 Triangle; Obtuse

One-angle measures between 90° and 180°.

Definition 5.1.22 Quadrilateral

A quadrilateral is a four-sided figure.

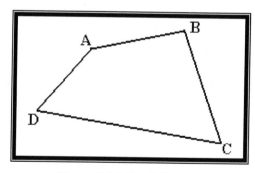

Figure 5.1.4 Quadrilateral

Definition 5.1.23 Square

A square has four congruent sides, with the four congruent angle vertices. Sometimes called a **_regular quadrilateral_**.

Definition 5.1.24 Rectangle

A rectangle has opposite sides congruent, with the four congruent angle vertices.

Definition 5.1.25 Rhombus

A rhombus has four congruent sides.

Definition 5.1.26 Trapezoid

Trapezoid has exactly one pair of parallel sides.

Definition 5.1.27 Congruent

Congruence is an undefined word alluded to in Hilbert's Axioms, Group IV on congruence; it is a relation of matching any two constructions (not the same or equal) component-by-component, such that the components are congruent. Any two matching geometrical objects, such as segments, angles or triangles are congruent if their sets of attributes correspond one-to-one then the relation is denoted by the symbol assignment operation [$\cong$]. The symbol [$\leftrightarrow$] shows a one-to-one correspondence between characterizing (vertices) or points of the two objects.

Definition 5.1.28 Concurrent

Any two geometrical objects having point-C $\leftrightarrow$ point C' are called concurrent if they intersect at or share their two points, represented by the symbol of identity operation [$\equiv$], such as two nonparallel lines having intersected at their respective points P $\equiv$ P'.

Definition 5.1.29 Similar Polygons

Similar is an undefined word used for any two geometric shapes that are alike. Similar polygon's have corresponding congruent angles and the corresponding opposing sides are proportional. Polygons are denoted as similar by the symbol of approximate operation of [~]. [MAR64, pg 131]

Definition 5.1.30 Transversal

[L. *transverses*, past part. of *transvertere*, to turn or direct across, fr. *trans-*, across + *vertere*, to turn.] A line; cutting across two or more lines. A transversal cutting two lines generates eight angles, four lying between the two lines, four external to the two lines. The external angles are called exterior angles; the others, interior angles. When two lines in the same plane are cut by a transversal, if the alternate interior (and alternate exterior) angles are equal, the lines are parallel. Conversely, if two parallel lines are cut by a transversal, the alternate interior (and alternate exterior) angles are equal. [The New Mathematics Dictionary and Handbook, pg 149]

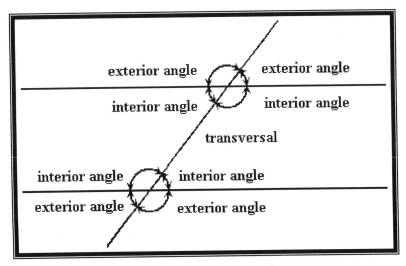

Figure 5.1.5 Transversal

Definition 5.1.31 Circle

A **circle** is a geometric plane figure having a locus of points equal distance [r], the radius, about a point of origin O.

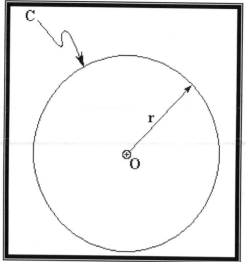

Figure 5.1.6 The Circle

Definition 5.1.32 Cord of a Circle

A **cord** is a line segment whose endpoint resides on the parameter of a circle and ends on the opposing side.

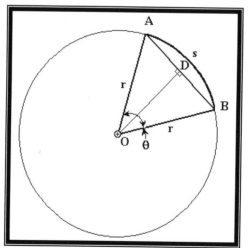

Figure 5.1.7 Cord of a Circle for Cord ≡ AB

Definition 5.1.33 Diameter of a Circle

The **diameter** of a circle is a cord-D_0 that passes through the center point-O of a circle.

Definition 5.1.34 Circumference of a Circle

The *circumference* of a circle C is the length, or complete distance around a circle. Having a constant ratio of the circumference with respect to the diameter of the circle D_0 represented by an irrational number π: [BEY88, pg 5]

$$\pi \equiv C / D_0$$
$$\pi \approx 3.14159\ 26535\ 89793\ 23846\ 26433\ 83279\ 50288\ 41971\ 69399\ 37511\ \ldots$$

Definition 5.1.35 Subtended Angle of a Cord

The *subtending angle*-θ (sometimes referred to as the *central angle*) of a cord spans the distance across the cord from the opposing vertex-O the center for the inscribing circle.

$$\text{Subtending Angle} \equiv \theta$$

Definition 5.1.36 Subtended Arc length of a Cord

The subtending cord sweeps a positive locus of length around a section of the circles' circumference this is called the *arc length-s* of the cord.

$$\text{Arc length} \equiv s \equiv \overset{\frown}{AB}.$$

A portion of an arc from a circle, or its length [s] is the fraction [f] of the circumference of a radian circle $[C_r]$. Let

f	$s = C_r x f = (2\pi f)r$	$\theta = 2\pi f$ angle in radians	θ angle in Degrees	Quadrature Angles CCW	
1	$(2\pi)r$	2π	360°	4*90°	QIV
3/2	$(3\pi/2)r$	$3\pi/2$	270°	3*90°	QIII
½	$(\pi)r$	π	180°	2*90°	QII
1/3	$(2\pi/3)r$	$2\pi/3$	120°		
¼	$(\pi/2)r$	$\pi/2$	90°	1*90°	QI
1/5	$(2\pi/5)r$	$2\pi/5$	75°		
1/6	$(\pi/3)r$	$\pi/3$	60°		
1/7	$(2\pi/7)r$	$2\pi/7$	51.43°		
1/8	$(\pi/4)r$	$\pi/4$	45°		
1/9	$(2\pi/9)r$	$2\pi/9$	40°		
1/10	$(\pi/5)r$	$\pi/5$	36°		
1/11	$(2\pi/11)r$	$2\pi/11$	32.<u>72</u>°		
1/12	$(\pi/6)r$	$\pi/6$	30°		
1/13	$(2\pi/13)r$	$2\pi/13$	27.69°		
1/14	$(\pi/7)r$	$\pi/7$	25.71°		
1/15	$(2\pi/15)r$	$2\pi/15$	24°		
$1/2\pi$	$(1)r$	1	57.30°		
...	...	...	...		

The angular measurement in radians is called a natural system of angular magnitude or measure, since it arises from the natural ratio of circumference to diameter. While degrees are an artificial artifice of measure, inherited from the Babylonians, several explanations have been put forward to account for the choice of these numbers, including the following, advocated by Otto Neugebauer. In early Sumerian times, there existed a large distance unit, a sort of Babylonian mile, equal to about seven of our miles. Since the Babylonian mile was used for measuring long distances, it was natural that it should also become a time unit, namely the time required to travel a Babylonian mile. People, in their slow–paced, non–technical, agrarian world, did not clearly understand the idea of rate so they would freely interchange time and distance to mean the same thing. Later on, sometime in the first millennium BCE, when Babylonian astronomy reached the stage with, which systematic records of celestial phenomena were kept, the Babylonian time-mile was adopted for measuring spans of time in a 360 day–year. Since a complete month was found to be equal to approximately twelve parts, and one complete month is closely equivalent to one-twelfth of a revolution of the sky in a year, so a complete circuit of a circle was divided into twelve equal parts. Thus, we arrive at 12 months * 30 days = 360 equal parts in a complete circuit. [EVE76, pg. 33]

An alternative explanation of why there is 360°, and more plausible, is that it comes from the first widely used positional numbering system in the Old World, which was the base-60 that approximated the number of days in a year, sexagesimal numbering system developed by the Babylonians around 2,000 BCE. This Babylonian number positional system is still in wide use today by virtually everyone in the form of hours, minutes, seconds used to represent time, and degrees using minutes and seconds for angular measure. Where minutes derive from the first fractional sexagesimal place; the second fractional place is the origin of the second. This was extremely convenient for Babylonians, because it simplified division by leaving no remainder, since 60 is a multiple of 2, 3, 4, 5, 6, 10, 12, 15, 20, 30 and 60. The sexagesimal system was really an early example of floating point. A zero symbol was added by Hellenistic times, after which the system spread over the entire civilized world, apart from China.

Users of the sexagesimal system, mostly astronomers and mathematicians performed computations such as long division by close analogy to what is done today using base-10. Ptolemy's famous "Almagest" of 140 AD, for example, utilized a zero symbol in conjunction with sexagesimal numerals and fractions exactly as we do today. Usage of sexagesimal by astronomers and mathematicians continued throughout the Byzantine and Islamic periods into the modern mathematical era allowing for advances in those areas [WORβ04] [NISβ04]. Sexagesimal systems ability to handle fractions as if they were rational numbers is

most likely why it was adapted to deal with a portion of a circle, an angle. The only apparent relationships between 360° to 365 days is that both systems are cyclic and of the same order of magnitude. Though 360°, as seen above may have had its roots in ancient Babylonian times when it was a time measurement of a year. They wanted an easy way to make calculations with multiples of 30 days per month verses the 30.4 days per month for 365 days, not very convenient.

Independently other cultures knew the length of a year, the picture to the right shows sunrise on the morning of summer solstice at Stonehenge (England). The builders, a semi-nomadic people that populated the Salisbury Plain, about 3500 BCE, could have very easily kept a tally stick

Figure 5.1.8 Stonehenge Summer Solstice

or bone and counted the days so would have known how long a year was to the day. And the Maya of South America had a 365-day Vague year, first century BCE, but solved their problem of making easy fractional calculations by having a second 260-day Tzolkin year using multiples of 13 lunar months or 20 days per month.

While degrees are artificial, they are often used because it is easer to write and say a nice whole integer number like 180° verses the irrational symbolic number π. For pure mathematical development, though, the natural system of angular measure is more meaningful. Both systems of measure can be used interchangeably and both placed into a one-to-one correspondence, hence countable and as seen from the above table can be placed into fractional parts as rational numbers. Even though the natural system of angular measure is based on multiples of a constant, that itself is the irrational number π. In summary, both systems can be categorized as belonging to the real number system and are interchangeable by the conversion ratio of π (radians) to 180° (degrees).

So as can be seen from the above table, if the fraction of arc were allowed being continuous for every point about the circle, the measure of arc length is directly proportional to the angle in radians with respect to the radius.

$s \propto \theta$ with respect to the radius-r, or	Equation A
$s \equiv \theta r$	Equation B
$\theta \equiv 2\pi f$ for [f] a fraction of the circumference of a circle	Equation C
$f_u \equiv 1/2\pi \approx 0.16 = 4\,/\,25 = (2/5)^2$	Equation D

        for 1 radian or one unit angular part of a circumference of a circle

Definition 5.1.37 Major-Minor arc and Semicircle

$0° \leq$ **Minor arc** $< 180°$, **Semicircle** $\equiv 180°$ and $180° <$ **Major arc** $< 360°$

Definition 5.1.38 Secant

In geometry, a **secant** is a straight line that intersects a circle at two points.

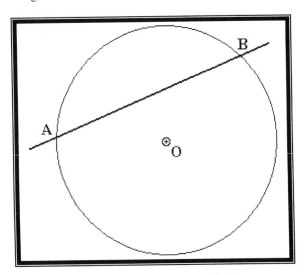

Figure 5.1.9 Secant to a Circle

Definition 5.1.39 Tangent

A line *tangent* to a circle is perpendicular to the radius of the circle drawn to the point of contact.

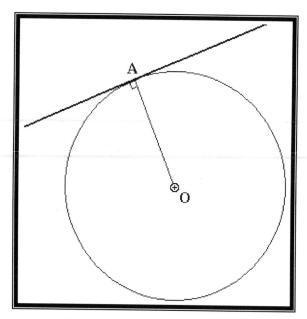

Figure 5.1.10 Tangent to a Circle

Definition 5.1.40 Circumcenter

The *circumcenter* point-O is the intersection point of the perpendicular bisectors inscribed in the triangle $\triangle ABC$. Where segments AO, BO and CO are equal distant specifying a circle inscribing the triangle $\triangle ABC$.

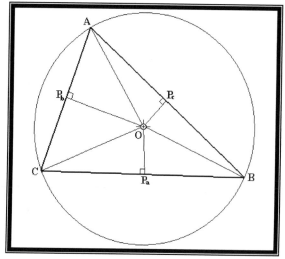

Figure 5.1.11 Circumcenter

Note that any pair of AO, BO and CO sides being equal distance constructs isosceles triangles $\triangle AOB$, $\triangle BOC$ and $\triangle COA$ assuring perpendiculars OP_a, OP_b and OP_c bisect the corresponding segments.

Definition 5.1.41 Orthocenter

The *orthocenter* point-H is where all altitudes of the triangle ΔABC are concurrent. When the four points A, B, C and H are distinct they constitute what is called an *orthocentric set* of four points, because each of the four points is the orthocenter of the triangle formed by the other three. For example, B is the orthocenter of the triangle with vertices A, C and H.

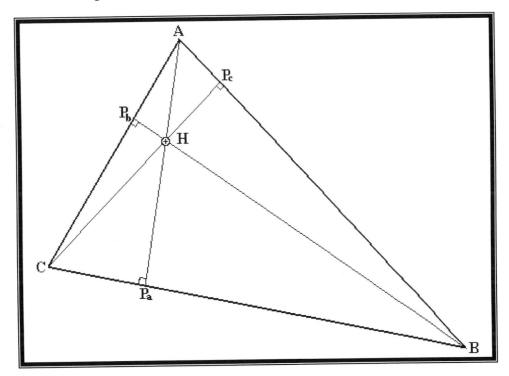

Figure 5.1.12 Orthocenter

Definition 5.1.42 Centroid

Point H is the *centroid* of ΔABC the intersection of the median vertices.

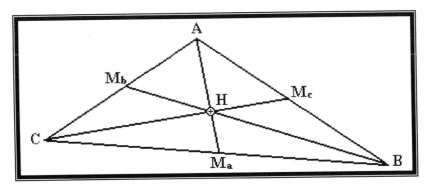

Figure 5.1.13 Centroid

Definition 5.1.43 Incenter and Incircle

Point O is the *incenter* of ΔABC and segments AB, BC and CA are tangent to the *incircle* or inscribing circle.

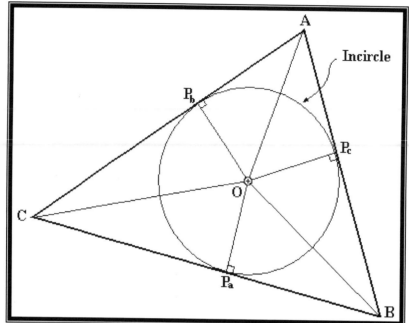

Figure 5.1.14 Inner Center and Circle

Definition 5.1.44 Excenter and Excircle

Point-E is the *excenter* equidistant from all three extended sides of the triangle characterizing a circle called the *excircle*. Where rays CR, CS and AB are tangent to the excircle.

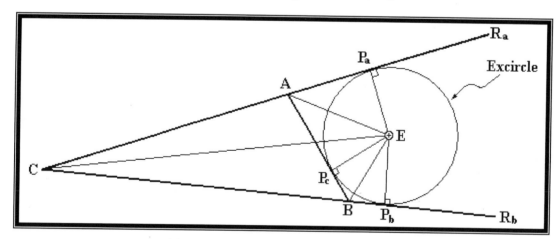

Figure 5.1.15 External Center and Circle

Definition 5.1.45 Medial Triangle

Mid points M_a, M_b and M_c constitute a triangle $\Delta M_a M_b M_c$ whose sides are parallel to opposing sides of the inscribing triangle ΔABC.

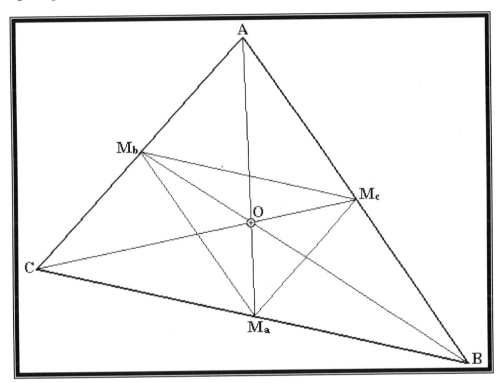

Figure 5.1.16 Medial Triangle

Definition 5.1.46 Cyclic Quadrilateral or Concyclic

A *cyclic quadrilateral* is a quadrilateral inscribed in a circle forcing opposing angles to be supplementary.

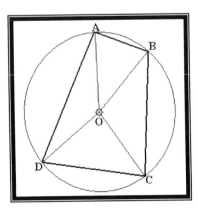

Figure 5.1.17 Cyclic Quadrilateral

Definition 5.1.47 Antiparallel

Opposite sides of a cyclic quadrilateral are called *antiparallel* with respect to the remaining pair of sides.

Definition 5.1.48 Cyclic Trapezoid

A cyclic quadrilateral with two sides parallel is called a *cyclic trapezoid*.

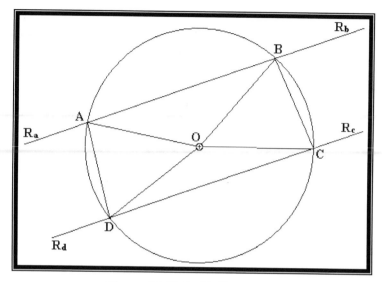

Figure 5.1.18 Cyclic Trapezoid

Definition 5.1.49 Area of a rectangle

Area of a rectangle is the width by height.

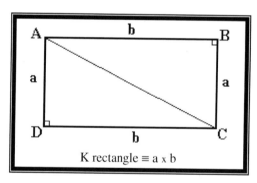

Figure 5.1.19 Area of a Rectangle

Definition 5.1.50 Congruent Triangles

Any two triangles, ΔABC and $\Delta A'B'C'$, are called congruent, $\Delta ABC \cong \Delta A'B'C'$, for all their parts, including vertices, having a one-to-one correspondence and are themselves congruent.

$$\text{vertices A} \longleftrightarrow \text{A' vertices}$$
$$\text{vertices B} \longleftrightarrow \text{B' vertices}$$
$$\text{vertices C} \longleftrightarrow \text{C' vertices}$$

(If axioms of construction are accepted, this definition needs to be replaced from Axiom group GVII Congruent Geometric Objects in equivalent axioms, see A5.18.9, 10, 11.)

Definition 5.1.51 External Apex Angle

For inscribed geometric objects having a constructed apex outside of the circle with the form of $\frac{1}{2}(\beta - \alpha)$ are called the **External Apex Angle**.

Definition 5.1.52 Internal Apex Angle

For inscribed geometric objects having a constructed apex inside of the circle with the form of $\frac{1}{2}(\alpha + \beta)$ are called the **Internal Apex Angle**.

Definition 5.1.53 Angle Bisector

Any angle divided by a line, ray or segment into two congruent sub-angles.

Definition 5.1.54 Perpendicular Bisector

Any segment or segment of a line divided into two congruent parts by a perpendicular line, ray or segment.

Table 5.1.1 Formal and Informal Naming of Polygons

Name[5.1..2]	Definition
Definition 5.1.55 Trigon or triangle	3-Sided Polygon
Definition 5.1.56 Quadragon or Quadrilateral or Trapezoid or Rhombus or Rectangle or Square	4-Sided Polygon
Definition 5.1.57 Pentagon or quintagon	5-Sided Polygon
Definition 5.1.58 Hexagon or sextagon	6-Sided Polygon
Definition 5.1.59 Heptagon	7-Sided Polygon
Definition 5.1.60 Octagon	8-Sided Polygon
Definition 5.1.61 Nonagon	9-Sided Polygon
Definition 5.1.62 Decagon	10-Sided Polygon
Definition 5.1.63 Hendecagon or Undecagon	11-Sided Polygon
Definition 5.1.64 Dodecagon or Duodecigon	12- Sided Polygon
Definition 5.1.65 n-gon	n- Sided Polygon

[5.1..2]Note: The origins of polygon names lead with a Greek prefix name.

Definition 5.1.66 Scaling Factor for Similar Polygons

The scaling factor is a constant of proportionality for similar polygons made from the ratio of the radii for two concentric circles inscribing two similar polygons.

$$p_r \equiv r' / r$$

This definition can be extended to inscribing spheres and hyperspheres in n-dimensional geometry.

Section 5.2 The Geometric Axiomatic System

Several different sets of axioms have been drawn up that are designed to eliminate the defects of Euclid's axioms. The best known of these are the axioms of David Hilbert (1862–1943), first presented by him in a series of lectures at the University of Gotteingen in 1898–1899. They are listed below in slightly modified form. [ADL66, pg 74]

Table 5.2.1 Group I Axioms of Connection Between Points-Lines and Lines-Planes

Axiom 5.2.1 GI Two Points on a $^\text{£}$SL	There is a straight line that contains two given distinct points.
Axiom 5.2.2 GI One $^\text{£}$SL with Two Points	There is at most one straight line that contains two given distinct points.
Axiom 5.2.3 GI Three Points in a Plane	There is a plane that contains three given points, which do not lie on the same straight line.
Axiom 5.2.4 GI One Plane with Three Points	There is at most one plane that contains three given points, which do not lie on the same straight line.
Axiom 5.2.5 GI $^\text{£}$SL in a Plane	If two points of a straight line lie in a plane, then every point of the straight line lies in the plane.
Axiom 5.2.6 GI Intersection of Two Planes	If two planes have one point in common, then they have at least a second point in common.
Axiom 5.2.7 GI Non-coplanar Points	On every straight line, there are least two points, in every plane, there are at least three points that are not on the same straight line, and in space, there are at least four points that are not on the same plane.

$^\text{£}$SL is straight line.

Group II Axioms of Order

The axioms of order introduce the undefined term between as a relation among points on a line. M. Pasch first formulated axioms of order in 1882.

Table 5.2.2 Group II Axioms of Order

Axiom 5.2.8 GII Order of Direction of Points on a $^£$SL	If A, B and C are points of a straight line and B is between A and C then B is also between C and A. $AC \cong AB + BC$
Axiom 5.2.9 GII Four Point Existence	If A and C are two points on a straight line then there is at least one point B that is between A and C, and there is at least one point D such that C is between A and D. $AC \cong AB + BC$ and $AD \cong AC + CD$
Axiom 5.2.10 GII Three Point Existence	Of any three points on a straight line, there is always one and only one, which are between the other two.
Axiom 5.2.11 GII Ordering of Points on a $^£$SL	Any four points on a straight line can be labeled A, B, C and D in such a way that B is between A and C and also between A and D, and that C is between A and D and also between B and D. On the basis of these axioms, a *segment* AB is defined as the set of points between the two points A and B. The term *segment* occurs in the next axiom of order:
Axiom 5.2.12 GII Intersection of a $^£$SL with a Triangle	Let A, B and C be three points that are not on the same straight line and let [a] be a straight line in the plane determined by these points and not passing through any of them. Then if [a] passes through a point of the segment AB, it will also pass through either a point of the segment BC or a point of the segment AC. (This axiom is know as the *Pasch Axiom*)

[ADL66, pg 75-76]

Figure 5.2.1 GII5.2.10: Order of Between any Three Points on a Line

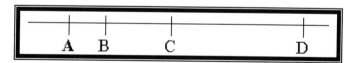

Figure 5.2.2 GII5.2.11: Existence of Points Between Points on a Line

Table 5.2.3 Group II Axioms of Parallels

Axiom 5.2.13 GIII Parallel Lines	In a plane containing a given straight line and a given point that is not on the line, there is one and only one straight line. That does not intersect the given line (Playfair Axiom).

[ADL66, pg 76]

The axioms of congruence introduce the undefined term ***congruent*** (represented by the usual symbol $\cong$) as a relation that may hold between two segments or two angles and characterized by the following properties:

Table 5.2.4 Group IV Axioms of Congruence

Axiom 5.2.14 GIV Symmetric-Reflexiveness of Segments	If AB is a segment and A' is a point on a line, then there is on that line, on a given side of A', one and only one point B' such that segment AB $\cong$ A'B' and segment A'B' $\cong$ AB[5.2.1]. segment AB $\cong$ AB Reflexive.	Equation A Equation B Equation C
Axiom 5.2.15 GIV Transitivity of Segments	If AB $\cong$ A'B' and AB $\cong$ A"B" then A'B' $\cong$ A"B".	
Axiom 5.2.16 GIV Congruence for Union of Segments	Let AB and BC be two segments of a straight line which lie on opposite sides of B, and let A'B' and B'C' be two segments of a straight line that lie on opposite sides of B'. Then if AB $\cong$ A'B' and BC $\cong$ B'C', we also have AC $\cong$ A'C'.	
Axiom 5.2.17 GIV Symmetric-Reflexiveness of Angles	If (h, k) is an angle and [a] is a line on a plane, and h' is a half-line on [a] with vertex O, then there is on the plane on a given side of [a] one and only one half-line k' with vertex O such that angle $\angle$[h, k] $\cong$ $\angle$[h', k'] and angle $\angle$[h', k'] $\cong$ $\angle$[h, k][5.2.1]. angle $\angle$[h, k] $\cong$ $\angle$[h, k] Reflexive.	Equation A Equation B Equation C
Axiom 5.2.18 GIV Transitivity of Angles	If angle $\angle$[h, k] $\cong$ $\angle$[h', k'] and $\angle$[h', k'] $\cong$ $\angle$[h", k"] then $\angle$[h', k'] $\cong$ $\angle$[h", k"].	
Axiom 5.2.19 GIV Congruence of Side-Side-Angle (SSA) and Opposite Angles	If in the two triangles ΔABC and ΔA'B'C', AB $\cong$ A'B', AC $\cong$ A'C' and $\angle$BAC $\cong$ $\angle$B'A'C' then BC $\cong$ B'C', $\angle$ABC $\cong$ $\angle$A'B'C' and $\angle$ACB $\cong$ $\angle$A'C'B'.	

[ADL66, pg 76-77]

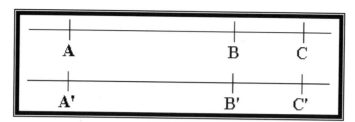

Figure 5.2.3 GIV5.1.16: Union of Segments

Table 5.2.5 Group V Axioms of Continuity

Axiom 5.2.20 GV Continuity of Points on a Straight Line	Let A_1 be any point on a straight line between the points A and B. Take the points A_2, A_3, A_4, ... , so that A_1 is between A and A_2, A_3 is between A_1 and A_2, A_3 is between A_3 and A_4, etc., and so that $$AA_1 \cong A_1A_2 \cong A_2A_3 \cong A_3A_4 \cong ....$$ Then there is a positive integer n such that B is between A and A_n.

[ADL66, pg 77]

The axiom of completeness says in effect that the space under consideration is the largest one that satisfies the other axioms.

Table 5.2.6 Group VI Axioms of Completeness

Axiom 5.2.21 GVI Closure of Axiom Set	No additional points, lines or planes can be added to the system without violating one of the axioms in groups I to V.

[ADL66, pg 77]

[5.2.1]Note: This modification of Hilbert's axioms adds this expression in order to complete the idea of symmetry of congruence, which Hilbert left out.

The geometry derived from these axioms or any equivalent set of axioms is now known as ***Euclidean Geometry***.

With the six groups of modified axioms of Hilbert as a foundation, the Euclid's geometry can now be developed without any logical flaws, yes? No! A systematic development has revealed other limitations of Hilbert's axioms such as symmetry for congruency of segments and angles as well as the notion of correspondence between vertices of geometric objects. As Kurt Gödel [SIN97, pg 139] pointed out, trying to create a complete and consistent mathematical system is impossible. Here even Hilbert has failed, though not do to his own fault, but because he was a mathematician of his time and held inherent misconception carried from Euclid on to current times. In the days of Euclid, nobody knew about negative numbers, only the positive counting integers. In today's modern math we can now consider negative numbers, so it is a possibility, and in some cases in modern fields of mathematics, such as complex numbers, to consider negative angles. This is somewhat disturbing, because what if one or more angles in a triangle were negative? The angles would not add to 180°. For non-Euclidean geometry, this is not a necessary criterion, but for Euclidean plan geometry, it's mandatory. It has always been a subconscious understanding in the way angles are used that they are a measure of magnitude for an arc and as such positive quantities. The specific fault for this case then is that neither by definition or axiom is it stated that angles are positive quantities; hence, all angles within a triangle are positive numbers. I clarify the problem by carefully crafting a definition for angle as magnitude, and to be backward compatible for geometric proof writing, I leave it as a theorem of proof that the angles within a triangle are positive.

Section 5.3 About Theorems from Euclid's Book I

Many new ideas in geometry have been discovered; since the days of Euclid, and new ways of constructing proofs devised, such as the revised axioms by Hilbert's. All of which, I've attempted to integrate within the following proofs. Hopefully this has lent itself to a cleaner and more clearly lucid set of proofs, which in tern should help to clarify Euclid's great work. The following sets of theorems are not, nor are they meant to entail, all of Euclid's theorems found in *The Elements*. These proofs are a systematic, sequential, deductive, coherent development, filling out the missing and incomplete parts of the logic. This has come about do to lack of thoroughness, or poor translation, or the fact that some of the books of the Elements were lost, and what might have been in them can now only be deduced.

These theorems are limited to the Euclidian flat plane no development into Euclid's solids is considered in the first part of this chapter. Forcing this study to stay strictly within Euclid's flat plane, sometimes these categories of proofs go by the name of **Euclidian Plane Geometry**.[5.3.1]

Since Euclidian geometry is not analytic, as in analytic geometry, the notion of measured distance and angle are simply not considered. Distance measured is done by the definition of a segment and marking out length with compass and straight edge. Measured length was done with rods or chains as used in land surveying and eventually with scales added for greater accuracy creating the first precession rulers. This in turn led to the concept of a, none zero, positive, number line, Descartes extended this idea by adding zero at the beginning and a negative scale on the other side of zero, then taking two number lines and crossing them perpendicular at zero to form what has become known as his Cartesian coordinate system. This allowed segment lengths embedded in such a system to be measured by using Pythagoras's theorem to calculate distance.

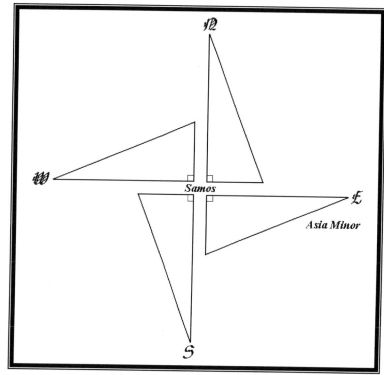

The early idea of angle was limited to the concept of multiples of perpendicular right angles[532] boxing the compass[533], known as **quadrature angles**. If we stand on Pythagoras's magical island of Samos in the Aegean Sea and looking across from the heights toward Asia Minor (now Turkey), we can layout a right triangle tile on the ground pointing in the direction, due east. The right triangle tile is chosen because it makes a good pointer and its asymmetry allows it to be laid, side–by–side, so it can be rotated CCW four times. From east to north, north to west, west to south and finally back again to Asia Minor due east. So, through our own experiences it can be seen

Figure 5.3.1 Greek's View of Boxing the Compass with Perpendicular Right Angles

that the natural world can be mapped by right angles.[532] These observations lead to perpendicular angles in Greek geometry, multiple *constructed* right angles, 90°s being a measured angle, measurement not clearly understood in those times. So, anytime quadrature angles 90°, 180°, 270° and 360° are seen in this treatises <u>they are not to be treated as measured angles</u>, but <u>constructed geometric constants</u> in this system of logic. [BRO73, pg 158]   It follows that any construction to a quadrature angle is congruent to these multiples of perpendicular angles.

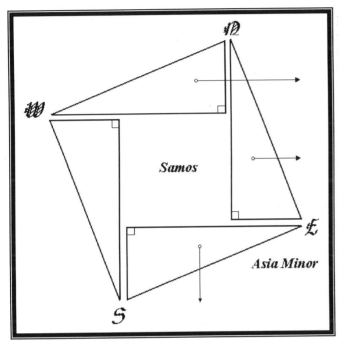

Notice in figure F5.3.1 if Pythagoras plays with the right triangles in a simple way as seen in F5.3.2, it would allow him to square the hypotenuse. Thus this line of thought; might be how he started thinking about this problem, maybe the other sides could be squared out of the same area? Which eventually lead to his famous theorem. For this development and other possible solutions see "Theorems on Right Triangles".

In Euclid's books, he gathers a great number of theorems that dealt with proportional ratios and numbers. This came about, because trigonometry had not yet been invented and a large number of theorems demonstrating every possibility, that might be needed, were required. Of course, every possibility is not practical and it took the general approach of using a simple triangle,

Figure 5.3.2 The Square of the Hypotenuse

the right triangle, which could relate angles and sides into proportional ratios to solve this problem. In this part of the book, I try to cover as many of these proportional problems as I could find in order to demonstrate his past philosophy.

[5.3.1]Note: Yet Euclid did a complete study of geometry in space, eventually would become known as Euclidean Solid Geometry in book XI [HEA56, Vol. 3 pg 272]. Those propositions will be summarized at the end of this chapter in a listing of those additional axioms.

[532]Note: $\kappa\alpha'\theta\varepsilon\tau o\zeta$, perpendicular, means literally let fall: The full phrase is perpendicular straight line, as we see from the enunciation of Euclid 1.II, and the notion is that of a straight line let fall at right angles to the surface of the earth, a plumb line, see Proclus (pg. 283, 9) alternate definition [HEA56, Vol. 1 pg 181]. So perpendicular angle may have its origins in some or all of these sources.

[533]Note: The term boxing the compass is a nautical phrase coming from the fact that originally the compass was mounted near the helm of the ship in a wooden square box. In order to test the compass's magnetic North against true North the ship would sail in a complete circle by the North Star spinning the compass needle around all four corners of the box, hence the phrase *boxing the compass*.

Section 5.4 Theorems on Intersection and Parallel Lines

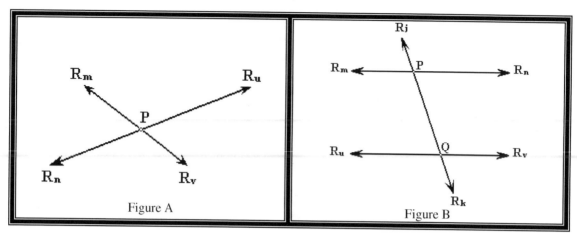

Figure A

Figure B

Figure 5.4.1 Intersection and Parallel Lines

Theorem 5.4.1 Intersection of Two Lines: Opposite Congruent Vertical Angles[5.4.1]

If two lines intersect, then they make the horizontal and vertical angles congruent.

1g 2g 3g	Given	the above geometry of Figure 5.4.1A $\angle R_mPR_n$, $\angle R_uPR_v$ are horizontal angles $\angle R_mPR_u$, $\angle R_nPR_v$ are vertical angles	
Steps	Hypothesis	$\angle R_mPR_n \cong \angle R_uPR_v$ are horizontal angles	$\angle R_mPR_u \cong \angle R_nPR_v$ are vertical angles
1	From 1g	Lines R_mR_v and R_uR_n are concurrent at P	
2	From 1g, 1 and D5.1.12 Supplementary Angles	$180° \equiv \angle R_uPR_v + \angle R_vPR_n$	$180° \equiv \angle R_mPR_u + \angle R_nPR_m$
3	From 1g, 1 and D5.1.12 Supplementary Angles	$180° \equiv \angle R_mPR_n + \angle R_vPR_n$	$180° \equiv \angle R_nPR_v + \angle R_nPR_m$
∴	From 2g, 3g, 2, 3, subtract, A4.2.7 Identity Add and T4.3.3 Equalities: Reversal of Right Cancellation by Addition	$\angle R_mPR_n \cong \angle R_uPR_v$ are horizontal angles	$\angle R_mPR_u \cong \angle R_nPR_v$ are vertical angles

[5.4.1]Note: This theorem also goes by the name of **The Bet Theorem**. Named after the Hebrew letter ב, which comes from the first letter and word in the book of Genesis ("origin") בְּרֵאשִׁית Pronounced (*bereshit*) out of the Hebrew Bible the Torah ("the five books of Moses"). Laterally interpreted as "When God began to create …", [PLU81, pg 18]. This letter is the first in the bible corresponding to the first theorem in this treatise's logical development of geometry. The reason this name was chosen was from a Midrash tale (Gensis Rabbah Chapter 1, verse 10), origin parable of why Bet is the first letter in the Torah. ". Just as the letter *bet* is enclosed on three sides but open to the front, we are not to speculate on the origins of God or what may have existed before creation." Hence, discourage efforts to prove the un-provable in circular arguments. Likewise, the logical development starts with this key theorem and everything else springs forth. However, it should be kept in mind that another author's development, might lead with another theorem depending on the objectives of the geometer, in which case that theorem would be the Bet Theorem.

Corollary 5.4.1.1 Intersection of Two Lines: Specific Solution for Flat line Angles

1g	Given	the above geometry of Figure 5.4.1A	
Steps	Hypothesis	$\angle R_uPR_v \equiv 180° - \angle R_vPR_n$ $\angle R_mPR_n \equiv 180° - \angle R_vPR_n$ $\angle R_vPR_n \equiv 180° - \angle R_uPR_v$ $\angle R_vPR_n \equiv 180° - \angle R_mPR_n$	$\angle R_mPR_u \equiv 180° - \angle R_nPR_m$ $\angle R_nPR_v \equiv 180° - \angle R_nPR_m$ $\angle R_nPR_m \equiv 180° - \angle R_mPR_u$ $\angle R_nPR_m \equiv 180° - \angle R_nPR_v$
1	From 1g	Lines R_mR_v and R_uR_n are concurrent at P	
2	From 1g, 1 and D5.1.12 Supplementary Angles	$180° \equiv \angle R_uPR_v + \angle R_vPR_n$	$180° \equiv \angle R_mPR_u + \angle R_nPR_m$
3	From 1g, 1 and D5.1.12 Supplementary Angles	$180° \equiv \angle R_mPR_n + \angle R_vPR_n$	$180° \equiv \angle R_nPR_v + \angle R_nPR_m$
∴	From 2, 3 and T4.3.3B Equalities: Right Cancellation by Addition	$\angle R_uPR_v \equiv 180° - \angle R_vPR_n$ $\angle R_mPR_n \equiv 180° - \angle R_vPR_n$ $\angle R_vPR_n \equiv 180° - \angle R_uPR_v$ $\angle R_vPR_n \equiv 180° - \angle R_mPR_n$	$\angle R_mPR_u \equiv 180° - \angle R_nPR_m$ $\angle R_nPR_v \equiv 180° - \angle R_nPR_m$ $\angle R_nPR_m \equiv 180° - \angle R_mPR_u$ $\angle R_nPR_m \equiv 180° - \angle R_nPR_v$

Theorem 5.4.2 Parallel Lines: Constructing a Congruent Line

1g	Given	line R_mR_n constructed through point-P on construction line R_iR_k with angle $\angle R_iPR_n$
2g		point-Q on line construction line R_iR_k
3g		such that P ~≅ Q
Steps	Hypothesis	$\angle R_iPR_n \cong \angle R_iQR_v$ for R_uR_v through point-Q with $\angle R_iQR_v$
∴	From 1g, 2g, 3g and A5.2.17 GIV Symmetric-Reflexiveness of Angles	$\angle R_iPR_n \cong \angle R_iQR_v$ for R_uR_v through point-Q with $\angle R_iQR_v$

Theorem 5.4.3 Parallel Lines: A Congruent Line is Parallel[5.4.2]

1g	Given	line R_mR_n constructed through point-P on construction line R_iR_k with angle $\angle R_iPR_n$
2g		point-Q on line construction line R_iR_k
3g		such that P ~≅ Q
Steps	Hypothesis	$R_uR_v \parallel R_mR_n$ for a congruent angle $\angle R_iQR_v$
1	From T5.4.2 Parallel Lines: Constructing a Congruent Line	$\angle R_iPR_n \cong \angle R_iQR_v$ for R_uR_v through point-Q with $\angle R_iQR_v$
2	From 1	Congruent angle $\angle R_iQR_v$ is singular and unique, hence line R_uR_v through point-Q is also, and does not intersect R_mR_n.
∴	From 2 and A5.2.13 GIII Parallel Lines	$R_uR_v \parallel R_mR_n$ for a congruent angle $\angle R_iQR_v$

[5.4.2]Note: This theorem is not necessary since the geometry, by this and inadvertently by Euclid's design (see axiom A5.2.17 "GIV Congruence of Angles"), stands on the concept of constructible congruent line, which in tern uses congruent alternate supplementary angles replacing the idea of parallel lines. This theorem is here for backwards compatibility for historical geometric logical development.

Theorem 5.4.4 Parallel Lines: Congruent Alternate Interior Angles

If two parallel lines are cut by a transversal, then the alternate interior angles are congruent and the corresponding angles are congruent.

1g	Given	the above geometry of Figure 5.4.1B	
Steps	Hypothesis	$\angle R_iQR_v \cong \angle R_mPR_k$	$\angle R_iQR_u \cong \angle R_nPR_k$
		for $R_uR_v \parallel R_mR_n$	
1	From 1g and T5.4.2 Parallel Lines: Constructing a Congruent Line	$\angle R_jPR_n \cong \angle R_jQR_v$	$\alpha \cong \beta$
2	From 1g, 1 and T5.4.1 Intersection of Two Lines: Opposite Congruent Vertical Angles	$\angle R_mPR_k \cong \angle R_jPR_n$	$\alpha \cong \alpha$
3	From 1, 2 and A5.2.18 GIV Transitivity of Angles, hence	$\angle R_jQR_v \cong \angle R_mPR_k$	$\beta \cong \alpha$
4	From 1g and D5.1.12 Supplementary Angles	$180° \cong \angle R_jPR_m + \angle R_jPR_n$	
5	From 4 and T4.3.3B Equalities: Right Cancellation by Addition	$\angle R_jPR_m \cong 180° - \angle R_jPR_n$	$\alpha' \cong 180° - \alpha$
6	From 5 and T5.4.1 Intersection of Two Lines: Opposite Congruent Vertical Angles	$\angle R_nPR_k \cong \angle R_jPR_m$	$\alpha' \cong \alpha'$
7	From 1g and D5.1.12 Supplementary Angles	$180° \cong \angle R_jQR_u + \angle R_jQR_v$	
8	From 7 and T4.3.3B Equalities: Right Cancellation by Addition	$\angle R_jQR_u \cong 180° - \angle R_jPR_n$	$\beta' \cong 180° - \beta$
9	From 1, 8 and A5.2.18 GIV Transitivity of Angles	$\angle R_jQR_u \cong 180° - \angle R_jQR_v$	$\beta' \cong 180° - \alpha$
10	From 5, 6, 9 and A5.2.18 GIV Transitivity of Angles, hence	$\angle R_jQR_u \cong \angle R_nPR_k$	$\beta' \cong \alpha'$
∴	From 3, 10 and T5.4.3 Parallel Lines: A Congruent Line is Parallel	$\angle R_iQR_v \cong \angle R_mPR_k : \beta \cong \alpha$	$\angle R_iQR_u \cong \angle R_nPR_k : \beta' \cong \alpha'$
		for $R_uR_v \parallel R_mR_n$	

Corollary 5.4.4.1 Parallel Lines: Congruent Alternate Supplementary Angles

1g	Given	the above geometry of Figure 5.4.1B	
Steps	Hypothesis	$\angle R_jPR_m \cong \angle R_jQR_u$ $\angle R_nPR_k \cong \angle R_vQR_k$	$\angle R_jPR_n \cong \angle R_jQR_v$ $\angle R_mPR_k \cong \angle R_uQR_k$
1	From T5.4.4 Parallel Lines: Congruent Alternate Interior Angles	$\angle R_nPR_k \cong \angle R_jQR_u$	$\angle R_mPR_k \cong \angle R_jQR_v$
2	From 1g and T5.4.1 Intersection of Two Lines: Opposite Congruent Vertical Angles	$\angle R_jPR_m \cong \angle R_nPR_k$	$\angle R_jPR_n \cong \angle R_mPR_k$
3	From 1, 2 and A5.2.18 GIV Transitivity of Angles hence	$\angle R_jPR_m \cong \angle R_jQR_u$	$\angle R_jPR_n \cong \angle R_jQR_v$
4	From 1g and T5.4.1 Intersection of Two Lines: Opposite Congruent Vertical Angles	$\angle R_jQR_u \cong \angle R_vQR_k$	$\angle R_jQR_v \cong \angle R_uQR_k$
5	From 1, 4 and A5.2.18 GIV Transitivity of Angles hence	$\angle R_nPR_k \cong \angle R_vQR_k$	$\angle R_mPR_k \cong \angle R_uQR_k$
∴	From 3 and 5	$\angle R_jPR_m \cong \angle R_jQR_u$ $\angle R_nPR_k \cong \angle R_vQR_k$	$\angle R_jPR_n \cong \angle R_jQR_v$ $\angle R_mPR_k \cong \angle R_uQR_k$

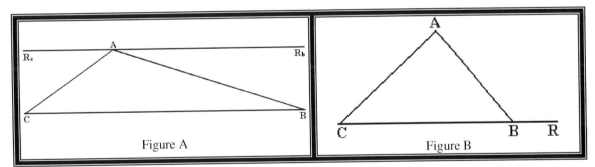

Figure 5.4.2 Triangular Internal and External Angles

Theorem 5.4.5 Sum of Internal Angles for a Triangle

1g 2g	Given	the above geometry from Figure 5.4.2A RcRb ∥ BC lines
Steps	Hypothesis	$180° \cong \angle ABC + \angle BAC + \angle BCA$
1	From 1g, 2g and T5.4.4 Parallel Lines: Congruent Alternate Interior Angles	$\angle R_bAB \cong \angle ABC$
2	From 1g, 2g and T5.4.4 Parallel Lines: Congruent Alternate Interior Angles	$\angle R_cAC \cong \angle BCA$
3	From 1g and D5.1.8 Flat Angle	$180° \cong \angle R_bAB + \angle BAC + \angle R_cAC$
∴	From 1, 2, 3 and A5.2.18 GIV Transitivity of Angles	$180° \cong \angle ABC + \angle BAC + \angle BCA$

Theorem 5.4.6 Solving for Third Angle of a Triangle

1g	Given	ΔABC
2g		$\angle ABC$ to solve for
Steps	Hypothesis	$\angle ABC \cong 180° - \angle BAC - \angle BCA$ for $(\Delta ABC, \angle ABC)$
1	From 1g and T5.4.5 Sum of Internal Angles for a Triangle	$180° \cong \angle ABC + \angle BCA + \angle CAB$
2	From 2g, 1 and A4.2.6 Associative Add	$180° \cong \angle ABC + (\angle BCA + \angle CAB)$
3	From 2, T4.3.3 Equalities: Right Cancellation by Addition	$180° - (\angle BCA + \angle CAB) \cong \angle ABC$
$\therefore$	From 3, A4.2.14 Distribution and A4.2.2B Equality	$\angle ABC \cong 180° - \angle BAC - \angle BCA$ for $(\Delta ABC, \angle ABC)$

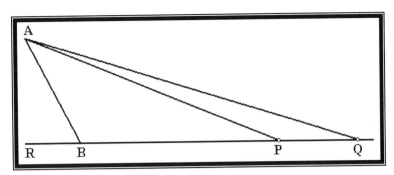

Figure 5.4.3 Limit of a Continuously Elongated Triangle

Theorem 5.4.7 Parallel Lines: Determined by a Continuously Elongated Triangle

1g	Given	the above geometry ABCD from Figure 5.4.3
2g		$\theta \equiv \angle BAP$
3g		$\alpha \equiv \angle ABR$
4g		$\Delta\theta \equiv \angle QAP$
5g		$\delta\alpha \equiv \alpha - \theta$
Steps	Hypothesis	$AP \parallel BP$ in the limit for $\angle ABR \equiv \angle BAP$
1	From 1g and D5.1.8 Flat Angle	$180° \cong \angle PAB + \angle ABR$
2	From 3g, 1, T4.3.3 Equalities: Right Cancellation by Addition and A5.2.18 GIV Transitivity of Angles	$\angle PAB \cong 180° - \angle ABR \cong 180° - \alpha$
3	From 1g and T5.4.6 Solving for a Specific Summation Angle of a Triangle	$\angle APB \cong 180° - \angle BAP - \angle PAB$ for $(\Delta APB, \angle APB)$
4	From 2g, 2, 3 and A5.2.18 GIV Transitivity of Angles	$\angle APB \cong 180° - \theta - (180° - \alpha)$
5	From 4, A4.2.14 Distribution, A4.2.8 Inverse Add, A4.2.7 Identity Add	$\angle APB \cong \alpha - \theta$
6	From 1g and D5.1.8 Flat Angle	$180° \cong \angle APQ + \angle APB$
7	From 5, 6, T4.3.3 Equalities: Right Cancellation by Addition and A5.2.18 GIV Transitivity of Angles	$\angle APQ \cong 180° - \angle APB \cong 180° - (\alpha - \theta)$
8	From 1g and T5.4.6 Solving for a Specific Summation Angle of a Triangle	$\angle AQP \cong 180° - \angle APQ - \angle QAP$ for $(\Delta AQP, \angle AQP)$

9	From 4g, 7, 8, and A5.2.18 GIV Transitivity of Angles	$\angle AQP \cong 180° - (180° - (\alpha - \theta)) - \Delta\theta$	
10	From 9, A4.2.14 Distribution, A4.2.8 Inverse Add, A4.2.7 Identity Add	$\angle AQP \cong \alpha - (\theta + \Delta\theta)$	
11	From 5, 10 and D5.1.6 Angle	$0 \leq \alpha - \theta$	$0 \leq \alpha - (\theta + \Delta\theta)$
12	From 11, A4.2.17 Continuity of Equality and Inequality, T4.3.4 Equalities: Reversal of Right Cancellation by Addition and A4.2.7 Identity Add	$\theta \leq \alpha$ no matter how large θ or $\theta + \Delta\theta$ become they cannot exceed α	$\theta + \Delta\theta \leq \alpha;$ hence,
13	From 4g and D5.1.6 Angle	$0 \leq \Delta\theta$	
14	From 13, A4.2.2 Equality and T4.7.2 Inequalities: Uniqueness of Addition by a Positive Number	$\theta \leq \theta + \Delta\theta$	
15	From 12, 14 and A5.2.18 GIV Transitivity of Angles	$\theta \leq \theta + \Delta\theta \leq \alpha$	
16	From 15, A4.2.17 Continuity of Equality and Inequality, A4.2.2 Equality and A4.2.8 Inverse Add	$\theta - \alpha \leq \theta + \Delta\theta - \alpha \leq 0$	
17	From 16, by −1, T4.7.6 Inequalities: Multiplication by Negative Number Reverses Order and A4.2.14 Distribution	$0 \leq \alpha - (\theta + \Delta\theta) \leq \alpha - \theta$	
18	From 5g, 10, 17 and A5.2.18 GIV Transitivity of Angles	$0 \leq \angle AQP \leq \delta\alpha$	
19	From 18 as ΔQAB elongates on the BQ base and $\delta\alpha$ decreases becoming ever smaller.	$\angle AQP$ decreases and becomes ever smaller as BP increases to a very long length.	
20	From 19 and in the limit	$0 \equiv \delta\alpha$ for BP $\equiv \infty$ in length	
21	From 2g, 3g, 5g, 20, T4.3.13 Equalities: Equality by Difference and A5.2.18 GIV Transitivity of Angles	$\alpha \equiv \theta$	or $\angle ABR \equiv \angle BAP$
22	From 1g	AB is a transversal intersecting AP and BP as a side to ΔABP	
∴	From 21, 22 and T5.4.4 Parallel Lines: Congruent Alternate Interior Angles	AP ∥ BP in the limit for $\angle ABR \equiv \angle BAP$	

Theorem 5.4.8 External Angle Supplementary of two Opposing Vertices

1g 2g 3g 4g 5g	Given	the above geometry from Figure 5.4.2B $\alpha \equiv \angle BAC$ $\beta \equiv \angle ABC$ $\delta \equiv \angle ACB$ $\gamma \equiv \angle ABR$
Steps	Hypothesis	$\gamma = \alpha + \delta$
1	From 1g and T5.4.5 Sum of Internal Angles for a Triangle	$180° = \angle BAC + \angle ABC + \angle ACB$
2	From 1g and D5.1.8 Flat Angle	$180° = \angle ABC + \angle ABR$
3	From 1, 2, T4.3.3 Equalities: Right Cancellation by Addition and A5.2.18 GIV Transitivity of Angles	$180° = \angle BAC + 180° - \angle ABR + \angle ACB$
4	From 3, T4.3.2 Equalities: Cancellation by Addition and T4.3.4 Equalities: Reversal of Right Cancellation by Addition	$\angle ABR = \angle BAC + \angle ACB$
∴	From 2g, 4g, 5g, 4 and A5.2.18 GIV Transitivity of Angles	$\gamma = \alpha + \delta$

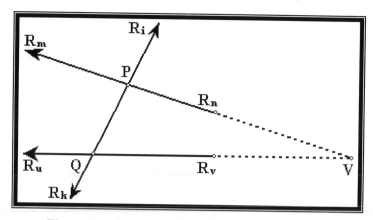

Figure 5.4.4 Converse of Implied Convergent Lines

The following two theorems determine parallel lines.

Theorem 5.4.9 Parallel Lines: Converse Congruent Alternate Interior Angles

If a line intersecting two other lines makes the alternate angle congruent to one another, then the lines will be parallel.

1g	Given	the above geometry from Figure F5.4.4
2g		$\angle PQR_v \cong \angle QPR_m$ constructed by PQ a <u>straight</u> transversal line cutting across R_mR_n, and R_uR_v at points P and Q.
3g		also assume R_mR_n, and R_uR_v to converge onto point-V
Steps	Hypothesis	$\angle QPR_m \cong \angle PQR_v$ than R_mR_n, $\parallel R_uR_v$
1	From 3g and T5.4.8 External Angle Supplementary of two Opposing Vertices	$\angle QPR_m \cong \angle PQR_v + \angle PVQ$
2	From 3g and 1	$\angle PQR_v < \angle PQR_v + \angle PVQ$
3	Form 1, 2 and A5.2.18 GIV Transitivity of Angles	$\angle PQR_v < \angle QPR_m$
4	From 1g, 2g and 3	the original assumption was $\angle PQR_v \cong \angle QPR_m$, hence $\angle PQR_v < \angle QPR_m$ can not be true
∴	From 4	$\angle QPR_m \cong \angle PQR_v$ than R_mR_n, $\parallel R_uR_v$

Corollary 5.4.9.1 Parallel Lines: Converse Non-Perpendicular Alternate Interior Angles

If a line intersecting two other lines makes the alternate angle not perpendicular to one another, then the lines will not be parallel.

1g	Given	the above geometry from Figure F5.4.4
2g		R_mR_n, and R_uR_v to converge onto point-V
3g		$90° \cong \angle PQV$ assume perpendicular angle
4g		$90° \cong \angle QPV$ assume perpendicular angle
Steps	Hypothesis	$180° > \angle PQV + \angle QPV$ than R_mR_n, are not parallel R_uR_v
1	Form 3g, 4g and D5.1.12 Supplementary Angles	$180° \cong \angle PQV + \angle QPV$
∴	From 1g, 2g, and 1 which is not true so we are forced to conclude	$180° > \angle PQV + \angle QPV$ than R_mR_n, are not parallel R_uR_v

Theorem 5.4.10 Parallel Lines: Angles Arising From the Transversal

If a line intersecting two other lines makes the exterior angle congruent to the interior and opposite angle on the same side, then the lines will be parallel.

1g	Given	the above geometry from Figure 5.4.4	
Steps	Hypothesis	R_mR_n, and R_uR_v must be parallel	
1	From 1g and T5.4.1 Intersection of Two Lines: Opposite Congruent Vertical Angles	$\angle R_jPR_m \cong \angle QPR_n$	$\angle R_jPR_n \cong \angle QPR_m$
2	From 1g and T5.4.1 Intersection of Two Lines: Opposite Congruent Vertical Angles	$\angle R_kQR_u \cong \angle PQR_v$	$\angle R_kQR_v \cong \angle PQR_u$
3	From 1g and T5.4.9 Parallel Lines: Converse Congruent Alternate Interior Angles	$\angle PQR_u \cong \angle QPR_n$ iff $R_mR_n \parallel R_uR_v$	$\angle QPR_m \cong \angle PQR_v$ iff $R_uR_v \parallel R_mR_n$
$\therefore$	From 1, 2 and 3	R_mR_n, and R_uR_v must be parallel	

Theorem 5.4.11 Parallel Lines: For Supplementary Angles

The sum of the interior angles on the same side equal to two right angles, then the lines will be parallel.

1g	Given	the above geometry from Figure 5.4.4	
Steps	Hypothesis	R_mR_n, and R_uR_v must be parallel for any supplementary angles on the same side	
1	From 1g, for line R_kR_i and D5.1.12 Supplementary Angles	$180° \cong \angle R_jPR_m + \angle QPR_m$	$180° \cong \angle R_jPR_n + \angle QPR_n$
2	From 1g, for line R_kR_i and D5.1.12 Supplementary Angles	$180° \cong \angle R_kQR_u + \angle PQR_u$	$180° \cong \angle R_kQR_v + \angle PQR_v$
3	From 1g, for line R_mR_n and D5.1.12 Supplementary Angles	$180° \cong \angle R_jPR_m + \angle R_jPR_n$	$180° \cong \angle QPR_m + \angle QPR_n$
4	From 1g, for line R_uR_v and D5.1.12 Supplementary Angles	$180° \cong \angle PQR_u + \angle PQR_v$	$180° \cong \angle R_uQR_k + \angle R_kQR_v$
$\therefore$	From 1, 2, 3, 4 and T5.4.10 Parallel Lines: Angles Arising From the Transversal	R_mR_n, and R_uR_v must be parallel for any supplementary angles on the same side	

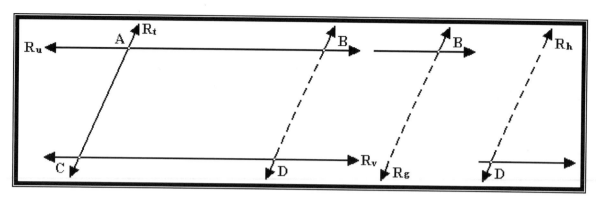

Figure 5.4.5 Translation of Parallel Lines

Theorem 5.4.12 Parallel Lines: Translation

1g	Given	the above geometry from Figure F5.4.5	
2g		AB ∥ CD	
3g		AB ≅ CD	
4g		AC transversal across AB ∥ CD	
Steps	Hypothesis	AC ∥ BD	A ←→ B and C ←→ D
1	From 2g, 3g and 4g	point-B is established at a distance AB from point-A spanning a distance AC	
2	From 2g, 3g and 4g	point-D is established at a distance CD from point-C spanning distance AC	
3	From 1g, 4g and A5.2.13 Axiom of Parallels	BR_g ∥ AC going through point-B	
4	From 1g, 4g and A5.2.13 Axiom of Parallels	DR_h ∥ AC going through point-D	
5	From 1, 2 and D5.1.2-3: Line	straight line BD passes through points-(B, D)	
6	From 3, 4 and 5	straight line BR_g ∥ BD and straight line DR_h ∥ BD	
∴	From 1, 3, 4, 6 and A5.2.15 GIV Transitivity of Segments	AC ∥ BD	A ←→ B and C ←→ D

Theorem 5.4.13 Parallel Lines: Congruence of Internal Base Angles

1g 2g 3g	Given	the above geometry for quadrilateral ABCD from Figure F5.4.5 AB ∥ CD AC ∥ BD
Steps	Hypothesis	∠BDR$_v$ ≅ ∠DCA
1	From 1g, 2g and T5.4.4 Parallel Lines: Congruent Alternate Interior Angles	∠CAR$_u$ ≅ ∠DCA for AC the transversal across AB ∥ CD
2	From 1g and T5.4.1 Intersection of Two Lines: Opposite Congruent Vertical Angles	∠R$_t$AB ≅ ∠CAR$_u$
3	From 1g, 3g and T5.4.4 Parallel Lines: Congruent Alternate Interior with Exterior Angles	∠ABD ≅ ∠R$_t$AB for AB the transversal across AC ∥ BD
4	From 1g, 2g and T5.4.2 Parallel Lines: Congruent Alternate Interior Angles	∠BDR$_v$ ≅ ∠ABD for BD the transversal across AB ∥ CD
5	From 2, 3 and A5.2.18 GIV Transitivity of Angles	∠ABD ≅ ∠CAR$_u$
∴	From 4, 5 and A5.2.18 GIV Transitivity of Angles	∠BDR$_v$ ≅ ∠DCA

Theorem 5.4.14 Parallel Lines: Relative Orientation of Parallel Lines

1g	Given	the above geometry for quadrilateral ABCD from Figure F5.4.5
Steps	Hypothesis	AC ∥ BC has the same orientation for AB ∥ CD
∴	From 1g and T5.4.13 Parallel Lines: Congruence of Internal Base Angles	AC ∥ BC has the same orientation for AB ∥ CD

Theorem 5.4.15 Parallel Lines: Congruency of Opposing Sides Along Parallel Rails

1g 2g 3g	Given	the above geometry for quadrilateral ABCD from Figure F5.4.5 AB ∥ CD AC ∥ BD
Steps	Hypothesis	AC ≅ BD along parallel rails AB ∥ CD
1	From 2g	Distance apart does not change from A to B
2	From 2g, 3g and T5.4.14 Parallel Lines: Relative Orientation of Parallel Lines	AC ∥ BC has the same orientation for AB ∥ CD
∴	From 1 and 2	AC ≅ BD along parallel rails AB ∥ CD

Theorem 5.4.16 Parallel Lines: Give Rise to Parallelograms

1g 2g 3g	Given	the above geometry for quadrilateral ABCD from Figure F5.4.4 AB ‖ CD AC ‖ BD
Steps	Hypothesis	quadrilateral ABCD is a parallelogram
∴	From 1g, 2g, 3g and D5.1.13 Parallelogram	quadrilateral ABCD is a parallelogram

Theorem 5.4.17 Parallel Lines: Parallelograms have Congruent Sides

1g 2g 3g	Given	the above geometry for quadrilateral ABCD from Figure F5.4.5 AB ‖ CD AC ‖ BD	
Steps	Hypothesis	AB ≅ CD	AC ≅ BD
1	From 1g, 2g, 3g and D5.1.13 Parallelogram	BDAC a parallelogram	ABCD a parallelogram
2	From 2g, 3g, 1 and T5.4.15 Parallel Lines: Congruency of Opposing Sides Along Parallel Rails	AB ≅ CD for parallelogram BDAC	
3	From 2g, 3g, 1 and T5.4.15 Parallel Lines: Congruency of Opposing Sides Along Parallel Rails	AC ≅ BD for parallelogram ABCD	
∴	From 2 and 3	AB ≅ CD	AC ≅ BD

Theorem 5.4.18 Parallel Lines: Parallelograms have Parallel Sides

1g	Given	Parallelogram ABCD	
Steps	Hypothesis	AB ‖ CD	AD ‖ BC
∴	1g and D5.1.13 Parallelogram	AB ‖ CD	AD ‖ BC

Theorem 5.4.19 Parallel Lines: Parallelograms have Corresponding Vertices

1g	Given	Parallelogram ABCD	
Steps	Hypothesis	A ←→ B and C ←→ D	A ←→ C and B ←→ D
1	1g and D5.1.13 Parallelogram	AB ‖ CD	AC ‖ BD
∴	From 1 and T5.4.12 Parallel Lines: Translation	A ←→ B and C ←→ D	A ←→ C and B ←→ D

Section 5.5 Theorems on Constructible Triangles

Many people have contributed to the understanding and proofs of the theorems on congruency and similarity, of course Euclid, Aristotle, Heiberg, Proclus and Savile after him, Thales, Simson, Talyor, Hilbert [HEA56, Vol. I, pg 229, 248, 262, 298, 304] the list goes on and now my meager contribution. In all the developing theorems, with bits and pieces of compelling proofs have been developed, but no coherent ordered method demonstrating congruence from first principles has been proposed. Even Hilbert attempts to prime the pump by creating an initializing axiom, Group IV, Axiom A5.2.19 GIV "Congruence of Side-Side-Angle (SSA) and Opposite Angles", is not very compelling in building a solid argument for congruency of triangles. In the following set of proofs, I attempt to do just that.

Hence starting from first principles two geometric objects, such as a triangle, are *congruent if-and only-if they can be placed into a one-to-one correspondence between components, or elements, comprising the objects, such that the components themselves are congruent.* This gives our first notion of what congruency is and how Euclid proved congruence for all his proofs.

What is meant here by component, or element, is a geometrical object constructed from base elements of construction. *Base parts, component, or elements of construction* mean that components themselves cannot be broken down any further into other discrete elements. Clearly, the line, constructed by the straightedge, and circle by the compass, can be broken down no further and as such are base elements. In most part, they are not very interesting, but in combination, all sorts of interesting things can be erected. As an example, an angle can only be constructed from two circles.

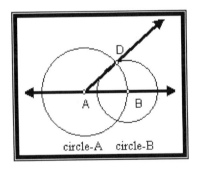

Figure 5.5.1 Constructible Angular Component

Where circle-A is called the construction circle, analogous to a construction line, and circle B the spanning circle defining the opening, or maw, straddling the angle and fixing its size. So, any angle is always comprised of two base elements that completely define that angle. The other aspect of this geometric object is the relationship between the two circles, where the spanning circle's center is someplace on the circumference of the construction circle along with a concurrent end-point straddling the radius thereby fixing the angle. The two circles and their relationship comprise a single geometric object the angle, which in tern can be used to build more sophisticated objects, and so on. So, a *geometric object* is defined here as any object of geometry that can be broken down into fundamental elements that have a unique relationship between them. With these new ideas about what we think constitutes a geometric object a modern definition can be devised based on Set Builder Notation.

Definition 5.5.1 Geometric Object

A ***geometric object*** is a special set G comprised of base geometric objects [g_i] and bound together by certain relationship rules of construction.

$g \equiv \{(_\bullet \, p_j \,_\bullet) \text{ for m-properties}\}$ base geometric object or component

$G \equiv \{(_\bullet \, g_i \,_\bullet) \text{ for n-base objects} \mid (_\bullet \, r_k \,_\bullet) \text{ for u–relationship rules of construction}\}$ geometric object

Correspondence and Constructible Geometric Objects:

It would follow that if two geometric objects can be shown to have a one-to-one correspondence between their base components that they are the same ***type of object***.

This might seem all that is needed to show congruency, between two objects, but there is one other all-encompassing property that must be considered. It has been alluded to, but is so basic that even Euclid and all the geometers before, and since, assumed it without ever bring it to the forefront and setting it apart from all other aspects of geometric proofs. It has always been considered self-evident and therefore taken for granted that construction is used to achieve geometric structures. In Euclid's book the elements construction is always there, but never setup in an axiomatic way using formal theorems, yet that is what a construction is a type of proof for propositions by using drawn components. Here I have discussed about constructing an angle and a congruent angle having congruent corresponding parts. So, it would seem to follow that if a geometric object is constructible and another geometric object of the same type where to be constructed from it, as a template, the two would be congruent. As a fundamental property of congruency, ***an object must be constructible; hence, its congruent counter part must likewise be constructible.*** Equally, the opposite would also have to be true; ***two congruent geometrical objects are constructible***.

Geometry is a unique discipline in mathematics in that there are two distinct types of proofs, <u>proofs by construction proving structure</u> and <u>proofs by deductive or inductive logic proving properties of structures</u>. Where construction being defined as a drafting method requiring a compass and a straight edge. Hence two geometric objects, such as a triangle, are ***congruent if-and only-if the second object can be constructed from the first, thereby establishing a one-to-one correspondence between components, or elements, comprising the objects***. This gives our second notion of congruency.

Before the second object can be constructed, we have to know that the first object is constructible. For a triangle, let's observe that it is made of six components or parts, three sides and three angles. The totality uniquely defines any given triangle. With the theorems of congruency and similarity, they require the following question to be answered, "What is the minimum number of components necessary to construct a triangle?" Without this question being answered congruency, or similarity, cannot be established, not even being able to determine a correspondence between two triangles. So, the following proofs try to answer the question what constitutes a constructible triangle?

- ***First notion of constructible triangles*** by using only a single element. This is self evident, if only one element is provided no triangle can be constructed. In other wards given just a single segment or angle, not enough information is available to construct a triangle.

- **Second notion of constructible triangles** by using only two elements. Given any two elements, is it possible to construct a triangle? Consequently, for all possible arrangements, there are 2^2 or 4 out of two elements that can be grouped for constructible triangle.

Given	Construction	Proof by Construction
Theorem 5.5.1 **Bi-Construction:** **Side-Side (SS)**	A ———————— B C —————————— D	Insufficient information to construct ΔABC. **Given:** AB and CD **Construction:** Clearly, no orientation of the segments or connection of endpoints would allow a triangle to be constructed. Without the information provided by angles, it just cannot happen.
Theorem 5.5.2 **Bi-Construction:** **Side-Angle (SA)**		Insufficient information to construct ΔABC. **Given:** AB and ∠ABC **Construction:** For any fixed segment joined at point-A, there are three possibilities either a rotation will fall short < AF, at point-E and no intersection established with the ray BC, hence no triangle defined Or congruent to AF establishing a triangle ΔABF as a special case, but providing no general solution to finding any triangle. More critically, this assumes priory knowledge of the length of AF, which is not given, so again no triangle can be established. Or it will be > AF and intersect at two points -D or C, but which is the correct one, again not enough information so no conclusion can be drawn and no triangle established. So, without a second segment being provided not enough information is given to construct a triangle.
Theorem 5.5.3 **Bi-Construction:** **Angle-Side (AS)**	See above construction	Same argument from Side-Angle

Given	Construction	Proof by Construction
Theorem 5.5.4 **Bi-Construction:** **Angle-Angle** **(AA)**		Insufficient information to construct ΔABC. **Given:** ∠ABC and ∠BCA **Construction:** Clearly, the third ray intersects AB determining ∠CAB and thereby establishing the shape of ΔABC. This is reaffirmed by Theorem T5.4.5 Sum of Internal Angles for a Triangle such that ∠CAB ≅ 180° – ∠ABC – ∠BCA establishing the third angle. Since no side is defined to scale, the size of the triangle a specific triangle is not established however by Definition D5.1.29 Similar Polygons alike triagon is established, which forms the foundation for the first theorem of similar triangles (AA).

Hence, with only two known elements it is possible to construct an infinite number of similar triangles, but no specific triangle. Only further criteria can establish which triangle should be used out of an infinite number unless only similarity is being considered in that case it does not matter. This is a geometric construction that is undecidable, because without a specific criterion the necessary information to contruct a triangle is simply is not present.

Tensor Calculus & Physics: A General Treatise

- ***Third notion of constructible triangles*** by using only three elements. Given the two basic elements side and angle than taking three elements at a time can a triangle be constructed without being inscribed within a circle? Inscription by circle is an advance geometry concept and out of sequence in the deductive geometry development. Also all constructions start on the constructor line BC. So, for all possible arrangements, there are 2^3 or 8 out of three elements that can be grouped for constructible triangle.

Given	Construction	Proof by Construction of the Triangle
Theorem 5.5.5 **Tri-Construction:** **Side-Side-Side** **(SSS)**		Sufficient information to construct ΔABC. **Given:** AB, BC and CA. **Construction:** With compass only striking intersecting arcs of appropriate length from either end of a given segment, completes the ΔABC with three angles, and all six components are accounted.
Theorem 5.5.6 **Tri-Construction:** **Side-Side-Angle** **(SSA)**		Sufficient information to construct ΔABC. **Given:** ∠A', AB and BC. **Construction:** With straightedge extend point-A' along angle ray to intersect line BC at point-B'. Mark off segment BB' from point-A' on constructed A'B" ∥ B'B constructing point-B". Construct line through points-B and B". Mark off given segment AB on BB" constructing point-A. Construct segment AC, which completes the third side and angle, and all six components are accounted.
Theorem 5.5.7 **Tri-Construction:** **Side-Angle-Side** **(SAS)**		Sufficient information to construct ΔABC. **Given:** AB, ∠B and BC. **Construction:** With straightedge extend point-A to point-C completes the second angle, the third side and angle, and all six components are accounted.
Theorem 5.5.8 **Tri-Construction:** **Angle-Side-Side** **(ASS)**		Sufficient information to construct ΔABC. **Given:** AB, BC and ∠C. **Construction:** With straightedge extend ray from point-C and with compass swing arc from point-B intersecting ray-C at point-A completing the third side and angle, and all six components are accounted.

Given	Construction	Determination of Construction
Theorem 5.5.9 Tri-Construction: Angle-Angle-Side (AAS)		Sufficient information to construct ∆ABC". **Given:** AB, ∠B and ∠C. **Construction:** With straightedge extend point-A and point-C to intersect at point A'. With compass and straightedge on A'C at point-C construct congruent, angles ∠AA'C ≅ ∠RCC' such that A'B ∥ CC' rays. From point-C construct extended congruent segment C'C ≅ A'A on parallel rays. Creating the parallelogram A'CC'A and mark off the intersection point-C" on CB. This completes the second, third side and angle, and all six components are accounted.
Theorem 5.5.10 Tri-Construction: Angle-Side-Angle (ASA)		Sufficient information to construct ∆ABC. **Given:** ∠A, AB and ∠B **Construction:** With straightedge extend ray from point-A to point-C completing the second, third side and angle, and all six components are accounted.
Theorem 5.5.11 Tri-Construction: Side-Angle-Angle (SAA)		Sufficient information to construct ∆A"BC. **Given:** ∠A, ∠B and BC. **Construction:** With straightedge extend point-B any where on ray construct at point A'. With compass and straightedge on A'B at point-A' construct congruent, angles ∠A ≅ ∠A' extending ray and constructing intersection point-C' on line BC. At point-C construct ∠A'C'B extending ray and constructing intersection point-A" on line A'B. This completes the second, third side and angle, and all six components are accounted.
Theorem 5.5.12 Tri-Construction: Angle-Angle-Angle (AAA)		Insufficient information to construct ∆ABC. **Given:** ∠A, ∠B and ∠C. **Construction:** Since no side is defined to scale, the size of the triangle is not established, however by definition D5.1.29 Similar Polygons such as a triagon can be established, which forms the foundation for the first theorem of similar triangles (AA).

A triangle requires a minimum of at lest three components are required in order to construct and properly define a complete triangle, except for the special case of angle-angle-angle.

From Hilbert's Axioms, the notion of congruence of lines, rays, segments and angles are established, but what of congruence of triangles? It would follow that in order for any two triangles to be congruent there must be a congruent one-to-one correspondence between all six components, see definition D5.1.27: "Congruent".

Logically it follows that if the initial triangle only has three components defined, the other elements can be found, and likewise for the second corresponding triangle, hence, all components are known and in one-to-one correspondence, therefore they are congruent triangles. However proving congruency of two triangles is a problem in topology as demonstrated by Euclid and confirmed by Hilbert's solution for their theorem (SAS), see the following two theorems:

For a formal proofs of these constructions see "Section 5.18 Modern Elements: Extending Hilbert's Axioms to Include Construction", theorems T5.19.1 through 7.

Theorem 5.5.13 Similar Triangles: Angle-Angle (AA)

If two triangles have two corresponding congruent angles, respectively, then the triangles are similar.

1g	Given	$\triangle ABC$, $\triangle A'B'C'$
2g		$\angle BAC \cong \angle B'A'C'$
3g		$\angle ABC \cong \angle A'B'C'$
Steps	Hypothesis	$\triangle ABC \sim \triangle A'B'C'$
1	From 1g, 2g, 3g and T5.4.6 Solving for a Specific Summation Angle of a Triangle	$\angle BCA \cong 180° - \angle ABC - \angle BAC$ for $(\triangle ABC, \angle BCA)$
2	From 1g, 2g, 3g and T5.4.6 Solving for a Specific Summation Angle of a Triangle	$\angle B'C'A' \cong 180° - \angle A'B'C' - \angle B'A'C'$ for $(\triangle A'B'C', \angle B'C'A')$
3	From 1g, 2g, 1 and A5.2.18 GIV Transitivity of Angles	$\angle BCA \cong 180° - \angle A'B'C' - \angle B'A'C'$
4	From 2, 3 and A5.2.18 GIV Transitivity of Angles	$\angle BCA \cong \angle B'C'A'$
$\therefore$	From 1g, 2g, 3g, 4 and D5.1.29 Similar Polygons	$\triangle ABC \sim \triangle A'B'C'$

Section 5.6 Theorems on Congruency of Two Triangles

Theorem 5.6.1 Congruent Triangles: Reflexive Property

1g	Given	$\triangle ABC$		
Steps	Hypothesis	$\triangle ABC \cong \triangle ABC$		
1	From 1g, A5.2.16 GIV Congruent-Reflexive Angles	$\angle CAB \cong \angle CAB$	$\angle ABC \cong \angle ABC$	$\angle BCA \cong \angle BCA$
2	From 1g, A5.2.14 GIV Symmetric-Reflexiveness of Segments	$AB \cong AB$	$BC \cong BC$	$CA \cong CA$
3	From 1g and D5.1.28 Concurrent vertices	point A $\leftrightarrow$ point A	point B $\leftrightarrow$ point B	point C $\leftrightarrow$ point C
∴	From 1, 2, 3 and D5.1.50 Congruent Triangles	$\triangle ABC \cong \triangle ABC$		

Theorem 5.6.2 Congruent Triangles: Symmetric Congruent Property

1g	Given	$\triangle ABC \cong \triangle A'B'C'$		
Steps	Hypothesis	$\triangle ABC \cong \triangle A'B'C'$	$\triangle A'B'C' \cong \triangle ABC$	
1	From 1g and D5.1.50 Congruent Triangles	$\angle CAB \cong \angle C'A'B'$	$\angle ABC \cong \angle A'B'C'$	$\angle BCA \cong \angle B'C'A'$
2	From 1g and D5.1.50 Congruent Triangles	$AB \cong A'B'$	$BC \cong B'C'$	$CA \cong C'A'$
3	From 1g and D5.1.50 Congruent Triangles	point A $\leftrightarrow$ point A'	point B $\leftrightarrow$ point B'	point C $\leftrightarrow$ point C'
4	From 1 and A5.2.17 GIV Symmetric-Reflexiveness of Angles	$\angle C'A'B' \cong \angle CAB$	$\angle A'B'C' \cong \angle ABC$	$\angle B'C'A' \cong \angle BCA$
5	From 2 and A5.2.14 GIV Symmetric-Reflexiveness of Segments	$A'B' \cong AB$	$B'C' \cong BC$	$C'A' \cong CA$
6	From 3 and D5.1.27 Congruent	point A' $\leftrightarrow$ point A	point B' $\leftrightarrow$ point B	point C' $\leftrightarrow$ point C
∴	From 1g, 4, 5, 6 and D5.1.50 Congruent Triangles	$\triangle ABC \cong \triangle A'B'C'$	$\triangle A'B'C' \cong \triangle ABC$	

Theorem 5.6.3 Congruent Triangles: Transitive Property

1g 2g	Given	$\triangle ABC \cong \triangle A'B'C'$ $\triangle A'B'C' \cong \triangle A''B''C''$		
Steps	Hypothesis	If $\triangle ABC \cong \triangle A'B'C'$ and $\triangle A'B'C' \cong \triangle A''B''C''$ then $\triangle ABC \cong \triangle A''B''C''$		
1	From 1g and D5.1.50 Congruent Triangles	$\angle CAB \cong \angle C'A'B'$	$\angle ABC \cong \angle A'B'C'$	$\angle BCA \cong \angle B'C'A'$
2	From 1g and D5.1.50 Congruent Triangles	$AB \cong A'B'$	$BC \cong B'C'$	$CA \cong C'A'$
3	From 1g and D5.1.50 Congruent Triangles	point A $\leftrightarrow$ point A'	point B $\leftrightarrow$ point B'	point C $\leftrightarrow$ point C'
4	From 2g and D5.1.50 Congruent Triangles	$\angle C'A'B' \cong \angle C''A''B''$	$\angle A'B'C' \cong \angle A''B''C''$	$\angle B'C'A' \cong \angle B''C''A''$
5	From 2g and D5.1.50 Congruent Triangles	$A'B' \cong A''B''$	$B'C' \cong B''C''$	$C'A' \cong C''A''$
6	From 2g and D5.1.50 Congruent Triangles	point A' $\leftrightarrow$ point A''	point B' $\leftrightarrow$ point B''	point C' $\leftrightarrow$ point C''
7	From 1, 4 and A5.2.18 GIV Transitivity of Angles	$\angle CAB \cong \angle C''A''B''$	$\angle ABC \cong \angle A''B''C''$	$\angle BCA \cong \angle B''C''A''$
8	From 2, 5 and A5.2.15 GIV Transitivity of Segments	$AB \cong A''B''$	$BC \cong B''C''$	$CA \cong C''A''$
9	From 3, 6 and D5.1.27 Congruent	point A $\leftrightarrow$ point A''	point B $\leftrightarrow$ point B''	point C $\leftrightarrow$ point C''
10	From 7, 8, 9 and D5.1.50 Congruent Triangles	$\triangle ABC \cong \triangle A''B''C''$		
$\therefore$	From 1g, 2g and 10	If $\triangle ABC \cong \triangle A'B'C'$ and $\triangle A'B'C' \cong \triangle A''B''C''$ then $\triangle ABC \cong \triangle A''B''C''$		

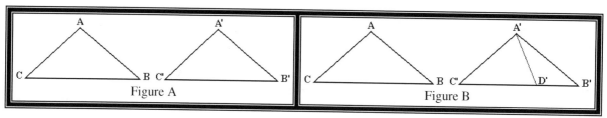

Figure 5.6.1 Euclid's and Hilbert's Proof of (SAS)

Theorem 5.6.4 Euclid's Proof of (SAS) [HEA56, Vol. I, pg 247]

1g	Given	the above geometry figure F5.6.1A
2g		ΔABC and ΔA'B'C'
3g		AB ≅ A'B' corresponding adjacent side
4g		∠CAB ≅ ∠C'A'B' corresponding adjacent angle
5g		AC ≅ A'C' corresponding adjacent side
6g		Corresponding points are equally spaced.

Steps	Hypothesis	ΔABC ≅ ΔA'B'C'	
1	From 3g, 4g, 5g and A5.2.19 GIV Congruence of Side-Side-Angle (SSA) and Opposite Angles	∠CBA ≅ ∠C'B'A'	∠ACB ≅ ∠A'C'B'
2	From 1g and 5g	point A → point A'	
3	From 5g and 3	point C → point C'	
4	Form 3g and 4g	point B → point B'	
5	From 3g, 6g, 3, 4 and D5.1.2-3	CB' as a straight line	
6	From 3, 4, 5 and A5.2.11 GII Ordering of Points on a SL	Points C, B and C', B' are collinear on straight line CB'	
7	From 6g, 6, D5.1.2-3 and A5.2.14 GIV Symmetric-Reflexiveness of Segments	CC' ≅ BB' every segment is congruent to itself.	
8	From 6, D5.1.2-3 and A5.2.14 GIV Symmetric-Reflexiveness of Segments	BC' ≅ CC' ∩ BB' ≅ <CB + BC' + C'B'> every segment is congruent to itself.	
9	From 1g, 6 and 8	CB ≅ CC' – BC'	
10	From 1g, 6 and 8	C'B' ≅ BB' – BC'	
11	From 9 and 10 take the difference	CB – C'B' ≅ CC' – BB' – BC' + BC'	
12	From 7, 11, A5.2.15 GIV Transitivity of Segments, A4.2.8 Inverse Add, A4.2.7 Identity Add and T4.3.13 Equalities: Equality by Difference	CB ≅ C'B' [Euclid proves this by reductio ad-absurdum with closure of the perimeter of congruent triangles. Hilbert's axioms, in steps 5 through 11 expedites the logical conclusion in a cleaner proof, but does the something.]	
∴	From 3g, 4g, 5g, 1, 12 and D5.1.50 Congruent Triangles[5.6.1]	ΔABC ≅ ΔA'B'C'	

Theorem 5.6.5 Hilbert's Proof of (SAS) [HEA56, Vol. I, pg 229]

1g	Given	the above geometry figure F5.6.1B
2g		ΔABC and $\Delta A'B'C'$
3g		$AB \cong A'B'$ corresponding adjacent side
4g		$\angle CAB \cong \angle C'A'B'$ corresponding adjacent angle
5g		$AC \cong A'C'$ corresponding adjacent side
6g		$BC \neq B'C'$ but $CB \cong C'D'$

Steps	Hypothesis	$\Delta ABC \cong \Delta A'B'C'$ by reductio ad-absurdum			
1	From 1g and A5.2.19 GIV Congruence of Side-Side-Angle (SSA) and Opposite Angles	$\angle ABC \cong \angle A'B'C'$		$\angle ACB \cong \angle A'C'B'$	
2	From 2g, 3g, 4g, 5g, 6g and A5.2.19 GIV Congruence of Side-Side-Angle (SSA) and Opposite Angles	$AB \cong A'D'$	$\angle CAB \cong \angle C'A'D'$	$AC \cong A'C'$ for $\Delta A'B'D'$	
3	From 3g, 2 and A5.2.18 GIV Transitivity of Angles	$\angle C'A'B' \cong \angle C'A'D'$			
4	From 1g, 3, and A5.2.17 GIV Symmetric-Reflexiveness of Angles	$\angle C'A'B' \neq \angle C'A'D'$ by geometry they are not congruent			
5	From 1g and 4	$CB \neq C'D'$			
$\therefore$	From 3g, 4g, 5g, 1 and 5	$\Delta ABC \cong \Delta A'B'C'$ by reductio ad-absurdum			

[5.6.1]Note: Definition of Congruent Triangles is an archaic definition that has been over used as if it were an axiom; it is only being kept here to be backwards compatible with Euclid's theorems. For use that is more modern, the three kinds of congruency theorems T5.6.15-on for triangles are used.

Of the two theorems, Euclid's is of most interest, because he takes advantage of the unique idea of topological coincident or correspondence of vertices to prove the congruence of triangles. This is a subconscious idea recognized by everybody through out the centuries, because geometric objects can be constructed anywhere in the Euclidian plane with any orientation. So it follows that if point-to-point or vertex-to-vertex can be matched the triangles are congruent and independent of orientation. This idea is so fundamental it could be restated as one of the axioms of construction, and is. See axiom A5.18.9 GVII "Oriented Congruent Geometric Objects".

These ideas on congruency lead to three fundamental ways of proving congruency for any two triangles.

- The first way is by placing the two triangles into a geometric frame; thereby establishing one-to-one correspondence of vertices, which systematically orients them.
- The second way is to parallel the construction of the two triangles on separate construction lines in the Euclidian Plane, thereby establishing the vertices into a one-to-one correspondence with their corresponding components.
- The third way is by projection of one triangle on top of the other, again establishing concurrency of vertices and congruency of sides.

The first method placing triangles in a frame limiting proofs of congruency to special cases, two cases will be considered. The second method is better because it leads to a greater generality by paralleling construction, which in turn leads to the more classical proofs used in geometry for testing congruency of triangles. Finally, the third technique appeals to our more intuitive understanding of being able to freely move geometric objects about and directly compare them.

Congruency of Triangles of the First Kind:

With our limited development of geometry, so far, two constructible triangles can be tested for congruency by placing them on a rail of parallel lines. This anchors the vertices and sides in proper relative orientation and vertically fixes them so that congruency of the triangles can be proven.

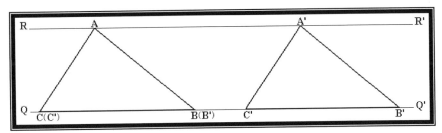

Figure 5.6.2 Congruent Triangles on Parallel Rails

Theorem 5.6.6 Congruent Triangles on Parallel Rails

1g	Given	the above geometry figure F5.6.2 "Congruent Triangles on Parallel Rails"			
2g		ΔABC is constructible			
3g		ΔABC and ΔA'B'C' are constructed between RR' $||$ QQ'			
Steps	Hypothesis	ΔABC $\cong$ ΔA'B'C'			
1	From 3g and T5.4.12 Parallel Lines: Translation	AC $||$ A'C'	A $\leftrightarrow$ A' C $\leftrightarrow$ C'	AB $||$ A'B'	A $\leftrightarrow$ A' B $\leftrightarrow$ B'
2	From 1 and T5.4.15 Parallel Lines: Congruency of Opposing Sides Along Parallel Rails	AC $\cong$ A'C'		AB $\cong$ A'B'	
3	From 3g, 1 and T5.4.17 Parallel Lines: Parallelograms have Congruent Sides	AA' $\cong$ CC' for parallelogram AA'CC'		AA' $\cong$ BB' for parallelogram AA'BB'	
4	From 3 and A5.2.15 GIV Transitivity of Segments	BB' $\cong$ CC'			
5	From 1g	CC' $\cong$ CB + BC'			
6	From 1g	BB' $\cong$ C'B' + BC'			
7	From 5, 6 and taking the difference	CC' $-$ BB' $\cong$ (CB + BC') $-$ (C'B' + BC')			
8	From 7, A4.2.14 Distribution and A4.2.5 Commutative Add	CC' $-$ BB' $\cong$ CB $-$ C'B' + BC' $-$ BC'			
9	From 8, A5.2.15 GIV Transitivity of Segments, A4.2.8 Inverse Add and A4.2.7 Identity Add	0 $\cong$ CB $-$ C'B'			
10	From 9 and T4.3.13 Equalities: Equality by Difference	CB $\cong$ C'B'			
11	From 3g, 1 and T5.4.13 Parallel Lines: Congruence of Internal Base Angles	$\angle$ACB $\cong$ $\angle$A'C'B'		$\angle$ABC $\cong$ $\angle$A'B'C'	

12	From 1g, 3g, 11 and T5.5.13 Similar Triangles: Angle-Angle (AA)	$\angle BAC \cong \angle B'A'C'$		
13	From 1, 2, 10, 11 and 12	A ⟷ A' B ⟷ B' C ⟷ C'	$AC \cong A'C'$ $AB \cong A'B'$ $BC \cong B'C'$	$\angle BAC \cong \angle B'A'C'$ $\angle ABC \cong \angle A'B'C'$ $\angle ACB \cong \angle A'C'B'$
∴	From 13 and D5.1.27 Congruent	$\Delta ABC \cong \Delta A'B'C'$		

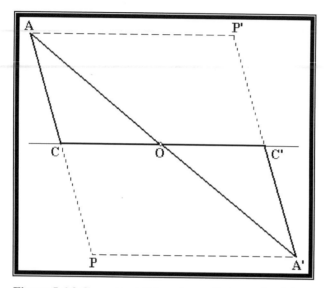

Figure 5.6.3 Congruent Triangles on Intersecting Lines

Theorem 5.6.7 Congruent Triangles on Intersecting Lines

1g	Given	the above geometry figure F5.6.3 "Congruent Triangles on Intersecting Lines"
2g		straight lines AA' and CC' bisects each other at point-O
3g		AP ‖ A'P'
4g		ΔABC and $\Delta A'B'C'$ are asymmetrical about straight line CC'
Steps	Hypothesis	$\Delta ABC \cong \Delta A'B'C'$
1	From 2g and T5.4.1 Intersection of Two Lines: Opposite Congruent Vertical Angles	$\angle AOC \cong \angle A'OC'$
2	From 2g, 3g and T5.4.4 Parallel Lines: Congruent Alternate Interior Angles	$\angle CAO \cong \angle C'A'O$ from the transversal AA' cutting AP ‖ A'P'

3	From 1, 2 and T5.5.13 Similar Triangles: Angle-Angle (AA)	$\angle ACO \cong \angle A'C'O$			
4	From 2 and T5.4.9 Parallel Lines: Converse Congruent Alternate Interior Angles	AP' ∥ A'P from the transversal AA'			
5	1g, 3g, 4 and D5.1.13 Parallelogram	AP'PA'			
6	From 5 and T5.4.19 Parallel Lines: Parallelograms have Corresponding Vertices	A ⟷ P and A' ⟷ P'		A ⟷ P' and A' ⟷ P	
7	From 6 and by correspondence	A ⟷ A'			
8	From 3g, 4 and D5.1.13 Parallelogram	AP'C'C			
9	From 8 and T5.4.19 Parallel Lines: Parallelograms have Corresponding Vertices	A ⟷ P' and C ⟷ C'			
10	From 2g and D5.1.28 Concurrent	O ≡ O			
11	From 3g, A5.2.14 GIV Symmetric-Reflexiveness of Segments	$CO \cong C'O$			
12	From 3g, A5.2.14 GIV Symmetric-Reflexiveness of Segments	$AO \cong A'O$			
13	From 4g and A5.2.14 GIV Symmetric-Reflexiveness of Segments	$A'C' \cong C'P'$			
14	From 8 and T5.4.17 Parallel Lines: Parallelograms have Congruent Sides	$C'P' \cong AC$			
15	From 13, 14 and A5.2.15 GIV Transitivity of Segments	$AC \cong A'C'$			
16	From 1, 2, 3, 7, 9, 10, 11, 12 and 15	A ⟷ A' O ≡ O' C ⟷ C'	$AC \cong A'C'$ $AO \cong A'O$ $CO \cong C'O$	$\angle CAO \cong \angle C'A'O$ $\angle AOC \cong \angle A'OC'$ $\angle ACO \cong \angle A'C'O$	
∴	From 16 and D5.1.27 Congruent	$\triangle ABC \cong \triangle A'B'C'$			

Congruency of Triangles of the Second Kind:

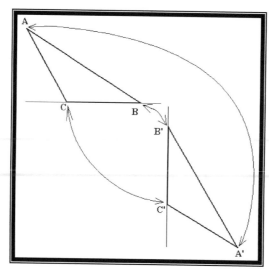

Figure 5.6.4 Congruency of Triangles of the Second Kind

Theorem 5.6.8 Congruency: by Side-Side-Side (SSS)

1g	Given	the above geometry figure F5.6.4 "Congruency of Triangles of the Second Kind"		
2g		construction lines CB and C'B'		
3g		B ←→ B' C ←→ C'	AC ≅ A'C' AB ≅ A'B' BC ≅ B'C'	
Steps	Hypothesis	ΔABC ≅ ΔA'B'C'		
1	From 2g, 3g and T5.5.5 Tri-Construction: Side-Side-Side (SSS)	A ←→ A' B ←→ B' C ←→ C'	AC ≅ A'C' AB ≅ A'B' BC ≅ B'C'	∠BAC ≅ ∠B'A'C' ∠ABC ≅ ∠A'B'C' ∠ACB ≅ ∠A'C'B'
∴	From 1 and D5.1.27 Congruent	ΔABC ≅ ΔA'B'C'		

Theorem 5.6.9 Congruency: by Side-Side-Angle (SSA)

1g	Given	the above geometry figure F5.6.4 "Congruency of Triangles of the Second Kind"		
2g		construction lines CB and C'B'		
3g		A ←→ A' B ←→ B' C ←→ C'	AB ≅ A'B' BC ≅ B'C'	∠BAC ≅ ∠B'A'C'
Steps	Hypothesis	ΔABC ≅ ΔA'B'C'		
1	From 2g, 3g and T5.5.6 Tri-Construction: Side-Side-Angle (SSA)	A ←→ A' B ←→ B' C ←→ C'	AC ≅ A'C' AB ≅ A'B' BC ≅ B'C'	∠BAC ≅ ∠B'A'C' ∠ABC ≅ ∠A'B'C' ∠ACB ≅ ∠A'C'B'
∴	From 1 and D5.1.27 Congruent	ΔABC ≅ ΔA'B'C'		

Tensor Calculus & Physics: A General Treatise

Theorem 5.6.10 Congruency: by Side-Angle-Side (SAS)

1g	Given	the above geometry figure F5.6.4 "Congruency of Triangles of the Second Kind"		
2g		construction lines CB and C'B'		
3g		A ←→ A' B ←→ B' C ←→ C'	AB ≅ A'B' BC ≅ B'C'	∠ABC ≅ ∠A'B'C'
Steps	Hypothesis	ΔABC ≅ ΔA'B'C'		
1	From 2g, 3g and T5.5.7 Side-Angle-Side (SAS)	A ←→ A' B ←→ B' C ←→ C'	AC ≅ A'C' AB ≅ A'B' BC ≅ B'C'	∠BAC ≅ ∠B'A'C' ∠ABC ≅ ∠A'B'C' ∠ACB ≅ ∠A'C'B'
∴	From 1 and D5.1.27 Congruent	ΔABC ≅ ΔA'B'C'		

Theorem 5.6.11 Congruency: by Angle-Side-Side (ASS)

1g	Given	the above geometry figure F5.6.4 "Congruency of Triangles of the Second Kind"		
2g		construction lines CB and C'B'		
3g		A ←→ A' B ←→ B' C ←→ C'	AB ≅ A'B' BC ≅ B'C'	∠ACB ≅ ∠A'C'B'
Steps	Hypothesis	ΔABC ≅ ΔA'B'C'		
1	From 2g, 3g and T5.5.8 Angle-Side-Side (ASS)	A ←→ A' B ←→ B' C ←→ C'	AC ≅ A'C' AB ≅ A'B' BC ≅ B'C'	∠BAC ≅ ∠B'A'C' ∠ABC ≅ ∠A'B'C' ∠ACB ≅ ∠A'C'B'
∴	From 1 and D5.1.27 Congruent	ΔABC ≅ ΔA'B'C'		

Theorem 5.6.12 Congruency: by Angle-Angle-Side (AAS)

1g	Given	the above geometry figure F5.6.4 "Congruency of Triangles of the Second Kind"		
2g		construction lines CB and C'B'		
3g		B ←→ B' C ←→ C'	AB ≅ A'B'	∠ABC ≅ ∠A'B'C' ∠ACB ≅ ∠A'C'B'
Steps	Hypothesis	ΔABC ≅ ΔA'B'C'		
1	From 2g, 3g and T5.5.9 Angle-Angle- Side (AAS)	A ←→ A' B ←→ B' C ←→ C'	AC ≅ A'C' AB ≅ A'B' BC ≅ B'C'	∠BAC ≅ ∠B'A'C' ∠ABC ≅ ∠A'B'C' ∠ACB ≅ ∠A'C'B'
∴	From 1 and D5.1.27 Congruent	ΔABC ≅ ΔA'B'C'		

Theorem 5.6.13 Congruency: by Angle-Side-Angle (ASA)

1g	Given	the above geometry figure F5.6.4 "Congruency of Triangles of the Second Kind"		
2g		construction lines CB and C'B'		
3g		A ←→ A' B ←→ B' C ←→ C'	AB ≅ A'B'	∠BAC ≅ ∠B'A'C' ∠ABC ≅ ∠A'B'C'
Steps	Hypothesis	ΔABC ≅ ΔA'B'C'		
1	From 2g, 3g and T5.5.10 Angle-Side-Angle (ASA)	A ←→ A' B ←→ B' C ←→ C'	AC ≅ A'C' AB ≅ A'B' BC ≅ B'C'	∠BAC ≅ ∠B'A'C' ∠ABC ≅ ∠A'B'C' ∠ACB ≅ ∠A'C'B'
∴	From 1 and D5.1.27 Congruent	ΔABC ≅ ΔA'B'C'		

Theorem 5.6.14 Congruency: by Side-Angle-Angle (SAA)

1g	Given	the above geometry figure F5.6.4 "Congruency of Triangles of the Second Kind"		
2g		construction lines CB and C'B'		
3g		B ←→ B' C ←→ C'	BC ≅ B'C'	∠BAC ≅ ∠B'A'C' ∠ABC ≅ ∠A'B'C'
Steps	Hypothesis	ΔABC ≅ ΔA'B'C'		
1	From 2g, 3g and T5.5.11 Side-Angle- Angle (SAA)	A ←→ A' B ←→ B' C ←→ C'	AC ≅ A'C' AB ≅ A'B' BC ≅ B'C'	∠BAC ≅ ∠B'A'C' ∠ABC ≅ ∠A'B'C' ∠ACB ≅ ∠A'C'B'
∴	From 1 and D5.1.27 Congruent	ΔABC ≅ ΔA'B'C'		

Congruency of Triangles of the Third Kind:

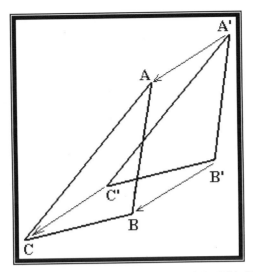

Figure 5.6.5 Congruency of Triangles of the Third Kind

Theorem 5.6.15 Congruent Triangles Projected on One Another

1g	Given	the above geometry figure F5.6.5 "Congruency of Triangles of the Third Kind"		
2g		A ≡ A'		
		B ≡ B'		
		C ≡ C'		
3g		ΔA'B'C' → ΔABC projected onto		
Steps	Hypothesis	ΔABC ≅ ΔA'B'C'		
1	From 2g and 3g	AB ≅ A'B' AC ≅ A'C BC ≅ B'C		
2	From 2g, 1 and T5.5.18 Congruency: by Side-Side-Side (SSS)	A ≡ A' B ≡ B' C ≡ C'	AB ≅ A'B' AC ≅ A'C BC ≅ B'C	∠CAB ≅ ∠C'A'B' ∠ABC ≅ ∠A'B'C' ∠BCA ≅ ∠B'C'A'
∴	From 2 and D5.1.27 Congruent	ΔABC ≅ ΔA'B'C'		

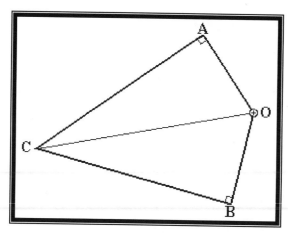

Figure 5.6.6 Common Hypotenuse-Leg (CHL)

Theorem 5.6.16 Congruency of Right Triangles by Common Hypotenuse-Leg (CHL)

1g	Given	the above geometry figure F5.6.6 "Common Hypotenuse-Leg (CHL)"		
2g		$90° \equiv \angle OAC \equiv \angle OBC$		
3g		$OA \equiv OB$		
Steps	Hypothesis	$\triangle AOC \cong \triangle BOC$	$\angle COA \cong \angle COB$	$\angle ACO \cong \angle BCO$
1	From 1g	OC common side to $\triangle AOC$ and $\triangle BOC$		
2	From 1, 2g, 3g and T5.6.9 Congruency: by Side-Side-Angle (SSA)	$\triangle AOC \cong \triangle BOC$		
3	From 2	$\angle COA \cong \angle COB$		
4	From 2	$\angle ACO \cong \angle BCO$		
∴	From 2, 3 and 4	$\triangle AOC \cong \triangle BOC$	$\angle COA \cong \angle COB$	$\angle ACO \cong \angle BCO$

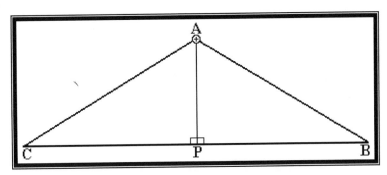

Figure 5.6.7 Common Leg-Leg (CLL)

Theorem 5.6.17 Congruency of Right Triangles by Common Leg-Leg (CLL)

1g	Given	the above geometry figure F5.6.7 "Common Leg-Leg (CLL)"		
2g		$90° \cong \angle APC \cong \angle APB$		
3g		$PC \cong PB$		
Steps	Hypothesis	$\triangle APC \cong \triangle APB$	$\angle ACP \cong \angle ABP$	$\angle CAP \cong \angle BAP$
1	From 1g	AP common side to $\triangle APC$ and $\triangle APB$		
2	From 1, 2g, 3g and T5.6.9 Congruency: by Side-Side-Angle (SSA)	$\triangle APC \cong \triangle APB$		
3	From 2	$\angle ACP \cong \angle ABP$		
4	From 2	$\angle CAP \cong \angle BAP$		
∴	From 2, 3 and 4	$\triangle APC \cong \triangle APB$	$\angle ACP \cong \angle ABP$	$\angle CAP \cong \angle BAP$

Tensor Calculus & Physics: A General Treatise

Section 5.7 Theorems on Similarity of Two Triangles

From the theorems on construction theorem T5.5.4 "Bi-Construction: Angle-Angle (AA)" leaves with an infinite possible family of triangles that have their internal angles invariant (constant) and from one triangle to another the angles are congruent. The best way to understand this particular geometry is to look at two concentric circles inscribing these triangles. Observe the relationships between corresponding sides as the isosceles triangles comprising the sides of the similar triangles, a long the radii, proportionally expand and contract relative to one another.

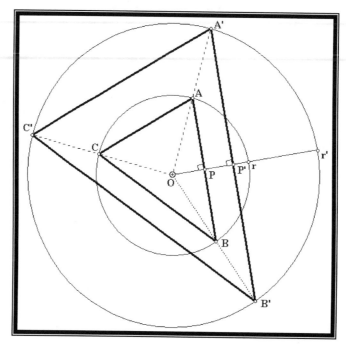

Figure 5.7.1 Proportional Triangles

Theorem T5.7.1 only proves sides are proportional for concentric triangles. It does not prove congruency of internal angles proving the triangles similar; hence, similar triangles have proportional sides!

Theorem 5.7.1 Similar Triangles: Proportional Triangles

1g	Given	the above geometry figure F5.7.1 "Proportional Triangles"		
2g		$\triangle ABC$ inscribed in circle-r		
3g		$\triangle A'B'C'$ inscribed in circle-r'		
4g		$\phi \cong \angle AOB \cong \angle A'OB'$		
5g		$k_\phi \cong r'/r = p_r$ scaling factor		
6g		$b \cong AP \cong PB$	$b' \cong A'P' \cong P'B'$	
7g		$\angle ABC \cong \angle A'B'C'$	$\angle BCA \cong \angle B'C'A'$	$\angle CAB \cong \angle C'A'B'$
Steps	Hypothesis	$k_\phi \equiv A'B' / AB \equiv B'C' / BC \equiv C'A' / CA$		$\triangle ABC \sim \triangle A'B'C'$
		$s'_\phi \equiv s_\phi$	$s'_\phi \equiv b'/r'$	$s_\phi \equiv b/r$
1	From 1g, 2g, 3g and D5.1.31 Circle	$r \cong AO \cong BO$		$r' \cong A'O \cong B'O$
2	From 4g, 1 and A5.2.15 GIV Transitivity of Segments	$k_\phi \equiv r'/r \equiv A'O / AO \equiv B'O / BO$		

3	From 1g and D5.1.29 Similar Polygons	$AB \longleftrightarrow \angle AOB$	$A'B' \longleftrightarrow \angle A'OB'$
4	From 4g, 3 and A5.2.18 GIV Transitivity of Angles	$AB \longleftrightarrow \angle A'OB'$	
5	From 3 and 4	$AB \longleftrightarrow A'B'$	
6	From 5 and D4.1.23C Ratio Proportional Quantities	$q_\phi \equiv A'B' / AB$	
7	From 1g and 6	$q_\phi \equiv (A'P' + P'B') / (AP + PB)$	
8	From 6g, 7, A5.2.15 GIV Transitivity of Segments and T4.3.10 Equalities: Summation of Repeated Terms by 2	$q_\phi \equiv 2b' / 2b$	
9	From 8, D4.1.4 Rational NumbersB, A4.2.10 Commutative Multp, T4.4.16A Equalities: Product by Division, Common Factor and T4.4.15 Equalities: Product by Division	$q_\phi \equiv b' / b$	
10	From 9 and T4.5.6 Equalities: Contraction of a Rational Chain	$q_\phi \equiv (b' / r') \, (r' / r) \, (r / b)$	
11	From 5g, 10, A4.2.10 Commutative Multp and A4.2.3 Substitution	$q_\phi \equiv (b' / r') \, (r / b) \, k_\phi$	
12	From 11 and T3.5.2 Equalities: Inverse of Reciprocal Products	$q_\phi \equiv (\, (b' / r') \, (r/r) / (b/r) \,) \, k_\phi$	
13	From 12, A4.2.13 Inverse Multp and T4.4.15A Equalities: Product by Division	$q_\phi \equiv (\, (b' / r') / (b / r) \,) \, k_\phi$	
14	From 13 and T4.4.4A Equalities: Right Cancellation by Multiplication	$q_\phi / k_\phi \equiv (b' / r') / (b / r)$	
15	From 14 and the right side is invariant with respect to ΔABC and $\Delta A'B'C'$ hence the left side is to:	$s'_\phi / s_\phi \equiv (b' / r') / (b / r)$ such that $\quad$ $s'_\phi \equiv b' / r'$ $\quad$ $s_\phi \equiv b / r$	
16	From 15 since s'_ϕ and s_ϕ are dependent only on ϕ and likewise invariant with respect to ΔABC and $\Delta A'B'C'$, hence	$s'_\phi \equiv s_\phi$	

17	From 16, A4.2.12 Identity Multp and T4.4.4 Equalities: Right Cancellation by Multiplication	$s'_\phi / s_\phi \equiv 1$		
18	From 14, 15, 17 and A4.2.3 Substitution	$q_\phi / k_\phi \equiv 1$		
19	From 18, T4.4.7 Equalities: Reversal of Left Cancellation by Multiplication and A4.2.12 Identity Multp	$q_\phi \equiv k_\phi$		
20	From 2, 6, 19 and A4.2.3 Substitution	$k_\phi \equiv r'/r \equiv A'O / AO \equiv B'O / BO \equiv A'B' / AB$		
21	From 1g and 20 it follows that if one side is proportional within the concentric circles that all sides must be:	$k_\phi \equiv A'B' / AB \equiv B'C' / BC \equiv C'A' / CA$		
∴	From 15, 16 and 21	$k_\phi \equiv A'B' / AB \equiv B'C' / BC \equiv C'A' / CA$ $s'_\phi \equiv s_\phi$	$s'_\phi \equiv b'/ r$	$\Delta ABC \sim \Delta A'B'C'$ $s_\phi \equiv b/ r$

Theorem 5.7.2 Similar Triangles: Converse Proportional Corresponding Sides

1g	Given	AB/A'B' = BC/B'C' = CA/C'A'		
Steps	Hypothesis	$\Delta ABC \sim \Delta A'B'C'$		
∴	From 1g and D5.1.29 Similar Polygons	$\Delta ABC \sim \Delta A'B'C'$		

Theorem 5.7.3 Similar Triangles: Corresponding Congruent Angles

1g	Given	$\Delta ABC \sim \Delta A'B'C'$		
Steps	Hypothesis	$\angle ABC \cong \angle A'B'C'$	$\angle BAC \cong \angle B'A'C'$	$\angle BCA \cong \angle B'C'A'$
∴	From 1g and D5.1.29 Similar Polygons	$\angle ABC \cong \angle A'B'C'$	$\angle BAC \cong \angle B'A'C'$	$\angle BCA \cong \angle B'C'A'$

Theorem 5.7.4 Similar Triangles: Converse Corresponding Congruent Angles

1g	Given	$\angle ABC \cong \angle A'B'C'$	$\angle BAC \cong \angle B'A'C'$	$\angle BCA \cong \angle B'C'A'$
Steps	Hypothesis	$\Delta ABC \sim \Delta A'B'C'$		
∴	From 1g and D5.1.29 Similar Polygons	$\Delta ABC \sim \Delta A'B'C'$		

Theorem 5.7.5 Similar Triangles: Proportional Sides Corresponding to Opposite Angles

1g	Given	$\triangle ABC \sim \triangle A'B'C'$
Steps	Hypothesis	AB/A'B' $\leftrightarrow$ $\angle BCA \cong \angle B'C'A'$
		BC/B'C' $\leftrightarrow$ $\angle BAC \cong \angle B'A'C'$
		CA/C'A' $\leftrightarrow$ $\angle ABC \cong \angle A'B'C'$
$\therefore$	From 1g, T5.7.1 Similar Triangles: Proportional Triangles and T5.7.3 Similar Triangles: Corresponding Congruent Angles	AB/A'B' $\leftrightarrow$ $\angle BCA \cong \angle B'C'A'$ BC/B'C' $\leftrightarrow$ $\angle BAC \cong \angle B'A'C'$ CA/C'A' $\leftrightarrow$ $\angle ABC \cong \angle A'B'C'$

Actually, theorem T5.7.5 is a property of the geometry of similar triangles and should be an axiom of the Hilbert's system.

When Euclid developed his theorems on congruency, specifically Theorem T5.6.13 "Euclid's Proof of (SAS) [HEA56, Vol. I, pg 247]", he sets vertex-to-vertex introducing the notion of correspondence. With his Definition D5.1.28 "Similar Polygons" the idea of corresponding ratio of sides to opposing angles is introduced. Following this, Hipparchus's at Alexandria (146 BCE) [EVE76, pg 143] introduces the idea of *sine* in his table of corresponding cord to the central angle, which is computed from a circularly inscribed isosceles triangle. This predates, but alludes to trigonometry and sets the stage for it to evolve into its formal form as found in the 17th century some 1800 years later and as we use it today.

The following theorem is not a true theorem of the geometric axiomatic system, because it uses trigonometry an external and later discipline of mathematics. Its use is to confer justification or instill confidence in setting up the correspondence in theorem T5.7.5 "Similar Triangles: Proportional Sides Corresponding to Opposite Angles" by interjecting techniques that clearly show the relationship between sides and angles. While the Law of Sine's is at its core, which in tern is based on the similarity of right triangles, in its general development it deals with the single triangle and never uses in its proof any of the theorems on similarity. Hence using it is not a circular digression, but a modern proof for this time demonstrating the proportional relationship of angle and side.

Theorem 5.7.6 Similar Triangles: Corresponding Congruent Angles by Trigonometry

1g	Given	$\triangle ABC \sim \triangle A'B'C'$		
2g	Law of Sins	$\sin(\angle ABC) / |AC| \equiv \sin(\angle BCA) / |AB| \equiv \sin(\angle CAB) / |BC|$		
Steps	Hypothesis	$\angle ABC \cong \angle A'B'C'$	$\angle BAC \cong \angle B'A'C'$	$\angle BCA \cong \angle B'C'A'$
1	From 1g and 2g	$\sin(\angle ABC) / |AC| \equiv \sin(\angle BCA) / |AB| \equiv$ $\sin(\angle CAB) / |CB|$		for $\triangle ABC$
2	From 1g and 2g	$\sin(\angle A'B'C') / |A'C'| \equiv \sin(\angle B'C'A') / |A'B'| \equiv$ $\sin(\angle C'A'B') / |C'B'|$		for $\triangle A'B'C'$
3	From 1 and 2 dividing equalities	$[\sin(\angle ABC) / \sin(\angle A'B'C')][|A'C'| / |AC|] \equiv$ $[\sin(\angle BCA) / \sin(\angle B'C'A')][|A'B'| / |AB|] \equiv$ $[\sin(\angle CAB) / \sin(\angle C'A'B')][|C'B'| / |CB|]$		
4	From 1g, 3 and T5.7.1 Similar Triangles: Proportional Triangles	$[\sin(\angle ABC) / \sin(\angle A'B'C')][|A'C'| / |AC|] \equiv |A'C'| / |AC| \equiv$ $[\sin(\angle BCA) / \sin(\angle B'C'A')][|A'B'| / |AB|] \equiv |A'B'| / |AB| \equiv$ $[\sin(\angle CAB) / \sin(\angle C'A'B')][|C'B'| / |CB|] \equiv |C'B'| / |CB|$		

| 5 | From 4 and T4.4.3 Equalities: Cancellation by Multiplication (dividing both sides by proportional segmented sides) | $\sin(\angle ABC) / \sin(\angle A'B'C') \equiv$ $\sin(\angle BCA) / \sin(\angle B'C'A') \equiv$ $\sin(\angle CAB) / \sin(\angle C'A'B') \equiv 1$ | | |
| ∴ | From 5 and reaffirming T5.7.3 Similar Triangles: Corresponding Congruent Angles | $\angle ABC \cong \angle A'B'C'$ | $\angle BAC \cong \angle B'A'C'$ | $\angle BCA \cong \angle B'C'A'$ |

Theorem 5.7.7 Similar Triangles: Proportional Side-Side-Angle (PSSA)

If two triangles have two proportional sides and a set of congruent angles, then the triangles are similar.

1g 2g 3g	Given	ΔABC and $\Delta A'B'C'$ $\angle ABC \cong \angle A'B'C'$ $C'A' / CA \equiv A'B' / AB$
Steps	Hypothesis	$\Delta ABC \sim \Delta A'B'C'$
1	From 2g and 3g	$CA/C'A' \longleftrightarrow \angle ABC \cong \angle A'B'C'$
2	From 3g and T5.7.5 Similar Triangles: Proportional Sides Corresponding to Opposite Angles	$AB/A'B' \longleftrightarrow \angle BCA \cong \angle B'C'A'$
∴	From 1, 2 and T5.5.13 Similar Triangles: Angle-Angle (AA)	$\Delta ABC \sim \Delta A'B'C'$

Theorem 5.7.8 Similar Triangles: Proportional Side-Side-Side (PSSS)

If two triangles have their corresponding sides proportional, then they are similar.

1g 2g	Given	ΔABC and $\Delta A'B'C'$ $AB/A'B' = BC/B'C' = CA/C'A'$
Steps	Hypothesis	$\Delta ABC \sim \Delta A'B'C'$
∴	From 1g, 2g, select any pair of proportional sides and by T5.7.2 Similar Triangles: Converse Proportional Corresponding Sides	$\Delta ABC \sim \Delta A'B'C'$

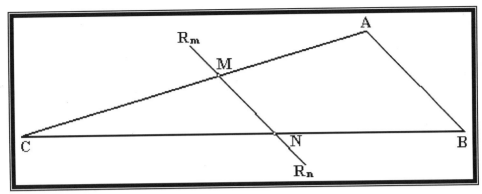

Figure 5.7.2 Similar Common Vertex with Two Sides Parallel (SCVTSP)

Theorem 5.7.9 Similar Common Vertex with Two Sides Parallel (SCVTSP)

1g	Given	the above geometry figure F5.7.2 "Similar Common Vertex with Two Sides Parallel (SCVTSP)"
2g		∠ACB ≅ ∠MCN
3g		∠BAC ≅ ∠NMC
4g		∠ABC ≅ ∠MNC
Steps	Hypothesis	AB ∥ MN for ΔABC ~ ΔMNC with common vertex-C
1	From 1g and T5.4.1 Intersection of Two Lines: Opposite Congruent Vertical Angles	∠NMC ≅ ∠AMR$_m$
2	From 1g and T5.4.1 Intersection of Two Lines: Opposite Congruent Vertical Angles	∠MNC ≅ ∠BNR$_n$
3	From 3g and 1	∠BAC ≅ ∠AMR$_m$
4	From 4g and 2	∠ABC ≅ ∠BNR$_n$
5	From 3, 4 and T5.4.4 Parallel Lines: Congruent Alternate Interior Angles	AB ∥ MN for ΔABC and ΔMNC
∴	From 2g, 3g and T5.5.13 Similar Triangles: Angle-Angle (AA)	AB ∥ MN for ΔABC ~ ΔMNC with common vertex-C

Theorem 5.7.10 Similar Common Vertex with Proportional Sides (STPS)

1g	Given	the above geometry figure F5.7.2 "Similar Common Vertex with Two Sides Parallel (SCVTSP)"
2g		AB ∥ MN for ΔABC ~ ΔMNC with common vertex-C
Steps	Hypothesis	AC / MC = AB / MN = BC / NC
∴	From 2g and T5.7.1 Similar Triangles: Proportional Triangles	AC / MC = AB / MN = BC / NC

Section 5.8 Theorems on Right Triangles

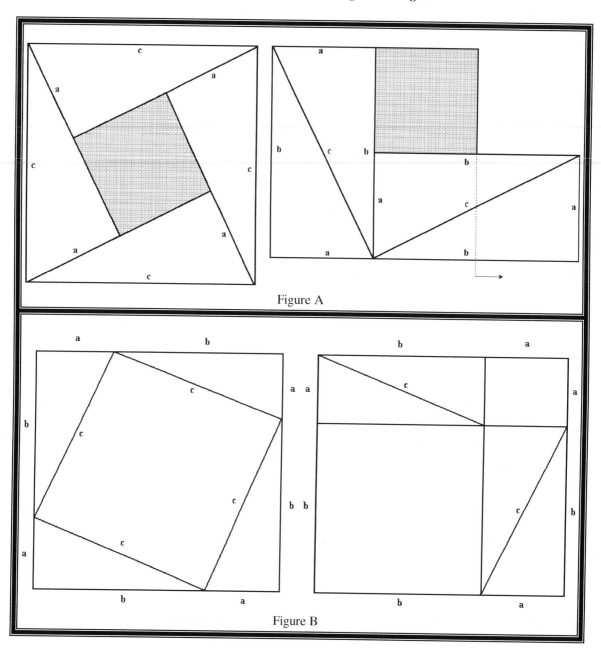

Figure A

Figure B

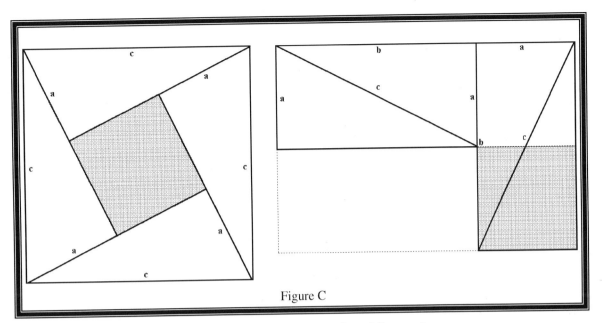

Figure C

Figure 5.8.1 Pythagorean Dissection of Square Area

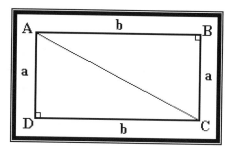

Figure 5.8.2 Area of a Right Triangle

Theorem 5.8.1 Right Triangle: Algebraic Area

1g	Given	the above geometry from figure F5.8.2 "Area of a Right Triangle"
Steps	Hypothesis	Area K of a right triangle $\equiv K_{rt} \equiv \frac{1}{2}ab$
1	From 1g and D5.1.23 Rectangle	$a \equiv AC \equiv BC$ $\qquad\qquad$ $b \equiv AB \equiv CD$
2	From 1g	AC the diagonal is common to $\triangle ABC$ and $\triangle ADC$
3	From 1, 2 and T5.5.8 Congruency: by Side-Side-Side (SSS)	$\triangle ABC \cong \triangle ADC$ dividing the rectangle ABCD into equal parts
∴	From 3 and D5.1.49 Area of a rectangle	Area K of a triangle $\equiv K_{rt} \equiv \frac{1}{2}ab$

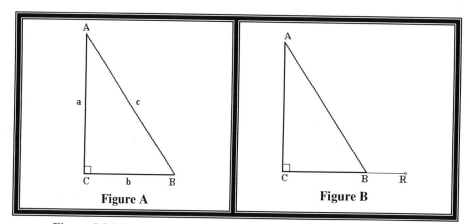

Figure 5.8.3 Trigonometric Area and External Angle Relationships

Theorem 5.8.2 Right Triangle: Trigonometric Area by Sine

1g	Given	the above geometry, figure F5.8.3A "Trigonometric Area and External Angle Relationships"
2g		$a \equiv |AC| \equiv c \sin \angle ABC$
3g		$b \equiv |BC|$
4g		$c \equiv |AB|$
Steps	Hypothesis	$K_{rt} \equiv \frac{1}{2}|AB||BC| \sin \angle ABC$
1	From 1g, 2g, and 4g	$a \equiv c \sin \angle ABC = |AB| \sin \angle ABC$
2	From 1, 3g and T5.8.1 Right Triangle: Algebraic Area	$K_{rt} \equiv \frac{1}{2}|AB| \sin \angle ABC |BC|$
∴	From 2	$K_{rt} \equiv \frac{1}{2}|AB||BC| \sin \angle ABC$

Theorem 5.8.3 Pythagorean Theorem A

For a right triangle with legs [a] and [b] the opposite side [c] is called the hypotenuse and related to the other legs by the formula $c^2 = a^2 + b^2$. While it is not known how Pythagoras proved his general theorem, it is thought it may have followed the dissection of square area type proof using ceramic tiles as demonstrated by the above two geometries in figures A and B, remember he did not know algebra. Here four identical right triangles of general type are used and than rearranged in various configurations to visually demonstrate the square area of the hypotenuse is the same as the area of the sum of the squares of the legs of the triangles. The three different proofs of dissection are possible ways Pythagoras might have solved the problem though theorem T5.8.3 "Pythagorean theorem A" is the simplest and easiest to divine.

1g	Given	the above geometry, figure F5.8.1A "Pythagorean Dissection of Square Area"[BRO73 pg 158]
Steps	Hypothesis	$c^2 = a^2 + b^2$
1	From 1g and D5.1.49 Area of a rectangle	area to the left = c x c
2	From 1g	area to the right = 4 area of right triangles + area of center square
3	From 2, T5.8.1 Right Triangle: Algebraic Area and D5.1.49 Area of a rectangle	area to the right = 4 (½a x b) + (b – a) x (b – a)
4	From 3, A4.2.14 Distribution T4.12.2 Polynomial Quadratic: The Perfect Square by Difference and A4.2.5 Commutative Add	area to the right = 2 a x b – 2a x b + b x b + a x a
5	From 4, A4.2.3 Inverse Add and A4.2.7 Identity Add	area to the right = b x b + a x a
6	From 1, 5, A4.2.3 Substitution and area conserved	c x c = b x b + a x a area has neither been added or subtracted hence it is conserved.
∴	From 8, D4.1.17 Exponential Notation and A4.2.5 Commutative Add	$c^2 = a^2 + b^2$

Tensor Calculus & Physics: A General Treatise

Theorem 5.8.4 Pythagorean Theorem B

1g	Given	the above geometry, figure F5.8.1B "Pythagorean Dissection of Square Area" [EVE76, pg 62]
Steps	Hypothesis	$c^2 = a^2 + b^2$
1	From 1g	Area to the left = c x c + 4 area of right triangles
2	From 1 and T5.8.1 Right Triangle: Algebraic Area	Area to the left = c x c + 4 (½a x b)
3	From 2, D4.1.19 Primitive Definition for Rational Arithmetic and A4.2.11 Associative Multp	Area to the left = c x c + 2 (2 (½a x b))
4	From 3 and T4.4.18A Equalities: Product by Division, Factorization by 2	Area to the left = c x c + 2 a x b
5	From 1g	Area to the right = a x a + b x b + 2 area of rectangle
6	From 5 and D5.1.49 Area of a rectangle	Area to the right = a x a + b x b + 2 a x b
7	From 4, 6, A4.2.3 Substitution and area conserved	c x c + 2 a x b = a x a + b x b + 2 a x b area has neither been added or subtracted hence it is conserved.
∴	From 7, D4.1.17 Exponential Notation and T4.3.2 Equalities: Cancellation by Addition	$c^2 = a^2 + b^2$

Theorem 5.8.5 Pythagorean Theorem C

1g	Given	the above geometry, figure F5.8.1C "Pythagorean Dissection of Square Area"
Steps	Hypothesis	$c^2 = a^2 + b^2$
1	From 1g	area to the left = c x c
2	From 1g	area to the right = total rectangle area – difference of center area
3	From 1, D5.1.49 Area of a rectangle	area to the right = b(a + b) – a(b – a)
4	From 3 and A4.2.14 Distribution	area to the right = b x a + b x b – a x b + a x a
5	From 4, A4.2.3 Inverse Add and A4.2.7 Identity Add	area to the right = b x b + a x a
6	From 1, 5 and A4.2.3 Substitution	c x c = b x b + a x a area has neither been added or subtracted hence it is conserved.
∴	From 7, D4.1.17 Exponential Notation and A4.2.5 Commutative Add	$c^2 = a^2 + b^2$

Tensor Calculus & Physics: A General Treatise

Theorem 5.8.6 Right Triangle: Non-Ninety Degree Angles are Complementary

1g	Given	the geometry of figure F5.8.3A "Trigonometric Area and External Angle Relationships"
Steps	Hypothesis	$90° \cong \angle CAB + \angle ABC$ are Complementary Angles
1	From 1g and D5.1.19 Triangle; Right	$90° \cong \angle BCA$
2	From 1g and T5.4.5 Sum of Internal Angles for a Triangle	$180° \cong \angle CAB + \angle ABC + \angle BCA$
3	From 1, 2 and A5.2.18 GIV Transitivity of Angles	$180° \cong \angle CAB + \angle ABC + 90°$
∴	From 3, T4.3.3 Equalities: Right Cancellation by Addition, D4.1.19 Primitive Definition for Rational Arithmetic and D5.1.11 Complementary Angles	$90° \cong \angle CAB + \angle ABC$ are Complementary Angles

Theorem 5.8.7 Right Triangle: External Angle Relationships

1g	Given	the above geometry, figure F5.8.3B "Trigonometric Area and External Angle Relationships"
Steps	Hypothesis	$\angle ABR \cong 90° + \angle CAB \cong 180° - \angle ABC$ $\cong 135° + ½(\angle CAB - \angle ABC)$
1	From 1g and T5.4.8 External Angle Supplementary of two Opposing Vertices	$\angle ABR \cong 90° + \angle CAB$
2	From 1g, D5.1.8 Flat Angle and T4.3.3 Equalities: Right Cancellation by Addition	$\angle ABR \cong 180° - \angle ABC$
3	From 1, 2, T4.3.1 Equalities: Uniqueness of Addition, T4.3.10 Equalities: Summation of Repeated Terms by 2 and D4.1.19 Primitive Definition for Rational Arithmetic	$2\angle ABR \cong 270° + \angle CAB - \angle ABC$
4	From 3, A4.2.2 Equality, T4.4.2 Equalities: Uniqueness of Multiplication, A4.2.14 Distribution and D4.1.19 Primitive Definition for Rational Arithmetic	$\angle ABD \cong 135° + ½(\angle CAB - \angle ABC)$
∴	From 1, 2 and 4	$\angle ABD \cong 90° + \angle CAB \cong 180° - \angle ABC$ $\cong 135° + ½(\angle CAB - \angle ABC)$

Theorem 5.8.8 Right Triangle: The Hypotenuse

Steps	Hypothesis	
	Hypothesis	$h = \pm\sqrt{a^2 + b^2}$
1	From T5.8.3 Pythagorean Theorem A	$h^2 = a^2 + b^2$
∴	From 1 and T4.10.7 Radicals: Reciprocal Exponent by Positive Square Root	$h = \pm\sqrt{a^2 + b^2}$

Theorem 5.8.9 Right Triangle: The Altitude

Steps	Hypothesis	
	Hypothesis	$a = \pm\sqrt{h^2 - b^2}$
1	From T5.8.3 Pythagorean Theorem A	$h^2 = a^2 + b^2$
2	From 1 and T4.3.3B Equalities: Right Cancellation by Addition	$a^2 = h^2 - b^2$
∴	From 2 and T4.10.7 Radicals: Reciprocal Exponent by Positive Square Root	$a = \pm\sqrt{h^2 - b^2}$

Theorem 5.8.10 Right Triangle: The Base

Steps	Hypothesis	
	Hypothesis	$b = \pm\sqrt{h^2 - a^2}$
1	From T5.8.3 Pythagorean Theorem A	$h^2 = a^2 + b^2$
2	From 1 and T4.3.3B Equalities: Right Cancellation by Addition	$b^2 = h^2 - a^2$
∴	From 2 and T4.10.7 Radicals: Reciprocal Exponent by Positive Square Root	$b = \pm\sqrt{h^2 - a^2}$

Theorem 5.8.11 Right Triangle: Ratio of Legs to the Hypotenuse

Steps	Hypothesis	
	Hypothesis	$1 = (a/h)^2 + (b/h)^2$
1	From T5.8.3 Pythagorean Theorem A	$h^2 = a^2 + b^2$
2	From 1, A4.2.2 Equality, T4.4.2 Equalities: Uniqueness of Multiplication and A4.2.14 Distribution	$(1/h^2)\, h^2 = (1/h^2)\, a^2 + (1/h^2)\, b^2$
3	From 2, T4.4.16B Equalities: Product by Division, Common Factor and A4.2.12 Identity Multp	$1 = a^2/h^2 + b^2/h^2$
∴	From 3 and T4.8.7 Integer Exponents: Distribution Across a Rational Number	$1 = (a/h)^2 + (b/h)^2$

Theorem 5.8.12 Right Triangle: Ratio of Altitude to the Hypotenuse

Steps	Hypothesis	$a/h = \pm\sqrt{1 - (b/h)^2}$
1	From T5.8.11 Right Triangle: Ratio of Legs to the Hypotenuse	$1 = (a/h)^2 + (b/h)^2$
2	From 1 and T4.3.3B Equalities: Right Cancellation by Addition	$(a/h)^2 = 1 - (b/h)^2$
∴	From 2 and T4.10.7 Radicals: Reciprocal Exponent by Positive Square Root	$a/h = \pm\sqrt{1 - (b/h)^2}$

Theorem 5.8.13 Right Triangle: Ratio of Base to the Hypotenuse

Steps	Hypothesis	$b/h = \pm\sqrt{1 - (a/h)^2}$
1	From T5.8.11 Right Triangle: Ratio of Legs to the Hypotenuse	$1 = (a/h)^2 + (b/h)^2$
2	From 1 and T4.3.3B Equalities: Right Cancellation by Addition	$(b/h)^2 = 1 - (a/h)^2$
∴	From 2 and T4.10.7 Radicals: Reciprocal Exponent by Positive Square Root	$b/h = \pm\sqrt{1 - (a/h)^2}$

Section 5.9 Theorems on Isosceles Triangles

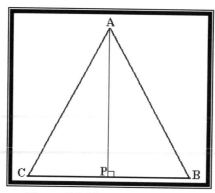

Figure 5.9.1 Isosceles Triangle

Given 5.9.1 Isosceles Triangle: Parameters

1g	Given	$\triangle ABC$ the above geometry figure F5.9.1 "Isosceles Triangle"	
2g		$r \equiv AC \cong AB$	hypotenuse of a right triangles
3g		$\theta \equiv \angle BAC \cong \angle CAB$	opposing vertex angle
4g		$\alpha \equiv \angle BAP \cong \angle CAP \equiv \tfrac{1}{2}\theta$	half opposing vertex angles
5g		$\beta \equiv \angle ABP \cong \angle ACP$	base vertex angles
6g		$\gamma \equiv \angle BPA$	perpendicular angle
7g		$a \equiv AP$	altitude of a right triangle
8g		$b \equiv BP \cong PC$	base of a right triangle
9g		$\triangle APB, \triangle APC$	right triangles

Theorem 5.9.1 Isosceles Triangle: Bisecting Vertex Between Congruent Sides Parses Equal Parts

Steps	Hypothesis	$\triangle APB \cong \triangle APC$	$\angle ABP \cong \angle ACP$	$\angle BPA \cong \angle CPA$
1	From G5.9.1-1g	AP common to $\triangle APB$ and $\triangle APC$		
∴	From G5.9.1-2g, 4g, 1 and T5.5.10 Congruency: by Side-Angle-Side (SAS)	$\triangle APB \cong \triangle APC$	$\angle ABP \cong \angle ACP$	$\angle BPA \cong \angle CPA$

Theorem 5.9.2 Isosceles Triangle: Bisected into Two Congruent Right Triangles

Steps	Hypothesis	$\Delta APB \cong \Delta APC$ Right Triangles	$\beta \equiv \angle ABP \cong \angle ACP \equiv 90° - \frac{1}{2}\theta$
1	From G5.9.1-5g, 6g and T5.9.1 Isosceles Triangle: Bisecting Vertex Between Congruent Sides Parses Equal Parts	$\beta \equiv \angle ABP \cong \angle ACP$	$\gamma \equiv \angle BPA \cong \angle CPA$
2	From G5.9.1-1g, T5.4.6 Solving for a Specific Summation Angle of a Triangle	$\angle BAC \cong 180° - (\angle ABP + \angle ACP)$ for $(\Delta ABC, \angle BAC)$	
3	From G5.9.1-3g, 2, A5.2.18 GIV Transitivity of Angles, T4.3.10 Equalities: Summation of Repeated Terms by 2	$\theta \equiv 180° - 2\beta$	
4	From 3, A4.2.2 Equality, A4.2.14 Distribution, T4.4.16B Equalities: Product by Division, Common Factor, T4.4.15B Equalities: Product by Division and D4.1.19 Primitive Definition for Rational Arithmetic	$\frac{1}{2}\theta \equiv 90° - \beta$	
5	From 4, T4.3.4 Equalities: Reversal of Right Cancellation by Addition and T4.3.5B Equalities: Left Cancellation by Addition	$\beta \equiv 90° - \frac{1}{2}\theta$	
6	From G5.9.1-1g and T5.4.5 Sum of Internal Angles for a Triangle	$180° \equiv \angle PAB + \angle ABP + \angle BPA$	
7	From G5.9.1-4g, 1 and A5.2.18 GIV Transitivity of Angles	$180° \equiv \frac{1}{2}\theta + \beta + \gamma$ for $\Delta APB \cong \Delta APC$	
8	From 5, 7, A5.2.18 GIV Transitivity of Angles, A4.2.8 Inverse Add and A4.2.7 Identity Add	$180° \equiv 90° + \gamma$	
9	From 8, T4.3.3 Equalities: Right Cancellation by Addition, and D4.1.19 Primitive Definition for Rational Arithmetic and A4.2.B Equality 2	$\gamma \equiv 90°$	
$\therefore$	From 5, 9 and D5.1.19 Triangle; Right	$\Delta APB \cong \Delta APC$ Right Triangles	$\beta \equiv \angle ABP \cong \angle ACP \equiv 90° - \frac{1}{2}\theta$

Theorem 5.9.3 Isosceles Triangle: Perpendicular Bisects Opposite Vertex

Steps	Hypothesis	$\perp$ to BC bisects the opposite vertex of $\triangle$ABC.
$\therefore$	From T5.9.2 Isosceles Triangle: Bisected into Two Congruent Right Triangles	hence base angles of right triangle are $\perp$ to BC, and the two opposite vertices are equal, thereby bisecting the opposite vertex $\triangle$ABC.

Theorem 5.9.4 Isosceles Triangle: Congruent Interior Angles and Sides

Steps	Hypothesis	$\triangle$ABC isosceles triangle	$\angle$BAP $\cong$ $\angle$CAP	AB $\cong$ AC
1	From G5.9.1-1g	AP common side to $\triangle$APB and $\triangle$APC		
2	From G5.9.1-4g	$\angle$BAP $\cong$ $\angle$CAP for $\triangle$ABC		
3	From G5.9.1-7g	AP altitude for AP $\perp$ BC, hence $\angle$APB $\cong$ $\angle$APC $\cong$ 90°		
4	From 1, 2, 3 and T5.6.12 Congruency: by Angle-Angle-Side (AAS)	$\triangle$APB $\cong$ $\triangle$APC	AB $\cong$ AC	
$\therefore$	From 5 and D5.1.16 Triangle; Isosceles	$\triangle$ABC isosceles triangle	$\angle$BAP $\cong$ $\angle$CAP	AB $\cong$ AC

Theorem 5.9.5 Isosceles Triangle: Right Triangles; Complementary Angles

Steps	Hypothesis	$\alpha + \beta \equiv \angle$BAP + $\angle$ABP $\equiv \angle$CAP + $\angle$ACP = 90° interior angles are complementary
1	From G5.9.1-1g and T5.9.2 Isosceles Triangle: Bisected into Two Congruent Right Triangles	$\alpha + \beta \equiv \angle$BAP + $\angle$ABP $\equiv \angle$CAP + $\angle$ACP = ½θ + 90° − ½θ
$\therefore$	From 1 and D5.1.11 Complementary Angle	$\alpha + \beta \equiv \angle$BAP + $\angle$ABP $\equiv \angle$CAP + $\angle$ACP = 90° interior angles are complementary

Theorem 5.9.6 Isosceles Triangle: Right Triangles Bisect Vertex Angle

Steps	Hypothesis	$\angle$CAP $\cong$ $\angle$BAP $\cong$ ½θ
1	From G5.9.1-1g and 2	$\angle$BAC $\cong$ $\angle$CAP + $\angle$BAP
2	From G5.9.1-3g, 4g and A5.2.18 GIV Transitivity of Angles	$\theta \cong \angle$BAP + $\angle$BAP
3	From 2 and T4.3.10 Equalities: Summation of Repeated Terms by 2	$\theta \cong 2\angle$BAP
4	From 3, A4.2.2A Equality, T4.4.16B Equalities: Product by Division, Common Factor and A4.2.2B Equality	$\angle$BAP $\cong$ ½θ
$\therefore$	From 4g, 4 and A5.2.18 GIV Transitivity of Angles	$\angle$CAP $\cong$ $\angle$BAP $\cong$ ½θ

Theorem 5.9.7 Isosceles Triangle: Constructed from Perpendicular Bisector

1g	Given	AP perpendicular bisector to BC	
Steps	Hypothesis	ΔABC isosceles triangle for perpendicular bisector AP	
1	From 1g and construct	ΔAPC and ΔAPB with AB and AC	
2	From 1g and 1	AP is a common side of ΔAPC and ΔAPB	
3	From 1g	BP ≅ CP	
4	From 1g	∠APC ≅ ∠APB ≅ 90°	
5	From 2, 3, 4 and T5.6.11 Congruency: by Angle-Side-Side (ASS)	ΔAPC ≅ ΔAPB	AB ≅ AC
∴	From 5 and D5.1.16 Triangle; Isosceles	ΔABC isosceles triangle for perpendicular bisector AP	

Theorem 5.9.8 Isosceles Triangle: Trigonometry; Base Length

Steps	Hypothesis	BC ≅ 2r sin ½θ
1	From G5.9.1-1g and T5.9.2 Isosceles Triangle: Bisected into Two Congruent Right Triangles	ΔAPB ≅ ΔAPC are right triangles
2	From 1 and D5.1.27 Congruent	PB ≅ PC
3	From 2 and DxE.1.1.1	AB sin ∠BAP ≡ AC sin ∠CAP
4	From G5.9.1-2g, G5.9.1-4g, 1, 2, 3, A5.2.16 GIV Transitivity of Segments and A5.2.18 GIV Transitivity of Angles	PB ≅ PC ≡ r sin ½θ
5	From G5.9.1-1g	BC ≅ PB + PC
∴	From 4, 5, A5.2.16 GIV Transitivity of Segments and T4.3.10 Equalities: Summation of Repeated Terms by 2	BC ≅ 2r sin ½θ

Theorem 5.9.9 Isosceles Triangle: Hypotenuse 2 Base with Base Angle

Steps	Hypothesis	$BC \cong 2r \cos(\beta)$
1	From T5.9.5 Isosceles Triangle: Right Triangles; Complementary Angles	$\alpha + \beta \cong 90°$
2	From G5.9.1-4g, 1 and A5.2.18 GIV Transitivity of Angles	$\tfrac{1}{2}\theta + \beta \cong 90°$
3	From 2 and T4.3.5B Equalities: Left Cancellation by Addition	$\tfrac{1}{2}\theta \cong 90° - \beta$
4	From T5.9.8 Isosceles Triangle: Trigonometry; Base Length and A5.2.18 GIV Transitivity of Angles	$BC \cong 2r \sin(90° - \beta)$
∴	From DxE.1.7.3	$BC \cong 2r \cos(\beta)$

Theorem 5.9.10 Isosceles Triangle: Base 2 Hypotenuse with Base Angle

Steps	Hypothesis	$r \cong BC / 2\cos(\beta)$
1	From T5.9.10 Isosceles Triangle: Hypotenuse 2 Base	$BC \cong 2r \cos(\beta)$
∴	From A4.2.10 Commutative Multp and T4.4.12 Equalities: Reversal of Cross Product of Proportions	$r \cong BC / 2\cos(\beta)$

Theorem 5.9.11 Isosceles Triangle: Pythagorean; Base Length

Steps	Hypothesis	$BC \cong 2\sqrt{r^2 - a^2}$
1	From G5.9.1-1g and T5.9.2 Isosceles Triangle: Bisected into Two Congruent Right Triangles	$\triangle APB \cong \triangle APC$ are right triangles
2	From G5.9.1-2g and 1	$r = AB \cong AC$
3	From 1 and D5.1.27 Congruent	$PB \cong PC$
4	From 1, G5.9.1-2g, G5.9.1-7g, T5.8.3 Pythagorean Theorem A and A5.2.15 GIV Transitivity of Segments	$PB \cong PC \cong \sqrt{r^2 - a^2}$
5	From G5.9.1-1g	$BC \cong PB + PC$
∴	From From 4, 5, A5.2.16 GIV Transitivity of Segments and T4.3.10 Equalities: Summation of Repeated Terms by 2	$BC \cong 2\sqrt{r^2 - a^2}$

Theorem 5.9.12 Isosceles Triangle: Area

Steps	Hypothesis	$K_{\Delta ABC} = ab$	
1	From T5.9.2 Isosceles Triangle: Bisected into Two Congruent Right Triangles	$\Delta APB \cong \Delta APC$ Right Triangles	
2	From G5.9.1-1g	$K_{\Delta ABC} = K_{\Delta APB} + K_{\Delta APC}$	
3	From 7g, 8g, T5.8.1 Right Triangle: Algebraic Area	$K_{\Delta APB} = \frac{1}{2}ab$	$K_{\Delta APC} = \frac{1}{2}ab$
∴	From 2, 3, A4.2.3 Substitution, T4.3.10 Equalities: Summation of Repeated Terms by 2 and T4.4.16A Equalities: Product by Division, Common Factor	$K_{\Delta ABC} = ab$	

Section 5.10 Theorems on Properties of Triangles

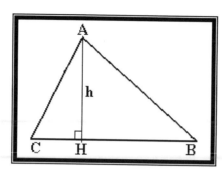

Figure 5.10.1 Trigonometric Area of a Triangle

Theorem 5.10.1 Trigonometric: Area of a Triangle

1g	Given	the above geometry figure F5.10.1 "Trigonometric Area of a Triangle"
2g		$h_a \cong AH_a$ perpendicular from $\angle BAC$ to BC
3g		$h_b \cong AH_b$ perpendicular from $\angle ABC$ to AC
4g		$h_c \cong AH_c$ perpendicular from $\angle ACB$ to AB
5g		$BC \cong BH_a + H_aC$
6g		$AC \cong AH_b + H_bC$
7g		$AB \cong AH_c + H_cB$
Steps	Hypothesis	$K = \tfrac{1}{2}\lvert AC\rvert\lvert CB\rvert \sin(\angle ACB) = \tfrac{1}{2}\lvert BA\rvert\lvert AC\rvert \sin(\angle BAC)$ $= \tfrac{1}{2}\lvert BC\rvert\lvert AB\rvert \sin(\angle ABC)$ for trigonometric area of $\triangle ABC$
1	From 1g	$K = K_{rt}(\triangle BH_aA) + K_{rt}(\triangle AH_aC)$
2	From 1g, 2g, T5.8.2 Right Triangle: Trigonometric Area by Sine, A4.2.3 Substitution and A5.2.15 GIV Transitivity of Segments	$K = \tfrac{1}{2}\lvert BH_a\rvert\lvert AC\rvert \sin(\angle ACB) + \tfrac{1}{2}\lvert H_aC\rvert\lvert AC\rvert \sin(\angle ACB)$
3	From 5g, 2, A4.2.14 Distribution and A5.2.15 GIV Transitivity of Segments	$K = \tfrac{1}{2}\lvert AC\rvert\lvert CB\rvert \sin(\angle ACB)$ for area of $\triangle ABC$
4	From 1g	$K = K_{rt}(\triangle AH_bB) + K_{rt}(\triangle BH_bC)$
5	From 1g, 3g, T5.8.2 Right Triangle: Trigonometric Area by Sine, A4.2.3 Substitution and A5.2.15 GIV Transitivity of Segments	$K = \tfrac{1}{2}\lvert AH_b\rvert\lvert AB\rvert \sin(\angle BAC) + \tfrac{1}{2}\lvert H_bC\rvert\lvert AB\rvert \sin(\angle BAC)$
6	From 6g, 5, A4.2.14 Distribution and A5.2.15 GIV Transitivity of Segments	$K = \tfrac{1}{2}\lvert BA\rvert\lvert AC\rvert \sin(\angle BAC)$ for area of $\triangle ABC$

7	From 1g	$K = K_{rt}(\triangle AH_cC) + K_{rt}(\triangle CH_cB)$
8	From 1g, 4g, T5.8.2 Right Triangle: Trigonometric Area by Sine, A4.2.3 Substitution and A5.2.15 GIV Transitivity of Segments	$K = \frac{1}{2}|AH_c||BC| \sin(\angle ABC) + \frac{1}{2}|H_cB||BC| \sin(\angle ABC)$
9	From 7g, 8, A4.2.14 Distribution and A5.2.15 GIV Transitivity of Segments	$K = \frac{1}{2}|AB||BC| \sin(\angle ABC)$ for area of $\triangle ABC$
$\therefore$	From 3, 6 and 9	$K = \frac{1}{2}|AC||CB| \sin(\angle ACB) = \frac{1}{2}|BA||AC| \sin(\angle BAC)$ $\frac{1}{2}|BC||AB| \sin(\angle ABC)$ for trigonometric area of $\triangle ABC$

Corollary 5.10.1.1 Algebraic: Area of a Triangle

1g	Given	$h_a \cong |AC| \sin(\angle ACB)$ apothem, altitude of a
2g		$h_b \cong |BA| \sin(\angle BAC)$ apothem, altitude of b
3g		$h_c \cong |AB| \sin(\angle ABC)$ apothem, altitude of s
4g		$a \cong BC$
5g		$b \cong AC$
6g		$c \cong AB$
Steps	Hypothesis	$K = \frac{1}{2} h_a a = \frac{1}{2} h_b b = \frac{1}{2} h_c c$ for trigonometric area of $\triangle ABC$
$\therefore$	From 1g, 2g, 3g, 4g, 5g, 6g, A4.2.3 Substitution and T5.10.1 Trigonometric: Area of a Triangle	$K = \frac{1}{2} h_a a = \frac{1}{2} h_b b = \frac{1}{2} h_c c$ for trigonometric area of $\triangle ABC$

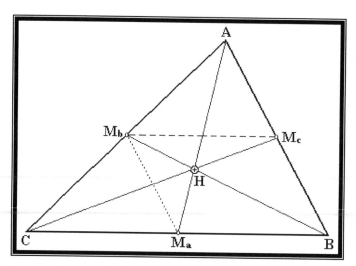

Figure 5.10.2 Parallel Median Lines to Opposite Triangle Sides

Theorem 5.10.2 Median Line is Parallel to the Opposite Side of a Triangle

1g	Given	the above geometry figure F5.10.2 "Parallel Median Lines to Opposite Triangle Sides"
2g		$AM_c \cong M_cB \cong \frac{1}{2}AB$
3g		$BM_a \cong M_aC \cong \frac{1}{2}BC$
4g		$CM_b \cong M_bA \cong \frac{1}{2}CA$
Steps	Hypothesis	$M_bM_c \parallel BC$ Median line is parallel to the Opposite Side of $\triangle ABC$
1	From 1g	$\angle BAC \cong \angle M_cAM_b$ for common vertex-A of $\triangle ABC$ and $\triangle AM_cM_b$
2	From 2g and T4.4.8 Equalities: Cross Product of Proportions	$AB \, / \, AM_c = 2$
3	From 4g and T4.4.8 Equalities: Cross Product of Proportions	$CA \, / \, CM_b = 2$
4	From 1, 2, 3 and T5.7.8 Similar Triangles: Proportional Side-Side-Angle (PSSA)	$\triangle ABC \sim \triangle AM_cM_b$
∴	From 1, 4 and T5.7.10 Similar Common Vertex with Two Sides Parallel (SCVTSP)	$M_bM_c \parallel BC$ Median line is parallel to the Opposite Side of $\triangle ABC$

Theorem 5.10.3 Median is Congruent to Half of the Opposite Side of a Triangle

1g	Given	the above geometry figure F5.10.2 "Parallel Median Lines to Opposite Triangle Sides"
2g		M_aM_b a median line
3g		M_bM_c a median line
4g		$AM_c \cong M_cB \cong \frac{1}{2}AB$
5g		$BM_a \cong M_aC \cong \frac{1}{2}BC$
6g		$CM_b \cong M_bA \cong \frac{1}{2}CA$
Steps	Hypothesis	$M_bM_c \cong BM_a \cong M_aC \cong \frac{1}{2}BC$
1	From 1g, 2g and T5.10.2 Median Line is Parallel to the Opposite Side of a Triangle	$M_aM_b \parallel AB$
2	From 1g, 3g and T5.10.2 Median Line is Parallel to the Opposite Side of a Triangle	$M_bM_c \parallel BC$
3	From 1g	BM_b is a common side to ΔBM_cM_b and ΔBM_aM_b
4	From 2 and T5.4.4 Parallel Lines: Congruent Alternate Interior Angles	$\angle M_aBM_b \cong \angle BM_bM_c$ for BM_b transverse line across $M_bM_c \parallel BC$
5	From 1 and T5.4.4 Parallel Lines: Congruent Alternate Interior Angles	$\angle M_cBM_b \cong \angle BM_bM_a$ for BM_b transverse line across $M_aM_b \parallel AB$
6	From 3, 4, 5 and T5.5.9 Tri-Construction: Angle-Angle-Side (AAS)	$\Delta BM_cM_b \cong \Delta BM_aM_b$
$\therefore$	From 3g, 5g, 6 and D5.1.27 Congruent	$M_bM_c \cong BM_a \cong M_aC \cong \frac{1}{2}BC$

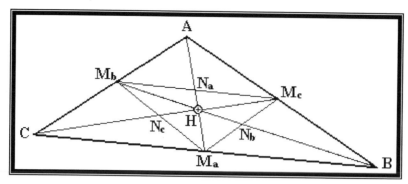

Figure 5.10.3 Median Similar to Triangle

Theorem 5.10.4 Any Given Triangle is Similar to its Medial Triangle

1g	Given	the above geometry figure F5.10.3 "Median Similar to Triangle"
2g		M_aM_b a median line
3g		M_bM_c a median line
4g		M_cM_a a median line
Steps	Hypothesis	$\triangle ABC \sim \triangle M_aM_bM_c$
1	From 2g and T5.10.3 Median is Congruent to Half of the Opposite Side of a Triangle	$M_aM_b \cong BM_c \cong M_cA \cong \tfrac{1}{2}AB$
2	From 3g and T5.10.3 Median is Congruent to Half of the Opposite Side of a Triangle	$M_bM_c \cong BM_a \cong M_aC \cong \tfrac{1}{2}BC$
3	From 4g and T5.10.3 Median is Congruent to Half of the Opposite Side of a Triangle	$M_cM_a \cong CM_b \cong M_bA \cong \tfrac{1}{2}CA$
4	From 1 and T4.4.8 Equalities: Cross Product of Proportions	$M_aM_b \,/\, AB = \tfrac{1}{2}$
5	From 2 and T4.4.8 Equalities: Cross Product of Proportions	$M_bM_c \,/\, BC = \tfrac{1}{2}$
6	From 3 and T4.4.8 Equalities: Cross Product of Proportions	$M_cM_a \,/\, CA = \tfrac{1}{2}$
∴	From 6 and T5.7.8 Similar Triangles: Proportional Side-Side-Angle (PSSA)	$\triangle ABC \sim \triangle M_aM_bM_c$

Section 5.11 Theorems on Circles, Cords and Arclength

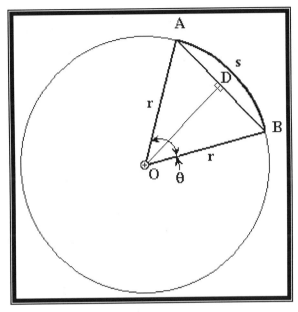

Figure 5.11.1 Anatomy of a Circle

Theorem 5.11.1 Cord Spanning Subtending Angle

1g	Given	the above geometry figure F5.11.1 "Anatomy of a Circle"
2g		point-O center of the circle
3g		$r \equiv OA$
4g		$\theta \equiv \angle AOB$
Steps	Hypothesis	$AB \equiv 2r \sin \tfrac{1}{2}\theta$
1	From 1g, 3g and D5.1.31 Circle	$r \equiv OA \equiv OB$
2	From 1 and D5.1.16 Isosceles Triangle	$\triangle AOB$ is isosceles
3	1g and D5.1.32 Cord of a Circle	AB cord of the circle of radius-r
∴	From 4g, 2, 3 and T5.9.8 Isosceles Triangle: Trigonometry; Base Length	$AB \equiv 2r \sin \tfrac{1}{2}\theta$

Theorem 5.11.2 Cord Spanning Arclength

1g	Given	the above geometry figure F5.11.1 "Anatomy of a Circle"
Steps	Hypothesis	$AB \equiv 2r \sin (s/2r)$
1	From 1g and T5.11.1 Cord Spanning Subtending Angle	$AB \equiv 2r \sin \frac{1}{2}\theta$
2	From D5.1.36 Subtended Arclength of a Cord, T4.4.4B Equalities: Right Cancellation by Multiplication	$\theta \equiv s/r$
3	From 2, A4.2.2 Equality, T4.4.15 Equalities: Product by Division	$\frac{1}{2}\theta \equiv s/2r$
∴	From 1, 3 and A4.2.3 Substitution	$AB \equiv 2r \sin (s/2r)$

Theorem 5.11.3 Cord Approximates Arclength for a Small Subtending Angle

1g	Given	the above geometry figure F5.11.1 "Anatomy of a Circle"
2g	$\delta\theta$ chosen such that; see OK.2.1A Best Approximate Limit of Sine and Cosine for a Small Angle	$(\frac{1}{2}\delta\theta) << 0.15$ radians or relatively smaller than 8.5°
Steps	Hypothesis	$AB \approx s$ or $AB \approx \overset{\frown}{AB}$
1	From T5.11.2 Cord Spanning Arclength	$AB \equiv 2r \sin (s/2r)$
2	From 2g, 1 and LxE.3.22 First Order Term	$AB \approx 2r (s/2r)$
∴	From 2, A4.2.10 Commutative Multp, A4.2.11 Associative Multp and T4.4.16A Equalities: Product by Division, Common Factor	$AB \approx s$ or $AB \approx \overset{\frown}{AB}$

Theorem 5.11.4 Radial Diameter of a Circle

1g	Given	the above geometry figure F5.11.1 "Anatomy of a Circle"
2g		$r \equiv OA$
3g		$\theta \equiv 180°$ a flat angle between OA and OB
Steps	Hypothesis	$D_0 = 2r$
1	From 1g and D5.1.31 Circle	$r \equiv OA \equiv OB$
2	From 1g, 3g and D5.1.33 Diameter of a Circle	$D_0 \equiv OA + OB$
∴	From 1, 2, A5.2.15 GIV Transitivity of Segments and T4.3.10 Equalities: Summation of Repeated Terms by 2	$D_0 = 2r$

Theorem 5.11.5 Congruency of any Two Circles yet Non-concentric

1g	Given	circle-O with radius-r
2g		circle-O' with radius-r'
3g		$r \cong r'$ for $O \not\equiv O'$
Steps	Hypothesis	circle-O $\cong$ circle-O' for $O \not\equiv O'$
$\therefore$	From 1g, 2g, 3g and D5.1.27 Congruent	circle-O $\cong$ circle-O' for $O \not\equiv O'$

Theorem 5.11.6 Congruency of any Two Circles yet Concentric

1g	Given	circle-O with radius-r
2g		circle-O' with radius-r'
3g		$r \cong r'$ for $O \equiv O'$
Steps	Hypothesis	circle-O $\cong$ circle-O' for $O \equiv O'$
$\therefore$	From 1g, 2g, 3g and D5.1.27 Congruent	circle-O $\cong$ circle-O' for $O \equiv O'$

Theorem 5.11.7 Non-congruency of any Two Circles yet Concentric

1g	Given	circle-O with radius-r
2g		circle-O' with radius-r'
3g		$r \not\cong r'$ for $O \equiv O'$
Steps	Hypothesis	circle-O $\not\cong$ circle-O' for $O \equiv O'$
$\therefore$	From 1g, 2g, 3g and D5.1.27 Congruent	circle-O $\not\cong$ circle-O' for $O \equiv O'$

Theorem 5.11.8 Non-congruency of any Two Circles yet Non-concentric

1g	Given	circle-O with radius-r
2g		circle-O' with radius-r'
3g		$r \not\cong r'$ for $O \not\equiv O'$
Steps	Hypothesis	circle-O $\not\cong$ circle-O' for $O \not\equiv O'$
$\therefore$	From 1g, 2g, 3g and D5.1.27 Congruent	circle-O $\not\cong$ circle-O' for $O \not\equiv O'$

Section 5.12 Theorems on Parallelograms

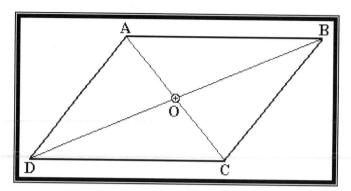

Figure 5.12.1 Parallelogram

Theorem 5.12.1 Opposing Angles of a Parallelogram are Congruent

1g	Given	the above geometry figure F5.12.1 "Parallelogram"	
Steps	Hypothesis	$\angle ABC \cong \angle ADC$	$\angle BAD \cong \angle BCD$
1	From 1g and D5.1.13 Parallelogram	$AB \parallel DC$	$AD \parallel BC$
2	From 1g, 1 and T5.4.4 Parallel Lines: Congruent Alternate Interior Angles	$\angle ABD \cong \angle BDC$, $\angle CBD \cong \angle ADB$	$\angle ACB \cong \angle CAD$, $\angle ACD \cong \angle BAC$
3	From 1g	$\angle ABC \cong \angle ABD + \angle CBD$, $\angle ADC \cong \angle ADB + \angle BDC$	$\angle BAD \cong \angle BAC + \angle CAD$, $\angle BCD \cong \angle ACB + \angle ACD$
4	From 2, 3 and A5.2.18 GIV Transitivity of Angles	$\angle ABC \cong \angle ABD + \angle CBD$, $\angle ADC \cong \angle CBD + \angle ABD$	$\angle BAD \cong \angle BAC + \angle CAD$, $\angle BCD \cong \angle CAD + \angle BAC$
$\therefore$	From 4 and A5.2.18 GIV Transitivity of Angles	$\angle ABC \cong \angle ADC$	$\angle BAD \cong \angle BCD$

Theorem 5.12.2 Diagonals of a Parallelogram Bisect One Another

1g	Given	the above geometry figure F5.12.1 "Parallelogram"	
Steps	Hypothesis	$AO \cong OC$ and $BO \cong OD$	
1	From 1g, D5.1.13 Parallelogram and T5.4.17 Parallel Lines: Parallelograms have Congruent Sides	$AB \parallel DC$, $AB \cong DC$	$AD \parallel BC$, $AD \cong BC$
2	From 1g, 1 and T5.4.4 Parallel Lines: Congruent Alternate Interior Angles	$\angle DCO \cong \angle BAO$, $\angle CDO \cong \angle ABO$	$\angle DAO \cong \angle BCO$, $\angle ADO \cong \angle CBO$
3	From 1, 2 and T5.6.13 Congruency: by Angle-Side-Angle (ASA)	$\triangle AOB \cong \triangle COD$	$\triangle AOD \cong \triangle BOC$
$\therefore$	From 3 and D5.1.27 Congruent	$AO \cong OC$	$BO \cong OD$

Theorem 5.12.3 Diagonals Partition a Parallelogram into Equal Parts

1g	Given	the above geometry figure F5.12.1 "Parallelogram"	
Steps	Hypothesis	$\triangle ABC \cong \triangle ADC$	$\triangle BAD \cong \triangle BCD$
1	From 1g	AC common side	BD common side
2	From 1g and D5.1.13 Parallelogram	AB ∥ DC, AB ≅ DC	AD ∥ BC, AD ≅ BC
3	From T5.12.1 Opposing Angels of a Parallelogram are Congruent	$\angle ABC \cong \angle ADC$	$\angle BAD \cong \angle BCD$
∴	From 1, 2, 3 and T5.6.11 Congruency: by Angle-Side-Side (ASS)	$\triangle ABC \cong \triangle ADC$	$\triangle BAD \cong \triangle BCD$

Theorem 5.12.4 Diagonals Partition a Parallelogram into Equal Areas

Steps	Hypothesis	$K_r = \frac{1}{2}\|AB\|\|BC\| \sin(\angle ABC)$ $= \frac{1}{2}\|AD\|\|DC\| \sin(\angle ADC)$	$K_l = \frac{1}{2}\|BA\|\|AD\| \sin(\angle BAD)$ $= \frac{1}{2}\|BC\|\|CD\| \sin(\angle BCD)$
1	From T5.12.3 Diagonals Partition a Parallelogram into Equal Parts	$\triangle ABC \cong \triangle ADC$	$\triangle BAD \cong \triangle BCD$
∴	From 1 and T5.10.1 Trigonometric Area of a Triangle	$K_r = \frac{1}{2}\|AB\|\|BC\| \sin(\angle ABC)$ $= \frac{1}{2}\|AD\|\|DC\| \sin(\angle ADC)$	$K_l = \frac{1}{2}\|BA\|\|AD\| \sin(\angle BAD)$ $= \frac{1}{2}\|BC\|\|CD\| \sin(\angle BCD)$

Theorem 5.12.5 Right Triangles Partition a Rectangle into Equal Areas

1g 2g 3g 4g	Given	from the geometry of a right triangle figure F5.8.1 "Parallelogram" $90° \cong \angle ABC \cong \angle ADC$ a = |AD| b = |AB|	$90° \cong \angle BAD \cong \angle BCD$ a = |BC| b = |BA|
Steps	Hypothesis	$K_{rtop} = \frac{1}{2}ab = K_{rbottom}$ Area of a Right Triangle	$K_{ltop} = \frac{1}{2}ab = K_{lbottom}$ Area of a Right Triangle
1	From 1g, 2g, 3g and 4g	a = |AD| = | BC | = |BC| = |AD|	b = |AB| = |DC| = |BA| = |CD|
2	From T5.12.4 Diagonals Partition a Parallelogram into Equal Areas	$K_r = \frac{1}{2}\|AB\|\|BC\| \sin(\angle ABC)$ $= \frac{1}{2}\|AD\|\|DC\| \sin(\angle ADC)$	$K_l = \frac{1}{2}\|BA\|\|AD\| \sin(\angle BAD)$ $= \frac{1}{2}\|BC\|\|CD\| \sin(\angle BCD)$
3	From 2g, 1, 2, A5.2.18 GIV Transitivity of Angles and A5.2.16 GIV Transitivity of Segments	$K_r = \frac{1}{2}ab \sin(90°)$ $= \frac{1}{2}ab \sin(90°)$	$K_l = \frac{1}{2}ab \sin(90°)$ $= \frac{1}{2}ab \sin(90°)$
∴	From 3, DxE.1.6.7 and A4.2.12 Identity Multp	$K_{rtop} = \frac{1}{2}ab = K_{rbottom}$ Area of a Right Triangle	$K_{ltop} = \frac{1}{2}ab = K_{lbottom}$ Area of a Right Triangle

Section 5.13 Theorems on Inscribed Cyclic Polygons

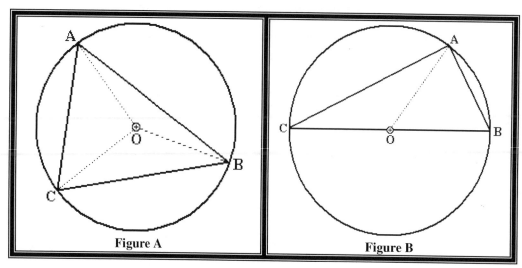

Figure 5.13.1 Inscribed Angles of a Triangle

Theorem 5.13.1 Inscribed Angle Theorem by Triangle Trisection of Circumference

1g	Given	the above geometry inscribed in circle-O see Figure F5.13.1A "Inscribed Angles of a Triangle"		
2g		$r \equiv OA$		
3g		$\theta \equiv \angle BOC$ for cord BC		
4g		$\alpha \equiv \angle AOB$ for cord AB		
5g		$\beta \equiv \angle AOC$ for cord AC		
Steps	Hypothesis	$\angle BAC \cong \tfrac{1}{2}\theta$ for cord BC	$\angle ABC \cong \tfrac{1}{2}\beta$ for cord AC	$\angle ACB \cong \tfrac{1}{2}\alpha$ for cord AB
1	From 1g, 2g and D5.1.31 Circle	$r \equiv OA \cong OB \cong OC$		
2	From 1g and adding angles about point-O	$360° \equiv \angle AOB + \angle AOC + \angle BOC$		
3	From 3g, 4g, 5g, 2 and A5.2.18 GIV Transitivity of Angles	$360° \equiv \alpha + \beta + \theta$		
4	From 3, A4.2.3 Equality, T4.4.2 Equalities: Uniqueness of Multiplication, T4.4.15 Equalities: Product by Division and D4.1.19 Primitive Definition for Rational Arithmetic	$180° \equiv \tfrac{1}{2}(\alpha + \beta + \theta)$		
5	From 1g, 1 and D5.1.16 Triangle; Isosceles	ΔBOC, ΔAOB and ΔAOC isosceles triangles		
6	From 3g, 4g, 5g, 5 and T5.9.2 Isosceles Triangle: Bisected into Two Congruent Right Triangles	$\angle BCO \cong \angle CBO = 90° - \tfrac{1}{2}\theta$	$\angle BAO \cong \angle ABO = 90° - \tfrac{1}{2}\alpha$	$\angle ACO \cong \angle CAO = 90° - \tfrac{1}{2}\beta$

7	From 1g	$\angle BAC \cong \angle BAO + \angle CAO$	$\angle ABC \cong \angle ABO + \angle CBO$	$\angle ACB \cong \angle BCO + \angle ACO$
8	From 6, 7 and A5.2.18 GIV Transitivity of Angles	$\angle BAC \cong 180° - \frac{1}{2}(\alpha + \beta)$	$\angle ABC \cong 180° - \frac{1}{2}(\alpha + \theta)$	$\angle ACB \cong 180° - \frac{1}{2}(\beta + \theta)$
9	From 8, A4.2.7 Identity Add, A4.2.8 Inverse Add, A4.2.14 Distribution and A4.2.6 Associative Add	$\angle BAC \cong 180° - \frac{1}{2}(\alpha + \beta + \theta) + \frac{1}{2}\theta$	$\angle ABC \cong 180° - \frac{1}{2}(\alpha + \beta + \theta) + \frac{1}{2}\beta$	$\angle ACB \cong 180° - \frac{1}{2}(\alpha + \beta + \theta) + \frac{1}{2}\alpha$
∴	From 4, 9, A5.2.18 GIV Transitivity of Angles, A4.2.8 Inverse Add and A4.2.7 Identity Add	$\angle BAC \cong \frac{1}{2}\theta$ for cord BC	$\angle ABC \cong \frac{1}{2}\beta$ for cord AC	$\angle ACB \cong \frac{1}{2}\alpha$ for cord AB

Theorem 5.13.2 Inscribed Triangle on the Diameter (Thale's Theorem) [5.13.1]

1g	Given	the above geometry inscribed in a circle-O see Figure F5.13.1B "Inscribed Angles of a Triangle"
2g		$D_0 \cong AB$ diameter
3g		$\theta \equiv \angle BOC$ for cord BC
Steps	Hypothesis	$\angle BAC \cong 90°$
1	From 1g, 2g, 3g and D5.1.8 Flat Angle	$\theta \equiv \angle BOC \cong 180°$
∴	From 1, T5.12.1 Inscribed Angle Theorem by Triangle Trisection of Circumference, A5.2.18 GIV Transitivity of Angles and T4.4.15 Equalities: Product by Division and D4.1.19 Primitive Definition for Rational Arithmetic	$\angle BAC \cong \frac{1}{2}(180°) \cong 90°$

[5.13.1]Note: As far as I can tell this theorem name is a misnomer no evidence shows that Thales (600 BCE) was responsible for the development of this theorem. Though it is included in his five propositions, probably why his name is associated with it, the Babylonians had recognized it some 1400 years earlier [EVE76, pg 55]. While Euclid acknowledges a number of theorems to Thales, in his book "The Elements", this theorem is found under the name of Hippocrate of Chios (440 BCE). Its use was for a quadratic solution found in a fragment of Hippocrate's "Quadrature of Lunes" (440 BCE), less than a hundred years before Euclid, preserved in a quotation by Simplicius from Eudemus' "History of Geometry" [HEA56, Vol. I, pg 387]. The modern use of it is accredited to the German Adam Riese (1489 – 1559) [EVE76, pg 216] and can be found in a humorous anecdote. It seems that one day Riese and a draftsman entered into a friendly contest to see which one of them could, with straightedge and compasses, draw more right angles in one minute. The draftsman drew a straight line, and then proceeded, by the standard construction of erecting a perpendicular to the line. Riese drew a semicircle and rapidly drew a large number of inscribed right angles. Easily wining the contest.

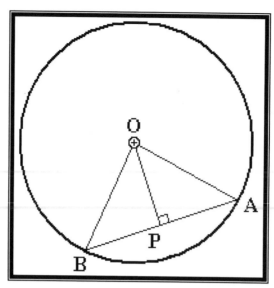

Figure 5.13.2 Inscribing an Isosceles Triangle on the Radius

Theorem 5.13.3 Inscribing an Isosceles Triangle: On the Radius

1g	Given	the above geometry figure F5.13.2 "Inscribing an Isosceles Triangle on the Radius"		
2g		$r \equiv OA$		
3g		$\theta \equiv \angle AOC$		
4g		$\alpha \equiv \angle AOP$		
5g		$\beta \equiv \angle OAP$		
Steps	Hypothesis	ΔAOC isosceles triangle	$\alpha \equiv \angle AOP \cong \angle COP = \tfrac{1}{2}\theta$	$\beta \equiv \angle OAP \cong \angle OCP = 90° - \tfrac{1}{2}\theta$
1	From 1g, 2g and D5.1.31 Circle	$r \equiv OA \cong OB \cong OC$		
2	From 1 and D5.1.16 Triangle; Isosceles	ΔAOC isosceles triangle		
∴	From 1g, 3g, 4g, 5g, 1 and T5.9.2 Isosceles Triangle: Bisected into Two Congruent Right Triangles	ΔAOC isosceles triangle	$\alpha \equiv \angle AOP \cong \angle COP = \tfrac{1}{2}\theta$	$\beta \equiv \angle OAP \cong \angle OCP = 90° - \tfrac{1}{2}\theta$

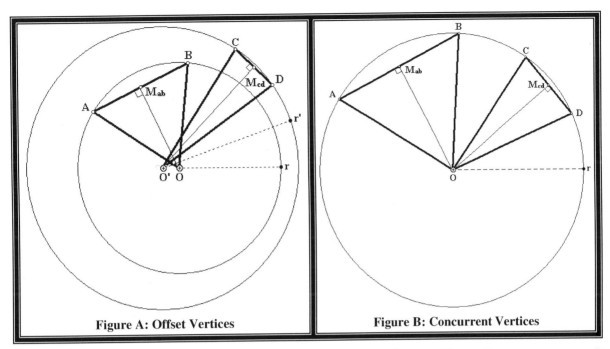

Figure 5.13.3 Inscribing Isosceles Triangles at a Common Vertex-O

Theorem 5.13.4 Inscribing Two Isosceles Triangles: Non-congruent Inscription

1g	Given	the above geometry figure F5.13.3A "Inscribing Isosceles Triangles at a Common Vertex-O"
2g		$M_{ab}O$ isosceles bisector for ΔAOB isosceles triangle
3g		$M_{cd}O'$ isosceles bisector for $\Delta CO'D$ isosceles triangle
4g		$M_{ab}O \ncong M_{cd}O'$ at Vertex-$O \neq O'$
5g		$BO \ncong CO'$
Steps	Hypothesis	circle-$O \ncong$ circle-O' and not concurrent, hence cannot be circumscribed by a single circle
1	From 1g, 2g, D5.1.16 Triangle; Isosceles and T5.12.3 Inscribing an Isosceles Triangle: on the Radius	$AO \cong BO \cong r$ inscribed by circle-O
2	From 1g, 3g, D5.1.16 Triangle; Isosceles; Isosceles and T5.12.3 Inscribing an Isosceles Triangle: on the Radius	$CO' \cong DO' \cong r'$ inscribed by circle-O'
3	From 5g, 1, 2 and A5.2.15 GIV Transitivity of Segments	$AO \ncong DO'$
4	From 4g, 1, 2, and A5.2.15 GIV Transitivity of Segments	$r \ncong r'$ for circle vertices $O \neq O'$ are not concurrent
∴	From 4 and D5.1.27 Congruent	circle-$O \ncong$ circle-O' and not concurrent, hence cannot be circumscribed by a single circle

Theorem 5.13.5 Inscribing Two Isosceles Triangles: Congruent Inscription

1g	Given	the above geometry figure F5.13.3B "Inscribing Isosceles Triangles at a Common Vertex-O"
2g		$M_{ab}O$ isosceles bisector for $\triangle AOB$ isosceles triangle
3g		$M_{cd}O'$ isosceles bisector for $\triangle CO'D$ isosceles triangle
4g		$M_{ab}O \cong M_{cd}O'$ at Vertex-$O \cong O'$
5g		$BO \cong CO'$
Steps	Hypothesis	circle-$O \cong$ circle-O' and are concurrent, hence can be circumscribed by a single circle
1	From 1g, 2g, D5.1.16 Triangle; Isosceles and T5.12.3 Inscribing an Isosceles Triangle: on the Radius	$AO \cong BO \cong r$ inscribed by circle-O
2	From 1g, 3g, D5.1.16 Triangle; Isosceles; Isosceles and T5.12.3 Inscribing an Isosceles Triangle: on the Radius	$CO' \cong DO' \cong r'$ inscribed by circle-O'
3	From 5g, 1, 2 and A5.2.15 GIV Transitivity of Segments	$AO \cong DO'$
4	From 4g, 1, 2, and A5.2.15 GIV Transitivity of Segments	$r \cong r'$ for circle vertices $O \cong O'$ are concurrent
$\therefore$	From 4 and D5.1.27 Congruent	circle-$O \cong$ circle-O' and are concurrent, hence can be circumscribed by a single circle

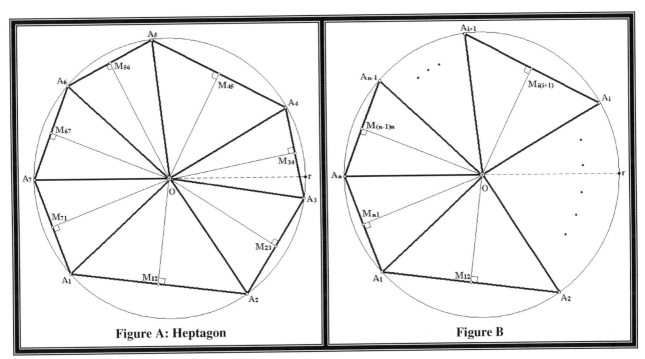

Figure A: Heptagon

Figure B

Figure 5.13.4 Inscribing an Irregular Polygon at a Common Vertex-O

Theorem 5.13.6 Inscribing n-Isosceles Triangles: Inscribed by a Circle

1g	Given	the above geometry figure F5.13.4B "Inscribing an Irregular Polygon at a Common Vertex-O"
2g		A_i are points shared by any two adjacent triangles $\Delta A_i O A_{(i+1)}$
3g		$M_{i(i+1)}O$ perpendicular bisector intersect at point-O for all i
4g		$r \cong A_1 O$
Steps	Hypothesis	circle-O at vertex-O inscribing any two isosceles triangles $\Delta A_i O A_{(i+1)}$, hence inscribing all n-isosceles triangles
1	From 3g and T5.9.7 Isosceles Triangle: Constructed from Perpendicular Bisector	$\Delta A_i O A_{(i+1)}$ isosceles triangle for any specific i
2	From 1 and D5.1.16 Triangle; Isosceles	$A_i O \cong A_{(i+1)} O$
3	From 2g and 2 over the finite set of [i] from 1 to n	$(\bullet \cong A_i O \bullet)$ for all i
4	From 4g, 3 and A5.2.18 GIV Transitivity of Angles	$(r \cong A_1 O \cong \bullet \cong A_k O \bullet)$ for all k = 2, …, n centered at vertex-O
∴	From 1, 4 and T5.12.5 Inscribing Two Isosceles Triangles: Congruent Inscription	circle-O at vertex-O inscribing any two isosceles triangles $\Delta A_i O A_{(i+1)}$, hence inscribing all n-isosceles triangles

Theorem 5.13.7 Inscribing n-Isosceles Triangles: Inscribed by a Circle have Closure

1g	Given	the above geometry figure F5.13.4B "Inscribing an Irregular Polygon at a Common Vertex-O"
2g		$\Delta A_iOA_{(i+1)}$ isosceles triangle for i = 1, …, n for n – 1 triangles so there is insufficient triangles to fill the circle all away around.
Steps	Hypothesis	Triangle can always be found to close and complete the set of n-inscribed triangles
1	From 2g and T5.12.8 Inscribing n-Isosceles Triangles: Inscribed by a Circle	$\Delta A_iOA_{(i+1)}$ isosceles triangle for i = 1, …, n for n – 1 triangles are inscribed in circle-O
2	From 1 and D5.1.31 Circle	$A_nO \cong A_1O$
3	From 1g and construct	A_nA_1 completing the triangle ΔA_nOA_1
∴	From 2, 3 and D5.1.16 Triangle; Isosceles	ΔA_nOA_1 is an isosceles triangle closing the circle of triangles, hence a triangle can always be found to close and complete the set of n-inscribed triangles.

Theorem 5.13.8 Inscribing n-Isosceles Triangles: Cyclic Vertices

1g	Given	the above geometry figure F5.13.4B "Inscribing an Irregular Polygon at a Common Vertex-O"
Steps	Hypothesis	vertex-A_{n+1} ←→ vertex-A_1 closure of vertices, hence cyclic
1	From 1g, T5.12.6 Inscribing n-Isosceles Triangles: Inscribed by a Circle and T5.12.7 Inscribing n-Isosceles Triangles: Inscribed by a Circle have Closure	$\Delta A_nOA_{(n+1)} \cong \Delta A_nOA_1$
∴	From 1 and D5.1.27 Congruent	vertex-$A_{(n+1)} \equiv$ vertex-A_1 are concurrent, hence cyclic

Theorem 5.13.9 Inscribing n-Isosceles Triangles: Tallied Area

1g	Given	the above geometry figure F5.13.4B "Inscribing an Irregular Polygon at a Common Vertex-O"
2g		$\Delta A_iOA_{(i+1)}$ isosceles triangle inscribed in circle-O
3g		$a_i \equiv M_{i(i+1)}O$
4g		$p_i \equiv A_iA_{(i+1)}$
5g		$K_T = \sum K_{\Delta AiOA(i+1)}$
Steps	Hypothesis	$K_T = \sum a_i\, p_i$ for cyclic vertices if closure applies
1	From 1g, 2g and T5.9.12 Isosceles Triangle: Area	$K_{\Delta AiOA(i+1)} = M_{i(i+1)}O\, A_iA_{(i+1)}$
2	From 3g, 4g, 1 and A5.2.15 GIV Transitivity of Segments	$K_{\Delta AiOA(i+1)} = a_i\, p_i$
∴	From 5g, 2, A4.2.3 Substitution and T5.12.8 Inscribing n-Isosceles Triangles: Cyclic Vertices	$K_T = \sum a_i\, p_i$ for cyclic vertices if closure applies

Theorem 5.13.10 Inscribing n-Isosceles Triangles: Parameter of a Regular Polygon

1g 2g	Given	$p \equiv p_i$ for all i when a regular polygon $P_T \equiv \sum p_i$ Total Parameter
Steps	Hypothesis	$P_T = np$ Total Parameter
1	From 1g, 2g, and A4.2.3 Substitution	$P_T = \sum p$
2	From 1, A4.2.12 Identity Multp and A4.2.14 Distribution	$P_T = p \sum 1$
∴	From 2 and LxE.3.29	$P_T = np$

Theorem 5.13.11 Inscribing n-Isosceles Triangles: Tallied Area of a Regular Polygon

1g 2g	Given	$a \equiv a_i$ for all i when a regular polygon $p \equiv p_i$ for all i when a regular polygon
Steps	Hypothesis	$K_T = \sum a_i p_i$ for cyclic vertices if closure applies
1	From T5.12.9 Inscribing n-Isosceles Triangles: Tallied Area	$K_T = \sum a_i p_i$
2	From 1g, 2g, 1, A5.2.15 GIV Transitivity of Segments, A4.2.12 Identity Multp and A4.2.14 Distribution	$K_T = a\, p \sum 1$
3	From 2, LxE.3.29 and A4.2.10 Commutative Multp	$K_T = a\,(n\,p)$
∴	From 3, T2.12.10 Inscribing n-Isosceles Triangles: Parameter of a Regular Polygon and A4.2.3 Substitution	$K_T = aP_T$

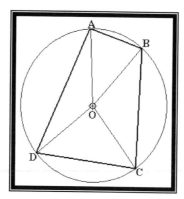

Figure 5.13.5 Cyclic Quadrilateral

Theorem 5.13.12 Cyclic Quadrilateral: Supplementary and Radial Base Vertex Angles

1g	Given	the above geometry figure F5.13.5 "Cyclic Quadrilateral" ∠DAB = ∠OAB + ∠OAD Cyclic Quadrilateral internal angle the sum of the base vertex angles of the radial isosceles triangles.ΔOAB and ΔOAD.
2g		$\alpha \equiv \angle AOB$
3g		$\beta \equiv \angle DOA$
Steps	Hypothesis	$\angle DAB = 180° - \frac{1}{2}(\alpha + \beta)$ EQA $\quad\mid\quad$ $\frac{1}{2}(\alpha + \beta) = 180° - \angle DAB$ EQB
1	From 1g, 2g, 3g, T5.9.2 Isosceles Triangle: Bisected into Two Congruent Right Triangles and A5.2.18 GIV Transitivity of Angles	$\angle DAB = 90° - \frac{1}{2}\alpha + 90° - \frac{1}{2}\beta$
2	From 1, T4.3.10 Equalities: Summation of Repeated Terms by 2, D4.1.19 Primitive Definition for Rational Arithmetic and A4.2.14 Distribution	$\angle DAB = 180° - \frac{1}{2}(\alpha + \beta)$
3	From 2, T4.3.4A Equalities: Reversal of Right Cancellation by Addition and T4.3.5B Equalities: Left Cancellation by Addition	$\frac{1}{2}(\alpha + \beta) = 180° - \angle DAB$
∴	From 2 and 3	$\angle DAB = 180° - \frac{1}{2}(\alpha + \beta)$ EQA $\quad\mid\quad$ $\frac{1}{2}(\alpha + \beta) = 180° - \angle DAB$ EQB

Theorem 5.13.13 Cyclic Quadrilateral: Opposing Angles are Supplementary

1g	Given	the above geometry figure F5.13.5 "Cyclic Quadrilateral"
2g		$\alpha \equiv \angle AOB$
3g		$\beta \equiv \angle BOC$
4g		$\delta \equiv \angle COD$
5g		$\gamma \equiv \angle DOA$
6g		$r \equiv OA \equiv OB \equiv OC \equiv OD$ inscribe in the circle
Steps	Hypothesis	$180° = \angle ABC + \angle CDA$
		$180° = \angle BCD + \angle DAB$ opposing angles are supplemental
1	1g and and adding angles about point-O	$360° \equiv \angle AOB + \angle BOC + \angle COD + \angle DOA$
2	From 1g, 2g, 3, 4g, 5g, 1 and A5.2.18 GIV Transitivity of Angles	$360° \equiv \alpha + \beta + \delta + \gamma$
3	From 1g, 6g and T5.12.2 Inscribing an Isosceles Triangle on the Radius	$\Delta AOB, \Delta BOC, \Delta COD$ and ΔDOA are isosceles triangles
4	From 1g	$\angle DAB = \angle OAB + \angle OAD$
5	From 2g, 5g, 4 and T5.12.12 Cyclic Quadrilateral: Supplementary and Radial Base Vertex Angles	$\frac{1}{2}(\alpha + \gamma) = 180° - \angle DAB$
6	From 1g	$\angle ABC = \angle OBA + \angle OBC$
7	From 2g, 3g, 6 and T5.12.12 Cyclic Quadrilateral: Supplementary and Radial Base Vertex Angles	$\frac{1}{2}(\alpha + \beta) = 180° - \angle ABC$
8	From 1g	$\angle BCD = \angle OCB + \angle OCD$
9	From 3g, 4g, 7 and T5.12.12 Cyclic Quadrilateral: Supplementary and Radial Base Vertex Angles	$\frac{1}{2}(\beta + \delta) = 180° - \angle BCD$
10	From 1g	$\angle CDA = \angle ODC + \angle ODA$
11	From 4g, 5g, 10 and T5.12.12 Cyclic Quadrilateral: Supplementary and Radial Base Vertex Angles	$\frac{1}{2}(\delta + \gamma) = 180° - \angle CDA$
12	From 2, A4.2.2 Equality, T4.4.15B Equalities: Product by Division and D4.1.19 Primitive Definition for Rational Arithmetic	$180° = \frac{1}{2}(\alpha + \beta + \delta + \gamma)$

13	From 12, A4.2.14 Distribution, A4.2.6 Associative Add	$180° = ½(α + β) + ½(δ + γ)$
14	From 7, 11, 13 and A5.2.18 GIV Transitivity of Angles	$180° = 180° − ∠ABC + 180° − ∠CDA$
20	From 14, A4.2.7 Identity Add, T4.3.2 Equalities: Cancellation by Addition, A4.2.6 Associative Add, A4.2.14 Distribution, T4.3.3A Equalities: Reversal of Right Cancellation by Addition and A4.2.2B Equality	$180° = ∠ABC + ∠CDA$
21	From 12, A4.2.5 Commutative Add, A4.2.14 Distribution, A4.2.6 Associative Add	$180° = ½(β +δ) + ½(α + γ)$
22	From 5, 9, 21 and A5.2.18 GIV Transitivity of Angles	$180° = 180° − ∠BCD + 180° − ∠DAB$
23	From 22, A4.2.7 Identity Add, T4.3.2 Equalities: Cancellation by Addition, A4.2.6 Associative Add, A4.2.14 Distribution, T4.3.3A Equalities: Reversal of Right Cancellation by Addition and A4.2.2B Equality	$180° = ∠BCD + ∠DAB$
∴	From 20 and 23	$180° = ∠ABC + ∠CDA$ $180° = ∠BCD + ∠DAB$

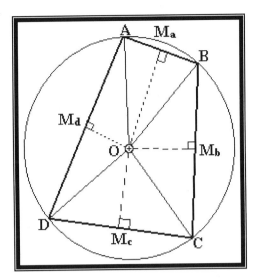

Figure 5.13.6 Cyclic Quadrilateral with Perpendiculars

Theorem 5.13.14 Cyclic Quadrilateral: Perpendiculars Intersect at Point-O

1g	Given	the above geometry figure F5.13.6 "Cyclic Quadrilateral with Perpendiculars"
2g		$r \equiv OA \equiv OB \equiv OC \equiv OD$ inscribe in the circle
Steps	Hypothesis	MO_a, MO_b, MO_c and MO_d $\perp$'s concurrent at Point-O
1	From 2g and D5.1.16 Triangle; Isosceles	$\triangle AOB$, $\triangle BOC$, $\triangle COD$ and $\triangle DOA$ isosceles triangles
2	From 1	$\angle AOB$, $\angle BOC$, $\angle COD$ and $\angle DOA$ vertices current at point-O
$\therefore$	From 1 and 2	MO_a, MO_b, MO_c and MO_d $\perp$'s concurrent at Point-O

Section 5.14 Theorems on Cyclic Transversal Secants and Tangents

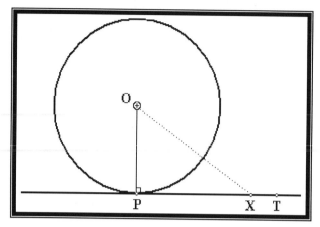

Figure 5.14.1 Perpendicular to Tangent

Theorem 5.14.1 Radius is Perpendicular to Tangent

1g	Given	the above geometry figure F5.14.1 "Perpendicular to Tangent"
2g		OP not perpendicular to PT
Steps	Hypothesis	OP perpendicular to PT
1	From D5.1.31 Circle	OP radius
2	From 2g	there must be another segment, say OX, perpendicular to PT
3	From 2	OX < OP for OX represents the shortest distance from O to PT
4	From 1g	X lies in the exterior of the circle-O
5	From 1g and 4	OX > OP because it exterior to circle-O
6	From 3 and 5	contradiction they both can not be greater and less than one another
7	From 2g and 6	assumption false
∴	From 7	OP perpendicular to PT

Theorem 5.14.2 Radius is Perpendicular Line Implies a Tangent

1g	Given	the above geometry figure F5.14.1 "Perpendicular to Tangent"
2g		OP perpendicular to PT at point-P
3g		PB is not a tangent line but intersects at another point-X
Steps	Hypothesis	PB tangent to circle-O
1	From 1g	∠OPX ≅ 90°
2	From 1 and D5.1.19 Triangle; Right	ΔOPX right triangle
3	From 2 and D5.1.19 Triangle; Right	OX is a hypotenuse of ΔOPX
4	From 3	OX > OP
5	From D5.1.31 Circle	OP radius
6	From 4 and 5	X must lie in the exterior of circle-O
7	From 5 and 6	contradiction for there is exactly one point at which PT intersects the circle-O
8	From 3g and 7	assumption false
∴	From 8	PT tangent to circle-O

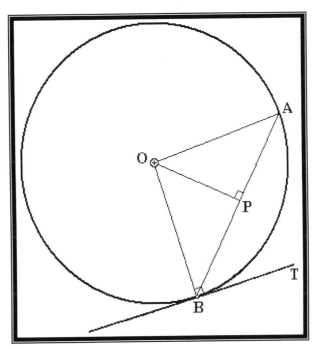

Figure 5.14.2 Cord-Tangent Angle

Theorem 5.14.3 Cord-Tangent Angle

1g	Given	the above geometry figure F5.14.2 "Cord-Tangent Angle"
2g		$r \equiv OA$
3g		$\theta \equiv \angle AOB$
4g		$\alpha \equiv \angle ABT$
5g		$\beta \equiv \angle OAP$
Steps	Hypothesis	$\alpha \equiv \tfrac{1}{2}\theta$
1	From 1g, 2g and D5.1.31 Circle	$r \equiv OA \cong OB$
2	From 1g, 1 and T5.12.2 Inscribing an Isosceles Triangle on the Radius	$\angle OBP = 90° - \tfrac{1}{2}\theta$
3	From 1g and D5.1.39 Tangent	$\angle OBT = 90°$
4	From 1g and 4g	$\alpha \equiv \angle ABT \cong \angle OBT - \angle OBP$
∴	From 2, 3, 4, A5.2.18 GIV Transitivity of Angles, A4.2.14 Distribution, A4.2.8 Inverse Add and A4.2.7 Identity Add	$\alpha \equiv \tfrac{1}{2}\theta$

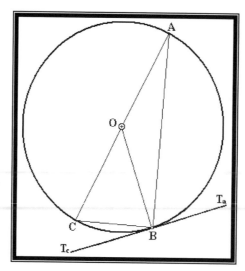

Figure 5.14.3 Radii-Tangent Angles

Theorem 5.14.4 Radii-Tangent Angles

1g	Given	the above geometry figure F5.14.3 "Radii-Tangent Angles"	
2g		$r \equiv OA$	
3g		$\theta \equiv \angle AOB$	
4g		$\phi \equiv \angle COB$	
5g		$\alpha \equiv \angle ABT_a$	
6g		$\beta \equiv \angle CBT_c$	
Steps	Hypothesis	$\alpha \equiv \tfrac{1}{2}\theta \cong \angle ABT_a$	$\beta \equiv \tfrac{1}{2}\phi \cong \angle CBT_c$
1	From 1g, 2g and D5.1.31 Circle	$r \equiv OA \cong OB \cong OC$	
2	From 1g, 1 and T5.12.2 Inscribing an Isosceles Triangle on the Radius	$\triangle AOB$ and $\triangle COB$ Isosceles	
∴	From 3g, 4g, 5g, 6g, 2 and T5.13.3 Cord-Tangent Angle	$\alpha \equiv \tfrac{1}{2}\theta \cong \angle ABT_a$	$\beta \equiv \tfrac{1}{2}\phi \cong \angle CBT_c$

Theorem 5.14.5 Diameter-Tangent Angles

1g	Given	the above geometry figure F5.14.3 "Radii-Tangent Angles"	
2g		$\theta \equiv \angle AOB$	
3g		$\phi \equiv \angle COB$	
4g		$\alpha \equiv \angle ABT_a$	
5g		$\beta \equiv \angle CBT_c$	
Steps	Hypothesis	$\alpha \equiv \tfrac{1}{2}\theta \cong \angle ABT_a$	$\beta \equiv \tfrac{1}{2}\phi \cong \angle CBT_c$
1	From 1g and D5.1.8 Flat Angle	$\angle COB \cong 180° - \angle AOB$	
2	From 2g, 3g, 1 and A5.2.18 GIV Transitivity of Angles	$\phi \equiv 180° - \theta$	
∴	From 2g, 3g, 4g, 5g, 2 and T5.13.4 Radii-Tangent Angles	$\alpha \equiv \tfrac{1}{2}\theta \cong \angle ABT_a$	$\beta \equiv \tfrac{1}{2}\phi \cong \angle CBT_c$

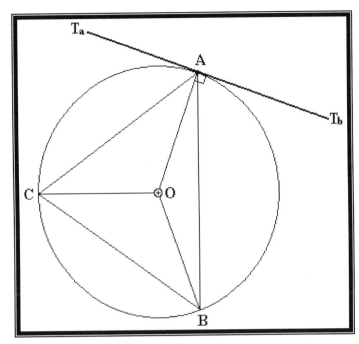

Figure 5.14.4 Inscribed Opposite Vertex of Congruent to Tangent Angle

Theorem 5.14.6 Opposite Vertex of Inscribed Triangle is Congruent to Tangent Angle

1g	Given	the above geometry figure F5.14.4 "$\triangle ABC$ inscribed in circle-O"	
2g		$2\theta \equiv \angle AOB$ for cord BC	
3g		$2\alpha \equiv \angle AOC$ for cord AB	
4g		$2\beta \equiv \angle BOC$ for cord AC	
5g		$\angle T_bAO \cong 90°$ for $T_aT_b \perp$ to AO and tangent to circle-O	
Steps	Hypothesis	$\angle T_bAB \cong \angle ACB \cong \theta$	$\angle T_aAC \cong \angle ABC \cong \alpha$
1	From 2g, 3g, 4g about point-O	$360° \equiv \angle AOC + \angle BOC + \angle AOB$	
2	From 2g, 3g, 4g, 1 and A5.2.18 GIV Transitivity of Angles	$360° \equiv 2\alpha + 2\beta + 2\theta$	
3	From 2, A4.2.2A Equality, T4.4.2 Equalities: Uniqueness of Multiplication, T4.4.15B Equalities: Product by Division, D4.1.19 Primitive Definition for Rational Arithmetic	$180° \equiv \frac{1}{2}(2\alpha + 2\beta + 2\theta)$	
4	From 3, A4.2.14 Distribution and T4.4.16B Equalities: Product by Division, Common Factor	$180° \equiv \alpha + \beta + \theta$	
5	From 1g, and D5.1.31 Circle	$OA \equiv OB \equiv OC$ radii of circle-O	

6	From 2g, 5 and T5.12.2 Inscribing an Isosceles Triangle on the Radius	$\angle BAO = 90° - \frac{1}{2}(\angle AOB)$		
7	From 2g, 6, A5.2.18 GIV Transitivity of Angles, T4.4.16B Equalities: Product by Division, Common Factor	$\angle BAO = 90° - \theta$		
8	From 1g, 5g, 7, A5.2.18 GIV Transitivity of Angles, A4.2.14 Distribution, A4.2.8 Inverse Add and A4.2.7 Identity Add	$\angle T_bAB \cong \angle T_bAO - \angle BAO \cong 90° - (90° - \theta) \cong \theta$		
9	From 2g, 3g, 4g, T5.12.1 Inscribed Angle Theorem by Triangle Trisection of Circumference and T4.4.16B Equalities: Product by Division, Common Factor	$\angle ACB \cong \theta$ for cord AB	$\angle ABC \cong \alpha$ for cord AC	$\angle BAC \cong \beta$ for cord BC
10	From 1g, 9, D5.1.8 Flat Angle and A5.2.18 GIV Transitivity of Angles	$\angle T_aAB \cong 180° - \angle ACB \cong 180° - \theta$		
11	From 1g, 9, 10 and A5.2.18 GIV Transitivity of Angles	$\angle T_aAC \cong \angle T_aAB - \angle BAC \cong 180° - \theta - \beta$		
12	From 4, A4.2.6 Associative Add, T4.3.3A Equalities: Right Cancellation by Addition and T4.2.14 Distribution	$180° - \theta - \beta \cong \alpha$		
13	From 11, 12 and A5.2.18 GIV Transitivity of Angles	$\angle T_aAC \cong \alpha$		
∴	From 8, 9 and 13	$\angle T_bAB \cong \angle ACB \cong \theta$	$\angle T_aAC \cong \angle ABC \cong \alpha$	

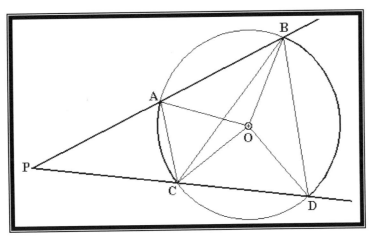

Figure 5.14.5 Secant Vertex Exterior to a Circle (VSS)

Theorem 5.14.7 Vertex Exterior to a Circle: by Secant-Secant (VSS)

1g	Given	the above geometry figure F5.14.5 "Secant Vertex Exterior to a Circle (VSS)"
2g		$\alpha \equiv \angle AOC$
3g		$\beta \equiv \angle BOD$
4g		$\theta \equiv \angle APC$
5g		$r \equiv OA$
Steps	Hypothesis	$\theta \equiv \frac{1}{2}(\beta - \alpha)$ external apex at point-P
1	From 1g, 5g and D5.1.31 Circle	$r \equiv OA \equiv OB \equiv OC \equiv OD$ inscribe in the circle
2	From 1g and T5.4.8 External Angle Supplementary of two Opposing Vertices	$\angle BCD \cong \angle APC + \angle ABC$
3	From 4g, 2, T4.3.3B Equalities: Right Cancellation by Addition and A5.2.18 GIV Transitivity of Angles	$\theta \equiv \angle BCD - \angle ABC$
4	From 1g and T5.12.1 Inscribed Angle Theorem by Triangle Trisection of Circumference	$\angle BCD \equiv \frac{1}{2}\angle BOD$
5	From 3g, 4 and A5.2.18 GIV Transitivity of Angles	$\angle BCD \equiv \frac{1}{2}\beta$
6	From 1g and T5.12.1 Inscribed Angle Theorem by Triangle Trisection of Circumference	$\angle ABC \equiv \frac{1}{2}\angle AOC$
7	From 2g, 6 and A5.2.18 GIV Transitivity of Angles	$\angle ABC \equiv \frac{1}{2}\alpha$
∴	From 3, 5, 7, A5.2.18 GIV Transitivity of Angles and A4.2.14 Distribution	$\theta \equiv \frac{1}{2}(\beta - \alpha)$

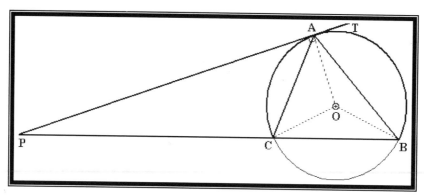

Figure 5.14.6 Secant-Tangent Vertex Exterior to a Circle (VST)

Theorem 5.14.8 Vertex Exterior to a Circle: by Secant-Tangent (VST)

1g	Given	the above geometry figure F5.14.6 "Secant-Tangent Vertex Exterior to a Circle (VST)"
2g		$\alpha \equiv \angle AOC$
3g		$\beta \equiv \angle AOB$
4g		$\theta \equiv \angle APC$
Steps	Hypothesis	$\theta \equiv \frac{1}{2}(\beta - \alpha)$ external apex at point-P
1	From 1g and T5.12.1 Inscribed Angle Theorem by Triangle Trisection of Circumference	$\angle ABC \cong \frac{1}{2}\angle AOC$ for $\triangle ABC$
2	From 2g, 1 and A5.2.18 GIV Transitivity of Angles	$\angle ABC \cong \frac{1}{2}\alpha$
3	From 1g and T5.12.1 Inscribed Angle Theorem by Triangle Trisection of Circumference	$\angle ACB \cong \frac{1}{2}\angle AOB$ for $\triangle ABC$
4	From 3g, 3 and A5.2.18 GIV Transitivity of Angles	$\angle ACB \cong \frac{1}{2}\beta$
5	From 1g, 4, D5.1.8 Flat Angle and A5.2.18 GIV Transitivity of Angles	$\angle PCA \cong 180° - \angle ACB \cong 180° - \frac{1}{2}\beta$
6	From 1g, 2g, 2, T5.13.6 Opposite Vertex of Inscribed Triangle is Congruent to Tangent Angle and A5.2.18 GIV Transitivity of Angles	$\angle CAP \cong \angle ABC \equiv \frac{1}{2}\alpha$
7	From 1g and T5.4.5 Sum of Internal Angles for a Triangle	$180° \equiv \angle APC + \angle PCA + \angle CAP$

| 8 | From 4g, 5, 6, 7 and A5.2.18 GIV Transitivity of Angles | $180° \equiv \theta + 180° - \tfrac{1}{2}\beta + \tfrac{1}{2}\alpha$ |
| $\therefore$ | From 8, T4.3.3A Equalities: Right Cancellation by Addition, A4.2.8 Inverse Add, A4.2.7 Identity Add and A4.2.14 Distribution | $\theta \equiv \tfrac{1}{2}(\beta - \alpha)$ |

Theorem 5.14.9 Vertex Exterior to a Circle: by Secant-Tangent Proportional (VSTP)

1g	Given	the above geometry figure F5.14.6 "Secant-Tangent Vertex Exterior to a Circle (VST)"
2g		$\alpha \equiv \angle AOC$
Steps	Hypothesis	$PA^2 = PC\,PB$ for $\angle ABC \cong \angle CAP \equiv \tfrac{1}{2}\alpha$
1	From 1g and T5.12.1 Inscribed Angle Theorem by Triangle Trisection of Circumference	$\angle ABC \cong \tfrac{1}{2}\angle AOC$ for ΔABC
2	From 2g, 1 and A5.2.18 GIV Transitivity of Angles	$\angle ABC \cong \tfrac{1}{2}\alpha$
3	From 1g, 2g, 2, T5.13.6 Opposite Vertex of Inscribed Triangle is Congruent to Tangent Angle and A5.2.18 GIV Transitivity of Angles	$\angle ABC \cong \angle CAP \equiv \tfrac{1}{2}\alpha$
4	From 1g	PA common to ΔCPA and ΔBPA
5	From 1g	$\angle APC \cong \angle APB$ common to ΔCPA and ΔBPA
6	From 3, 5 and T5.5.13 Similar Triangles: Angle-Angle (AA)	$\Delta CPA \sim \Delta BPA$
7	From 6	PA/ PC = PB / PA
$\therefore$	From 7, T4.4.8 Equalities: Cross Product of Proportions and D4.1.17 Exponential Notation	$PA^2 = PC\,PB$ for $\angle ABC \cong \angle CAP \equiv \tfrac{1}{2}\alpha$

Theorem 5.14.10 Vertex Exterior to a Circle: Corresponding to (Parallel Cord)-Tangent (VPCT)

1g	Given	the above geometry figure F5.14.6 "Secant-Tangent Vertex Exterior to a Circle (VST)"		
2g		$\theta \equiv \angle ABC$ for cord AC		
3g		$\alpha \equiv \angle ACB$ for cord AB		
4g		$\angle PAO \cong 90°$ for PT $\perp$ to AO and tangent to circle-O		
5g		Tangent line PT intersects cord PC or PB at point-P		
Steps	Hypothesis	For inscribed isosceles triangle and parallel Cord-Tangent than $\angle$Tangent $\cong$ $\angle$Complementary		
1	From 2g, 3g, T5.13.6 Opposite Vertex of Inscribed Triangle is Congruent to Tangent Angle and A5.2.18 GIV Transitivity of Angles	$\angle CAP \cong \theta$		$\angle BAT \cong \alpha$
2	From 4g, 5g, 1 and T5.4.7 Parallel Lines: Determined by a Continuously Elongated Triangle	PT ‖ PB and $\angle CAP \cong \angle BAT$		
3	From 1, 2 and T5.4.4 Parallel Lines: Congruent Alternate Interior Angles	$\angle ACB \cong \angle CAP \cong \theta$		$\angle ABC \cong \angle BAT \cong \alpha$
4	From 2, 3 and A5.2.18 GIV Transitivity of Angles	$\angle ACB \cong \angle ABC$ for $\triangle ABC$		$\theta \cong \alpha$
$\therefore$	From 4 and T5.9.4 Isosceles Triangle: Congruent Interior Angles and Sides	For inscribed isosceles triangle and parallel Cord-Tangent than $\angle$Tangent $\cong$ $\angle$Isosceles Complementary		

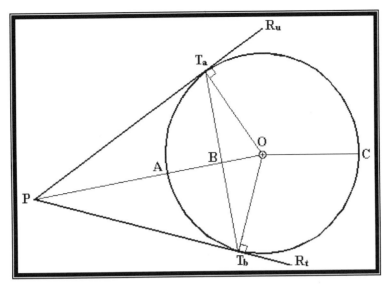

Figure 5.14.7 Tangent-Tangent Vertex Exterior to a Circle (VTT)

Theorem 5.14.11 Vertex Exterior to a Circle: by Tangent-Tangent (VTT)

1g	Given	the above geometry figure F5.14.7 "Tangent-Tangent Vertex Exterior to a Circle (VTT)"	
2g		$\alpha \equiv \angle T_aOT_b$	
3g		$\beta \equiv \angle T_aCT_b$	
4g		$\theta \equiv \angle T_aPT_b$	
Steps	Hypothesis	$\theta \equiv \frac{1}{2}(\beta - \alpha)$ external apex at point-P	
1	From 1g	$90° \equiv \angle PT_aO$	$90° \equiv \angle PT_bO$
2	From 1g, 1 and D5.1.19 Triangle; Right	ΔPT_aO right triangle	ΔPT_bO right triangle
3	From 2	OP hypotenuse for ΔPT_aO	OP hypotenuse for ΔPT_bO
4	From 2, 3 and T5.6.16 Congruency of Right Triangles by Common Hypotenuse-Leg (CHL)	$\Delta PT_aO \cong \Delta PT_bO$ $\quad \angle T_aOP \cong \angle T_bOP$	$\angle T_aPO \cong \angle T_bPO$
5	From 1g	$\angle T_aOT_b \cong \angle T_aOP + \angle T_bOP$	
6	From 2g, 4, 5, A5.2.18 GIV Transitivity of Angles, T4.3.10 Equalities: Summation of Repeated Terms by 2, A4.2.2A Equality, T4.4.2 Equalities: Uniqueness of Multiplication and T4.4.16B Equalities: Product by Division, Common Factor	$\angle T_aOP \cong \angle T_bOP \equiv \frac{1}{2}\alpha$	

7	From 1g	$\angle T_aPT_b \cong \angle T_aPO + \angle T_bPO$	
8	From 4g, 4, 7, A5.2.18 GIV Transitivity of Angles, T4.3.10 Equalities: Summation of Repeated Terms by 2, A4.2.2A Equality, T4.4.2 Equalities: Uniqueness of Multiplication and T4.4.16B Equalities: Product by Division, Common Factor	$\angle T_aPO \cong \angle T_bPO \cong \tfrac{1}{2}\theta$	
9	From 1g, 1 and T5.8.6 Right Triangle: Non-Ninety Degree Angles are Complementary	$90° \equiv \angle T_aOP + \angle T_aPO$	$90° \equiv \angle T_bOP + \angle T_bPO$
10	From 9, 6, 8 and A5.2.18 GIV Transitivity of Angles	$90° \equiv \tfrac{1}{2}\alpha + \tfrac{1}{2}\theta$	$90° \equiv \tfrac{1}{2}\alpha + \tfrac{1}{2}\theta$
11	From 10, A4.2.2A Equality, D4.1.19 Primitive Definition for Rational Arithmetic, A4.2.14 Distribution and T4.4.16A Equalities: Product by Division, Common Factor and A4.2.12 Identity Multp	$180° \equiv \alpha + \theta$	
12	From 1g	$360° \equiv \angle T_aOT_b + \angle T_aCT_b$	
13	From 2g, 3g, 12 and A5.2.18 GIV Transitivity of Angles	$360° \equiv \alpha + \beta$	
14	From 13, A4.2.2A Equality and D4.1.19 Primitive Definition for Rational Arithmetic	$180° \equiv \tfrac{1}{2}(\alpha + \beta)$	
15	From 11 and 14	$\alpha + \theta \equiv \tfrac{1}{2}(\alpha + \beta)$	
∴	From 15, A4.2.14 Distribution, T4.3.5B Equalities: Left Cancellation by Addition, D4.1.19 Primitive Definition for Rational Arithmetic	$\theta \equiv \tfrac{1}{2}(\beta - \alpha)$	

Theorem 5.14.12 Vertex Exterior to a Circle: by Tangent-Tangent Proportional (VTTP)

1g	Given	the above geometry figure F5.14.7 "Tangent-Tangent Vertex Exterior to a Circle (VTT)"
2g		$r \equiv OA$
Steps	Hypothesis	$r^2 = PO\ BO$
1	From 1g, 2g and D5.1.31 Circle	$r \equiv OA \equiv OT_a \equiv OT_b \equiv OC$ inscribe in the circle	
2	From 1g	$90° \equiv \angle PT_aO$	$90° \equiv \angle PT_bO$
3	From 1g, 2 and D5.1.19 Triangle; Right	ΔPT_aO right triangle	ΔPT_bO right triangle
4	From 1g and 3	OP hypotenuse common for ΔPT_aO	OP hypotenuse common for ΔPT_bO
5	From 1, 2, 3 and T5.6.11 Congruency: by Angle-Side-Side (ASS)	$\Delta PT_aO \cong \Delta PT_bO$	
6	From 1g, 1 and D5.1.16 Triangle; Isosceles	ΔT_aOT_b isosceles triangle	
7	From 1, 2, 6, D5.1.16 Triangle; Isosceles and T5.9.2 Isosceles Triangle: Bisected into Two Congruent Right Triangles	$\Delta BT_aO \cong \Delta BT_bO$ since ΔPT_aO, ΔPT_bO are in equal parts	
8	From 1g, 6, 7 and T5.9.2 Isosceles Triangle: Bisected into Two Congruent Right Triangles	$\angle T_aOB \cong \frac{1}{2}\theta$ $\angle BT_aO \cong 90° - \frac{1}{2}\theta$	$\angle T_bOB \cong \frac{1}{2}\theta$ $\angle BT_bO \cong 90° - \frac{1}{2}\theta$
9	From 1g and T5.4.6 Solving for Third Angle of a Triangle	$\angle T_aBO \cong 180° - \angle BT_aO - \angle T_aOB$ for $(\Delta T_aBO, \angle T_aBO)$	$\angle T_bBO \cong 180° - \angle BT_bO - \angle T_bOB$ for $(\Delta T_bBO, \angle T_bBO)$
10	From 1g, 8, 9, A5.2.18 GIV Transitivity of Angles, A4.2.8 Inverse Add and A4.2.7 Identity Add	$\angle T_aBO \cong \angle T_bBO \cong 90°$	
11	From 7, 10 and D5.1.19 Triangle; Right	ΔBT_aO and ΔBT_bO are right triangles	
12	From 1g, 2, T5.8.6 Right Triangle: Non-Ninety Degree Angles are Complementary and T4.3.5B Equalities: Left Cancellation by Addition	$\angle BT_aO \cong 90° - \angle BT_aP$	$\angle BT_bO \cong 90° - \angle BT_bP$
13	From 2, 12, T5.8.6 Right Triangle: Non-Ninety Degree Angles are Complementary and A5.2.18 GIV Transitivity of Angles	$\angle BT_aO \cong 90° - (90° - \angle T_aPO)$	$\angle BT_bO \cong 90° - (90° - \angle T_bPO)$

14	From 13, T4.3.4A Equalities: Reversal of Right Cancellation by Addition and T4.3.5B Equalities: Left Cancellation by Addition	$90° - \angle BT_aO \cong 90° - \angle T_aPO$	$90° - \angle BT_bO \cong 90° - \angle T_bPO$
15	From 14, T4.3.2 Equalities: Cancellation by Addition and T4.4.3 Equalities: Cancellation by Multiplication	$\angle BOT_a \cong \angle T_aOP$	$\angle BOT_b \cong \angle T_bOP$
16	From 2, 10, 15 and T5.5.13 Similar Triangles: Angle-Angle (AA)	$\Delta T_aPO \sim \Delta BT_aO$	$\Delta T_aPO \sim \Delta BT_bO$
17	From 16	$PO / T_aO = PT_a / T_aB = T_aO / BO$	$PO / T_bO = PT_b / T_bB = T_bO / BO$
$\therefore$	From 1, 17, A5.2.15 GIV Transitivity of Segments and T4.4.8 Equalities: Cross Product of Proportions	$r^2 = PO\ BO$	

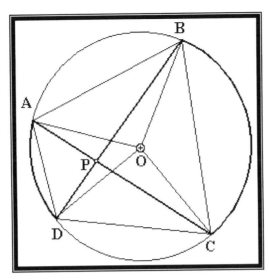

Figure 5.14.8 Vertex Interior to a Circle (VIC)

Theorem 5.14.13 Vertex Interior to a Circle (VIC)

1g	Given	the above geometry figure F5.14.8 "Vertex Interior to a Circle (VIC)"
2g		$\alpha \equiv \angle AOD$
3g		$\beta \equiv \angle BOC$
4g		$\theta \equiv \angle BPC$
Steps	Hypothesis	$\theta \equiv \tfrac{1}{2}(\alpha + \beta)$ internal apex at point-P
1	From 1g and T5.12.1 Inscribed Angle Theorem by Triangle Trisection of Circumference	$\angle ACD \cong \tfrac{1}{2}\angle AOD$ for $\triangle ACD$
2	From 2g, 1and A5.2.18 GIV Transitivity of Angles	$\angle ACD \cong \tfrac{1}{2}\alpha$
3	From 1g and T5.12.1 Inscribed Angle Theorem by Triangle Trisection of Circumference	$\angle BDC \cong \tfrac{1}{2}\angle BOC$ for $\triangle BCD$
4	From 3g, 31and A5.2.18 GIV Transitivity of Angles	$\angle BDC \cong \tfrac{1}{2}\beta$
5	From 1g and T5.4.8 External Angle Supplementary of two Opposing Vertices	$\angle BPC \cong \angle PCD + \angle PDC$
6	From 1g share common angle	$\angle PCD \cong \angle ACD$ $\Big|$ $\angle PDC \cong \angle BDC$
7	From 2, 4, 5, 631and A5.2.18 GIV Transitivity of Angles	$\angle BPC \cong \tfrac{1}{2}\alpha + \tfrac{1}{2}\beta$
∴	From 4g, 7, A5.2.18 GIV Transitivity of Angles and A4.2.14 Distribution	$\theta \equiv \tfrac{1}{2}(\alpha + \beta)$

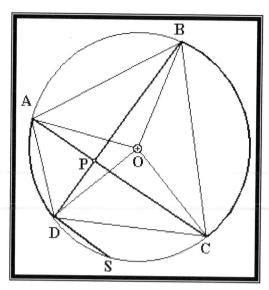

Figure 5.14.9 Vertex and Parallel Lines Interior to a Circle (VPLIC)

Theorem 5.14.14 Vertex and Parallel Lines Interior to a Circle (VPLIC)

1g	Given	the above geometry figure F5.14.9 "Vertex and Parallel Lines Interior to a Circle (VPLIC)"
2g		AC ∥ DS
3g		$\alpha \equiv \angle AOD$
4g		$\beta \equiv \angle BOC$
5g		$\theta \equiv \angle BPC$
Steps	Hypothesis	$\angle BDS \cong \theta \equiv \frac{1}{2}(\alpha + \beta)$ internal apex at point-P for AC ∥ DS
1	From 1g, 5g and point-P inscribed in a circle by T5.13.13 Vertex Interior to a Circle (VIC)	$\theta \equiv \angle BPC \equiv \frac{1}{2}(\alpha + \beta)$
2	From 1g, 2g and T5.4.4 Parallel Lines: Congruent Alternate Interior Angles	$\angle BDS \cong \angle BPC$ for DB as transversal
∴	From 1, 2 and A5.2.18 GIV Transitivity of Angles	$\angle BDS \cong \theta \equiv \frac{1}{2}(\alpha + \beta)$

Section 5.15 Theorems on Cyclic Trapezoids

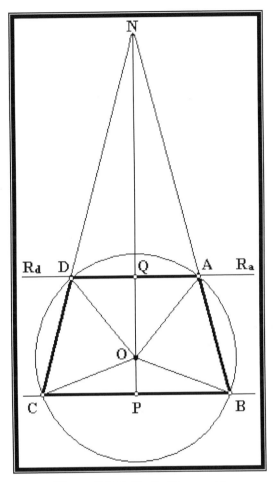

Figure 5.15.1 Cyclic Trapezoid

Theorem 5.15.1 Cyclic Trapezoid: Congruency of Externally Inscribed Isosceles Triangles

1g	Given	the above geometry figure F5.15.1 " ABCD Cyclic Trapezoid inscribed in circle-O"
2g		$R_a R_d \parallel BC$
3g		$r \equiv OA$
Steps	Hypothesis	$\Delta BOC \cong \Delta COD$ are isosceles
1	From 1g, 4g and D5.1.31 Circle	$r = OA = OB = OC = OD$
2	From 1g, 1 and D5.1.16 Isosceles Triangle	$\Delta AOD, \Delta AOB, \Delta BOC, \Delta COD$ are isosceles
3	From 2 and T5.9.2 Isosceles Triangle: Bisected into Two Congruent Right Triangles	$\Delta OPC \cong \Delta OPB$ are right triangles $\Delta OQD \cong \Delta OQA$ are right triangles

4	From 3	$\angle OQD \cong 90°$ for ΔOQD	$\angle OQA \cong 90°$ for ΔOQA
5	From 3 and T5.4.1 Intersection of Two Lines: Opposite Congruent Vertical Angles	$\angle NQD \cong 90°$	$\angle NQA \cong 90°$
6	From 3	$AQ \cong DQ$	
7	From 1g	NQ a common side	
8	From 5, 6, 7, T5.6.11 Congruency: by Angle-Side-Side (ASS) and D5.1.19 Triangle; Right	$\Delta NQD \cong \Delta NQA$ right triangle	
9	From 8 and D5.1.27 Congruent	$\angle NDQ \cong \angle NAQ$	$NA \cong ND$
10	From 9 and D5.1.16 Triangle; Isosceles	ΔAND isosceles triangle	
11	From 1g, 9 and T5.4.1 Intersection of Two Lines: Opposite Congruent Vertical Angles	$\angle R_dDC \cong \angle NAQ$	$\angle R_aAB \cong \angle NAQ$
12	From 2g, 11 and T5.4.4 Parallel Lines: Congruent Alternate Interior Angles	$\angle DCP \cong \angle R_dDC$	$\angle ABP \cong \angle R_aAB$
13	From 9, 11, 12 and A5.2.18 GIV Transitivity of Angles	$\angle DCP \cong \angle ABP$	
14	From 2 and T5.9.1 Isosceles Triangle: Bisecting Vertex Between Congruent Sides Parses Equal Parts	$\angle OCP \cong \angle OBP \equiv \alpha$	
15	From 1g	$\angle DCP \cong \angle OCP + \angle OCD$	$\angle ABP \cong \angle OBP + \angle OBA$
16	From 13, 15 and A5.2.18 GIV Transitivity of Angles	$\angle OCP + \angle OCD \cong \angle OBP + \angle OBA$	
17	From 14, 16 and A5.2.18 GIV Transitivity of Angles	$\alpha + \angle OCD \cong \alpha + \angle OBA$	
18	From 17 and T4.3.2 Equalities: Cancellation by Addition	$\angle OCD \cong \angle OBA$	
∴	From 1, 2, 18 and T5.6.9 Congruency: by Side-Side-Angle (SSA)	$\Delta BOC \cong \Delta COD$ are isosceles	

Theorem 5.15.2 Cyclic Trapezoid: Congruency of Internal Arclengths

1g	Given	the above geometry figure F5.15.1 " ABCD Cyclic Trapezoid inscribed in circle-O"	
2g		$r \equiv OA$	
3g		$\alpha \equiv \angle AOD$	
4g		$\delta \equiv \angle BOC$	
Steps	Hypothesis	$\widehat{AD} \cong \alpha r$	$\widehat{BC} \cong \delta r$ internal arclengths
1	From 2g and D5.1.31 Circle	$r \equiv OA = OB = OC = OD$	
2	From T5.14.1 Cyclic Trapezoid: Congruency of Inscribed Isosceles Triangles	ΔAOD inscribed isosceles triangles and not necessarily congruent	ΔBOC
2	From 2, 3g, 4g and D5.1.35 Subtending Angle of the Cord	$\alpha \equiv \angle AOD$ subtending angle resulting from the parallel sides of the cyclic trapezoid transecting the circle	$\delta \equiv \angle BOC$
∴	From 1, 2 and D5.1.35 Subtending Arclength of a Cord	$\widehat{AD} \cong \alpha r$	$\widehat{BC} \cong \delta r$ internal arclengths

Theorem 5.15.3 Cyclic Trapezoid: External Arclengths

1g	Given	the above geometry figure F5.15.1 " ABCD Cyclic Trapezoid inscribed in circle-O"		
2g		$\beta \equiv \angle AOB$		
3g		$\gamma \equiv \angle DOC$		
4g		$r \equiv OA$		
Steps	Hypothesis	$\widehat{AB} \cong r\beta \cong \widehat{CD} \cong r\gamma$ external arclength		
1	From 4g and D5.1.31 Circle	$r \equiv OA = OB = OC = OD$		
2	From 1g and T5.14.1 Cyclic Trapezoid: Congruency of Inscribed Isosceles Triangles	$\Delta AOB \cong \Delta COD$		
3	From 2g, 3g, 2, A5.2.18 GIV Transitivity of Angles and D5.1.27 Congruent	$AB \cong CD$	$\angle AOB \cong \angle COD$	$\beta \cong \gamma$
4	From 3, A4.2.2 Equality and T4.4.2 Equalities: Uniqueness of Multiplication	$r\beta \cong r\gamma$		
∴	From 2, 4, and D5.1.36 Subtending Arclength of a Cord	$\widehat{AB} \cong r\beta \cong \widehat{CD} \cong r\gamma$ external arclength		

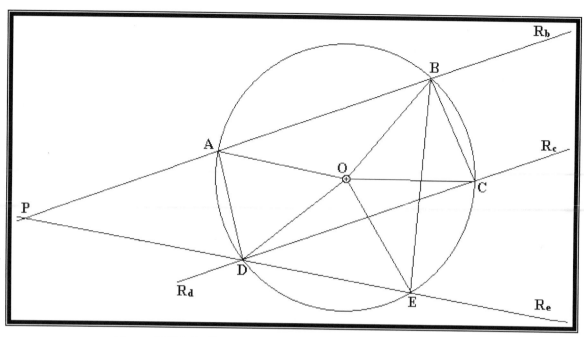

Figure 5.15.2 Cyclic Trapezoid Congruent External Apex Angles

Theorem 5.15.4 Cyclic Trapezoid: Congruency of External Apex Angles

1g	Given	the above geometry figure F5.15.2 "Cyclic Trapezoid Congruent External Apex Angles"	
2g		$PR_b \parallel R_dR_c$	
3g		$r \equiv OA$	
4g		$\alpha \equiv \angle AOD$	
5g		$\beta \equiv \angle BOE$	
6g		$\theta \equiv \angle R_bPR_e$	
Steps	Hypothesis	$\theta \equiv \angle R_bPR_e \cong \frac{1}{2}(\beta - \alpha)$	$\theta \equiv \angle CDR_e \cong \frac{1}{2}(\beta - \alpha)$
1	From 3g and D5.1.31 Circle	$r \equiv OA = OB = OC = OD = OE$	
2	From 1g, 4g, 5g and T5.13.7 Vertex Exterior to a Circle: by Secant-Secant (VSS)	$\theta \equiv \frac{1}{2}(\beta - \alpha)$	
3	From 1g	PR_e transverse line across $PR_b \parallel R_dR_c$ intersecting at points P and D	
4	From 3 and T5.4.4 Parallel Lines: Congruent Alternate Interior Angles	$\theta \equiv \angle R_bPR_e \equiv \angle CDR_e$	
∴	From 2 and 4	$\theta \equiv \angle R_bPR_e \cong \frac{1}{2}(\beta - \alpha)$	$\theta \equiv \angle CDR_e \cong \frac{1}{2}(\beta - \alpha)$

Section 5.16 Theorems on Specific Origins

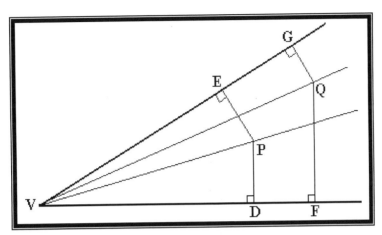

Figure 5.16.1 Internal Orthoangle Bisector

Theorem 5.16.1 Internal Orthoangle Bisector

1g 2g 3g	Given	the above geometry figure F5.16.1 "Internal Orthoangle Bisector" ∠EVP ≅ ∠DVP 90° ≡ ∠VEP ≅ ∠VDP construct perpendicular from angle bisector	
Steps	Hypothesis	EP ≅ PD, hence VP is the internal orthoangle bisector of ∠GVF	
1	From 3g, D5.1.19 Triangle; Right	ΔVPE	ΔVPD are right triangles
2	From 1g	VP ≅ VP common hypotenuse for ΔVPE and ΔVPD	
3	From 2g, 2 and T5.6.16 Congruency of Right Triangles by Common Hypotenuse-Leg (CHL)	ΔVPE ≅ ΔVPD	
∴	From 3 and D5.1.27 Congruent	EP ≅ PD, hence VP is the internal orthoangle bisector of ∠GVF	

Theorem 5.16.2 Converse of Internal Orthoangle Bisector

1g 2g 3g	Given	the above geometry figure F5.16.1 "Internal Orthoangle Bisector" QF ≅ QG 90° ≡ ∠VGQ ≅ ∠VFQ construct perpendicular from angle bisector	
Steps	Hypothesis	∠QVG ≅ ∠QVF hence VQ is the internal orthoangle bisector and lies on VP	
1	From 3g, D5.1.19 Triangle; Right	ΔQGV	ΔQFV are right triangles
2	From 1g	VQ ≅ VQ common hypotenuse for ΔQGV and ΔQFV	
3	From 2g, 2 and T5.6.16 Congruency of Right Triangles by Common Hypotenuse-Leg (CHL)	ΔQGV ≅ ΔQFV	
∴	From 3 and D5.1.27 Congruent	∠QVG ≅ ∠QVF hence VQ is the internal orthoangle bisector and lies on VP	

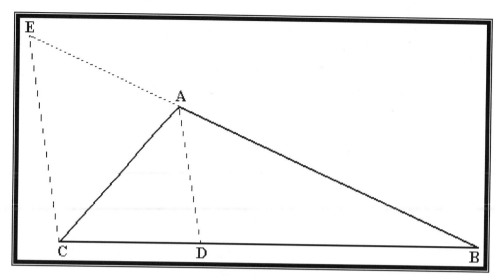

Figure 5.16.2 Proportions of Internal Angle Bisector

Theorem 5.16.3 Proportions of Internal Angle Bisector

1g	Given	the above geometry figure F5.16.2 "Proportions of Internal Angle Bisector"
2g		$\angle CAD \cong \angle DAB \cong \frac{1}{2}\angle CAB$ by segment AD
3g		EC $\parallel$ AD
Steps	Hypothesis	AC / BA = CD / BD
1	From 1g, D5.1.8 Flat Angle and T4.3.3B Equalities: Right Cancellation by Addition	$\angle EAC \cong 180° - \angle CAB$
2	From 3g and T5.4.4 Parallel Lines: Congruent Alternate Interior Angles	$\angle ACE \cong \angle CAD$
3	From 1, 2, T5.4.6 Solving for Third Angle of a Triangle and A5.2.18 GIV Transitivity of Angles	$\angle AEC \cong 180° - (180° - \angle CAB) - \angle CAD$ for ($\triangle AEC, \angle AEC$)
4	From 3, A4.2.14 Distribution, A4.2.8 Inverse Add and A4.2.7 Identity Add	$\angle AEC \cong \angle CAB - \angle CAD$
5	From 1g and T4.3.3B Equalities: Right Cancellation by Addition	$\angle DAB \cong \angle CAB - \angle CAD$
6	From 4, 5 and A5.2.18 GIV Transitivity of Angles	$\angle AEC \cong \angle DAB$
7	From 2g, 6 and A5.2.18 GIV Transitivity of Angles	$\angle AEC \cong \angle CAD$

8	From 2, 7 and A5.2.18 GIV Transitivity of Angles	$\angle ACE \cong \angle AEC$
9	From 8 and T5.9.4 Isosceles Triangle: Congruent Interior Angles and Sides	$\triangle ACE$ isosceles triangle
10	From 9 and D5.1.16 Triangle; Isosceles	$AE \cong AC$
11	From 1g	$\triangle EBC$ and $\triangle ABD$ have common vertex-B, hence share $\angle ABD$
12	From 3g and T5.4.4 Parallel Lines: Congruent Alternate Interior Angles	$\angle ADB \cong \angle ECB$ $\angle BEC \cong \angle BAD$ for EB and BC transversals cutting EC ∥ AD
13	From 11, 12 and D5.1.29 Similar Polygons	$\triangle EBC \sim \triangle ABD$
14	From 3g, 13 and T5.7.10 Similar Common Vertex with Proportional Sides (STPS)	$BE / BA = CE / DA = BC / BD$
15	From 14	$BE / BA = BC / BD$
16	From 1g, 12 and A5.2.15 GIV Transitivity of Segments	$(BA + AE) / BA = (BD + CD) / BD$
17	From 16, A4.2.14 Distribution, A4.2.13 Inverse Multp and T4.3.2 Equalities: Cancellation by Addition	$AE / BA = CD / BD$
∴	From 10, 17 and A5.2.15 GIV Transitivity of Segments	$AC / BA = CD / BD$

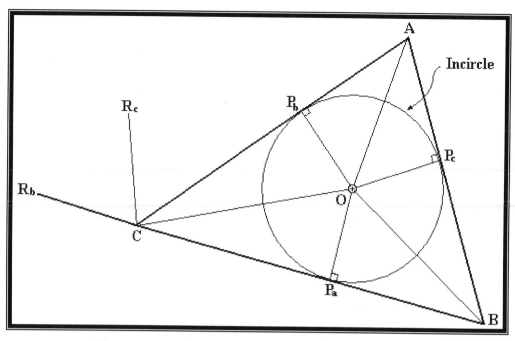

Figure 5.16.3 Concurrent Orthocenter and Angle Bisectors

Theorem 5.16.4 Concurrent Orthocenter and Angle Bisectors

1g	Given	the above geometry figure F5.16.3 "Concurrent Orthocenter and Angle Bisectors"		
2g		$r \equiv OP_a$		
3g		BC Tangent to P_a, AC Tangent to P_b and AB Tangent to P_c		
Steps	Hypothesis	OA bisects $\angle P_bAP_c$	OB bisects $\angle P_aAP_c$	OC bisects $\angle P_aAP_b$
		Hence concurrent point of angle bisectors coincides with the Orthocenter		
1	From 1g and 2g	$r \equiv OP_a \equiv OP_b \equiv OP_c$		
2	From 1g	OA common side to $\triangle AOP_b$ and $\triangle AOP_c$		
3	From 1g	OB common side to $\triangle BOP_a$ and $\triangle BOP_c$		
4	From 1g	OC common side to $\triangle COP_a$ and $\triangle COP_b$		
5	From 3g, 1, 2 and T5.6.16 Congruency of Right Triangles by Common Hypotenuse-Leg (CHL)	$\triangle AOP_b \cong \triangle AOP_c$		$\angle P_bAO \cong \angle P_cAO$
6	From 3g, 1, 3 and T5.6.16 Congruency of Right Triangles by Common Hypotenuse-Leg (CHL)	$\triangle BOP_a \cong \triangle BOP_c$		$\angle P_aBO \cong \angle P_cBO$
7	From 3g, 1, 4 and T5.6.16 Congruency of Right Triangles by Common Hypotenuse-Leg (CHL)	$\triangle COP_a \cong \triangle COP_b$		$\angle P_aCO \cong \angle P_bCO$
$\therefore$	From 5, 6 and 7	OA bisects $\angle P_bAP_c$	OB bisects $\angle P_aAP_c$	OC bisects $\angle P_aAP_b$

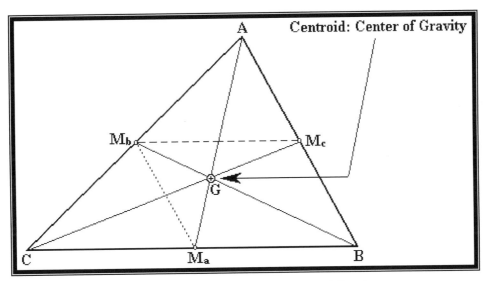

Figure 5.16.4 Centroid as the Center of Gravity

Theorem 5.16.5 Centroid: Center of Gravity

1g	Given	the above geometry figure F5.16.4 "Centroid as the Center of Gravity"
2g		M_aM_b a median line
3g		M_bM_c a median line
4g		$AM_c \cong M_cB \cong \frac{1}{2}AB$
Steps	Hypothesis	$AG / AMa = 2/3$
1	From 2g and T5.10.2 Median Line is Parallel to the Opposite Side of a Triangle	$M_aM_b \parallel AB$ Median line is parallel to the opposite side of $\triangle ABC$
2	From 3g and T5.10.2 Median Line is Parallel to the Opposite Side of a Triangle	$M_bM_c \parallel BC$ Median line is parallel to the opposite side of $\triangle ABC$
3	From 1g and T5.4.1 Intersection of Two Lines: Opposite and Equal Angles	$\angle M_aGM_b \cong \angle AGB$ for $\triangle M_aGM_b$ and $\triangle AGB$
4	From 1g, 1 and T5.4.4 Parallel Lines: Congruent Alternate Interior Angles	$\angle M_aM_bG \cong \angle ABG$
5	From 1g, 2 and T5.4.4 Parallel Lines: Congruent Alternate Interior Angles	$\angle M_bM_aG \cong \angle BAG$
6	From 3, 4, 5 and T5.5.13 Similar Triangles: Angle-Angle (AA)	$\triangle AGB \sim \triangle M_aGM_b$

7	From 1g	$AM_a \cong AG + GM_a$
8	From A4.2.2A Equality	$AG / AM_a = AG / AM_a$
9	From 7, 8 and A5.2.15 GIV Transitivity of Segments	$AG / AM_a = AG / (AG + GM_a)$
10	From 9, A4.2.12 Identity Multp, A4.2.13 Inverse Multp, A4.2.14 Distribution and A4.2.13 Inverse Multp	$AG / AM_a = 1 / (1 + GM_a / AG)$
11	From 6 and T5.7.1 Similar Triangles: Proportional Triangles	$GM_a / AG = M_aM_b / AB$
12	From 4g, A4.2.2A Equality, T4.4.2 Equalities: Uniqueness of Multiplication, T4.4.15 Equalities: Product by Division, Common Factor, A4.2.12 Identity Multp and A4.2.2B Equality	$AB = 2\, M_cB$
13	From 2g, 12, T5.10.3 Median is Congruent to Half of the Opposite Side of a Triangle and A5.2.15 GIV Transitivity of Segments	$AB = 2\, M_aM_b$
14	From 11, 13, A5.2.15 GIV Transitivity of Segments, A4.2.10 Commutative Multp, A4.2.11 Associative Multp, D4.1.4A Rational Numbers, A4.2.13 Inverse Multp and A4.2.12 Identity Multp	$GMa / AG = MaMb / 2\, MaMb = \tfrac{1}{2}$
15	From 10, 14 and A4.2.3 Substitution	$AG / AMa = 1 / (1 + \tfrac{1}{2})$
16	From 15, T4.5.5 Equalities: Addition of Two Rational Fractions and D4.1.19 Primitive Definition for Rational Arithmetic	$AG / AMa = 1 / (3 / 2)$ for $a = b = c = 1$ and $d = 2$
∴	From 16 and T4.5.3 Equalities: Compound Reciprocal Products	$AG / AMa = 2/3$ for $a = c = 1$ and $b = 3$ and $d = 2$

Section 5.17 Signed Distance (Sense Magnitudes by Proportions)

Through out this treatise the notation of a directed-signed quantity-giving rise to a sense of ordination is a theme that constantly arises. So in Chapter 1 we see the orientation of a surface has a meaning. In the following proof, orientation of magnitude has to be considered before understanding how Menelau's notation of a Directed Sense applies to his theorems. This proof was proposed and written, as is, by Dr. Richard E. Pfiefer, Department of Mathematics, California State University of San Jose, in order to explain the property of Menelau's signed distance.

F5.18.1-A

Where $0 \le AB$ is defined as distance or magnitude of displacement and reordering points gives $BA < 0$ or rewriting reordered point notation

$$AB = -BA.$$

In absolute positional notation | AB | is the regular position direction.

F5.18.1-B

With a freely positional [x] between A and B a simple ratio can be constructed dividing A and B by x:

$$\text{Ratio} \equiv \frac{AX}{XB}$$

As an example:

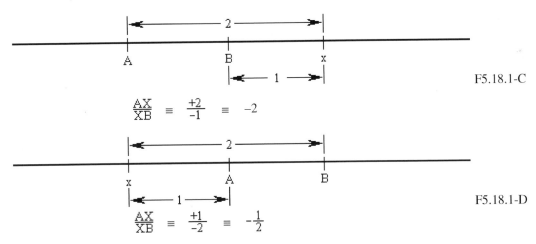

F5.18.1-C

$$\frac{AX}{XB} \equiv \frac{+2}{-1} \equiv -2$$

F5.18.1-D

$$\frac{AX}{XB} \equiv \frac{+1}{-2} \equiv -\frac{1}{2}$$

Graph geometric points as real number variables

F5.18.1-E

A free non-fixed point such as [x] can be represented as a function in the classical way,

$$f(x) = \frac{AX}{XB}$$

As an analytic geometric function,

$$f(x) \;=\; \frac{x-a}{b-x}$$

This meets our notion of how [x] my move between [a] and [b], however such a function must account for analytic anomalies, which can be resolved by the idea of a limit:

$$\frac{AX}{XB} \;\equiv\; \lim_{x\to\infty} \frac{x-a}{b-x} = -1$$

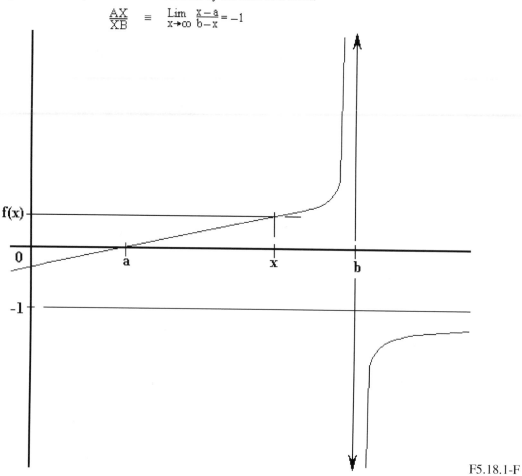

Figure 5.17.1 Directed Sense

F5.18.1-F

Here the hyperbola function f(x) has a
Domain of x: all reals ≠ b
Range of f(x):   all reals ≠ –1
The function f(x) is 1–1 crossing in y only at one point.

Finally, we can see that the ratio of magnitudes can have a negative sense under this notation of direction in the limit:

$$\frac{AX}{XB} \;=\; -1$$

Using Menelau's Theorem, select a path CWs or CCWs (see arrows about the triangle ΔABC) to establish a sense of positive direction. Whatever course is selected for continuity in order to maintain that sense of a positive direction it must be preserved, hence BA = –AB for a CCW or AB = –BA CW direction.

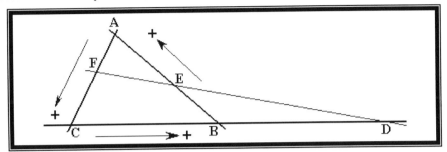

Figure 5.17.2 Menelau's Directed Sense

Determining the Menelau's Product

Rule 1) <u>By the Finger Rule</u>

Weather extended side or not, take your pointing index finger, starting at vertex-A, trace counter clockwise about the edge of the triangle without <u>ever</u> leaving the parameter path. Make sure the path traced alternates by vertex than transversal point. Moving backwards over the same line, and through a vertex to get to another transversal point, or vertex is permitted. However, this <u>must always</u> be done in the same <u>sense of direction</u>.

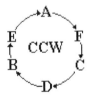

Rule 2) <u>By the CCW Menelau's Cycle</u>

Also, notice that the product of Menelau's Theorem is cyclic, starting with the triangle ΔACB in the Counter-Clock-Wise direction. Interleaving with the transversal points. Starting with the internal endpoint partitioning the first two vertices in the CCW direction, the external endpoint and finally the midpoint.

Theorem 5.17.1 Menelau's Absolute Sense of Direction

1g	Given	AB ≡ –BA Menelau's Sense of Direction
Steps	Hypothesis	| AB | = | BA |
∴	From 1g, A5.2.15 GIV Transitivity of Segments and D4.1.11 Absolute Value | a |	| AB | = | BA |

Tensor Calculus & Physics: A General Treatise

Theorem 5.17.2 Menelau's Theorem-CCW

1g	Given	the above geometry figure F5.17.2 "Menelau's Directed Sense; ΔACB and transversal F, E and D cutting FD"
2g		$\alpha \equiv \angle CDF$
3g		$\beta \equiv \angle DEB$
4g		$\gamma \equiv \angle AFE$
Steps	Hypothesis	$(AF / FC)\ (CD / DB)\ (BE / EA) = -1$
1	From 1g and LxE.2.1A	$\sin(\angle DEB) / |BD| \equiv \sin(\angle BDE) / |EB|$ for ΔBDE
2	From 1g and LxE.2.1A	$\sin(\angle FEA) / |AF| \equiv \sin(\angle AFE) / |AE|$ for ΔAFE
3	From 1g and LxE.2.1A	$\sin(\angle CDF) / |FC| \equiv \sin(\angle DFC) / |CD|$ for ΔDFC
4	From 2g, 3g, 1 and A5.2.18 GIV Transitivity of Angles	$\sin(\beta) / |BD| \equiv \sin(\alpha) / |EB|$ for ΔEBD
5	From 3g, 4g, 2, T5.4.4 Parallel Lines: Congruent Alternate Interior Angles and A5.2.18 GIV Transitivity of Angles	$\sin(\beta) / |AF| \equiv \sin(\gamma) / |AE|$ for ΔAFE
6	From 1g, 2g, 4g, 3, C4.4.1.1 Intersection of Two Lines: Specific Solution for Flat line Angles and A5.2.18 GIV Transitivity of Angles	$\sin(\alpha) / |FC| \equiv \sin(180° - \gamma) / |CD|$ for ΔDFC
7	From 4 and T4.4.14 Equalities: Formal Cross Product	$|EB| / |BD| = \sin(\alpha) / \sin(\beta)$ following CCW
8	From 5 and T4.4.14 Equalities: Formal Cross Product	$|AF| / |AE| = \sin(\beta) / \sin(\gamma)$ following CCW
9	From 6, DxE.1.7.5 and T4.4.14 Equalities: Formal Cross Product	$|CD| / |FC| = \sin(\gamma) / \sin(\alpha)$ following CCW
10	From 7, 8, 9, T4.4.2 Equalities: Uniqueness of Multiplication, A4.2.10 Commutative Multp, A4.2.11 Associative Multp, A4.2.13 Inverse Multp and A4.2.12 Identity Multp	$(|EB| / |BD|)\ (|AF| / |AE|)\ (|CD| / |FC|) = 1$
11	From 10, D4.4.1B Rational Numbers, A4.2.11 Associative Multp and T5.16.1 Menelau's Absolute Sense of Direction	$(|BE|\ |AF|\ |CD|) / (|EA|\ |DB|\ |FC|) = 1$

12	From 11 and T4.6.3 Equalities: Absolute Product is the Absolute of the Products	|(BE AF CD) / (DB EA FC)| = 1
13	From 12, A4.2.10 Commutative Multp and D4.1.11 Absolute Value |a|	(AF CD BE) / (FC DB EA) = ±1
14	From 1g, 13 and Menelau's Sense of Direction by sign	(+1 +1 +1) / (+1 −1 +1) = −1
∴	From 13 and 14	(AF / FC) (CD / DB) (BE / EA) = −1

Rule 3) By the CW Menelau's Cycle

Also, notice for the same geometry that Menelau's Theorem is symmetrical starting with the triangle ΔABC in the Counter-Clock-Wise direction. Again interleaving with the transversal points. However, this time starting with the midpoint partitioning the first two vertices in the CW direction, the external endpoint and finally the internal endpoint.

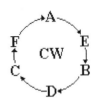

Theorem 5.17.3 Menelau's Theorem-CW

1g	Given	the above geometry figure F5.17.2 "Menelau's Directed Sense; ΔACB and transversal F, E and D cutting FD"
2g		$\alpha \equiv \angle CDF$
3g		$\beta \equiv \angle DEB$
4g		$\gamma \equiv \angle AFE$
Steps	Hypothesis	(AF / FC) (CD / DB) (BE / EA) = −1
1	From 1g and LxE.2.1A	$\sin(\angle EFA) / \|AE\| \equiv \sin(\angle AEF) / \|FA\|$ for ΔEFA
2	From 1g and LxE.2.1A	$\sin(\angle BDE) / \|EB\| \equiv \sin(\angle DEB) / \|BD\|$ for ΔBDE
3	From 1g and LxE.2.1A	$\sin(\angle CFD) / \|DC\| \equiv \sin(\angle FDC) / \|CF\|$ for ΔCFD
4	From 1g, 3g, 4g, 1, T5.4.1 Intersection of Two Lines: Opposite Congruent Vertical Angles and A5.2.18 GIV Transitivity of Angles	$\sin(\gamma) / \|AE\| \equiv \sin(\beta) / \|FA\|$ for ΔEFA
5	From 1g, 3g, 2 and A5.2.18 GIV Transitivity of Angles	$\sin(\alpha) / \|EB\| \equiv \sin(\beta) / \|BD\|$ for ΔBDE
6	From 1g, 2g, 4g, 3, C4.4.1.1 Intersection of Two Lines: Specific Solution for Flat line Angles and A5.2.18 GIV Transitivity of Angles	$\sin(180° − \gamma) / \|DC\| \equiv \sin(\alpha) / \|CF\|$ for ΔCFD
7	From 4 and T4.4.14 Equalities: Formal Cross Product	$\|FA\| / \|AE\| = \sin(\beta) / \sin(\gamma)$ following CW
8	From 5 and T4.4.14 Equalities: Formal Cross Product	$\|EB\| / \|BD\| = \sin(\alpha) / \sin(\beta)$ following CW
9	From 6, DxE.1.7.5 and T4.4.14 Equalities: Formal Cross Product	$\|DC\| / \|CF\| = \sin(\gamma) / \sin(\alpha)$ following CW

10	From 7, 8, 9, T4.4.2 Equalities: Uniqueness of Multiplication, A4.2.10 Commutative Multp, A4.2.11 Associative Multp, A4.2.13 Inverse Multp and A4.2.12 Identity Multp	(|FA| / |AE|) (|EB| / |BD|) (|CD| / |CF|) = 1
11	From 10, D4.4.1B Rational Numbers, A4.2.11 Associative Multp, T5.16.1 Menelau's Absolute Sense of Direction and A4.2.10 Commutative Multp	(|BE| |AF| |CD|) / (|EA| |DB| |FC|) = 1
12	From 11 and T4.6.3 Equalities: Absolute Product is the Absolute of the Products	|(BE AF CD) / (DB EA FC)| = 1
13	From 12, A4.2.10 Commutative Multp and D4.1.11 Absolute Value |a|	(AF CD BE) / (FC DB EA) = ±1
14	From 1g, 13 and Menelau's Sense of Direction by sign	[(+1) (+1) (+1)] / [(+1) (−1) (+1)] = −1
∴	From 13 and 14	(AF / FC) (CD / DB) (BE / EA) = −1

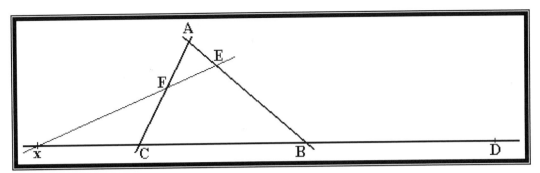

Figure 5.17.3 Converse of Menelau's Theorem

Theorem 5.17.4 Converse of Menelau's Theorem

1g	Given	the above geometry figure F5.17.3 "Converse of Menelau's Theorem; ΔABC with points X, F and E on AB, BC and CA"
2g		the above geometry figure F4.16.2 of ΔABC with points D, E and F on AB, BC and CA
Steps	Hypothesis	D, E, and F lie on a collinear line the transversal
1	From 1g	EF cuts AB through [x], hence XEF is a transversal
2	From 1 and T5.16.2 Menelau's Theorem-CCW	$(AF / FC)(CX / XB)(BE / EA) = -1$
3	From 1g	EF cuts AB through D, hence DEF is a transversal
4	From 3 and T5.16.2 Menelau's Theorem-CCW	$(AF / FC)(CD / DB)(BE / EA) = -1$
5	From 2, 4 and A4.2.3 Substitution	$(AF / FC)(CX / XB)(BE / EA) = (AF / FC)(CD / DB)(BE / EA)$
6	From 5 and T4.4.3 Equalities: Cancellation by Multiplication	$CX / XB = CD / DB$
7	From 6 and T4.4.11 Equalities: Uniqueness of Division	$CX = CD$ $XB = DB$
8	From 7	$X \longleftrightarrow D$ vertices
∴	From 8	It follows D, E, and F lie on a collinear line the transversal

The Ceva's triangle is constructed using Ceva's line segments from any given vertex in a triangle to any point on the opposing side. They are constructed in just such away that they are concurrent at a point-P. In the use of Ceva's Theorem, the path for the product is CCWs (see arrows about the triangle ΔABC) following the segments point-by-point about the triangle. This creates a sense of positive direction, hence AB = –BA for a CCW direction.

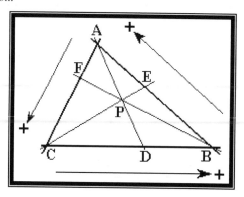

Figure 5.17.4 Ceva's Directed Sense

Theorem 5.17.5 Ceva's Theorem

1g 2g 3g	Given	the above geometry figure F5.17.4 "Ceva's Directed Sense" Ceva's segments AD, CE and BF concurrent at point-P AB = –BA for a CCW sense of direction
Steps	Hypothesis	(AF / FC) (CD / DB) (BE / EA) = +1
1	From 1g and T5.16.2 Menelau's Theorem-CCW	(AF / FC) (CB / BD) (DP / PA) = –1 for F, P and B transversal line cutting ΔACD
2	From 1g and T5.16.2 Menelau's Theorem-CCW	(AP / PD) (DC / CB) (BE / EA) = –1 for C, P and E transversal line cutting ΔABD
3	From 1, 2 and T4.4.2 Equalities: Uniqueness of Multiplication	(AF / FC) (DC / C̶B̶) (C̶B̶ / BD) (DP / PA) (AP / PD) (BE / EA) = +1
4	From 3, A4.2.11 Associative Multp, A4.2.13 Inverse Multp and A4.2.12 Identity Multp	(AF / FC) (DC / BD) (BE / EA) (DP / PA) (AP / PD) = +1
5	From 1g and 3g	CD = –DC, DB = –BD, DP = –PD and PA = –AP
6	From 4, 5 and A5.2.15 GIV Transitivity of Segments	(AF / FC) (–CD / –DB) (BE / EA) (–P̶D̶ / –A̶P̶) (AP / P̶D̶) = +1
∴	From 6, A4.2.11 Associative Multp, A4.2.13 Inverse Multp and A4.2.12 Identity Multp	(AF / FC) (CD / DB) (BE / EA) = +1

Notice that the product of Ceva's Theorem cycles are

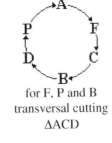

for F, P and B
transversal cutting
ΔACD

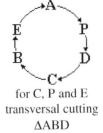

for C, P and E
transversal cutting
ΔABD

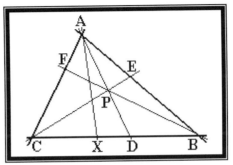

Figure 5.17.5 Converse of Ceva's Theorem

Theorem 5.17.6 Converse of Ceva's Theorem

1g 2g 3g	Given	the above geometry figure F5.17.5 "Converse of Ceva's Theorem" Ceva's segments AD, CE and BF for ΔABC concurrent at point-P Ceva's segments AX for ΔABC not concurrent at point-P
Steps	Hypothesis	Given Ceva's segments AD, CE and BF than they are concurrent at point-P
1	From 2g and T4.16.4 Ceva's Theorem	(AF / FC) (CD / DB) (BE / EA) = +1 for AD, CE and BF concurrent at point-P
2	From 3g and T4.16.4 Ceva's Theorem	(AF / FC) (CX / XB) (BE / EA) = +1 for AX, CE and BF not concurrent at point-P
3	From 1, 2 and A4.2.3 Substitution	(AF / FC) (CD / DB) (BE / EA) = (AF / FC) (CX / XB) (BE / EA)
4	From 3 and T4.4.3 Equalities: Cancellation by Multiplication	CD / DB = CX / XB
5	From 4 and T4.4.11 Equalities: Uniqueness of Division	CD = CX | DB = XB
6	From 5	X ←→ D vertices
∴	From 6	Given Ceva's segments AD, CE and BF than they are concurrent at point-P

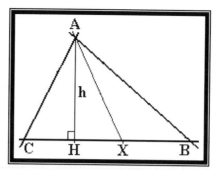

Figure 5.17.6 Ceva's Area

Theorem 5.17.7 Ceva's Area

1g	Given	the above geometry figure F5.17.6 "Ceva's Area"
Steps	Hypothesis	CX / XB = (|CA| sin(∠CAX)) / (|AB| sin(∠XAB))
1	From 1g and A4.2.2A Equality	CX / XB = CX / XB
2	From 1 and T4.5.2 Equalities: Inverse of Reciprocal Products	CX / XB = (½h x CX) / (½h x XB)
3	From 2 and C4.9.1.1 Algebraic: Area of a Triangle	CX / XB = K$_{CX}$ / K$_{XB}$
4	From 1 and T4.9.1 Trigonometric: Area of a Triangle	CX / XB = (½|CA||AX| sin(∠CAX)) / (½|XA||AB| sin(∠XAB))
∴	From 4, T5.16.1 Menelau's Absolute Sense of Direction, A4.2.10 Commutative Multp, and T4.5.6 Equalities: Contraction of a Rational Chain for n = 2	CX / XB = (|CA| sin(∠CAX)) / (|AB| sin(∠XAB))

Theorem 5.17.8 Ceva's Theorem by Trigonometry

1g 2g	Given	the above geometry figure F5.17.6 "Ceva's Area" AD, BF and CE are Ceva's segments for $\triangle ABC$
Steps	Hypothesis	$+1 = (\sin(\angle ABF)\ \sin(\angle CAD)\ \sin(\angle BCE))\ /$ $(\sin(\angle FBC)\ \sin(\angle DAB)\ \sin(\angle ECA))$
1	From 1g, 2g and T4.16.7 Ceva's Area	$CD/DB = (|CA|\sin(\angle CAD)\ /\ (|AB)\ \sin(\angle DAB))$
2	From 1g, 2g and T4.16.7 Ceva's Area	$BE/EA = (|CB|\sin(\angle BCE)\ /\ (|CA|\ \sin(\angle ECA))$
3	From 1g, 2g and T4.16.7 Ceva's Area	$AF/FC = (|BA|\sin(\angle ABF)\ /\ (|BC|\ \sin(\angle FBC))$
4	From 1, 2, 3, T4.4.2 Equalities: Uniqueness of Multiplication, T5.16.1 Menelau's Absolute Sense of Direction and T4.5.6 Equalities: Contraction of a Rational Chain n = 3	$(AF\ /\ FC)\ (CD\ /\ DB)\ (BE\ /\ EA) =$ $(\sin(\angle ABF)\ \sin(\angle CAD)\ \sin(\angle BCE))\ /$ $(\sin(\angle FBC)\ \sin(\angle DAB)\ \sin(\angle ECA))$
$\therefore$	From 4, A4.2.3 Substitution and T4.16.4 Ceva's Theorem	$+1 = (\sin(\angle ABF)\ \sin(\angle CAD)\ \sin(\angle BCE))\ /$ $(\sin(\angle FBC)\ \sin(\angle DAB)\ \sin(\angle ECA))$

Section 5.18 Modern Elements: Extend Hilbert's Axioms to Include Construction

Table 5.18.1 Group VII Axioms of Construction

Axiom 5.18.1 GVII Constructible Point	A point can be constructed at the intersection of two or more geometric objects marking a singular position.
Axiom 5.18.2 GVII Construction-Line	Any constructed datum line that shares at least one or more points with one or more constructions. Establishes relative relationships of distance and orientation between geometric objects.
Axiom 5.18.3 GVII Union of Collinear Lines	Let AB be one segment of a straight line and B'C' be one segment of a straight line that B corresponds to B', and let A"B" and B"C" be two segments of a straight line that lie on opposite sides of B". Then if AB $\cong$ A"B" and B'C' $\cong$ B"C", we also have A"C" $\cong$ AB + B'C'.
Axiom 5.18.4 GVII Intersection of Collinear Lines	Let AB and BC be two segments of a straight line which lie on opposite sides of B, and let B'C' and C'D' be two segments of a straight line that lie on opposite sides of C', where B corresponds to B' and C corresponds to C', If BC $\cong$ B'C', as the intersecting segment. Also let A"B" and B"C" be two segments of a straight line which lie on opposite sides of B", where A corresponds to A", B' corresponds to B", B" corresponds to C' and C" corresponds to D', such that $$0 \leq |\,AC - B'D'\,| \text{ and}$$ $$A"C" \cong AB + C'D' \cong AC - B'D'.$$
Axiom 5.18.5 GVII Intersecting Non-collinear Lines	Let a non-collinear line [a] with line [b] have relative, respective angles $\angle A$ and $\angle B$ on a construction line, such that they are not congruent, then they construct one and only one concurrent point-P.
Axiom 5.18.6 GVII Intersecting Circles (Arcs)	Two circles C and C' having centers O and O' on a construction line have radius [r'] and [r] respectively and are distanced apart by OO', then they have: No constructible intersection points when $r' + r < OO'$ or $$(r'/OO') + (r/OO') < 1 \qquad \text{EQA}$$ One constructible intersection point-P when $r' + r \cong OO'$ or $$(r'/OO') + (r/OO') = 1 \qquad \text{EQB}$$ Two constructible intersection points-P and Q when $$OO' + 2r > r' + r > OO' \text{ or}$$ $$1 + 2(r/OO') > (r'/OO') + (r/OO') > 1 \qquad \text{EQC}$$ One constructible intersection point-P' when $r' - r \cong OO'$ or $$(r'/OO') - (r/OO') = 1 \qquad \text{EQD}$$ C-inscribed in C' for $r < r'$ when $r' - r > OO'$ or $$(r'/OO') - (r/OO') > 1 \qquad \text{EQE}$$

Axiom 5.18.7 GVII **Radius on a Construction Line**	Let a circle-C having a center concurrent at point-B on construction line AB and passing though point-A have a segment radius AB.
Axiom 5.18.8 GVII **Intersecting Two Sides of an Angle with a Circle**	Let a circle-A with center concurrent to vertex-A of angle $\angle A$ intersect point-B on angle ray-b and point-C on opposite angle ray-c, such that distances AB and AC are the radius of the circle-A.
Axiom 5.18.9 GVII **Oriented Congruent Geometric Objects**[5.18.1]	A geometric object's elementary components can be identically constructed any place on the geometric plane by translation and / or orientation, such that characterizing points (vertices) maintain a one-to-one correspondence, thereby constructing a congruent geometric object. Geometric object $\cong$ Geometric object' for all V $\longleftrightarrow$ V'
Axiom 5.18.10 GVII **Constructibility of Congruent Objects**	If one geometric object G is constructible then any congruent object G' can also be constructed. G $\cong$ G' is constructible
Axiom 5.18.11 GVII **Converse of Constructibility of Congruent Objects**	If two geometric objects are congruent and G is not constructible, then G' is not constructible as well. G $\cong$ G' are not constructible
Axiom 5.18.12 GVII **Absolute Magnitude of an Angle**[5.18.2]	If (h, k) is a constructed angle than it is a positive quantity $0 \leq$ (h, k). It would follow since a polygon is a constructed geometric object that all angles within that polygon are positive magnitudes.
Axiom 5.18.13 GVII **Sides of Similar Polygons Correspond to Opposite Angles**[5.18.3]	Similar polygons $\rho \sim \rho'$ have sides that correspond to ($\longleftrightarrow$) opposite and congruent angles, where AB $\longleftrightarrow \angle C$ and A'B' $\longleftrightarrow \angle C'$ such that $\angle C \cong \angle C'$ for all angles.
Axiom 5.18.14 GVII **Reflectivity of Correspondence**	If the quantity A $\longleftrightarrow$ A then A $\longleftrightarrow$ A it is *concurrent*
Axiom 5.18.15 GVII **Symmetry of Correspondence**	If the quantities A $\longleftrightarrow$ B then B $\longleftrightarrow$ A they are *congruent*
Axiom 5.18.16 GVII **Transitivity of Correspondence**	If the quantities A $\longleftrightarrow$ B and B $\longleftrightarrow$ C then A $\longleftrightarrow$ C.

From chapter 1, section 1.0, a seashell is seen to be asymmetric. When an image is reflected about a line, such as RR', that line is called the **line of symmetry** if symmetrical and **line of asymmetry** if otherwise and the resulting reflection is called the **anti- or mirror image** of the original object. If such an axis can be found through an object, that object is said to be **symmetrical** about the line of symmetry. If the axis of symmetry splits the object into two apparent independent objects, each object is said to be **asymmetrical** by them selves, which can be seen in the following figure of counter spiral seashells.

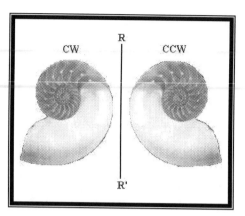

Figure 5.18.1 Asymmetrical Spiral Seashells

Table 5.18.2 Group VIII Axioms of Constructed Symmetries

Axiom 5.18.17 GVIII Symmetric Construction	Any two symmetric geometrical objects have a one-to-one correspondence about their line of symmetry in their space of construction.
Axiom 5.18.18 GVIII Asymmetric Construction	If any two asymmetric geometrical objects have a one-to-one correspondence about their line of asymmetry, in their space of construction, then there exists a hyperspace above the space of construction allowing the geometric object to be rotated through.
Axiom 5.18.19 GVIII Converse of Asymmetric Construction	If there exists a hyperspace above the space of construction that allows the geometric object to be rotated through, then there exists a hyper-axis of symmetry and the two asymmetric geometrical objects have a one-to-one correspondence.

The point as defined here is not to be confused with an analytic point, but the classical Euclidean point. In construction, the concern is, does the intersection of two geometric objects exist, so the point can be constructed. The nature of these axioms is just that, they allow us to establish constructible points. These points are not to be confused with precisely located points, such as found in analytic geometry, referenced to some coordinate grid, but relatively located to or on the construction.

Axiom 5.18.6 "GVII Intersecting Circles (Arcs)" is conditional and limited. One it is based on the ability that a triangle can be constructed and guarantees a certain number of erected intersection points. For other geometric objects it requires reevaluation to see which equation applies.

From these axioms constructible proofs can be developed, however caution needs to be used, this set may not be optimized, yet they may seem to cover all the possibilities, however they are only proposals, and need time and further research to acquire acceptable validity.

Axiom A5.18.6 "GVII Intersecting Circles (Arcs)" was designed only for triangles, hence any new construction, not a triangle, must be evaluated to verify compatibility and assigned the appropriate equation A through E. Example:

Table 5.18.3 Number of Intersections for A5.18.6 "GVII Intersecting Circles (Arcs)"

	A	B	C	D	E
Triangle	0	1	2	1	0
Adding Angles			2		

[5.18.1]Note: Congruency of segments, the other fundamental geometric element, is given by Hilbert's axiom group IV, A5.2.14 "GIV Symmetric-Reflexiveness of Segments" so does not need to be made part of the set of axioms on construction.

[5.18.2]Note: Axiom A5.18.12 GVII "Absolute Magnitude of an Angle" is one of those pivot axioms, such as Euclid's Fifth Postulate. It's a true axiom and stands on its own, for if it is modified, new and independent geometries arise forking off onto new paths. For example, Signed Distances necessary to evaluate Menelau's Theorem in projective geometry, or quadrature sign of an angle for analytic geometry, or tensor's sense of orientation in a manifold space.

[5.18.3]Note: Extended Hilbert's group VII Axioms of Construction; Axiom A5.18.13 "GVII Sides of Similar Polygons Correspond to Opposite Angles" replaces definition D5.1.29 Similar Polygons and the use of the following theorems where appropriate.

Section 5.19 Construction of Basic Congruent Geometric Objects

Theorem 5.19.1 Construction: Congruent Segments

1g	Given	circle-O has radius-r	
Step	Hypothesis	circle-O $\cong$ circle-O' for O $\longleftrightarrow$ O'	$r \cong r'$
$\therefore$	From 1g, and A5.18.9 GVII Congruent Geometric Objects[5.18.1]	circle-O $\cong$ circle-O' for O $\longleftrightarrow$ O'	$r \cong r'$

Theorem 5.19.2 Construction: Congruent Circles

1g	Given	circle-O has radius-r	
Step	Hypothesis	circle-O $\cong$ circle-O' for O $\longleftrightarrow$ O'	$r \cong r'$
$\therefore$	From 1g, and A5.18.9 GVII Congruent Geometric Objects[5.18.1]	circle-O $\cong$ circle-O' for O $\longleftrightarrow$ O'	$r \cong r'$

Theorem 5.19.3 Construction: Congruent Triangles

1g	Given	$\triangle ABC$		
Step	Hypothesis	$\triangle ABC \cong \triangle A'B'C'$ for A $\longleftrightarrow$ A' B $\longleftrightarrow$ B' C $\longleftrightarrow$ C'	$AB \cong A'B'$ $BC \cong B'C'$ $CA \cong C'A'$	$\angle CAB \cong \angle C'A'B'$ $\angle ABC \cong \angle A'B'C'$ $\angle BCA \cong \angle B'C'A'$
$\therefore$	From 1g and A5.18.9 GVII Congruent Geometric Objects[5.18.1]	$\triangle ABC \cong \triangle A'B'C'$ for A $\longleftrightarrow$ A' B $\longleftrightarrow$ B' C $\longleftrightarrow$ C'	$AB \cong A'B'$ $BC \cong B'C'$ $CA \cong C'A'$	$\angle CAB \cong \angle C'A'B'$ $\angle ABC \cong \angle A'B'C'$ $\angle BCA \cong \angle B'C'A'$

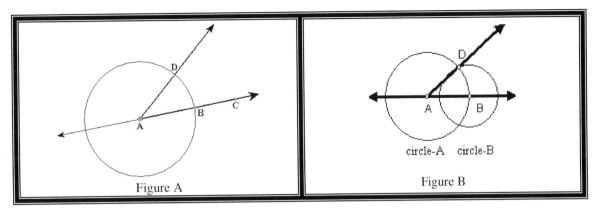

Figure 5.19.1 Component Construction of an Angle

Theorem 5.19.4 Construction: Segment Radius Spanning an Angle

1g	Given	the above geometry, figure F5.19.1A "Component Construction of an Angle"
2g		∠A at vertex-A
3g		construction line AC
4g		circle-A has radius-r < AC
5g		center of circle-A concurrent with vertex-A
Step	Hypothesis	segment radii AB and AD spans angle ∠BAD
1	From 4g, 5g and A5.18.8 GVII Intersecting Two Sides of an Angle with a Circle-C	construct intersecting point-B on AC and point-D on opposite ray and radius ≡ AB ≅ AD for circle-A.
2	From 4g, 1, A5.2.10 GII Three Point Existence and A5.2.8 GII Order of Direction of Points on a SL	A to B and B to C and C to B and B to A
∴	From 1, 2 and A5.18.7 GVII Radius on a Construction Line	construct circle-A with center concurrent with point-A passing through point-D, such that segment radii AB and AD spans angle ∠BAD

Theorem 5.19.5 Construction: Angle

1g	Given	the above geometry, figure F5.19.1B "Component Construction of an Angle"
2g		construction line AB
3g		∠BAD concurrent with vertex-A
4g		AB/AB + DB/AB > 1 for circle-A and circle-B
Step	Hypothesis	construct rays AD and AB bounding any ∠BAD, hence constructible
1	From 1g, 2g, 3g and T5.18.4 Construction: Segment Radius Spanning an Angle	construct intersection point-D with segment radius AB and AD spanning angle ∠BAD
2	From 4g, 1 and A.22.5 GVII Intersecting Circles (Arcs) EQC	DB is the radius of circle-B AB > 0 and DB > 0 bounding any ∠BAD
∴	From 2g, 1, 2 and D5.1.3 Ray	construct ray AD, such that with ray AB, they bound any ∠BAD, hence constructible

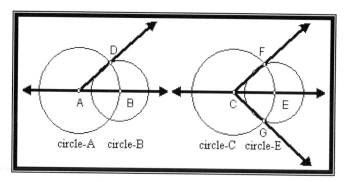

Figure 5.19.2 Constructing Congruent Angles

Theorem 5.19.6 Construction: Congruent Angles

1g	Given	the above geometry figure F5.19.2 "Constructing Congruent Angles"
2g		construction line AC
3g		construction line CE
4g		∠BAD concurrent with vertex-A
5g		AB/AB + DB/AB > 1 for circle-A and circle-B
6g		circle-A ≅ circle-C for construction circles
7g		circle-B ≅ circle-E for spanning circles
Step	Hypothesis	∠BAD ≅ ∠ECF ≅ ∠ECG for all parts congruent
1	From 1g, 2g, 4g and T5.18.5 Construction: Angle	construct rays AD and AB bounding ∠BAD
2	From 5g, 1 and A.22.5 GVII Intersecting Circles (Arcs) EQC	DB is the radius of spanning circle-B
3	From 6g and T5.18.1 Construction: Congruency of Circles	radii AB ≅ CE
4	From 1g, 3g and 3	construction circle-C concurrent with point-C
5	From 7g and T5.18.1 Construction: Congruency of Circles	radii DB ≅ FE ≅ GE
6	From 5g, 6g, 7g and A5.18.6 GVII Intersecting Circles (Arcs) EQC	construct intersecting point-F and point-G
7	From 4, 6 and D5.1.3 Ray	construct rays CF and CG bounding ∠ECF and ∠ECG
∴	From 6g, 7g, 1 and 7	∠BAD ≅ ∠ECF ≅ ∠ECG for all parts congruent

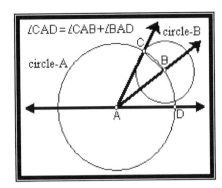

Figure 5.19.3 Adding Angles by Construction

Theorem 5.19.7 Construction: Addition of Angles

1g	Given	the above geometry figure F5.19.3 "Adding Angles by Construction"
2g		construction line AD along ray-a
3g		∠DAB concurrent with vertex-A and radius AD
4g		BC spanning radius of ∠BAC
5g		circle-A for construction circle
Step	Hypothesis	∠DAB + ∠BAC ≅ ∠DAC
1	From 1g, 2g, 3g, 4g and A5.18.8 GVII Intersecting Two Sides of an Angle with a Circle	construct ray AB spanning ∠DAB which intersects circle-A at point B with ray-b
2	From 3g, 4g, and A5.18.8C GVII Intersecting Two Sides of an Angle	AD/AB + BC/AB > 1 for circle-B intersecting circle-A at point-C
3	From 2 and A5.18.8 GVII Intersecting Two Sides of an Angle with a Circle	constructs a ray-c passing through point-C
4	From 1, 3 and D5.1.6 Angle	constructs ∠BAC
5	From 2g, 3 and D5.1.6 Angle	constructs ∠DAC
∴	From 3g, 4 and 5 have a common ray-b so it follows that	∠DAB + ∠BAC ≅ ∠DAC

Section 5.20 Constructibility of a Triangle Given Only Three Elements

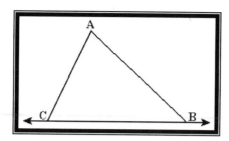

Figure 5.20.1 Constructing Side-Side-Side (SSS)

Theorem 5.20.1 Tri-Construction: Side-Side-Side (SSS)

1g	Given	the above geometry figure F5.20.1 "Constructing Side-Side-Side (SSS)"
2g		BC construction line
3g		AB and CA radii, Э (AB /BC) + (CA / BC) > 1 A5.18.6C GVII

Step	Hypothesis	Sufficient information to construct ΔABC implying		
		∠CAB	∠ABC	∠BCA
1	From 1g, 2g, 3g and A5.18.6 GVII Intersecting Circles (Arcs) EQC	Construct point-A at intersection where AB + CA > BC		
2	From 2g, 1 and D5.1.4-3 Line Segment and A5.2.1 GI Two Points on a SL	construct segments AB and CA		
∴	From 2g, 1, 2 and D5.1.13 Triangle	Sufficient information to construct ΔABC implying		
		∠CAB	∠ABC	∠BCA

Corollary 5.20.1.1 Tri-Construction: Side-Side-Side (SSS), Constructible Inequality

1g	Given	the above geometry figure F5.20.1 "Constructing Side-Side-Side (SSS)"
2g		BC construction line
3g		AB and CA radii, Э (AB /BC) + (CA / BC) > 1 A5.18.6C GVII

Step	Hypothesis	AB + CA > BC implying ΔABC is constructible
1	From 2g, 1 and D5.1.4-3 Line Segment and A5.2.1 GI Two Points on a SL	construct segments AB and CA
2	From 1g, 2g, 3g, 1 and A5.18.6 GVII Intersecting Circles (Arcs) EQC	Construct point-A at intersection where AB + CA > BC
∴	From 2g, 1, 2 and D5.1.13 Triangle	AB + CA > BC implying ΔABC is constructible

Corollary 5.20.1.2 Tri-Construction: Side-Side-Side (SSS), Non-constructible

1g	Given	the above geometry figure F5.20.1 "Constructing Side-Side-Side (SSS)"
2g		BC construction line
3g	To small never intersects	AB and CA radii, ∋ (AB /BC) + (CA / BC) < 1 A5.18.6A
4g	On the edge one intersection	AB and CA radii, ∋ (AB /BC) + (CA / BC) = 1 A5.18.6B
5g	On the edge one intersection	AB and CA radii, ∋ (AB /BC) – (CA / BC) = 1 A5.18.6D
6g	To large never intersects	AB and CA radii, ∋ (AB /BC) – (CA / BC) > 1 A5.18.6E
Step	Hypothesis	AB + CA < BC implying ΔABC cannot be constructed
		AB ± CA = BC implying ΔABC cannot be constructed
		AB – CA > BC implying ΔABC cannot be constructed
1	From 2g, 1 and D5.1.4-3 Line Segment and A5.2.1 GI Two Points on a SL	construct segments AB and CA
2	From 1g, 2g, 3g, 1 and A5.18.6 GVII Intersecting Circles (Arcs) EQA	Cannot construct intersecting point-A where AB + CA < BC
3	From 1g, 2g, 4g, 1 and A5.18.6 GVII Intersecting Circles (Arcs) EQB	Can construct intersecting point-A on BC only where AB + CA = BC
4	From 1g, 2g, 5g, 1 and A5.18.6 GVII Intersecting Circles (Arcs) EQD	Can construct intersecting point-A on BC only where AB + CA = BC
5	From 1g, 2g, 6g, 1 and A5.18.6 GVII Intersecting Circles (Arcs) EQE	Cannot construct intersecting point-A where AB – CA > BC AB >> CA inside one another
∴	From 2	AB + CA < BC implying ΔABC cannot be constructed
	From 3 and 4	AB ± CA = BC implying ΔABC cannot be constructed
	From 5	AB – CA > BC implying ΔABC cannot be constructed

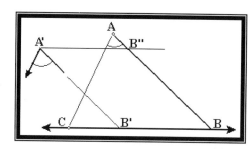

Figure 5.20.2 Constructing Side-Side-Angle (SSA)

Theorem 5.20.2 Tri-Construction: Side-Side-Angle (SSA)

1g	Given	the above geometry figure F5.20.2 "Constructing Side-Side-Angle (SSA)"
2g		∠A'
3g		BC construction line
4g		segment AB

Step	Hypothesis	Sufficient information to construct ΔABC implying		
		∠CAB	∠ABC	∠BCA
1	From 2g, 4g, D5.1.6 Angle, D5.1.3 Ray and A5.18.5 GVII Intersecting Non-collinear Lines	construct extended ray-∠A' and point-B' at intersection		
2	From 1	∠CB'A' is constructible		
3	From 2 and T5.18.6 Construction: Congruent Angles	construct ∠B'A'B" ≅ ∠CB'A'		
4	From 3 and D5.1.3 Ray	construct ray A'B" bounding ∠B'A'B"		
5	From 3 and T5.4.4 Parallel Lines: Converse Congruent Alternate Interior Angles	segment BB' ∥ A'B"		
6	From 1g and 5	construct point-B" at a distance BB'		
7	From 1g, 6 and A5.2.14 GIV Symmetric-Reflexiveness of Angles	A'B" ≅ BB'		
8	From 5, 7 and T5.4.12 Parallel Lines: Translation	A'B' ∥ B"B		
9	From 3g, 6 and D5.1.3 Ray	construct ray BB"		
10	From 4g and 9	construct point-A on ray BB"		
11	From 3g, 10 and D5.1.4-3 Line Segment and A5.2.1 GI Two Points on a SL	construct segment AC		
∴	From 3g, 4g, 11 and D5.1.13 Triangle	Sufficient information to construct ΔABC implying		
		∠CAB	∠ABC	∠BCA

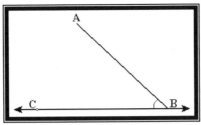

Figure 5.20.3 Constructing Side-Angle-Side (SAS)

Theorem 5.20.3 Tri-Construction: Side-Angle-Side (SAS)

1g	Given	the above geometry figure F5.20.3 "Constructing Side-Angle-Side (SAS)"
2g		segment AB
3g		∠B
4g		BC construction line

Step	Hypothesis	Sufficient information to construct ΔABC implying		
		∠CAB	∠ABC	∠BCA
1	From 2g, 4g and D5.1.4-3 Line Segment and A5.2.1 GI Two Points on a SL	construct segment AC		
2	From 4g and D5.1.4-3 Line Segment and A5.2.1 GI Two Points on a SL	construct segment BC		
∴	From 2g, 1, 2 and D5.1.13 Triangle	Sufficient information to construct ΔABC implying		
		∠CAB	∠ABC	∠BCA

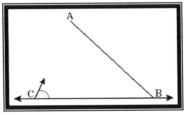

Figure 5.20.4 Constructing Angle-Side-Side (ASS)

Theorem 5.20.4 Tri-Construction: Angle-Side-Side (ASS)

1g	Given	the above geometry figure F5.20.4 "Constructing Angle-Side-Side (ASS)"
2g		segment AB
3g		BC construction line
4g		∠C with extended ray-c
5g		AB > BC so line AB is a secant through circle-B
6g		center of circle-B concurrent with vertex-B
7g		radius AB for circle-B

Step	Hypothesis	Sufficient information to construct ΔABC implying		
		∠CAB	∠ABC	∠BCA
1	From 4g, 5g, 6g and 7g	construct point-A at intersection of ray-c and circle-B		
2	From 1 and D5.1.4-3 Line Segment and A5.2.1 GI Two Points on a SL	construct segment AC		
∴	From 2g, 3g, 2 and D5.1.13 Triangle	Sufficient information to construct ΔABC implying		
		∠CAB	∠ABC	∠BCA

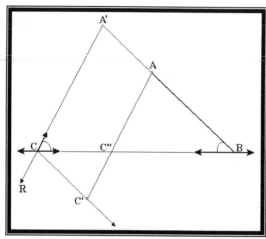

Figure 5.20.5 Constructing Angle-Angle-Side (AAS)

Theorem 5.20.5 Tri-Construction: Angle-Angle-Side (AAS)

	Given	the above geometry figure F5.20.5 "Constructing Angle-Angle-Side (AAS)"
1g		
2g		segment AB
3g		BC construction line
4g		∠B with extended ray-b
5g		∠C with extended ray-c
6g		circle-B has radius-r ≡ AB
7g		center of circle-B concurrent with vertex-B
8g		circle-C has radius-r ≡ A'A
9g		center of circle-C concurrent with vertex-C

Step	Hypothesis	Sufficient information to construct ΔABC" implying		
		∠C"AB	∠ABC"	∠BC"A
1	From 4g, 5g and A5.18.5 GVII Intersecting Non-collinear Lines	construct point-A' at intersection of ray-b and ray-c		
2	From 1	∠BA'C and ray-A'R		
3	From 2 and T5.18.6 Construction: Congruent Angles	∠RCC' and ray-c'		
4	From 2, 3 and T5.4.10 Parallel Lines: Angles Arising From the Transversal	ray-b ‖ ray-c'		
5	From 4g, 6g, 7g and T5.18.4 Construction: Segment Radius Spanning an Angle	construct point-A where AB lies on ray-b		

6	From 1, 5 and A5.18.4 GVII Intersection of Collinear Lines	constructed segment A'A
7	From 8g, 9g, 6 and A5.18.7 GVII Radius on a Construction Line	construct point-C' such that A'A ≅ C'C
8	From 4, 7 and T5.4.12 Parallel Lines: Translation	A'C ∥ AC'
9	From 3g, 8 and A5.18.5 GVII Intersecting Non-collinear Lines	construct point-C" at intersection of BC and AC'
10	From 5, 9, D5.1.4-3 Line Segment and A5.2.1 GI Two Points on a SL	construct segment AC"
11	From 3g, 9, D5.1.4-3 Line Segment and A5.2.1 GI Two Points on a SL	construct segment BC"

∴	From 5, 10, 11 and D5.1.13 Triangle	Sufficient information to construct ΔABC" implying
		∠C"AB $\quad\mid\quad$ ∠ABC" $\quad\mid\quad$ ∠BC"A

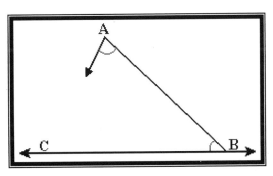

Figure 5.20.6 Constructing Angle-Side-Angle (ASA)

Theorem 5.20.6 Tri-Construction: Angle-Side-Angle (ASA)

1g	Given	the above geometry figure F5.20.6 "Constructing Angle-Side-Angle (ASA)"
2g		∠A with extended ray-a
3g		segment AB
4g		∠B with extended ray-b
5g		BC construction line

Step	Hypothesis	Sufficient information to construct ΔABC implying		
		∠CAB	∠ABC	∠BCA
1	From 1g, 3g, 4g and A5.18.5 GVII Intersecting Non-collinear Lines	at intersection of ray-a with BC construct point-C		
2	From 2g, 1, D5.1.4-3 Line Segment and A5.2.1 GI Two Points on a SL	construct segment AC		
3	From 4g, 1, D5.1.4-3 Line Segment and A5.2.1 GI Two Points on a SL	construct segment BC		
∴	From 3g, 2, 3 and D5.1.13 Triangle	Sufficient information to construct ΔABC implying		
		∠CAB	∠ABC	∠BCA

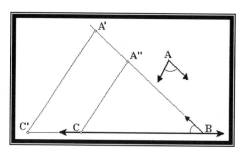

Figure 5.20.7 Constructing Side-Angle-Angle (SAA)

Theorem 5.20.7 Tri-Construction: Side-Angle-Angle (SAA)

1g	Given	the above geometry figure F5.20.7 "Constructing Side-Angle-Angle (SAA)"
2g		∠A with extended rays
3g		∠B with extended ray-b
4g		segment BC
5g		BC construction line

Step	Hypothesis	Sufficient information to construct ΔA"BC implying		
		∠CA"B	∠A"BC	∠BCA"
1	From 3g	any where on ray-b construct point-A'		
2	From 2g, 1 and T5.18.6 Construction: Congruent Angles	∠A ≅ ∠A' with extended ray-a'		
3	From 5g, 2 and A5.18.5 GVII Intersecting Non-collinear Lines	at intersection of ray-a' with BC construct point-C'		
4	From 3 and T5.18.5 Construction: Angle	construct ∠C'		
5	From 4, 5g and T5.18.3 Construction: Congruent Angles	∠C ≅ ∠C' with extended ray-c		
6	From 3g, 5 and A5.18.5 GVII Intersecting Non-collinear Lines	at intersection of ray-b with ray-c construct point-A"		
7	From 4g, 6, D5.1.4-3 Line Segment and A5.2.1 GI Two Points on a SL	construct segment A"C		
8	From 4g, 7, D5.1.4-3 Line Segment and A5.2.1 GI Two Points on a SL	construct segment A"B		
∴	From 4g, 7, 8 and D5.1.13 Triangle	Sufficient information to construct ΔA"BC implying		
		∠CA"B	∠A"BC	∠BCA"

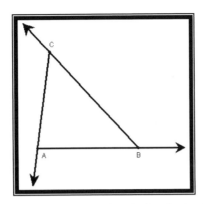

Figure 5.20.8 Constructing Angle-Angle-Angle (AAA)

Theorem 5.20.8 Tri-Construction: Angle-Angle-Angle (AAA)

1g	Given	the above geometry figure F5.20.8 "Constructing Angle-Angle-Angle (AAA)"		
2g		∠A with extended line-a		
3g		∠B with extended line-b		
4g		∠C with extended line-c		
Step	Hypothesis	Insufficient information to construct ∆ABC size undetermined		
		∠C	∠A	∠B
1	From 4g, 2g and A5.18.5 GVII Intersecting Non-collinear Lines	at intersection of line-c with line-a construct point-A		
2	From 2g, 3g and A5.18.5 GVII Intersecting Non-collinear Lines	at intersection of line-a with line-b construct point-B		
3	From 3g, 4g and A5.18.5 GVII Intersecting Non-collinear Lines	at intersection of line-b with line-c construct point-C		
4	From 1 and T5.18.5 Construction: Angle	construct ∠A		
5	From 2 and T5.18.5 Construction: Angle	construct ∠B		
6	From 3 and T5.18.5 Construction: Angle	construct ∠C		
∴	From 4, 5, 6 and D5.1.13 Triangle	Without a constructible side, there is insufficient information to construct ∆ABC for a specific size.		
		∠C	∠A	∠B

Tensor Calculus & Physics: A General Treatise

Section 5.21 Theorems: Constructibility of Similar Polygons

Definition D5.1.29 "Similar Polygons" is the classical definition used by Euclid, but it has been used more as an axiom than a definition to prove various theorems. Not only is it taken for granted that the sides are always proportional to one another. It is for these reasons that the definition has been replaced with Axiom A5.18.13 "GVII Sides of Similar Polygons Correspond to Opposite Angles". In this section on similarity, the axiom on construction is used and proportional sides are left as a proof.

Given 5.21.1 Similar Polygons: Parameters of Isosceles and Their Right Triangles

1g	Given	the above geometry figure F5.20.1 "Similar Isosceles Triangles with Common Vertex"
2g		$r \cong OA \cong OB$
3g		$r' \cong OA' \cong OB'$
4g		$a \cong OP$
5g		$a' \cong OP'$
6g		$b \cong AP \cong PB$
7g		$b' \cong A'P' \cong P'B'$
8g		$\theta \cong \angle AOB \cong \angle A'OB'$
9g		$\phi \cong \angle OAP$
10g		$r' / r \equiv k$ as the radii contract and expand

Theorem 5.21.1 Similar Polygons: Similarity of Isosceles and Their Right Triangles

Step	Hypothesis	$\Delta OAP \sim \Delta OA'P'$	$\Delta OBP \sim \Delta OB'P'$
		$\Delta AOB \sim \Delta A'OB'$ isosceles triangles for $R_aR_b \parallel A'B'$	
1	From G5.20.1-(1g, 2g, 3g) and D5.1.16 Triangle; Isosceles	ΔAOB and $\Delta A'OB'$ isosceles triangles	
2	From G5.20.1-9g, 1 and T5.9.1 Isosceles Triangle: Bisecting Vertex Between Congruent Sides Parses Equal Parts	$\phi \cong \angle OAP \cong \angle OBP \cong \angle OA'P' \cong \angle OB'P'$	
3	From 1 and T5.8.6 Isosceles Triangle: Right Triangles Bisect Vertex Angle	$\angle AOP \cong \angle BOP \cong \angle A'OP \cong \angle B'OP \cong \frac{1}{2}\theta$	
4	From 2 and T5.4.1 Intersection of Two Lines: Opposite Congruent Vertical Angles	$\angle A'AR_a \cong \angle OAP \cong \phi$	
5	From G5.20.1-3g, 2, 4 and T5.4.10 Parallel Lines: Angles Arising From the Transversal	$R_aR_b \parallel A'B'$ for OA' transversal cutting across lines R_aR_b and A'B'	
∴	From 1, 2, 3 and A5.18.13 GVII Sides of Similar Polygons Correspond to Opposite Angles	$\Delta OAP \sim \Delta OA'P'$	$\Delta OBP \sim \Delta OB'P'$
		$\Delta AOB \sim \Delta A'OB'$ isosceles triangles for $R_aR_b \parallel A'B'$	

Theorem 5.21.2 Similar Polygons: Proportionality of Right Triangles

Step	Hypothesis	$a / r = a' / r' = s_\phi$	$b/r = b'/r' = c_\phi$
1	From T5.7.1 Similar Triangles: Proportional Triangles	$\phi \rightarrow a / r = a' / r' = s_\phi$ as a function of ϕ	
2	From 1, T5.9.2 Isosceles Triangle: Bisected into Two Congruent Right Triangles and T5.7.11 Right Triangle: Ratio of Altitude to the Hypotenuse	$\pm\sqrt{1 - (b/r)^2} = \pm\sqrt{1 - (b'/r')^2}$	
3	From 2, T4.8.6 Integer Exponents: Uniqueness of Exponents, T4.10.2 Radicals: Commutative Product and T4.10 3 Radicals: Identity	$1 - (b/r)^2 = 1 - (b'/r')^2$	
4	From 3, T4.3.2 Equalities: Cancellation by Addition	$(b/r)^2 = (b'/r')^2$	
5	From 1, 4, T4.9.5 Rational Exponent: Uniqueness of Exponents, T4.10.3 Radicals: Identity and D4.1.23C Ratio of Quantities	$\phi \rightarrow b/r = b'/r' = c_\phi$ as a function of ϕ	
∴	From 1 and 5	$\phi \rightarrow a / r = a' / r' = s_\phi$	$\phi \rightarrow b/r = b'/r' = c_\phi$

From this theorem springs all of trigonometry, hence its other name ***The Fundamental Theorem of Trigonometry***. It clearly shows why right triangles are so useful, because for any size of a right triangle by similarity they have constants of proportionality for their ratios of legs (s_ϕ, c_ϕ). The history of trigonometry demonstrates many such uses.

Theorem 5.21.3 Similar Polygons: Constant of Proportionality for Sine and Cosine

Step	Hypothesis	$\sin(\phi) = a / r = s_\phi$	$\cos(\phi) = b / r = c_\phi$
1	From T5.20.2 Fundamental Theorem of Trigonometry	$\phi \rightarrow a / r = s_\phi$	$\phi \rightarrow b/r = c_\phi$
2	From 1, DK.1.4 Image of a Domain Element, than the image of the domain of ϕ, DxE.1.1.1 and DxE.1.1.2	$sine(\phi) = a / r = s_\phi$[5.20.1]	$cosine(\phi) = b / r = c_\phi$
∴	From 2 and abbreviating to three letters	$\sin(\phi) = a / r = s_\phi$	$\cos(\phi) = b / r = c_\phi$

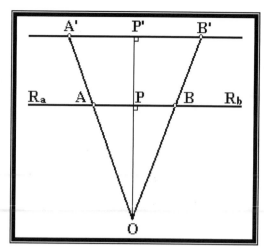

Figure 5.21.1 Similar Isosceles Triangles with Common Vertex

Theorem 5.21.4 Similar Polygons: Proportionality Bases of Right Triangles

1g	Given	the above geometry figure F5.21.1 "Similar Isosceles Triangles with Common Vertex"	
2g		$b \cong AP \cong PB$	
3g		$b' \cong A'P' \cong P'B'$	
Step	Hypothesis	$(\frac{1}{2}AB) / r = (\frac{1}{2}A'B') / r'$	$AB / 2r = A'B' / 2r'$
1	From 1g	$AB \cong AP + PB$	$A'B' \cong A'P' + P'B'$
2	From G5.20.1-(6g, 7g), A5.2.15 GIV Transitivity of Segments and T4.3.10 Equalities: Summation of Repeated Terms by 2	$AB \cong 2b$	$A'B' \cong 2b'$
3	From 2, A4.2.2 Equality, T4.4.2 Equalities: Uniqueness of Multiplication, T4.4.16B Equalities: Product by Division, Common Factor, A4.2.12 Identity Multp and A4.2.2B Equality	$b \cong \frac{1}{2}AB$	$b' \cong \frac{1}{2}A'B'$
∴	From 3, T5.20.2 Similar Polygons: Proportionality of Right Triangles, A5.2.15 GIV Transitivity of Segments and D4.1.4C Rational Numbers	$(\frac{1}{2}AB) / r = (\frac{1}{2}A'B') / r'$	$AB / 2r = A'B' / 2r'$

Theorem 5.21.5 Similar Polygons: Constant Proportionality for Sides

1g	Given	the above geometry figure F5.21.1 "Similar Isosceles Triangles with Common Vertex"
2g		$b \cong AP \cong PB$
3g		$b' \cong A'P' \cong P'B'$
Step	Hypothesis	$(\frac{1}{2}AB) / r = (\frac{1}{2}A'B') / r'$
1	From T5.20.4 Similar Polygons: Proportionality Bases of Right Triangles	$AB / 2r = A'B' / 2r'$
2	From 1 and T4.5.11 Equalities: Formal Cross Product	$A'B' / AB = 2r' / 2r$
∴	From 2, A4.2.11 Associative Multp, A4.2.13 Inverse Multp and A4.2.12 Identity Multp	$A'B' / AB = r' / r$

[5.20.1]Note: In the 12th century the Arabic word for the half chord of the double arc, found by Hipparchus's at Alexandria (146 BCE) [EVE76, pg 143], was confused with another word and translated sinus (sine). [BRI60, Vol. 22, pg 475]

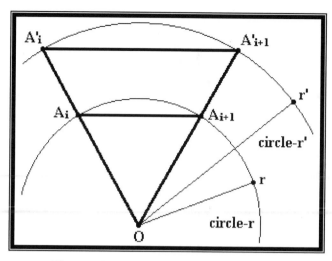

Figure 5.21.2 Side of Concentric Polygons

Theorem 5.21.6 Similar Polygons: Have Proportional Sides

1g	Given	the above geometry figure F5.21.2 "Side of Concentric Polygons"
2g		$\rho \equiv A_1 A_2 \dots A_n$ n-points for irregular polygon
3g		$\rho' \equiv A'_1 A'_2 \dots A'_n$ for $\rho \sim \rho'$
4g		vertices of ρ lies on circle-r
5g		vertices of ρ' lies on circle-r'
6g		vertices A_i and A_{i+1} lie on a radial line as they expand or contract relative to one another for all i.
Step	Hypothesis	$A'_i A'_{i+1} / A_i A_{i+1} = r' / r = p_r$ for all i
1	1g and A5.18.9 GVII Oriented Congruent Geometric Objects	ρ is centered in ρ' of concentric circle-r and circle-r' sharing center-O
2	From 4g, 6g and 1	$r \cong OA_i \cong OA_{i+1}$
3	From 5g, 6g and 1	$r' \cong OA'_i \cong OA'_{i+1}$
4	From 2, 3 and D5.1.16 Triangle; Isosceles	$\Delta A_{i+1} O A_i$ and $\Delta A'_{i+1} O A'_i$ are isosceles of common vertex-O
∴	From 4, D5.1.66 Scaling Factor for Similar Polygons and T5.20.5 Similar Polygons: Constant Proportionality Sides	$A'_i A'_{i+1} / A_i A_{i+1} = r' / r = p_r$ for all i

Notice that in order to prove congruence of triangles they have to be oriented by placing them in a parallel frame, likewise in order to prove similarity polygons have to be placed in a circularly concentric frame of two inscribing circles. Also for backwards compatibility Axiom A5.18.13 GVII "Sides of Similar Polygons Correspond to Opposite Angles" and Theorem T5.20.5 "Similar Polygons: Constant Proportionality for Sides" simply replace definition D5.1.29 "Similar Polygons".

Theorem 5.21.7 Similar Polygons: Converse Corresponding Congruent Angles

1g	Given	$\rho \equiv A_1A_2 \dots A_n$	n-points for an n-sided polygon
2g		$\rho' \equiv A'_1A'_2 \dots A'_n$	such that $\rho \sim \rho'$
3g		$A'_iA'_{i+1} / A_iA_{i+1} = p_r$	
Step	Hypothesis	$\angle C_i \cong \angle C_i'$ for all i	
∴	From 2g and A5.18.13 GVII Sides of Similar Polygons Correspond to Opposite Angles	$\angle C_i \cong \angle C_i'$ for all i	

This theorem replaces Theorem T5.7.4 "Similar Triangles: Converse Corresponding Congruent Angles".

Theorem 5.21.8 Similar Polygons: Proportional Perimeter of Similar Polygons

1g	Given	$\rho \equiv A_1A_2 \dots A_n$	n-points for an n-sided polygon
2g		$\rho' \equiv A'_1A'_2 \dots A'_n$	
3g		$\rho \sim \rho'$	
4g		$P_T \equiv \sum A_iA_{(i+1)}$ total perimeter of polygon	$P_T \in R$
5g		$P'_T \equiv \sum A'_iA'_{(i+1)}$	$P_T \in R$
Step	Hypothesis	$P'_T / P_T \equiv p_r$	
1	From 1g, 2g, 3g and T5.20.6 Similar Polygons: Have Proportional Sides	$A'_iA'_{(i+1)} / A_iA_{(i+1)} \equiv p_r$	such that $k \in R$ for all i
2	From 1 and T4.4.7 Equalities: Reversal of Left Cancellation by Multiplication	$A'_iA'_{(i+1)} \equiv p_r A_iA_{(i+1)}$	
3	From 2, summing over all terms and A4.2.14 Distribution	$\sum A'_iA'_{(i+1)} \equiv p_r \sum A_iA_{(i+1)}$	
4	From 4g, 5g, 3 and A5.2.15 GIV Transitivity of Segments	$P'_T \equiv p_r P_T$	
∴	From 4 and T4.4.4 Equalities: Right Cancellation by Multiplication	$P'_T / P_T \equiv p_r$	

Theorem 5.21.9 Similar Polygons: Proportional Area of Similar Polygons

1g	Given	$\rho \equiv A_1 A_2 \ldots A_n$ n-points for an n-sided polygon
2g		$\rho' \equiv A'_1 A'_2 \ldots A'_n$
3g		$\rho \sim \rho'$
4g		$K_T \equiv \sum a_i A_i A_{(i+1)}$ total area of polygon $K_T \in R$
5g		$K'_T \equiv \sum a_i' A'_i A'_{(i+1)}$ $K_T \in R$
Step	Hypothesis	$K'_T / K_T \equiv p_r^2$
1	From 1g, 2g, 3g and T5.20.6 Similar Polygons: Have Proportional Sides	$A'_i A'_{(i+1)} / A_i A_{(i+1)} \equiv p_r$ such that $k \in R$ for all i
2	From 1 and T4.4.7 Equalities: Reversal of Left Cancellation by Multiplication	$A'_i A'_{(i+1)} \equiv p_r A_i A_{(i+1)}$
3	From T5.20.2 Similar Polygons: Proportionality of Right Triangles	$a_i / r = a_i' / r'$
4	From 3 and T4.4.14 Equalities: Formal Cross Product	$a_i' / a_i = r' / r$
5	From 4, D5.1.66 Scaling Factor for Similar Polygons and A4.2.3 Substitution	$a_i' / a_i = p_r$
6	From 5 and T4.4.7 Equalities: Reversal of Left Cancellation by Multiplication	$a_i' = p_r a_i$
7	From 2, 6, T4.4.2 Equalities: Uniqueness of Multiplication D4.1.17 Exponential Notation and T5.9.12 Isosceles Triangle: Area	$a_i' A'_i A'_{(i+1)} \equiv p_r^2 a_i A_i A_{(i+1)}$ area of similar isosceles triangles
8	From 7, summing over all i, 5g, 6g and A4.2.3 Substitution	$K'_T \equiv p_r^2 K_T$
∴	From 8 and T4.4.4A Equalities: Right Cancellation by Multiplication	$K'_T / K_T \equiv p_r^2$

Section 5.22 Constructibles

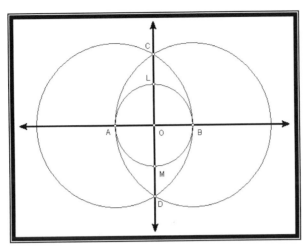

Figure 5.22.1 Constructing Perpendicular Lines

Theorem 5.22.1 Construction: Perpendicular Lines and Corresponding Right Angles

1g	Given	the above geometry figure F5.22.1 "Constructing Perpendicular Lines"
2g		intersecting circles-A and B concurrent at points A and B AB/AB + BA/AB > 1
3g		construction line AB
4g		construct circle-O having diameter AB ≡ 2r
Step	Hypothesis	constructed perpendicular or right angle
1	From 2g	construct intersecting points-C and D
2	From 1 and D5.1.4-3 Line Segment and A5.2.1 GI Two Points on a SL	construct line CD
3	From 3g, 2 and A5.18.5 GVII Intersecting Non-collinear Lines	construct point-O at intersection of AB and CD
4	From 1g and T5.4.1 Intersection of Two Lines: Opposite Congruent Vertical Angles	∠BOL ≅ ∠AOM ∠BOM ≅ ∠AOL
5	From 1g and 3g	AO ≅ BO ≅ LO ≅ MO ≅ ½AB ≡ r
6	From 4 and 5	r∠BOL ≅ r∠AOM r∠BOM ≅ r∠AOL
7	From 6 and D5.1.36 Subtended Arclength of a Cord	arc BL ≅ arc AM arc BM ≅ arc AL
8	From 4g and 7 arc of circle-O	arc BL ≅ arc BM arc AM ≅ arc AL
9	From 8 and D5.1.36 Subtended Arclength of a Cord	r∠BOL ≅ r∠BOM r∠AOM ≅ r∠AOL
10	From 9	∠BOL ≅ ∠BOM ∠AOM ≅ ∠AOL
11	From 4 and 10	∠BOL ≅ ∠BOM ≅ ∠AOM ≅ ∠AOL
12	From 1g	360° ≡ ∠BOL + ∠BOM + ∠AOM + ∠AOL
13	From 11 and 12	360° ≡ ∠BOL + ∠BOL + ∠BOL + ∠BOL
14	From 13	360° ≡ 4∠BOL
15	From 11 and 14	∠BOL ≅ ∠BOM ≅ ∠AOM ≅ ∠AOL ≡ 90°
∴	From 15 and D5.1.7 Right Angle	constructed perpendicular or right angle

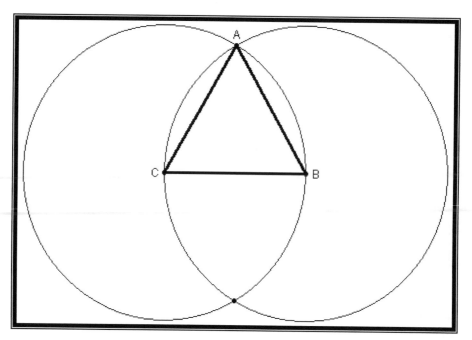

Figure 5.22.2 Constructing an Equilateral Triangle

Theorem 5.22.2 Constructible: Equilateral Triangle

Construct two interlocking circles such that the centers are on the circumference points B and C.
Construct intersection point-A.
Construct segments AB, BC and CA.
Given: BC = r the radius of either circle.
Prove: Both circles are congruent
Prove: AB ≅ BC ≅ CA
Prove: ∠CAB ≅ ∠ABC ≅ ∠BCA ≡ 60°

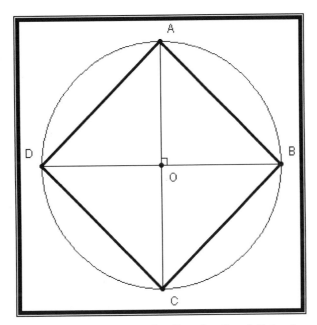

Figure 5.22.3 Constructing Regular Quadrilaterals

Theorem 5.22.3 Constructible: Regular Quadrilateral (Euclidian Square)

Construct two lines AC ⊥ BD.
Construct at the center-O a circle with a given radius AO.
Construct intersection points-A, B, C and D.
Construct segments AB, BC, CD and DA.
Given: AO = r the circle radius.
Prove: AO ≅ BO ≅ CO ≅ DO
Prove: AB ≅ BC ≅ CD ≅ DA
Prove: ∠DAB ≅ ∠ABC ≅ ∠BCD ≅ ∠CDA ≡ 90°

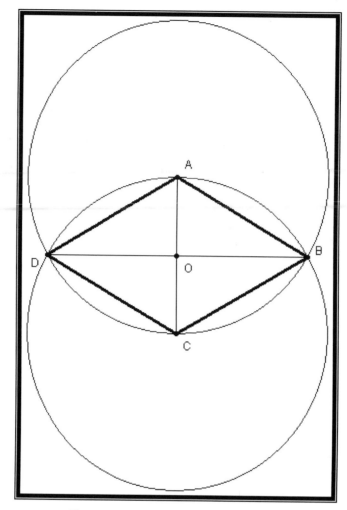

Figure 5.22.4 Constructing a Rhombus

Theorem 5.22.4 Constructible: Rhombus

Construct two interlocking circles such that the centers are on the circumference points A and C.
Construct intersection points-B and D.
Construct segments AB, BC, CD and DA.
Given: AC = r the radius of either circle.
Prove: Both circles are congruent
Prove: AB ≅ BC ≅ CD ≅ DA
Prove: ∠CAB≅ ∠BCD ≡ 120° ∠ABC ≅ ∠CDA ≡ 60°
Prove: Diagonals AC and BC are perpendicular

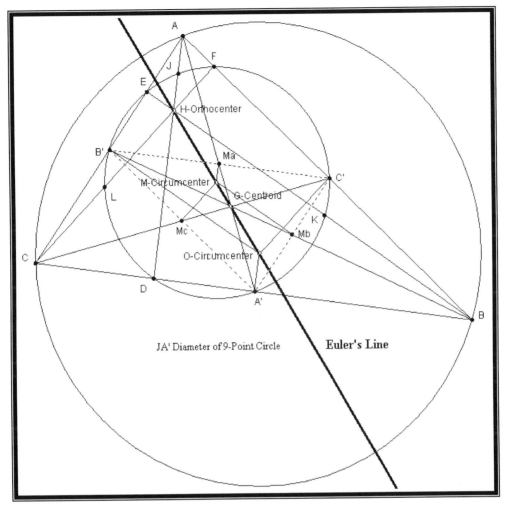

Figure 5.22.5 Constructing a 9-Point Circle

Theorem 5.22.5 9-Point Circle

Construct triangle ΔABC.

Construct midpoints A', B', C' on segments AB, BC and CA.

Construct medium segments B'C', C'A' and A'B'.

Construct medium midpoints Ma, Mb and Mc.

Construct perpendicular lines intersecting at point-M Circumcenter.

Construct circle through points A', B' and C' at center-M.

Construct perpendicular altitudes to points A, B and C of ΔABC and label concurrent point-H Orthocenter.

Construct intersecting points-(D, J), (E, K), (F, L) completing the 9-points.

Construct medium intersecting point-O Circumcenter from angle bisectors.

Construct circle through points A, B and C at center-O.

Construct intersection point-G Centroid from lines C'McC, A'MaA and B'MbB.

Construct Euler's Line through points H, M, G and O.

Given: JA' = d the diameter of the 9-Point Circle.

Prove: JA' is the diameter of the 9-Point Circle

Prove: H, M, G and O are collinear points on the Euler Line.

Section 5.23 Introductory Principles of Irregular Polygons

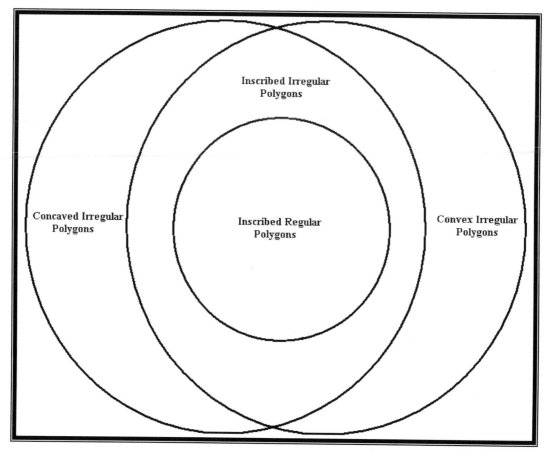

Figure 5.23.1 Polygon Hierarchy

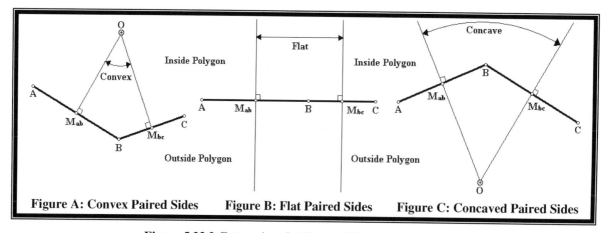

Figure A: Convex Paired Sides   **Figure B: Flat Paired Sides**   **Figure C: Concaved Paired Sides**

Figure 5.23.2 Categories of Adjacent Sides from a Polygon

Definition 5.23.1 Convex Sides

For any two adjacent sides external angles are on the outside of the polygon and intersection of their perpendiculars point inward. (See Figure 5.19.2A)

Definition 5.23.2 Flat Sides

For any two adjacent sides external angles have a flat angle between them and their perpendiculars point vertically. (See Figure 5.19.2B)

Definition 5.23.3 Concaved Sides

For any two adjacent sides external angles are on the inside of the polygon and intersection of their perpendiculars point outward. (See Figure 5.19.2C)

Definition 5.23.4 Convex Irregular Polygons

Convex Irregular Polygon is a contiguously connected closed polygon where all adjacent sides of the polygon are convex.

Definition 5.23.5 Concaved Irregular Polygons

Concaved Irregular Polygon is a contiguously connected closed polygon where at least one or more adjacent sides are concaved.

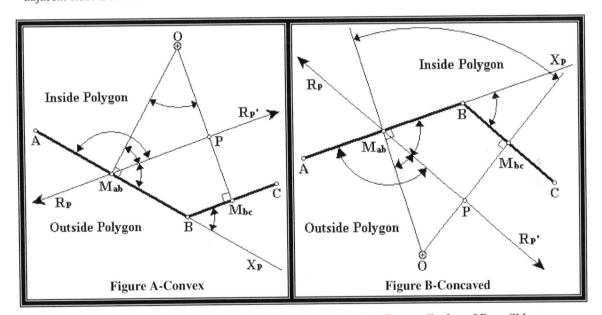

Figure 5.23.3 Relationship of Polygon External Angle to Perpendicular of Base-Side

Observation 5.23.1: Direction of External Angles

From Figure 5.23.3 "Relationship of Polygon External Angle to Perpendicular of Base-Side" it can be seen that convex sides have their external angle $\angle CBX_p$ outside the closed bounds of the polygon, while concave sides have their external angle $\angle CBX_p$ inside, like wise for the inside angle $\angle CBA$. Since all angles are positive for convex polygons, it follows that the external angle goes negative for the concave polygon relative to the convex polygon.

Theorem 5.23.1 Irregular Polygons: Internal and External Angles are Supplemental

1g	Given	the above geometry figure F5.23.3A "Relationship of Polygon External Angle to Perpendicular of Base-Sid"convex polygon
2g		the above geometry figure F5.23.3B "Relationship of Polygon External Angle to Perpendicular of Base-Sid"concaved polygon
Steps	Hypothesis	$180° \cong \angle CBA + \angle CBX_p$
$\therefore$	From 1g, 2g and D5.1.12 Supplementary Angles	$180° \cong \angle CBA + \angle CBX_p$

Theorem 5.23.2 Irregular Polygons: Perpendicular and External Angles are Complementary

1g	Given	the above geometry figure F5.23.3A "Relationship of Polygon External Angle to Perpendicular of Base-Sid"convex polygon	
2g		the above geometry figure F5.23.3B "Relationship of Polygon External Angle to Perpendicular of Base-Sid"concaved polygon	
3g		$R_p R_{p'} \parallel BC$	
Steps	Hypothesis	$\angle CBX_p \cong 90° - \angle R_{p'} M_{ab} O$	$\angle R_{p'} M_{ab} O \cong 90° - \angle CBX_p$
1	From 1g, 2g, 3g and C5.4.2.1 Parallel Lines: Congruent Alternate Supplementary Angles	$\angle R_{p'} M_{ab} X_p \cong \angle CBX_p$	
2	From 1g and D5.1.11 Complementary Angles	$90° \cong \angle R_{p'} M_{ab} O + \angle R_{p'} M_{ab} X_p$	
3	From 1, 2 and A5.2.18 GIV Transitivity of Angles	$90° \cong \angle R_{p'} M_{ab} O + \angle CBX_p$	
4	From 3 and T4.3.3B Equalities: Right Cancellation by Addition	$\angle CBX_p \cong 90° - \angle R_{p'} M_{ab} O$	
5	From 3 and T4.3.3B Equalities: Right Cancellation by Addition	$\angle R_{p'} M_{ab} O \cong 90° - \angle CBX_p$	
$\therefore$	From 4 and 5	$\angle CBX_p \cong 90° - \angle R_{p'} M_{ab} O$	$\angle R_{p'} M_{ab} O \cong 90° - \angle CBX_p$

Theorem 5.23.3 Irregular Polygons: Perpendicular and Internal Angles are Complementary

Steps	Hypothesis	$\angle CBA \cong 90° + \angle R_{p'} M_{ab} O$	$\angle R_{p'} M_{ab} O \cong \angle CBA - 90°$
1	From T5.19.2 Irregular Polygons: Perpendicular and External Angles are Complementary	$90° \cong \angle R_{p'} M_{ab} O + \angle CBX_p$	
2	From T5.19.1 and T4.3.3B Equalities: Right Cancellation by Addition	$\angle CBX_p \cong 180° - \angle CBA$	
3	From 1, 2 and A5.2.18 GIV Transitivity of Angles	$90° \cong \angle R_{p'} M_{ab} O + 180° - \angle CBA$	

4	From 3, T4.3.3A Equalities: Right Cancellation by Addition, T4.3.4A Equalities: Reversal of Right Cancellation by Addition and T4.3.5B Equalities: Left Cancellation by Addition	$180° - 90° \cong \angle CBA - \angle R_p \cdot M_{ab}O$	
5	From 4 and D4.1.19 Primitive Definition for Rational Arithmetic	$90° \cong \angle CBA - \angle R_p \cdot M_{ab}O$	
6	From 5 and T4.3.4B Equalities: Reversal of Right Cancellation by Addition	$\angle CBA \cong 90° + \angle R_p \cdot M_{ab}O$	
7	From 6 and T4.3.2B Equalities: Right Cancellation by Addition	$\angle R_p \cdot M_{ab}O \cong \angle CBA - 90°$	
∴	From 6 and 7	$\angle CBA \cong 90° + \angle\ R_p \cdot M_{ab}O$	$\angle R_p \cdot M_{ab}O \cong \angle CBA - 90°$

Theorem 5.23.4 Irregular Polygons: Congruent External and Perpendicular Intersecting Angles

1g 2g 3g	Given	convex polygon see above figure F5.19.3A concaved polygon see above figure F5.19.3B $R_pR_{p'} \parallel BC$
Steps	Hypothesis	$\angle M_{ab}OP \cong \angle CBX_p$
1	From 1g, 2g, 3g and C5.4.2.1 Parallel Lines: Congruent Alternate Supplementary Angles	$90° \cong \angle BM_{bc}P \cong \angle M_{ab}PO$
2	From 1 and D5.1.19 Triangle; Right	$\Delta M_{ab}PO$ is a right triangle
3	From 2 and T5.7.6 Right Triangle: Non-Ninety Degree Angles are Complementary	$90° \cong \angle M_{ab}OP + \angle R_p \cdot M_{ab}O$
4	From 3 and T4.3.3B Equalities: Right Cancellation by Addition	$\angle M_{ab}OP \cong 90° - \angle R_p \cdot M_{ab}O$
5	From 4, T5.19.2 Irregular Polygons: Perpendicular and External Angles are Complementary and A5.2.18 GIV Transitivity of Angles	$\angle M_{ab}OP \cong 90° - (90° - \angle CBX_p)$
∴	From 5, A4.2.14 Distribution, D4.1.19 Primitive Definition for Rational Arithmetic, A4.2.8 Inverse Add and A4.2.7 Identity Add	$\angle M_{ab}OP \cong \angle CBX_p$

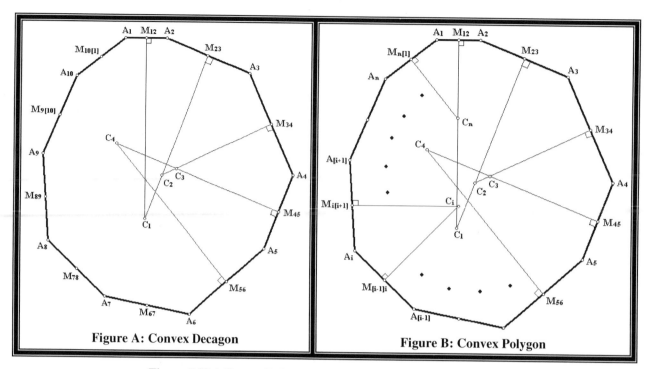

Figure A: Convex Decagon | **Figure B: Convex Polygon**

Figure 5.23.4 Convex Polygon: Accumulated Intersecting Angles

Theorem 5.23.5 Convex and Concave Polygons: Accumulating Intersecting Angles

1g	Given	the above geometry figure F5.23.4B "Convex Polygon: Accumulated Intersecting Angles"
2g		$\angle A_{[i+1]}A_iX_{pi}$ external angle for the ith angle
Steps	Hypothesis	$360° \cong \sum \angle A_{[i+1]}A_iX_{pi}$
1	From 1g	$\Delta M_{[i-2][i-1]}C_iM_{[i-1]i}$, $\Delta M_{[i-1]i}C_iM_{i[i+1]}$ share the common side $M_{[i-1]i}$ for all i = 1, 2, ..., n
2	From 1g and D5.19.4 Convex Irregular Polygons	$\Delta M_{n[i]}C_nM_{12}$, $\Delta M_{12}C_1M_{23}$ share the common side M_{12} hence have closure about 360°
3	From 1 and 2	$360° \cong \sum \angle M_{[i-1]i}C_iM_{i[i+1]}$ accumulated intersecting angles about 360°
∴	From 2g, 3, T5.19.4 Irregular Polygons: Congruent External and Perpendicular Intersecting Angles and A5.2.18 GIV Transitivity of Angles	$360° \cong \sum \angle A_{[i+1]}A_iX_{pi}$

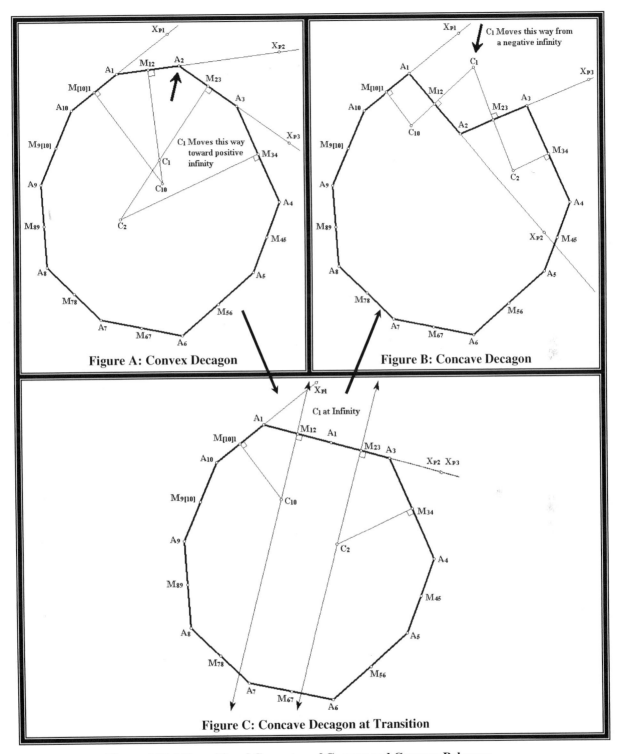

Figure A: Convex Decagon

Figure B: Concave Decagon

Figure C: Concave Decagon at Transition

Figure 5.23.5 Transitional Geometry of Convex and Concave Polygons

Observation 5.23.2 Convex and Concave Polygons: Accumulating Intersecting Angles

	∠Intersection ≅ ∠External		Summation of External Angles
Convex	83° ≅ $\angle M_{12}C_1M_{23}$ ≅ $\angle A_2A_1X_{p1}$ ≅ 83°		360° ≅ $\sum \angle M_{[i-1]i}C_iM_{i[i+1]}$ for n = 10
Transition	0° ≅ $\angle M_{12}C_1M_{23}$ ≅ $\angle A_2A_1X_{p1}$ ≅ 0°		360° ≅ $\sum \angle M_{[i-1]i}C_iM_{i[i+1]}$ for n = 10
Concave	140.53° ≅ $\angle M_{12}C_1M_{23}$ ≅ $\angle A_2A_1X_{p1}$ ≅ 140.53°		360° ≅ $\sum \angle M_{[i-1]i}C_iM_{i[i+1]}$ for n = 10

From the above table two observations can be made:

A) All convex polygon external angles sum to 360° about the circumference of the polygon F5.23.5 "Transitional Geometry of Convex and Concave Polygons". For a formal proof, see theorem T5.23.5 "Convex and Concave Polygons: Accumulating Intersecting Angles".

B) The intersection point of any two adjacent perpendiculars in moving from convex to concave moves toward the positive side of infinity, at a flat angle is at infinity and from the negative side of infinity becoming concave, likewise in reverse. See Figure F5.23.5A, C and B.

Section 5.24 Theorems on Mapping Irregular to Regular Polygons

In the next three sections, proofs of regular and irregular polygons are investigated. Also a specific proof is considered that may lead to a general proof of the *zero property* of polygons. The zero property of polygons is the net perceived perimeter path about the polygon projected onto a construction line that accumulates to a net magnitude of zero. This comes about from vector analysis when a sequence of vectors circularly close on themselves forming an irregular polygon having a directed sense allowing it to add up to net of zero.

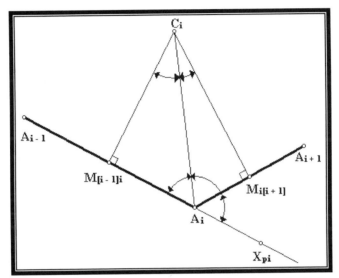

Figure 5.24.1 ith Intersecting Component of an Irregular Polygon

Given 5.24.1: Parameters of an Irregular Polygon

1g	Given	the above geometry figure F5.24.1 "ith Intersecting Component of an Irregular Polygon"
2g		geometry of a polygon has closer in the sense that such that $\Delta A_n C_n A_{n+1}$ is adjacent to $\Delta A_1 C_1 A_2$ around vertex-C
3g		component $\Delta A_i C_i A_{i+1}$ for i = 1, 2, …, n of sides to the polygon
4g		$a_i \equiv M_{[i-1]i} C_i$ apothem, altitude of $\Delta A_i C_i A_{i+1}$
5g		$p_i \equiv A_i A_{i+1}$ parameter base
6g		$r_i \equiv C_i A_i$ radial side
7g		$r_{i-1} \cong r_i$ sharing the common vertex C_i
8g		$\psi_i \equiv \angle A_i C_i M_{[i-1]i}$ complementary intersection (CI) angle
9g		$\psi_{i+1} \equiv \angle A_i C_i M_{i[i+1]}$ complementary intersection (CI) angle
10g		$\xi_i \equiv \angle M_{i[i+1]} C_i M_{[i-1]i}$ intersection angle
11g		$\phi_i \equiv \angle A_i C_i A_{i-1}$ central angle
12g		$\chi_i \equiv \angle A_{[i+1]} A_i X_{pi}$ external angle
13g		$\eta_i \equiv \angle M_{[i-1]i} A_i C_i$ complementary interior (CN) angle
14g		$\eta_{i+1} \equiv \angle M_{i[i+1]} A_i C_i$ complementary interior (CN) angle
15g		$\kappa_i \equiv \angle M_{i[i+1]} A_i M_{[i-1]i}$ interior angle

16g	$p \equiv p_i$ for all [i] when polygon is regular
17g	$a \equiv a_i$ for all [i] when polygon is regular
18g	$r \equiv r_i$ for all [i] when polygon is regular
19g	$\phi \equiv \phi_i$ for all [i] when polygon is regular
20g	$\psi \equiv \psi_i = \psi_{i+1}$ for all [i] when polygon is regular
21g	$\xi \equiv \xi_i$ for all [i] when polygon is regular
22g	$\eta \equiv \eta_i \equiv \eta_{i+1}$ for all [i] when polygon is regular
23g	$\kappa \equiv \kappa_i \equiv \kappa_{i+1}$ for all [i] when polygon is regular
24g	$(\bullet \equiv C_i \bullet)$ for all [i] when polygon is regular
25g	$(\bullet \cong A_iC_i \bullet)$ for all [i] when polygon is regular
26g	$(\bullet \cong M_{i[i+1]}C_i \bullet)$ for all [i] when polygon is regular
27g	$(\bullet \cong M_{i[i+1]}A_i \bullet)$ for all [i] when polygon is regular
28g	$(\bullet \cong A_iA_{i+1} \bullet)$ for all [i] when polygon is regular
29g	$\kappa_{av} \equiv 180° (n-2)/n$ average internal angle
30g	$P_T \equiv \sum p_i$ total perimeter of polygon
31g	$P_{av} \equiv (1/n)\, P_T$ average parameter base
32g	K_i area of component triangle
33g	$K_T \equiv \sum K_i$ total area of polygon

Theorem 5.24.1 Irregular Polygons: Accumulated External Angle

Steps	Hypothesis	$360° \cong \sum \chi_i$ Equation A	$360° \cong \sum \xi_i$ Equation B
1	From G5.19.1-12g, T5.19.5 Convex and Concave Polygons: Accumulating Intersecting Angles and A5.2.18 GIV Transitivity of Angles	$360° \cong \sum \chi_i$	
2	From G5.19.1-12g, T5.19.4 Irregular Polygons: Congruent External and Perpendicular Intersecting Angles and A5.2.18 GIV Transitivity of Angles	$360° \cong \sum \xi_i$	
∴	From 1 and 2	$360° \cong \sum \chi_i$ Equation A	$360° \cong \sum \xi_i$ Equation B

Theorem 5.24.2 Irregular Polygons: Intersection and Complementary Angle Relationship

Steps	Hypothesis	$\xi_i \cong \psi_i + \psi_{i+1}$
1	From G5.19.1-1g	$\angle M_{i[i+1]}C_iM_{[i-1]i} \cong \angle A_iC_iM_{[i-1]i} + \angle A_iC_iM_{i[i+1]}$
∴	From G5.19.1-8g, 9g, 10g, 1 and A5.2.18 GIV Transitivity of Angles	$\xi_i \cong \psi_i + \psi_{i+1}$

Theorem 5.24.3 Irregular Polygons: Interior and Complementary Angle Relationship

Steps	Hypothesis	$\kappa_i \cong \eta_i + \eta_{i+1}$
1	From G5.19.1-1g	$\angle M_{i[i+1]}A_iM_{[i-1]i} \cong \angle M_{[i-1]i}A_iC_i + \angle M_{i[i+1]}A_iC_i$
∴	From G5.19.1-13g, 14g, 15g, 1 and A5.2.18 GIV Transitivity of Angles	$\kappa_i \cong \eta_i + \eta_{i+1}$

Theorem 5.24.4 Irregular Polygons: Complementary Parsed Angles

Steps	Hypothesis	$90° \cong \psi_i + \eta_i$	$90° \cong \psi_{i+1} + \eta_{i+1}$
1	From G5.19.1-1g	$90° \cong \angle A_i M_{[i-1]i} C_i$	$90° \cong \angle A_i M_{i[i+1]} C_i$
2	From G5.19.1-1g, 1 and D5.1.19 Triangle; Right	$\Delta A_i M_{[i-1]i} C_i$	$\Delta A_i M_{i[i+1]} C_i$
3	From 2 and T5.7.6 Right Triangle: Non-Ninety Degree Angles are Complementary	$90° \cong \angle A_i C_i M_{[i-1]i} + \angle M_{[i-1]i} A_i C_i$	$90° \cong \angle A_i C_i M_{i[i+1]} + \angle M_{i[i+1]} A_i C_i$
∴	From G5.19.1-8g, 9g, 13g, 14g, 3 and A5.2.18 GIV Transitivity of Angles	$90° \cong \psi_i + \eta_i$	$90° \cong \psi_{i+1} + \eta_{i+1}$

Theorem 5.24.5 Irregular Polygons: Supplementary Relation of External to Interior Angle

Steps	Hypothesis	$180° \cong \kappa_i + \chi_i$
1	From G5.19.1-1g and D5.1.12 Supplementary Angles	$180° \cong \angle M_{i[i+1]} A_i M_{[i-1]i} + \angle A_{[i+1]} A_i X_{pi}$
∴	From G5.19.1-12g, 15g, 1 and A5.2.18 GIV Transitivity of Angles	$180° \cong \kappa_i + \chi_i$

Theorem 5.24.6 Irregular Polygons: Accumulated Internal Angle

Steps	Hypothesis	$180°(n-2) \cong \sum \kappa_i$
1	From T5.20.5 and summing over all n-component angles	$\sum 180° \cong \sum \kappa_i + \sum \chi_i$
2	From 1, A4.2.12 Identity Multp, A4.2.14 Distribution and LxE.3.29	$180°n \cong \sum \kappa_i + \sum \chi_i$
3	From 2, T5.20.1 Irregular Polygons: Accumulated External Angle and A5.2.18 GIV Transitivity of Angles	$180°n \cong \sum \kappa_i + 360°$
4	From 3, T4.3.3A Equalities: Right Cancellation by Addition	$180°n - 360° \cong \sum \kappa_i$
5	From 4 and T4.2.15 Equalities: Product by Division, Factorization by 2	$180°n - 2 \cdot 180° \cong \sum \kappa_i$
∴	From 5 and A4.2.14 Distribution	$180°(n-2) \cong \sum \kappa_i$

Theorem 5.24.7 Irregular to Regular Polygons: Total Central Angle

Steps	Hypothesis	$\phi = 360° / n$
1	From G5.19.1-19g, T5.20.1 Irregular Polygons: Accumulated External Angle and A5.2.18 GIV Transitivity of Angles	$360° = \sum \phi$
2	From 1, A4.2.12 and A4.2.14	$360° = \phi \sum 1$
∴	From 2, LxE.3.29 and T4.4.5 Equalities: Left Cancellation by Multiplication	$\phi = 360° / n$

Theorem 5.24.8 Irregular to Regular Polygons: Relation Intersecting to CI Angles

Steps	Hypothesis	$\xi \cong 2\psi$
1	From G5.19.1-20g, 21g, T5.20.2 Irregular Polygons: Intersection and Complementary Angle Relationship and A5.2.18 GIV Transitivity of Angles	$\xi_i \cong \psi_i + \psi_{i+1}$
∴	From 1 and T4.3.10 Equalities: Summation of Repeated Terms by 2	$\xi \cong 2\psi$

Theorem 5.24.9 Irregular to Regular Polygons: Relation Central to CI Angles

Steps	Hypothesis	$\phi \cong 2\psi$
1	From G5.19.1-27g	$\angle A_{i-1}C_iM_{[i-1]i} \cong \angle A_iC_iM_{[i-1]i} \cong \angle A_iC_iM_{i[i+1]} \cong \angle A_{i+1}C_iM_{i[i+1]}$ for all [i] when polygon is regular
2	From G5.19.1-8g, 9g, 1 and A5.2.18 GIV Transitivity of Angles	$\psi \cong \angle A_{i-1}C_iM_{[i-1]i} \cong \angle A_iC_iM_{[i-1]i}$ $\cong \angle A_iC_iM_{i[i+1]} \cong \angle A_{i+1}C_iM_{i[i+1]}$ for all [i] when polygon is regular
3	From G5.19.1-1g, 11g and 1	$\angle A_{i-1}C_iM_{[i-1]i} \cong \angle A_iC_iM_{[i-1]i}$ $\cong \angle A_iC_iM_{i[i+1]} \cong \angle A_{i+1}C_iM_{i[i+1]} \cong \frac{1}{2} \angle A_iC_iA_{i-1}$
4	From G5.19.1-11g, 2, 3 and A5.2.18 GIV Transitivity of Angles	$\psi \cong \frac{1}{2}\phi$
∴	From 4 and T4.4.4B Equalities: Reversal of Right Cancellation by Multiplication	$\phi \cong 2\psi$

Theorem 5.24.10 Irregular to Regular Polygons: Congruent Central to Intersecting Angle

Steps	Hypothesis	$\phi \cong \xi$
$\therefore$	From T5.20.8 Irregular to Regular Polygons: Relation Intersecting to CI Angles, T5.20.9 Irregular to Regular Polygons: Relation Central to CI Angles and A5.2.17 GIV Congruence of Angles	$\phi \cong \xi$

Theorem 5.24.11 Irregular to Regular Polygons: Relation of CI to CN Angle

Steps	Hypothesis	$90° \cong \psi + \kappa$	
1	From G5.19.1-20g, 22g, T5.20.4 Irregular Polygons: Complementary Parsed Angles and A5.2.18 GIV Transitivity of Angles	$90° \cong \psi_i + \eta_i$	$90° \cong \psi_{i+1} + \eta_{i+1}$
$\therefore$	From 1	$90° \cong \psi + \kappa$	

Theorem 5.24.12 Irregular to Regular Polygons: Supplementary Relation Central to Internal Angle

Steps	Hypothesis	$180° \cong \phi + 2\kappa$
1	From A4.2.2 Equality T5.20.11 Irregular to Regular Polygons: Relation of CI to CN Angle, T4.4.2 Equalities: Uniqueness of Multiplication and A4.2.14 Distribution and D4.1.19 Primitive Definition for Rational Arithmetic	$2 = 2$
2		$180° \cong 2\psi + 2\kappa$
$\therefore$	From G5.19.1-20g, 2, T5.20.9 Irregular to Regular Polygons: Relation Central to CI Angles and A5.2.18 GIV Transitivity of Angles	$180° \cong \phi + 2\kappa$

Theorem 5.24.13 Irregular to Regular Polygons: Supplementary Central Angle

Steps	Hypothesis	$\phi \cong 180° - 2\kappa$
$\therefore$	From T5.20.12 Irregular to Regular Polygons: Supplementary Relation Central to Internal Angle and T4.3.3A Equalities: Right Cancellation by Addition	$\phi \cong 180° - 2\kappa$

Theorem 5.24.14 Irregular to Regular Polygons: Supplementary Internal Angle

Steps	Hypothesis	$\kappa \cong 90° - \frac{1}{2}\phi$
1	From A4.2.2 Equality	$\frac{1}{2} = \frac{1}{2}$
2	From T5.19.13 Irregular Polygons: Supplementary Central Angle and T4.3.3A Equalities: Right Cancellation by Addition	$2\kappa \cong 180° - \phi$
∴	From 1, T4.4.2 Equalities: Uniqueness of Multiplication, D4.1.4A Rational Numbers, A4.2.13 Inverse Multp, A4.2.12 Identity Multp, A4.2.14 Distribution and D4.1.19 Primitive Definition for Rational Arithmetic	$\kappa \cong 90° - \frac{1}{2}\phi$

Theorem 5.24.15 Irregular to Regular Polygons: Internal Angle, Function (n-Sides)

Steps	Hypothesis	$\kappa \cong 90° - (180° / n)$
1	From T5.20.14 Irregular to Regular Polygons: Internal Angle, Function (n-Sides), T5.20.7 Irregular to Regular Polygons: Total Central Angle and A4.2.3 Substitution	$\kappa \cong 90° - \frac{1}{2}(360° / n)$
∴	From 1, A4.2.11 Associative Multp and D4.1.19 Primitive Definition for Rational Arithmetic	$\kappa \cong 90° - (180° / n)$

Tensor Calculus & Physics: A General Treatise

Theorem 5.24.16 Irregular to Regular Polygons: Double Internal Angle, Function (n-Sides)

Steps	Hypothesis	$2\kappa \cong 180° (1 - (2 / n))$ EQA	$2\kappa \cong 180° - (360° / n)$ EQB
1	From A4.2.2 Equality	$2 = 2$	
1	From 1, T5.20.15 Irregular to Regular Polygons: Internal Angle, Function (n-Sides), T4.4.2 Equalities: Uniqueness of Multiplication and D4.1.19 Primitive Definition for Rational Arithmetic	$2\kappa \cong 180° - (360° / n)$	
2	From 1, T4.4.18 Equalities: Product by Division, Factorization by 2, A4.2.10 Commutative Multp and A4.2.14 Distribution	$2\kappa \cong 180° (1 - (2 / n))$	
∴	From 1 and 2	$2\kappa \cong 180° (1 - (2 / n))$ EQA	$2\kappa \cong 180° - (360° / n)$ EQB

Theorem 5.24.17 Irregular Polygons: Averaging Internal Angles

Steps	Hypothesis	$\kappa_{av} \cong (1/n) \sum \kappa_i$ Equation A	$\sum \kappa_i \cong 180° (n - 2)$ Equation B
1	From T5.20.1B Irregular Polygons: Accumulated External Angle	$360° \cong \sum \xi_i$	
2	From 1, T5.20.2 Irregular Polygons: Intersection and Complementary Angle Relationship and A5.2.18 GIV Transitivity of Angles	$360° \cong \sum \psi_i + \psi_{i+1}$	
3	From T5.20.4 Irregular Polygons: Complementary Parsed Angles and T4.3.3B Equalities: Right Cancellation by Addition	$\psi_i \cong 90° - \eta_i$	$\psi_{i+1} \cong 90° - \eta_{i+1}$
4	From 2, 3 and A5.2.18 GIV Transitivity of Angles	$360° \cong \sum 90° - \eta_i + 90° - \eta_{i+1}$	
5	From 4, D4.1.19 Primitive Definition for Rational Arithmetic, D4.1.20 Negative Coefficient, A4.2.14 Distribution	$360° \cong \sum 180° - (\eta_i + \eta_{i+1})$	
6	From 5, T5.20.3 Irregular Polygons: Interior and Complementary Angle Relationship, A5.2.18 GIV Transitivity of Angles, A4.2.12 Identity Multp, A4.2.14 Distribution and LxE.3.29	$360° \cong \sum 180° - \kappa_i$	
5		$360° \cong 180° n - \sum \kappa_i$	

Steps		
7	From 5 and T4.3.4A Equalities: Reversal of Right Cancellation by Addition and T4.3.5B Equalities: Left Cancellation by Addition	$\sum \kappa_i \cong 180° \, n - 360°$
8	From 6, T4.4.18 Equalities: Product by Division, Factorization by 2, A4.2.10 Commutative Multp and A4.2.14 Distribution	$\sum \kappa_i \cong 180° \, (n - 2)$
9	From A4.2.2 Equality	$1/n = 1/n$
10	From 8, 9 and T4.4.2 Equalities: Uniqueness of Multiplication	$(1/n) \sum \kappa_i \cong (1/n) \, (180° \, (n - 2))$
11	From 10, D4.1.19 Primitive Definition for Rational Arithmetic, A4.2.10 Commutative Multp, D4.1.4 Rational Numbers	$(1/n) \sum \kappa_i \cong 180°(n - 2) / n$
∴	From 29g, 8, 11, A5.2.18 GIV Transitivity of Angles and A5.2.17 GIV Congruence of Angles	$\kappa_{av} \cong (1/n) \sum \kappa_i$ Equation A $\sum \kappa_i \cong 180° \, (n - 2)$ Equation B

Theorem 5.24.18 Irregular to Regular Polygons: Averaging Internal Angles

Steps	Hypothesis	$\kappa_{av} = \kappa = 180° \, (n - 2) / n$
1	From G5.19.1-23g, T5.20.17 Irregular Polygons: Averaging Internal Angles, A5.2.18 GIV Transitivity of Angles, A4.2.12 Identity Multp, and A4.2.14 Distribution	$\kappa_{av} \cong (1/n) \sum \kappa$
2	From 1, A4.2.14 Distribution, LxE.3.29, A4.2.10 Commutative Multp and A4.2.11 Associative Multp	$\kappa_{av} \cong (\, (1/n) \, n) \, \kappa$
∴	From G5.19.1-29g, 2, T4.4.16B Equalities: Product by Division, Common Factor and A4.2.12 Identity Multp	$\kappa_{av} = \kappa = 180° \, (n - 2) / n$

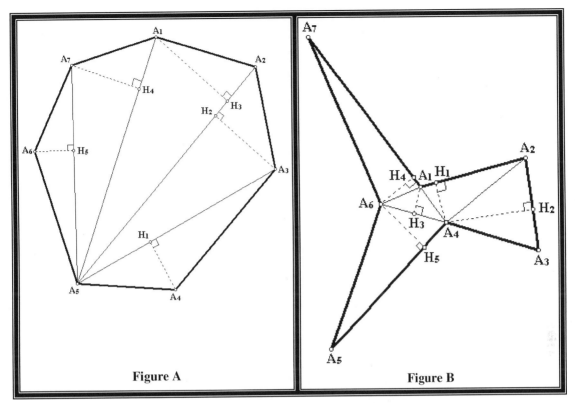

Figure 5.24.2 Parsing Convex and Concave Polygons into Triangles

Observation 5.24.1 Parsing Convex and Concave Polygons into Triangles

1) Notice in Figure 5.24.2A the polygon, from any vertices can be parsed into a fan of triangles, also that except for the first and last triangles only one side of the polygon is used by the other triangles to make them up. This means that n – 2 points are left over to parse the polygon into exactly n – 2 triangles.

2) The Concave Polygon there is not a nice systemic way of parsing the polygon as with the convex, but it still can be divided into n – 2 triangles.

3) Also notice that in Figure 5.24.2B the central triangle $\Delta A_1A_4A_6$ shares no side of the polygon, the argument made with observation 5.24.2A does not hold, yet it still works.

4) If sides are counted for all triangles in Figure 5.24.2 A and B a table can be compiled:

Table 5.24.1 Convex-Concave Sides of a Polygon

Convex Polygon		Concave Polygon	
Parsed Triangles	**N-Polygon Sides**	**Parsed Triangles**	**N-Polygon Sides**
$\Delta A_5A_4A_3$	2	$\Delta A_1A_4A_6$	0
$\Delta A_5A_3A_2$	1	$\Delta A_1A_7A_6$	2
$\Delta A_5A_2A_1$	1	$\Delta A_1A_4A_2$	1
$\Delta A_5A_1A_7$	1	$\Delta A_2A_4A_3$	2
$\Delta A_5A_7A_6$	2	$\Delta A_5A_4A_6$	2
Total Sides	7	Total Sides	7

E) From Table 5.24.2 there is conservation of the sides of a polygon of n-sides regardless whether it's convex or concave.

F) The sides used by the parsing triangles can be no greater than two, or $0 \leq \Delta sides \leq 2$. Consider a polygon of three sides, a triangle, since the triangle, the fundamental space filling shape of the polygon, this is the only case were $\Delta sides = 3$. Consider if it is four the max $\Delta sides = 2$, because the triangles have to share just so many side of the polygon. In fact, for any polygon greater than 3, a triangle having only three sides, this would be the maximum number of sides a triangle can possible have, so this rule would only apply for $n > 3$.

G) The number of triangles is exactly the number of the sum measure of the internal angle normalized with respect to 180°, so from theorem T5.20.17B "Irregular Polygons: Averaging Internal Angles"

$$(n - 2) \equiv (1 / 180°) \sum \kappa_i \qquad \text{Equation G}$$

So the maximum number of triangles that can fill a polygon, whether convex or concave, by parsing is $(n - 2)$. This in tern unifies conservation of sides with the sum measure of the internal angle. This and the specific observation of Figure 5.20.2B would seem reinforces the validity of Observation 5.19.2A.

Theorem 5.24.19 Irregular Polygons: Area of a Polygon by Triangulation

1g	Given	From Observation 5.24.1G
2g		From Figure 5.24.2
3g		$a_i = A_{\beta(i)}H_{\sigma(i)}$ for $\Delta A_{\alpha(i)}A_{\beta(i)}A_{\gamma(i)}$ for all [i]
4g		$p_i = A_{\alpha(i)}A_{\gamma(i)}$ for $\Delta A_{\alpha(i)}A_{\beta(i)}A_{\gamma(i)}$ for all [i]
Steps	Hypothesis	$K_T = \sum \frac{1}{2}a_i\, p_i$ Total Area of a polygon
1	Form G5.19.1-32g; 2g, 3g, 4g and C5.9.1.1 Algebraic: Area of a Triangle	$K_i = \frac{1}{2}a_i\, p_i$ area of parsed triangle i
∴	From G5.19.1-33g; 1g, 1 summing over all [i]	$K_T = \sum \frac{1}{2}a_i\, p_i$ Total Area of a polygon

Theorem 5.24.20 Irregular to Regular Polygons: Area

Steps	Hypothesis	$K_T = \frac{1}{2}a\, P_T$
1	From G5.19.1-16g, 17g, 26g, 28g, T5.20.19 Irregular Polygons: Area of a Polygon by Triangulation and A5.2.15 GIV Transitivity of Segments	$K_T = \sum \frac{1}{2}a\, p$
2	From 1, A4.2.12 Identity Multp and A4.2.14 Distribution	$K_T = \frac{1}{2}a\, p \sum 1$
3	From 2 and LxE.3.29	$K_T = \frac{1}{2}a\, np$
∴	From 3 and T5.12.10 Inscribing n-Isosceles Triangles: Parameter of a Regular Polygon	$K_T = \frac{1}{2}a\, P_T$

Section 5.25 Theorems on the Zero Property of Regular Polygons Order 3 to 11

n-gon	Polygon	Polar Coord-Pts	Geometric Figure: Regular Polygon
3	Triagon	A: $(1, -150°)$ B: $(1, -30°)$ C: $(1, +90°)$	
4	Quadragon Quadrilateral Square	A: $(1, -135°)$ B: $(1, -45°)$ C: $(1, +45°)$ D: $(1, +135°)$	
5	Pentagon	A: $(1, -126°)$ B: $(1, -54°)$ C: $(1, +18°)$ D: $(1, +90°)$ E: $(1, +162°)$	

n-gon	Polygon	Polar Coord-Pts	Geometric Figure: Regular Polygon
6	Hexagon	A: (1, −120°) B: (1, −60°) C: (1, +0°) D: (1, +60°) E: (1, +120°) F: (1, −180°)	
7	Heptagon	A: (1, −115.71°) B: (1, −64.28°) C: (1, +12.86°) D: (1, +38.57°) E: (1, +90.00°) F: (1, +141.43°) G: (1, −167.14°)	
8	Octagon	A: (1, −112.5°) B: (1, −67.5°) C: (1, −22.5°) D: (1, +22.5°) E: (1, +67.5°) F: (1, +112.5°) G: (1, +157.5°) H: (1, −157.5°)	

n-gon	Polygon	Polar Coord-Pts	Geometric Figure: Regular Polygon
9	Nonagon	A: (1, −110°) B: (1, −70°) C: (1, −30°) D: (1, +10°) E: (1, +50°) F: (1, +90°) G: (1, +130°) H: (1, +170°) I: (1, −150°)	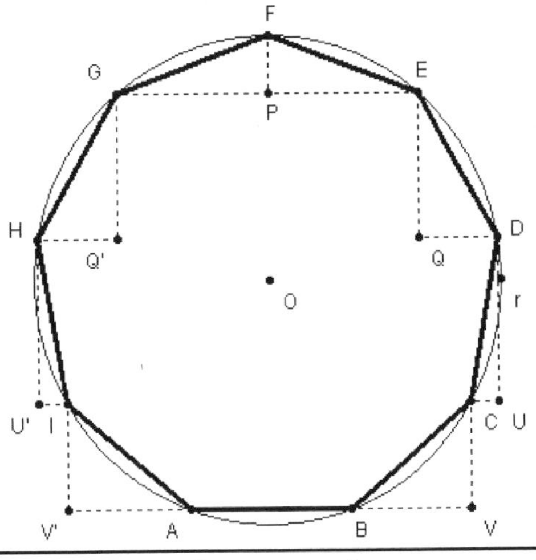
10	Decagon	A: (1, −108°) B: (1, −72°) C: (1, −36°) D: (1, +0°) E: (1, +36°) F: (1, +72°) G: (1, +108°) H: (1, +144°) I: (1, −180°) J: (1, −144°)	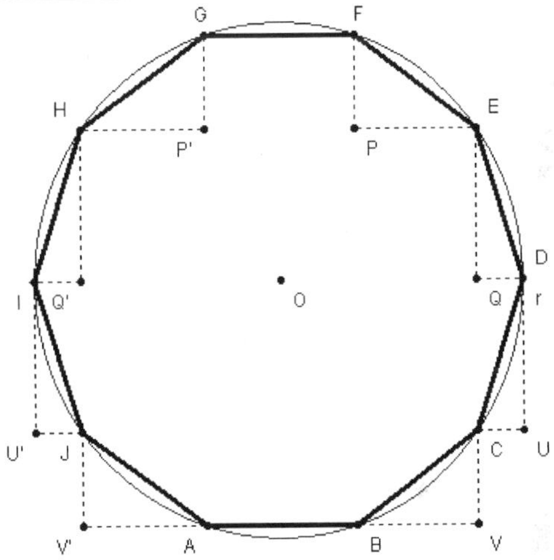
11	Endlagon	A: (1, −106.36°) B: (1, −73.63°) C: (1, −40.91°) D: (1, −8.18°) E: (1, +24.55°) F: (1, +57.27°) G: (1, +90.00°) H: (1, +122.73°) I: (1, +155.46°) J: (1, −171.82°) K: (1, −139.09°)	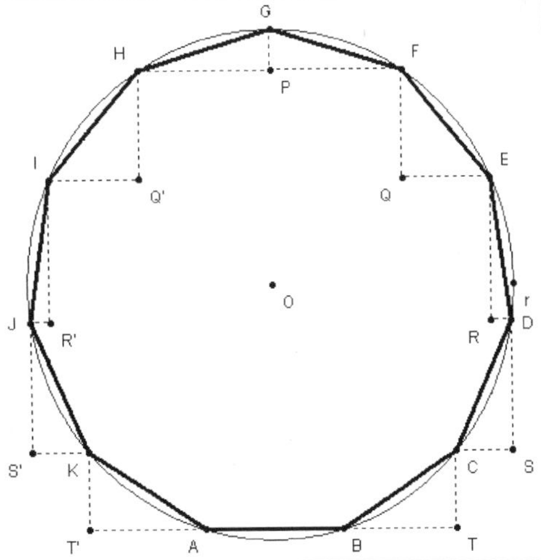

Figure 5.25.1 n-Gon

From T5.24.7 "Irregular Polygons: Total Central Angle" and T5.24.17 "Irregular Polygons: Averaging Internal Angles" the following table can be constructed.

Given 5.25.1: N-Gon Parameters

n-gon	φ Central Angle	κ Internal Angle	Polygon Internal Angles	Regular Polygon Sides
3	120°	30°	∠CBA ≅ ∠BAC ≅ ∠ACB ≅ 2κ	AB ≅ AC ≅ CB ≅ R
4	90°	45°	∠CBA ≅ ∠BAD ≅ ∠ADC ≅ ∠DCB ≅ 2κ	AB ≅ AD ≅ DC ≅ CB ≅ R
5	72°	54°	∠CBA ≅ ∠BAE ≅ ∠AED ≅ ∠EDC ≅ ∠DCB ≅ 2κ	AB ≅ AE ≅ ED ≅ DC ≅ CB ≅ R
6	60°	60°	∠CBA ≅ ∠BAF ≅ ∠AFE ≅ ∠FED ≅ ∠EDC ≅ ∠DCB ≅ 2κ	AB ≅ AF ≅ FE ≅ ED ≅ DC ≅ CB ≅ R
7	51.43°	64.29°	∠CBA ≅ ∠BAG ≅ ∠AGF ≅ ∠GFE ≅ ∠FED ≅ ∠EDC ≅ ∠DCB ≅ 2κ	AB ≅ AG ≅ GF ≅ FE ≅ ED ≅ DC ≅ CB ≅ R
8	45°	67.5°	∠CBA ≅ ∠BAH ≅ ∠AHG ≅ ∠HGF ≅ ∠GFE ≅ ∠FED ≅ ∠EDC ≅ ∠DCB ≅ 2κ	AB ≅ AH ≅ HG ≅ GF ≅ FE ≅ ED ≅ DC ≅ CB ≅ R
9	40°	70°	∠CBA ≅ ∠BAI ≅ ∠AIH ≅ ∠IHG ≅ ∠HGF ≅ ∠GFE ≅ ∠FED ≅ ∠EDC ≅ ∠DCB ≅ 2κ	AB ≅ AI ≅ IH ≅ HG ≅ GF ≅ FE ≅ ED ≅ DC ≅ CB ≅ R
10	36°	72°	∠CBA ≅ ∠BAJ ≅ ∠AJI ≅ ∠JIH ≅ ∠IHG ≅ ∠HGF ≅ ∠GFE ≅ ∠FED ≅ ∠EDC ≅ ∠DCB ≅ 2κ	AB ≅ AJ ≅ JI ≅ IH ≅ HG ≅ GF ≅ FE ≅ ED ≅ DC ≅ CB ≅ R
11	32.72°	73.63°	∠CBA ≅ ∠BAK ≅ ∠AKJ ≅ ∠KJI ≅ ∠JIH ≅ ∠IHG ≅ ∠HGF ≅ ∠GFE ≅ ∠FED ≅ ∠EDC ≅ ∠DCB ≅ 2κ	AB ≅ AK ≅ KJ ≅ JI ≅ IH ≅ HG ≅ GF ≅ FE ≅ ED ≅ DC ≅ CB ≅ R

Tensor Calculus & Physics: A General Treatise

n-gon	Projected Sides by Right Triangles	Projected Isosceles Triangles	Orient able Sense of Direction	Directed Sign
3	$\triangle APC \cong \triangle BPC$ Right Triangles	$\triangle ABC$	CW	+
4				
5	$\triangle EPD \cong \triangle CPD$ Right Triangles $\triangle EAS' \cong \triangle CBS$ Right Triangles	$\triangle ECD,$ $\triangle BA[CE]$	CW, CCW	+, −
6	$\triangle EP'F \cong \triangle CPD$ Right Triangles $\triangle FS'A \cong \triangle CSB$ Right Triangles	$\triangle FC[ED],$ $\triangle BA[CF]$	CW, CCW	+, −
7	$\triangle EPF \cong \triangle EPD$ Right Triangles $\triangle FQ'G \cong \triangle DQC$ Right Triangles $\triangle GS'A \cong \triangle CSB$ Right Triangles	$\triangle FDE,$ $\triangle GC[FD],$ $\triangle BA[CG]$	CW, CW, CCW	+, +, −
8	$\triangle FP'G \cong \triangle EPD$ Right Triangles $\triangle HQ'A \cong \triangle CQB$ Right Triangles	$\triangle GD[FE],$ $\triangle BA[CH]$	CW, CCW	+, −
9	$\triangle FPG \cong \triangle FPE$ Right Triangles $\triangle GQ'H \cong \triangle EQD$ Right Triangles $\triangle HU'I \cong \triangle DUC$ Right Triangles $\triangle IV'A \cong \triangle CVB$ Right Triangles	$\triangle GEF,$ $\triangle HD[GE],$ $\triangle CI[DH],$ $\triangle BA[CI]$	CW, CW, CCW, CCW	+, +, −, −
10	$\triangle GP'H \cong \triangle FPE$ Right Triangles $\triangle HQ'I \cong \triangle EQD$ Right Triangles $\triangle IU'J \cong \triangle QUC$ Right Triangles $\triangle JV'A \cong \triangle CVB$ Right Triangles	$\triangle HE[GF],$ $\triangle ID[HE],$ $\triangle CJ[DI],$ $\triangle BA[CJ]$	CW, CW, CCW, CCW	+, +, −, −
11	$\triangle GPH \cong \triangle GPF$ Right Triangles $\triangle HQ'I \cong \triangle FQE$ Right Triangles $\triangle IR'J \cong \triangle ERD$ Right Triangles $\triangle JS'K \cong \triangle DSC$ Right Triangles $\triangle KT'A \cong \triangle CTB$ Right Triangles	$\triangle HFG,$ $\triangle IE[HF],$ $\triangle JD[IE],$ $\triangle KC[JD]$ $\triangle AB[KC]$	CW, CW, CW, CCW, CCW	+, +, +, −, −

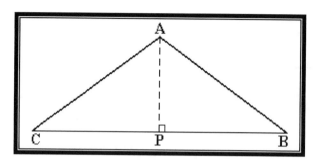

Figure 5.25.2 Zero Harmonic; Isosceles Projection Angle

Theorem 5.25.1 Regular Polygons: Zero Harmonic; Isosceles Projection Angle

1g 2g 3g	Given	$\triangle ABC$ Isosceles Triangle see above figure F5.20.2 $\angle CAB \cong 2\kappa$ $AC \cong AB \cong R$	
Steps	Hypothesis	$\angle BCA \cong \angle CBA \cong 90° - \kappa$	$BP \cong CP \cong R \sin(\kappa)$
1	From 1g, 2g and T5.8.2 Isosceles Triangle: Bisected into Two Congruent Right Triangles	$\angle BCA \cong \angle CBA \cong 90° - ½(2\kappa)$	
2	From 1, A4.2.13 and A4.2.12	$\angle BCA \cong \angle CBP \cong 90° - \kappa$	
3	From 1g, 3g, DxE.1.2.2, A5.2.15 GIV Transitivity of Segments, 2, A5.2.18 GIV Transitivity of Angles and DxE.1.7.3	$BP \cong AB \cos(\angle CBP)$ $CP \cong AC \cos(\angle BCA)$ <td>$BP \cong CP$ $\cong R \cos(90° - \kappa)$</td> <td>$BP \cong CP$ $\cong R \sin(\kappa)$</td>
∴	From 2 and 3	$\angle CBP \cong 90° - \kappa$	$BP \cong CP \cong R \sin(\kappa)$

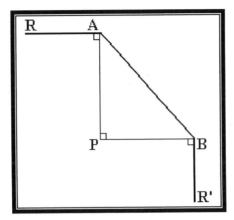

Figure 5.25.3 Square Projection Angle

Theorem 5.25.2 Regular Polygons: Square Projection Angle

1g 2g 3g	Given	$\triangle$APB Right Triangle see above figure F5.20.3 $\angle RAB \cong \angle ABR' \cong 2\kappa$ $AB \cong R$		
Steps	Hypothesis	$\angle PAB \cong \angle PBA \cong 2\kappa - 90°$		$AP \cong BP \cong R \sin(2\kappa)$
1	From 1g	$\angle PAB \cong \angle PBA \cong \angle RAB - 90° \cong \angle ABR' - 90°$		
2	From 2g, 1 and A5.2.18 GIV Transitivity of Angles	$\angle PAB \cong \angle PBA \cong 2\kappa - 90°$		
3	From 1g, 3g, DxE.1.2.2, A5.2.15 GIV Transitivity of Segments, 2, A5.2.18 GIV Transitivity of Angles and DxE.1.7.3	$AP \cong AB \cos(\angle PAB)$ $BP \cong AB \cos(\angle PBA)$	$AP \cong BP$ $\cong R \cos(2\kappa - 90°)$	$AP \cong BP$ $\cong R \sin(2\kappa)$
$\therefore$	From 2 and 3	$\angle PAB \cong \angle PBA \cong 2\kappa - 90°$	$AP \cong BP \cong R \sin(2\kappa)$	

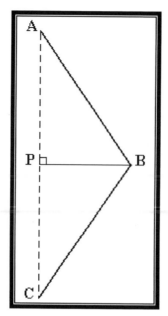

Figure 5.25.4 Vertical Projection Angle

Theorem 5.25.3 Regular Polygons: Vertical Projection Angle

1g	Given	ΔABC Isosceles Triangle see above figure F5.20.4	
2g		∠ABC ≅ 2κ	
3g		AB ≅ R	
Steps	Hypothesis	∠ABP ≅ κ	BP ≅ R cos(κ)
1	From 1g and T5.8.2 Isosceles Triangle: Bisected into Two Congruent Right Triangles	ΔAPB ≅ ΔCPB Right Triangles	
2	From 2g, 1 and T5.8.6 Isosceles Triangle: Right Triangles Bisecting Vertex Angle	∠ABP ≅ ½(2κ)	
3	From 2, D4.1.4-C, A4.2.13 and A4.2.212	∠ABP ≅ κ	
4	From 1g, 3g, DxE.1.2.2, A5.2.15 GIV Transitivity of Segments, 3 and A5.2.18 GIV Transitivity of Angles	BP ≅ AB cos(∠ABP)	BP ≅ R cos(κ)
∴	From 3 and 4	∠ABP ≅ κ	BP ≅ R cos(κ)

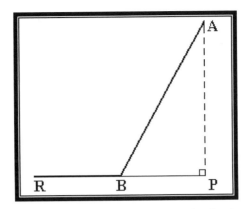

Figure 5.25.5 First Harmonic; External Projection Angle

Theorem 5.25.4 Regular Polygons: First Harmonic; External Projection Angle

1g	Given	ΔAPB Right Triangle see above figure F5.20.5	
2g		∠RBA ≅ 2κ	
3g		AB ≅ R	
Steps	Hypothesis	∠ABP ≅ 180° − 2κ	BP ≅ –R cos(2κ)
1	From 1g, D5.1.8: Flat Angle and D5.1.12: Supplementary Angles	180° ≅ ∠RBA + ∠ABP	
2	From 1 and T4.3.3 Equalities: Right Cancellation by Addition	∠ABP ≅ 180° − ∠RBA	
3	From 2g, 2 and A5.2.18 GIV Transitivity of Angles	∠ABP ≅ 180° − 2κ	
4	From 1g, 3g, DxE.1.2.2, A5.2.15 GIV Transitivity of Segments, 3, A5.2.18 GIV Transitivity of Angles and DxE.1.7.5	BP ≅ AB cos(∠ABP) BP ≅ R cos(180° − 2κ) BP ≅ –R cos(2κ)	
∴	From 3 and 4	∠ABP ≅ 180° − 2κ	BP ≅ –R cos(2κ)

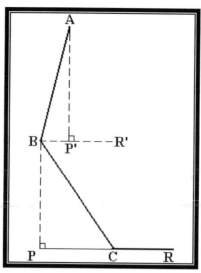

Figure 5.25.6 Second Harmonic; Septa Vertical Projection Angle

Theorem 5.25.5 Regular Polygons: Second Harmonic; Septa Vertical Projection Angle

1g	Given	see above figure		
2g		$\angle ABC \cong \angle BCR \cong 2\kappa$		
3g		$AB \cong BC \cong R$		
4g		PR ∥ BR'		
Steps	Hypothesis	$\angle ABP' \cong 4\kappa - 180°$		$BP \cong -R\cos(4\kappa)$
1	From 1g, D5.1.8: Flat Angle and D5.1.12: Supplementary Angles	$180° \cong \angle RCB + \angle PCB$		
2	From 1 and T4.3.3 Equalities: Right Cancellation by Addition	$\angle PCB \cong 180° - \angle RCB$		
3	From 2g, 2 and A5.2.18 GIV Transitivity of Angles	$\angle PCB \cong 180° - 2\kappa$		
4	From 1g, 4g, 3 and T5.4.4 Parallel Lines: Congruent Alternate Interior Angles	$\angle R'BC \cong \angle PCB \cong 180° - 2\kappa$		
5	From 1g	$\angle ABP' \cong \angle CBA - \angle R'BC$		
6	From 2g, 4, 5 and A5.2.18 GIV Transitivity of Angles	$\angle ABP' \cong 2\kappa - (180° - 2\kappa)$		
7	From 6, A4.2.4, D4.1.19 Primitive Definition For Rational Arithmetic and T4.3.9 Equalities: Summation of Repeated Terms	$\angle ABP' \cong 4\kappa - 180°$ for n =2 terms		
8	From 1g, 3g, DxE.1.2.2, A5.2.15 GIV Transitivity of Segments, 7, A5.2.18 GIV Transitivity of Angles and DxE.1.7.5	$BP \cong AB \cos(\angle ABP')$	$BP \cong R\cos(4\kappa - 180°)$	$BP \cong -R\cos(4\kappa)$
∴	From 7 and 8	$\angle ABP' \cong 4\kappa - 180°$		$BP \cong -R\cos(4\kappa)$

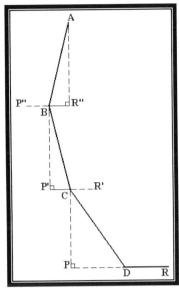

Figure 5.25.7 Second Harmonic; Nona Vertical First Projection Angle

Theorem 5.25.6 Regular Polygons: Second Harmonic; Nona Vertical First Projection Angle

1g 2g 3g 4g	Given	see above figure F5.20.7 $\angle ABC \cong \angle BCD \cong \angle CDE \cong 2\kappa$ $AB \cong BC \cong CD \cong R$ P"R" ‖ P'R' ‖ PR	
Steps	Hypothesis	$\angle BCP' \cong 360° - 4\kappa$	$CP' \cong R\cos(4\kappa)$
1	From 1g and T4.20.5 Regular Polygons: Second Harmonic; Septa Vertical Projection Angle	$\angle BCR' \cong 4\kappa - 180°$	
2	From 1g, D5.1.8: Flat Angle and D5.1.12: Supplementary Angles	$180° \cong \angle BCP' + \angle BCR'$	
3	From 2 and T4.3.3 Equalities: Right Cancellation by Addition	$\angle BCP' \cong 180° - \angle BCR'$	
4	From 1, 3 and A5.2.18 GIV Transitivity of Angles	$\angle BCP' \cong 180° - (4\kappa - 180°)$	
5	From 4, A4.2.14, D4.1.19 Primitive Definition For Rational Arithmetic and T4.3.9 Equalities: Summation of Repeated Terms	$\angle BCP' \cong 360° - 4\kappa$ for n = 2 terms	
6	From 1g, 3g, DxE.1.2.2, A5.2.15 GIV Transitivity of Segments, 5, A5.2.18 GIV Transitivity of Angles and DxE.1.7.9	$CP' \cong BC$ $\cos(\angle BCP')$	$CP' \cong R\cos(360° - 4\kappa)$ $\quad$ $CP' \cong R\cos(4\kappa)$
∴	From 5 and 6	$\angle BCP' \cong 360° - 4\kappa$	$CP' \cong R\cos(4\kappa)$

Theorem 5.25.7 Regular Polygons: Third Harmonic; Nona Vertical Second Projection Angle

	Given	see above figure F5.20.7	
1g			
2g		$\angle ABC \cong \angle BCD \cong 2\kappa$	
3g		$AB \cong BC \cong CD \cong R$	
4g		P"R" ‖ P'R' ‖ PR	
Steps	Hypothesis	$\angle ABR" \cong 6\kappa - 360°$	$BR" \cong R\cos(6\kappa)$
1	From 1g and T4.20.6 Regular Polygons: Second Harmonic; Nona Vertical First Projection Angle	$\angle BCP' \cong 360° - 4\kappa$	
2	From 1g, 4g, 1 and T5.4.4 Parallel Lines: Congruent Alternate Interior Angles	$\angle R"BC \cong \angle BCP' \cong 360° - 4\kappa$	
3	From 1g	$\angle ABR" \cong \angle ABC - \angle R"BC$	
4	From 2g, 2 and A5.2.18 GIV Transitivity of Angles	$\angle ABR" \cong 2\kappa - (360° - 4\kappa)$	
5	From 4, A4.2.14 and D4.1.19 Primitive Definition For Rational Arithmetic	$\angle ABR" \cong 6\kappa - 360°$	
6	From 1g, 3g, DxE.1.2.2, A5.2.15 GIV Transitivity of Segments, 5, A5.2.18 GIV Transitivity of Angles and DxE.1.7.9	$BR" \cong AB \cos(\angle ABR")$ $\quad$ $BR" \cong R\cos(6\kappa - 360°)$ $\quad$ $BR" \cong R \cos(6\kappa)$	
∴	From 5 and 6	$\angle ABR" \cong 6\kappa - 360°$	$BR" \cong R\cos(6\kappa)$

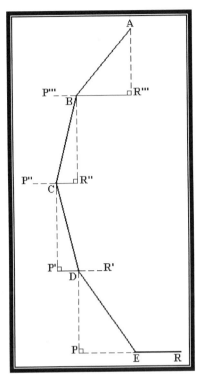

Figure 5.25.8 Fourth Harmonic; Endla Vertical First Projection Angle

Theorem 5.25.8 Regular Polygons: Fourth Harmonic; Endla Vertical First Projection Angle

1g 2g 3g 4g	Given	see above figure F5.20.8 $\angle ABC \cong \angle BCD \cong \angle CDE \cong \angle DER \cong 2\kappa$ $AB \cong BC \cong CD \cong DE \cong R$ P'''R''' ‖ P''R'' ‖ P'R' ‖ PR	
Steps	Hypothesis	$\angle ABR''' \cong 8\kappa - 540°$	$BR''' \cong -R\cos(8\kappa)$
1	From 1g and T4.20.7 Regular Polygons: Third Harmonic; Nona Vertical Second Projection Angle	$\angle BCR'' \cong 6\kappa - 360°$	
2	From 1g, 4g, 1 and T5.4.4 Parallel Lines: Congruent Alternate Interior Angles	$\angle P'''BC \cong \angle BCR'' \cong 6\kappa - 360°$	
3	From 1g, D5.1.8: Flat Angle and D5.1.12: Supplementary Angles	$180° \cong \angle CBR''' + \angle P'''BC$	
4	From 3 and T4.3.3 Equalities: Right Cancellation by Addition	$\angle CBR''' \cong 180° - \angle P'''BC$	
5	From 1, 2 and A5.2.18 GIV Transitivity of Angles	$\angle CBR''' \cong 180° - (6\kappa - 360°)$	
6	From 5, A4.2.14 and D4.1.19 Primitive Definition For Rational Arithmetic	$\angle CBR''' \cong 540° - 6\kappa$	

7	From 1g	$\angle ABC \cong \angle CBR''' + \angle ABR'''$		
8	From 7 and T4.3.3 Equalities: Right Cancellation by Addition	$\angle ABR''' \cong \angle ABC - \angle CBR'''$		
9	From 2g, 8, 6 and A5.2.18 GIV Transitivity of Angles	$\angle ABR''' \cong 2\kappa - (540° - 6\kappa)$		
10	From 9, A4.2.14 and D4.1.19 Primitive Definition For Rational Arithmetic	$\angle ABR''' \cong 8\kappa - 540°$		
11	From 10 and parts of 360° cycles	$\angle ABR''' \cong 8\kappa - 180° - 360°$		
12	From 1g, 3g, DxE.1.2.2, A5.2.15 GIV Transitivity of Segments, 11, A5.2.18 GIV Transitivity of Angles, DxE.1.7.9 and DxE.1.7.5	$BR''' \cong AB \cos(\angle ABR''')$	$BR''' \cong R \cos(8\kappa - 180° - 360°)$	$BR''' \cong -R \cos(8\kappa)$
∴	From 10 and 12	$\angle ABR''' \cong 8\kappa - 540°$	$BR''' \cong -R \cos(8\kappa)$	

Theorem 5.25.9 Regular Polygons: 3-gon Projection

1g 2g	Given	see F5.20.1 3-Gon see G5.20.1 3-Gon Parameters	assume
Steps	Hypothesis	$AB \cong AP + PB$	$1 \cong 2\cos(2\kappa)$
1	From 1g and DxE.1.1.2	$AP \cong AC \cos(\angle BAC)$	
2	From 1g and DxE.1.1.2	$BP \cong BC \cos(\angle CBA)$	
3	From 2g, 1, A5.2.15 GIV Transitivity of Segments and A5.2.18 GIV Transitivity of Angles	$AP \cong R \cos(2\kappa)$	
4	From 2g, 2, A5.2.15 GIV Transitivity of Segments and A5.2.18 GIV Transitivity of Angles	$BP \cong R \cos(2\kappa)$	
5	From 2g, Hypothesis, A5.2.15 GIV Transitivity of Segments	$R \cong R \cos(2\kappa) + R \cos(2\kappa)$	
6	From A4.2.14, T4.4.3 Equalities: Cancellation by Multiplication and T4.3.9 Equalities: Summation of Repeated Terms	$1 \cong 2\cos(2\kappa)$ for n = 2 terms	
7	From 2g, 6 and A5.2.18 GIV Transitivity of Angles	$1 \cong 2\cos(2\,(\,30°\,))$	
8	From 7 and D4.1.19 Primitive Definition for Rational Arithmetic	$1 \cong 2\cos(60°)$	
9	From 8, DxE.1.6.5 and A4.2.10	$1 \cong 2\,(½)$	
10	From 9 and T4.4.18A Equalities: Product by Division, Factorization by 2	$1 \equiv 1$	
∴	From 2g, 6, 10 and by identity	$AB \cong AP + PB$	$1 \cong 2\cos(2\kappa)$

Theorem 5.25.10 Regular Polygons: 4-gon Projection

1g	Given	see F5.20.1 4-Gon	
2g		see G5.20.1 4-Gon Parameters	
3g		$AB \cong AD - DC + CB$ assume	$1 \cong 1$ assume
Steps	Hypothesis	$AB \cong AD - DC + CB$	$1 \cong 1$
1	From 1g, 2g, A5.2.15 GIV Transitivity of Segments, DxE.1.1.2, and DxE.1.6.14	$AD \cong R \cos(270°)$	$AD \cong 0$
2	From 1g, 2g, A5.2.15 GIV Transitivity of Segments, DxE.1.1.2 and DxE.1.6	$DC \cong R \cos(180°)$	$DC \cong -R$
3	From 1g, 2g, A5.2.15 GIV Transitivity of Segments, DxE.1.1.2 and DxE.1.6.7	$CB \cong R \cos(90°)$	$CB \cong 0$
4	From Hypothesis, 1, 2, 3 and A5.2.15 GIV Transitivity of Segments	$AB \cong 0 + R + 0$	
5	From 2g, 4 and A5.2.15 GIV Transitivity of Segments	$R \cong R$	
6	From 5, A4.2.12 and T4.4.3 Equalities: Cancellation by Multiplication	$1 \cong 1$	
∴	From 3g, 6 and by identity	$AB \cong AD - DC + CB$?	$1 \cong 1$?

Theorem 5.25.11 Regular Polygons: 5-gon Projection

1g 2g 3g	Given	see F5.20.1 5-Gon see G5.20.1 5-Gon Parameters $AB \cong -AS' + EP + PC - BS$ Directed Sense assume			$1 \cong 2(\sin(\kappa) + \cos(2\kappa))$ assume
Steps	Hypothesis	$AB \cong -AS' + EP + PC - BS$ Directed Sense			$1 \cong 2(\sin(\kappa) + \cos(2\kappa))$
1	From 1g, 2g and T5.20.4 Regular Polygons: First Harmonic; External Projection Angle T5.20.2 Regular Polygons: Square Projection Angle	$AS' \cong -R \cos(2\kappa)$	$EP \cong R \sin(\kappa)$	$PC \cong R \sin(\kappa)$	$SB \cong -R \cos(2\kappa)$
2	From 2g, Hypothesis, 1 and A5.2.15 GIV Transitivity of Segments	$R \cong +R \cos(2\kappa) + R \sin(\kappa) + R \sin(\kappa) + R \cos(2\kappa)$			
3	From 2, A4.2.14, A4.2.12 and T4.4.3 Equalities: Cancellation by Multiplication	$1 \cong \cos(2\kappa) + \sin(\kappa) + \sin(\kappa) + \cos(2\kappa)$			
4	From T4.3.9 Equalities: Summation of Repeated Terms, A4.2.14 and A4.2.5	$1 \cong 2(\sin(\kappa) + \cos(2\kappa))$			
5	From 2g, 4 and A5.2.18 GIV Transitivity of Angles	$1 \cong 2(\sin(54°) + \cos(2 (54°)))$			
6	From 5 and D4.1.19 Primitive Definition for Rational Arithmetic	$1 \cong 2(\sin(54°) + \cos(108°))$			
7	From 6, Evaluate Trig functions and A4.2.10	$1 \cong 2 (½)$			
8	From 7, D4.1.4-A-B Rational Numbers and A4.2.13 Inverse Multp	$1 \equiv 1$			
∴	From 3g, 8 and by identity	$AB \cong -AS' + EP + PC - BS$ Directed Sense			$1 \cong 2(\sin(\kappa) + \cos(2\kappa))$

Theorem 5.25.12 Regular Polygons: 6-gon Projection

1g 2g 3g	Given	see F5.20.1 6-Gon see G5.20.1 6-Gon Parameters $AB \cong + AS' + FP'$ $+ AB - ED$ $-PC\ -BS$ Directed Sense assume		$1 \cong 1$ assume	
Steps	Hypothesis	$AB \cong + AS' + FP'$ $+ AB - ED$ $-PC\ -BS$ Directed Sense		$1 \cong 1$	
1	From 1g, 2g and T5.20.4 Regular Polygons: First Harmonic; External Projection Angle T5.20.2 Regular Polygons: Square Projection Angle	$AS' \cong -R$ $\cos(2\kappa)$	$FP' \cong R \sin(\kappa)$	$ED \cong R \cos(180°)$	
2	From 1g, 1 and symmetry of regular polygons	$BS \cong -R$ $\cos(2\kappa)$	$PC \cong R \sin(\kappa)$	$AB \cong R \cos(0°)$	
3	From 2g, Hypothesis, 1, 2 and A5.2.15 GIV Transitivity of Segments	$R \cos(0°) \cong -R \cos(2\kappa) + R \sin(\kappa) - R \cos(180°)$ $- R \sin(\kappa) + R \cos(2\kappa)$			
4	From 3, DxE.1.6.1, DxE.1.6.13, A4.2.14, A4.2.12 and T4.4.3 Equalities: Cancellation by Multiplication	$1 \cong -\cos(2\kappa) + \sin(\kappa) - (-1) - \sin(\kappa) + \cos(2\kappa)$			
5	From 4, A4.2.5, A4.2.8, D4.1.20 Negative Coefficient and D4.1.19 Primitive Definition for Rational Arithmetic	$1 \cong 1$			
∴	From 3g, 5 and by identity	$AB \cong + AS' + FP'$ $+ AB - ED$ $-PC\ -BS$ Directed Sense		$1 \cong 1$	

Theorem 5.25.13 Regular Polygons: 7-gon Projection

1g 2g 3g	Given	see F5.20.1 7-Gon see G5.20.1 7-Gon Parameters $AB \cong -AS' + GQ'$ $+ FP + PD$ $+ QC - BS$ Directed Sense assume		$1 \cong 2(\sin(\kappa) - \cos(4\kappa) + \cos(2\kappa))$ assume
Steps	Hypothesis	$AB \cong -AS' + GQ'$ $+ FP + PD$ $+ QC - BS$ Directed Sense		$1 \cong 2(\sin(\kappa) - \cos(4\kappa) + \cos(2\kappa))$
1	From 1g, 2g, T5.20.4 Regular Polygons: First Harmonic; External Projection Angle, T5.20.5 Regular Polygons: Second Harmonic; Septa Vertical Projection Angle and T5.20.1 Regular Polygons: Zero Harmonic; Isosceles Projection Angle	$AS' \cong -R \cos(2\kappa)$	$GQ' \cong -R \cos(4\kappa)$	$FP \cong R \sin(\kappa)$
2	From 1g, 1 and symmetry of regular polygons	$PD \cong -R \cos(2\kappa)$	$QC \cong -R \cos(4\kappa)$	$SB \cong R \sin(\kappa)$
3	From 2g, Hypothesis, 1, 2 and A5.2.15 GIV Transitivity of Segments	$R \cong +R \cos(2\kappa) - R \cos(4\kappa) + R \sin(\kappa)$ $+R \sin(\kappa) - R \cos(4\kappa) + R \cos(2\kappa)$		
4	From 3, DxE.1.6.1, DxE.1.6.13, A4.2.14, A4.2.12 and T4.4.3 Equalities: Cancellation by Multiplication	$1 \cong \cos(2\kappa) - \cos(4\kappa) + \sin(\kappa) + \sin(\kappa) - \cos(4\kappa) + \cos(2\kappa)$		
5	From 4, A4.2.5, T4.3.9 Equalities: Summation of Repeated Terms and A4.2.14	$1 \cong 2(\sin(\kappa) - \cos(4\kappa) + \cos(2\kappa))$ for n = 2 terms		
6	From 2g, 5 and A5.2.18 GIV Transitivity of Angles	$1 \cong 2(\sin(64.29°) + \cos(4(64.29°)) - \cos(2(64.29°)))$		
7	From 5 and D4.1.19 Primitive Definition for Rational Arithmetic	$1 \cong 2(\sin(64.29°) + \cos(257.16°) - \cos(128.58°))$		
8	From 6, Evaluate Trig functions and A4.2.10	$1 \cong (½) 2$		
9	From 5, D4.1.4-A-B Rational Numbers and A4.2.13 Inverse Multp	$1 \equiv 1$ by identity		
∴	From 3g, 9 and by identity	$AB \cong -AS' + GQ'$ $+ FP + PD$ $+ QC - BS$ Directed Sense		$1 \cong 2(\sin(\kappa) - \cos(4\kappa) + \cos(2\kappa))$

Theorem 5.25.14 Regular Polygons: 8-gon Projection

1g 2g 3g	Given	see F5.20.1 8-Gon see G5.20.1 8-Gon Parameters $AB \cong\ + AQ' + HH$ $+ GP' - FE$ $- DP - CC - BQ$ Directed Sense		$1 \cong 1$	
Steps	Hypothesis	$AB \cong\ + AQ' + HH$ $+ GP' - FE$ $- DP - CC - BQ$ Directed Sense assume		$1 \cong 1$ assume	
1	From 1g, 2g, T5.20.4 Regular Polygons: First Harmonic; External Projection Angle and T5.20.2 Regular Polygons: Square Projection Angle	$AQ' \cong -R$ $\cos(2\kappa)$	$HH \cong R$ $\cos(270°)$	$GP' \cong R \sin(2\kappa)$	$FE \cong R$ $\cos(180°)$
2	From 1g, 1 and symmetry of regular polygons	$PD \cong -R$ $\cos(2\kappa)$	$CC \cong R$ $\cos(90°)$	$DP \cong R \sin(2\kappa)$	
3	From 2g, Hypothesis, 1, 2 and A5.2.15 GIV Transitivity of Segments	$R \cong -R \cos(2\kappa) + R \cos(270°) + R \sin(2\kappa) - R \cos(180°)$ $- R \sin(2\kappa) - R \cos(90°) - (-R \cos(2\kappa))$			
4	From 3, DxE.1.6.14, DxE.1.6.13, DxE.1.6.7, A4.2.14, A4.2.12 and T4.4.3 Equalities: Cancellation by Multiplication	$1 \cong -\cos(2\kappa) + 0 + \sin(2\kappa) - (-1) - \sin(2\kappa) - 0 + \cos(2\kappa)$			
5	From 4, A4.2.5, A4.2.8, A4.2.7, D4.1.20 Negative Coefficient and D4.1.19 Primitive Definition for Rational Arithmetic	$1 \equiv 1$			
∴	From 3g, 5 and by identity	$AB \cong\ + AQ' + HH$ $+ GP' - FE$ $- DP - CC - BQ$ Directed Sense		$1 \cong 1$	

Theorem 5.25.15 Regular Polygons: 9-gon Projection

1g 2g	Given	see F5.20.1 9-Gon see G5.20.1 9-Gon Parameters			
Steps	Hypothesis	$AB \cong -AV' + IU'$ $+ HQ' + GP$ $+ PE + QD$ $+ UC - BV$ Directed Sense		$1 \cong 2(\cos(2\kappa) - \cos(4\kappa) +$ $\cos(6\kappa) + \sin(\kappa))$	
1	From 1g, 2g, T5.20.4 Regular Polygons: First Harmonic; External Projection Angle, T5.20.6 Regular Polygons: Second Harmonic; Nona Vertical First Projection Angle, T5.20.7 Regular Polygons: Third Harmonic; Nona Vertical Second Projection Angle, T5.20.5 Regular Polygons: Second Harmonic; Septa Vertical Projection Angle and T5.20.1 Regular Polygons: Zero Harmonic; Isosceles Projection Angle	$AV' \cong -R$ $\cos(2\kappa)$	$IU' \cong -R$ $\cos(4\kappa)$	$HQ' \cong R$ $\cos(6\kappa)$	$GP \cong R \sin(\kappa)$
2	From 1g, 1 and symmetry of regular polygons	$BV \cong -R$ $\cos(2\kappa)$	$UC \cong -R$ $\cos(4\kappa)$	$QD \cong R$ $\cos(6\kappa)$	$PE \cong R \sin(\kappa)$
3	From 2g, Hypothesis, 1, 2 and A5.2.15 GIV Transitivity of Segments	$R \cong +R \cos(2\kappa) - R \cos(4\kappa) + R \cos(6\kappa) + R \sin(\kappa)$ $+R \cos(2\kappa) - R \cos(4\kappa) + R \cos(6\kappa) + R \sin(\kappa)$			
4	From 3, DxE.1.6.1, DxE.1.6.13, A4.2.14, A4.2.12 and T4.4.3 Equalities: Cancellation by Multiplication	$1 \cong \cos(2\kappa) - \cos(4\kappa) + \cos(6\kappa) + \sin(\kappa) +$ $\cos(2\kappa) - \cos(4\kappa) + \cos(6\kappa) + \sin(\kappa)$			
5	From 4, A4.2.5, T4.3.9 Equalities: Summation of Repeated Terms and A4.2.14	$1 \cong 2(\cos(2\kappa) - \cos(4\kappa) + \cos(6\kappa) + \sin(\kappa))$ for n = 2 terms			
6	From 2g, 5 and A5.2.18 GIV Transitivity of Angles	$1 \cong 2(\cos(2(70°)) - \cos(4(70°)) + \cos(6(70°)) + \sin(70°))$			
7	From 6 and D4.1.19 Primitive Definition for Rational Arithmetic	$1 \cong 2(\cos(140°) - \cos(280°) + \cos(420°) + \sin(70°))$			
8	From 7, Evaluate Trig functions and A4.2.10	$1 \cong (½) 2$			

9	From 8, D4.1.4-A-B Rational Numbers and A4.2.13 Inverse Multp	$1 \equiv 1$ by identity	
$\therefore$	From 2g and 9; by identity	$AB \cong -AV' + IU'$ $+ HQ' + GP$ $+ PE + QD$ $+ UC - BV$ Directed Sense EQ A	$1 \cong 2(\,\cos(2\kappa) - \cos(4\kappa) +$ $\cos(6\kappa) + \sin(\kappa)\,)$ EQ B

Theorem 5.25.16 Regular Polygons: 11-gon Projection

1g 2g	Given	see F5.20.1 11-Gon see G5.20.1 11-Gon Parameters				
Steps	Hypothesis	$AB \cong -AT' + KS'$ $+ JR' + IQ'$ $+ HP + PF$ $+ QE + RD$ $+ SC - TB$ Directed Sense		$1 \cong 2(\,\cos(2\kappa) - \cos(4\kappa) + \cos(6\kappa)$ $- \cos(8\kappa) + \sin(\kappa)$		
1	From 1g, 2g, T5.20.4 Regular Polygons: First Harmonic; External Projection Angle, T5.20.6 Regular Polygons: Second Harmonic; Nona Vertical First Projection Angle, T5.20.7 Regular Polygons: Nona Vertical Second Projection Angle T5.20.8 Regular Polygons: Fourth Harmonic; Endla Vertical First Projection Angle and T5.20.1 Regular Polygons: Zero Harmonic; Isosceles Projection Angle	$AT' \cong -R$ $\cos(2\kappa)$	$KS' \cong -R$ $\cos(4\kappa)$	$JR' \cong R$ $\cos(6\kappa)$	$IQ' \cong -R$ $\cos(8\kappa)$	$HP \cong R$ $\sin(\kappa)$
2	From 1g, 1 and symmetry of regular polygons	$TB \cong -R$ $\cos(2\kappa)$	$SC \cong -R$ $\cos(4\kappa)$	$RD \cong R$ $\cos(6\kappa)$	$QE \cong -R$ $\cos(8\kappa)$	$PF \cong R$ $\sin(\kappa)$
3	From 2g, Hypothesis, 1, 2 and A5.2.15 GIV Transitivity of Segments	$R \cong +R\cos(2\kappa) - R\cos(4\kappa) + R\cos(6\kappa) - R\cos(8\kappa) + R\sin(\kappa)$ $+R\cos(2\kappa) - R\cos(4\kappa) + R\cos(6\kappa) - R\cos(8\kappa) + R\sin(\kappa)$				
4	From 3, DxE.1.6.1, DxE.1.6.13, A4.2.14, A4.2.12 and T4.4.3 Equalities: Cancellation by Multiplication	$1 \cong \cos(2\kappa) - \cos(4\kappa) + \cos(6\kappa) - \cos(8\kappa) + \sin(\kappa) +$ $\cos(2\kappa) - \cos(4\kappa) + \cos(6\kappa) - \cos(8\kappa) + \sin(\kappa)$				
5	From 4, A4.2.5, T4.3.9 Equalities: Summation of Repeated Terms and A4.2.14	$1 \cong 2(\,\cos(2\kappa) - \cos(4\kappa) + \cos(6\kappa) - \cos(8\kappa) + \sin(\kappa)$ for n = 2 terms				
6	From 2g, 5 and A5.2.18 GIV Transitivity of Angles	$1 \cong 2(\,\cos(2(73.\underline{63}°)) - \cos(4(73.\underline{63}°)) + \cos(6(73.\underline{63}°))$ $- \cos(8(73.\underline{63}°)) + \sin(73.\underline{63}°)\,)$				

7	From 6 and D4.1.19 Primitive Definition for Rational Arithmetic	$1 \cong 2(\cos(147.\underline{27}°) - \cos(294.\underline{54}°) + \cos(441.\underline{81}°) - \cos(589.\underline{09}°)$ $+ \sin(73.\underline{63}°))$	
8	From 7, Evaluate Trig functions and A4.2.10	$1 \cong (½) 2$	
9	From 8, D4.1.4-A-B Rational Numbers and A4.2.13 Inverse Multp	$1 \equiv 1$ by identity	
∴	From 9 and ???	$AB \cong -AT' + KS'$ $+ JR' + IQ'$ $+ HP + PF$ $+ QE + RD$ $+ SC - TB$ Directed Sense	$1 \cong 2(\cos(2\kappa) - \cos(4\kappa) +$ $\cos(6\kappa)$ $- \cos(8\kappa) +$ $\sin(\kappa)$

Summary of Unit Net Sides for N-Gons

n-gon	unit net side
3	$1 \cong 2\cos(2\kappa)$
4	$1 \cong -\cos(180°)$
5	$1 \cong 2(\cos(2\kappa) + \sin(\kappa))$
6	$1 \cong -\cos(180°)$
7	$1 \cong 2(\cos(2\kappa) - \cos(4\kappa) + \sin(\kappa))$
8	$1 \cong -\cos(180°)$
9	$1 \cong 2(\cos(2\kappa) - \cos(4\kappa) + \cos(6\kappa) + \sin(\kappa))$
10	$1 \cong -\cos(180°)$
11	$1 \cong 2(\cos(2\kappa) - \cos(4\kappa) + \cos(6\kappa) - \cos(8\kappa) + \sin(\kappa))$

Because the net perceived perimeter path must have accumulative directional sense of zero, and then do to the bilateral symmetry of even polygons, opposite sides subtract each other out leaving only the top opposing side, see 4, 6, 8, 10 from the above table.

Theorem 5.25.17 Regular Polygons: Even Number of Sides Integrate to Equal Opposing Sides

1g	Given	$AB/R \equiv \sum_{i=1}^{m-1} C_iD_i/R - C_mD_m/R - \sum_{j=m+1}^{2m} C_jD_j/R$
		Directed Sense
2g		Opposing sides are equal
3g		$C_mD_m/R \equiv \cos(180°)$ directed angle opposing resultant
4g		$-\cos(180°) \equiv AB/R$ assume
Steps	Hypothesis	$AB/R \equiv -\cos(180°)$
1	From 4g	$-\cos(180°) \equiv AB/R$
2	From 1, DxE.1.6.13 cosine of 180° and T4.8.2 Integer Exponents: Negative One Squared	$1 = AB/R$
3	From 1g, 2, A4.2.3 Substitution and A4.2.5 Commutative Add	$1 = \sum_{i=1}^{m-1} C_iD_i/R - \sum_{j=m+1}^{2m} C_jD_j/R - C_mD_m/R$
4	From 2g, 3, A4.2.14 Distribution and A4.2.6 Associative Add	$1 = \sum_{i=1}^{m-1} (C_iD_i/R - C_iD_i/R) - C_mD_m/R$
5	From 4, A4.2.8 Inverse Add and A4.2.7 Identity Add	$1 = -C_mD_m/R$
6	From 3g, 5 and A4.2.3 Substitution	$1 = -\cos(180°)$
7	From 5, DxE.1.6.13 cosine of 180° and T4.8.2 Integer Exponents: Negative One Squared	$1 \equiv 1$
∴	From 1, 7 and by identity	$AB/R \equiv -\cos(180°)$

However, odd numbered sides are another matter all together. From the above table odd sided polygons require an alternate directional sense of harmonic progression and can be easily generalized, see 3, 5, 7, 9, 11 from the above table and the given by following intermediate theorem "Regular Polygons: Harmonic Geometric Progression at the Central Angle". The question than arises is this a true generalization and would it satisfy all odd sided polygons? The following proof will show exactly that and for any value of [n] will always contract to ½ being evaluated at the central angle allowing the net path to integrate to zero.

Theorem 5.25.18 Regular Polygons: Harmonic Geometric Progression at the Central Angle

1g 2g	Given	$X = \sin(\kappa) - X_0$ for $M = \frac{1}{2}(n - 3)$ general harmonic progression $X_0 = \sum_{m=1}^{M} (-1)^m \cos(m\, 2\kappa)$ kernel harmonic progression
Steps	Hypothesis	$X = \frac{1}{2}$ for all odd numbers of $n > 1$
1	From 1g, 2g, A4.2.3 Substitution and LxE.3.72	$X = \sin(\kappa) - (-\frac{1}{2} + (-1)^M \cos((2M + 1)\kappa) / 2\cos(\kappa))$
2	From 1, A4.2.14 Distribution, D4.1.20 Negative Coefficient and D4.1.19 Primitive Definition for Rational Arithmetic	$X = \sin(\kappa) + \frac{1}{2} - ((-1)^M \cos((2M + 1)\kappa) / 2\cos(\kappa))$
3	From 2 and A4.2.5 Commutative Add	$X = \frac{1}{2} + \sin(\kappa) - ((-1)^M \cos((2M + 1)\kappa) / 2\cos(\kappa))$
4	From 3, A4.2.12 Identity Multp, A4.2.13 Inverse Multp and A4.2.14 Distribution	$X = \frac{1}{2} + (2\cos(\kappa)\sin(\kappa) - (-1)^M \cos((2M + 1)\kappa) / 2\cos(\kappa))$
5	From 4 and LxE.1.34	$X = \frac{1}{2} + (\sin(2\kappa) - (-1)^M \cos((2M + 1)\kappa) / 2\cos(\kappa))$
6	From 1g, 5, T5.19.15 Irregular Polygons: Internal Angle, Function of Number of Sides and A4.2.3 Substitution	$X = \frac{1}{2} + (\sin(2\kappa) - (-1)^M \cos((2M + 1)\kappa) / 2\cos(\kappa))$
7	From 1g, 6, T5.19.16 Irregular Polygons: Double Internal Angle, Function of Number of Sides, T5.19.15 Irregular Polygons: Internal Angle, Function of Number of Sides and A4.2.3 Substitution	$X = \frac{1}{2} + (\sin(180° - (360° / n))$ $- (-1)^M \cos((2(\frac{1}{2}(n - 3)) + 1)(90° - (180° / n))) /$ $2\cos(90° - (180° / n)))$
8	From 7, DxE.1.7.5, DxE.1.7.3, D4.1.4 Rational Numbers-A, A4.2.13 Inverse Multp, A4.2.12 Identity Multp and D4.1.19 Primitive Definition for Rational Arithmetic	$X = \frac{1}{2} + (\sin(360° / n)$ $- (-1)^M \cos((n - 2)(90° - (180° / n))) / 2\sin(180° / n))$
9	From 8 and A4.2.14 Distribution	$X = \frac{1}{2} + (\sin(360° / n)$ $- (-1)^M \cos(n\, 90° + n(- (180° / n) - 2(90°)$ $+ (- 2)(- (180° / n))) / 2\sin(180° / n))$

10	From 9, A4.2.10 Commutative Multp, A4.2.13 Inverse Multp, A4.2.12 Identity Multp, D4.1.20 Negative Coefficient and D4.1.19 Primitive Definition for Rational Arithmetic	$X = \frac{1}{2} + (\sin(360° / n)$ $- (-1)^M \cos(90°n - 180° - 180° + 360° / n)) / 2\sin(180° / n))$
11	From 10, T4.3.9 Equalities: Summation of Repeated Terms and D4.1.19 Primitive Definition for Rational Arithmetic	$X = \frac{1}{2} + (\sin(360° / n)$ $- (-1)^M \cos(90°n - 360° + 360° / n)) / 2\sin(180° / n))$ for 2 phase terms
12	From 11, DxE.1.7.9	$X = \frac{1}{2} + (\sin(360° / n)$ $- (-1)^M \cos(90°n + 360° / n)) / 2\sin(180° / n))$
13	From 12, and AE.1.1 for n odd and multiples of 90° the cofunction is sine	$X = \frac{1}{2} + (\sin(360° / n) - (-1)^M(-1)^M \sin(360° / n))/2\sin(180° / n))$
14	From 12, 15, A4.2.18 Summation Exp and T4.3.9 Equalities: Summation of Repeated Terms	$X = \frac{1}{2} + (\sin(360° / n) - (-1)^{2M} \sin(360° / n)) / 2\sin(180° / n))$ for 2 exponent terms
15	From 14 and T4.8.3 Integer Exponents: Negative One Raised to an Even Number	$X = \frac{1}{2} + (\sin(360° / n) - (+1) \sin(360° / n)) / 2\sin(180° / n))$
16	From 15 and A4.2.12 Identity Multp	$X = \frac{1}{2} + (\sin(360° / n) - \sin(360° / n)) / 2\sin(180° / n))$
17	From 16 and A4.2.8 Inverse Add	$X = \frac{1}{2} + (0) / 2\sin(180° / n))$
∴	From 17, DxE.1.6.13 and T4.3.1 Equalities: Any Quantity Multiplied by Zero is Zero	$X = \frac{1}{2}$ for all odd numbers of n > 1

Notice that the proof does not require the Induction Property of Integers [$\flat_2$] the fact that it evaluates to [½] proves that the harmonic progression is a unique solution around the integrated path, hence is valid for all n-odd sided polygons.

Theorem 5.25.19 Regular Polygons: Odd Number of Sides Integrate to Equal Opposing Sides

1g	Given	$AB/R \equiv \sum_{i=1}^{\frac{1}{2}(n-1)} C_i D_i/R + \sum_{j=\frac{1}{2}(n+1)}^{n-1} C_j D_j/R$
		Directed Sense assume
2g		Opposing sides are equal
3g		$C_1 D_1/R \equiv C_{n-1} D_{n-1}/R \equiv -C_1 D_1/R$
4g		$X \equiv \sum_{i=1}^{\frac{1}{2}(n-1)} C_i D_i/R$
5g		$1 \equiv AB/R$
Steps	Hypothesis	$AB/R \equiv 2\,(\sin(\kappa) - \sum_{m=1}^{M} (-1)^m \cos(m\,2\kappa))$
		for $\kappa \equiv 90° - (180°/n)$ evaluated at the central angle
1	From 5g	$1 = AB/R$
2	From 1g, 2g, 1 and A4.2.3 Substitution	$1 = \sum_{i=1}^{\frac{1}{2}(n-1)} C_i D_i/R + \sum_{i=1}^{\frac{1}{2}(n-1)} C_i D_i/R$
3	From 4g, 2 and A4.2.3 Substitution	$1 = X + X$
4	From 3 and T4.3.9 Equalities: Summation of Repeated Terms	$1 = 2X$ for 2 harmonic progressions
5	Form 4, T5.20.18 Regular Polygons: Harmonic Geometric Progression at the Central Angle and A4.2.3 Substitution	$1 = 2\,(\frac{1}{2})$
6	From 5, D4.1.4-A-B Rational Numbers and A4.2.13 Inverse Multp	$1 = 1$
∴	From 1, 6 and by identity	$AB/R \equiv 2\,(\sin(\kappa) - \sum_{m=1}^{M} (-1)^m \cos(m\,2\kappa))$
		for $\kappa \equiv 90° - (180°/n)$ evaluated at the central angle

Section 5.26 Coplanar Proof of Validity of the Zero Property

Theorems "Irregular Polygons: Averaging Internal Angles" and "Irregular Polygons: Averaging Internal Angles for Regular Polygon" coupled with the given G5.20.1 "Parameters of an Irregular Polygon"-28g, the average parameter base, completely describes an averaged polygon topologically mapped from an irregular polygon. This is a very powerful notion that any irregular polygon always has a uniquely defined regular polygon, hence corresponds and maps into the other. This would imply that any property of a regular polygon would also have to be a property of an irregular polygon. From theorems "Regular Polygons: Even Number of Sides Integrate to Equal Opposing Sides" and "Regular Polygons: Odd Number of Sides Integrate to Equal Opposing Sides" implies the property of the integral projected path about a regular polygon must also be true for any number of sides of an irregular polygon, QED.

Theorem 5.26.1 Irregular Polygons: Have the same Properties as Regular Polygons

1g	Given	T_{aia}	Proposition relationship between internal averages of irregular polygon angles
2g		T_{ara}	Proposition relationship between internal averages of regular polygon angles
3g		P_{reg}	Property of regular polygon
4g		Av_{reg}	Set of all average regular polygons
5g		I_{reg}	Set of all irregular polygons
6g		$f< T_{aia} \wedge T_{ara} >: I_{reg} \rightarrow Av_{reg}$	
7g		$f^{-1}< T_{ara} \wedge T_{aia} >: Av_{reg} \rightarrow I_{reg}$	
8g		$P_{reg} \in Av_{reg}$	
Steps	Hypothesis	$P_{reg} \in I_{reg}$	for $f^{-1}< T_{ara} \wedge T_{aia} >: Av_{reg} \rightarrow I_{reg}$
1	From 6g, (T5.20.17 $\in T_{aia}$) Irregular Polygons: Averaging Internal Angles and (T5.20.18 $\in T_{ara}$) Irregular to Regular Polygons: Averaging Internal Angles	$f< T_{aia} \wedge T_{ara} >: I_{reg} \rightarrow Av_{reg}$	valid [True]
2	From 7g, 1 by Av_{reg} (1-1 and onto) I_{reg}	$f^{-1}< T_{ara} \wedge T_{aia} >: Av_{reg} \rightarrow I_{reg}$	valid [True]
$\therefore$	From 8g and 2	$P_{reg} \in I_{reg}$	

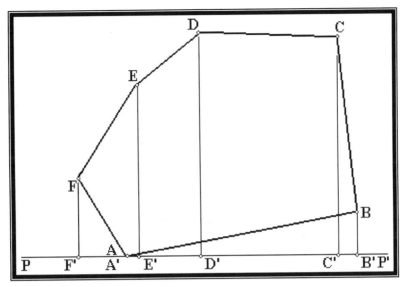

Figure 5.26.1 Projected Integral Path about a Polygon

Theorem 5.26.2 Irregular Polygons: Zero property of the Integral Projected Path

1g	Given	see above geometry figure F5.22.1	
2g		T_{aia}	Proposition relationship between internal averages of irregular polygon angles
3g		T_{ara}	Proposition relationship between internal averages of regular polygon angles
4g		P_{zro}	Zero property of the integral projected path about a regular polygon sums to zero
5g		Av_{reg}	Set of all regular polygons
6g		I_{reg}	Set of all irregular polygons
Steps	Hypothesis	$P_{zro} \in I_{reg}$	for $f^{-1} < T_{ara} \wedge T_{aia} >: Av_{reg} \to I_{reg}$
1	From 1g, 3g, 4g, 5g by (T5.21.17$\in T_{ara}$) Regular Polygons: Even Number of Sides Integrate to Equal Opposing Sides and (T5.21.19 $\in T_{ara}$) Regular Polygons: Odd Number of Sides Integrate to Equal Opposing Sides	$P_{zro} \in Av_{reg}$	
$\therefore$	From 3g, 1 and T5.22.1 Irregular Polygons: Have the same Properties as Regular Polygons	$P_{zro} \in I_{reg}$	

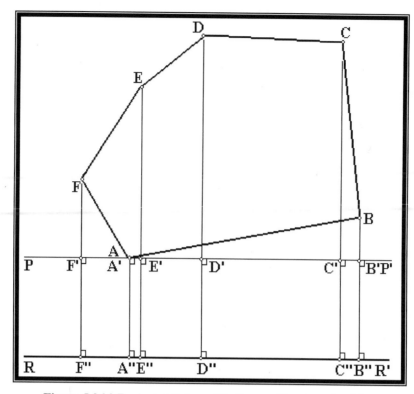

Figure 5.26.2 Projected Integral Path on a Construction Line

Theorem 5.26.3 Irregular Polygons: Zero Property for Integral Path on a Construction Line

1g	Given	above geometry figure F5.22.2
2g		PP' ‖ RR'
3g		$P_{zro} \in I_{reg}$ of irregular polygons projected vertices onto PP'
4g		T_{ireg} Set of irregular polygons projected vertices onto RR'
5g		$I_{reg} \rightarrow T_{ireg}$ PP' maps onto construction line RR'
Steps	Hypothesis	$P_{zro} \in T_{ireg}$
1	From 2g, 3g and 5g	$I_{reg} \equiv T_{ireg}$
∴	From 3g, 1 and A4.2.3 Substitution	$P_{zro} \in T_{ireg}$

Section 5.27 Non-Coplanar Proof of Validity of the Zero Property

This chapter has been building up to irregular polygons of general order-n, however it has been on the assumption that the polygons have been constructed on a flat Euclidian plane, but the finial proof is yet to be considered. The objective is to model a sequence of vectors that circularly close on themselves. In n-dimensions vectors are not constrained to a plane, but have the freedom to angle in any direction, certainly not confined to a 2-dimensional plane embedded in n-space. The only conditions are that they be contiguous (connect end-to-end) and close on themselves (last vector connecting with the tail end of the first vector).

While all of the proofs where constructed on a plane that is not a necessary requirement for a general proof because the proof really requires construction of the projection any line segment onto a construction line and then sum all quantities independent of any other line segment.

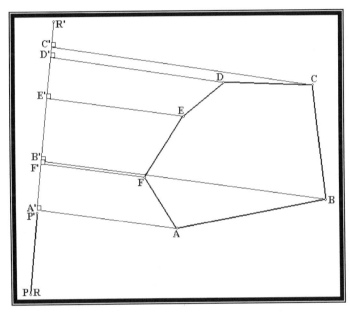

Figure 5.27.1 Non-Coplanar Polygon Projected onto a Co-linear Plane

Theorem 5.27.1 Irregular Polygons: Non-Coplanar Polygon's Zero Property for Integral Path

1g	Given	above geometry figure F5.23.1
2g		Polygon any spatial polygon
3g		Polygon' projected polygon
4g		Polygon $\rightarrow$ Polygon' maps onto plane-RR'
5g		construction line PP' co-planar on plane-RR'
6g		S_{ireg} Set of non-coplanar irregular polygons
Step	Hypothesis	$P_{zro} \in S_{ireg}$
1	From T5.22.3 Irregular Polygons: Zero Property for Integral Path on a Construction Line	$P_{zro} \in T_{ireg}$
2	From 1g and 4g	$T_{reg} \equiv S_{ireg}$
∴	From 2g, 1 and A4.2.3 Substitution	$P_{zro} \in S_{ireg}$

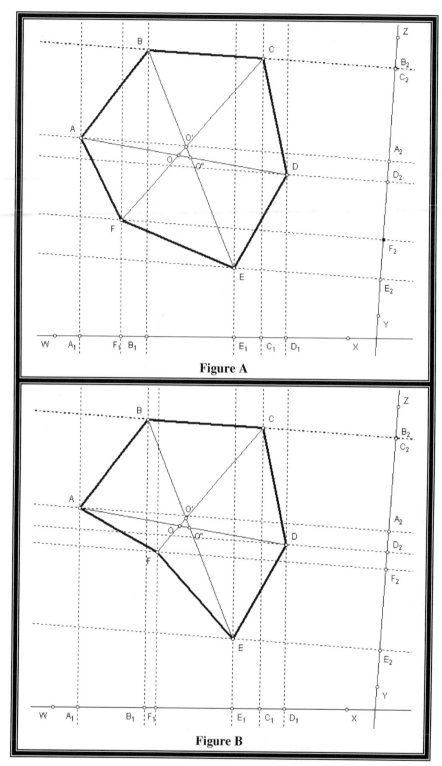

Figure A

Figure B

Figure 5.27.2 Zero Property of Convex and Concave Polygons

Observation 5.27.1 Zero Property of Convex and Concave Polygons

A) From Figure 5.23.2A and B these specific examples maintain and hold the zero property. This is true as long as the concave points do not cross one another disorganizing the sense of direction. The proofs for the zero Property hold for ether case of convex or concave polygons this good indicator of their validity.

$$A_1F_1 + F_1E_1 + E_1D_1 - D_1C_1 - C_1B_1 - B_1A_1 = 0.00 \text{ cm}$$
$$-A_2F_2 - F_2E_2 + E_2D_2 + D_2C_2 - C_1B_1 - B_1A_1 = 0.00 \text{ cm}$$

Section 5.28 Theorems on Irregular Polygons of Order 3 to 7

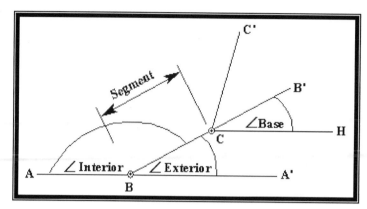

Figure 5.28.1 Geometry of an Irregular Polygon

Definition 5.28.1 Irregular Polygon

An irregular polygon is a polygon that has n-irregular lengths for sides and n-interior (base) angles.

Definition 5.28.2 Regular Polygon

A regular polygon is a polygon that has alternate equal n-sides and n-interior (base) angles.

Definition 5.28.3 Regular Equilateral Polygon

A *regular equilateral polygon* or simply known as an *equilateral polygon* is the special case of a regular polygon, were all sides are equal.

Definition 5.28.4 Polygon Critical Angle

The critical angle is the angle to delineating positive from negative going slopes and having the form $-180^\circ - (\gamma - \alpha_i)$ for $2 < i < n - 1$ and $n > 3$.

Definition 5.28.5 Projected Angle

The projected angle is the angle at the vertex and the base of the right triangle on the construction line being projected onto.

Definition 5.28.6 Polygon Base (Resultant) side

The base side of a polygon is the resultant side of a polygon that all other projected sides add up to its projection. Also, when its base angle is zero it is the side that the polygon rests on, having a zero slope.

Definition 5.28.7 Polygon Base Angle

The base angle is made from the projection onto a construction line that parallels the base frame of orientation for the polygon.

Definition 5.28.8 Polygon Interior Angle

The interior angle lies between any two adjacent sides on the inside of the polygon.

Definition 5.28.9 Polygon Exterior Angle

The exterior angle is a supplementary angle ($180° \equiv \angle$Exterior $+ \angle$Interior) lying on the outside of the polygon from the external projection of a side.

Definition 5.28.10 Simply Closed Polygon

A simple polygon is a polygon that is contiguously connected and having one-and-only one critical angle.

Definition 5.28.11 Multifaceted Closed Polygon

A multifaceted polygon has more than one critical angle.

This next section is reaffirmation of the Zero Property and that it real does work for irregular polygons. So here a general analysis of irregular polygons is directly developed is taken to a logical conclusion if possible. Certainly, even number of sides is proven valid for the Zero Property but as can be seen for odd number of sides this is not the case. This analysis gives some, not all, confidence that the zero property of polygons does work.

Irregular Polygons are of interest because they come up with limits of Line Integrals for closed paths and in this treatise. Most importantly, they play a major role in Tensor Geometry, specifically with the distribution of tensor operators. So, the following theorems investigate irregular and regular polygons. Polygons arise in vector addition. So if n-vectors lie in a plane and are added together and their path is closed with a resulting vector the geometric shape that arises is an n-sided polygon sometimes called an ***n-gon***. Since the vectors will most likely be of irregular length and direction than we are interested in irregular polygons. The distribution of the dot product operating on a system of summed vectors depends on the projected resultant of the sides of irregular polygons, which is the concern of these geometric developments.

Given 5.28.1: Lines of Construction

1g	Given	$\overline{H_nH_n'}$ is a line of construction parallel to the x-axis. for n = 1, 2, 3, …

Lets start with the simplest polygon of irregular shape a triangle or trigon.

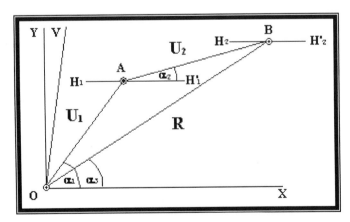

Figure 5.28.2 Geometry of a Trigon

Theorem 5.28.1 Irregular Trigon: Sum of All Interior Angles Add to 180°

1g	Given	$\angle XOA = \alpha_1$
2g		$\angle BAH'_1 = \alpha_2$
3g		$\angle XOB = \alpha_3$
4g		TIA (Total Internal Angle) = $\angle AOB + \angle OAB + \angle OBA$
Steps	Hypothesis	TIA = 180° for any irregular Trigon (Triangle)
1	From F5.24.2	$\angle AOB = \angle XOA - \angle XOB$
2	From F5.24.2 and 1g	$\angle OAB = \angle BAH'_1 + 180° - \angle OAH_1$
3	From F5.24.2 and 1g	$\angle OBA = \angle OBH_2 - \angle ABH_2$
4	From F5.24.2 by T5.4.4 Parallel Lines: Congruent Alternate Interior Angles	$\angle XOA = \angle OAH_1$
5	From F5.24.2 by T5.4.4 Parallel Lines: Congruent Alternate Interior Angles	$\angle BAH'_1 = \angle ABH_2$
6	From F5.24.2 by T5.4.4 Parallel Lines: Congruent Alternate Interior Angles	$\angle XOB = \angle OBH_2$
7	From 1, 1g, 3g and A4.2.3	$\angle AOB = \alpha_1 - \alpha_3$
8	From 2, 2g, 4, 1g and A4.1.3	$\angle OAB = \alpha_2 + 180° - \alpha_1$
9	From 3, 6, 3g, 5, 2g and A4.2.3	$\angle OBA = \alpha_3 - \alpha_2$
10	From 7, 8, 9, 4g and A4.2.3	TIA = $\alpha_1 - \alpha_3 + \alpha_2 + 180° - \alpha_1 + \alpha_3 - \alpha_2$
11	From 10 and A4.2.5	TIA = $\alpha_1 - \alpha_1 + \alpha_2 - \alpha_2 + \alpha_3 - \alpha_3 + 180°$
12	From 11 and A4.2.8	TIA = $0 + 0 + 0 + 180°$
∴	From 10 and A4.2.7	TIA = $180°$

Two things maybe observed here:

Observation 5.28.1 The Sum of the Internal Angles for a Trigon is always 180°

Based on T5.3.1 the Total Internal Angle of a Trigon always sums to 180° regardless weather it is irregular or the special case of a regular.

Observation 5.28.2 Indirect Proof of Irregular Trigon

Also from T5.3.1 the logic is symmetrical, and shows that if the proof had been done just for a regular trigon that would also have been a proof for a irregular trigon, since all constructed angles and sides change proportionately, hence obey the same principles. So, a proof for regular trigon would be a valid proof for any irregular trigon.

Classically this proof is usual proven by constructing the interior angles on a straight line.

Projective Geometry for a Trigon

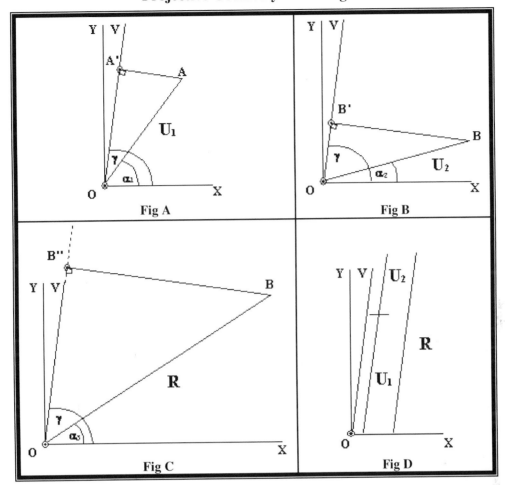

Figure 5.28.3 Irregular Trigon: Sum of the Projective Sides Equal the Resultant

Lets see what can be gleaned from the above figures?

Theorem 5.28.2 Irregular Trigon: Summation of Projected Sides

1g	Given	$\angle AOA' = \gamma - \alpha_1$ from F5.24.3A.
2g		$\angle BOB' = \gamma - \alpha_2$ from F5.24.3B.
3g		$\angle BOB'' = \gamma - \alpha_3$ from F5.24.3C.
4g		$OA = U_1$ from F5.24.3A
5g		$OB = U2$ from F5.24.3B
6g		$OB = R$ from F5.24.3C
Steps	Hypothesis	$R \cos(\gamma - \alpha_3) = U_1 \cos(\gamma - \alpha_1) + U_2 \cos(\gamma - \alpha_2)$
1	From F5.24.3A	$OA' = OA \cos(\angle AOA')$
2	From F5.24.3B	$OB' = OB \cos(\angle BOB')$
3	From F5.24.3C	$OB'' = OB \cos(\angle BOB'')$
4	From 1, 1g and A4.2.3	$OA' = OA \cos(\gamma - \alpha_1)$
5	From 2, 2g and A4.2.3	$OB' = OB \cos(\gamma - \alpha_2)$
6	From 3, 3g and A4.2.3	$OB'' = OB \cos(\gamma - \alpha_3)$
7	From F5.24.3D	$OB'' = OA' + OB'$
8	From 7, 4, 5, 6 and A4.2.3	$OB \cos(\gamma - \alpha_3) = OA \cos(\gamma - \alpha_1) + OB \cos(\gamma - \alpha_2)$
∴	From 9, 4g, 5g, 6g and A4.2.3	$R \cos(\gamma - \alpha_3) = U_1 \cos(\gamma - \alpha_1) + U_2 \cos(\gamma - \alpha_2)$

What are some ways that this expression can be simplified without losing geometric generality? One way is to let $\alpha_3 = 0$ this only tilts the trigon to the horizontal axis.

Corollary 5.28.2.1 Irregular Trigon: Rotated to the Horizontal

1g	Given	$\alpha_3 = 0$
Steps	Hypothesis	$R \cos(\gamma) = U_1 \cos(\gamma - \alpha_1) + U_2 \cos(\gamma - \alpha_2)$
1	From T5.24.2 Irregular Trigon: Summation of Projected Sides	$R \cos(\gamma - \alpha_3) = U_1 \cos(\gamma - \alpha_1) + U_2 \cos(\gamma - \alpha_2)$
∴	From 1, 1g and A4.2.3	$R \cos(\gamma) = U_1 \cos(\gamma - \alpha_1) + U_2 \cos(\gamma - \alpha_2)$

Tensor Calculus & Physics: A General Treatise

Now lets look at the special case of a regular trigon where $U = U_1 = U_2$ and $\alpha = \alpha_1 = -\alpha_2$. This creates an isosceles triangle with a further relationship for the base R.

Theorem 5.28.3 Regular Trigon: Summation of Projected Sides

1g 2g 3g	Given	$U = U_1 = U_2$ $\alpha = \alpha_1 = -\alpha_2$ $R = 2\,U\cos(\alpha)$ from an isosceles triangle
Steps	Hypothesis	$R\cos(\gamma - \alpha_3) = U_1\cos(\gamma - \alpha_1) + U_2\cos(\gamma - \alpha_2)$
1	From C5.24.2.1 Irregular Trigon: Rotated to the Horizontal	$R\cos(\gamma) = U_1\cos(\gamma - \alpha_1) + U_2\cos(\gamma - \alpha_2)$
2	From 1, 1g, 2g, 3g and A4.2.3	$2\,U\cos(\alpha)\cos(\gamma) = U\cos(\gamma + \alpha) + U\cos(\gamma - \alpha)$
3	From 2 and A4.2.12	$2\,U\cos(\alpha)\cos(\gamma) = U(\cos(\gamma + \alpha) + \cos(\gamma - \alpha))$
4	From 3 and T4.4.3 Equalities: Cancellation by Multiplication	$2\cos(\alpha)\cos(\gamma) = \cos(\gamma + \alpha) + \cos(\gamma - \alpha)$
5	From 4, LxE.1.24 and LxE.1.25	$2\cos(\alpha)\cos(\gamma) = \cos(\gamma)\cos(\alpha) - \sin(\gamma)\sin(\alpha) + \cos(\gamma)\cos(\alpha) + \sin(\gamma)\sin(\alpha)$
6	From 5 and A4.2.5	$2\cos(\alpha)\cos(\gamma) = \cos(\gamma)\cos(\alpha) + \cos(\gamma)\cos(\alpha) - \sin(\gamma)\sin(\alpha) + \sin(\gamma)\sin(\alpha)$
7	From 6, A4.2.8 and A4.2.14	$2\cos(\alpha)\cos(\gamma) = (1 + 1)\cos(\gamma)\cos(\alpha) + 0$
8	From 7, A4.2.7, D4.1.19 Primitive Definition For Rational Arithmetic and A4.1.2 (Identity Proof)	$2\cos(\alpha)\cos(\gamma) = 2\cos(\alpha)\cos(\gamma)$
∴	From 8 and O5.24.2	$R\cos(\gamma - \alpha_3) = U_1\cos(\gamma - \alpha_1) + U_2\cos(\gamma - \alpha_2)$

From Observation O5.24.2 since a regular trigon is proven than it is also a proof for any irregular trigon. So this has been confirmed by the above analytical proof and geometrical construction in F5.24.3D.

Now will graduate to the next higher polygon a Quadragon.

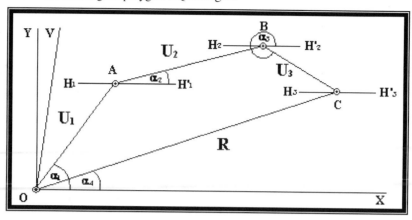

Figure 5.28.4 Geometry of a Quadragon

Theorem 5.28.4 Irregular Quadragon: Sum of All Interior Angles Add to 360°

1g	Given	$\angle XOA = \alpha_1$
2g		$\angle BAH'_1 = \alpha_2$
3g		$\angle H'_2BC = \alpha_3$
4g		$\angle XOC = \alpha_4$
5g		TIA (Total Internal Angle) $= \angle AOC + \angle OAB + \angle ABC + \angle BCO$
Steps	Hypothesis	TIA $= 360°$ for any irregular Quadragon (Quadrilateral)
1	From F5.24.4, 1g and 4g	$\angle AOC = \angle XOA - \angle XOC$
2	From F5.24.4 and 2g	$\angle OAB = \angle BAH'_1 + 180° - \angle OAH_1$
3	From F5.24.4	$\angle ABC = 180° - \angle ABH_2 - \angle CBH'_2$
4	From F5.24.4	$\angle BCO = \angle OCH_3 + \angle BCH_3$
5	From F5.24.4 and 3g	$\angle CBH'_2 = 360° - \angle H'_2BC$
6	From F5.24.4 by T5.4.4 Parallel Lines: Congruent Alternate Interior Angles	$\angle OAH_1 = \angle XOA$
7	From F5.24.4 by T5.4.4 Parallel Lines: Congruent Alternate Interior Angles	$\angle ABH_2 = \angle BAH'_1$
8	From F5.24.4 by T5.4.4 Parallel Lines: Congruent Alternate Interior Angles	$\angle BCH_3 = \angle CBH'_2$
9	From F5.24.4 by T5.4.4 Parallel Lines: Congruent Alternate Interior Angles	$\angle OCH_3 = \angle XOC$
10	From 1, 1g, 4g and A4.2.3	$\angle AOC = \alpha_1 - \alpha_4$
11	From 2, 2g, 6, 1g, A4.2.3 and A4.2.5	$\angle OAB = \alpha_2 - \alpha_1 + 180°$
12	From 2, 7, 2g, 5, 3g and A4.2.3	$\angle ABC = 180° - \alpha_2 - (360° - \alpha_3)$
13	From 4, 9, 4g, 8, 5, 3g and A4.2.3	$\angle BCO = \alpha_4 + 360° - \alpha_3$

14	From 5g, 10, 11, 12, 13 and A4.1.3	$TIA = \alpha_1 - \alpha_4 + \alpha_2 - \alpha_1 + 180^\circ + 180^\circ - \alpha_2 - (360^\circ - \alpha_3) + \alpha_4 + 360^\circ - \alpha_3$
15	From 14, A4.2.14 and D4.1.19 Primitive Definition For Rational Arithmetic	$TIA = \alpha_1 - \alpha_4 + \alpha_2 - \alpha_1 + 180^\circ + 180^\circ - \alpha_2 - 360^\circ + \alpha_3 + \alpha_4 + 360^\circ - \alpha_3$
16	From 15 and A4.2.5	$TIA = 180^\circ + 180^\circ + 360^\circ - 360^\circ + \alpha_1 - \alpha_1 + \alpha_2 - \alpha_2 + \alpha_3 - \alpha_3 + \alpha_4 - \alpha_4$
17	From 16, A4.2.8, and A4.2.14	$TIA = (1 + 1)\,180^\circ + 0 + 0 + 0 + 0 + 0$
18	From 17 and D4.1.19 Primitive Definition For Rational Arithmetic	$TIA = 2\,180^\circ$
$\therefore$	From 10 and A4.1.8	$TIA = 360^\circ$

Again two similar observations can be made for Quadragon in the same way they where made for a Trigon:

Observation 5.28.3 The Sum of the Internal Angles for a Quadragon is always 360°

Based on T5.3.4 the Total Internal Angle of a Quadragon always sums to 360° regardless weather it is irregular or the special case of a regular.

Observation 5.28.4 Indirect Proof of Irregular Quadragon

Also from T5.3.4 the logic is symmetrical, and shows that if the proof had been done just for a regular Quadragon that would also have been a proof for a irregular Quadragon, since all constructed angles and sides change proportionately, hence obey the same principles. So a proof for regular Quadragon would be a valid proof for any irregular Quadragon.

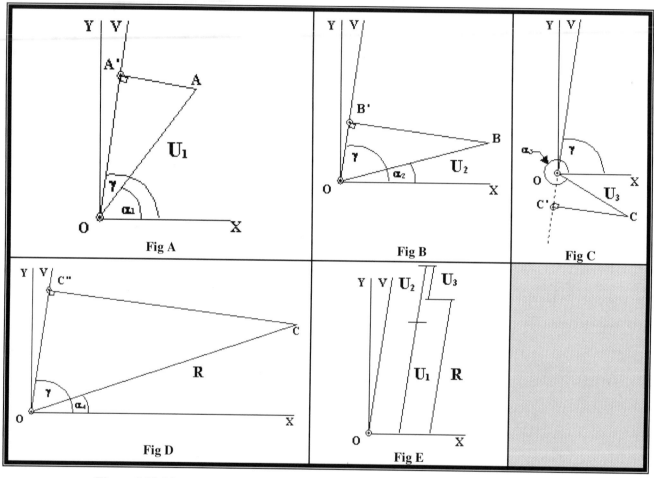

Figure 5.28.5 Irregular Quadragon: Sum of the Projective Sides Equal the Resultant

Theorem 5.28.5 Irregular Quadragon: Summation of Projected Sides

1g	Given	$\angle AOA' = \gamma - \alpha_1$	from F5.24.5A.
2g		$\angle BOB' = \gamma - \alpha_2$	from F5.24.5B.
3g		$\angle COC' = -180° - (\gamma - \alpha_3)$	from F5.24.5C.
4g		$\angle COC'' = \gamma - \alpha_4$	from F5.24.5D.
5g		$OA = U_1$	from F5.24.5A
6g		$OB = U_2$	from F5.24.5B
7g		$OC = U_3$	from F5.24.5C
8g		$OB = R$	from F5.24.5D
Steps	Hypothesis	$R \cos(\gamma - \alpha_3) = U_1 \cos(\gamma - \alpha_1) + U_2 \cos(\gamma - \alpha_2) - U_3 \cos(\gamma - \alpha_3)$	
1	From F5.24.5A	$OA' = OA \cos(\angle AOA')$	
2	From F5.24.5B	$OB' = OB \cos(\angle BOB')$	
3	From F5.24.5C	$OC' = OC \cos(\angle COC')$	
4	From F5.24.5D	$OC'' = OC \cos(\angle COC'')$	
5	From 1, 1g and A4.2.3	$OA' = OA \cos(\gamma - \alpha_1)$	
6	From 2, 2g and A4.2.3	$OB' = OB \cos(\gamma - \alpha_2)$	
7	From 3, 3g and A4.2.3	$OC' = OC \cos(-180° - (\gamma - \alpha_3))$	
8	From 7 and DxE.1.7.4	$OC' = -OC \cos(\gamma - \alpha_3)$	
9	From 4, 4g and A4.2.3	$OC'' = OC \cos(\gamma - \alpha_4)$	
10	From F5.24.5E	$OC'' = OA' + OB' + OC'$	
11	From 10, 5, 6, 8, 9 and A4.1.3	$OC \cos(\gamma - \alpha_4) = OA \cos(\gamma - \alpha_1) + OB \cos(\gamma - \alpha_2) - OC \cos(\gamma - \alpha_3)$	
∴	From 11, 5g, 6g, 7g, 8g and A4.2.3	$R \cos(\gamma - \alpha_4) = U_1 \cos(\gamma - \alpha_1) + U_2 \cos(\gamma - \alpha_2) - U_3 \cos(\gamma - \alpha_3)$	

What are some ways that this expression can be simplified without losing geometric generality? One way is to let $\alpha_4 = 0$ this only tilts the Quadragon to the horizontal axis.

Corollary 5.28.5.1 Irregular Quadragon: Rotated to the Horizontal

1g	Given	$\alpha_4 = 0$
Steps	Hypothesis	$R \cos(\gamma) = U_1 \cos(\gamma - \alpha_1) + U_2 \cos(\gamma - \alpha_2) - U_3 \cos(\gamma - \alpha_3)$
1	From T5.24.5 Irregular Quadragon: Summation of Projected Sides	$R \cos(\gamma - \alpha_4) = U_1 \cos(\gamma - \alpha_1) + U_2 \cos(\gamma - \alpha_2) - U_3 \cos(\gamma - \alpha_3)$
∴	From 1, 1g and A4.2.3	$R \cos(\gamma) = U_1 \cos(\gamma - \alpha_1) + U_2 \cos(\gamma - \alpha_2) - U_3 \cos(\gamma - \alpha_3)$

Now lets look at the special case of a regular Quadragon where $U = U_1 = U_3$.and $\alpha = \alpha_1$, $\alpha_2 = 0$ and $\alpha_3 = \alpha_1$. This creates a parallelogram with a further relationship for the base $R = U_2$.

Theorem 5.28.6 Regular Quadragon: Summation of Projected Sides

	Given	$U = U_1 = U_3$
1g		
2g		$\alpha = \alpha_1$
3g		$\alpha_2 = 0$
4g		$\alpha_3 = \alpha_1, \neq 180^\circ + \alpha_1$ has already been accounted for in TA5.5, 3g.
5g		$R = U_2$ from parallelogram
Steps	Hypothesis	$R \cos(\gamma) = U_1 \cos(\gamma - \alpha_1) + U_2 \cos(\gamma - \alpha_2) - U_3 \cos(\gamma - \alpha_3)$
1	From C5.24.5.1 Irregular Quadragon: Rotated to the Horizontal	$R \cos(\gamma) = U_1 \cos(\gamma - \alpha_1) + U_2 \cos(\gamma - \alpha_2) - U_3 \cos(\gamma - \alpha_3)$
2	From 1, 1g, 2g, 3g, 4g, 5g and A4.2.3 Substitution	$R \cos(\gamma) = U \cos(\gamma - \alpha) + R \cos(\gamma - 0) - U \cos(\gamma - \alpha)$
3	From 2, A4.2.7 Identity Add	$R \cos(\gamma) = U \cos(\gamma - \alpha) + R \cos(\gamma) - U \cos(\gamma - \alpha)$
4	From 3 and T4.3.7 Uniqueness of Subtraction	$0 = U \cos(\gamma - \alpha) - U \cos(\gamma - \alpha)$
5	From 4 and A4.2.8 Inverse Add	$0 = 0$
$\therefore$	From 5 and O5.24.4 Indirect Proof of Irregular Quadragon	$R \cos(\gamma) = U_1 \cos(\gamma - \alpha_1) + U_2 \cos(\gamma - \alpha_2) - U_3 \cos(\gamma - \alpha_3)$

From Observation 5.24.4 "Indirect Proof of Irregular Quadragon" since a regular Quadragon is proven than it is also a proof for any irregular Quadragon. So this has been confirmed by the above analytical proof and geometrical construction in F5.24.5E.

Observation 5.28.5 Focus on Summation of Projected Sides

The Trigon and Quadragon show what is of critical interest, where the sides of an n-gon summing to its projected resultant. So the following proofs on pentagons, hexagons and heptagons start with the geometrical setup and go directly to the construction of the resulting projections, side stepping the sum of interior angles.

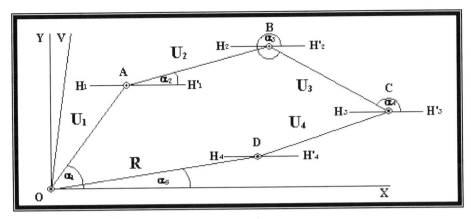

Figure 5.28.6 Geometry of a Pentagon

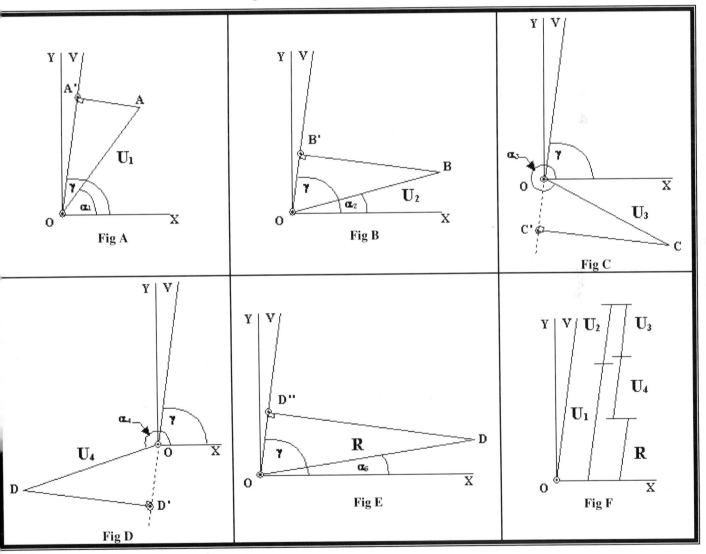

Figure 5.28.7 Irregular Pentagon: Sum of the Projective Sides Equal the Resultant

Theorem 5.28.7 Irregular Pentagon: Summation of Projected Sides

1g	Given	$\angle AOA' = \gamma - \alpha_1$	from F5.24.7A.
2g		$\angle BOB' = \gamma - \alpha_2$	from F5.24.7B.
3g		$\angle COC' = -180^\circ - (\gamma - \alpha_3)$	from F5.24.7C.
4g		$\angle DOD' = 180^\circ + \gamma - \alpha_4$	from F5.24.7D.
5g		$\angle DOD'' = \gamma - \alpha_5$	from F5.24.7E.
6g		$OA = U_1$	from F5.24.7A
7g		$OB = U_2$	from F5.24.7B
8g		$OC = U_3$	from F5.24.7C
9g		$OD = U_4$	from F5.24.7D
10g		$OD = R$	from F5.24.7E
Steps	Hypothesis	$R \cos(\gamma - \alpha_5) = U_1 \cos(\gamma - \alpha_1) + U_2 \cos(\gamma - \alpha_2)$ $- U_3 \cos(\gamma - \alpha_3) - U_4 \cos(\gamma - \alpha_4)$	
1	From F5.24.7A	$OA' = OA \cos(\angle AOA')$	
2	From F5.24.7B	$OB' = OB \cos(\angle BOB')$	
3	From F5.24.7C	$OC' = OC \cos(\angle COC')$	
4	From F5.24.7D	$OD' = OD \cos(\angle DOD')$	
5	From F5.24.7E	$OD'' = OD \cos(\angle DOD'')$	
6	From 1, 1g and A4.2.3	$OA' = OA \cos(\gamma - \alpha_1)$	
7	From 2, 2g and A4.2.3	$OB' = OB \cos(\gamma - \alpha_2)$	
8	From 3, 3g and A4.2.3	$OC' = OC \cos(-180^\circ - (\gamma - \alpha_3))$	
9	From 8 and DxE.1.7.4	$OC' = -OC \cos(\gamma - \alpha_3)$	
10	From 4, 4g and A4.2.3	$OD' = OD \cos(180^\circ + \gamma - \alpha_4)$	
11	From 10 and DxE.1.7.4	$OD' = -OD \cos(\gamma - \alpha_4)$	
12	From 4, 4g and A4.2.3	$OD'' = OD \cos(\gamma - \alpha_5)$	
13	From F5.24.7F	$OD'' = OA' + OB' + OC' + OD'$	
14	From 13, 7, 9, 11, 12 and A4.2.3	$OD \cos(\gamma - \alpha_5) = OA \cos(\gamma - \alpha_1) + OB \cos(\gamma - \alpha_2) - OC \cos(\gamma - \alpha_3)$ $- OD \cos(\gamma - \alpha_4)$	
∴	From 14, 6g, 7g, 8g, 9g, 10g and A4.2.3	$R \cos(\gamma - \alpha_5) = U_1 \cos(\gamma - \alpha_1) + U_2 \cos(\gamma - \alpha_2)$ $- U_3 \cos(\gamma - \alpha_3) - U_4 \cos(\gamma - \alpha_4)$	

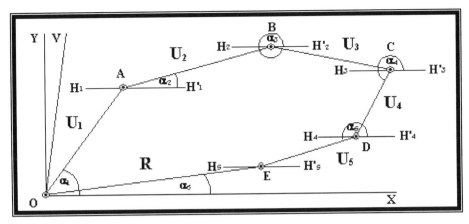

Figure 5.28.8 Geometry of a Hexagon

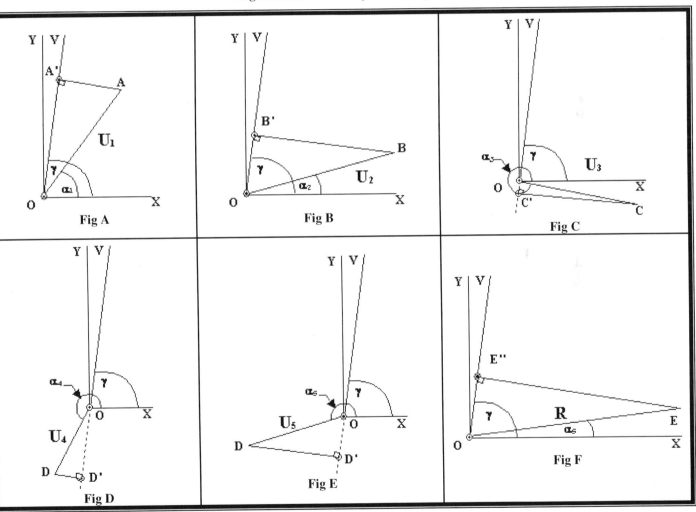

Figure 5.28.9 Irregular Hexagon: Sum of the Projective Sides Equal the Resultant

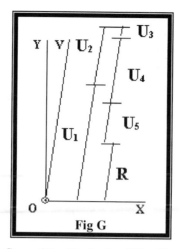

Fig G

Figure 5.28.10 Irregular Hexagon: Sum of the Projective Sides Equal the Resultant (Continued)

Theorem 5.28.8 Irregular Hexagon: Summation of Projected Sides

1g	Given	$\angle AOA' = \gamma - \alpha_1$	from F5.24.9A.
2g		$\angle BOB' = \gamma - \alpha_2$	from F5.24.9B.
3g		$\angle COC' = -180^\circ - (\gamma - \alpha_3)$	from F5.24.9C.
4g		$\angle DOD' = 180^\circ + \gamma - \alpha_4$	from F5.24.9D.
5g		$\angle EOE' = 180^\circ + \gamma - \alpha_5$	from F5.24.9E.
6g		$\angle EOE'' = \gamma - \alpha_6$	from F5.24.9F.
7g		$OA = U_1$	from F5.24.9A
8g		$OB = U_2$	from F5.24.9B
9g		$OC = U_3$	from F5.24.9C
10g		$OD = U_4$	from F5.24.9D
11g		$OE = U_5$	from F5.24.9E
12g		$OE = R$	from F5.24.9F
Steps	Hypothesis	$R \cos(\gamma - \alpha_6) = U_1 \cos(\gamma - \alpha_1) + U_2 \cos(\gamma - \alpha_2) - U_3 \cos(\gamma - \alpha_3)$ $\quad - U_4 \cos(\gamma - \alpha_4) - U_5 \cos(\gamma - \alpha_5)$	
1	From F5.24.9A	$OA' = OA \cos(\angle AOA')$	
2	From F5.24.9B	$OB' = OB \cos(\angle BOB')$	
3	From F5.24.9C	$OC' = OC \cos(\angle COC')$	
4	From F5.24.9D	$OD' = OD \cos(\angle DOD')$	
5	From F5.24.9E	$OE' = OE \cos(\angle EOE')$	
6	From F5.24.9F	$OE'' = OE \cos(\angle EOE'')$	
7	From 1, 1g and A4.2.3	$OA' = OA \cos(\gamma - \alpha_1)$	
8	From 2, 2g and A4.2.3	$OB' = OB \cos(\gamma - \alpha_2)$	
9	From 3, 3g and A4.2.3	$OC' = OC \cos(-180^\circ - (\gamma - \alpha_3))$	
10	From 9 and DxE.1.7.4	$OC' = -OC \cos(\gamma - \alpha_3)$	
11	From 4, 4g and A4.2.3	$OD' = OD \cos(180^\circ + \gamma - \alpha_4)$	
12	From 10 and DxE.1.7.4	$OD' = -OD \cos(\gamma - \alpha_4)$	
13	From 5, 5g and A4.2.3	$OE' = OE \cos(180^\circ + \gamma - \alpha_5)$	
14	From 13 and DxE.1.7.4	$OE' = -OE \cos(\gamma - \alpha_5)$	
15	From 6, 6g and A4.2.3	$OE'' = OE \cos(\gamma - \alpha_6)$	
16	From F5.24.9G	$OD'' = OA' + OB' + OC' + OD' + OE'$	
17	From 16, 7, 8, 10, 12, 14, 15 and A4.2.3	$OD \cos(\gamma - \alpha_6) = OA \cos(\gamma - \alpha_1) + OB \cos(\gamma - \alpha_2) - OC \cos(\gamma - \alpha_3)$ $\quad - OD \cos(\gamma - \alpha_4) - OE \cos(\gamma - \alpha_5)$	
$\therefore$	From 17, 7g, 8g, 9g, 10g, 11g, 12g and A4.2.3	$R \cos(\gamma - \alpha_6) = U_1 \cos(\gamma - \alpha_1) + U_2 \cos(\gamma - \alpha_2)$ $\quad - U_3 \cos(\gamma - \alpha_3) - U_4 \cos(\gamma - \alpha_4) - U_5 \cos(\gamma - \alpha_5)$	

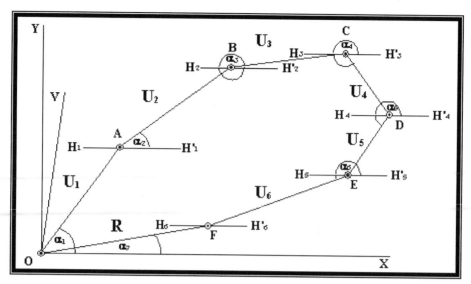

Figure 5.28.11 Geometry of a Heptagon

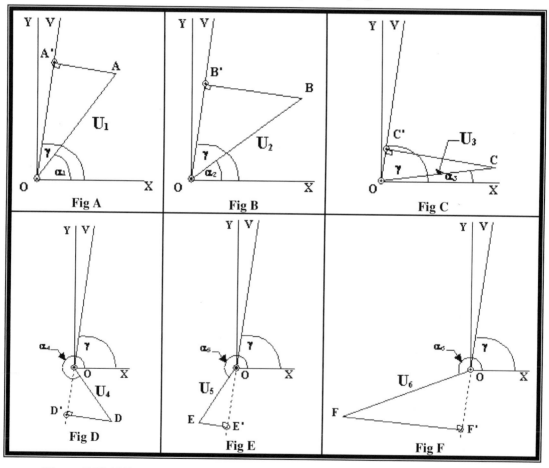

Figure 5.28.12 Irregular Heptagon: Sum of the Projective Sides Equal the Resultant

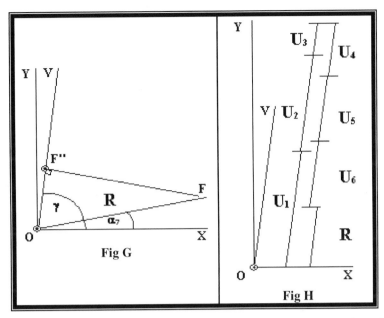

Figure 5.28.13 Irregular Heptagon: Sum of the Projective Sides Equal the Resultant (Continued)

Theorem 5.28.9 Irregular Heptagon: Summation of Projected Sides

1g	Given	$\angle AOA' = \gamma - \alpha_1$	from F5.24.11A.
2g		$\angle BOB' = \gamma - \alpha_2$	from F5.24.11B.
3g		$\angle COC' = \gamma - \alpha_3$	from F5.24.11C.
4g		$\angle DOD' = -180° - (\gamma - \alpha_4)$	from F5.24.11D.
5g		$\angle EOE' = 180° + \gamma - \alpha_5$	from F5.24.11E.
6g		$\angle FOF' = 180° + \gamma - \alpha_6$	from F5.24.11F.
7g		$\angle FOF'' = \gamma - \alpha_7$	from F5.24.11G.
8g		$OA = U_1$	from F5.24.11A
9g		$OB = U_2$	from F5.24.11B
10g		$OC = U_3$	from F5.24.11C
11g		$OD = U_4$	from F5.24.11D
12g		$OE = U_5$	from F5.24.11E
13g		$OF = U_6$	from F5.24.11F
14g		$OF = R$	from F5.24.11G
Steps	Hypothesis	$R \cos(\gamma - \alpha_7) = U_1 \cos(\gamma - \alpha_1) + U_2 \cos(\gamma - \alpha_2) + U_3 \cos(\gamma - \alpha_3)$ $- U_4 \cos(\gamma - \alpha_4) - U_5 \cos(\gamma - \alpha_5) - U_6 \cos(\gamma - \alpha_6)$	
1	From F5.24.11A	$OA' = OA \cos(\angle AOA')$	
2	From F5.24.11B	$OB' = OB \cos(\angle BOB')$	
3	From F5.24.11C	$OC' = OC \cos(\angle COC')$	
4	From F5.24.11D	$OD' = OD \cos(\angle DOD')$	
5	From F5.24.11E	$OE' = OE \cos(\angle EOE')$	
6	From F5.24.11F	$OF' = OF \cos(\angle FOF')$	
7	From F5.24.11G	$OF'' = OF \cos(\angle FOF'')$	
8	From 1, 1g and A4.2.3	$OA' = OA \cos(\gamma - \alpha_1)$	
9	From 2, 2g and A4.2.3	$OB' = OB \cos(\gamma - \alpha_2)$	
10	From 3, 3g and A4.2.3	$OC' = OC \cos(\gamma - \alpha_3)$	
11	From 4, 4g and A4.2.3	$OD' = OD \cos(-180° - (\gamma - \alpha_4))$	
12	From 11 and DxE.1.7.4	$OD' = -OD \cos(\gamma - \alpha_4)$	
13	From 5, 5g and A4.2.3	$OE' = OE \cos(180° + \gamma - \alpha_5)$	
14	From 13 and DxE.1.7.4	$OE' = -OE \cos(\gamma - \alpha_5)$	
15	From 5, 5g and A4.2.3	$OF' = OF \cos(180° + \gamma - \alpha_6)$	
16	From 15 and DxE.1.7.4	$OF' = -OF \cos(\gamma - \alpha_6)$	
17	From 6, 6g and A4.2.3	$OF'' = OF \cos(\gamma - \alpha_7)$	
18	From F5.24.11G	$OF'' = OA' + OB' + OC' + OD' + OE' + OF'$	
19	From 18, 8, 9, 10, 12, 14, 16, 17 and A4.2.3	$OF \cos(\gamma - \alpha_7) = OA \cos(\gamma - \alpha_1) + OB \cos(\gamma - \alpha_2)$ $+ OC \cos(\gamma - \alpha_3) - OD \cos(\gamma - \alpha_4)$ $- OE \cos(\gamma - \alpha_5) - OF \cos(\gamma - \alpha_6)$	
∴	From 19, 8g, 9g, 10g, 11g, 12g, 13g, 14g and A4.2.3	$R \cos(\gamma - \alpha_7) = U_1 \cos(\gamma - \alpha_1) + U_2 \cos(\gamma - \alpha_2) + U_3 \cos(\gamma - \alpha_3)$ $- U_4 \cos(\gamma - \alpha_4) - U_5 \cos(\gamma - \alpha_5) - U_6 \cos(\gamma - \alpha_6)$	

Table 5.28.1 Summary Irregular Polygons 3 to 7: Summation of Projected Sides

Poly-gon	Constructed Projected Bases	Irregular Projected Angles	$\cos(\gamma - \alpha_i)$	Slope	Equilateral $\angle$Base	E-Slop
3	$|R| = |U'_1| + |U'_2|$	$\angle AOA' = \gamma - \alpha_1$	+	+	$\alpha_1 = \angle$Interior	+s
		$\angle BOB' = \gamma - \alpha_2$	+	+	$\alpha_2 = +180° + 2\alpha_1$	−s
		$\angle BOB'' = \gamma - \alpha_3$	+	+	$\alpha_3 = 0$	+b
4	$|R| = |U'_1| + |U'_2| - |U'_3|$	$\angle AOA' = \gamma - \alpha_1$	+	+	$\alpha_1 = \angle$Interior $= 90°$	+s
		$\angle BOB' = \gamma - \alpha_2$	+	+	$\alpha_2 = +360°$	−t
		$\angle COC' = -180° - (\gamma - \alpha_3)$	−	−	$\alpha_3 = +180° + \alpha_1$	−s
		$\angle COC'' = \gamma - \alpha_4$	+	+	$\alpha_4 = 0°$	+b
5	$|R| = |U'_1| + |U'_2| - |U'_3| - |U'_4|$	$\angle AOA' = \gamma - \alpha_1$	+	+	$\alpha_1 = \angle$Interior	−s1l
		$\angle BOB' = \gamma - \alpha_2$	+	+	$\alpha_2 = -180° + 2\alpha_1$	+s2l
		$\angle COC' = -180° - (\gamma - \alpha_3)$	−	−	$\alpha_3 = +3\alpha_1$	−s3r
		$\angle DOD' = +180° + (\gamma - \alpha_4)$	−	−	$\alpha_4 = +360° - \alpha_1$	+s4r
		$\angle DOD'' = \gamma - \alpha_5$	+	+	$\alpha_5 = 0°$	+b
6	$|R| = |U'_1| + |U'_2| - |U'_3| - |U'_4| - |U'_5|$	$\angle AOA' = \gamma - \alpha_1$	+	+	$\alpha_1 = \angle$Interior	−s1l
		$\angle BOB' = \gamma - \alpha_2$	+	+	$\alpha_2 = +180° - 2\alpha_1$	+s2l
		$\angle COC' = -180° - (\gamma - \alpha_3)$	−	−	$\alpha_3 = +0°$	−t
		$\angle DOD' = +180° + (\gamma - \alpha_4)$	−	−	$\alpha_4 = +180° + \alpha_1$	−s3r
		$\angle EOE' = +180° + (\gamma - \alpha_5)$	−	−	$\alpha_5 = +360° - \alpha_1$	+s4r
		$\angle EOE'' = \gamma - \alpha_6$	+	+	$\alpha_6 = 0°$	+b
7	$|R| = |U'_1| + |U'_2| + |U'_3| - |U'_4| - |U'_5| - |U'_6|$	$\angle AOA' = \gamma - \alpha_1$	+	+	$\alpha_1 = \angle$Interior	−s1l
		$\angle BOB' = \gamma - \alpha_2$	+	+	$\alpha_2 = -180° + 2\alpha_1$	+s2l
		$\angle COC' = \gamma - \alpha_3$	+	+	$\alpha_3 = -360° + 3\alpha_1$	+s3l
		$\angle DOD' = -180° - (\gamma - \alpha_4)$	−	−	$\alpha_4 = -180° + 4\alpha_1$	−s4r
		$\angle EOE' = +180° + (\gamma - \alpha_5)$	−	−	$\alpha_5 = +540° - 2\alpha_1$	−s5r
		$\angle FOF' = +180° + (\gamma - \alpha_6)$	−	−	$\alpha_6 = +360° - \alpha_1$	+s6r
		$\angle FOF'' = \gamma - \alpha_7$	+	+	$\alpha_7 = 0°$	+b

Where $|U'_i| = |U_i| \cos(\gamma - \alpha_i)$ and equilaterals have a fixed or constant $\gamma \equiv 90°$.

Observation 5.28.6 Irregular Polygons: Correspondence Between Projected Angle and Slope
What can be gleaned from all of this is there exists a one-to-one correspondence between projected angle and slope of the polygon line segment.

Observation 5.28.7 Irregular Polygons: Closer
That there is closer, the loop made of contiguously straight line segments, has at least one critical angle of the form [$-180° - (\gamma - \alpha_i)$] for [2 < i < n-1]. Notice that the triangle is the exception to the rule being precluded, because n = 3 is simple and degenerate.

Observation 5.28.8 Irregular Polygons: Demarcation for Positive and Negative Slopes
The critical angle constitutes the demarcation between positive going and negative slops returning the sequenced lines to the base segment, thereby closing the loop.

Observation 5.28.9 Irregular Polygons: Existence of Projected Sides

While polygons have been studied for thousand of years in their purest form as equilateral space filling tiles these studies have not included projected geometry of irregular polygons. In the context of projective geometry, the construction-line must be considered as integral part of its structure.

The construction-line was designed to allow its angel [γ] to be variable in a general way, but apparently a special set of polygons can be found that require a fixed value for [γ] otherwise the projective relationship of its sides cannot exist. This gets into something that is extremely critical to tensors and that is the existences of distributed operators. Maybe it should be said that in the projective geometry of irregular polygons one can always find a unique [γ] that will satisfy the relationship, hence demonstrate its existence. The use of such a proposition should be used carefully otherwise; it could lead to a curricular argument for existence for such an operator. Case in point with the dot-product operator being distributed across the addition of a set of vectors, most text simply define the dot operator and than go back and prove distribution using base vectors in conjunction with it, which of course is circular. This proof goes directly to the heart of the geometry lying behind what actually makes it possible, the projective geometry of irregular polygons.

Let's summarize the resulting irregular polygon projections.

Observation 5.28.10 Generalization of Irregular Polygon Projection Equation

n-gon	Odd	Even
3	$R \cos(\gamma - \alpha_3) = U_1 \cos(\gamma - \alpha_1) + U_2 \cos(\gamma - \alpha_2)$	
4		$R \cos(\gamma - \alpha_3) = U_1 \cos(\gamma - \alpha_1) + U_2 \cos(\gamma - \alpha_2)$ $- U_3 \cos(\gamma - \alpha_3)$
5	$R \cos(\gamma - \alpha_5) = U_1 \cos(\gamma - \alpha_1) + U_2 \cos(\gamma - \alpha_2)$ $- U_3 \cos(\gamma - \alpha_3) - U_4 \cos(\gamma - \alpha_4)$	
6		$R \cos(\gamma - \alpha_6) = U_1 \cos(\gamma - \alpha_1) + U_2 \cos(\gamma - \alpha_2)$ $- U_3 \cos(\gamma - \alpha_3) - U_4 \cos(\gamma - \alpha_4)$ $- U_5 \cos(\gamma - \alpha_5)$
7	$R \cos(\gamma - \alpha_7) = U_1 \cos(\gamma - \alpha_1) + U_2 \cos(\gamma - \alpha_2)$ $+ U_3 \cos(\gamma - \alpha_3) - U_4 \cos(\gamma - \alpha_4)$ $- U_5 \cos(\gamma - \alpha_5) - U_6 \cos(\gamma - \alpha_6)$	
odd	$R \cos(\gamma - \alpha_n) = \sum_{i=1}^{n\backslash2} U_i \cos(\gamma - \alpha_i)$ $- \sum_{j=n\backslash2+1}^{n-1} U_j \cos(\gamma - \alpha_j)$	
even		$R \cos(\gamma - \alpha_n) = \sum_{i=1}^{(n-1)\backslash2} U_i \cos(\gamma - \alpha_i)$ $- \sum_{j=(n-1)\backslash2+1}^{n-1} U_j \cos(\gamma - \alpha_j)$
in gen	$R \cos(\gamma - \alpha_n) = \sum_{i=1}^{(n-1)\backslash2} U_i \cos(\gamma - \alpha_i) - \sum_{j=(n-1)\backslash2+1}^{n-1} U_j \cos(\gamma - \alpha_j)$	

Observation 5.28.11 Regular Polygons Limiting Special Case of Irregular Polygons

From theorems, T5.24.3 and 6 on irregular polygons demonstrate in these cases that regular polygons are a limiting case of irregular polygons. From this notion, the following proof demonstrates conclusively that for at least even number of sides up hold that the projection must sum to the result.

Theorem 5.28.10 Irregular Polygons: Summation of Projected Even Number of Sides

1g	Given	Assume equilateral n-gon for n even and n > 4
2g		$R = U_n$
3g		$U = U_k$ $\qquad$ for all k
4g		$\alpha = \alpha_k$ $\qquad$ for all k
5g		$\gamma = 90^\circ$
Steps	Hypothesis	$R\cos(\gamma - \alpha_n) = \sum^{n-4} U_k\cos(\gamma - \alpha_k) - \sum^{n-1} U_k \cos(\gamma - \alpha_k)$
1	From A4.2.2 Equality	$0 = 0$
2	From 1 and A4.2.8 Inverse Add	$0 = \frac{1}{2} n - \frac{1}{2} n$
3	From 2 and sum of n one's	$0 = \sum^{\frac{1}{2} n} 1 - \sum^{\frac{1}{2} n} 1$
4	From A4.2.2 Equality	$U \sin(\alpha) = U \sin(\alpha)$
5	From 3, 4 and T4.4.1 Equalities: Any Quantity Multiplied by Zero is Zero	$U \sin(\alpha)\, 0 = U \sin(\alpha)\, (\sum_{\frac{1}{2} n} 1 - \sum_{\frac{1}{2} n} 1\,)$
6	From 5, T3.3.2 Equalities: Any Quantity Multiplied by Zero is Zero and A4.2.14 Distribution	$0 = \sum^{\frac{1}{2} n} U \sin(\alpha) - \sum^{\frac{1}{2} n} U \sin(\alpha)$
7	From 6 and DxE.1.7.4 Minus Sine Reduction	$0 = \sum^{\frac{1}{2} n} U \sin(\alpha) + \sum^{\frac{1}{2} n} U \cos(180^\circ - \alpha)$
8	From 3g, 4g, 5g, 6, A5.2.15 GIV Transitivity of Segments, A5.2.18 GIV Transitivity of Angles and summing over all k	$0 = -U_n \cos(\gamma - \alpha_n) + \sum^{n-4} U_k\cos(\gamma - \alpha_k) - \sum^{n-1} U_k \cos(\gamma - \alpha_k)$
9	From 8, A4.2.5 Commutative Add and T4.3.4A Equalities: Reversal of Right Cancellation by Addition	$U_n \cos(\gamma - \alpha_n) = \sum^{n-4} U_k\cos(\gamma - \alpha_k) - \sum^{n-1} U_k \cos(\gamma - \alpha_k)$
$\therefore$	From 9, 2g and A4.2.3 Substitution	$R\cos(\gamma - \alpha_n) = \sum^{n-4} U_k\cos(\gamma - \alpha_k) - \sum^{n-1} U_k \cos(\gamma - \alpha_k)$

From stratagem, $\natural_2$ an attempt is made to prove observation 5.24.10 by extrapolating the projection equation.

Theorem 5.28.11 Irregular Polygons: Summation of Projected Sides by Induction

1g	Given	Let $\exists$ a m + 1 for any irregular m + 1-gon
Step	Hypothesis	hence only even elements in S can be counted by positive integers.
1	From T5.24.2, 5, 7, 8, 9	n = 3, 4, 5, 6 and 7 polygons have been proven valid.
2	From 1 and O5.24.10: Generalization of Irregular Polygon Projection Equation	$R\cos(\gamma - \alpha_n) = \sum_{i=1}^{(n-1)\backslash 2} U_i \cos(\gamma - \alpha_i)$ $- \sum_{j=(n-1)\backslash 2+1}^{n-1} U_j \cos(\gamma - \alpha_j)$
3	From 1g, 2 and let $\exists$ n = m + 1	$R\cos(\gamma - \alpha_{m+1}) = \sum_{i=1}^{(m)\backslash 2} U_i \cos(\gamma - \alpha_i)$ $- \sum_{j=(m)\backslash 2+1}^{m} U_j \cos(\gamma - \alpha_j)$
4	From 3 and T5.24.10 Irregular Polygons: Summation of Projected Even Number of Sides	$R\cos(\gamma - \alpha_{m+1}) = \sum_{i=1}^{(m)\backslash 2} U_i \cos(\gamma - \alpha_i)$ $- \sum_{j=(m)\backslash 2+1}^{m} U_j \cos(\gamma - \alpha_j)$ for m even
5	From m odd no generalization is found, hence	not all m + 1 elements can be placed in S do to the asymmetry geometry of odd side polygons.
$\therefore$	From 4 shows only **n** without limit for even sides.	Hence, only even elements in S can be counted by positive integers.

This is an important theorem, because the ***inductive properties of integers*** cannot apply to odd number of sides to a polygon. The proof requires an actual construction of the projective geometry for odd number sided polygon. This type of geometric construction does not lend itself to generalization, each problem being specific, than an alternate method has to be found. It is interesting to note that a general proof for even number of sides for equilateral polygons can be established do to perfect symmetry, but do to the asymmetry of the odd number of sides this cannot be done.

However, using the proof based on Langlands's Hypothesis based on $\natural_4$ stratification can circumvent this limitation.

Theorem 5.28.12 Irregular Polygons: Summation of Projected Sides by Langlands's Hypothesis

Step	Hypothesis	$R\cos(\gamma - \alpha_n) = \sum_{i=1}^{(n-1)\backslash 2} U_i \cos(\gamma - \alpha_i)$ $- \sum_{j=(n-1)\backslash 2+1}^{n-1} U_j \cos(\gamma - \alpha_j)$
1	From 1 and O5.24.10: Generalization of Irregular Polygon Projection Equation	$R\cos(\gamma - \alpha_n) = \sum_{i=1}^{(n-1)\backslash 2} U_i \cos(\gamma - \alpha_i)$ $- \sum_{j=(n-1)\backslash 2+1}^{n-1} U_j \cos(\gamma - \alpha_j)$
$\therefore$	From 1 and T5.23.1 Irregular Polygons: Non-Coplanar Polygon's Zero Property for Integral Path	$R\cos(\gamma - \alpha_n) = \sum_{i=1}^{(n-1)\backslash 2} U_i \cos(\gamma - \alpha_i)$ $- \sum_{j=(n-1)\backslash 2+1}^{n-1} U_j \cos(\gamma - \alpha_j)$

Section 5.29 Observation on Regular Polygon Symmetry

The next n-gon would be a 5-gon or Pentagon and it can be constructed just as the Trigon and Quadragon demonstrating irregular TIA = 540° than a Hexagon of 720° and so on, however as was seen with the previous proofs this is a very complicated processes and the size of the proof increases exponentially with [n]. In order to get around such laborious proofs it is fortunate that polygons have some nice properties working in the proof writer's favor.

These fundamental notions and properties of polygons can be rendered into pseudo axioms, or conjectures because they may be provable.

Observation 5.29.1 On Regular Polygon Symmetry

A)
        Regular Polygon: Mirror Symmetry Odd Number of Sides: There are n-lines of symmetry bisecting every vertex and opposite side of a regular polygon.

B)
        Regular Polygon: Mirror Symmetry Even Number of Sides: There are 2n-lines of symmetry bisecting every vertex and its opposite and every side and its opposite.

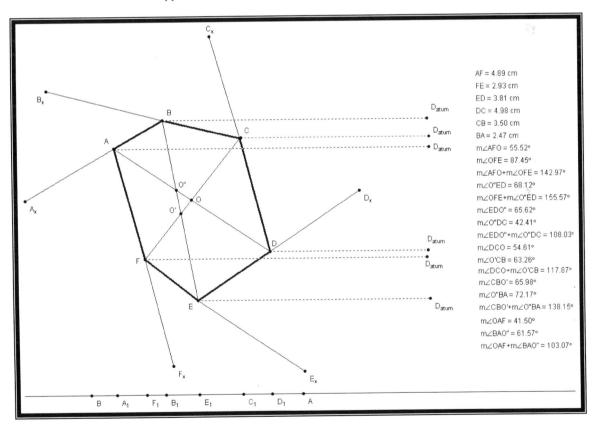

Figure 5.29.1 Irregular Hexagon

Section 5.30 Circles

A simple thing one would think finding the area of a circle. Finding the circumference only took 3,000 years starting with the written history of the Egyptian's. Archimedes 240 B.C.E. and others attempted to approximate it by bounding it with regular polygons. By this method, Archimedes was able to approximate $\pi \approx 3.14$. These approximation techniques that Archimedes was so found of using allowed him to do certain incredible feats in mathematics that predated calculus. These early attempts in calculus could not have happened without Eudoxus', 370 B.C.E., discovery of his "Method of Exhaustion", which eventually found its way into Euclid's Books "The Elements". Let's consider how to calculate the area of a circle following this method, a practice that Archimedes might have used.

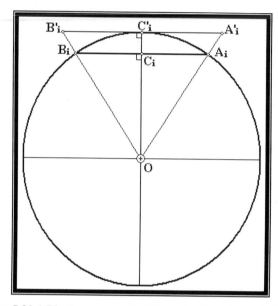

Figure 5.30.1 Finding Area of a Circle by Inscribing Polygons

Theorem 5.30.1 Finding Area of a Circle by Bounding with Regular Polygons

1g	Given	the above geometry of Figure F5.18.1
2g		$R = OA_i = OB_i = OC'_i$ radius of the circle being inscribed
3g		$A'_i B'_i$ Outerside of the polygon
4g		$A_i B_i$ Inside of the polygon
5g		$\theta \cong \angle A_i OB_i \cong \angle A'_i OB'_i$
6g		$\rho \sim \rho'$ regular polygons sharing center point-O and having n-sides
7g		A_c area of circle

Step	Hypothesis	$A_c = \pi R^2$ in the bounding limit	
1	From 1g, 2g and D5.1.16 Triangle; Isosceles	$\Delta A_i OB_i$ is an isosceles triangles	
2	From 6g and 1	$\Delta A_i OB_i \sim \Delta A'_i OB'_i$	
3	From 1 and 2	$\Delta A'_i OB'_i$ is an isosceles triangles	
4	From 4g, 6g and T5.20.7 Irregular to Regular Polygons: Total Central Angle	$\theta \cong 2\pi / n$	
5	From 3 and D5.1.16 Triangle; Isosceles	$OA'_i = OB'_i$	
6	From 5g, 1, 3 and T5.8.6 Isosceles Triangle: Right Triangles Bisect Vertex Angle	$\angle A_i OC_i \cong \angle A'_i OC'_i \cong \frac{1}{2}\theta$	
7	From 1g, 1, 3 and T5.8.8 Isosceles Triangle: Trigonometry; Base Length	$A_i C_i = OA_i \sin \angle A_i OC_i$	$OC_i = OA_i \cos \angle A_i OC_i$
8	From 1g, 1, 3 and T5.8.8 Isosceles Triangle: Trigonometry; Base Length	$A'_i C'_i = OA'_i \sin \angle A'_i OC'_i$	$OC'_i = OA'_i \cos \angle A'_i OC'_i$
9	From 2g, 6, 7, A5.2.15 GIV Transitivity of Segments and A5.2.18 GIV Transitivity of Angles	$A_i C_i = R \sin \frac{1}{2}\theta$	$OC_i = R \cos \frac{1}{2}\theta$
10	From 2g, 8, A5.2.15 GIV Transitivity of Segments and A5.2.18 GIV Transitivity of Angles	$A'_i C'_i = OA'_i \sin \frac{1}{2}\theta$	$OC'_i = R = OA'_i \cos \frac{1}{2}\theta$
11	From 10 and T4.4.4B Equalities: Right Cancellation by Multiplication	$OA'_i = R / \cos \frac{1}{2}\theta$	
12	From 1g and T5.8.12 Isosceles Triangle: Area	$K_i = A_i C_i \, OC_i$	$K'i = A'_i C'_i \, OC'_i$
13	From 9, 10, 12 and A5.2.15 GIV Transitivity of Segments	$K_i = R \sin \frac{1}{2}\theta \, R \cos \frac{1}{2}\theta$	$K'_i = OA'_i \sin \frac{1}{2}\theta \, r$
14	From 11, 13 and A5.2.15 GIV Transitivity of Segments	$K_i = R \sin \frac{1}{2}\theta \, R \cos \frac{1}{2}\theta$	$K'_i = (R / \cos \frac{1}{2}\theta) \sin \frac{1}{2}\theta \, R$
15	From 14, A4.2.10 Commutative Multp, D4.1.17 Exponential Notation, A4.2.12 Identity Multp, A4.2.13 Inverse Multp, LxE.1.34, A4.2.11 Associative Multp and LxE.1.15	$K_i = R^2 \, \frac{1}{2}\sin \theta$	$K'_i = R^2 \tan \frac{1}{2}\theta$

16	From 1g and summing all of the outer and inner polygon areas	$K = \sum^n K_i$	$K' = \sum^n K'_i$
17	From 15, 16 and A4.2.3 Substitution	$K = \sum^n R^2 \, \tfrac{1}{2}\sin \theta$	$K' = \sum^n R^2 \tan \tfrac{1}{2}\theta$
18	From 17, A4.2.12 Identity Multp and A4.2.14 Distribution	$K = R^2 \, \tfrac{1}{2}\sin \theta \sum^n 1$	$K' = R^2 \tan \tfrac{1}{2}\theta \sum^n 1$
19	From 18, LxE.3.29 and A4.2.10 Commutative Multp	$K = R^2 \, n \, \tfrac{1}{2}\sin \theta$	$K' = R^2 \, n \tan \tfrac{1}{2}\theta$
20	From 4, 19 and A5.2.18 GIV Transitivity of Angles	$K = R^2 \, n \, \tfrac{1}{2}\sin 2\pi / n$	$K' = R^2 \, n \tan \tfrac{1}{2} \, 2\pi / n$
21	From 20, A4.2.13 Inverse Multp, T4.4.18 Equalities: Product by Division, Factorization by 2 T4.5.7 Equalities: Product and Reciprocal of a Product and A4.2.10 Commutative Multp	$K = \pi R^2 \dfrac{(\sin 2\pi / n \,)}{(2\pi / n \,)}$	$K' = \pi R^2 \dfrac{(\tan \pi / n \,)}{(\pi / n \,)}$
22	From TK.2.2 Reciprocal Numbers Tend to Zero as Denominator Increases without Limit	$\lim_{n \to \infty} (2\pi / n) = 0$	
23	From TK.2.2 Reciprocal Numbers Tend to Zero as Denominator Increases without Limit	$\lim_{n \to \infty} (\pi / n) = 0$	
24	From 1g, 7g and area- ρ, area- ρ' bound the area of the circle	$K \leq A_c \leq K'$ in the bounding area	
25	From 21, 24 and A4.2.3 Substitution	$\pi R^2 (\sin 2\pi / n \,) / (2\pi / n) \leq A_c \leq \pi R^2 (\tan \pi / n) / (\pi / n)$	
26	From LxK.2.2.3 Obvious Limit, LxK.2.2.5 Limit of the Products and LxK.2.2.2 Limit of a Constant	$\pi R^2 \lim_{n \to \infty} (\sin 2\pi / n \,) / (2\pi / n) \leq A_c$ $\leq \pi R^2 \lim_{n \to \infty} (\tan \pi / n) / (\pi / n)$	
27	From 22, 23, 26, TK.2.5 Removing Singularity at Zero for Sine(X) / X, TK.2.7 Removing Singularity at Zero for Tangent(X) / X and A4.2.12 Identity Multp	$\pi R^2 \leq A_c \leq \pi R^2$	
$\therefore$	From 27 and LxK.2.2.10 Sandwiching Theorem	$A_c = \pi R^2$ in the bounding limit	

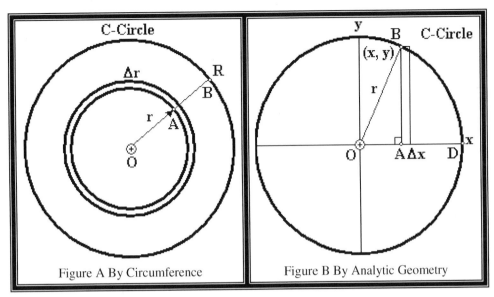

Figure 5.30.2 Finding Area of a Circle

Theorem 5.30.2 Finding Area of a Circle by Integrating Rings of Circumference

1g 2g 3g 4g	Given	the above geometry of Figure F5.18.2A r = OA radius of the circle-C R = OB radius of the bounding circle A_c area of circle
Step	Hypothesis	$A_c = \pi R^2$
1	From 1g, 2g and D5.1.36 Subtended Arclength of a Cord	$c = 2\pi r$ for f = 1 angle in radians
2	From 1g, 4g, 1 and the difference in area	$\Delta A_c \approx c\, \Delta r$
3	From 2 and TK.2.1 Exact Differential from First Principles and Separation of Variable	$dA_c = c\, dr$
4	From 1, 3 and A4.2.3 Substitution	$dA_c = 2\pi r\, dr$
5	From 1g, 3g, 4 and integrating over the length of the radii	$A_c = \int_0^R 2\pi r\, dr$
6	From 5 and LxK.3.2.1	$A_c = 2\pi \int_0^R r\, dr$
7	From 6, LxK.3.1.1 and A4.2.3 Substitution of bounding limits	$A_c = 2\pi\, \tfrac{1}{2}(R^2 - 0)$
∴	From 7, A4.2.7 Identity Add, A4.2.10 Commutative Multp and T4.4.18B Equalities: Product by Division, Factorization by 2	$A_c = \pi R^2$

Theorem 5.30.3 Finding Area of a Circle by Analytic Geometry

1g	Given	the above geometry of Figure F5.18.2B
2g		$r = OB$ radius of the circle-C
3g		$R = OD$ radius of the bounding position
4g		A_c area of circle
Step	Hypothesis	$A_c = \pi R^2$
1	From 1g, 2g, D5.1.19 Triangle; Right Subtended Arclength of a Cord and T5.7.9 Right Triangle: The Altitude	$y = \sqrt{r^2 - x^2}$
2	From 1g, 4g, 1 and the difference in area	$\Delta A_c \approx y\, \Delta x$
3	From 2 and TK.2.1 Exact Differential from First Principles and Separation of Variable	$dA_c = y\, dx$
4	From 1, 3 and A4.2.3 Substitution	$dA_c = (\sqrt{r^2 - x^2})\, dx$
5	From 1g, 3g, 4 and integrating over the length of [x]	$A_c = 4 \int_0^R (\sqrt{R^2 - x^2})\, dx$
6	From 5 and LxK.3.3.1	$A_c = 4\, \tfrac{1}{2}[\, x(\sqrt{R^2 - x^2}) + R^2 \sin^{-1} x/|R|\,]\,\big|_0^R$
7	From 5, D4.1.19 Primitive Definition for Rational Arithmetic and A4.2.3 Substitution of bounding limits	$A_c = 2[\, R\,(\sqrt{R^2 - R^2}) + R^2 \sin^{-1} R/|R|\,]$ $\quad - 2[\, 0\,(\sqrt{R^2 - 0^2}) + R^2 \sin^{-1} 0/|R|\,]$
8	From 7, A4.2.8 Inverse Add, A4.2.7 Identity Add, T4.10.3 Radicals: Identity	$A_c = 2[\, R\,0 + R^2 \sin^{-1} 1\,]$ $\quad - 2[\, 0\,R + R^2 \sin^{-1} 0/|R|\,]$
9	From 8, T4.4.1 Equalities: Any Quantity Multiplied by Zero is Zero and A4.2.7 Identity Add	$A_c = 2R^2 \sin^{-1} 1$ $\quad - 2R^2 \sin^{-1} 0/|R|$
10	From 9, A4.2.10 Commutative Multp, DxE.1.6.7 and DxE.1.6.1	$A_c = 2\, \tfrac{1}{2}\pi R^2$ $\quad - 2\,0\,R^2$
$\therefore$	From 7, T4.4.18B Equalities: Product by Division, Factorization by 2, T4.4.1 Equalities: Any Quantity Multiplied by Zero is Zero and A4.2.7 Identity Add	$A_c = \pi R^2$

Section 5.31 Spheres, Cones and Trihedral Pyramids

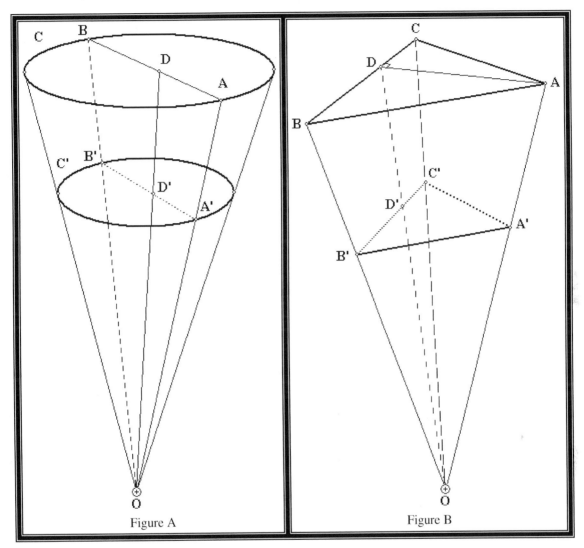

Figure A   Figure B

Figure 5.31.1 Inscribed Cone and Trihedral Pyramids in a Sphere

Theorem 5.31.1 Volume of a Cone by Integrating Circular Disks (Theorems of Pappus)

1g	Given	the above geometry of Figure F5.18.1A
2g		R = OA = OB of inscribing sphere
3g		R' = OA' = OB' of inscribing sphere
4g		r = AD = DB radius of circle-C
Step	Hypothesis	$V = (1/3 \; \pi \sin^2 \tfrac{1}{2}\theta) \; R^3$
1	From 1g	AB = AD + DB
2	From 3g, 1, A5.2.15 GIV Transitivity of Segments and T4.3.10 Equalities: Summation of Repeated Terms by 2	AB = 2r
3	From 2g and D5.1.16 Triangle; Isosceles	$\triangle$AOB Isosceles triangle
4	From 3 and T5.8.8 Isosceles Triangle: Trigonometry; Base Length	$AB = 2R \sin \tfrac{1}{2}\theta$
5	From 2, 4, A5.2.14 GIV Congruence of Segments and T4.4.3 Equalities: Cancellation by Multiplication	$r = R \sin \tfrac{1}{2}\theta$
6	From 5, T5.18.1 Finding Area of a Circle by Bounding with Regular Polygons and A5.2.15 GIV Transitivity of Segments	$A_c = \pi \, (R \sin \tfrac{1}{2}\theta)^2$
7	From 6, A4.2.21 Distribution Exp and A4.2.10 Commutative Multp	$A_c = \pi \sin^2 \tfrac{1}{2}\theta \; R^2$
8	From 2g, 3g and taking the difference	$\Delta R = R - R'$
9	From 7, 8 and the difference in volume	$\Delta V \approx A_c \, \Delta R$
10	From 9 and TK.2.1 Exact Differential from First Principles and Separation of Variable	$dV = A_c \, d\rho$
11	From 6, 10, A4.2.3 Substitution and A4.2.10 Commutative Multp	$dV = (\pi \sin^2 \tfrac{1}{2}\theta) \; \rho^2 \, d\rho$
12	From 11 and integrating over the length of the cone	$V = \int_0^R (\pi \sin^2 \tfrac{1}{2}\theta) \; \rho^2 \, d\rho$
13	From 12 and LxK.3.2.1	$V = (\pi \sin^2 \tfrac{1}{2}\theta) \int_0^R \rho^2 \, d\rho$
14	From 13, LxK.3.1.1 and A4.2.3 Substitution of bounding limits	$V = (1/3 \; \pi \sin^2 \tfrac{1}{2}\theta) \; (R^3 - 0)$
$\therefore$	From 14 and A4.2.7 Identity Add	$V = (1/3 \; \pi \sin^2 \tfrac{1}{2}\theta) \; R^3$

Theorem 5.31.2 Volume of Trihedral Pyramid Inscribed by a Sphere

1g	Given	the above geometry of Figure F5.18.1B
2g		$R = OA = OB = OC$ of inscribing sphere
3g		$R' = OA' = OB' = OC'$ of inscribing sphere
4g		$\varphi \cong \angle COA$
5g		$\phi \cong \angle AOB$
6g		$\theta \cong \angle BOC$
7g		$\sigma = s/R$
Step	Hypothesis	$V = (1/3 \ (4\sqrt{ \ \tfrac{1}{2}\sigma \ (\tfrac{1}{2}\sigma - \sin \tfrac{1}{2}\varphi)(\tfrac{1}{2}\sigma - \sin \tfrac{1}{2}\theta)}$ $(\tfrac{1}{2}\sigma - \sin \tfrac{1}{2}\varphi))) \ R^3$
1	From 2g and D5.1.16 Triangle; Isosceles	$\Delta COA, \Delta AOB, \Delta BOC$ are isosceles triangles
2	From 2g, 4g, 5g, 6g, 1, A5.2.15 GIV Transitivity of Segments, A5.2.18 Transitivity of Angles and T5.8.8 Isosceles Triangle: Trigonometry; Base Length	$AC \cong 2R \sin \tfrac{1}{2}\varphi$ $\quad\big|\quad$ $BC \cong 2R \sin \tfrac{1}{2}\theta$ $\quad\big|\quad$ $AC \cong 2R \sin \tfrac{1}{2}\varphi$
3	From LxE.2.13	$A_t = \sqrt{s(s - AC)(s - BC)(s - AC)}$ for $s = \tfrac{1}{2} (AC + BC + AC)$
4	From 2, 3 and A5.2.15 GIV Transitivity of Segments	$A_t = \sqrt{s(s - 2R \sin \tfrac{1}{2}\varphi)(s - 2R \sin \tfrac{1}{2}\theta)(s - 2R \sin \tfrac{1}{2}\varphi)}$ for $s = \tfrac{1}{2} (2R \sin \tfrac{1}{2}\varphi + 2R \sin \tfrac{1}{2}\theta + 2R \sin \tfrac{1}{2}\varphi)$
5	From 4 and A4.2.14 Distribution	$A_t = \sqrt{s(s - 2R \sin \tfrac{1}{2}\varphi)(s - 2R \sin \tfrac{1}{2}\theta)(s - 2R \sin \tfrac{1}{2}\varphi)}$ for $s = \tfrac{1}{2} 2R (\sin \tfrac{1}{2}\varphi + \sin \tfrac{1}{2}\theta + \sin \tfrac{1}{2}\varphi)$
6	From 5, A4.2.13 Inverse Multp, D4.1.17 Exponential Notation, A4.2.11 Associative Multp, A4.2.10 Commutative Multp and D4.1.4(B, C) Rational Numbers	$A_t = \sqrt{(2R)^4 \ s/2R \ (s/2R - \sin \tfrac{1}{2}\varphi)(s/2R - \sin \tfrac{1}{2}\theta)}$ $(s/2R - \sin \tfrac{1}{2}\varphi)$ for $s = R (\sin \tfrac{1}{2}\varphi + \sin \tfrac{1}{2}\theta + \sin \tfrac{1}{2}\varphi)$
7	From 7g, 6, D4.1.19 Primitive Definition for Rational Arithmetic, A4.2.22 Product Exp, T4.10.4 Radicals: Distribution Across a Product, T4.10.3 Radicals: Identity, T4.4.4A Equalities: Right Cancellation by Multiplication, D4.1.4(B, C) Rational Numbers, A4.2.10 Commutative Multp and A4.2.3 Substitution	$A_t = (2R)^2\sqrt{ \ \tfrac{1}{2}\sigma \ (\tfrac{1}{2}\sigma - \sin \tfrac{1}{2}\varphi)(\tfrac{1}{2}\sigma - \sin \tfrac{1}{2}\theta)}$ $(\tfrac{1}{2}\sigma - \sin \tfrac{1}{2}\varphi)$ for $\sigma = \sin \tfrac{1}{2}\varphi + \sin \tfrac{1}{2}\theta + \sin \tfrac{1}{2}\varphi$
8	From 7, A4.2.21 Distribution Exp, D4.1.19 Primitive Definition for Rational Arithmetic and A4.2.10 Commutative Multp	$A_t = R^2 4\sqrt{ \ \tfrac{1}{2}\sigma \ (\tfrac{1}{2}\sigma - \sin \tfrac{1}{2}\varphi)(\tfrac{1}{2}\sigma - \sin \tfrac{1}{2}\theta)}$ $(\tfrac{1}{2}\sigma - \sin \tfrac{1}{2}\varphi)$ for $\sigma = \sin \tfrac{1}{2}\varphi + \sin \tfrac{1}{2}\theta + \sin \tfrac{1}{2}\varphi$
9	From 2g, 3g and taking the difference	$\Delta R = R - R'$
10	From 8, 9 and the difference in volume	$\Delta V \approx A_t \ \Delta R$

11	From 10 and TK.2.1 Exact Differential from First Principles and Separation of Variable	$dV = A_t\, d\rho$
12	From 8, 11, A4.2.3 Substitution and A4.2.10 Commutative Multp	$dV = (4\sqrt{\tfrac{1}{2}\sigma}\,(\tfrac{1}{2}\sigma - \sin\tfrac{1}{2}\varphi)(\tfrac{1}{2}\sigma - \sin\tfrac{1}{2}\theta)(\tfrac{1}{2}\sigma - \sin\tfrac{1}{2}\varphi))\ \rho^2\, d\rho$
13	From 12 and integrating over the length of the cone	$V = \int_0^R (4\sqrt{\tfrac{1}{2}\sigma}\,(\tfrac{1}{2}\sigma - \sin\tfrac{1}{2}\varphi)(\tfrac{1}{2}\sigma - \sin\tfrac{1}{2}\theta)(\tfrac{1}{2}\sigma - \sin\tfrac{1}{2}\varphi))\ \rho^2\, d\rho$
14	From 13 and LxK.3.2.1	$V = (4\sqrt{\tfrac{1}{2}\sigma}\,(\tfrac{1}{2}\sigma - \sin\tfrac{1}{2}\varphi)(\tfrac{1}{2}\sigma - \sin\tfrac{1}{2}\theta)(\tfrac{1}{2}\sigma - \sin\tfrac{1}{2}\varphi))\ \int_0^R \rho^2\, d\rho$
15	From 14, LxK.3.1.1 and A4.2.3 Substitution of bounding limits	$V = (1/3\ (4\sqrt{\tfrac{1}{2}\sigma}\,(\tfrac{1}{2}\sigma - \sin\tfrac{1}{2}\varphi)(\tfrac{1}{2}\sigma - \sin\tfrac{1}{2}\theta)(\tfrac{1}{2}\sigma - \sin\tfrac{1}{2}\varphi)))\ (R^3 - 0)$
∴	From 15, A4.2.7 Identity Add, A4.2.11 Associative Multp, A4.2.10 Commutative Multp and D4.1.4(A, B) Rational Numbers	$V = (4/3\sqrt{\tfrac{1}{2}\sigma}\,(\tfrac{1}{2}\sigma - \sin\tfrac{1}{2}\varphi)(\tfrac{1}{2}\sigma - \sin\tfrac{1}{2}\theta)(\tfrac{1}{2}\sigma - \sin\tfrac{1}{2}\varphi)\,)\ R^3$

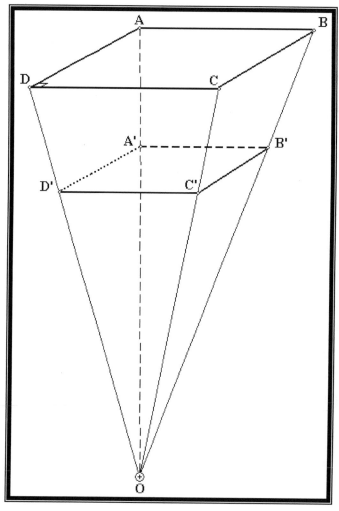

Figure 5.31.2 Inscribed Rectangle Pyramid in a Sphere

Theorem 5.31.3 Volume of Rectangular Pyramid Inscribed by a Sphere

1g	Given	the above geometry of Figure F5.18.2
2g		$R = OA = OB = OC = OD$ of inscribing sphere
3g		$R' = OA' = OB' = OC' = OD'$ of inscribing sphere
4g		$\theta \cong \angle AOB \cong \angle COD$
5g		$\phi \cong \angle BOC \cong \angle AOD$
Step	Hypothesis	$V = (\tfrac{4}{3} \sin \tfrac{1}{2}\theta \sin \tfrac{1}{2}\phi) R^3$
1	From 2g and D5.1.16 Triangle; Isosceles	$\Delta AOB, \Delta BOC, \Delta COD, \Delta DOB$ are isosceles triangles
2	From 2g, 4g, 1, A5.2.15 GIV Transitivity of Segments, A5.2.18 Transitivity of Angles and T5.8.8 Isosceles Triangle: Trigonometry; Base Length	$AB \cong 2R \sin \tfrac{1}{2}\theta$ $\qquad\qquad DC \cong 2R \sin \tfrac{1}{2}\theta$
3	From 2g, 5g, 1, A5.2.15 GIV Transitivity of Segments, A5.2.18 Transitivity of Angles and T5.8.8 Isosceles Triangle: Trigonometry; Base Length	$BC \cong 2R \sin \tfrac{1}{2}\phi$ $\qquad\qquad AD \cong 2R \sin \tfrac{1}{2}\phi$
4	From 1g and D5.1.49 Area of a rectangle	$A_r = AB\ BC$
5	From 4 and A5.2.15 GIV Transitivity of Segments	$A_r = (2R \sin \tfrac{1}{2}\theta)\ (2R \sin \tfrac{1}{2}\phi)$
6	From 5, A4.2.10 Commutative Multp, A4.2.11 Associative Multp, D4.1.19 Primitive Definition for Rational Arithmetic and D4.1.17 Exponential Notation	$A_r = R^2\ 4 \sin \tfrac{1}{2}\theta \sin \tfrac{1}{2}\phi$
7	From 2g, 3g and taking the difference	$\Delta R = R - R'$
8	From 6, 7 and the difference in volume	$\Delta V \approx A_r\ \Delta R$
9	From 8 and TK.2.1 Exact Differential from First Principles and Separation of Variable	$dV = A_r\ d\rho$
10	From 6, 9, A4.2.3 Substitution and A4.2.10 Commutative Multp	$dV = (4\sin \tfrac{1}{2}\theta \sin \tfrac{1}{2}\phi)\ \rho^2\ d\rho$
11	From 10 and integrating over the length of the cone	$V = \int_0^R (4\sin \tfrac{1}{2}\theta \sin \tfrac{1}{2}\phi)\ \rho^2\ d\rho$
12	From 12 and LxK.3.2.1	$V = (4\sin \tfrac{1}{2}\theta \sin \tfrac{1}{2}\phi) \int_0^R \rho^2\ d\rho$
13	From 13, LxK.3.1.1 and A4.2.3 Substitution of bounding limits	$V = \tfrac{1}{3} (4\sin \tfrac{1}{2}\theta \sin \tfrac{1}{2}\phi)\ (R^3 - 0)$
$\therefore$	From 15, A4.2.7 Identity Add, A4.2.11 Associative Multp, A4.2.10 Commutative Multp and D4.1.4(A, B) Rational Numbers	$V = (\tfrac{4}{3} \sin \tfrac{1}{2}\theta \sin \tfrac{1}{2}\phi) R^3$

Tensor Calculus & Physics: A General Treatise

Section 5.32 Trihedral Pyramid

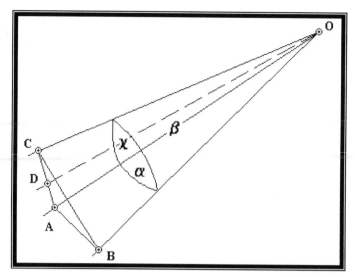

Figure 5.32.1 A Trihedral Pyramid

A Trihedral Pyramid frequently manifest itself in many manifolds and relationships so its only appropriate to provide a study on this geometric construction. [AYR54 pg 146, 150]

Definition 5.32.1 Trihedral Pyramid

Is when three planes have one and only one point in common **O** called the **vertex** of the trihedral.

Definition 5.32.2 Trihedral angle (Euclid's Solid Angle)

Is a **solid angle** constructed from ∠O-ABC.

Definition 5.32.3 Trihedral Face Angles

Are constructed in F4.21.1 from common vertices at O.
$\alpha \equiv \angle AOB$
$\beta \equiv \angle BOC$
$\chi \equiv \angle AOC$

Definition 5.32.4 Trihedral Faces (Planes)

Are the planes intersecting at the common vertex point O.
ΔAOB plane
ΔBOC plane
ΔAOC plane

Definition 5.32.5 Trihedral Edges

Are the common intersections of the Trihedral Faces

$a_o \equiv \overline{AO}$ edge

$b_o \equiv \overline{BO}$ edge

$c_o \equiv \overline{CO}$ edge

Definition 5.32.6 Trihedral Base Face

Is the plane that cuts cross the Trihedral Faces defining the base of the pyramid, $\triangle ABC$.

Definition 5.32.7 Trihedral Base Edges

Are the common edges with the face and base of the pyramid.

$a \equiv \overline{AB}$ edge

$b \equiv \overline{BC}$ edge

$c \equiv \overline{AC}$ edge

Definition 5.32.8 Trihedral Base Angles

Are constructed from F4.21.1 based on $\triangle ABC$.

$\beta_a \equiv \angle BAC$

$\beta_b \equiv \angle ABC$

$\beta_c \equiv \angle ACB$

Theorem 5.32.1 Trihedral: The Sum of Two Face Angles are Greater than the Third

Step	Hypothesis	$\angle AOB + \angle BOC > \angle AOC$
1	From F4.21.1 construct $\angle AOD \ni$	$\angle AOB \cong \angle AOD$
2	From 1, $\overline{AO}$ common between $\triangle AOD$ and $\triangle AOB$ than by ET4.2.26	$\overline{BO} \cong \overline{DO}$ and $\overline{AB} \cong \overline{DA}$ ← Fix proof wrong!!!!!!!!
3	From F4.21.1 and $\triangle ABC$	$\overline{AB} + \overline{BC} > \overline{CA}$
4	From F4.21.1 and edge $\overline{CA}$	$\overline{CA} \cong \overline{CD} + \overline{DA}$
5	From 3 and 4	$\overline{AB} + \overline{BC} > \overline{CD} + \overline{DA}$
6	From 2 and A3.1.3	$\overline{AB} + \overline{BC} > \overline{CD} + \overline{AB}$
7	From 6 and T3.4.5	$\overline{BC} > \overline{CD}$
8	From 7 and $\overline{CO}$ a common side than $\triangle BOC$ and $\triangle DOC$ are non-similar	$\angle BOC > \angle DOC$
9	From 1, 8, A3.2.17 Continuity of Equality and Inequality and T3.3.1	$\angle AOB + \angle BOC > \angle AOD + \angle DOC$
10	From F4.21.1 by construction	$\angle AOC \cong \angle AOD + \angle DOC$
∴	From 9, 10 and A3.2.3	$\angle AOB + \angle BOC > \angle AOC$

Corollary 5.32.1.1 Trihedral: The Sum of any Two Face Angles are Greater than the Third

Step	Hypothesis	
		$\angle AOB + \angle BOC > \angle AOC$ $\angle AOC + \angle BOC > \angle AOB$
∴	From T4.21.1 Trihedral: The Sum of Two Face Angles are Greater than the Third and since no face angle is indistinguishable from any other	$\angle AOB + \angle AOC > \angle BOC$ $\angle AOC + \angle BOC > \angle AOB$

Corollary 5.32.1.2 Trihedral: Sum of any Two Vertex Face Angles are Greater than the Third

Step	Hypothesis	
		$\angle OAB + \angle OAC > \angle BAC$ $\angle OBA + \angle OBC > \angle ABC$ $\angle OCA + \angle OCB > \angle ACB$
∴	From T4.21.1 Trihedral: The Sum of Two Face Angles are Greater than the Third and any co-joining point can be selected as the vertex	$\angle OAB + \angle OAC > \angle BAC$ $\angle OBA + \angle OBC > \angle ABC$ $\angle OCA + \angle OCB > \angle ACB$

Theorem 5.32.2 Trihedral: The Sum all Face Angles are Less than 360°

Step	Hypothesis	$360^\circ > \angle AOB + \angle AOC + \angle BOC$
1	From ΔBOC, ΔAOB and ΔAOC the Trihedral faces have total angle of	$540^\circ = 180^\circ + 180^\circ + 180^\circ$
2	From 1 and T4.18.1 Irregular Trigon: Sum of All Interior Angles Add to 180°	$540^\circ = (\angle AOB + \angle OAB + \angle OBA) +$ $(\angle AOC + \angle OAC + \angle OCA) +$ $(\angle BOC + \angle OBC + \angle OCB)$
3	From 2, A3.2.5 and A3.2.5	$540^\circ = (\angle AOB + \angle AOC + \angle BOC) +$ $(\angle OAB + \angle OAC) +$ $(\angle OBC + \angle OBA) +$ $(\angle OCA + \angle OCB)$
4	From 3, C4.21.1.2 Trihedral: Sum of any Two Vertex Face Angles are Greater than the Third and A3.2.3	$540^\circ > (\angle AOB + \angle AOC + \angle BOC) +$ $(\angle BAC + \angle ABC + \angle ACB)$
5	From 4 and T4.18.1 Irregular Trigon: Sum of All Interior Angles Add to 180° for ΔABC	$540^\circ > (\angle AOB + \angle AOC + \angle BOC) + 180^\circ$
6	From 5, A3.2.8 and T3.7.3 Inequalities: Uniqueness of Subtraction by a Positive Number	$540^\circ - 180^\circ > \angle AOB + \angle AOC + \angle BOC$
∴	From 6 and D3.1.18 Primitive Definition for Rational Arithmetic	$360^\circ > \angle AOB + \angle AOC + \angle BOC$

Theorem 5.32.3 Trihedral: Law of Sine for Faces

Step	Hypothesis	
	Hypothesis	$a / \sin(\alpha) = b_0 / \sin(\angle BAO) = a_0 / \sin(\angle ABO)$
		$b / \sin(\beta) = b_0 / \sin(\angle BCO) = c_0 / \sin(\angle CBO)$
		$c / \sin(\chi) = a_0 / \sin(\angle ACO) = c_0 / \sin(\angle CAO)$
1	From F4.21.1 and LxD.2.1	$\overline{AB} / \sin(\angle AOB) = \overline{BO} / \sin(\angle BAO) = \overline{AO} / \sin(\angle ABO)$
		$\overline{BC} / \sin(\angle BOC) = \overline{BO} / \sin(\angle BCO) = \overline{CO} / \sin(\angle CBO)$
		$\overline{AC} / \sin(\angle AOC) = \overline{AO} / \sin(\angle ACO) = \overline{CO} / \sin(\angle CAO)$
∴	From 1, D4.21.3 Trihedral Face Angles, D4.21.5 Trihedral Edges, D4.21.7 Trihedral Base Edges and A3.2.3	$a / \sin(\alpha) = b_0 / \sin(\angle BAO) = a_0 / \sin(\angle ABO)$ $b / \sin(\beta) = b_0 / \sin(\angle BCO) = c_0 / \sin(\angle CBO)$ $c / \sin(\chi) = a_0 / \sin(\angle ACO) = c_0 / \sin(\angle CAO)$

Theorem 5.32.4 Trihedral: Weak[5.32.1] Unification Law of Sine for Faces and Base

	Given	
1g	Given	$\angle ACO = \angle ABO$ and $\angle BAO = \angle BCO$
2g		$\angle ACO = \angle ABO$ and $\angle CBO = \angle CAO$
3g		$\angle BAO = \angle BCO$ and $\angle CBO = \angle CAO$
Step	Hypothesis	$a / \sin(\alpha) = b / \sin(\beta) = c / \sin(\chi)$
1	From F4.21.1 and LxD.2.1	$a / \sin(\alpha) = b_0 / \sin(\angle BAO) = a_0 / \sin(\angle ABO)$ $b / \sin(\beta) = b_0 / \sin(\angle BCO) = c_0 / \sin(\angle CBO)$ $c / \sin(\chi) = a_0 / \sin(\angle ACO) = c_0 / \sin(\angle CAO)$
∴	From 1 by 1g, or 2g, or 3g and A3.2.3	$a / \sin(\alpha) = b / \sin(\beta) = c / \sin(\chi)$

[5.32.1]Note: Weak verses Strong in this sense means equality holds with conditions.

Theorem 5.32.5 Trihedral: Law of Sine for Base

Step	Hypothesis	$a / \sin(\beta_c) = b / \sin(\beta_a) = c / \sin(\beta_b)$
1	From F4.21.1, LxD.2.1	$\overline{AB} / \sin(\angle ACB) = \overline{BC} / \sin(\angle BAC) = \overline{AC} / \sin(\angle ABC)$
∴	From 1, D4.21.7 Trihedral Base Edges, D4.21.8 Trihedral Base Angles and A3.2.3	$a / \sin(\beta_c) = b / \sin(\beta_a) = c / \sin(\beta_b)$

Theorem 5.32.6 Trihedral: Weak Unification Law of Sine

Step	Hypothesis	$\sin(\beta_c) / \sin(\alpha) = \sin(\beta_a) / \sin(\beta) = \sin(\beta_b) / \sin(\chi)$
1	From T4.21.5 Trihedral: Law of Sine for Base and T3.4.5	$a \sin(\beta_a) = b \sin(\beta_b) = c \sin(\beta_c)$
2	From 1 and T3.4.6	$\sin(\beta_c) / a = \sin(\beta_a) / b = \sin(\beta_b) / c$
3	From 2, T4.21.4 Trihedral: Weak[5.32.1] Unification Law of Sine for Faces and Base and T3.3.9	$(\sin(\beta_c) / a)(a / \sin(\alpha)) = (\sin(\beta_a) / b)(b / \sin(\beta)) = (\sin(\beta_b) / c)(c / \sin(\chi))$
∴	From 3, A3.2.11, A3.2.13 and A3.2.12	$\sin(\beta_c) / \sin(\alpha) = \sin(\beta_a) / \sin(\beta) = \sin(\beta_b) / \sin(\chi)$

Theorem 5.32.7 Trihedral: Law of Cosine for Faces

Step	Hypothesis	
		$a^2 = a_o{}^2 + b_o{}^2 - 2\,a_o\,b_o\,\cos(\alpha)$ $c^2 = a_o{}^2 + c_o{}^2 - 2\,a_o\,c_o\,\cos(\chi)$ $b^2 = b_o{}^2 + c_o{}^2 - 2\,b_o\,c_o\,\cos(\beta)$
1	From F4.21.1, LxD.2.2, LxD.2.3 and LxD.2.4	$\overline{AB}^2 = \overline{AO}^2 + \overline{BO}^2 - 2\,\overline{AO}\,\overline{BO}\,\cos(\angle AOB)$ $\overline{AC}^2 = \overline{AO}^2 + \overline{CO}^2 - 2\,\overline{AO}\,\overline{CO}\,\cos(\angle AOC)$ $\overline{BC}^2 = \overline{BO}^2 + \overline{CO}^2 - 2\,\overline{BO}\,\overline{CO}\,\cos(\angle BOC)$
∴	From 1, D4.21.7 Trihedral Base Edges, D4.21.5 Trihedral Edges, D4.21.3 Trihedral Face Angles and A3.2.3	$a^2 = a_o{}^2 + b_o{}^2 - 2\,a_o\,b_o\,\cos(\alpha)$ $c^2 = a_o{}^2 + c_o{}^2 - 2\,a_o\,c_o\,\cos(\chi)$ $b^2 = b_o{}^2 + c_o{}^2 - 2\,b_o\,c_o\,\cos(\beta)$

Theorem 5.32.8 Trihedral: Law of Cosine for Base

Step	Hypothesis	
		$a^2 = c^2 + b^2 - 2\,c\,b\,\cos(\beta_c)$ $c^2 = a^2 + b^2 - 2\,a\,b\,\cos(\beta_b)$ $b^2 = a^2 + c^2 - 2\,a\,c\,\cos(\beta_a)$
1	From F4.21.1, LxD.2.2, LxD.2.3 and LxD.2.4	$\overline{AB}^2 = \overline{AC}^2 + \overline{BC}^2 - 2\,\overline{AC}\,\overline{BC}\,\cos(\angle ACB)$ $\overline{AC}^2 = \overline{AB}^2 + \overline{BC}^2 - 2\,\overline{AB}\,\overline{BC}\,\cos(\angle ABC)$ $\overline{BC}^2 = \overline{AB}^2 + \overline{AC}^2 - 2\,\overline{AB}\,\overline{AC}\,\cos(\angle BAC)$
∴	From 1, D4.21.8 Trihedral Base Angles, D4.21.7 Trihedral Base Edges and A3.2.3	$a^2 = c^2 + b^2 - 2\,c\,b\,\cos(\beta_c)$ $c^2 = a^2 + b^2 - 2\,a\,b\,\cos(\beta_b)$ $b^2 = a^2 + c^2 - 2\,a\,c\,\cos(\beta_a)$

Theorem 5.32.9 Trihedral: Unification for Law of Cosine

Step	Hypothesis	
		$a^2 = c^2 + b^2 - 2\,c\,b\,\cos(\beta_c) = a_o{}^2 + b_o{}^2 - 2\,a_o\,b_o\,\cos(\alpha)$ $c^2 = a^2 + b^2 - 2\,a\,b\,\cos(\beta_b) = a_o{}^2 + c_o{}^2 - 2\,a_o\,c_o\,\cos(\chi)$ $b^2 = a^2 + c^2 - 2\,a\,c\,\cos(\beta_a) = b_o{}^2 + c_o{}^2 - 2\,b_o\,c_o\,\cos(\beta)$
∴	From T4.21.7 Trihedral: Law of Cosine for Faces, T4.21.8 Trihedral: Law of Cosine for Base and A3.2.2	$a^2 = c^2 + b^2 - 2\,c\,b\,\cos(\beta_c) = a_o{}^2 + b_o{}^2 - 2\,a_o\,b_o\,\cos(\alpha)$ $c^2 = a^2 + b^2 - 2\,a\,b\,\cos(\beta_b) = a_o{}^2 + c_o{}^2 - 2\,a_o\,c_o\,\cos(\chi)$ $b^2 = a^2 + c^2 - 2\,a\,c\,\cos(\beta_a) = b_o{}^2 + c_o{}^2 - 2\,b_o\,c_o\,\cos(\beta)$

Theorem 5.32.10 Trihedral: Weak Unification for Law of Sine and Cosine

Step	Hypothesis	
		$\sqrt{c^2 + b^2 - 2\,c\,b\,\cos(\beta_c)}\,/\sin(\alpha) =$ $\sqrt{a^2 + c^2 - 2\,a\,c\,\cos(\beta_a)}\,/\sin(\beta) =$ $\sqrt{a^2 + b^2 - 2\,a\,b\,\cos(\beta_b)}\,/\sin(\chi)$
1	From T4.21.8 Trihedral: Law of Cosine for Base and A3.2.25	$a = \sqrt{c^2 + b^2 - 2\,c\,b\,\cos(\beta_c)}$ $c = \sqrt{a^2 + b^2 - 2\,a\,b\,\cos(\beta_b)}$ $b = \sqrt{a^2 + c^2 - 2\,a\,c\,\cos(\beta_a)}$
∴	From 1, T4.21.4 Trihedral: Weak[5.32.1] Unification Law of Sine for Faces and Base and A3.2.3	$\sqrt{c^2 + b^2 - 2\,c\,b\,\cos(\beta_c)}\,/\sin(\alpha) =$ $\sqrt{a^2 + c^2 - 2\,a\,c\,\cos(\beta_a)}\,/\sin(\beta) =$ $\sqrt{a^2 + b^2 - 2\,a\,b\,\cos(\beta_b)}\,/\sin(\chi)$

Theorem 5.32.11 Equilateral Trihedral: Unification for Law of Cosine for Faces

1g	Given	$h_o = a_o = b_o = c_o$
Step	Hypothesis	$a^2 = h_o^2\, 2\,(1 - \cos(\alpha))$ $c^2 = h_o^2\, 2\,(1 - \cos(\chi))$ $b^2 = h_o^2\, 2\,(1 - \cos(\beta))$
1	From 1g, T4.21.7 Trihedral: Law of Cosine for Faces and A3.13	$a^2 = h_o^2 + h_o^2 - 2\, h_o\, h_o \cos(\alpha)$ $c^2 = h_o^2 + h_o^2 - 2\, h_o\, h_o \cos(\chi)$ $b^2 = h_o^2 + h_o^2 - 2\, h_o\, h_o \cos(\beta)$
2	From 1, A3.2.12 and A3.2.14	$a^2 = h_o^2\, (1 + 1 - 2\,1\,\cos(\alpha))$ $c^2 = h_o^2\, (1 + 1 - 2\,1\,\cos(\chi))$ $b^2 = h_o^2\, (1 + 1 - 2\,1\,\cos(\beta))$
3	From 2, A3.2.12 and D3.2.16	$a^2 = h_o^2\, (2 - 2\,\cos(\alpha))$ $c^2 = h_o^2\, (2 - 2\,\cos(\chi))$ $b^2 = h_o^2\, (2 - 2\,\cos(\beta))$
$\therefore$	From 3 and A3.2.14	$a^2 = h_o^2\, 2\,(1 - \cos(\alpha))$ $c^2 = h_o^2\, 2\,(1 - \cos(\chi))$ $b^2 = h_o^2\, 2\,(1 - \cos(\beta))$

Theorem 5.32.12 Equilateral Trihedral: Unification for Law of Half-Sine for Faces

Step	Hypothesis	$a = h_o\, 2\,\sin(\tfrac{1}{2}\alpha)$ $c = h_o\, 2\,\sin(\tfrac{1}{2}\chi)$ $b = h_o\, 2\,\sin(\tfrac{1}{2}\beta)$
1	From T4.21.11 Equilateral Trihedral: Unification for Law of Cosine for Faces, A3.2.12, A3.2.13 and A3.2.11	$a^2 = h_o^2\, (2\,2)\,\tfrac{1}{2}(1 - \cos(\alpha))$ $c^2 = h_o^2\, (2\,2)\,\tfrac{1}{2}(1 - \cos(\chi))$ $b^2 = h_o^2\, (2\,2)\,\tfrac{1}{2}\,(1 - \cos(\beta))$
2	From 1, D3.1.16, LxD.1.68, T3.8.5 and T3.10.3	$a^2 = h_o^2\, 2^2\, \sin^2(\tfrac{1}{2}\alpha)$ $c^2 = h_o^2\, 2^2\, \sin^2(\tfrac{1}{2}\chi)$ $b^2 = h_o^2\, 2^2\, \sin^2(\tfrac{1}{2}\beta)$
3	From 2 and A3.2.21	$a^2 = (\, h_o\, 2\,\sin(\tfrac{1}{2}\alpha)\,)^2$ $c^2 = (\, h_o\, 2\,\sin(\tfrac{1}{2}\chi)\,)^2$ $b^2 = (\, h_o\, 2\,\sin(\tfrac{1}{2}\beta)\,)^2$
$\therefore$	From 3 and T3.8.5	$a = h_o\, 2\,\sin(\tfrac{1}{2}\alpha)$ $c = h_o\, 2\,\sin(\tfrac{1}{2}\chi)$ $b = h_o\, 2\,\sin(\tfrac{1}{2}\beta)$

Theorem 5.32.13 Equilateral Trihedral: Face Triangles are Isosceles

1g	Given	$h_o = a_o = b_o = c_o$
2g		$a_n \equiv a / h_o$
3g		$b_n \equiv b / h_o$
4g		$c_n \equiv c / h_o$
5g		$\tfrac{1}{2}\alpha \equiv \angle BAO = \angle ABO$
6g		$\tfrac{1}{2}\beta \equiv \angle BCO = \angle CBO$
7g		$\tfrac{1}{2}\chi \equiv \angle ACO = \angle CAO$
Step	Hypothesis	$a_n = \sin(\alpha) / \sin(\tfrac{1}{2}\alpha)$
		$b_n = \sin(\beta) / \sin(\tfrac{1}{2}\beta)$
		$c_n = \sin(\chi) / \sin(\tfrac{1}{2}\chi)$ All face triangles are isosceles.
1	From 1g, 2g, 3g, T4.21.3 Trihedral: Law of Sine for Faces and A3.2.3	$a / \sin(\alpha) = h_o / \sin(\angle BAO) = h_o / \sin(\angle ABO)$
		$b / \sin(\beta) = h_o / \sin(\angle BCO) = h_o / \sin(\angle CBO)$
		$c / \sin(\chi) = h_o / \sin(\angle ACO) = h_o / \sin(\angle CAO)$
2	From 1, T2.3.9, 2g, 3g, 4g and A3.2.3	$a_n / \sin(\alpha) = 1 / \sin(\angle BAO) = 1 / \sin(\angle ABO)$
		$b_n / \sin(\beta) = 1 / \sin(\angle BCO) = 1 / \sin(\angle CBO)$
		$c_n / \sin(\chi) = 1 / \sin(\angle ACO) = 1 / \sin(\angle CAO)$
3	From 2 and T3.4.5	$\sin(\alpha) / a_n = \sin(\angle BAO) = \sin(\angle ABO)$
		$\sin(\beta) / b_n = \sin(\angle BCO) = \sin(\angle CBO)$
		$\sin(\chi) / c_n = \sin(\angle ACO) = \sin(\angle CAO)$
4	From 3 by sine's of the same angular argument.	$\angle BAO = \angle ABO$
		$\angle BCO = \angle CBO$
		$\angle ACO = \angle CAO$
5	From 3, 4, 5g, 6g, 7g, A3.2.3 and DxD.2.1	$\sin(\alpha) / a_n = \sin(\tfrac{1}{2}\alpha)$
		$\sin(\beta) / b_n = \sin(\tfrac{1}{2}\beta)$
		$\sin(\chi) / c_n = \sin(\tfrac{1}{2}\chi)$ All face triangles are isosceles.
∴	From 5, T3.3.3 and T3.4.5	$a_n = \sin(\alpha) / \sin(\tfrac{1}{2}\alpha)$
		$b_n = \sin(\beta) / \sin(\tfrac{1}{2}\beta)$
		$c_n = \sin(\chi) / \sin(\tfrac{1}{2}\chi)$ All face triangles are isosceles.

Theorem 5.32.14 Equilateral Trihedral: Weak Unification for Law of Half-Sine

Step	Hypothesis	$\sin(\tfrac{1}{2}\alpha) / \sin(\beta_c) = \sin(\tfrac{1}{2}\beta) / \sin(\beta_a) = \sin(\tfrac{1}{2}\chi) / \sin(\beta_b)$
1	From T4.21.4 Trihedral: Weak[5.32.1] Unification Law of Sine for Faces and Base, T4.21.12 Equilateral Trihedral: Unification for Law of Half-Sine for Faces and A3.2.3	$h\,2 \sin(\tfrac{1}{2}\alpha) / \sin(\beta_c) = h\,2 \sin(\tfrac{1}{2}\beta) / \sin(\beta_a) = h\,2 \sin(\tfrac{1}{2}\chi) / \sin(\beta_b)$
∴	From 1 and T3.4.8	$\sin(\tfrac{1}{2}\alpha) / \sin(\beta_c) = \sin(\tfrac{1}{2}\beta) / \sin(\beta_a) = \sin(\tfrac{1}{2}\chi) / \sin(\beta_b)$

Theorem 5.32.15 Equilateral Trihedral: Reciprocal Unification for Law of Half-Sine

Step	Hypothesis	$\sin(\tfrac{1}{2}\alpha) \sin(\beta_a) = \sin(\tfrac{1}{2}\beta) \sin(\beta_b) = \sin(\tfrac{1}{2}\chi) \sin(\beta_c)$
∴	From T4.21.14 Equilateral Trihedral: Weak Unification for Law of Half-Sine and T3.4.5	$\sin(\tfrac{1}{2}\alpha) \sin(\beta_a) = \sin(\tfrac{1}{2}\beta) \sin(\beta_b) = \sin(\tfrac{1}{2}\chi) \sin(\beta_c)$

Theorem 5.32.16 Equilateral Trihedral: Isosceles Base for Faces

1g	Given	$h = c = b$
2g		$b_n \equiv b / h_o$
3g		$c_n \equiv c / h_o$
4g		$h_n = h / h_0$
Step	Hypothesis	$a_n = \sin(\alpha) / \sin(\tfrac{1}{2}\alpha)$
		$h_n = \sin(\beta) / \sin(\tfrac{1}{2}\beta) = \sin(\chi) / \sin(\tfrac{1}{2}\chi)$ for $\beta = \chi$
1	T4.21.13 Equilateral Trihedral: Face Triangles are Isosceles, 1g, 2g, 3g, 4g and A3.2.3	$a_n = \sin(\alpha) / \sin(\tfrac{1}{2}\alpha)$ $h_n = \sin(\beta) / \sin(\tfrac{1}{2}\beta)$ $h_n = \sin(\chi) / \sin(\tfrac{1}{2}\chi)$
2	From 1 and A3.2.2	$a_n = \sin(\alpha) / \sin(\tfrac{1}{2}\alpha)$ $h_n = \sin(\beta) / \sin(\tfrac{1}{2}\beta) = \sin(\chi) / \sin(\tfrac{1}{2}\chi)$
3	From 2 by sine's of the same angular argument.	$\beta = \chi$
∴	From 2 and 3	$a_n = \sin(\alpha) / \sin(\tfrac{1}{2}\alpha)$ $h_n = \sin(\beta) / \sin(\tfrac{1}{2}\beta) = \sin(\chi) / \sin(\tfrac{1}{2}\chi)$ for $\beta = \chi$

Theorem 5.32.17 Equilateral Trihedral: Isosceles Base

1g	Given	$\beta_c = \theta$
2g		$\beta_a = \beta_b = \tfrac{1}{2}\theta$
3g		$h = c = b$
4g		$a_b = a / h$
Step	Hypothesis	$a_b = \sin(\theta) / \sin(\tfrac{1}{2}\theta)$ for $\beta_a = \beta_b = \tfrac{1}{2}\theta$
1	From 3g, T4.21.5 Trihedral: Law of Sine for Base and A3.2.3	$a / \sin(\beta_c) = h / \sin(\beta_a) = h / \sin(\beta_b)$
2	From 1, T3.4.1A, 4g, D3.1.4 and A3.1.3	$a_b / \sin(\beta_c) = 1 / \sin(\beta_a) = 1 / \sin(\beta_b)$
3	From 2 and T3.4.5	$\sin(\beta_c) / a_b = \sin(\beta_a) = \sin(\beta_b)$
4	From 3 by sine's of the same angular argument.	$\beta_a = \beta_b$
5	From 4, 1g, 2g and A3.2.3	$\sin(\theta) / a_b = \sin(\tfrac{1}{2}\theta)$
∴	From 4, 5 and T3.4.5	$a_b = \sin(\theta) / \sin(\tfrac{1}{2}\theta)$ for $\beta_a = \beta_b = \tfrac{1}{2}\theta$

Theorem 5.32.18 Equilateral Trihedral: Unified Isosceles Base for Faces

1g	Given	$a_n \equiv a / h_o$
2g		$a_b = a / h$
Step	Hypothesis	$h \sin(\theta) / \sin(\tfrac{1}{2}\theta) = h_0 \sin(\alpha) / \sin(\tfrac{1}{2}\alpha)$
1	T4.21.16 Equilateral Trihedral: Isosceles Base for Faces, 1g and T3.4.4B	$a = h_0 \sin(\alpha) / \sin(\tfrac{1}{2}\alpha)$
2	T4.21.17 Equilateral Trihedral: Isosceles Base, 2g and T3.4.4B	$a = h \sin(\theta) / \sin(\tfrac{1}{2}\theta)$
∴	From 1, 2 and A3.2.2	$h \sin(\theta) / \sin(\tfrac{1}{2}\theta) = h_0 \sin(\alpha) / \sin(\tfrac{1}{2}\alpha)$

Theorem 5.32.19 Equilateral Trihedral: Right Triangular Base

1g	Given	$\beta_c = 90^\circ$
2g		$\beta_b = 90^\circ - \theta$
3g		$\beta_a = \theta$
Step	Hypothesis	$a = \sin(\theta) / b = \cos(\theta) / c$
1	From 1g, 2g, 3g, T4.21.5 Trihedral: Law of Sine for Base and A3.2.3	$a / \sin(90^\circ) = b / \sin(\theta) = c / \sin(90^\circ - \theta)$
2	From 1, DxTD.1.6 and DxTD.1.7	$a / 1 = b / \sin(\theta) = c / \cos(\theta)$
$\therefore$	From 2 and T3.4.5	$1 / a = \sin(\theta) / b = \cos(\theta) / c$

Theorem 5.32.20 Equilateral Trihedral: Right Triangular Base Trigonometry

Step	Hypothesis	$\sin(\theta) = b / a$
		$\cos(\theta) = c / a$
		$\tan(\theta) = b / c$
1	From T4.21.19 Equilateral Trihedral: Right Triangular Base, T3.3.3 and T3.4.5	$\sin(\theta) = b / a$
2	From T4.21.19 Equilateral Trihedral: Right Triangular Base, T3.3.3 and T3.4.5	$\cos(\theta) = c / a$
3	From 1, 2, T3.4.7, D3.1.4, A3.2.13 and A3.2.12	$\tan(\theta) = b / c$
$\therefore$	From 1, 2 and 3	$\sin(\theta) = b / a$
		$\cos(\theta) = c / a$
		$\tan(\theta) = b / c$

Theorem 5.32.21 Equilateral Trihedral: Right Base to Faces

1g	Given	$a_n \equiv a / h_o$
2g		$b_n \equiv b / h_o$
3g		$c_n \equiv c / h_o$
Step	Hypothesis	$\sin(\theta) = (\sin(\tfrac{1}{2}\alpha) \sin(\beta)) / (\sin(\alpha) \sin(\tfrac{1}{2}\beta))$
		$\cos(\theta) = (\sin(\tfrac{1}{2}\alpha) \sin(\chi)) / (\sin(\alpha) \sin(\tfrac{1}{2}\chi))$
		$\tan(\theta) = (\sin(\beta) \sin(\tfrac{1}{2}\chi)) / (\sin(\tfrac{1}{2}\beta) \sin(\chi))$
1	From T4.21.20 and T3.4.10	$\sin(\theta) = (b / h_0) / (a / h_0)$
		$\cos(\theta) = (c / h_0) / (a / h_0)$
		$\tan(\theta) = (b / h_0) / (c / h_0)$
2	From 1, 1g, 2g, 3g and A3.2.3	$\sin(\theta) = b_n / a_n$
		$\cos(\theta) = c_n / a_n$
		$\tan(\theta) = b_n / c_n$
3	From 2, T4.21.13 Equilateral Trihedral: Face Triangles are Isosceles and A3.2.3	$\sin(\theta) = (\sin(\beta) / \sin(\tfrac{1}{2}\beta)) / (\sin(\alpha) / \sin(\tfrac{1}{2}\alpha))$
		$\cos(\theta) = (\sin(\chi) / \sin(\tfrac{1}{2}\chi)) / (\sin(\alpha) / \sin(\tfrac{1}{2}\alpha))$
		$\tan(\theta) = (\sin(\beta) / \sin(\tfrac{1}{2}\beta)) / (\sin(\chi) / \sin(\tfrac{1}{2}\chi))$
$\therefore$	From 3, T3.5.1 Equilateral Trihedral: Right Triangular Base Trigonometry and A3.2.10	$\sin(\theta) = (\sin(\tfrac{1}{2}\alpha) \sin(\beta)) / (\sin(\alpha) \sin(\tfrac{1}{2}\beta))$
		$\cos(\theta) = (\sin(\tfrac{1}{2}\alpha) \sin(\chi)) / (\sin(\alpha) \sin(\tfrac{1}{2}\chi))$
		$\tan(\theta) = (\sin(\beta) \sin(\tfrac{1}{2}\chi)) / (\sin(\tfrac{1}{2}\beta) \sin(\chi))$

Theorem 5.32.22 Equilateral Trihedral: Equilateral Base, Equal Face Angles

1g	Given	$a_n = b_n = c_n = h_n$
Step	Hypothesis	$\alpha = \beta = \chi$
1		$h_n = \sin(\alpha) / \sin(\tfrac{1}{2}\alpha)$
		$h_n = \sin(\beta) / \sin(\tfrac{1}{2}\beta)$
		$h_n = \sin(\chi) / \sin(\tfrac{1}{2}\chi)$
2		$\sin(\alpha) / \sin(\tfrac{1}{2}\alpha) = \sin(\beta) / \sin(\tfrac{1}{2}\beta) = \sin(\chi) / \sin(\tfrac{1}{2}\chi)$
$\therefore$	From 2 by sine's of the same angular argument.	$\alpha = \beta = \chi$

Section 5.33 Euclid's Book XI: Exploring the Geometry of Space, Definitions

This treatise exclusively deals with the development of plane geometry; however there are proofs that require the solid geometry of Euclid. So in this section a summary of Euclid's work on solid geometry is reviewed here for a quick lookup reference to the first twenty theorems [HEA56, Vol. 3 pg 260].

Definition 5.33.1 Solid
A solid is that which has length, breadth, and depth.

Definition 5.33.2 Surface
An extremity of a solid is a surface.

Definition 5.33.3 Intersection of Two Planes at Right Angles to One Another
A straight line is at right angles to a plane, when it makes right angles with all the straight lines, which meet it and are in the plane.

Definition 5.33.4 A Plane at Right Angles to Another Plane
A plane is at right angles to a plane when the straight lines drawn, in one of the planes, at right angles to the common section of the planes are at right angles to the remaining plane.

Definition 5.33.5 Inclination of a Straight Line to a Plane
The inclination of a straight line to a plane is, assuming a perpendicular drawn from the extremity of the straight line which is elevated above the plane to the plane, and a straight line joined from the point thus arising to the extremity of the straight line which is in the plane, the angle contained by the straight line so drawn and the straight line standing up.

Definition 5.33.6 Inclination of a Plane to a Plane
The inclination of a plane to a plane is the acute angle contained by the straight lines drawn at right angles to the common section at the same point, one in a each of the planes.

Definition 5.33.7 Similarly Inclined Planes
A plane is said to be similarly inclined to a plane as another is to another when the said angles of the inclinations are equal to one another.

Definition 5.33.8 Parallel Planes
Parallel planes are those, which do not meet.

Definition 5.33.9 Similar Solid Figures
Similar solid figures are those contained by similar planes equal in multitude.

Definition 5.33.10 Equal and Similar Solid Figures
Equal and similar solid figures are those contained by similar planes equal in multitude and in magnitude.

Definition 5.33.11 Solid Angle

1. A solid angle is the inclination constituted by more than two lines, which meet one another and are not in the same surface, towards all the lines.

   Otherwise: A solid angle is that which is contained by more than two plane angles which are not in the same plane and are constructed to one point. [HEA56, Vol. 3 pg 261]

A more modern definition would be:
2. A solid angle is bounded by three plane angles at a common point and no two plane angles are coplanar.

Definition 5.33.12 Pyramid

A pyramid is a solid figure, contained by planes, which is constructed from one plane to one point.

Definition 5.33.13 Prism

A prism is a solid figure contained by planes two of which, namely those, which are opposite, are equal, similar and parallel, while the rest are parallelograms.

Definition 5.33.14 Sphere

When, the diameter of a semicircle remaining fixed, the semicircle is carried round and restored again to the same position from which it began the move, the figure comprehended is a sphere.

Definition 5.33.15 Axis of a Sphere

The axis of the sphere is the straight line which remains fixed and about which the semicircle is turned.

Definition 5.33.16 Center of the Sphere

The center of a sphere is the same as that of the semicircle.

Definition 5.33.17 Diameter of the Sphere

A diameter of the sphere is any straight line drawn through the center and terminated in both directions by the surface of the sphere.

Definition 5.33.18 Cone: Right, Obtuse and Acute-Angle

When, one sided of those about the right angle in a tight-angled triangle remaining fixed, the triangle is carried round and restored again to the same position from which it began to be moved, the figure so comprehended is a cone.

And, if the straight line which remains fixed be equal to the remaining side about the right angle which is carried round, the cone will be right-angled; if less, obtuse-angled; and if greater, acute-angled.

Definition 5.33.19 Axis of a Cone

The axis of the cone is the straight line which remains fixed and about which the triangle is turned.

Definition 5.33.20 Base of a Cone

The base of a cone is the circle described by the straight line, which is carried round.

Definition 5.33.21 Cylinder

When, one side of those about the right angle in a rectangular parallelogram remaining fixed, the parallelogram is carried round and restored again to the same position from which it began to be moved, the figure so comprehended is a cylinder.

Definition 5.33.22 Axis of a Cylinder

Axis of the cylinder is the straight line, which remains fixed, and about which the parallelogram is turned.

Definition 5.33.23 Bases of a Cylinder

The bases of a cylinder are the circles described by the two sides opposite to one another, which are carried round.

Definition 5.33.24 Similar Cones and Cylinders

Similar cones and cylinders are those in which the axes and the diameters of the bases are proportional.

Definition 5.33.25 Cube

A cube is a solid figure contained by six equal squares.

Definition 5.33.26 Octahedron

An octahedron is a solid figure contained by eight equal and equilateral triangles.

Definition 5.33.27 Icosahedron

An icosahedron is a solid figure contained by twenty equal and equilateral triangles.

Definition 5.33.28 Dodecahedron

A dodecahedron is a solid figure contained by twelve equal, equilateral, and equiangular pentagons.

Mathematicians have learned so much in these last two thousand years that while I was writing down these definitions I had a strong urge to rewrite them in a more modern analytic form for accuracy. However, the thought occurred to me that if I never had seen a sphere, cylinder, cone or dodecahedron I might write them in the same descriptive way. I think this might have been the situation in Euclid's day trying to convey these ideas to people who knew nothing about geometry, but who wanted to learn it for the first time.

Some phrases were archaic, such as, "carried round and restored again", which could be replaced with "turned around about its axis full circle" but I left them in because they had already been just translated from Greek and as the translator had pointed out some changes had already been made do to the lack of an equivalent word or unclear meaning so I didn't want to move them any further away from their original interpretation.

Section 5.34 *Euclid's Book XI: Theorems, Exploring the Geometry of Space*

Euclid explored the geometry of space by using perpendicular lines to a plane, planes, pyramids and parallelepipedal solids. [HEA56, Vol. 3 pg 272] All of these can be found in his thirty-nine theorems or as he called them propositions. The following table is a quick lookup reference for the first twenty Euclidean theorems based on lines and planes.

Table 5.34.1 Book XI: Theorems on Space Geometry

Theorem	Proposition	Diagram
Theorem 5.34.1 Elevated Line of a Plane	A part of a straight line cannot be in the plane of reference and a part in a plane more elevated	
Theorem 5.34.2 Two Intersecting Lines Define a Plane	If two straight lines cut one another, they are in one plane and every triangle is in one plane.	
Theorem 5.34.3 Intersection of Two Planes is a Line	If two planes cut one another, their common section is a straight line.	
Theorem 5.34.4 Line Perpendicular to a Plane, Intersects Two Lines is at Right Angles	If a straight line be set up at right angles to two straight lines which cut one another, at their common point of intersection, it will be at right angles to the plane through them.	
Theorem 5.34.5 Line Perpendicular to a Plane, Intersects Three Lines is at Right Angles	If a straight line be set up at right angles to three straight which meet one another, at their common point of intersection, the three straight lines are in one plane	
Theorem 5.34.6 Two Right Angle Lines to a Plane are Parallel	If two straight lines be at right angles to the same plane, the straight lines will be parallel.	
Theorem 5.34.7 Joined Parallel Lines Lie in the same Plane	If two straight lines be parallel and points be taken at random on each of them, the straight lines joining the points is in the same plane with the parallel straight lines.	

Theorem 5.34.8 **Parallel Lines, One Perpendicular to Plane then so to is the other**	If two straight lines be parallel, and one of them be at right angles to any plane, the remaining one will also be at right angles to the same plane.	
Theorem 5.34.9 **Parallel Lines between Planes Mark Out Parallel Lines**	Straight lines, which are parallel to the same straight line and are not in the same plane with it, are also parallel to one another. It is implied that AD ‖ BC then AD ‖ EF and BC ‖ EF.	
Theorem 5.34.10 **Parallel Lines between Planes have Congruent Angles**	If two straight lines meeting one another be parallel to two straight lines meeting one another not in the same plane, they will contain equal angles. It is implied AB ‖ DF and AC ‖ DE then ∠BAC ≅ ∠FDE and AD ‖ BF ‖ CE.	
Theorem 5.34.11 **Point above a Plane is Perpendicular to the Plane**	From a given elevated point to draw a straight line perpendicular to a given plane. It is implied that Point-A ⊥ Plane-BCF	
Theorem 5.34.12 **Point Above a Plane a Perpendicular Line can be Construct to the Plane**	To set up a straight line at right angles to a given plane from a given point in it. It is implied 1 select B, 2 drop a ⊥to Plane at C, 3 construct AD ‖ BC, hence AD ⊥ Plane.	
Theorem 5.34.13 **Angles in the same Plane cannot be Congruent**	From the same point two straight lines cannot be setup at right angles to the same plane on the same side. It is implied that ∠BAE and ∠CAE in the same plane cannot be the same ⊥ angle.	
Theorem 5.34.14 **Parallel Planes Never Meet**	Planes to which the same straight line is at right angles will be parallel. It is implied that ΔAKB cannot have two right angles hence does not exist and planes CD and FE can never meet thereby they are parallel. Perpendicular planes are parallel.	
Theorem 5.34.15 **Planes are Parallel if Parallel Lines are in Different Planes**	If two straight lines meeting one another be parallel to two straight lines meeting one another, not being in the same plane, the planes through them are parallel. It is implied that BG ⊥ plane-HGK, AB ‖ HG and BC ‖ GK, HG ‖ DE and GK ‖ EF, since all lines are ⊥ to BG then the two planes are parallel.	

Theorem 5.34.16 Intersection Lines between Two Planes and a Third are Parallel	If two parallel planes be cut by any plane, their common intersection are parallel. It is implied EFK on the same straight line GHK as well K cannot be shared by both lines, hence intersection lines GH and EF must be parallel.	
Theorem 5.34.17 Parallel Planes Cutting Adjacent Triangles have Proportional Sides	If two straight lines be cut by parallel planes, they will be cut in the same ratios. It is implied planes GH ‖ KL ‖ MN, and the bases of CA ‖ FO and BD ‖ ED for ΔDAB and ΔOAE, hence ΔDAB ~ ΔOAE likewise ΔADC ~ ΔODF, therefore all sides of ΔDAB and ΔOAE are proportional ratios likewise for ΔDAB and ΔOAE.	
Theorem 5.34.18 All Planes and their Axis Line are at Right Angles to a Reference Plane	If a straight line be at right angles to any plane, all the planes through it will also be at right angles to the same plane. It is implied AB $\perp$ reference-REC then all through plane-AB will be $\perp$ reference-REC. It follows any $\perp$ line to a plane of reference will have all planes through it $\perp$ to the reference plane.	
Theorem 5.34.19 There is only one Right Angle Intersection Line for Two Intersecting Planes	If two planes which cut one another be at right angles to any plane, their common intersection will also be at right angles to the same plane It implies that plane-AB and BC are $\perp$ reference-RCA, hence the line of intersection BD is $\perp$. Now assume BD is not perpendicular, but DE and DF are by implication of being in planes-AB and BC, there cannot be two $\perp$ lines, therefore only the common intersection can be $\perp$.	
Theorem 5.34.20 The Sum of any Two Solid Angles, are Greater than the Remaining One	If a solid angle be contained by three plane angles, any two, taken together in any manner, are greater than the remaining one. In a simplistic way let all three plane angles be equal then the sum of any two have to be two times greater than the remaining angle. For more general proof see Note[5.22.1].	

Euclid does not have the following theorem in his list of spatial constructions so in order to prove the analytic equation of a plane the following theorem is required:

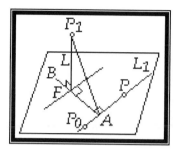

Figure 5.34.1 Line L_1 through P_0 and P is perpendicular to L and in a Plane

Theorem 5.34.21 Line L_1 through P_0 and P is perpendicular to L in a Plane

1g	Given	the above geometry of Figure F5.22.1
Step	Hypothesis	If Line-L_1 passes through P_0 and P while $\perp$ to L, then P is on plane-AB
1	From 1g and T5.22.11 A Point above a Plane is Perpendicular to the Plane	There exists line-L drawn from P_1 intersecting $\perp$ to plane-AB at point-F
2	From 1g, 2 and A5.2.3 GI Three Points in a Plane	There exists a point-F and another pair of points, P_0 and P, on plane-AB
3	From 3 and A5.2.2 GI One $^£$SL with Two Points	There exists one line-L_1 containing points P_0 and P
$\therefore$	From 1, 2 and 3 it can be concluded	If Line-L_1 passes through P_0 and P while $\perp$ to L, then P is on plane-AB

Phrasing the Theorem:

From Euclidean geometry we recall that if a line L_1 through P_0 and P is perpendicular to L, then P must be in the desired plane.

If P_1 is perpendicular to a plane-AB and the resulting perpendicular line-L_1 intersects at point-F, then any two points, lets say P_0 and P on line-L are also on the plane-AB.

[5.22.1] Notes: General Proof for T5.22.20

1g	Given	From Figure T5.22.20
2g		s∠A is a solid angle
3g		AE ≅ AD
4g		AB common
Step	Hypothesis	m∠BAC < m∠DAB + m∠DAC for s∠A
		m∠DBC < m∠DBA + m∠CBA for s∠B
		m∠DCB < m∠ACB + m∠ACD for s∠C
1	From 3g, 5g, 6g and T5.6.10 Congruency: by Side-Angle-Side (SAS)	ΔDBA ≅ ΔEBA
2	From 1 and D5.1.50 Congruent Triangles	BD ≅ BE
3	From 1 and D5.1.50 Congruent Triangles	∠BAE ≅ ∠DAB
4	From 1g	∠BAE + ∠EAC ≅ ∠BAC for ΔABC
5	From 1g, C5.19.1.1 Tri-Construction: Side-Side-Side (SSS), Constructible Inequality and geometry of ΔDBC	BD + DC > BC for ΔDBC
6	From 1g and geometry of ΔDBC	BC ≅ BE + EC
7	From 5, 6 and A5.2.15 GIV Transitivity of Segments	BD + DC > BE + EC
8	From 2, 7 and A5.2.15 GIV Transitivity of Segments	BE + DC > BE + EC
9	From 8, A5.18.4 GVII Intersection of Collinear Lines	DC > EC BE being the intersection, thereby leaving the inequality balanced.
10	From 3g, 9, ΔEAC and ΔDAC	∠EAC < ∠DAC
11	From 3, 10 and T5.18.6 Construction: Addition of Angles	∠BAE + ∠EAC < ∠DAB + ∠DAC
∴	From 4, 11 and A5.2.18 GIV Transitivity of Angles	∠BAC < ∠DAB + ∠DAC for s∠A
∴	Same above proof	∠DBC < ∠DBA + ∠CBA for s∠B
∴	Same above proof	∠DCB < ∠ACB + ∠ACD for s∠C

Section 5.35 Modern Elements: A Proposal for Restructuring Euclid's Axioms

As it will be found out there are other issues that have been left unsaid, these issues have been brought about by the problem of congruency, which shows the need for axioms to support proofs on construction. Classically the compass and straight edge have been used to prove geometry by construction and it's always been understood that all proofs have their origins and are validated by such methods. However, construction allows certain short cuts, and external or overview observations to be made that rigorous logic cannot extrapolate to as seen in building geometric objects from elementary components. This has lead unwittingly to a divergence between logical theorems and construction. In the previous part of this chapter, there is a proposed list of axioms Section 5.17 "Modern Elements: Extend Hilbert's Axioms to Include Construction" on construction, which might be used to supplement Hilbert's original postulates and hopefully reunify this discrepancy between these two methods of proof writing. These axioms achieved this by eliminating the need for the mechanical compass and straight edge, thereby creating a complete deductive system of geometrical logic.

The Greeks had a third geometry that they invented based on a collapsible compass, since the mechanics of their day could not build a compass that stayed set to a fix distance. With the elimination of the compass by these new axioms, the two geometric systems can now be amalgamated into one.

It is also of interest, and practical use, that Euclid originally laid out his axioms in what seemed to follow a formal Galois Group. These axioms were for the most part abandoned by Archimedes, Playfair, Hilbert and others as they started identifying weakness in Euclid's development of geometry and tried to repair the holes in a strict logical way. After all Euclid was attempting something that had never been done before and how he possible could have had the insight, and a backward view, that two thousand plus years of experience can bring. By working out a number of his theorems it has become evident that Euclid was a very practical man relying on experimentation with the use of construction to validate his work. For example if in a theorem something was needed to make it work he constantly went back and revaluated his axioms to see if one, they were truly fundamental and two, what else might be needed. As a result of this ongoing cycle, evaluation of his axioms started to take on the form of a Galois Group.

What so special about a Galois Group, so far in the experience of mathematics every new system of math seems to start out in some way or another to take on this structure when listing axioms. If generally true, then this would provide away of standardizing a new system, thereby providing <u>completeness</u> for a set of axioms, and to some extent <u>validating</u> and <u>building confidence</u> in those postulates and propositions.

Completeness is very important in a set of axioms without it you cannot move from point-A to point-B in a proof. For example if the notion of equality were to be removed that would stop the development of most proofs in math, then what would our confidence be in that system? This is what is brought into question with the Playfair and Hilbert's axioms, certainly they are very logical and most of the holes have been filled in, but in doing so have new omissions been created, was something left out, are they complete? Maybe in away Euclid was correct all along.

This is why I am proposing going back to the original axioms and writing them in the frame of a Galois Group, improve the wording, symbolism and placing them in a more analytical modern structure, including the old corrected axioms were appropriate. One thing for sure using these improved set of axioms it would now be possible to prove Euclid's theorems, because they would now be backwards compatible.

Euclid's Original Initiating Axioms

These postulates and common notions are actually axioms, because they are fundamental statements of basic observations for properties of geometric objects.

Postulates
1)   To draw a straight line from any point to any point.
2)   To produce a finite straight line continuously is a straight line.
3)   To describe a circle with any centre and distance.
4)   That all right angles are equal to one another.
5)   That, if a straight line falling on two straight lines make the interior angles on the same side less than two right angles, the two straight lines, if produced indefinitely, meet on the side on which are the angles less than the two right angles. [HEA56, Vol. 1 pg 154]

Postulates 1 to 4 are written in a way that they could be rewritten as descriptions making them definitions, but it might be argued that they are elementary axioms on how to construct geometric objects. This brings up the point Euclid and his peers thought in terms of constructing geometric objects in order to prove their ideas; hence his axioms should also be based on constructions. Also Postulate 5 is actually provable, see Corollary 5.4.9.1 "Parallel Lines: Converse Non-Perpendicular Alternate Interior Angles", therefore cannot be an axiom.

Common Notions
1)   Things which are equal to the same thing are also equal to one another.
2)   If equals be added to equals, the wholes are equal.
3)   If equals be subtracted from equals, the remainders are equal.
4)   Things which coincide with one another are equal to one another.
5)   The whole is greater than the part. [HEA56, Vol. 1 pg 155]

Common notions, such as 1 through 4, deal with a very important property of geometry the ability to add and subtract from a set of objects and establish an equal balance between object groups. Also (5) alludes to geometric objects being less than or greater than in their assembled construction. These properties Euclid freely used in his proofs, however they are not to be found in the set of Hilbert's axioms making the axioms discoordant, hence not backwards compatible, and so cannot prove some of his theorems.

Frame for a Galois Group

Axioms by Group Quantities	Description
$\exists$ a field G $\ni$, quantities where $(e, a, a^{-1}) \in G$	Existence of a Group Field where [e] is called the identity element and has the unique identity property of leaving quantities unaltered under the operation of the group system.
$a = b$ Equation A $b = a$ Equation B	Equality of quantities
$b = a$ and $c = b$ $\therefore c = a$	Substitution of quantities
Axioms by Group Operator	**Description**
$(a \circ b \wedge b \circ a) \in G$	Closure with respect to Group Operator
$a \circ b = b \circ c$	Commutative by Group Operator
$a \circ (b \circ c) = (a \circ b) \circ c$	Associative by Group Operator
$a \circ e = a$	Identity by Group Operator
$a \circ a^{-1} = e$	Inverse Group Operator by reciprocal Quantity

Also see D2.2.17 Group.

Table 5.35.1 Group IX Axioms on Constructions for Lines and Planes

Axiom	Axioms on Construction of Lines and Planes
Axiom 5.35.1 GIX Two Points in a Line	Given point-A and point-B a straight line-AB can be constructed through the two points.
Axiom 5.35.2 GIX One Line through Two Points	Given a straight line-AB then there is a point-A and point-B constructed on the straight line.
Axiom 5.35.3 GIX Three Points in a Plane	Given point-A, point-B and point-C that are not collinear, a plane-ABC can be constructed through the three points.
Axiom 5.35.4 GIX One Point and a Line in a Plane	Given point-A and point-B on a straight line-AB, and point-C not on line-AB a plane-ABC can be constructed through the point and straight line.
Axiom 5.35.5 GIX Every Point in a Line are in a Plane	Given point-A and point-B in a straight line-AB and straight line-AB lies in a plane then every point on line-AB lies in the plane-AB.
Axiom 5.35.6 GIX Intersection of Two Planes	Intersection of two planes-A, B is a straight line-AB.
Axiom 5.35.7 GIX Progressive Ordering of Space	In every straight line-AB, there are at least two points, in every plane-ABC, there are at least three points that are not in the same straight line, and in space-ABCD, there are at least four points that are not in the same plane.

Table 5.35.2 Group X Axiom for Geometric Objects Comprised of Components

Axioms	Axioms by Component of Objects[5.35.1]		Description
Axiom 5.35.8 GX Existence	$\exists$ a field G $\ni$, quantities $(\ \bullet\ g_i\ \bullet\ ) \in$ G, see D5.5.1 Geometric Object		Existence of geometric constructions
Axiom 5.35.9 GX Congruency	$g_i \cong h_i$ EQ A $h_i \cong g_i$ EQ B	$G \cong H$ EQ A $H \cong G$ EQ B	Congruency of constructible components for all $i^{[5.35.2]}$
Axiom 5.35.10 GX Transitivity	$f_i \cong g_i$ and $g_i \cong h_i$ $\therefore f_i \cong h_i$	$F \cong G$ and $G \cong H$ $\therefore F \cong H$	Transitivity of constructible components for all i

Axioms by Group Quantities	Description
$\exists$ a field G $\ni$, quantities where $(e, a, a^{-1}) \in$ G	Existence of a Group Field where [e] is called the identity element and has the unique identity property of leaving quantities unaltered under the operation of the group system.
a = b Equation A b = a Equation B	Equality of quantities
b = a and c = b $\therefore$ c = a	Substitution of quantities

If one goes back to Euclid's original theorems and meticulously proves them it can be found that Euclid subconsciously used, but never formalized these ideas. For example Euclid's proof for Proposition 4 of (SAS) [HEA56, Vol. I, pg 247] demonstrating the use of correspondence of points. Also in his proof Proposition 20 on "Summation of Regular Angles from a Solid Angle", [HEA56, Vol. 3, pg 307] he uses the idea of unequal segment lengths and angles and removes equal quantities form both sides maintaining the proportion or balance of the inequality in order to prove his theorem. Nowhere in his work does he sanctify the notion of inequality, thereby legitimatising its use. Now consider that geometric length, while they can be equal, one length can be greater than the other, which can be demonstrated by constructing segments of unequal lengths, hence AB:length-a > GH:length-g and the inequality signs can be used, so AB $\sim\cong$ GH. For simplicity it can be rewritten as AB > GH meaning the property of one length can be greater than another.

Table 5.35.3 Group XI Axioms Trichotomy of Geometric Properties

Axioms	Axioms Trichotomy of Properties		Description
Axiom 5.35.11 GXI Inequality: Properties	g_i: property > h_i: property	G $\sim\cong$ H	Inequality of components and their unequal properties for all i
Axiom 5.35.12 GXI Inequality: Transitivity	$f_i > g_i$ and $g_i > h_i$ $\therefore f_i > h_i$	F > G and G > H $\therefore$F > H	Transitivity for inequality of components and their properties for all i
Axiom 5.35.13 GXI Correspondence of Trichotomy	[>, $\geq$, <, $\leq$]: Regardless of the inequality operator they hold true for any axiom or theorem as long as the order is left unchanged between properties.		Correspondence of inequality operations for axioms and theorems.

Table 5.35.4 Group XII Axioms on Segments

Axiom	Axioms by Addition of Segments	Description
Axiom 5.35.14 GXII Continuity of Straight Line	Given a straight line-AB, with point-A = A_1 and point-B = A_n, such that point-A_{i-1} and point-A_{i+1}, lie between point-A and point-B, then any point-A_i can be found between point-A_{i-1} and point-A_{i+1}, such that the points are ordered and the following segments are constructed: $A_{i-1}A_i$, A_iA_{i+1} Repeat process and increase [n] until the line-AB is completely constructed with no gaps. If true than the line is continuous without gaps or holes.	
Axiom 5.35.15 GXII Segment of Symmetric-Reflexiveness	If AB is a segment and A' is a point on a line, then there is on that line, on a given side of A', one and only one point B' such that $AB \cong A'B'$ and Equation A $A'B' \cong AB$. Equation B	
Axiom 5.35.16 GXII Segment Addition	If a line segment-AC has a point B between them, than $AC \cong AB + BC$ is the addition of the segments.	
Axiom 5.35.17 GXII Segment Addition Converse	If $AC \cong AB + BC$, the addition of the segments, than the corresponding line segment-AC has a point B between A and C.	
Axiom 5.35.18 GXII Closure Seg	$(AB + BC \wedge BC + AB) \in S$	Closure with respect to addition of segments
Axiom 5.35.19 GXII Commutative Seg	$AB + BC \cong BC + AB$	Commutative for segments
Axiom 5.35.20 GXII Associative Seg	$AB + (BC + CD) \cong (AB + BC) + CD$	Associative for segments
Axiom 5.35.21 GXII Identity Seg	$AB + 0 \cong AB$ [5.35.3]	Identity for segments
Axiom 5.35.22 GXII Inverse Seg	$AB - AB \cong 0$ [5.35.4]	Inverse for segments

Axioms by Group Operator	Description
$(a \circ b \wedge b \circ a) \in G$	Closure with respect to Group Operator
$a \circ b = b \circ c$	Commutative by Group Operator
$a \circ (b \circ c) = (a \circ b) \circ c$	Associative by Group Operator
$a \circ e = a$	Identity by Group Operator
$a \circ a^{-1} = e$	Inverse Group Operator by reciprocal Quantity

Table 5.35.5 Group XIII Axioms on Angles

Axiom	Axioms by Addition of Angles	Description
Axiom 5.35.23 GXIII Continuity of Angle of Arc	Given an $\angle AB$, with point-A = A_1 and point-B = A_n, such that point-A_{i-1} and point-A_{i+1}, lie between point-A and point-B, then any point-A_i can be found between point-A_{i-1} and point-A_{i+1}, such that the points are ordered and the following angle of arcs are constructed: $\angle A_{i-1}A_i$, $\angle A_iA_{i+1}$ Repeat process and increase [n] until the $\angle AB$ is completely constructed with no gaps. If true than the angle of arc is continuous without gaps or holes.	
Axiom 5.35.24 GXIII Angle of Symmetric-Reflexiveness	If $\angle AOB$ is an angle built on a construction line [a] and [a] is on a plane, and OA' is a half-line on [a] with vertex O, then there is on the plane on a given side of [a] one and only one half-line OB' with vertex O such that $\angle AB \cong \angle A'B'$ and $\angle A'B' \cong \angle AB.$ Equation A / Equation B	
Axiom 5.35.25 GXIII Addition of Angles	If a $\angle AC$ has a point B between them, than $\angle AC \cong \angle AB + \angle BC$ is the addition of angles.	
Axiom 5.35.26 GXIII Addition of Angles Converse	If $\angle AC \cong \angle AB + \angle BC$, the addition of angles, than the corresponding $\angle AC$ has a point B between A and C.	
Axiom 5.35.27 GXIII Closure Ang	$(\angle A + \angle B \wedge \angle B + \angle A) \in A$	Closure with respect to Angles
Axiom 5.35.28 GXIII Commutative Ang	$\angle A + \angle B \cong \angle B + \angle A$	Commutative for Angles
Axiom 5.35.29 GXIII Associative Ang	$\angle A + (\angle B + \angle C) \cong (\angle A + \angle B) + \angle C$	Associative for Angles
Axiom 5.35.30 GXIII Identity Ang	$\angle A + 0 \cong \angle A$ [5.35.3]	Identity for Angles
Axiom 5.35.31 GXIII Inverse Ang	$\angle A - \angle A \cong 0$ [5.35.4]	Inverse for Angles

Axioms by Group Operator	Description
$(a \circ b \wedge b \circ a) \in G$	Closure with respect to Group Operator
$a \circ b = b \circ c$	Commutative by Group Operator
$a \circ (b \circ c) = (a \circ b) \circ c$	Associative by Group Operator
$a \circ e = a$	Identity by Group Operator
$a \circ a^{-1} = e$	Inverse Group Operator by reciprocal Quantity

Table 5.35.6 Group XIV Axiom on Parallel Lines

Axiom	Axiom on Parallel Lines
Axiom 5.35.32 GXIV Parallel Lines	Given a plane-ABC containing a given straight line-AB and a given point-C that is not on the line-AB, there is one and only one straight line-c that can be constructed. That does not intersect the given line-AB (Modified Playfair Axiom).

[5.35.1]Note: Group XIII is written in a general way applying to segments as well as angles and any other object that fits into this category.

[5.35.2]Note: Congruency is defined with a more specific definition here the construction of a component to match another includes associated properties. For example a segment length can be reconstructed on any construction line, but to be congruent the property of length must be equal in order to make the components identical. Likewise a reconstructed circle to be congruent must have equal radii. So $AB \cong GH$ if and only if their properties AB:length-a = GH:length-g. It is important to know that geometric length is being defined here as equal in order not to confuse it with a measured equal length, which would complicate things, since a measured length is not precise. Geometric constructions however have always been understood to be exact structures.

[5.35.3]Note: Using definition 4.1.2.2 "The Number Zero" it implies either no object present, empty, or only an intersection point is left.

[5.35.4]Note: The "Inverse Segment" axiom heralds back to Euclid's Common Notion (3), establishing the ability to remove equal proportional components from two congruent objects while leaving the resulting constructions congruent. Not to be found in the set of Hilbert's axioms making it an incomplete axiomatic system. Hilbert had it wrong, Euclid had it right all along.

Table 5.35.7 The Modern Elements: Proposal Comparing New and Old Axioms

The Modern Elements: Axioms	The Elements: Axioms	Comments
A5.35.1 GIX Two Points in a Line	A5.2.1 GI Two Points on a Straight Line	
A5.35.2 GIX One Line through Two Points	A5.2.2 GI One Straight Line with Two Points	
A5.35.3 GIX Three Points in a Plane	A5.2.3 GI Three Points in a Plane	
A5.35.4 GIX One Point and a Line in a Plane	A5.2.4 GI One Plane with Three Points	
A5.35.5 GIX Every Point in a Line are in a Plane	A5.2.5 GI Straight Line in a Plane	
A5.35.6 GIX Intersection of Two Planes	A5.2.6 GI Intersection of Planes	
A5.35.7 GIX Progressive Ordering of Space	A5.2.7 GI Non-coplanar Points	
A5.35.8 GX Existence	A5.2.10 GII Three Point Existence	
A5.35.9 GX Congruency	A5.2.16 GIV Congruence for Union of Segments	
A5.35.10 GX Transitivity	A5.2.15 GIV Transitivity of Segments	A5.2.18 GIV Transitivity of Angles
A5.35.11 GXI Inequality: Properties		
A5.35.12 GXI Inequality: Transitivity		
A5.35.13 GXI Correspondence of Trichotomy		
A5.35.14 GXII Continuity of Straight Line	A5.2.20 GV Continuity of Points on a Straight Line	
A5.35.15 GXII Segment Symmetric-Reflexiveness	A5.2.14 GIV Symmetric-Reflexiveness of Segments	
A5.35.16 GXII Segment Addition	A5.2.8 GII Order of Direction of Points on a straight line	
A5.35.17 GXII Segment Addition Converse	A5.2.9 GII Four Point Existence	
A5.35.18 GXII Closure Seg	A5.2.21 GIV GVI Closure of Axiom Set	
A5.35.19 GXII Commutative Seg		
A5.35.20 GXII Associative Seg		
A5.35.21 GXII Identity Seg		
A5.35.22 GXII Inverse Seg		

A5.35.23 GXIII Continuity of Angle of Arc		
A5.35.24 GXIII Angle Symmetric-Reflexiveness	A5.2.17 GIV Symmetric-Reflexiveness of Angles	
A5.35.25 GXIII Addition of Angles		T5.18.7 Construction: Addition of Angles
A5.35.26 GXIII Addition of Angles Converse		
A5.35.27 GXIII Closure Ang	A5.2.21 GIV GVI Closure of Axiom Set	
A5.35.28 GXIII Commutative Ang		
A5.35.29 GXIII Associative Ang		
A5.35.30 GXIII Identity Ang		
A5.35.31 GXIII Inverse Ang		
A5.35.32 GXIV Parallel Lines	A5.2.13 GIII Parallel Lines	
	A5.2.11 GII Ordering of Points on a Straight Line	Redundant: A5.23.13 GXII Continuity of Straight Line embodies the ordering of points
	A5.2.12 GII Intersection of a Straight Line with a Triangle	Not desirable to be included in the list of axioms, because possibly an unlimited combination of shape intersections could be found that would be just as fundamental and they would have to be included in the list of axioms making it cumbersome, not only that none of them would be required in the original Euclidian proofs.
	A5.2.19 GIV Congruence of Side-Side-Angle (SSA) and Opposite Angles	T5.19.2 Tri-Construction: Side-Side-Angle (SSA)

This table shows huge gaps or incompleteness of the old Playfair-Hilbert axioms to the new modern set.

Section 5.36 Modern Elements: Proposal for Axioms on Making Measurements

Now, the notation of magnitude (measured distance) and a measured angle are none existent in Greek geometry. It wouldn't be till 1637 that Descartes published his famous work on calculated distance and angles was it possible to analytically make a computed measure. However, it would be from another area of applied mathematics, the mathematics of land surveying that measured distance and angle would be made possible with the invention of special instrumentation.

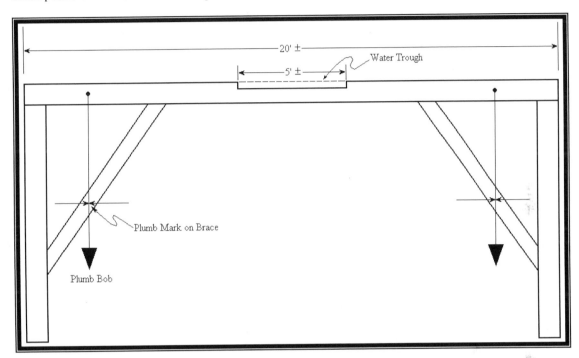

Figure 5.36.1 Roman Leveler (Chorobate)

The invention of these instruments was a slow evolutionary process, the Romans picking up the Greek's ideas of measurement, which came to them from the Egyptians, in tern the Babylonians and so on. The Romans made good use of practical surveying techniques for many centuries to construct roadways, aqueducts, military camps, towns, cities and coliseums. Two thousand years later some Roman roads, such as the ***apian way***, and aqueducts still exist and some parts are still functional. For leveling surfaces and measuring accurate angles, the Romans used a ***chorobate***, an approximate 20–foot wooden structure with plumbed end braces and an approximate 5–foot groove for a water trough. Linear measurements were often made with wooden poles 10 to 17 ft long. With the fall of the Roman Empire, surveying and most other intellectual endeavors became lost arts to the Western World.

Renewed interest in intellectual pursuits may have been fostered by the explorers' need for effective navigational skills. The lodestone, a naturally magnetized rock (magnetite), was used to locate magnetic north. The compass would later be used for navigation on both land and water allowing an angled direction to be measured against a fixed frame of reference. In the mid–1500s, the surveyors' chain was first used in the Netherlands. An Englishman, Thomas Digges, in 1571, first used the term ***theodolite*** to describe an instrument, which has a circle that was graduated in 360° and used to measure horizontal and vertical angles from a small, leveled platform or table. [KAVβ04, pg 626]

In 1658, Phillips picked up on Digges instrument and made it out of brass consisting of a scale on a semicircle and uses it in surveying of land. The instrument now known as a ***protractor***, eventually works its way into geometry for lying down and measuring angles in drawings. [OXF71, Vol. II, protractor, pg 2338] So, from this point on a new idea that distances and angles can be measured starts to enter into geometry, now coupled with Descartes coordinate system a new discipline is created called analytic geometry unifying algebra, geometry and measurement.

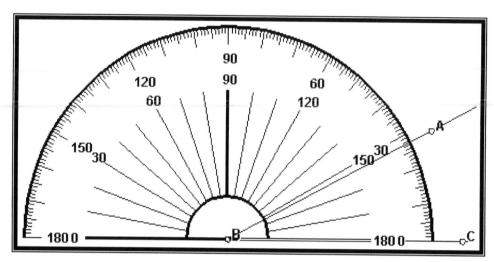

Figure 5.36.2 Modern Protractor Measuring ∠ABC

The notion of measuring distance and angles was left to reside in analytic geometry and geometry was left unsullied from measurement until the 1990's when in order to more easily teach students, congruency, which is difficult to prove, was replaced with the more algebraic intuitive concept of measurement.

In text books measurement of distance or magnitude and angle are now stated as axioms of Euclidian geometry, which of course they are not. Since measurement converts, the magnitude and angle to a number the equal sign can now be used without worrying about congruency and what is a geometric object? Now we lose track of what and which object we are dealing with, resulting in confusion and non-rigorous proofs. It is for those reasons that with great reluctance I introduce these axioms and warn the proof writer not to use them as a foundation of any Euclidian geometric proof.

Definition 5.36.1 Measured Magnitude

Measured magnitude is symbolically represented in the following ways:
$$a \equiv AB \equiv m\overline{AB} \equiv |\overline{AB}| \equiv |\overleftrightarrow{AB}| \equiv |\overrightarrow{AB}|$$
for measured distance between points (A, B).

Definition 5.36.2 Measurement of Angle

Measurement of angle is symbolically represented in the following way:
$$\alpha \equiv m\angle ABC \text{ and for simplicity the vertex point can be dropped, then } \alpha \equiv m\angle AC$$
for measured angle through the arc $\overset{\frown}{ABC}$ likewise can be written $\overset{\frown}{AC}$.

Definition 5.36.3 Proportional Error

If a quantity being measured and the true value [Q] is given, but there is an error of an amount $[\mu \equiv \pm\mu_0]$ then the Proportional Error $\equiv (\mu/Q) \times 100$ is read in percent.

Table 5.36.1 Group XV Axioms on Making Measurements

Axiom 5.36.1 GXV Ruler Postulate	The set of points (A ≤ . P_i . ≤ B for all countable numbers {i} ∈ **M** on any line segment the points have a one-to-one correspondence to the set of real numbers (0 ≤ p_i ≤ b for all countable numbers {i} ∈ **R** such that, given any two points P and Q on the line, P correspond to zero, and Q corresponds to any positive number.
Axiom 5.36.2 GXV Segment Addition Postulate	If a line segment $\overline{AC}$ has a point B between them, than AC = AB + BC the addition of the measured segments.
Axiom 5.36.3 GXV Segment Addition Converse Postulate	If AC = AB + BC the addition of the measured segments than the corresponding line segment $\overline{AC}$ has a point B between A and C.
Axiom 5.36.4 GXV Protractor Postulate	The set of points on the arc of a semicircle (A ≤ . P_i . ≤ B for all countable numbers {i} ∈ **Pr** have a one-to-one correspondence to the set of real numbers (0 ≤ p_i ≤ b for all countable numbers {i} ∈ **R** so that, given any two points P and Q on the arc, P correspond to zero, and Q corresponds to any positive number having units of angle of arc, such as degrees or radians.
Axiom 5.36.5 GXV Angle Addition Postulate	If an angle of arc $\overset{\frown}{AC}$ or ∠AC has an angle point B between them, than m∠AC = m∠AB + m∠BC the addition of the measured angles.
Axiom 5.36.6 GXV Angle Addition Converse Postulate	If m∠AC = m∠AB + m∠BC the addition of the measured angles than the angle of arc $\overset{\frown}{AC}$ or ∠AC has a point B between them.
Axiom 5.36.7 GXV Measure Numbers are Algebraic Axiomatic	All measured numbers are a field of numbers [a] where (0 ≤ a) ∈ R and being real numbers they follow all the axioms and theorems the algebraic real number system imposes.
Axiom 5.36.8 GXV Limit of Accuracy for a Measured Quantity	Any measured quantity [mQ] cannot be measured better than an ideal quantity [Q] within an error of measurement [μ] that is limited to the accuracy of the instrument being used to measure with. $Q - \mu_0 < mQ < Q + \mu_0$ Equation A or $\lvert mQ - Q \rvert < \mu_0.$ Equation B
Axiom 5.36.9 GXV Greater Precision	As the precision improves the error of measurement will decrease toward a very small value of a plus or minus number for: $0 < \lvert \mu_0 \rvert$ as Equation A $\mu_0 \rightarrow 0$ Equation B

Observation 5.36.1 |μ_0| Gets in the way: Can't get there from here!

From axiom A5.24.9 "GXV Greater Precision" |μ_0| can never be zero only approaches it, hence no measurement can ever truly be known.

Theorem 5.36.1 Accumulated Error for a Measured Series

1g	Given	$i = 1, 2, \ldots , n$ for $0 < n$
2g		$mQ_{av} = (1/n) \, (\sum mQ_i \,)$ average of all measured quantities
3g		$\mu_{av} = (1/n)\sum \mu_i$ average of all errors of measurement
Step	Hypothesis	$\mid mQ_{av} - Q \mid < \mu_{av}$
1	From 1g and A5.24.8A	$Q - \mu_i < mQ_i < Q + \mu_i$
2	From 1 and summing over all i	$\sum Q - \sum \mu_i < \sum mQ_i < \sum Q + \sum \mu_i$
3	From 2 and T4.3.9 Equalities: Summation of Repeated Terms	$n \, Q - \sum \mu_i < \sum mQ_i < n \, Q + \sum \mu_i$
4	From 1g, 3 and T4.7.5 Inequalities: Multiplication by Positive Number	$(1/n)(\, n \, Q - \sum \mu_i \,) < (1/n) \, (\sum mQ_i \,) < (1/n) \, (n \, Q + \sum \mu_i)$
5	From 4 and A4.2.14 Distribution	$(1/n) \, n \, Q - (1/n) \sum \mu_i < (1/n) \, (\sum mQ_i \,)$ $< (1/n) \, n \, Q + (1/n)\sum \mu_i$
6	From 5, T4.4.16 Equalities: Product by Division, Common Factor and A4.2.12 Identity Multp	$Q - (1/n) \sum \mu_i < (1/n) \, (\sum mQ_i \,) < Q + (1/n)\sum \mu_i$
7	From 2g, 3g, 6 and A4.2.3 Substitution	$Q - \mu_{av} < mQ_{av} < Q + \mu_{av}$
8	From 7 and T4.7.37 Inequalities: Subtraction Across Ternary Inequality	$- \mu_{av} < \pm(mQ_{av} - Q) < + \mu_{av}$ for $Q < mQ_{av}$ or $mQ_{av} < Q$
∴	From 9 and T4.7.29 Inequalities: Equal Values About a Lesser Absolute Quantity	$\mid mQ_{av} - Q \mid < \mu_{av}$

Observation 5.36.2 Weighted Deviation

Building an aggregate system theorem T5.24.1 "Accumulated Error for a Measured Series" finds the average error of measurement. Finding the deviation of the over all composites requires the constant of error to be considered. This can be done by weighting deviation with the errors of measurement, which is given by the following special formula:

$$\sigma^2 \equiv (1/n) \sum \mu_i(mQ_i - mQ_{av})^2 ./ \mu_{av} \qquad \text{Equation A}$$

The accumulated error arises in such measurement situation for finding the accumulated center of gravity of a missile system or the precision length of a fabricated Traveling Wave-Guide.

Observation 5.36.3 Nonlinear Accumulated Errors

The majority of physical formulas are products of system parameters, hence nonlinear. How can accumulated errors for products be calculated?

Suppose Q is a quantity measured at [m], the value of [m] to be determined by the instrument of measure. This measurement is not precise (measurements never are) and there is an error [μ] in measuring the value [m]. The error in the measurement [Q] is $\Delta Q \equiv Q(m + \mu) - Q(m)$ and the ***proportional error*** in the measurement is simply

$$\Delta Q \equiv Q(m + \mu) - Q(m) \qquad \text{Equation A}$$

Theorem 5.36.2 Differential Approximate Error

1g	Given	$\mu = \Delta m$	
2g		$\delta\mu \approx \mu$	approximate error as $\mu \to 0$
Step	Hypothesis	$\dfrac{\delta Q}{Q} \approx \dfrac{Q'(m)}{Q(m)}\, \delta\mu$	
1	From 1g and O5.21.2A	$\Delta Q = Q(m + \mu) - Q(m)$	
2	From 1 and T4.5.8 Equalities: Division of a Constant and T4.5.9 Equalities: Multiplication of Unity	$\dfrac{\Delta Q}{Q} = \dfrac{[Q(m + \mu) - Q(m)]\,(\mu / \mu)}{Q(m)}$	
3	From 2, D4.1.4 Rational Numbers and A4.2.10 Commutative Multp	$\dfrac{\Delta Q}{Q} = \dfrac{[Q(m + \mu) - Q(m)] / \mu}{Q(m)}\,\mu$	
4	From 2, LxK.2.1.1 Uniqueness of Limits and A5.21.9 GX Greater Precision	$\lim_{\mu \to 0} \dfrac{\Delta Q}{Q} = \lim_{\mu \to 0} \dfrac{[Q(m + \mu) - Q(m)] / \mu}{Q(m)}\,\mu$	
$\therefore$	From 2g, 4, A4.2.3 Substitution, DxK.3.1.3 Definition of a Derivative as a Limit, DxK.3.2.1B Definitions on Differential Notation and A4.2.3 Substitution	$\dfrac{\delta Q}{Q} \approx \dfrac{Q'(m)}{Q(m)}\, \delta\mu$ [5.22.1]	

Theorem 5.36.3 Accumulated Error for a Product

1g	Given	$Q = \prod X_i(\bullet\, m_i\, \bullet)$ each component is dependent on the uncertainty m_i and the other quantities.
Step	Hypothesis	$\delta\langle Q/Q\rangle_{[\approx 1]} \sim \sum_i DCF_i\, \delta\mu_i$ [5.36.1]
1	From 1g and LxK.3.2.1A Differential Operators (exact differential)	$dQ = \sum_j \prod_i^{j-1} X_i(\bullet\, m_i\, \bullet)\, [X_{j,j}(\bullet\, m_i\, \bullet)\, dm_j]\, \prod_{i=j-1}^{n} X_i(\bullet\, m_i\, \bullet)$
2	From 1g, 1, T4.5.8 Equalities: Division of a Constant, A4.2.14 Distribution and expand product	$\dfrac{dQ}{Q} = \dfrac{\sum_j \prod_i^{j-1} X_i(\bullet\, m_i\, \bullet)\, [X_{j,j}(\bullet\, m_i\, \bullet)\, dm_j]\, \prod_{i=j-1}^{n} X_i(\bullet\, m_i\, \bullet)}{\prod_i^{j-1} X_i(\bullet\, m_i\, \bullet)\, [X_j(\bullet\, m_i\, \bullet)]\, \prod_{i=j-1}^{n} X_i(\bullet\, m_i\, \bullet)}$
3	From 2, T4.4.22 Equalities: Reversal of Identity and Inverse of Multiplication	$\delta Q/Q \approx \sum_j DCF_i\, [X_{j,j}(\bullet\, m_i\, \bullet) / X_j(\bullet\, m_i\, \bullet)]\, \delta\mu_j$ [5.22.1]
$\therefore$	From 4 and; The normalization of	$\delta\langle Q/Q\rangle_{[\approx 1]} \sim \sum_i DCF_i\, \delta\mu_i$
	the coefficients moves them toward unity, hence the delta variations are directly domentate and their straight differential summation [DCF_i] alown is what counts to find the accumulated error of the product.	

In random measurements of aggregate systems the above three theorems are probably the most important and useful. However measurement is a vast topic and a number of extensive disciplines are devoted to the study of this issue, they are Statistics, Probability and Stochastic Processes.

Observation 5.36.4 The Quantum Uncertainty in Measuring Precision

What is the greatest accuracy that one could ever hope to attain? In Quantum Mechanics the error in measuring position of an electron is Δx and the error in its momentum Δp the net precision between the two errors can only be $\Delta x \, \Delta p \geq h/2\pi$ or $\Delta x \, \Delta p \geq \hbar$ the normalized Plank's Constant. If measuring itself introduces a certain impreciseness or error, then as the precision increases that error must go toward zero, but $[\hbar]$ is in the way so the certainty of ever knowing the position of the electron can never truly be known, hence $|\mu_0|$ can never equal zero, likewise the procession of a measurement can never truly be known.

The reason for the uncertainty again is due to the instrument or method of measurement. Case in point for finding the position of an electron the situation is quite unique. We can only hope to "see" the electron if we reflect light, or another particle, from it. In this case the recoil that the electron experiences when the light (photon) bounces, by absorption and emission, from it completely alters the electron's motion in a way that cannot be avoided or even corrected for. Even if the smallest amount of photon energy could be used motion would still occur and the certainty of its position would still be in question.

Our inherent inability to describe the position and motion of the electron in a classical way finds expression in the above ***uncertainty principle***, enunciated by Werner Heisenberg in 1927. [HAL66 pg 1210] This idea of quantum uncertainty effects other physical measurements. The geometry of measurement is far more intricate than this treatises is prepared to go into so it is left to the reader to explore these other topics.

[5.36.1]Note: The exact differential is the resulting proportional error, δQ from the $\delta\mu$'s, see O5.22.2A, hence can only be an accumulative approximation from the other perimeters as well.

Section 5.37 Analytic Geometry

Definition 5.37.1 Analytic Geometry

1) Analytic geometry is the unification of geometry and algebra allowing for the development of analytic formulas to represent geometric objects.

2) Analytic Geometry is constructed on the foundation of Euclidian n-space.

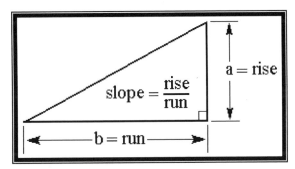

Figure 5.37.1 Slope of a Right Triangle

Definition 5.37.2 Slope of a Right Triangle

> **Slope of a Right Triangle** is the ratio of rise over run for a right triangle. See figure F5.25.1 "Slope".

$$slope \equiv rise / run \qquad \text{Equation A}$$
$$slope \equiv a / b \qquad \text{Equation B}$$

Definition 5.37.3 Slope of a Straight Line

If a right triangle's hypotenuse is collinear to a straight line, then the slope of the right triangle is the slope of the straight line.

Section 5.38 Analytic Geometry: Cartesian Coordinates in a Euclidian Plane

The manifold upon which analytic geometry is constructed, is an n-dimensional cartesian[2.16.2] space.

Axiom 5.38.1 A Cartesian or Euclidian Coordinate Space $\mathfrak{C}^n$

Let $\exists$ a set of n-tuple quantities $(\bullet\, x_i\, \bullet)$ called **coordinates** and $(\bullet A_i \bullet)$ called an **axial set**, than the set S_n is called a **Cartesian or Euclidian Coordinate Space or Simple n-Space**, having the following relationships and properties:

$$S^n \equiv \{\textstyle\prod_n \times A_i\} = \{(\bullet\, x_i\, \bullet) \mid x_i \in A_i,\ \text{for all } i = 1, 2, 3, \ldots n\}$$
cartesian coordinate space or **cartesian coordinate system**

A point in that field is represented by an order group of numbers or n-tuples,

$P(x_1, x_2, \ldots, x_i, \ldots, x_n)$ an n-tuple or simply for any i-listed quantities, abbreviated as $P_n(\bullet\, x_i\, \bullet)$ called the **coordinate or point of P_n**.

By its very definition of being an ordered set:

$A_i \times A_j \neq A_i \times A_j$, for any $i \neq j$, hence is non-commutative.

Where the black dots bounding the quantity $(\bullet\, x_i\, \bullet)$ of [i] means that it is understood there are n-elements as defined by the n subscript on the point and $i = 1, 2, 3, \ldots n$. Where [n] is not necessary to be stated because it is understood to be n-quantities unless otherwise specified.

Section 5.39 Magnitude in Cartesian Space

In the world of tensors distance is measured in a more classical manner with the Pythagoras's Theorem. So, Pythagoras will define magnitude and it will be read as such:

Definition 5.39.1 Pythagorean Magnitude

$$D = \sqrt{a^2 + b^2}$$

(see T5.8.8 "Right Triangle: The Hypotenuse")

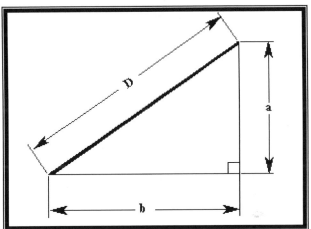

Figure 5.39.1 Pythagorean Magnitude

The next step is to build a general version of Pythagoras's Theorem in an n-space coordinate system.

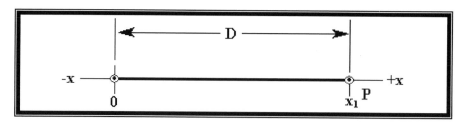

Figure 5.39.2 Magnitude D in 1-Dimension

Theorem 5.39.1 Pythagorean Theorem for Magnitude D in One Dimension

1g	Given	the above geometry of Figure F5.27.2 "Magnitude D in 1-Dimension"for $0 \le D$
2g		$OP = x_1$
Steps	Hypothesis	$D = \sqrt{x_1^2} = |x_1|$
1	From 1g, 2g and A5.26.1 A Cartesian or Euclidian Coordinate Space $\mathfrak{C}^1$	$D = \sqrt{x_1^2 + 0}$
2	From 1 and A4.2.7 Identity Add	$D = \sqrt{x_1^2}$
∴	From 2 and D4.1.11 Absolute Value	$D = \sqrt{x_1^2} = |x_1|$

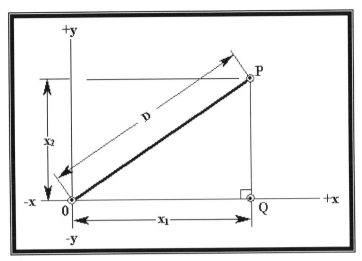

Figure 5.39.3 Magnitude D in 2-Dimensions

Theorem 5.39.2 Pythagorean Theorem for Magnitude D in Two Dimensions

1g	Given	the above geometry of Figure F5.27.3 "Magnitude D in 2-Dimensions" for $0 \le D$
2g		$OQ = x_1$
3g		$QP = x_2$
Steps	Hypothesis	$D = \sqrt{x_1^2 + x_2^2}$
1	From 1g, and D5.27.1 Pythagoras's Theorem for Magnitude; hypotenuse of ΔOPQ	$D = \sqrt{OQ^2 + QP^2}$
$\therefore$	From 2g, 3g, 1, A5.26.1 A Cartesian or Euclidian Coordinate Space $\mathbb{C}^2$ and A4.2.3 Substitution	$D = \sqrt{x_1^2 + x_2^2}$

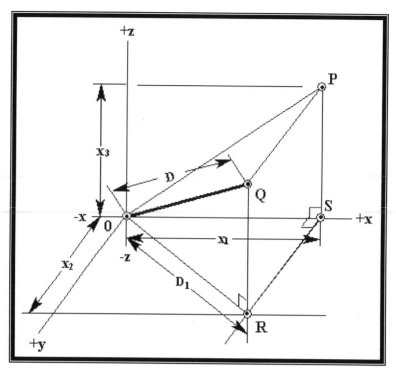

Figure 5.39.4 Magnitude D in 3-Dimensions

Theorem 5.39.3 Pythagorean Theorem for Magnitude D in Three Dimensions

1g	Given	the above geometry of Figure F5.27.4 "Magnitude D in 3-Dimensions" for $0 \le D$
2g		$OS = x_1$
3g		$SR = x_2$
4g		$OR = D_1$
5g		$RQ = x_3$
Steps	Hypothesis	$D = \sqrt{x_1{}^2 + x_2{}^2 + x_3{}^2}$
1	From 1g, ΔOSR and D8.2.1	$D_1 = \sqrt{OS^2 + SR^2}$
2	From 1, 2g, 3g and A4.2.3 Substitution	$D_1 = \sqrt{x_1{}^2 + x_2{}^2}$
3	From F8.2.4, ΔORQ and D8.2.1	$D = \sqrt{OR^2 + RQ^2}$
4	From 3, 4g, 5g and A4.2.3 Substitution	$D = \sqrt{D_1{}^2 + x_3{}^2}$
$\therefore$	From 2, 4, 1, A5.26.1 A Cartesian or Euclidian Coordinate Space $\mathbb{C}^3$ and A4.2.3 Substitution	$D = \sqrt{x_1{}^2 + x_2{}^2 + x_3{}^2}$

There is a clear pattern of building one dimension upon another from right triangle to right triangle. So a count from 1 to 3-dimensions has been established as correct. Let's assume that this dimensional progress continues for D_i and on to D_{i+1} as follows:

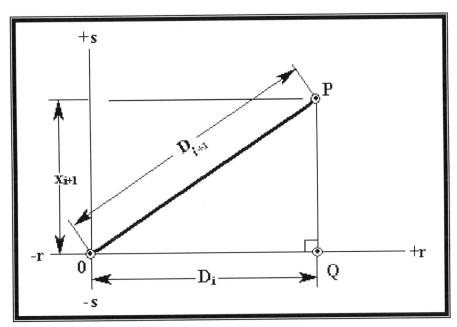

Figure 5.39.5 Magnitude D in (i + 1)-Dimensions

Theorem 5.39.4 Pythagorean Theorem for Magnitude D in n-Dimensions

1g	Given	the above geometry of Figure F5.27.5 "Magnitude D in (i + 1)-Dimensions" for $0 \le D$
2g		$OQ^2 = D_i^2 = \sum_i x_j^2$
3g		$QP = x_{i+1}$
Steps	Hypothesis	$D_n = \sqrt{\sum_n x_i^2}$
1	From 1g; $\triangle OQP$	$D_{i+1} = \sqrt{OQ^2 + QP^2}$
2	From 1, 2g, 3g, A5.26.1 A Cartesian or Euclidian Coordinate Space $\mathfrak{C}^n$ and A4.2.3 Substitution	$D_{i+1} = \sqrt{x_{i+1}^2 + D_i^2}$
3	From 2, 2g and A4.2.3 Substitution	$D_{i+1} = \sqrt{x_{i+1}^2 + \sum_i x_j^2}$
4	From 3 and summing over j to i+1	$D_{i+1} = \sqrt{\sum_{i+1} x_j^2}$
$\therefore$	From 4 and A3.11.1 Induction Property of Integers	$D_n = \sqrt{\sum_n x_i^2}$

Section 5.40 Three Point Straight Line

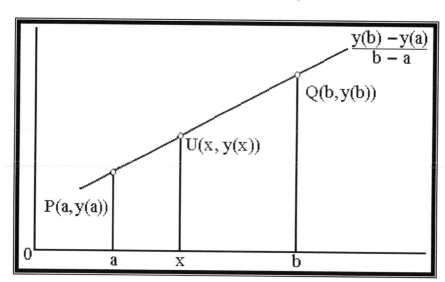

Figure 5.40.1 Three Point Straight Line

Theorem 5.40.1 Constant Slope of the Three Point Straight Line

1g 2g	Given	the above geometry, F5.28.2 "Three Point Straight Line" [x] a variable between $a \le x \le b$
Steps	Hypothesis	$[y(x) - y(a)] / (x - a) = [y(b) - y(a)] / (b - a)$
1	From 1g	rise PU = $y(x) - y(a)$
2	From 1g	run PU = $x - a$
3	From 1, 2, D5.25.2 Slope of a Straight line; between points PU and A4.2.3 Substitution	$[y(x) - y(a)] / (x - a)$ = slope between two points PU
4	From 1g	rise = $y(b) - y(a)$
5	From 1g	run = $b - a$
6	From 4, 5, D5.25.2 Slope of a Straight line; between points UQ and A4.2.3 Substitution	slope between two points UQ = $[y(b) - y(a)] / (b - a)$
7	From 1g, 2g and A4.2.2 Equality	slope between two points PU = slope between two points UQ
∴	From 3, 6, 7 and A4.2.3 Substitution	$[y(x) - y(a)] / (x - a) = [y(b) - y(a)] / (b - a)$

Observation 5.40.1 In the Interval the Three Point Straight Line has an Infinite Number of Slopes

In theorem T5.28.1 "Constant Slope of the Three Point Straight Line", [x] is a variable between $a \le x \le b$ and as such can freely move between [a, b], since the slope is the same for any value of [x] then all infinite slopes must have the same in the interval between [a, b];

$$slope = [y(b) - y(a)] / (b - a) \qquad \text{Equation A}$$

Section 5.41 Theorems on Straight Hyperline Manifolds

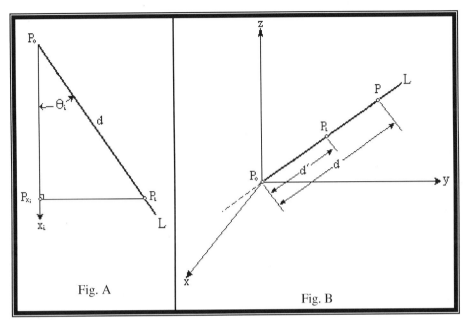

Figure 5.41.1 Three Points Lying on a Line

Theorem 5.41.1 Direction Cosines of L

1g	Given	the above geometry from figure F5.41.1A "Three Points Lying on a Line"
2g		the above geometry from figure F5.41.1B "Three Points Lying on a Line"
3g		A set of angular parameters ($\bullet \theta_i \bullet$ for all i) referenced to the directed line-L
4g		$g_{\alpha i}{}^n \equiv \{ \ \Delta P_L P_0 P_{xi}$ a right triangle for $P_L(\bullet x_{iL} \bullet), P_0(\bullet x_{i0} \bullet), P_{xi}(\bullet x_{(i-1)0}, x_i, x_{(i+1)0} \bullet) \}$
Step	Hypothesis	$\cos(\theta_i) = (x_i - x_{i0}) / d$ for all i between points $P_0 P_{xi}$
1	From 1g, 4g and D4.1.16C Distance or Magnitude on a Number Line	$|P_0 P_{xi}| = x_i - x_{i0}$ for $x_{i0} < x_i$
2	From 1g, 3g and DxE.1.2.2	$\cos(\theta_i) = |P_0 P_{xi}| / d$
3	From 1, 2 and A4.2.3 Substitution	$\cos(\theta_i) = (x_i - x_{i0}) / d$ is true for $g_{\alpha i}{}^n$
4	From 3 and i = 1	$\cos(\theta_1) = (x_1 - x_{1(0)}) / d$ is true for $g_{\alpha i}{}^n$
$\therefore$	From 3, 4 and A3.12.1 Equivalency Property of Constructible Geometric Components	$\cos(\theta_i) = (x_i - x_{i0}) / d$ for all i between points $P_0 P_{xi}$

Theorem 5.41.2 Two Point Parametric Line-L

1g	Given	the above geometry from figure F5.41.1A "Three Points Lying on a Line"
2g		A set of angular parameters $(\bullet, \theta_i, \bullet$ for all i) referenced to the directed line-L
3g		$g_{\alpha\mathcal{L}^n} \equiv \{$ $P_L P_0 P$ three collinear points on a line-L $P_L(\bullet\, x_{iL}\, \bullet),\ P_0(\bullet\, x_{i0}\, \bullet),\ P(\bullet x_i\, \bullet)\}$
4g		$v_m\, t \equiv d\,/\,d'$ conversion to parametric variable with v_m the ***frequency of parametric transformation*** having dimension $[T^{-1}]$ [5.41.1]
Step	Hypothesis	$v_m\, t = (x_i - x_{i0})\,/\,(x_{iL} - x_{i0})$ for all i
1	From 1g, 2g and T9.3.1 Direction Cosines of L	$\cos(\theta_i) = (x_i - x_{i0})\,/\,d$ for $g_{\alpha\mathcal{L}^n}$ and all i
2	From 1g, 2g and T9.3.1 Direction Cosines of L	$\cos(\theta_i) = (x_{iL} - x_{i0})\,/\,d'$ for $g_{\alpha\mathcal{L}^n}$ and all i
3	From 1, 2 and A4.2.3 Substitution	$\cos(\theta_i) = (x_i - x_{i0})\,/\,d = (x_{iL} - x_{i0})\,/\,d'$
4	From 3, T4.4.8 Equalities: Cross Product of Proportions and D4.1.35 Constant of Proportional Quantities	$(x_i - x_{i0})\,/\,(x_{iL} - x_{i0}) = d\,/\,d'$ constant of proportionality for all i
∴	From 4g, 4 and A4.2.3 Substitution	$v_m\, t = (x_i - x_{i0})\,/\,(x_{iL} - x_{i0})$ for all i

Theorem 5.41.3 Two Point Parametric Equation of Slope for a Line-L

1g	Given	$m_i \equiv [v_m\,(x_{iL} - x_{i0})]$ with dimensions $[L][T^{-1}]$	
Step	Hypothesis	$x_i = x_{i0} + m_i\, t$ EQ A	$(x_i - x_{i0})\,/\,m_i = t$ EQ B
1	From T5.29.2 Two Point Parametric Line-L	$(x_i - x_{i0})\,/\,(x_{iL} - x_{i0}) = v_m\, t$ for all i	
2	From 1 and T4.4.7B Equalities: Reversal of Left Cancellation by Multiplication	$x_i - x_{i0} = t\,[v_m\,(x_{iL} - x_{i0})]$	
3	From 1g, 2 and A4.2.3 Substitution;	$x_i - x_{i0} = t\, m_i$	
4	From 3 and T4.4.4 Equalities: Right Cancellation by Multiplication; independent of [i]	$(x_i - x_{i0})\,/\,m_i = t$	
5	From 1g, 3, A4.2.3 Substitution, A4.2.10 Commutative Multp and T4.3.6A Equalities: Reversal of Left Cancellation by Addition; for all [i]	$x_i = m_i\, t + x_{i0}$	
∴	From 3 and A4.2.5 Commutative Add	$x_i = x_{i0} + m_i\, t$ EQ A	$(x_i - x_{i0})\,/\,m_i = t$ EQ B

Theorem 5.41.4 Two Point Parametric Equations and their Directed Numbers

Step	Hypothesis	$x_i = x_{i0} + m_i t$ all i [5.41.2] EQ A	$t = (x_i - x_{i0}) / m_i$ all i EQ B [5.41.3]
			$m_i = (x_i - x_{i0}) / t$ all i EQ C
1	From T5.29.3A Two Point Parametric Equation of Slope for a Line-L	$x_i = x_{i0} + m_i t$ for all i	
2	From 1, A4.2.25 Commutative Add and T4.3.3A Equalities: Right Cancellation by Addition	$x_i - x_{i0} = m_i t$	
∴	From 1, 2 and T4.4.4A Equalities: Right Cancellation by Multiplication	$x_i = x_{i0} + m_i t$ all i EQ A	$t = (x_i - x_{i0}) / m_i$ all i EQ B
			$m_i = (x_i - x_{i0}) / t$ all i EQ C

Theorem 5.41.5 Directed Numbers are Slopes for Parametric Line-L

Step	Hypothesis	$x'_i = m_i$ for all i [5.41.4]
1	From T5.29.4A Two Point Parametric Equations and their Directed Numbers	$x_{1i} = x_{i0} + m_i t_1$ for all i and point $P_1(\bullet x_{1i} \bullet)$ at t_1 on line-L
2	From T5.29.4A Two Point Parametric Equations and their Directed Numbers	$x_{2i} = x_{i0} + m_i t_2$ for all i and point $P_2(\bullet x_{2i} \bullet)$ at t on line-L
3	From 1, 2 and taking the difference	$x_{2i} - x_{1i} = x_{i0} + m_i t_2 - (x_{i0} + m_i t_1)$
4	From 3, D4.1.20 Negative Coefficient, A4.2.14 Distribution and A4.2.5 Commutative Add	$x_{2i} - x_{1i} = x_{i0} - x_{i0} + m_i t_2 - m_i t_1$
5	From 4, A4.2.8 Inverse Add, A4.2.7 Identity Add and A4.2.14 Distribution	$x_{2i} - x_{1i} = m_i (t_2 - t_1)$
6	From 5 and DxK.3.1.1 Delta: Increment by Difference	$\Delta x_i = m_i \Delta t$
7	From 6 and T4.4.4A Equalities: Right Cancellation by Multiplication	$\Delta x_i / \Delta t = m_i$
8	From 7 and as P_1 is selected closer and closer toward P_2 then in the limit $\Delta t \to 0$	$\lim_{\Delta t \to 0} \dfrac{\Delta x_i}{\Delta t} = m_i$
∴	From 8 and DxK.3.1.3 Definition of a Derivative	$x'_i = m_i$ for all i

Theorem 5.41.6 Bounded Parametric Magnitude Between Two Points

1g	Given	$d^2 \equiv \sum^n [\, x_i(t) - a_i(t)\,]^2$
2g	for	$0 < d$
3g		$|\, X(b) - A(a)\,| \equiv \sum^n [\, x_i(t) - a_i(t)\,]^2$ in the interval I_t
Steps	Hypothesis	$0 < |\, X(b) - A(a)\,| / \sqrt{n} < k$
		in the normalized interval δI_t [5.41.5]
1	From A11.6.1P3 Parametric Hypercurve Manifold $\mathfrak{P}^n$	$|x_i(t) - a_i(t)| < k$
2	From 1, A4.2.17 Correspondence of Equality and Inequality, T4.6.2 Equalities: Square of an Absolute Value and T4.8.6 Integer Exponents: Uniqueness of Exponents	$[\, x_i(t) - a_i(t)\,]^2 < k^2$
3	From 2 and summing over both sides of the inequality for all coordinate [i]	$\sum^n [\, x_i(t) - a_i(t)\,]^2 < \sum^n k^2$
4	From 3, T4.3.9 Equalities: Summation of Repeated Terms	$\sum^n [\, x_i(t) - a_i(t)\,]^2 < nk^2$
5	From 1g, 4 and A4.2.3 Substitution	$d^2 < nk^2$
6	From 5 and T4.10.4 Radicals: Identity Radical Raised to a Power	$d^2 < (\sqrt{n})^2 k^2$
7	From 6, A4.2.21 Distribution and A4.2.10 Commutative Multp	$d^2 < (\, k\, \sqrt{n}\,)^2$
8	From 2g, 7, T4.8.6 Integer Exponents: Uniqueness of Exponents, A4.2.16 The Trichotomy Law of Ordered Numbers	$0 < d < k\sqrt{n}$
9	From 1g, 2g, T4.10.1 Radicals: Uniqueness of Radicals and T4.10.3 Radicals: Identity Power Raised to a Radical	$d = \sqrt{\sum^n [\, x_i(t) - a_i(t)\,]^2}$ for $0 < d$
10	From 8, 9 and A4.2.3 Substitution	$0 < \sqrt{\sum^n [\, x_i(t) - a_i(t)\,]^2} < k\sqrt{n}$
11	From 3g, 10 and A4.2.3 Substitution	$0 < |\, X(b) - A(a)\,| < k\sqrt{n}$
∴	From 11, A4.2.12 Identity Multp, and T4.7.31 Inequalities: Cross Multiplication with Positive Reciprocal Product	$0 < |\, X(b) - A(a)\,| / \sqrt{n} < k$ in the normalized interval δI_t

[5.41.1]Note: The dimension [T] in this case does not necessarily mean time, but any parametric dimension as specified.

[5.41.2]Note: Clearly from the inverse equations of T5.41.4B the m_i's are the directed numbers from Definition 11.6.4 "Directed Numbers".

[5.41.3]Note: From T5.41.4B "Two Point Parametric Equations and their Directed Numbers", gives rise to the possibility of dividing by zero, because it is permissible to have a line lying horizontal to the base axis with a slope of zero. Since this is a geometric real possibility, the singularity is removed by not dividing, but it is understood to mean the slope is horizontal.

[5.41.4]Note: It is seen that the directed numbers are nothing more than the derivative or slope for the variable $[x_i]$ at an instant of the parametric parameter [t].

[5.41.5]Note: $(\, |\, X(b) - A(a)\,| / \sqrt{n} < k\,) \rightarrow (\, |t - \alpha| < \kappa\,)$ within the interval δI_t so it can now be said that the magnitude is continuous within the interval [a, b], hence differentiable.

Tensor Calculus & Physics: A General Treatise

Table of Contents

Tensor Calculus & Physics: A General Treatise

List of Tables

List of Observations

Chapter 6 Matrix Algebra

This chapter takes those parts of Linear Algebra that lay down the key definitions and theorems that form the foundation for tensors. Other theorems will be drawn upon but they will not have the immediate importance of the theorems in this section, therefore they will not be proven but just refereed to from S. Lipschutz book on *Linear Algebra* [LIP68].

Section 6.1 Matrix Definitions

When n-elements of the matrix span a real R or complex number field C^2 the elements will be indexed to mark their position:

$A_n = (a_i)$ for i = 1, 2, …, n a row vector following ***covariant notation*** for lower index.

or

$B^m = (b^j)$ for j = 1, 2, …, m a column vector following ***contavariant notation*** for upper index.

Since this symbolism is confusing as to what is a row or column matrix vector superscripts will be used to represent rows and subscripts columns:

Definition 6.1.1 Vector Matrix Notation Row and Column

$A_n \equiv (a_i)$ for i = 1, 2, …, n   a row   vector a (n x 1) matrix

or

$B^m \equiv (b^j)$ for j = 1, 2, …, m   a column vector a (1 x m) matrix

This follows the contra and covariant tensor notation, which will be discussed later.

Definition 6.1.2 Rectangular Matrix

Is an n-row by m-column (n x m) array of elements. It is represented as follows:

$$A_n^{\ m} \equiv (a_i^{\ j}) \text{ for i = 1, 2, …, n row and j = 1, 2, …, m column position.}$$

Where (n x m) is called the ***order*** of the rows and columns of the matrix or simply the order of the matrix. It is interesting to note that different authors use different terminology the most common are to read them straight as **n-by-m**, or the numbers as a pair the *size* or *shape* of the matrix.

Definition 6.1.3 Square Matrix

Is an (n x n) matrix having equal number of rows and columns. It is represented as follows:

$A_n^{\ n} \equiv A$ where the super and subscripts are dropped since there the same. Referred to as an ***order-n*** square matrix.

Definition 6.1.4 Zero Matrix

Is where all elements are set to zero (0). It is represented as follows $0 \equiv (0)$.

Definition 6.1.5 Diagonal Matrix

Is a square matrix with all elements set to zero except the diagonal elements. It is represented as follows:

$$\text{Diagonal}(d) \equiv \text{Diag }(d) \equiv (d_i^{\;j}) \equiv (0\backslash d^i\backslash 0) \text{ for i = 1, 2, ..., n}$$

where

$$d_i^{\;j} \equiv \begin{cases} d_i \text{ or } d^j \text{ for } i = j \\ 0 \qquad\quad \text{for } i \neq j \end{cases} \qquad \text{Equation A}$$

$$d_i^{\;j} \equiv \delta_i^{\;j} d^j \qquad\qquad \text{Equation B}$$

$$d_i^{\;j} \equiv \delta_i^{\;j} d_i \qquad\qquad \text{Equation C}$$

Diagonal(d) is read **Diagonal of d** as a function operating on a row or column vector d.

$$d = \begin{pmatrix} d^1 \\ d^2 \\ \vdots \\ d^j \\ \vdots \\ d^n \end{pmatrix} \text{ or } (d_1 \; d_2 \; ... \; d_i \; ... \; d_n)$$

Definition 6.1.6 Identity Matrix

Is a diagonal matrix where the diagonal elements are set to one (1). It is represented as follows:

$$I \equiv (\delta_i^{\;j}) \equiv (0\backslash 1\backslash 0) \qquad\qquad\qquad\qquad\qquad \text{Equation A}$$

where

$$\delta_i^{\;j} \equiv \begin{cases} 1 \text{ for } i = j \\ 0 \text{ for } i \neq j \end{cases} \qquad\qquad\qquad\qquad \text{Equation B}$$

and $\delta_i^{\;j}$ is called the **Kronecker Delta**.

Definition 6.1.7 Lower Triangular Matrix

Is a matrix with elements above the diagonal zeroed.

$$T_L = (t_i^{\;j}) \text{ for } t_i^{\;j} = 0 \text{ and } i < j \qquad \text{Equation A}$$
or
$$T_L = (t_j^{\;i}) \text{ for } t_j^{\;i} = 0 \text{ and } i > j \qquad \text{Equation B relative to a } T_U \text{ matrix}$$

Definition 6.1.8 Upper Triangular Matrix

Is a matrix with elements low the diagonal zeroed.

$$T_U = (t_i^{\;j}) \text{ for } t_i^{\;j} = 0 \text{ and } i > j \qquad \text{Equation A}$$
or
$$T_U = (t_j^{\;i}) \text{ for } t_j^{\;i} = 0 \text{ and } i < j \qquad \text{Equation B relative to a } T_L \text{ matrix}$$

Definition 6.1.9 Lower Identity Triangule Matrix

Is a triangular matrix where the lower elements are set to one (1). It is represented as follows:

$$I_L \equiv (\tau_{Li}{}^j) \equiv (0\backslash0\backslash1)$$

where

$$\tau_{Li}{}^j \equiv \begin{cases} 0 \text{ for } i \geq j \\ 1 \text{ for } i < j \end{cases}$$

and $\tau_{Li}{}^j$ is called the **Lower Tau Matrix**.

Definition 6.1.10 Upper Identity Triangule Matrix

Is a triangular matrix where the upper elements are set to one (1). It is represented as follows:

$$I_U \equiv (\tau_{Ui}{}^j) \equiv (1\backslash0\backslash0)$$

where

$$\tau_{Ui}{}^j \equiv \begin{cases} 1 \text{ for } i < j \\ 0 \text{ for } i \geq j \end{cases}$$

and $\tau_{Ui}{}^j$ is called the **Upper Tau Matrix**.

Definition 6.1.11 Elementary Matrices

Have unique properties that can be used to construct general matrices, such matrices are based on either an upper or lower Triangular Matrix, represented generally as T:

$$T \in {}_\varepsilon E \quad ({}_\varepsilon E \text{ being the set of elementary matrices}).$$

An Elementary Matrix, $E \equiv (e_i{}^j)$, is any Triangular Matrix that has been or not been modified in the following way:

a)   Any two rows or columns have been interchanged, $\qquad T \times (R_i \rightarrow R_r \text{ or } C^j \rightarrow C^s)$.

b)   Any row or column has been multiplied by a nonzero constant, $kT \times (R_i \text{ or } C^j)$.

c)   Any row or column has been multiplied by a nonzero constant and then added to another row or column, $\qquad kT \times (R_i \text{ or } C^j) + T \times (R_r \text{ or } C^s)$

$$\text{for } i \neq r \text{ or } j \neq s \text{ necessarily.}$$

These actions are calculated operations so that when the determinate of an elementary matrix is taken simple results occur, given E an elementary matrix and a nonzero constant [k] that comprises it, then:

$$|E| = \pm k \, \Pi e_i{}^i$$

Where the plus and minus sign reflect any change of rows or columns. Of course, any operation such as (c) leave the determinate unaltered.

Definition 6.1.12 Matrix Addition Expediential Notation

$$(A_n{}^m)^{+p+q} \equiv (A_n{}^m)^{+p} (A_n{}^m)^{+q} \text{ Equation A}$$
$$(A_n{}^m)^{-q+p} \equiv (A_n{}^m)^{+q} (A_n{}^m)^{+p} \text{ Equation B}$$

Definition 6.1.13 Matrix Subtraction Expediential Notation

$$(A_n{}^m)^{+p-q} \equiv (A_n{}^m)^{+p} (A_n{}^m)^{-q} \quad \text{Equation A}$$
$$(A_n{}^m)^{-q+p} \equiv (A_n{}^m)^{-q} (A_n{}^m)^{+p} \quad \text{Equation B}$$

<h2 style="text-align:center">Section 6.2 Matrix Arithmetic</h2>

Axiom 6.2.1 Closure Addition

$[(A_n{}^m + B_n{}^m \wedge B_n{}^m + A_n{}^m) \wedge (A_n{}^m - B_n{}^m \wedge B_n{}^m - A_n{}^m)] \in M$
Closure with respect to Addition and Subtraction

Axiom 6.2.2 Closure Multiplication

$[(A_n{}^m B_p{}^q \wedge B_p{}^q A_n{}^m)] \in M$
Closure with respect to Multiplication

Axiom 6.2.3 Equality of Vectors

$C_n \equiv A_n$	for i = 1, 2, …, n a row vector	Equation A
$A^m \equiv C^m$	for j = 1, 2, …, m a column vector	Equation B
$C^m \equiv A^m$	for j = 1, 2, …, m a column vector	Equation C
$A_n \equiv C_n$	for i = 1, 2, …, n a row vector	Equation D

Axiom 6.2.4 Vector Addition

$C_n \equiv A_n + B_n \equiv (a_i) + (b_i) \equiv (a_i + b_i) \equiv (c_i)$ for i = 1, 2, …, n a row vector Equation A
or
$C^m \equiv A^m + B^m \equiv (a^j) + (b^j) \equiv (a^j + b^j) \equiv (c^j)$ for j = 1, 2, …, m a column vector Equation B

Axiom 6.2.5 Scalar-Vector Multiplication

$k * A_n \equiv (ka_i)$ for i = 1, 2, …, n a row vector   Equation A
or
$k * B^m \equiv (kb^j)$ for j = 1, 2, …, m a column vector Equation B
for k a scalar number.

Axiom 6.2.6 Column Vector Multiplication

$a \equiv A_{vn} * B_v{}^n = \sum_{k=1}^{n} a_k b^k$ (1 x n) = (1 x n)(n x 1) column to row
Now let there be n-column vectors multiplying one row vector such that:
$C^n{}_n = A_v{}^n * B_{vn} = (a^i b_j)$ (n x n) = (n x 1)(1 x n) column to row
In general matrix multiplication is defined as follows in A6.2.10 "Matrix Multiplication".

Axiom 6.2.7 Equality of Matrices

$C_n{}^m \equiv A_n{}^m$ for order conserved (n x m)   Equation A
$A_n{}^m \equiv C_n{}^m$ for order conserved (n x m)   Equation B

Axiom 6.2.8 Matrix Addition

$C_n{}^m \equiv A_p{}^m + B_n{}^p$ Equation A
$(c_i{}^j) \equiv (a_k{}^j) + (b_i{}^k)$ for order conserved (n x m) = (n x m) + (n x m) Equation B
where
$c_i{}^j \equiv a_i{}^j + b_i{}^j$ for order conserved (n x m) = (n x m) + (n x m) Equation C

Axiom 6.2.9 Matrix Multiplication by a Scalar

In order to be constant with scalar multiplication of vectors scalar multiplication of matrices is as follows:

$$k * A_p^{\ m} \equiv (k \ a_k^{\ j}) \qquad\qquad \text{Equation A}$$
$$A_p^{\ m} * k \equiv (a_k^{\ j} \ k) \qquad\qquad \text{Equation B}$$

for all i and j (n x m) and for k a scalar number.

Axiom 6.2.10 Matrix Multiplication

$$C_n^{\ m} \equiv A_p^{\ m} * B_n^{\ p} \qquad\qquad\qquad\qquad\qquad \text{Equation A}$$
$$(c_i^{\ j}) \equiv (a_k^{\ j}) * (b_i^{\ k}) \qquad \text{for } (n \times m) = (p \times m) * (n \times p) \qquad \text{Equation B}$$

where

$$c_i^{\ j} \equiv \sum_{k=1}^{p} a_i^{\ k} b_k^{\ j} \qquad \text{for } (n \times m) = (p \times m) * (n \times p) \qquad \text{Equation C}$$

Axiom 6.2.11 Matrix Transposition

Transposition of a matrix transposes rows for columns.

$$^tA_n^{\ m} \equiv {}^t(a_i^{\ j}) \equiv (a_j^{\ i}) \equiv A_m^{\ n} \qquad\qquad \text{for } {}^t(n \times m) = (m \times n)$$

Transposition has the effect of transposing the elements of the matrix about its diagonal.

Axiom 6.2.12 Matrix Decomposition into Elementary Matrices

Any given square matrix that is invertible, | A | ≠ 0 otherwise [A] independent, can be broken down into a set of elementary matrices.

$$A \equiv \prod E^i \qquad \text{for } E^i \in {}_\varepsilon E \text{ for all i such that } | A | \neq 0.$$

The above property of a matrix is represented as an axiom to prevent a recursive argument; however there is a proof for this in Lipschutz [LIP68] but it has the potential in this system to become circular.

Section 6.3 Properties of Matrix Operations Under Addition and Multiplication

To have a true arithmetic of matrices the uniqueness of transposition, addition and multiplication is established. That is the balance of quantities on either side of the equality must be maintained to keep the logic consistent. What one does to one side of the equality then the something must be done to the other side to maintain balance.

Theorem 6.3.1 Substitution of Matrices

1g	Given	Matrices $A_n{}^m$, $B_n{}^m$, $C_n{}^m \in (R^n$ or $C^n)$
2g		For $A_n{}^m \equiv B_n{}^m = (r_i^j)$
3g		$B_n{}^m \equiv C_n{}^m = (s_i^j)$
Steps	Hypothesis	$A_n{}^m \equiv B_n{}^m$
		$\underline{B_n{}^m \equiv C_n{}^m}$
		$\therefore A_n{}^m \equiv C_n{}^m$
1	From 1g and A6.2.7 Equality of Matrices	$B_n{}^m = B_n{}^m$
2	From 1 by 2g and 3g	$(r_i^j) = (s_i^j)$ element by element
$\therefore$	From 2 by 2g and 3g	$A_n{}^m = C_n{}^m$

Theorem 6.3.2 Uniqueness of Matrix Addition

What one does to one side of the equality then the something must be done to the other side to maintain balance.

1g	Given	Matrices $A_n{}^m$, $B_n{}^m$, $C_n{}^m$, $D_n{}^m \in (R^n$ or $C^n)$	
2g		For $A_n{}^m \equiv B_n{}^m = (r_i^j)$	
3g		$C_n{}^m \equiv D_n{}^m = (s_i^j)$	
4g		$q_i^j \equiv r_i^j + s_i^j$	
Steps	Hypothesis	$A_n{}^m \equiv B_n{}^m$	
		$\underline{+ C_n{}^m \equiv D_n{}^m}$	
		$\therefore A_n{}^m + C_n{}^m \equiv B_n{}^m + D_n{}^m$	
1	From 2g and 3g	$A_n{}^m + C_n{}^m$	$B_n{}^m + D_n{}^m$
2	From 1 by 2g and 3g	$(r_i^j) + (s_i^j)$	$(r_i^j) + (s_i^j)$
3	From 2 by A6.2.8 Matrix Addition	$(r_i^j + s_i^j)$	$(r_i^j + s_i^j)$
4	From 3 by 4g	(q_i^j)	(q_i^j)
$\therefore$	From 4 and A6.2.7 Equality of Matrices	$A_n{}^m \equiv B_n{}^m$ $\underline{+ C_n{}^m \equiv D_n{}^m}$ $\therefore A_n{}^m + C_n{}^m \equiv B_n{}^m + D_n{}^m$	

Theorem 6.3.3 Uniqueness of Scalar Multiplication

What one does to one side of the equality then the something must be done to the other side to maintain balance.

1g 2g 3g	Given	Matrices $A_n^m, B_n^m \in (R^n$ or $C^n)$ For $A_n^m \equiv B_n^m = (r_i^j)$ $k = k$ a scalar number.
Steps	Hypothesis	$k * A_n^m \equiv k * B_n^m$
1	From 2g	$A_n^m = B_n^m$
2	From 1 and by 2g	$(r_i^j) = (r_i^j)$
3	From 2, 3g and T4.4.2 Equalities: Uniqueness of Multiplication	$k * (r_i^j) = k * (r_i^j)$
4	From 3 by A6.2.9 Matrix Multiplication by a Scalar	$(k * r_i^j) = (k * r_i^j)$ element by element
$\therefore$	From 4 the equality left unaltered.	$k * A_n^m \equiv k * B_n^m$

Theorem 6.3.4 Uniqueness of Matrix Multiplication

What one does to one side of the equality then the something must be done to the other side to maintain balance.

1g 2g 3g 4g	Given	Matrices $A_n^m, B_n^m, C_n^m, D_n^m \in (R^n$ or $C^n)$ For $A_n^m \equiv B_n^m = (r_i^j)$ $C_n^m \equiv D_n^m = (s_i^j)$ $q_i^j \equiv \sum_{k=1}^{p} r_k^j * s_i^k$
Steps	Hypothesis	$A_n^m \equiv C_n^m$ $\underline{* C_n^m \equiv D_n^m}$ $\therefore A_n^m * C_n^m \equiv C_n^m * D_n^m$
1	From 2g and 3g	$A_n^m * C_n^m$ $\qquad$ $B_n^m * D_n^m$
2	From 1 by 2g and 3g	$(r_j^i) * (s_i^j)$ $\qquad$ $(r_j^i) * (s_i^j)$
3	From 2 by A6.2.10 Matrix Multiplication	$(\sum_{k=1}^{p} r_k^j * s_i^k)$ $\qquad$ $(\sum_{k=1}^{p} r_k^j * s_i^k)$
4	From 3 by 4g	(q_i^j) $\qquad$ (q_i^j)
$\therefore$	From 4 and A6.2.7 Equality of Matrices	$A_n^m \equiv B_n^m$ $\underline{* C_n^m \equiv D_n^m}$ $\therefore A_n^m * C_n^m \equiv B_n^m * D_n^m$

Theorem 6.3.5 Matrix Association for Addition

Steps	Hypothesis	$(A_n^m + B_n^m) + C_n^m$	$\equiv$	$A_n^m + (B_n^m + C_n^m)$
1	From hypothesis by D6.1.2 Rectangular Matrix	$((a_i^j) + (b_i^j)) + (c_i^j)$	$=$	$(a_i^j) + ((b_i^j) + (c_i^j))$
2	From 1 by A6.2.8 Matrix Addition	$(a_i^j + b_i^j) + (c_i^j)$	$=$	$(a_i^j) + (b_i^j + c_i^j)$
3	From 2 by A6.2.8 Matrix Addition	$(a_i^j + b_i^j + c_i^j)$	$=$	$(a_i^j + b_i^j + c_i^j)$
$\therefore$	From 4 and A6.2.7 Equality of Matrices	$(A_n^m + B_n^m) + C_n^m$	$\equiv$	$A_n^m + (B_n^m + C_n^m)$

Theorem 6.3.6 Matrix Identity for Addition

Steps	Hypothesis	$A_n^m + \mathbf{0}$	$\equiv$	A_n^m
1	From hypothesis by D6.1.2 Rectangular Matrix	$(a_i^j) + (0)$	$=$	(a_i^j)
2	From 1 by A6.2.8 Matrix Addition	$(a_i^j + 0)$	$=$	(a_i^j)
3	From 2 and A4.2.9 Identity Add	(a_i^j)	$=$	(a_i^j)
$\therefore$	From 1, 3 and A6.2.7 Equality of Matrices	$A_n^m + \mathbf{0}$	$\equiv$	A_n^m

Theorem 6.3.7 Matrix Inverse for Addition

Steps	Hypothesis	$A_n^m + (-A_n^m)$	$\equiv$	$\mathbf{0}$
1	From hypothesis by D6.1.2 Rectangular Matrix	$(a_i^j) + (-1 * (a_i^j)) =$		(0)
2	From 1 by A6.2.9 Matrix Multiplication by a Scalar	$(a_i^j) + (-a_i^j)$	$=$	(0)
3	From 2 by A6.2.8 Matrix Addition	$(a_i^j - a_i^j)$	$=$	(0)
4	From 3 by A4.2.10 Inverse Add	(0)	$=$	(0)
$\therefore$	From 1, 4 and A6.2.7 Equality of Matrices	$A_n^m + (-A_n^m)$	$\equiv$	$\mathbf{0}$

Theorem 6.3.8 Matrix Commutative for Addition

Steps	Hypothesis	$A_n^m + B_n^m$	$\equiv$	$B_n^m + A_n^m$
1	From hypothesis by D6.1.2 Rectangular Matrix	$(a_i^j) + (b_i^j)$	$=$	$(b_i^j) + (a_i^j)$
2	From 1 by A6.2.8 Matrix Addition	$(a_i^j + b_i^j)$	$=$	$(b_i^j + a_i^j)$
3	From 2 by A4.2.7 Commutative Add	$(b_i^j + a_i^j)$	$=$	$(b_i^j + a_i^j)$
$\therefore$	From 3 by A6.2.7 Equality of Matrices	$A_n^m + B_n^m$	$\equiv$	$B_n^m + A_n^m$

Theorem 6.3.9 Distribution of a Scalar for Matrix Addition

1g 2g	Give	Matrices A_n^m, $B_n^m \in (R^n$ or $C^n)$ k a scalar number		
Steps	Hypothesis	$k * (A_n^m + B_n^m) \equiv$		$k * B_m^n + k * A_n^m$
1	From Hypothesis, (1 & 2)g and D6.1.2 Rectangular Matrix	$k * ((a_i^j) + (b_i^j))$	$=$	$k * (b_i^j) + k * (a_i^j)$
2	From 1 by A6.2.9 Matrix Multiplication by a Scalar	$k * ((a_i^j) + (b_i^j))$	$=$	$(k * b_i^j) + (k * a_i^j)$
3	From 2 by A6.2.8 Matrix Addition	$k * (a_i^j + b_i^j)$	$=$	$(k * b_i^j + k * a_i^j)$
4	From 3 by A6.2.9 Matrix Multiplication by a Scalar	$(k * (a_i^j + b_i^j))$	$=$	$(k * b_i^j + k * a_i^j)$
5	From 4 by A4.2.14 Distribution	$(k * a_i^j + k * b_i^j)$	$=$	$(k * b_i^j + k * a_i^j)$
$\therefore$	From 5 by A6.2.7 Equality of Matrices	$k * (A_n^m + B_m^n) \equiv$		$k * B_m^n + k * A_n^m$

Theorem 6.3.10 Matrix Distribution for Scalar Addition

1g 2g	Give	Matrices $A_n^{\,m}, B_n^{\,m} \in (R^n$ or $C^n)$ $k1$ and $k2$ a scalar numbers	
Steps	Hypothesis	$(k1 + k2) * A_n^{\,m} \equiv$	$k1 * A_m^{\,n} + k2 * A_n^{\,m}$
1	From hypothesis, (1&2)g and D6.1.2 Rectangular Matrix	$(k1 + k2) * (a_i^{\,j}) =$	$k1 * (a_i^{\,j}) + k2 * (a_i^{\,j})$
2	From 1 by A6.2.9 Matrix Multiplication by a Scalar	$((k1 + k2) * a_i^{\,j}) =$	$(k1 * a_i^{\,j}) + (k2 * a_i^{\,j})$
3	From 2 by A4.2.14 Distribution	$(k1 * a_i^{\,j} + k2 * a_i^{\,j}) =$	$(k1 * a_i^{\,j}) + (k2 * a_i^{\,j})$
4	From 3 by A6.2.8 Matrix Addition	$(k1 * a_i^{\,j} + k2 * a_i^{\,j}) =$	$(k1 * a_i^{\,j} + k2 * a_i^{\,j})$
$\therefore$	From 4 by A6.2.7 Equality of Matrices	$(k1 + k2) * A_n^{\,m} \equiv$	$k1 * A_m^{\,n} + k2 * A_n^{\,m}$

Theorem 6.3.11 Matrix Association for Multiplication

Steps	Hypothesis	$(A_p^{\,m} * B_r^{\,p}) * C_n^{\,r} \equiv$	$A_p^{\,m} * (B_r^{\,p} * C_n^{\,r})$
1	From hypothesis by D6.1.2 Rectangular Matrix	$((a_k^{\,j}) * (b_l^{\,k})) * (c_i^{\,l}) =$	$(a_k^{\,j}) * ((b_l^{\,k}) * (c_i^{\,l}))$
2	From 1 by A6.2.10 Matrix Multiplication	$(\sum\limits_{k=1}^{p} a_k^{\,j} * b_l^{\,k}) * (c_i^{\,l}) =$	$(a_k^{\,j}) * (\sum\limits_{l=1}^{r} b_l^{\,k} * c_i^{\,l})$
3	From 2 by A6.2.10 Matrix Multiplication	$(\sum\limits_{l=1}^{r} \sum\limits_{k=1}^{p} a_k^{\,j} * b_l^{\,k} * c_i^{\,l}) =$	$(\sum\limits_{l=1}^{r} \sum\limits_{k=1}^{p} a_k^{\,j} * b_l^{\,k} * c_i^{\,l})$
$\therefore$	From 4 by A6.2.7 Equality of Matrices	$(A_p^{\,m} * B_r^{\,p}) * C_n^{\,r} \equiv$	$A_p^{\,m} * (B_r^{\,p} * C_n^{\,r})$

Theorem 6.3.12 Matrix Identity right of Multiplication

Steps	Hypothesis	$A * I \equiv$	A
1	From hypothesis by D6.1.2 Rectangular Matrix	$(a_k^{\,j}) * (\delta_i^{\,k}) =$	$(a_i^{\,j})$
2	From 1 by A6.2.10 Matrix Multiplication	$(\sum\limits_{k=1}^{p} a_k^{\,j} * \delta_i^{\,k}) =$	$(a_i^{\,j})$
3	From 2 by evaluation of the Kronecker Delta	$(a_i^{\,j}) =$	$(a_i^{\,j})$
$\therefore$	From 3 by A6.2.7 Equality of Matrices	$A * I \equiv$	A

Theorem 6.3.13 Matrix Identity left of Multiplication

Steps	Hypothesis	$I * A \equiv$	A
1	From hypothesis by D6.1.2 Rectangular Matrix	$(\delta_k^{\,j}) * (a_i^{\,k}) =$	$(a_i^{\,j})$
2	From 1 by A6.2.10 Matrix Multiplication	$(\sum\limits_{k=1}^{p} \delta_k^{\,j} * a_i^{\,k}) =$	$(a_i^{\,j})$
3	From 2 by evaluation of the Kronecker Delta	$(a_i^{\,j}) =$	$(a_i^{\,j})$
$\therefore$	From 3 by A6.2.7 Equality of Matrices	$I * A \equiv$	A

Theorem 6.3.14 Matrix Zero Right of Multiplication

Steps	Hypothesis	$A * \mathbf{0}$	$\equiv$	$\mathbf{0}$
1	From hypothesis by D6.1.2 Rectangular Matrix	$(a_k^{\ j}) * (0)$	$=$	(0)
2	From 1 by A6.2.10 Matrix Multiplication	$(\sum\limits_{k=1}^{p} a_k^{\ j} * 0)$	$=$	(0)
3	From 2 by summing over zero, zeros the matrix	(0)	$=$	(0)
$\therefore$	From 3 by A6.2.7 Equality of Matrices	$A * \mathbf{0}$	$\equiv$	$\mathbf{0}$

Theorem 6.3.15 Matrix Zero Left of Multiplication

Steps	Hypothesis	$\mathbf{0} * A$	$\equiv$	$\mathbf{0}$
1	From hypothesis by D6.1.2 Rectangular Matrix	$(0) * (a_k^{\ j})$	$=$	(0)
2	From 1 by A6.2.10 Matrix Multiplication	$(\sum\limits_{k=1}^{p} 0 * a_k^{\ j})$	$=$	(0)
3	From 2 by summing over zero, zeros the matrix	(0)	$=$	(0)
$\therefore$	From 3 by A6.2.7 Equality of Matrices	$\mathbf{0} * A$	$\equiv$	$\mathbf{0}$

Theorem 6.3.16 Matrix Non-Commutative for Multiplication

Steps	Hypothesis	$A_p^{\ m} * B_n^{\ p}$	$\neq$	$B_p^{\ m} * A_n^{\ p}$
1	From hypothesis by D6.1.2 Rectangular Matrix	$(a_k^{\ j}) * (b_i^{\ k})$	$=$	$(b_k^{\ j}) * (a_i^{\ k})$
2	From 1 by A6.2.10 Matrix Multiplication	$(\sum\limits_{k=1}^{p} a_k^{\ j} * b_i^{\ k})$	$=$	$(\sum\limits_{k=1}^{p} b_k^{\ j} * a_i^{\ k})$
3	From 2 by the order of summation is different on either side of the equality	$(\sum\limits_{k=1}^{p} a_k^{\ j} * b_i^{\ k})$	$\neq$	$(\sum\limits_{k=1}^{p} b_k^{\ j} * a_i^{\ k})$
$\therefore$	From 3 by A6.2.7 Equality of Matrices	$A_p^{\ m} * B_n^{\ p}$	$\neq$	$B_p^{\ m} * A_n^{\ p}$

Theorem 6.3.17 Left Distribution for Matrices

Steps	Hypothesis	$A_p^{\ m} * (B_n^{\ p} + C_n^{\ p})$	$\equiv$	$A_p^{\ m} * B_n^{\ p} + A_p^{\ m} * C_n^{\ p}$
1	From hypothesis by D6.1.2 Rectangular Matrix	$(a_k^{\ j}) * ((b_i^{\ k}) + (c_i^{\ k}))$	$=$	$(a_k^{\ j}) * (b_i^{\ k}) + (a_i^{\ k}) * (c_i^{\ k})$
2	From 1 by A6.2.10 Matrix Multiplication	$(a_k^{\ j}) * (b_i^{\ k} + c_i^{\ k})$	$=$	$(\sum\limits_{k=1}^{p} a_k^{\ j} * b_i^{\ k}) + (\sum\limits_{k=1}^{p} a_k^{\ j} * c_i^{\ k})$
3	From 2 by A6.2.10 Matrix Multiplication	$(\sum\limits_{k=1}^{p} a_k^{\ j} * (b_i^{\ k} + c_i^{\ k}))$	$=$	$(\sum\limits_{k=1}^{p} a_k^{\ j} * b_i^{\ k}) + (\sum\limits_{k=1}^{p} a_k^{\ j} * c_i^{\ k})$
4	From 3 by A4.2.14 Distribution	$(\sum\limits_{k=1}^{p} a_k^{\ j} * b_i^{\ k}) + (\sum\limits_{k=1}^{p} a_k^{\ j} * c_i^{\ k})$		
			$=$	$(\sum\limits_{k=1}^{p} a_k^{\ j} * b_i^{\ k}) + (\sum\limits_{k=1}^{p} a_k^{\ j} * c_i^{\ k})$
$\therefore$	From 4 by A6.2.7 Equality of Matrices	$A_p^{\ m} * (B_n^{\ p} + C_n^{\ p})$	$\equiv$	$A_p^{\ m} * B_n^{\ p} + A_p^{\ m} * C_n^{\ p}$

Theorem 6.3.18 Right Distribution for Matrices

Steps	Hypothesis	$(B_n^p + C_n^p) * A_p^m$	$\equiv$	$B_n^p * A_p^m + C_n^p * A_p^m$
1	From hypothesis by D6.1.2 Rectangular Matrix	$((b_i^k) + (c_i^k)) * (a_k^j)$	$=$	$(b_i^k) * (a_k^j) + (c_i^k) * (a_k^j)$
2	From 1 by A6.2.10 Matrix Multiplication	$((b_i^k) + (c_i^k)) * (a_k^j)$	$=$	$(\sum\limits_{k=1}^{p} b_i^k * a_k^j) + (\sum\limits_{k=1}^{p} c_i^k * a_k^j)$
3	From 2 by A6.2.8 Matrix Addition	$(b_i^k + c_i^k) * (a_k^j)$	$=$	$(\sum\limits_{k=1}^{p} b_i^k * a_k^j) + (\sum\limits_{k=1}^{p} c_i^k * a_k^j)$
4	From 3 by A6.2.10 Matrix Multiplication	$(\sum\limits_{k=1}^{p} ((b_i^k + c_i^k) * a_k^j))$	$=$	$(\sum\limits_{k=1}^{p} b_i^k * a_k^j) + (\sum\limits_{k=1}^{p} c_i^k * a_k^j)$
5	From 4 by A4.2.14 Distribution	$(\sum\limits_{k=1}^{p} b_i^k * a_k^j) + (\sum\limits_{k=1}^{p} c_i^k * a_k^j)$		
			$=$	$(\sum\limits_{k=1}^{p} b_i^k * a_k^j) + (\sum\limits_{k=1}^{p} c_i^k * a_k^j)$
$\therefore$	From 5 by A6.2.7 Equality of Matrices	$(B_n^p + C_n^p) * A_p^m$	$\equiv$	$B_n^p * A_p^m + C_n^p * A_p^m$

Theorem 6.3.19 Association of a Scalar with Matrix Multiplication

1g	Give	c a scalar number				
Steps	Hypothesis	$c * (A_p^m * B_n^p)$	$\equiv$	$(c * A_p^m) * B_n^p$	$\equiv$	$A_p^m * (c * B_n^p)$
1	From hypothesis, 1g and D6.1.2 Rectangular Matrix	$c * ((a_k^j) * (b_i^k))$	$=$	$(c * (a_k^j)) * (b_i^k)$	$=$	$(a_j^k) * (c * (b_i^k))$
2	From 1 by A6.2.10 Matrix Multiplication	$c * (\sum\limits_{k=1}^{p} a_k^j * b_i^k) =$ $(a_k^j) * (c * (b_i^k))$		$(c * (a_k^j)) * (b_i^k)$	$=$	
3	From 2 by A6.2.9 Matrix Multiplication by a Scalar	$(c * \sum\limits_{k=1}^{p} a_k^j * b_i^k) =$ $(a_k^j) * (c * b_i^k)$		$(c * a_k^j) * (b_i^k)$	$=$	
4	From 3 by A6.2.10 Matrix Multiplication	$(c * \sum\limits_{k=1}^{p} a_k^j * b_i^k) =$ $(\sum\limits_{k=1}^{p} a_k^j * c * b_i^k)$		$(\sum\limits_{k=1}^{p} c * a_k^j * b_i^k) =$		
5	From 4 by A4.2.14 Distribution	$(c * \sum\limits_{k=1}^{p} a_k^j * b_i^k) =$ $(c * \sum\limits_{k=1}^{p} a_k^j * b_i^k)$		$(c * \sum\limits_{k=1}^{p} a_k^j * b_i^k) =$		
$\therefore$	From 5 by A6.2.7 Equality of Matrices	$c * (A_p^m * B_n^p)$	$\equiv$	$(c * A_p^m) * B_n^p$	$\equiv$	$A_p^m * (c * B_n^p)$

Theorem 6.3.20 Cancellation Law for Matrix Addition

1g	Given	Matrices $A_n^{\ m}, B_n^{\ m}, C_n^{\ m} \in (R^n \text{ or } C^n)$			
Steps	Hypothesis	If	$A_n^{\ m} + C_n^{\ m}$ $\equiv$ $A_n^{\ m}$	$B_n^{\ m} + C_n^{\ m}$ $\equiv$ $B_n^{\ m}$	then
1	From hypothesis	$A_n^{\ m} + C_n^{\ m}$	$\equiv$	$B_n^{\ m} + C_n^{\ m}$	
2	From 1, 1g and T6.3.3 Uniqueness of Scalar Multiplication	$- C_n^{\ m}$	$\equiv$	$- C_n^{\ m}$	
3	From 2 by T6.3.2 Uniqueness of Matrix Addition	$A_n^{\ m} + C_n^{\ m} - C_n^{\ m}$	$\equiv$	$B_n^{\ m} + C_n^{\ m} - C_n^{\ m}$	
4	From 3 by T6.3.5 Matrix Association for Addition	$A_n^{\ m} + (C_n^{\ m} - C_n^{\ m})$	$\equiv$	$B_n^{\ m} + (C_n^{\ m} - C_n^{\ m})$	
5	From 4 by T6.3.7 Matrix Inverse for Addition	$A_n^{\ m} + \mathbf{0}$	$\equiv$	$B_n^{\ m} + \mathbf{0}$	
∴	From 5 by T6.3.6 Matrix Identity for Addition	$A_n^{\ m}$	$\equiv$	$B_n^{\ m}$	

Theorem 6.3.21 Matrix Exponential Notation

1g	Given	for $0 < p$ an integer
Steps	Hypothesis	$(A_n^{\ m})^p = \prod^p *(A_n^{\ m})$
∴	From D4.1.17 Exponential Notation	$(A_n^{\ m})^p = \prod^p *(A_n^{\ m})$

Theorem 6.3.22 Matrix Uniqueness of Exponents

1g	Given	for $0 < p$ an integer
Steps	Hypothesis	$(A_n^{\ m})^p = (B_n^{\ m})^p$
1	From A6.2.7A Equality of Matrices and T6.3.21 Matrix Exponential Notation	$(A_n^{\ m})^1 = (B_n^{\ m})^1$
2	From 1, T6.3.4 Uniqueness of Matrix Multiplication and T6.3.21 Matrix Exponential Notation	$(A_n^{\ m})^2 = (B_n^{\ m})^2$
∴	From Repeat Steps 1 and 2 $p - 1$ times	$(A_n^{\ m})^p = (B_n^{\ m})^p$

Section 6.4 Properties of Matrix Transposition

Theorem 6.4.1 Uniqueness of Transposition of Matrices

1g	Given	Matrices $A_n{}^m$, $B_n{}^m \in (R^n$ or $C^n)$
2g		For $A_n{}^m \equiv B_n{}^m = (r_i{}^j)$
Steps	Hypothesis	${}^t A_n{}^m \equiv {}^t B_n{}^m$
1	From 2g and A6.2.9 Matrix Multiplication by a Scalar	${}^t A_n{}^m = (r_j{}^i)$
∴	From 1, 2g and A6.2.9 Matrix Multiplication by a Scalar	${}^t A_n{}^m = {}^t B_n{}^m$

Theorem 6.4.2 Distribution of the transpose operation for sums

Steps	Hypothesis	${}^t(A_n{}^m + B_n{}^m)$	$\equiv$	${}^t A_n{}^m + {}^t B_n{}^m$
1	From hypothesis by D6.1.2 Rectangular Matrix	${}^t((a_i{}^j) + (b_i{}^j))$	$=$	$({}^t(a_i{}^j) + {}^t(b_i{}^j))$
2	From 1 by A6.2.9 Matrix Multiplication by a Scalar	${}^t((a_i{}^j) + (b_i{}^j))$	$=$	$((a_j{}^i) + (b_j{}^i))$
3	From 2 by A6.2.8 Matrix Addition	${}^t((a_i{}^j) + (b_i{}^j))$	$=$	$(a_j{}^i + b_j{}^i)$
4	From 3 by A6.2.8 Matrix Addition	${}^t(a_i{}^j + b_i{}^j)$	$=$	$(a_j{}^i + b_j{}^i)$
5	From 4 by A6.2.9 Matrix Multiplication by a Scalar	$(a_j{}^i + b_j{}^i)$	$=$	$(a_j{}^i + b_j{}^i)$
∴	From 5 by A6.2.7 Equality of Matrices	${}^t(A_n{}^m + B_n{}^m)$	$\equiv$	${}^t A_n{}^m + {}^t B_n{}^m$

Theorem 6.4.3 Invariance of the transpose operation

Steps	Hypothesis	${}^t({}^t A_n{}^m)$	$\equiv$	$A_n{}^m$
1	From hypothesis by D6.1.2 Rectangular Matrix	${}^t({}^t(a_i{}^j))$	$=$	$(a_i{}^j)$
2	From 1 by A6.2.9 Matrix Multiplication by a Scalar	${}^t(a_j{}^i)$	$=$	$(a_i{}^j)$
3	From 2 by A6.2.9 Matrix Multiplication by a Scalar	$(a_i{}^j)$	$=$	$(a_i{}^j)$
∴	From 3 by A6.2.7 Equality of Matrices	${}^t({}^t A_n{}^m)$	$\equiv$	$A_n{}^m$

Theorem 6.4.4 Transpose of a constant

1g	Given	k a scalar constant		
Steps	Hypothesis	$^t(k*A_n^m)$	$\equiv$	$k*^tA_n^m$
1	From hypothesis, 1g and D6.1.2 Rectangular Matrix	$^t(k*(a_i^j))$	$=$	$k*^t(a_i^j)$
2	From 1 by A6.2.9 Matrix Multiplication by a Scalar	$^t(k*a_j^i)$	$\equiv$	$k*^t(a_j^i)$
3	From 2 by A6.2.9 Matrix Multiplication by a Scalar	$(k*a_j^i)$	$=$	$k*(a_j^i)$
4	From 2 by A6.2.9 Matrix Multiplication by a Scalar	$(k*a_i^j)$	$=$	$(k*a_i^j)$
$\therefore$	From 4 by A6.2.7 Equality of Matrices	$^t(k*A_n^m)$	$\equiv$	$k*^tA_n^m$

Theorem 6.4.5 Distribution of the transpose operation for products

1g	Given	$c_i^j \equiv \sum\limits_{k=1}^{p} a_k^j\, b_i^k$		
Steps	Hypothesis	$^t(A_n^m * B_n^m)$	$\equiv$	$^tA_n^m * {}^tB_n^m$
1	From hypothesis by D6.1.2 Rectangular Matrix	$^t((a_i^j) * (b_i^j))$	$=$	$(^t(a_i^j) * {}^t(b_i^j))$
2	From 1 by A6.2.9 Matrix Multiplication by a Scalar	$^t((a_i^j) * (b_i^j))$	$=$	$((a_j^i) * (b_j^i))$
3	From 2 by A6.2.10 Matrix Multiplication	$^t(\sum\limits_{k=1}^{p} a_k^j\, b_i^k)$	$=$	$(\sum\limits_{k=1}^{p} a_k^i\, b_j^k)$
4	From 3 and 1g	$^t(c_i^j)$	$=$	(c_i^i)
5	From 3 by A6.2.9 Matrix Multiplication by a Scalar	(c_j^i)	$=$	(c_j^i)
$\therefore$	From 5 by A6.2.7 Equality of Matrices	$^t(A_p^m * B_n^p)$	$\equiv$	$^tA_p^m * {}^tB_n^p$

Theorem 6.4.6 Transpose of the Identity Matrix

Steps	Hypothesis	$^tI \equiv I$
1	From D6.1.6 Identity Matrix	$I = (\delta_i^j)$
2	From 1 and T6.4.1 Uniqueness of Transposition of Matrices	$^tI = (\delta_i^j)^\tau$
3	From 2 and A6.2.9 Matrix Multiplication by a Scalar	$^tI = (\delta_j^i)$
4	From 3 transposing of off elements leaves the matrix unaltered being all zeros.	$^tI = (\delta_i^j)$
$\therefore$	From 4 and D6.1.6 Identity Matrix	$^tI = I$

Theorem 6.4.7 Transpose of Triangular Matrix T_L to T_U

Steps	Hypothesis	${}^tT_L \equiv T_U$
1	From D6.1.9 Lower Triangular Matrix	$T_L = (t_i^j)$ for $t_i^j = 0$ and $i < j$
2	From 1 and T6.4.1 Uniqueness of Transposition of Matrices	${}^tT_L = {}^\tau(t_i^j)$
3	From 2 and A6.2.9 Matrix Multiplication by a Scalar	${}^tT_L = (t_j^i)$ for $t_j^i = 0$ and $i < j$
∴	From 4 and D6.1.10B Upper Triangular Matrix	${}^tT_L = T_U$

Theorem 6.4.8 Transpose of Triangular Matrix T_U to T_L

Steps	Hypothesis	${}^tT_U$	$\equiv T_L$
1	From T6.4.6 Transpose of the Identity Matrix	${}^tT_L$	$= T_U$
2	From 1 and T6.4.1 Uniqueness of Transposition of Matrices	${}^t({}^tT_L)$	$= {}^tT_U$
3	From 2 and T6.4.2 Distribution of the transpose operation for sums	T_L	$= {}^tT_U$
∴	From 3 reordering	${}^tT_U$	$= T_L$

Section 6.5 Permutation

Definition 6.5.1 Notation of Factorials

For each positive interger [n], the product of the first n-positivae intergers is callled n-*factorial* and is denoted by [n!]. [JOH69 pg 278 and 279]

$$n! \equiv \prod_{i=1}^{n} i \qquad \text{Equation A}$$

$$0! \equiv 1 \qquad \text{Equation B}$$

Definition 6.5.2 General definition of Permutation

Permutation takes place when m-objects from a set of n-objects are arranged without repetition of the m-objects. [JOH69 pg 280] The count of the number of permutations is given by:

$$_mP_n \equiv \frac{n!}{(n-m)!} \qquad \text{Equation A}$$

For this discussion into permutations, as applied to determinants, the full set of objects is used, hence m = n and

$$_nP_n \equiv n! \qquad \text{Equation B}$$

In terms of determinants, the development of the topic of permutation is based on the reordered set of n-integers. A more formal definition is required in order to handle the complexity of this topic.

Definition 6.5.3 Permutation

Permutation is a finite ordered set of integers that have the following properties:

Given		a set of positive integers; $\sigma(1) \equiv \{i \mid 1, 2, 3, \ldots, i, \ldots, n\}$
Then	Reordering the set $\sigma(1)$ gives a unique ordered permutated set represented by the index [j]. The countable collections of integers having indices $[j_i]$ represent unique one to one mappings from $\sigma(1)$ without repetition.	$\sigma(j) \equiv \{i \mid j_1, j_2, j_3, \ldots, j_i, \ldots, j_n,$ where $j_i \equiv \gamma[i]\}$
Where	Every ordering being unique can only have combinations that do not repeat the indices, hence are only ***countable*** from 1 to n!. See D6.5.2 General definition of Permutation; above.	$j \equiv 1, 2, 3, \ldots, n!$ ***permutation count index***
Such that	The collections of permutated integers form a set;	$S_n \equiv \{j \mid s:\sigma(1) \rightarrow \sigma(j)$ for $j \equiv 1, 2, 3, \ldots, n!\}$

Definition 6.5.4 The Permutated index $j_i \equiv \gamma[i]$

It is a set in its own right containing one integer from the permutation.

Definition 6.5.5 The identity permutation

It is $\varepsilon \equiv \sigma(1)$ where $\varepsilon \in S_n$, $\qquad \gamma_\varepsilon[i] \in \varepsilon$ for $\gamma_\varepsilon[i] \equiv i$, such that $\qquad i \equiv 1, 2, 3, \ldots, n$.

Definition 6.5.6 The identity transposition

It is $\varepsilon^t \equiv \sigma(n!)$ where $\varepsilon^t \in S_n$ for $\gamma^t_\varepsilon[i] \equiv n - i + 1$ $\qquad$ and $\qquad i \equiv 1, 2, 3, \ldots, n$.

Definition 6.5.7 The inverse identity transposition

It is $\varepsilon^{-t} \equiv \varepsilon^t$.

Definition 6.5.8 The transposition of permutation indices

It is where $\alpha[i]$, $\alpha[j] \in \tau(k)$ and $i < j$ than if i and j are reordered such that $i > j$ than $\alpha[i]$ and $\alpha[j]$ are transposed and all other ordered numbers are left unchanged within the permutation.

Composition mapping performs a multiplication type of operation; as such, it will follow the exponent notation from algebra.

Definition 6.5.9 Composition exponent

It is a short hand notation for n-factors $\sigma \circ \sigma \circ \sigma \circ \sigma \circ \sigma \circ \sigma \equiv \sigma^n$.

Definition 6.5.10 Composition zero exponent *for* $\sigma^0 \equiv \varepsilon$.

Permutation has unique properties associated with operations on it. Since permutations have typical properties of groups than the following axioms follow the basic propositions of groups. [LIP68]

Axiom 6.5.1 Closure with Respect To The Composition Operator

The composition mapping operator always yields a new permutation, such that
$(\tau \equiv \eta \circ \sigma) \in S_n$
where the operator transforms as follows;
$$\alpha[i] \equiv \gamma[\ \beta[i]\ ] \text{ such that } \alpha[i] \in \tau, \beta[i] \in \eta, \gamma[i] \in \sigma \qquad \text{for } i \equiv 1, 2, 3, \ldots, n$$

Axiom 6.5.2 Inverse Composition Operator

For each $\sigma \in S_n$ there exists a permutation $\sigma^{-1} \in S_n$, called the inverse permutation, such that
$$\varepsilon \equiv \sigma^{-1} \circ \sigma \equiv \sigma \circ \sigma^{-1}$$
where
$$i \equiv \gamma[\ \gamma^{-1}[i]\ ] \equiv \gamma^{-1}[\ \gamma[i]\ ] \text{ such that } i \in \varepsilon, \gamma^{-1}[i] \in \sigma^{-1}, \gamma[i] \in \sigma \text{ for } i \equiv 1, 2, 3, \ldots, n$$

Theorem 6.5.1 Substitution for Composition

1g	Given	For any permutation $(\tau, \eta, \sigma) \in S_n$
2g		where $i \equiv 1, 2, 3, ..., n$
3g		$\alpha[i] \in \tau, \beta[i] \in \eta, \gamma[i] \in \sigma$
4g		$\tau = \eta$
5g		$\eta = \sigma$
Steps	Hypothesis	$\tau = \sigma$ by substitution
1	From (2,3,4)g, by equating all indices.	$\alpha[i] = \beta[i]$
2	From (2,3,5)g, by equating all indices.	$\beta[i] = \gamma[i]$
3	From 1 and 2, and equating all indices	$\alpha[i] = \gamma[i]$
$\therefore$	From 4g	$\tau = \eta$
	From 5g	$\underline{\eta = \sigma}$
	From 3 it is implied	$\therefore \tau = \sigma$

Theorem 6.5.2 Uniqueness of Left Composition for Permutations

1g	Given	For any permutation $(\tau, \eta, \mu, \sigma) \in S_n$
2g		$i \equiv 1, 2, 3, ..., n$
3g		$\tau = \eta$ for $\alpha[i] \in (\tau, \eta)$
4g		$\mu = \sigma$ for $\gamma[i] \in (\mu, \sigma)$
Steps	Hypothesis	$\tau \circ \mu = \eta \circ \sigma$
1	From hypothesis, (1,3,4)g by A6.5.1 Closure with Respect To The Composition Operator.	$\gamma[\, \alpha[i]\,] = \gamma[\, \alpha[i]\,]$
$\therefore$	From 1 and (3 & 4)g with left composition.	$\tau = \eta$
		$\underline{\circ\ \mu = \sigma}$
		$\tau \circ \mu = \eta \circ \sigma$

Since composition is non-commutative order is important, composition moves from left to right, hence the name left composition.

Theorem 6.5.3 Non-Commutative Property for Permutations

1g	Given	For any permutation $(\tau, \sigma) \in S_n$
2g		$i \equiv 1, 2, 3, ..., n$
3g		$\alpha[i] \in \tau, \gamma[i] \in \sigma$
Steps	Hypothesis	$\tau \circ \sigma \neq \sigma \circ \tau$
1	Assume opposite of hypothesis, by the (1 & 3)g and A6.5.1 Closure with Respect To The Composition Operator	$\gamma[\, \alpha[i]\,] = \alpha[\, \gamma[i]\,]$
2	From 1 by D6.5.3 Permutation; every permutation is unique and non-repetitive, hence cannot be equal.	$\gamma[\, \alpha[i]\,] \neq \alpha[\, \gamma[i]\,]$
$\therefore$	From 2 by A6.5.1 Closure with Respect To The Composition Operator	$\tau \circ \sigma \neq \sigma \circ \tau$

Theorem 6.5.4 Non-Commutative for Uniqueness of Right Composition for Permutations

1g	Given	For any permutation $(\tau, \eta, \mu, \sigma) \in S_n$
2g		$i \equiv 1, 2, 3, \ldots, n$
3g		$\tau = \eta$ for $\alpha[i] \in (\tau, \eta)$
4g		$\mu = \sigma$ for $\gamma[i] \in (\mu, \sigma)$
5g	Assume right composition	$\tau \circ \mu = \sigma \circ \eta$
Steps	Hypothesis	$\tau \circ \mu \neq \sigma \circ \eta$
1	From 5g and substituting (3 & 4)g.	$\tau \circ \sigma = \sigma \circ \tau$
2	From 1 by T6.5.3 The Permutated index $j_i \equiv \gamma[i]$	$\tau \circ \sigma \neq \sigma \circ \tau$
$\therefore$	From 3 and (3 & 4)g with right composition.	$\tau = \eta$ $\underline{\circ \; \mu = \sigma}$ $\tau \circ \mu \neq \sigma \circ \eta$

Theorem 6.5.5 Identity Commutativity for Composition

1g	Given	For permutations $(\varepsilon, \sigma) \in S_n$
2g		$\gamma[i] \in \sigma$
3g		$i \equiv 1, 2, 3, \ldots, n$
4g	Assume	$\sigma \equiv \sigma \circ \varepsilon$
5g	Assume	$\sigma \equiv \varepsilon \circ \sigma$
Steps	Hypothesis	$\sigma \equiv \sigma \circ \varepsilon \equiv \varepsilon \circ \sigma$
1	From 4g	$\sigma = \sigma \circ \varepsilon$
2	From 1 by 2g, D6.5.5 The identity permutation and A6.5.1 Closure with Respect To The Composition Operator	$\gamma[i] = \gamma_\varepsilon[\, \gamma[i]\,]$
3	From 2 and as $\gamma[i]$ selects indices from the ascending ordered identity index $\gamma_\varepsilon[i]$, its order is left unaltered.	$\gamma[i] = \gamma[i]$
4	From 3 by D6.5.3 Permutation	$\sigma = \sigma \circ \varepsilon$
5	From 5g	$\sigma = \varepsilon \circ \sigma$
6	From 5 by 2g, D6.5.5 The identity permutation and A6.5.1 Closure with Respect To The Composition Operator	$\gamma[i] = \gamma[\, \gamma_\varepsilon[i]\,]$
7	From 6 and as the ascending ordered identity index $\gamma_\varepsilon[i]$, select indices from $\gamma[i]$, $\gamma[i]$'s indices are left unaltered.	$\gamma[i] = \gamma[i]$
8	From 7 by D6.5.3 Permutation	$\sigma = \varepsilon \circ \sigma$
$\therefore$	From 4, 8 Identity for composition is valid and commutative	$\sigma \equiv \sigma \circ \varepsilon \equiv \varepsilon \circ \sigma$

Tensor Calculus & Physics: A General Treatise

Theorem 6.5.6 Association for Composition

1g 2g 3g	Given	For permutations $(\tau, \eta, \sigma) \in S_n$ $\alpha[i] \in \tau, \beta[i] \in \eta, \gamma[i] \in \sigma$ $i \equiv 1, 2, 3, \ldots, n$
Steps	Hypothesis	$(\tau \circ \eta) \circ \sigma \equiv \tau \circ (\eta \circ \sigma)$
1	From hypothesis.	$(\tau \circ \eta) \circ \sigma \qquad = \tau \circ (\eta \circ \sigma)$
2	From 1 by 2g and A6.5.1 Closure with Respect To The Composition Operator.	$(\beta[\,\alpha[i]\,]) \circ \sigma \qquad = \tau \circ (\gamma[\,\beta[i]\,])$
3	From 2 by 2g and A6.5.1 Closure with Respect To The Composition Operator.	$\gamma[\,\beta[\,\alpha[i]\,]\,] \qquad = \gamma[\,\beta[\,\alpha[i]\,]\,]$
∴	From 3 by D6.5.3 Permutation	$(\tau \circ \eta) \circ \sigma \equiv \tau \circ (\eta \circ \sigma)$

Composition mapping has a single operation as such it has no reciprocal operator. The following theorem develops such a represented composition operation.

Theorem 6.5.7 Reciprocal Property for Permutation

1g 2g	Given	For permutations $(\tau, \sigma^{-1}) \in S_n$ $\sigma \equiv \sigma$
Steps	Hypothesis	$\sigma \equiv (\sigma^{-1})^{-1} \equiv \sigma^{-1} \circ \sigma^2$
1	From the hypothesis	$\sigma = (\sigma^{-1})^{-1}$
2	From 1 and 1g than by T6.5.1 Substitution for Composition and using inverse permutation as an intermediate operator, there will be some permutation τ that will satisfy the equality.	$\sigma = \sigma^{-1} \circ \tau$
3	From 2 by 2g and T6.5.2 Uniqueness of Left Composition for Permutations	$\sigma \circ \sigma = \sigma \circ (\sigma^{-1} \circ \tau)$
4	From 3 by T6.5.6 Association for Composition, D6.5.9 Composition exponent and D4.1.17 Exponential Notation	$\sigma^2 = (\sigma \circ \sigma^{-1}) \circ \tau$
5	From 4 by A6.5.2 Closure with respect to the identity permutation	$\sigma^2 = \varepsilon \circ \tau$
6	From 5 by T6.5.5 Identity Commutativity for Composition	$\sigma^2 = \tau$
∴	From 2 and 6 by T6.5.1 Substitution for Composition	$\sigma \equiv (\sigma^{-1})^{-1} \equiv \sigma^{-1} \circ \sigma^2$

Theorem 6.5.8 Closure of identity transposition with respect to identity permutation

1g	Given	$\varepsilon \equiv \varepsilon^t \circ \varepsilon^{-t}$
2g		$\varepsilon \equiv \varepsilon^{-t} \circ \varepsilon^t$
Steps	Hypothesis	$\varepsilon \equiv \varepsilon^t \circ \varepsilon^{-t} \equiv \varepsilon^{-t} \circ \varepsilon^t$
1	From (1 or 2)g and D6.5.7 The inverse identity transposition.	$\varepsilon \equiv \varepsilon^t \circ \varepsilon^t$
2	From 1 and D6.5.5 The identity permutation, D6.5.6 The identity transposition and A6.5.1 Closure with Respect To The Composition Operator	$\gamma_\varepsilon[i] = \gamma^t_\varepsilon[\ \gamma^t_\varepsilon[i]\ ]$
3	From 2 by D6.5.6 The identity transposition	$\gamma_\varepsilon[i] = \gamma^t_\varepsilon[\ n - i + 1\ ]$
4	From 3 by the reverse selection of $\gamma^t_\varepsilon()$ indices the permutation is reordered in an ascending progression, which is nothing more than the identity permutation indices.	$\gamma_\varepsilon[i] = \gamma_\varepsilon[i]$
$\therefore$	From 4 by D6.5.3 Permutation	$\varepsilon \equiv \varepsilon^t \circ \varepsilon^{-t} \equiv \varepsilon^{-t} \circ \varepsilon^t$

Theorem 6.5.9 Transposition and Commutativity for Composition

1g	Given	For permutations $(\sigma^t, \sigma) \in S_n$
2g		$\gamma^t[i] \in \sigma^t$
3g		$\gamma[i] \in \sigma$
4g		$i \equiv 1, 2, 3, \ldots, n$
5g	Assume	$\sigma^t \equiv \sigma \circ \varepsilon^t$
6g	Assume	$\sigma^t \equiv \varepsilon^t \circ \sigma$
Steps	Hypothesis	$\sigma^t \equiv \sigma \circ \varepsilon^t \equiv \varepsilon^t \circ \sigma$
1	From 5g.	$\sigma^t = \sigma \circ \varepsilon^t$
2	From 1 by (2&3)g D6.5.6 The identity transposition and A6.5.1 Closure with Respect To The Composition Operator	$\gamma^t[i] = \gamma^t_\varepsilon[\ \gamma[i]\ ]$
3	From 2 and as $\gamma[i]$ selects indices from the descending ordered transpose index $\gamma^t_\varepsilon[i]$, its order is transposed.	$\gamma^t[i] = \gamma^t[i]$
4	From 3 by D6.5.3 Permutation	$\sigma^t = \sigma \circ \varepsilon^t$
5	From 6g	$\sigma = \varepsilon^t \circ \sigma$
6	From 5 by (2&3)g D6.5.6 The identity transposition and A6.5.1 Closure with Respect To The Composition Operator	$\gamma^t[i] = \gamma[\ \gamma^t_\varepsilon[i]\ ]$

7	From 6 and as the descending ordered transpose index $\gamma^t_\varepsilon[i]$, selects indices from $\gamma[i]$ its order is transposed.	$\gamma^t[i] = \gamma^t[i]$
8	From 7 by D6.5.3 Permutation	$\sigma^t = \varepsilon^t \circ \sigma$
$\therefore$	From 4, 8 Transposition of composition is valid and commutative	$\sigma^t \equiv \sigma \circ \varepsilon^t \equiv \varepsilon^t \circ \sigma$

Theorem 6.5.10 Closure for identity transposition, right & left handed inverse transposition

1g	Given	For permutations $(\sigma, \sigma^{-1}, \sigma^{-t}_R, \sigma^{-t}_L, \varepsilon^t) \in S_n$
2g		$\sigma \circ \sigma^{-t}_R \equiv \varepsilon^t$ assume
3g		$\sigma^{-t}_L \circ \sigma \equiv \varepsilon^t$
4g		$\sigma^{-1} \equiv \sigma^{-1}$
Steps	Hypothesis	$\sigma \circ \sigma^{-t}_R \equiv \sigma^{-t}_L \circ \sigma \equiv \varepsilon^t$
1	From 2g.	$\sigma \circ \sigma^{-t}_R \qquad = \varepsilon^t$
2	From 1 by 4g and T6.5.2 Uniqueness of Left Composition for Permutations	$\sigma^{-1} \circ (\sigma \circ \sigma^{-t}_R) \quad = \sigma^{-1} \circ \varepsilon^t$
3	From 2 by T6.5.6 Association for Composition	$(\sigma^{-1} \circ \sigma) \circ \sigma^{-t}_R \quad = \sigma^{-1} \circ \varepsilon^t$
4	From 3 by A6.5.2 Closure with respect to the identity permutation	$\varepsilon \circ \sigma^{-t}_R \qquad = \sigma^{-1} \circ \varepsilon^t$
5	From 4 by T6.5.5 Identity Commutativity for Composition	$\sigma^{-t}_R \qquad = \sigma^{-1} \circ \varepsilon^t$
6	From 5 Since σ^{-1} and ε^t are valid than by T6.5.1 Substitution for Composition, σ^{-t}_R must also be a valid permutation within set S_n.	$\sigma \circ \sigma^{-t}_R \qquad = \varepsilon^t$ Hence the original assumption in step 1 must be valid.
7	From 3g.	$\sigma^{-t}_L \circ \sigma \qquad = \varepsilon^t$
8	From 1 by 4g and T6.5.2 Uniqueness of Left Composition for Permutations	$(\sigma^{-t}_L \circ \sigma) \circ \sigma^{-1} \quad = \varepsilon^t \circ \sigma^{-1}$
9	From 2 by T6.5.6 Association for Composition	$\sigma^{-t}_L \circ (\sigma \circ \sigma^{-1}) \quad = \varepsilon^t \circ \sigma^{-1}$
10	From 3 by A6.5.2 Closure with respect to the identity permutation	$\sigma^{-t}_L \circ \varepsilon \qquad = \varepsilon^t \circ \sigma^{-1}$
11	From 4 by T6.5.5 Identity Commutativity for Composition	$\sigma^{-t}_L \qquad = \varepsilon^t \circ \sigma^{-1}$
12	From 5 Since σ^{-1} and ε^t are valid than by T6.5.1 Substitution for Composition, σ^{-t}_L must also be a valid permutation within set S_n.	$\sigma^{-t}_L \circ \sigma \qquad = \varepsilon^t$ Hence the original assumption in step 7 must be valid.
$\therefore$	From 2g and 12 by identity	$\sigma \circ \sigma^{-t}_R \equiv \sigma^{-t}_L \circ \sigma \equiv \varepsilon^t$

Theorem 6.5.11 Non-Commutative Property of right and left handed transposition

1g	Given	For permutations $(\sigma, \sigma^{-1}, \sigma^{-t}_R, \sigma^{-t}_L, \varepsilon^t) \in S_n$
2g	Let	$\sigma^{-t}_R = \sigma^{-1} \circ \varepsilon^t$
3g	Let	$\sigma^{-t}_L = \varepsilon^t \circ \sigma^{-1}$
4g	Assume	$\sigma^{-t}_R = \sigma^{-t}_L$
Steps	Hypothesis	$\sigma^{-t}_R \neq \sigma^{-t}_L$
1	From 4g.	$\sigma^{-t}_R \quad = \sigma^{-t}_L$
2	From 1, 2g, 3g and T6.5.1 Substitution for Composition	$\sigma^{-1} \circ \varepsilon^t \quad = \varepsilon^t \circ \sigma^{-1}$
3	From 2 by T6.5.2 Uniqueness of Left Composition for Permutations	$\sigma^{-1} \circ \varepsilon^t \quad \neq \varepsilon^t \circ \sigma^{-1}$, since it is non-commutative.
$\therefore$	From 3 the assumption in step 1 was incorrect, implying the opposite is true.	$\sigma^{-t}_R \neq \sigma^{-t}_L$

Theorem 6.5.12 Closure of right & left handed inverse transposition for identity permutation

1g	Given	For permutations $(\sigma, \sigma^{-1}, \sigma^{-t}_R, \sigma^{-t}_L, \varepsilon^t) \in S_n$
2g		$\varepsilon^{-t} \equiv \varepsilon^{-t}$
3g		$\sigma \circ \sigma^{-t}_R \equiv \varepsilon^t \qquad$ assume
4g		$\sigma^{-t}_L \circ \sigma \equiv \varepsilon^t$
Steps	Hypothesis	$\varepsilon^{-t} \circ (\sigma \circ \sigma^{-t}_R) \equiv (\sigma^{-t}_L \circ \sigma) \circ \varepsilon^{-t} \equiv \varepsilon$
1	From 3g.	$\sigma \circ \sigma^{-t}_R \quad = \varepsilon^t$
2	From 1, 2g and T6.5.2 Uniqueness of Left Composition for Permutations	$\varepsilon^{-t} \circ (\sigma \circ \sigma^{-t}_R) \quad = \varepsilon^{-t} \circ \varepsilon^t$
3	From 2 by T6.5.8 Closure of identity transposition with respect to identity permutation	$\varepsilon^{-t} \circ (\sigma \circ \sigma^{-t}_R) \quad = \varepsilon$
4	From 4g.	$\sigma^{-t}_L \circ \sigma \quad = \varepsilon^t$
5	From 4, 2g and T6.5.2 Uniqueness of Left Composition for Permutations	$(\sigma^{-t}_L \circ \sigma) \circ \varepsilon^{-t} \quad = \varepsilon^t \circ \varepsilon^{-t}$
6	From 5 by T6.5.8 Closure of identity transposition with respect to identity permutation	$(\sigma^{-t}_L \circ \sigma) \circ \varepsilon^{-t} \quad = \varepsilon$
$\therefore$	From 3g, 3 and 6 by identity	$\varepsilon^{-t} \circ (\sigma \circ \sigma^{-t}_R) \equiv (\sigma^{-t}_L \circ \sigma) \circ \varepsilon^{-t} \equiv \varepsilon$

Section 6.6 Parity

Definition 6.6.1 Association of Parity to Integer Numbers

Parity is the fact of being an even and not odd number. [OXF73] This is not a satisfying definition for what are even or odd numbers? However, a more quantifiable numerical definition comes from the word even, divisible integrally into two equal parts. [OXF73] So a number is even if divisible by two with no modulo remainder, than the following numerical definition can be written for integer numbers:

$$\text{if } (r \equiv n / 2 = 0) \quad \text{than n is an even number with a modulus remainder of zero,} \quad \text{Equation A}$$

likewise;

$$\text{if } (r \equiv n / 2 = 1) \quad \text{than n is an odd number with a modulus remainder of one,} \quad \text{Equation B.}$$

where $n \equiv 0, \pm 1, \ldots$ for all integers.

Definition 6.6.2 Association of Parity to Odd and Even Numbers

Parity can now be abstracted to represent odd or even giving the formal relations called parity.

$$\text{parity}(n) \equiv \begin{cases} \text{odd} & \text{for n/2 with modulo remainder} \\ \text{even} & \text{for n/2 with no modulo remainder} \end{cases}, \quad \text{Equation A}$$

or representing odd and even numerically by one and zero,

$$\text{parity}(n) \equiv \begin{cases} 1 & \text{for n/2 with modulo remainder} \\ 0 & \text{for n/2 with no modulo remainder} \end{cases}, \quad \text{Equation B}$$

Definition 6.6.3 Association of Sign to Parity

Parity raised to a negative number is called the **sign to parity** or **sign of parity** (see equation B). Associating a plus or minus sign with parity.

$$\text{sgn}(n) \qquad \equiv (-1)^n \qquad\qquad \text{Equation A}$$

$$\text{sgn}(\text{parity}(n)) \quad \equiv (-1)^{\text{parity}(n)} \qquad \text{Equation B,}$$
using D6.6.2.B Association of Parity to Odd And Even Numbers

The concept of odd and even integers can be quantified into an operator by a table of additions:

Table 6.6.1 Parity of Integer Addition

K	L	K + L
even	even	even
even	odd	odd
odd	even	odd
odd	odd	even

Its not to surprising that the table falls into the form of a truth table, since parity is binary dealing in two quantities odd and even. The logical operator that this table fits is the logical equivalence [$\equiv$] or negated exclusive-OR [$\sim(a \oplus b)$] operation. In order to avoid confusion between [$\equiv$] as a logical operator and [$\equiv$] as, equivalence by identity or by definition, the logical operator will be redefined as [✿]. Now by Definition 6.6.2.A a logical operator can be defined:

Definition 6.6.4 Parity Operator

parity(K) ✿ parity(L) ≡ parity(K + L)

Equation A by D6.6.2A
Association of Parity to Odd
And Even Numbers (state)

or

parity(K) ✿ parity(L) ≡ parity(K) + parity(L)

Equation B by D6.6.2B
Association of Parity to Odd
And Even Numbers (number)

for

Table 6.6.2 Parity Operator

parity(K)	parity(L)	parity(K) ✿ parity(L)
even	even	even
even	odd	odd
odd	even	odd
odd	odd	even

Theorem 6.6.1 Uniqueness of the parity operator

1g	Given	$n = m$ integer numbers.
Steps	Hypothesis	parity(n) ≡ parity(m) for $n = m$
1	From D6.6.2B Association of Parity to Odd And Even Numbers	parity(n) = parity(n) = some number of parity
∴	From 1 and 1g	parity (n) = parity (m)

Theorem 6.6.2 Uniqueness of the sign operator

1g	Given	$n = m$ integer numbers.
Steps	Hypothesis	sgn(n) ≡ sgn(m) for $n = m$
1	From 1g and D6.6.3A Association of Sign to Parity	$sgn(n) = (-1)^n$
2	From 1 and 1g	$sgn(n) = (-1)^m$
∴	From 2 and D6.6.3A Association of Sign to Parity	sgn(n) = sgn(m) for $n = m$

Playing with odd and even integers reveal certain properties. The following proofs are general within the system of integer numbers.

Theorem 6.6.3 For all sequences of integer numbers odd alternates with even

1g	Given	$n = m$ any integer
Steps	Hypothesis	$L = K + 1$ sequentially even alternates with odd.
1	From K is even for any m.	$K = 2*m$ K an even number
2	From L is odd for any n.	$L = 2* n + 1$ L an odd number
∴	From 1, 2 and 1g by A4.2.3 Substitution gives	$L = K + 1$

Theorem 6.6.4 Addition of an even and odd integer number always results in an odd number

1g	Given	$n \neq m$ for any integer other than itself.	
Steps	Hypothesis	$K + L = L + K = 2 * (m + n) + 1$ always odd.	
1	From K is even for any m.	$K = 2*m$	K an even number
2	From L is odd for any n.	$L = 2* n + 1$	L an odd number
∴	From 1, 2 and 1g by addition gives	$K + L = L + K = 2 * (m + n) + 1$ odd number	

Theorem 6.6.5 Addition of an even and even integer number always results in an even number

1g	Given	$n \neq m$ for any integer other than itself.	
Steps	Hypothesis	$K + L = L + K = 2 * (m + n)$ always even.	
1	From K is even for any m.	$K = 2*m$	K an even number
2	From L is even for any n.	$L = 2* n$	L an even number
∴	From 1, 2 and 1g by addition gives	$K + L = L + K = 2 * (m + n)$ even number	

Theorem 6.6.6 Addition of an odd and odd integer number always results in an even number

1g	Given	$n \neq m$ for any integer other than itself.	
Steps	Hypothesis	$K + L = L + K = 2 * (m + n + 1)$ always even.	
1	From K is odd for any m.	$K = 2*m + 1$	K an odd number
2	From L is odd for any n.	$L = 2* n + 1$	L an odd number
∴	From 1, 2 and 1g by addition gives	$K + L = L + K = 2 * (m + n + 1)$ even number	

So, the operation of parity can be thought of in two ways and used interchangeable, as the need requires. A sort of summary of these tables can be restated as a proof by using the proof of exhaustion Definition 3.1.15.

Theorem 6.6.7 Distribution of the Parity Operator

parity(K)	parity(L)	K+L	parity(K+L)	parity(K) + parity(L)	parity(K) ✿ parity(L)
even	even	even	even	even	even
even	odd	odd	odd	odd	odd
odd	even	odd	odd	odd	odd
odd	odd	even	even	even	even

Theorem 6.6.8 Sign of parity distributed over the sum of integers

1g	Given	integers n and m
2g	Let	$sgn(n+m) = sgn(n+m)$
Steps	Hypothesis	$sgn(n+m) = sgn(n) * sgn(m)$
1	From 2g.	$sgn(n+m) = sgn(n+m)$
2	From 1 by D6.6.3A Association of Sign to Parity	$sgn(n+m) = (-1)^{n+m}$
3	From 2 by A4.2.18 Summation Exp	$sgn(n+m) = (-1)^{n} * (-1)^{m}$
∴	From 3 by D6.6.3A Association of Sign to Parity	$sgn(n+m) = sgn(n) * sgn(m)$

Tensor Calculus & Physics: A General Treatise

Theorem 6.6.9 Sign of parity distributed over the sum of parity

1g	Given	integers m and n
2g	Let	$sgn(parity(n)+parity(m)) = sgn(parity(n) + parity(m))$
Steps	Hypothesis	$sgn(parity(n)+parity(m)) = sgn(parity(n)) * sgn(parity(m))$
1	From 2g.	$sgn(parity(n)+parity(m)) = sgn(parity(n) + parity(m))$
2	From 1 by D6.6.3B Association of Sign to Parity	$sgn(n+m) = (-1)^{parity(n) + parity(m)}$
3	From 2 by A4.2.18 Summation Exp	$sgn(n+m) = (-1)^{parity(n)} * (-1)^{parity(m)}$
∴	From 3 by D6.6.3B Association of Sign to Parity	$sgn(parity(n)+parity(m)) = sgn(parity(n)) * sgn(parity(m))$

Corollary 6.6.9.1 Sign of parity distributed over the Parity Operator

Steps	Hypothesis	$sgn(parity(n) \maltese parity(m)) = sgn(parity(n)) * sgn(parity(m))$
1	From T6.6.8 Sign of parity distributed over the sum of integers	$sgn(parity(n) + parity(m)) = sgn(parity(n)) * sgn(parity(m))$
∴	From 1 by D6.6.4B Parity Operator	$sgn(parity(n) \maltese parity(m)) = sgn(parity(n)) * sgn(parity(m))$

Section 6.7 Parity and Permutation

Definition 6.7.1 Association of parity to permutated indices

For any pair of permutated indices $\gamma[i]$ and $\gamma[j]$, there exists a number associated with the pair called parity by association.

$$\text{paritya}(\gamma[i], \gamma[j]) \equiv \begin{cases} 1 & \text{for} \quad \gamma[i] > \gamma[j] \\ 0 & \text{for} \quad \gamma[i] < \gamma[j] \end{cases} \quad \text{such that} \qquad (i, j) \leq n \text{ and } i \neq j.$$

Definition 6.7.2 Association of a parity to permutation

For the sum of all permutated, transposition of paired indices is called the parity number of the permutation.

$$\text{parityn}(\sigma(k)) \equiv \sum_{i=1}^{n-1} \sum_{j=i+1}^{n} \text{paritya}(\gamma[i], \gamma[j]) \qquad \text{for } k \equiv 1, 2, 3, \ldots, n!$$

Definition 6.7.3 Conditional Constraint on Parity

$$\text{psn}(\eta, \sigma) \equiv \text{parityn}(\eta) + \text{parityn}(\sigma)$$

Where the constraint is $\text{psn}(\eta, \sigma) < \frac{1}{2}n(n-1)$ or $\text{psn}(\eta, \sigma) \geq \frac{1}{2}n(n-1)$, see Observation I and II and summarized in Axiom 6.7.1.

The above definitions tell how to calculate parity from a permutation, but they say nothing on how the parity operator is related to the composition operator. Looking at examples for composition and parity a certain relationship does appear in four typical cases demonstrating this property.

Given the following relation for a composition of $\eta(k) \circ \sigma(l) = \tau(m)$ than;

Table 6.7.1 Permutation and Parity, Case I: $\eta(k) \leq \sigma(l)$ for $k \leq l$ and $l < m < n!$

	5!	4!	3!	2!	$\gamma[4]$	$\gamma[5]$	Parity Number	Parity
p_r	120	24	6	2	τ_i	τ_j	p_n	p
i	=	1	2	3	4	5		
$\eta(k)$	10	1	3	4	5	2	3	odd
$\circ \, \sigma(l)$	77	4	1	5	2	3	5	odd
$\tau(m)$	94	4	5	2	3	1	8	even

$\beta[1] = 1 \rightarrow \{i = 1\} \rightarrow \gamma[1] = 4 \rightarrow \alpha[1] \equiv 4; \quad \beta[2] = 3 \rightarrow \{i = 3\} \rightarrow \gamma[3] = 5 \rightarrow \alpha[2] \equiv 5;$
$\beta[3] = 4 \rightarrow \{i = 4\} \rightarrow \gamma[4] = 2 \rightarrow \alpha[3] \equiv 2; \quad \beta[4] = 5 \rightarrow \{i = 5\} \rightarrow \gamma[5] = 3 \rightarrow \alpha[4] \equiv 3;$
$\beta[5] = 2 \rightarrow \{i = 2\} \rightarrow \gamma[2] = 1 \rightarrow \alpha[5] \equiv 1;$

Table 6.7.2 Permutation and Parity, Case II: $\eta(k) \leq \sigma(l)$ for $k \leq l$ and $m < k$ cycles around.

	5!	4!	3!	2!	$\gamma[4]$	$\gamma[5]$	Parity Number	Parity
p_r	120	24	6	2	τ_i	τ_j	p_n	p
$\eta(k)$	79	4	2	1	3	5	4	even
$\circ \, \sigma(l)$	115	5	4	1	2	3	7	odd
$\tau(m)$	41	2	4	5	1	3	5	odd

Table 6.7.3 Permutation and Parity, Case III: $\eta(k) > \sigma(l)$ for $k > l$ and $k < m < n!$

	5!	4!	3!	2!	$\gamma[4]$	$\gamma[5]$	Parity Number	Parity
p_r	120	24	6	2	τ_i	τ_j	p_n	p
$\eta(k)$	55	3	2	1	4	5	3	odd
$\circ\ \sigma(l)$	12	1	3	5	4	2	4	even
$\tau(m)$	110	5	3	1	4	2	7	odd

Table 6.7.4 Permutation and Parity, Case IV: $\eta(k) > \sigma(l)$ for $k > l$ and $m < l$ cycles around

	5!	4!	3!	2!	$\gamma[4]$	$\gamma[5]$	Parity Number	Parity
p_r	120	24	6	2	τ_i	τ_j	p_n	p
$\eta(k)$	100	5	1	3	4	2	6	even
$\circ\ \sigma(l)$	34	2	3	4	5	1	4	even
$\tau(m)$	4	1	2	4	5	3	2	even

Table 6.7.5 Permutation and Parity, Case V: $\eta(k) > \sigma(l)$ for $k > l$ and $l < m < k$ cycles around

	5!	4!	3!	2!	$\gamma[4]$	$\gamma[5]$	Parity Number	Parity
p_r	120	24	6	2	τ_i	τ_j	p_n	p
$\eta(k)$	100	5	1	3	4	2	6	even
$\circ\ \sigma(l)$	38	2	4	1	5	3	4	even
$\tau(m)$	56	3	2	1	5	4	4	even

When does a conjecture become an axiom or proven theorem? Answer if it is not provable then it is not a theorem and if it is a fundamental property of the deductive system than it becomes an axiom. Specifically the difference between a conjecture and an axiom is the degree of faith one has in it. While axioms are never to be proven, in a deductive system they do have validity outside of the system, which in turn gives confidence in its use. Experimentation or some other method of proof that can validate it can give this confidence. In this way, it can be used and everything that is based on it can carry that notion of validity with it.

Case in point. Euclid's famous postulate for parallel lines had been thought of as immutable for centuries. However, as time went by and more was learned about geometry people's confidence eroded in it and finally other axioms where tried and non-Euclidean geometry was born.

Newton's axiomatic system of physical, flat, absolute space like wise had a similar fate when Einsteinain relativistic, curved, non-absolute space replaced it.

So, is it a good idea to develop an axiomatic system with axioms that might possible be wrong? Yes. The mathematician has to start some place, some time. If the axiom is wrong or as it is in the Newtonian system, not broad enough to accurately model the system, in time these weaknesses are revealed. The axioms are than replaced with better ones and through an iterative learning process; the accuracy of the system is improved.

A similar situation exists here these example problems demonstrate a possible unification of permutation and parity. Yet, I've looked at a number of possible proofs none of which are rigorous enough to prove the relationships. Though having done a large number of examples, as shown in Appendix-F.1 and above patterns have been established giving a high degree of confidence to these conjectures.

Observation 6.7.1 The Summation of Parity Numbers

In Case I and III, the sums of the parity numbers give the correct odd or even result for the end composition product. This is to be expected for a counting system that is infinite. However this is not the kind of system that is being dealt with, the system is finite and the count as defined in Theorem 6.7.2 cannot exceed [½n(n - 1)]. So the conditional constraint on the above summation is

$$\text{parityn}(\eta°\sigma) = \text{parityn}(\eta) + \text{parityn}(\sigma) \text{ for psn}(\eta, \sigma) < \tfrac{1}{2}n(n-1).$$

Observation 6.7.2 The Summation of Parity Numbers as an Upper Bound

However beyond [½n(n-1)], the system being finite, the process becomes cyclic Cases II, IV, and V show this breaks down. The best that can be said is that

$$\text{parityn}(\eta°\sigma) < \text{parityn}(\eta) + \text{parityn}(\sigma) \qquad \text{for psn}(\eta, \sigma) \geq \tfrac{1}{2}n(n-1).$$

So with a very small apology I move on establishing the following axioms. Now by combining Observations I and II:

Axiom 6.7.1 Association of Parity Number with the cyclic Composition Operator

$$\text{parityn}(\eta°\sigma) \leq \text{parityn}(\eta) + \text{parityn}(\sigma) \qquad \text{for all parity numbers of } [\eta, \sigma].$$

Observation 6.7.3 The Summation of Parity

Cases II, IV, and V show the resulting cyclic permutation number still gives the correct parity as if the sum is without bound. parity((parityn(η°σ)) ≡ parity(parityn(η) + parityn(σ)) for any parity numbers of [η, σ].

Axiom 6.7.2 Association of Parity with the cyclic Composition Operator

parityn(η)	parityn(σ)	parityn(η) + parityn(σ)	parityn(η°σ)
even	even	even	even
even	odd	odd	odd
odd	even	odd	odd
odd	odd	even	even

for all parity numbers of [η, σ].

In summary this axiom states that even though psn(η, σ) may not always be equal to parityn(η°σ) it will always be correct adding up to an odd or even number.

Observe that the operation can be represented as (even, odd) or numerically (0, 1), see Definition 6.6.2B.

Observation 6.7.4 Upper Bound for the Difference in Parity Numbers

Cases II, IV, and V show the permutation number of the composition has a bounding constraint where

$$\text{parityn}(\eta°\sigma) \geq |\,\text{parityn}(\eta) - \text{parityn}(\sigma)\,|.$$

This holds for composition on all permutations, where the absolute value is taken about the difference to account for the situation of $\text{parityn}(\eta) < \text{parityn}(\sigma)$, since the parity number is always positive. Of course the situation where $\text{parityn}(\eta) = \text{parityn}(\sigma)$ is accounted for with the lower limit of zero.

$$|\,\text{parityn}(\eta) - \text{parityn}(\sigma)\,| \geq 0 \text{ for composition on all permutations.}$$

Axiom 6.7.3 Constraint on permutation number for composition

$$\text{parityn}(\eta°\sigma) \geq |\,\text{parityn}(\eta) - \text{parityn}(\sigma)\,| \geq 0 \qquad \text{for composition on all permutations.}$$

While Axioms 6.7.1,2 and 3 are not conclusively proven from Appendix F.1 they are perfectly valid for permutations n = 2, 3, 4, and 5. To extrapolate them to any order through the ***Induction Property of Integers*** Theorem 3.11.1 is not within the scope of this work, it is simply sufficient and expedient to start here with these axioms.

Theorem 6.7.1 Uniqueness of parity number operator

1g	Given	$\eta \equiv \sigma$
2g		$parityn(\eta) = parityn(\eta)$
Steps	Hypothesis	$parityn(\eta) \equiv parityn(\sigma)$ for all permutations
1	From 2g	$parityn(\eta) = parityn(\eta)$
∴	From 1, 1g by T6.5.1 Substitution for Composition	$parityn(\eta) = parityn(\sigma)$ for all permutations

Theorem 6.7.2 Maximum value of a Parity Number

1g	Given	$k = n!$
2g		$\sigma(n!) = n\,(n-1)\,(n-2)\,\ldots\,5\,4\,3\,2\,1$
Steps	Hypothesis	$parityn(\sigma(n!)) \equiv \tfrac{1}{2}n\,(n-1)$
1	From 1g and 2g	permutated indices are $\gamma[i] > \gamma[j]$ for all i
2	From 1 and D6.7.1 Association of parity to permutated indices	$paritya(\gamma[i], \gamma[j]) = 1$
3	From 2 and D6.7.2 Association of a parity to permutation	$parityn(\sigma(n!)) \equiv \sum\limits_{i=1}^{n-1} \sum\limits_{j=i+1}^{n} (1)$
4	From 3 and expanding summation	$parityn(\sigma(n!)) \equiv \sum\limits_{i=2}^{n} (1) + \sum\limits_{i=3}^{n} (1) + \sum\limits_{i=4}^{n} (1) + \ldots + \sum\limits_{i=n-1}^{n} (1) + \sum\limits_{i=n}^{n} (1)$
5	From 4 and counting ones over the summation	$parityn(\sigma(n!)) \equiv (n-1) + (n-2) + (n-3) + \ldots + (n - (n-1)) + (n - (n-0))$
6	From 5 and grouping	$parityn(\sigma(n!)) \equiv \sum\limits_{i=1}^{n} (n) - \sum\limits_{i=1}^{n} (i)$
7	From 6 and sum of powers of the first n integers	$parityn(\sigma(n!)) \equiv n\,n - \tfrac{1}{2}n\,(n+1)$
∴	From 7 and carrying out the indicated algebra	$parityn(\sigma(n!)) \equiv \tfrac{1}{2}n\,(n-1)$

Corollary 6.7.2.1 Parity Number for the Identity Transposition

Steps	Hypothesis	$parityn(\varepsilon^t) = 0$
1	From D6.5.6 The identity transposition	$\sigma(n!) \equiv \varepsilon^t$
∴	From 1, 2 and T6.7.1 Uniqueness of parity number operator	$parityn(\varepsilon^t) \equiv \tfrac{1}{2}n\,(n-1)$

Theorem 6.7.3 Minimum value of a Parity Number

1g	Given	$k = 1$
2g		$\sigma(1) = 1\ 2\ 3\ 4\ 5\ \dots\ (n-2)\ (n-1)\ n$
Steps	Hypothesis	$\text{parityn}(\sigma(1)) = 0$
1	From 1g and 2g	permutated indices are $\gamma[i] < \gamma[j]$ for all i
2	From 1 and D6.7.1 Association of parity to permutated indices	$\text{paritya}(\gamma[i], \gamma[j]) = 0$
3	From 2 and D6.7.2 Association of a parity to permutation	$\text{parityn}(\sigma(1)) \equiv \sum_{i=1}^{n-1} \sum_{j=i+1}^{n} (0)$
∴	From 3 and A4.2.7 Identity Add for all zeros	$\text{parityn}(\sigma(1)) \equiv 0$

Corollary 6.7.3.1 Parity Number of the Identity Permutation

Steps	Hypothesis	$\text{parityn}(\varepsilon) = 0$
1	From T6.5.4 Non-Commutative for Uniqueness of Right Composition for Permutations	$\sigma(1) \equiv \varepsilon$
∴	From 1, 3 and T6.7.1 Association of parity to permutated indices	$\text{parityn}(\varepsilon) \equiv 0$

Corollary 6.7.3.2 Sign of the Identity Permutation

Steps	Hypothesis	$\text{sgn}(\text{parityn}(\varepsilon)) \equiv 1$
1	From C6.7.3.1 Parity Number of the Identity Permutation	$\text{parityn}(\varepsilon) = 0$
2	From 1 and T6.6.2 Uniqueness of the sign operator	$\text{sgn}(\text{parityn}(\varepsilon)) = \text{sgn}(0)$
3	From 2 and D6.6.3A Association of Sign to Parity; for all sequences of integer numbers odd alternates with even	$\text{sgn}(\text{parityn}(\varepsilon)) = (-1)^0$
∴	From 3	$\text{sgn}(\text{parityn}(\varepsilon)) = 1$

Tensor Calculus & Physics: A General Treatise

Theorem 6.7.4 Bounding all values of a Parity Number

Steps	Hypothesis	$0 \leq \text{parityn}(\sigma(k)) \leq \frac{1}{2}n(n-1)$
1	Since $\sigma(1)$ has the simplest ordering of permutated indices than any $\sigma(k)$ for $k \neq 1$ will have a more complex ordering and contribute ones to the summation by D6.7.2 Association of a parity to permutation, hence	$\text{parityn}(\sigma(1)) \leq \text{parityn}(\sigma(k))$
2	Since $\sigma(n!)$ has the highest ordering of permutated indices than any $\sigma(k)$ for $k \neq n!$ will have a less complex ordering hence contribute fewer ones to the summation by D6.7.2 Association of a parity to permutation	$\text{parityn}(\sigma(k)) \leq \text{parityn}(\sigma(n!))$
3	From 1 and 2	$\text{parityn}(\sigma(1)) \leq \text{parityn}(\sigma(k)) \leq \text{parityn}(\sigma(n!))$
$\therefore$	From 3 and substituting the results from T6.7.2 Maximum value of a Parity Number and T6.7.3 Minimum value of a Parity Number	$0 \leq \text{parityn}(\sigma(k)) \leq \frac{1}{2}n(n-1)$ for all $k = 1, ..., n!$

Theorem 6.7.5 Parity and Addition of the Parity Numbers

1g	Given	$\text{parityn}(\eta°\sigma) = \text{parityn}(\eta°\sigma)$
Steps	Hypothesis	$\text{parity}((\text{parityn}(\eta°\sigma)) = \text{parity}(\text{parityn}(\eta)) + \text{parity}(\text{parityn}(\sigma)))$ for all parity numbers of $[\eta, \sigma]$
1	From 1g	$\text{parityn}(\eta°\sigma) = \text{parityn}(\eta°\sigma)$
2	From 1 and T6.6.1 Uniqueness of the parity operator	$\text{parity}((\text{parityn}(\eta°\sigma)) = \text{parity}((\text{parityn}(\eta°\sigma)))$
3	From 2 and A6.7.1 Association of Parity Number with the cyclic Composition Operator	$\text{parity}((\text{parityn}(\eta°\sigma)) = \text{parity}(\text{parityn}(\eta) + \text{parityn}(\sigma))$ for all parity numbers of $[\eta, \sigma]$
4	From 3 and T6.6.7 Distribution of the Parity Operator	$\text{parity}((\text{parityn}(\eta°\sigma)) = \text{parity}(\text{parityn}(\eta)) + \text{parity}(\text{parityn}(\sigma)))$
∴	From 4	$\text{parity}((\text{parityn}(\eta°\sigma)) = \text{parity}(\text{parityn}(\eta)) + \text{parity}(\text{parityn}(\sigma)))$ for all parity numbers of $[\eta, \sigma]$

Corollary 6.7.5.1 Parity and the Parity Number Operator

Steps	Hypothesis	$\text{parity}((\text{parityn}(\eta°\sigma)) = \text{parity}(\text{parityn}(\eta)) ✿ \text{parity}(\text{parityn}(\sigma)))$ for all parity numbers of $[\eta, \sigma]$
1	From T6.7.5 Parity and Addition of the Parity Numbers	$\text{parity}((\text{parityn}(\eta°\sigma)) = \text{parity}(\text{parityn}(\eta)) + \text{parity}(\text{parityn}(\sigma)))$ for all parity numbers of $[\eta, \sigma]$
∴	From 1 and D6.6.4B Parity Operator; Addition of an even and odd integer number always results in an odd number	$\text{parity}((\text{parityn}(\eta°\sigma)) = \text{parity}(\text{parityn}(\eta)) ✿ \text{parity}(\text{parityn}(\sigma)))$ for all parity numbers of $[\eta, \sigma]$

Theorem 6.7.6 Sign of parity number distributed over the sum of parity numbers

1g	Given	$parityn(\eta°\sigma) = parityn(\eta°\sigma)$
Steps	Hypothesis	$sgn((parityn(\eta°\sigma)) = sgn(parityn(\eta)) * sgn(parityn(\sigma)))$ for all parity numbers of $[\eta, \sigma]$
1	From 1g	$parityn(\eta°\sigma) = parityn(\eta°\sigma)$
2	From 1 and T6.6.2 Uniqueness of the sign operator	$sgn((parityn(\eta°\sigma)) = sgn((parityn(\eta°\sigma))$
3	From 2 and A6.7.1 Association of Parity Number with the cyclic Composition Operator	$sgn((parityn(\eta°\sigma)) = sgn(parityn(\eta) + parityn(\sigma))$ for all parity numbers of $[\eta, \sigma]$
4	From 3 and T6.6.8 Sign of parity distributed over the sum of integers	$sgn((parityn(\eta°\sigma)) = sgn(parityn(\eta)) * sgn(parityn(\sigma)))$
∴	From 4	$sgn((parityn(\eta°\sigma)) = sgn(parityn(\eta)) * sgn(parityn(\sigma)))$ for all parity numbers of $[\eta, \sigma]$

Theorem 6.7.7 Permutation and Inverse Permutation have equivalent parity numbers

Steps	Hypothesis	$parityn(\sigma^{-1}) \equiv parityn(\sigma)$
1	From A6.5.2 Closure with respect to the identity permutation and T6.7.1 Association of parity to permutated indices	$parityn(\epsilon) = parityn(\sigma^{-1} ° \sigma)$
2	Applying A6.7.3 Conditional Constraint on Parity to the composition $\sigma^{-1} ° \sigma$	$parityn(\sigma^{-1} ° \sigma) \geq | parityn(\sigma^{-1}) - parityn(\sigma) | \geq 0$
3	From 1, 2 and A4.2.3 Substitution.	$parityn(\epsilon) \geq | parityn(\sigma^{-1}) - parityn(\sigma) | \geq 0$
4	From 3 by C6.7.3.1 Parity Number of the Identity Permutation	$0 \geq | parityn(\sigma^{-1}) - parityn(\sigma) | \geq 0$
∴	From 4	$parityn(\sigma^{-1}) = parityn(\sigma)$ for all permutations

Theorem 6.7.8 Inverse permutation has the same parity as its permutation

Steps	Hypothesis	$parity(parityn(\sigma)) \equiv parity(parityn(\sigma^{-1}))$
1	From T6.7.7 Permutation and Inverse Permutation have equivalent parity numbers	$parityn(\sigma^{-1})$ $= parityn(\sigma)$
∴	From 1 and T6.6.1 Uniqueness of the parity operator	$parity(parityn(\sigma)) = parity(parityn(\sigma^{-1}))$

For examples see Tables, Appendix F, Section F.2 n = 2, 3, and 4 are validated.

Theorem 6.7.9 The sign of the inverse permutation has the same sign as its permutation

Steps	Hypothesis	$\text{sgn}(\text{parityn}(\sigma))$	$\equiv \text{sgn}(\text{parityn}(\sigma^{-1}))$
1	From T6.7.7 Permutation and Inverse Permutation have equivalent parity numbers	$\text{parityn}(\sigma^{-1})$	$= \text{parityn}(\sigma)$
$\therefore$	From 1 by T6.6.2 Uniqueness of the sign operator	$\text{sgn}(\text{parityn}(\sigma))$	$= \text{sgn}(\text{parityn}(\sigma^{-1}))$

Tensor Calculus & Physics: A General Treatise

Section 6.8 Predicting Composite Parity Numbers

Predicting the composite parity number can be done without carrying out the operation by realizing certain characteristics of permutations. First a geometric representation of permutation needs to be established in order to test for patterns of number counting that are necessary as well as sufficient. Simply making calculations and looking at the results is not thorough and may not establish the generality that is needed for insight into these principles.

To establish a geometry let's create a coordinate system for permutations. This can be done by plotting the permutation cycles against the permutation index, see Appendix F. Something nice happens when this is done not only has a coordinate point been defined P(cycle, index), but also between points when a line is drawn the composition operator provides a preferred direction; hence the coordinate permutation system is actually a vector coordinate system. So, for $\eta°\sigma$, a preferred direction is established from $[\eta] \xrightarrow{\text{to}} [\sigma]$ providing a new permutation either high or low.

Definition 6.8.1 Cross Parity Number

Given by parityn($\eta x\sigma$) is the term that compensates for the absolute parity number sum so it adds correctly to the composite parity number. In most algebras this extra term is usually the cross term between two quantities, hence the name cross, in this case the quantities are the parity numbers between the permutations beginning operated on $[\eta, \sigma]$.

Theorem 6.8.1 Permutation Coordinate System n > 2

Proof is done by the method of exhaustion see Appendix F Section F.5.

Theorem 6.8.2 The sum of parity numbers predict composition parity

1g 2g 3g	Given	$\text{parityn}(\eta) + \text{parityn}(\sigma) < \frac{1}{2}\,n(n-1)$ $\text{parityn}(\eta) \neq \text{parityn}(\sigma)$ $\text{parityn}(\eta \times \sigma) \equiv \mid \text{parityn}(\eta) + \text{parityn}(\sigma) \mid - \text{parityn}(\eta \circ \sigma)$
Steps	Hypothesis	$\exists\! \mid \text{parityn}(\eta \times \sigma) \ni \text{parityn}(\eta \circ \sigma) = \mid \text{parityn}(\eta) + \text{parityn}(\sigma) \mid - \text{parityn}(\eta \times \sigma)$ for all parity numbers of $[\eta, \sigma]$
1	From 1 and A6.7.1 Association of Parity Number with the cyclic Composition Operator	$\text{parityn}(\eta \circ \sigma) < \text{parityn}(\eta) + \text{parityn}(\sigma)$ ignoring the trivial case of equality.
2	From T6.7.4 Bounding all values of a Parity Number	$0 \leq \text{parityn}(\eta)$ and $0 \leq \text{parityn}(\sigma)$
3	From 2 and D4.1.11 Absolute Value $\mid a \mid$	$\text{parityn}(\eta) + \text{parityn}(\sigma) = \mid \text{parityn}(\eta) + \text{parityn}(\sigma) \mid$
4	From 1g and 3	$\mid \text{parityn}(\eta) + \text{parityn}(\sigma) \mid < \frac{1}{2}\,n(n-1)$
5	From 1 and 3 by A4.2.3 Substitution	$\text{parityn}(\eta \circ \sigma) < \mid \text{parityn}(\eta) + \text{parityn}(\sigma) \mid$
6	From 4, 5 and A4.2.16 The Trichotomy Law of Ordered Numbers	$\text{parityn}(\eta \circ \sigma) < \mid \text{parityn}(\eta) + \text{parityn}(\sigma) \mid < \frac{1}{2}\,n(n-1)$
7	From A4.2.2 Equality	$\text{parityn}(\eta \circ \sigma) = \text{parityn}(\eta \circ \sigma)$
8	From 6, 7, A4.2.17 Correspondence of Equality and Inequality and T4.3.7 Equalities: Uniqueness of Subtraction	$\text{parityn}(\eta \circ \sigma) - \text{parityn}(\eta \circ \sigma) < \mid \text{parityn}(\eta) + \text{parityn}(\sigma) \mid - \text{parityn}(\eta \circ \sigma)$ $< \frac{1}{2}\,n(n-1) - \text{parityn}(\eta \circ \sigma)$
9	From 8 and A4.2.8 Inverse Add	$0 < \mid \text{parityn}(\eta) + \text{parityn}(\sigma) \mid - \text{parityn}(\eta \circ \sigma) < \frac{1}{2}\,n(n-1) - \text{parityn}(\eta \circ \sigma)$
10	From 3g, 9 and A4.2.3 Substitution	$0 < \text{parityn}(\eta \times \sigma) < \frac{1}{2}\,n(n-1) - \text{parityn}(\eta \circ \sigma)$ So a $\text{parityn}(\eta \times \sigma)$ can always be found, because it is bounded and uniquely exists.
11	From 3g and 10	$\text{parityn}(\eta \times \sigma) = \mid \text{parityn}(\eta) + \text{parityn}(\sigma) \mid - \text{parityn}(\eta \circ \sigma)$
12	From A4.2.2 Equality	$-\text{parityn}(\eta \times \sigma) + \text{parityn}(\eta \circ \sigma) = -\text{parityn}(\eta \times \sigma) + \text{parityn}(\eta \circ \sigma)$
13	From 11, 12 and T4.3.1 Equalities: Uniqueness of Addition	$\text{parityn}(\eta \times \sigma) - \text{parityn}(\eta \times \sigma) + \text{parityn}(\eta \circ \sigma) = \mid \text{parityn}(\eta) + \text{parityn}(\sigma) \mid - \text{parityn}(\eta \circ \sigma) - \text{parityn}(\eta \times \sigma) + \text{parityn}(\eta \circ \sigma)$
14	From 13 and A4.2.5 Commutative Add	$\text{parityn}(\eta \times \sigma) - \text{parityn}(\eta \times \sigma) + \text{parityn}(\eta \circ \sigma) = \mid \text{parityn}(\eta) + \text{parityn}(\sigma) \mid - \text{parityn}(\eta \circ \sigma) + \text{parityn}(\eta \circ \sigma) - \text{parityn}(\eta \times \sigma)$
15	From 14 and A4.2.8 Inverse Add	$0 + \text{parityn}(\eta \circ \sigma) = \mid \text{parityn}(\eta) + \text{parityn}(\sigma) \mid 0 - \text{parityn}(\eta \times \sigma)$
$\therefore$	From 15 and A4.2.7 Identity Add	$\text{parityn}(\eta \circ \sigma) = \mid \text{parityn}(\eta) + \text{parityn}(\sigma) \mid - \text{parityn}(\eta \times \sigma)$ for all parity numbers of $[\eta, \sigma]$

Theorem 6.8.3 The difference of parity numbers predict composition parity

1g 2g 3g	Given	$\text{parityn}(\eta°\sigma) < \frac{1}{2}\,n(n-1)$ $\text{parityn}(\eta) \neq \text{parityn}(\sigma)$ $\text{parityn}(\eta\times\sigma) \equiv -\mid \text{parityn}(\eta) - \text{parityn}(\sigma) \mid + \text{parityn}(\eta°\sigma)$
Steps	Hypothesis	$\exists\mid \text{parityn}(\eta\times\sigma) \ni \text{parityn}(\eta°\sigma) = \mid \text{parityn}(\eta) - \text{parityn}(\sigma) \mid + \text{parityn}(\eta\times\sigma)$ for all parity numbers of $[\eta, \sigma]$
1	From 1, A6.7.3 and 2g ignoring the trivial case of equality.	$\text{parityn}(\eta°\sigma) > \mid \text{parityn}(\eta) - \text{parityn}(\sigma) \mid > 0$
2	From A4.2.2 Equality	$\text{parityn}(\eta°\sigma) = \text{parityn}(\eta°\sigma)$
3	From 1, 2 and T4.3.7 Equalities: Uniqueness of Subtraction	$\text{parityn}(\eta°\sigma) - \text{parityn}(\eta°\sigma) > \mid \text{parityn}(\eta) - \text{parityn}(\sigma) \mid - \text{parityn}(\eta°\sigma) > -\text{parityn}(\eta°\sigma)$
4	From 3 and A4.2.8 Inverse Add	$0 > \mid \text{parityn}(\eta) - \text{parityn}(\sigma) \mid - \text{parityn}(\eta°\sigma) > -\text{parityn}(\eta°\sigma)$
5	From 4, T4.4.2 Equalities: Uniqueness of Multiplication and T4.7.6 Inequalities: Multiplication by Negative Number Reverses Order and A4.2.14 Distribution	$0 < -\mid \text{parityn}(\eta) - \text{parityn}(\sigma) \mid + \text{parityn}(\eta°\sigma) < \text{parityn}(\eta°\sigma)$
6	From 5, 3g and A4.2.3 Substitution	$0 < \text{parityn}(\eta\times\sigma) < \text{parityn}(\eta°\sigma)$ So a $\text{parityn}(\eta\times\sigma)$ can always be found, because it is bounded and uniquely exists.
7	From 3g and 6	$\text{parityn}(\eta\times\sigma) = -\mid \text{parityn}(\eta) - \text{parityn}(\sigma) \mid + \text{parityn}(\eta°\sigma)$
8	From A4.2.2 Equality	$\mid \text{parityn}(\eta) - \text{parityn}(\sigma) \mid = \mid \text{parityn}(\eta) - \text{parityn}(\sigma) \mid$
9	From 7, 8 and T4.3.1 Equalities: Uniqueness of Addition	$\text{parityn}(\eta\times\sigma) + \mid \text{parityn}(\eta) - \text{parityn}(\sigma) \mid = -\mid \text{parityn}(\eta) - \text{parityn}(\sigma) \mid + \mid \text{parityn}(\eta) - \text{parityn}(\sigma) \mid + \text{parityn}(\eta°\sigma)$
10	From 9 and A4.2.8 Inverse Add	$\text{parityn}(\eta\times\sigma) + \mid \text{parityn}(\eta) - \text{parityn}(\sigma) \mid = 0 + \text{parityn}(\eta°\sigma)$
$\therefore$	From 10 and A4.2.7 Identity Add	$\text{parityn}(\eta°\sigma) = \mid \text{parityn}(\eta) - \text{parityn}(\sigma) \mid + \text{parityn}(\eta\times\sigma)$ for all parity numbers of $[\eta, \sigma]$

Corollary 6.8.3.1 The Arithmetic of parity numbers predict composition parity

Steps	Hypothesis	$\text{parityn}(\eta°\sigma) = \mid \text{parityn}(\eta) \pm \text{parityn}(\sigma) \mid \mp \text{parityn}(\eta\times\sigma)$ for all parity numbers of $[\eta, \sigma]$
$\therefore$	From T6.8.2 The sum of parity numbers predict composition parity and T6.8.3 The difference of parity numbers predict composition parity	$\text{parityn}(\eta°\sigma) = \mid \text{parityn}(\eta) \pm \text{parityn}(\sigma) \mid \mp \text{parityn}(\eta\times\sigma)$ for all parity numbers of $[\eta, \sigma]$

Tensor Calculus & Physics: A General Treatise

From Corollary 6.8.3.1 "The Arithmetic of parity number predicts composition parity" the question needs to be asked how and when is the sign chosen? Fortunately this turns out to be very simple because the signs have a direct geometric interpretation. The ± signs are simply the sign of the slope between the parity numbers as demonstrated in the examples in Appendix A Section 4 and follow accordingly for the inverted plus and minus sign in the last term.

Now finding the cross parity number is a little more difficult of course one way is by inspection and a process of logical elimination. On the other hand this can be a little more systematic if a number independent of the outcome can be considered such as the sum of the composition parity numbers.

Theorem 6.8.4 Finding Cross Parity Number by Rational Decomposition

1g	Given	$pny \equiv parityn(\eta°\sigma)$ composite with extended upper limit
Steps	Hypothesis	$parityn(\eta x\sigma) + pny = parityn(\eta) + parityn(\sigma) < n^2$
1	From T6.8.2 The sum of parity numbers predict composition parity	$parityn(\eta°\sigma) = \mid parityn(\eta) + parityn(\sigma) \mid - parityn(\eta x\sigma)$
2	From A4.2.2 Equality	$parityn(\eta x\sigma) = parityn(\eta x\sigma)$
3	From 1, 2 and T4.3.1 Equalities: Uniqueness of Addition	$parityn(\eta°\sigma) + parityn(\eta x\sigma) = \mid parityn(\eta) + parityn(\sigma) \mid - parityn(\eta x\sigma) + parityn(\eta x\sigma)$
4	From 3 and A4.2.8 Inverse Add	$parityn(\eta°\sigma) + parityn(\eta x\sigma) = \mid parityn(\eta) + parityn(\sigma) \mid + 0$
5	From 4 and A4.2.7 Identity Add	$parityn(\eta°\sigma) + parityn(\eta x\sigma) = \mid parityn(\eta) + parityn(\sigma) \mid$
6	From 5, 1g and A4.2.3 Substitution	$pny + parityn(\eta x\sigma) = \mid parityn(\eta) + parityn(\sigma) \mid$
7	From observation II section 6.7	$\mid parityn(\eta) + parityn(\sigma) \mid < \frac{1}{2} n(n-1) = \frac{1}{2} n^2 - \frac{1}{2} n < n^2$ worst possible case
∴	From 6, 7 and A4.2.3 Substitution	$parityn(\eta x\sigma) + pny = parityn(\eta) + parityn(\sigma) < n^2$ thereby extending the limit for the worst possible case of $parityn(\eta) = parityn(\sigma) = \frac{1}{2} n(n-1)$.

What Theorem 6.8.4 means is that the sum of the parity numbers can always be broken apart into the cross parity number and some other parity number [pny]. As an example if the parity number sum where 11 than it could be broken apart as follows:

```
0 + 11 = 11
1 + 10 = 11
2 +  9 = 11
3 +  8 = 11
4 +  7 = 11
5 +  6 = 11
```

Now the possibilities are finite and small so a process of logical elimination will reveal the cross parity number. For a complete example, see Appendix F Table F.4.7.

There are ways of finding the cross parity number by immediate inspection such as the case where it is zero. As seen from Appendix F Table F.4.1 if the slope is ether vertical or horizontal the cross parity number is zero. This is only construed as a conjecture here, but this would indicated that somehow the cross parity number is closely tied to the geometrical slope so in the future it might be proven and the number immediately calculated.

Section 6.9 Determinants

Definition 6.9.1 Determinant of A

To every square matrix A of order [n] over a field K there is assigned a specific scalar denoted by:

$$\det(A) \equiv |A| \qquad \text{Equation A}$$

$$|A| \equiv |a_i^{\ j}| \qquad \text{Equation B}$$

$$|A| \equiv \begin{vmatrix} a_1^{\ 1} & a_2^{\ 1} & \cdots & a_i^{\ 1} & \cdots & a_n^{\ 1} \\ a_1^{\ 2} & a_2^{\ 2} & \cdots & a_i^{\ 2} & \cdots & a_n^{\ 2} \\ \vdots & \vdots & & \vdots & & \vdots \\ a_1^{\ j} & a_2^{\ j} & \cdots & a_i^{\ j} & \cdots & a_n^{\ j} \\ \vdots & \vdots & & \vdots & & \vdots \\ a_1^{\ n} & a_2^{\ n} & \cdots & a_i^{\ n} & \cdots & a_1^{\ n} \end{vmatrix} \qquad \text{Equation C}$$

The evaluation of the determinant is done as follows:

$$|A| \equiv \sum_{j=1}^{n!} \text{sgn}(\text{parityn}(\sigma(j))) \prod_{i=1}^{n} a_i^{\ j}i \ . \qquad \text{Equation D}$$

For ease of manipulation $\sigma(j)$ is reduced to just σ and sgn(parityn($\sigma(j)$)) to sgn(σ) it than is implied that summing is done over all permutations of σ:

$$|A| \equiv \sum_{\sigma \in S_r} \text{sgn}(\sigma) \prod_{i=1}^{n} a_i^{\ \gamma[i]} \ . \qquad \text{Equation E}$$

Unification of determinant elements with permutation requires a redefinition of permutation in order to model the way the factors handle the indices.

$$a \equiv |A| \text{ reperesented by lower case letter} \qquad \text{Equation F}$$

Definition 6.9.2 The Permutation Matrix

Let;

$$_m\sigma \equiv \begin{pmatrix} 1 & 2 & 3 & \dots & n \\ \gamma[1] & \gamma[2] & \gamma[3] & \dots & \gamma[n] \end{pmatrix} \qquad \text{Equation A}$$

Notice that row-1 represents the row indices for the matrix factors and row-2 the column indices. The permutation matrix is a special type of matrix having specific manipulation rules corresponding to manipulation of factors:

         Rule A   Columns can be transposed or permutated.

         Rule B   Rows are fixed and cannot be transposed or permutated.

Symbolically the permutation matrix can be represented as:

$$_m\sigma \equiv \begin{pmatrix} i \\ \gamma[i] \end{pmatrix} \qquad \text{for i = 1, 2 ..., n such that } \gamma[i] \in \sigma$$

From Appendix F, OF.1.7 it's observed that the parity about the center permutation is symmetrical.

From Definition 6.9.1 Rule B must hold, reordering the columns can transpose rows, such that the $\gamma[i]$'s are transposed to $[i]$'s in row-2 yielding a new set of permutated indices in row-1. This new permutation matrix $_m\tau$ is the elements in a determinant that have been transposed.

$$_m\tau \equiv \begin{pmatrix} \alpha[1]\,\alpha[2]\,\alpha[3] & \ldots & \alpha[n] \\ 1 \quad 2 \quad 3 & \ldots & n \end{pmatrix}$$

or

$$_m\tau \equiv \begin{pmatrix} \alpha[i] \\ i \end{pmatrix}$$

Definition 6.9.3 Minor of a Square Matrix A

Is M_i^j of order $[n-1]$ with the ith row and j^{th} column deleted from A.

Definition 6.9.4 Cofactor of a Square Matrix A

Is the determinant A_i^j is the derived from the minor M_i^j [6..9.1]
$$A_i^j \equiv sgn(i + j)\ |M_i^j|$$

Definition 6.9.5 Adjoint of a Square Matrix A

Is derived from the cofactor A_i^j:
$$adj\ A \equiv {}^t(A_i^j)$$

Definition 6.9.6 Inverse Matrix of a Square Matrix A

$$A^{-1} \equiv adj\ A\ /\ |A|\ for\ |\,A\,| \neq 0$$

Definition 6.9.7 Matrix 2 x 2

$$A_{2x2} = \begin{pmatrix} a & b \\ c & d \end{pmatrix}$$

[6..9.1]Note: The reason why the Cofactor and Minor determinates have both raised indices is later on we will see in tensor calculus that they are related to the contravariant metric, which is a tensor of second rank.

Theorem 6.9.1 Uniqueness of a Determinate

1g	Given	$A \equiv B$
Steps	Hypothesis	$\mid A \mid \equiv \mid B \mid$
1	From D6.9.1A Determinant of A	$\det(A) = \mid A \mid$
2	From 1 and 2g	$\det(A) = \mid B \mid$
$\therefore$	From 1 and 2	$\mid A \mid = \mid B \mid$

Theorem 6.9.2 Transpose of a Determinate leaves it unaltered

1g	Given	For permutations $(\sigma, \tau) \in S_n$
2g		$\gamma[i] \in \sigma$ and $\beta[i] \in \tau$
3g		${}^t A = B$
4g		$A = (a_i^{\,j})$
5g		$B = (b_i^{\,j})$
Steps	Hypothesis	$\mid {}^t A \mid = \mid A \mid$
1	From 3g,4g and 5g	$b_i^{\,j} = a_j^{\,i}$ equating element by element.
2	From 1 and D6.9.1E Determinant of A	$\mid B \mid = \sum_{\tau \in S_r} \text{sgn}(\tau) \prod_{i=1}^{n} b_i^{\,\beta[i]}$
3	From 2 substitute 1 into the product of factors.	$\mid B \mid = \sum_{\tau \in S_r} \text{sgn}(\tau) \prod_{i=1}^{n} a_{\beta[i]}^{\,i}$
4	From 3 substitute 3g into determinate.	$\mid {}^t A \mid = \sum_{\tau \in S_r} \text{sgn}(\tau) \prod_{i=1}^{n} a_{\beta[i]}^{\,i}$
5	From 4 the a-factors can be transposed to obtain the σ ordering	$\mid {}^t A \mid = \sum_{\tau \in S_r} \text{sgn}(\tau) \prod_{i=1}^{n} a_{\beta[\gamma[i]]}^{\,\gamma[i]}$
6	From 5 and 3g since this is ordered over σ it must be concluded	$\tau = \sigma^{-1}$ the inverse permutation, hence $\beta = \gamma^{-1}$
7	From 6 and 5 and summing over σ	$\mid {}^t A \mid = \sum_{\sigma \in S_r} \text{sgn}(\sigma^{-1}) \prod_{i=1}^{n} a_{\gamma^{-1}[\gamma[i]]}^{\,\gamma[i]}$
8	From 7 and A6.5.1 Closure with Respect To The Composition Operator	$\mid {}^t A \mid = \sum_{\sigma \in S_r} \text{sgn}(\sigma^{-1}) \prod_{i=1}^{n} a_i^{\,\gamma[i]}$
9	From 8 and T6.7.9 The sign of the inverse permutation has the same sign as its permutation	$\mid {}^t A \mid = \sum_{\sigma \in S_r} \text{sgn}(\sigma) \prod_{i=1}^{n} a_i^{\,\gamma[i]}$
$\therefore$	From 9, 4g and D6.9.1E Determinant of A	$\mid {}^t A \mid = \mid A \mid$

Theorem 6.9.3 The sign of the cofactor on the diagonal is always positive

1g	Given	$\text{sgn}(i+j)$ for $|A| \equiv |a^i_j|$
2g		$i \equiv j$ for the diagonal
Steps	Hypothesis	$\text{sgn}(i+j) \equiv +1$
1	From 1g	$\text{sgn}(i+j) = \text{sgn}(i+j)$
2	From 1 by 2g for the diagonal	$\text{sgn}(i+j) = \text{sgn}(2*i)$
3	From 2 by D6.6.3A Association of Sign to Parity; for all sequences of integer numbers odd alternates with even	$\text{sgn}(i+j) = (-1)^{2*i}$
4	From 3 by A4.2.22 Product Exp	$\text{sgn}(i+j) = ((-1)^2)^i$
5	From 4 by T4.8.2 Integer Exponents: Negative One Squared	$\text{sgn}(i+j) = (+1)^i$
∴	From 5 by T4.8.1 Integer Exponents: Unity Raised to any Integer Value	$\text{sgn}(i+j) \equiv +1$

From Appendix F.6.3 the following property is observed in expanding a determinate about a row (column):

Theorem 6.9.4 Parity of an expanded determinate about a row (column)

1g	Given	For permutations $(\sigma, \tau) \in S_n$ and
2g		$(\mu, \kappa) \in Q_{n-1}$
3g		$_rp_{\sigma n} \equiv \text{parity}(\text{parityn}(\sigma))$
4g		$_rp^i_{\sigma[n-1]} \equiv \text{parity}(\text{parity}[n-1](\sigma^i))$ where the ith permutation index is removed
5g		$p_{\sigma n} \equiv \text{parityn}(\sigma)$
6g		$p^i_{\sigma[n-1]} \equiv \text{parity}[n-1](\sigma^i)$
7g		$_rp(N) \equiv \text{parity}(N)$
8g		$_rp(p_{\sigma n}) = {_rp}(p^i_{\mu[n-1]} + (i+j))$ proven for
9g		$n = 3, 4,$ and 5 Appendix F.5.3
Steps	Hypothesis	$_rp(p_{\sigma n}) \equiv {_rp}(p^i_{\mu[n-1]} + (i+j))$ valid for all permutations $\in S_n$
1	From (1&2)g by transpose any two permutation indices not i.	$\sigma \to \tau$, and since μ is derived from σ than $\mu \to \kappa$ transposes as well.
2	From 1 and (5,6,7,8)g with i and j constant	$_rp(p_{\tau n}) = {_rp}(p^i_{\kappa[n-1]} + (i+j))$
∴	From 2 by equivalence, parity is consistent, hence	$_rp(p_{\sigma n}) = {_rp}(p^i_{\mu[n-1]} + (i+j))$

For any ordered square matrix, the proof follows since no inconsistencies arise for any transposition in any permutation. The equation is general for any row or column.

Corollary 6.9.4.1 Sign of an expanded determinate about a row (column)

1g	Given	$_rp(p_{\sigma n}) = {_rp}(p^i_{\mu[n-1]} + (i+j))$
2g		For permutations $\sigma \in S_n$ and $\mu \in Q_{n-1}$
Steps	Hypothesis	$\text{sgn}(p_{\sigma n}) = \text{sgn}(p^i_{\mu[n-1]} + (i+j))$
1	From 1g	$_rp(p_{\sigma n}) = {_rp}(p^i_{\mu[n-1]} + (i+j))$
∴	From 1 by T6.6.2 Uniqueness of the sign operator	$\text{sgn}(p_{\sigma n}) = \text{sgn}(p^i_{\mu[n-1]} + (i+j))$

Theorem 6.9.5 Expanding Determinant [A] of order [n] about a Row or Column

Is equal to the sum of the products obtained by multiplying the elements of any row (column) by their respective cofactors:

1g 2g	Given	For permutations $\sigma \in S_n$ and $\mu \in Q_{n-1}$ $A \equiv (a_i^j)$	
Steps	Hypothesis	$\|A\| = \sum\limits_{j=1}^{n} a_i^{\,j} A_i^{\,j}$ the ith row Equation A	$\|A\| = \sum\limits_{i=1}^{n} a_i^{\,j} A_i^{\,j}$ the jth column Equation B
1	From D6.9.4 Cofactor of a square matrix A	$A_i^{\,j} = sgn(i+j)\,\|M_i^j\|$	
2	From D6.9.1D Determinant of A	$\|M_i^j\| = \sum\limits_{q=1}^{(n-1)!} sgn(\,parity_{[n-1]}(\mu(q))\,) \prod\limits_{p=1}^{n-1} m_p^{\,q}{}_p$	
3	From 1g, 2g, 1, 2 and A4.2.3 Substitution	$\|A\| = \sum\limits_{j=1}^{n} sgn(i+j)\,a^i{}_j (\,\sum\limits_{q=1}^{(n-1)!} sgn(\,parity_{[n-1]}(\mu(q))\,) \prod\limits_{p=1}^{n-1} m_p^{\,q}{}_p\,)$	
4	From 3 by A4.2.14 Distribution	$\|A\| = \sum\limits_{j=1}^{n!} sgn(\,parity_{[n-1]}(\mu(j))\,)*sgn(i+j_i)*a^i{}_{j_i} \prod\limits_{p=1}^{n-1} m_p^{\,j}{}_p$	
5	From 4 by completing the product.	$\|A\| = \sum\limits_{j=1}^{n!} sgn(\,parity_{[n-1]}(\mu(j))\,)*sgn(i+j_i) \prod\limits_{i=1}^{n} a_i^{\,j}{}_i$	
6	From 5 by T6.6.8 Sign of parity distributed over the sum of integers	$\|A\| = \sum\limits_{j=1}^{n!} sgn(\,parity_{[n-1]}(\mu(j)) + i + j_i) \prod\limits_{i=1}^{n} a_i^{\,j}{}_i$	
7	From 6 by C6.9.4.1 Sign of an expanded determinate about a row (column)	$\|A\| = \sum\limits_{j=1}^{n!} sgn(\,parity_n(\sigma(j))\,) \prod\limits_{i=1}^{n} a_i^{\,j}{}_i$	
8	From 7 and D6.9.1D Determinant of A	$\|A\| = \|A\|$	
∴	From 8 and A6.2.7 Equality of Matrices	$\|A\| = \sum\limits_{j=1}^{n} a_i^{\,j} A_i^{\,j}$ the ith row Equation A	$\|A\| = \sum\limits_{i=1}^{n} a_i^{\,j} A_i^{\,j}$ the jth column Equation B

The following theorems look at certain cases where determinants can evaluate immediately.

Theorem 6.9.6 The determinant of a matrix having a row or column of zeros is zero

1g 2g	Given	$A \equiv (a_i^j)$ $a_i^j \equiv 0$ for any row or column
Steps	Hypothesis	$\|A\| \equiv 0$
1	From T6.9.5 Expanding Determinant [A] of order [n] about a Row or Column, Adjoint of a square matrix A	$\|A\| = \sum\limits_{j=1}^{n} a_i^{\,j} A_i^{\,j}$
∴	From 1 and 2g, A4.2.3 Substitution and summing.	$\|A\| = 0$

Theorem 6.9.7 Interchanging of any two rows in a determinant

1g	Given	$A \equiv (a_i^{\ j})$
2g		$B \equiv (b_i^{\ j})$
3g		$b_i^{\ j} \equiv a_{p_i}^{\ j}$ for $p_i \in \tau$ where any two indices are transposed out of ε. In this case any two rows.
4g		$\sigma \in S_n$ and $\mu \in Q_n$
Steps	Hypothesis	$|B| \equiv \sum\limits_{j=1}^{n!} \text{sgn}(\text{parityn}(\sigma(j)) + \text{parityn}(\tau)) \prod\limits_{i=1}^{n} a_i^{\ j_i}$
1	From 2g, 4g by D6.9.1D Determinant of A	$|B| = \sum\limits_{k=1}^{n!} \text{sgn}(\text{parityn}(\mu(k))) \prod\limits_{i=1}^{n} b_i^{\ k_i}$
2	From 1, 3g and A4.2.3 Substitution	$|B| = \sum\limits_{k=1}^{n!} \text{sgn}(\text{parityn}(\mu(k))) \prod\limits_{i=1}^{n} a_{p_i}^{\ k_i}$
3	From 2 by transposing τ back to ε, maps p_i back to i and shifts parity to odd for every term since τ is odd by definition.	$|B| = \sum\limits_{k=1}^{n!} \text{sgn}(\text{parityn}(\sigma(k)) + \text{parityn}(\tau)) \prod\limits_{i=1}^{n} a_i^{\ k_i}$
$\therefore$	From 3 summing over all j	$|B| = \sum\limits_{j=1}^{n!} \text{sgn}(\text{parityn}(\sigma(j)) + \text{parityn}(\tau)) \prod\limits_{i=1}^{n} a_i^{\ j_i}$

Theorem 6.9.8 Interchanging of any two columns in a determinant

1g	Given	$A \equiv (a_i^{\ j})$
2g		$B \equiv (b_i^{\ j})$
3g		$b_i^{\ j} \equiv a_i^{\ p_i}$ for $p_i \in \tau$ where any two indices are transposed out of ε. In this case any two columns.
4g		$\sigma \in S_n$ and $\mu \in Q_n$
Steps	Hypothesis	$|B| \equiv \sum\limits_{j=1}^{n!} \text{sgn}(\text{parityn}(\sigma(j)) + \text{parityn}(\tau)) \prod\limits_{i=1}^{n} a_i^{\ j_i}$
1	From 2 and 4g by D6.9.1D Determinant of A	$|B| = \sum\limits_{k=1}^{n!} \text{sgn}(\text{parityn}(\mu(k))) \prod\limits_{i=1}^{n} b_i^{\ k_i}$
2	From 1, 3g and A4.2.3 Substitution	$|B| = \sum\limits_{k=1}^{n!} \text{sgn}(\text{parityn}(\mu(k))) \prod\limits_{i=1}^{n} a_i^{\ k_i}$
3	From 2 by transposing τ back to ε, maps p_i back to i and shifts parity to odd for every term since τ is odd by definition.	$|B| = \sum\limits_{k=1}^{n!} \text{sgn}(\text{parityn}(\sigma(k)) + \text{parityn}(\tau)) \prod\limits_{i=1}^{n} a_i^{\ k_i}$
$\therefore$	From 3 summing over all j	$|B| = \sum\limits_{j=1}^{n!} \text{sgn}(\text{parityn}(\sigma(j)) + \text{parityn}(\tau)) \prod\limits_{i=1}^{n} a_i^{\ j_i}$

Theorem 6.9.9 Interchanging of any two rows or columns changes the sign in a determinant

1g	Given	$A \equiv (a_i^j)$
2g		$B \equiv (b_i^j)$
3g		For $p_i \in \tau$ where any two indices are transposed out of ε, hence is odd.
4g		$\sigma \in S_n$
Steps	Hypothesis	$|B| \equiv -|A|$
1	From T6.9.7 Interchanging of any two rows in a determinant or T6.9.8 Interchanging of any two columns in a determinant	$|B| = \sum_{j=1}^{n!} sgn(\, parityn(\sigma(j)) + parityn(\tau)\,) \prod_{i=1}^{n} a_i^{j_i}$
2	From 1 by T6.6.8 Sign of parity distributed over the sum of integers	$|B| = \sum_{j=1}^{n!} sgn(\, parityn(\sigma(j))\,) * sgn(\, parityn(\tau)\,) \prod_{i=1}^{n} a_i^{j_i}$
3	From 2 by 3g τ being odd.	$|B| = - \sum_{j=1}^{n!} sgn(\, parityn(\sigma(j))\,) \prod_{i=1}^{n} a_i^{j_i}$
$\therefore$	From 3 D6.9.1D Determinant of A	$|B| = -|A|$

Theorem 6.9.10 A determinant with two identical rows or columns is zero

1g	Given	$A \equiv (a_i^j)$
2g		Any two rows or columns are identical
Steps	Hypothesis	$|A| \equiv 0$
1	From 1g and 2g	$|A| = |A|$
2	From 1 by T6.9.9 Interchanging of any two rows or columns changes the sign in a determinant, interchanging the identical rows or columns:	$|A| = -|A|$ leaves the determinant unaltered except for sign.
$\therefore$	From 3 by adding to both sides $|A|$ and dividing by 2:	$|A| = 0$

Theorem 6.9.11 Multiply a row or column of a determinant [A] by a scalar k

1g	Given	k constant for $k \in (R^n$ or $C^n)$
2g		$A \equiv (a_i^j)$
3g		$B \equiv (b_i^j)$
4g		$b_i^j \equiv ka_i^j$ for single row or column
Steps	Hypothesis	$|B| \equiv k\,|A|$
1	From 3g by D6.9.1D Determinant of A	$|B| = \sum_{j=1}^{n!} sgn(\, parityn(\sigma(j))\,) \prod_{i=1}^{n} b_i^{j_i}$
2	From 1, 4g, A4.2.3 Substitution and A4.2.14 Distribution	$|B| = k\,(\, \sum_{j=1}^{n!} sgn(\, parityn(\sigma(j))\,) \prod_{i=1}^{n} a_i^{j_i}\,)$
$\therefore$	From 2 by D6.9.1D Determinant of A	$|B| \equiv k\,|A|$

Theorem 6.9.12 Addition of determinants by Row or Column

1g	Given	$A \equiv (a_i^j)$
2g		$B \equiv (b_i^j)$
3g		$C \equiv (c_i^j)$
4g		$a_i^j \equiv b_i^j + c_i^j$ for single row or column
Steps	Hypothesis	$|A| \equiv |B| + |C|$
1	From 1g by D6.9.1D Determinant of A	$|A| = \sum_{j=1}^{n!} \text{sgn}(\text{parityn}(\sigma(j))) \prod_{i=1}^{n} a_i^{j_i}$
2	From 1 by 4g substituting terms for either the ith or j^{th} row or column:	$|A| = \sum_{j=1}^{n!} \text{sgn}(\text{parityn}(\sigma(j))) \prod_{i=1}^{n} (b_i^{j_i} + c_i^{j_i})$
3	From 2 and A4.2.14 Distribution	$|A| = \sum_{j=1}^{n!} \text{sgn}(\text{parityn}(\sigma(j)))\prod_{i=1}^{n} b_i^{j_i} + \sum_{j=1}^{n!} \text{sgn}(\text{parityn}(\sigma(j)))\prod_{i=1}^{n} c_i^{j_i}$
∴	From 3 by D6.9.1D Determinant of A	$|A| = |B| + |C|$

Theorem 6.9.13 Adding a row inside a determinate [A] leaves it unaltered

1g	Given	$A \equiv (a_i^j)$
2g		$B \equiv (b_i^j)$
3g	where	$C \equiv (c_i^j)$ for $c_k^j = c_i^j$ row k and i are the same.
4g	and	$a_k^j \equiv b_k^j + c_k^j$ for k^{th} row.
Steps	Hypothesis	$|A| \equiv |B|$ adding the ith row in B to the kth row.
1	From 1g and T6.9.1 Uniqueness of a Determinate	$|A| = |A|$
2	From 1, 4g by T6.9.12 Addition of Determinants by Row or Column	$|A| = |B| + |C|$
3	From 2 and 3g and T6.9.10 A determinant with two identical rows or columns is zero	$|C| = 0$
∴	From 2, 3 by A4.2.3 Substitution and A4.2.7 Identity Add	$|A| = |B|$

Theorem 6.9.14 Adding a column inside a determinate [A] leaves it unaltered

1g	Given	$A \equiv (a_i^j)$
2g		$B \equiv (b_i^j)$
3g	where	$C \equiv (c_i^j)$ for $c_i^k = c_i^j$ column k and j are the same.
4g	and	$a_i^k \equiv b_i^k + c_i^k$ for k^{th} column.
Steps	Hypothesis	$|A| \equiv |B|$ adding the jth column in B to the kth column.
1	From 1g and T6.9.1 Uniqueness of a Determinate	$|A| = |A|$
2	From 1, 4g by T6.9.12 Addition of Determinants by Row or Column	$|A| = |B| + |C|$
3	From 2 and 3g and T6.9.9 Interchanging of any two rows or columns changes the sign in a determinant	$|C| = 0$
∴	From 2, 3 by A4.2.3 Substitution and A4.2.7 Identity Add	$|A| = |B|$

Theorem 6.9.15 Closer of Square Matrix Multiplication, by row

1g	Given	$A \equiv (a_i^j)$ any square matrix.
2g	and for	$|A| \neq 0$ independent
Steps	Hypothesis	$I \equiv A * ((adj\ A)\ /\ |A|)$ for $|A| \neq 0$
1	From T6.9.5 Expanding Determinant [A] of order [n] about a Row or Column, expanding about the ith row.	$|A| = \sum_{k=1}^{n} a_i^{\cdot k} A_i^{\cdot k}$
2	From 1 can be cast into another form by allowing i = j.	$|A| = \sum_{k=1}^{n} a_i^{\cdot k} A_j^{\cdot k}$ for i = j
3	From T6.9.5 Expanding Determinant [A] of order [n] about a Row or Column and T6.9.10 A determinant with two identical rows or columns is zero, identical row.	$0 = \sum_{k=1}^{n} a_i^{\cdot k} A_j^{\cdot k}$ for $i \neq j$
4	Combining steps 2 and 3 with the Kronecker Delta notation D6.1.6 Identity Matrix	$|A|\delta_i^j = \sum_{k=1}^{n} a_i^{\cdot k} A_j^{\cdot k}$
5	From 4 the summation can be placed into matrix form D6.1.2 Rectangular Matrix.	$(|A|\delta_i^j) = (\sum_{k=1}^{n} a_i^{\cdot k} A_j^{\cdot k})$
6	From 5 by factoring out $|A|$ T6.3.3 Uniqueness of Scalar Multiplication, from the Identity Matrix left side and factoring matrix A and cofactor right side of equality by A6.2.10 Matrix Multiplication.	$|A|*I = A * (A_j^{\cdot i})$
7	From 6 by A6.2.9 Matrix Multiplication by a Scalar	$|A|*I = A * {}^t(A_i^j)$
8	From 7 by D6.9.5 Adjoint of a square matrix A	$|A|*I = A * (adj\ A)$
9	From A4.2.2 Equality	$1/|A| = 1/|A|$
10	From 8, 9 and T4.4.2 Equalities: Uniqueness of Multiplication	$|A|(1/|A|)*I = A * (1/|A|) *(adj\ A)$
11	From 10, T4.4.16 Equalities: Product by Division, Common Factor and A4.2.13 Inverse Multp	$1 * I = A * (1/|A|) *(adj\ A)$
$\therefore$	From 11, A6.2.9 Matrix Multiplication by a Scalar and A4.2.12 Identity Multp	$I = A * ((adj\ A)\ /\ |A|)$ for $|A| \neq 0$

Theorem 6.9.16 Closer of Square Matrix Multiplication, by column

1g	Given	$A \equiv (a_i^j)$ any square matrix.
2g	and for	$|A| \neq 0$ independent
Steps	Hypothesis	$I \equiv A * ((adj\ A) / |A|)$ for $| A | \neq 0$
1	From T6.9.5 Expanding Determinant [A] of order [n] about a Row or Column, expanding about the jth column.	$|A| = \sum_{k=1}^{n} a_k^{\ j} A_k^{\ j}$
2	From 1 can be cast into another form by allowing i = j.	$|A| = \sum_{k=1}^{n} a_k^{\ i} A_k^{\ j}$ for i = j
3	From T6.9.5 Expanding Determinant [A] of order [n] about a Row or Column, Expanding Determinant [A] of order [n] about a Row or Column and T6.9.10 A determinant with two identical rows or columns is zero, identical row.	$0 = \sum_{k=1}^{n} a_k^{\ i} A_k^{\ j}$ for i ≠ j
4	Combining steps 2 and 3 with the Kronecker Delta notation D6.1.6 Identity Matrix	$|A|\delta_j^{\ i} = \sum_{k=1}^{n} a_k^{\ i} A_k^{\ j}$
5	From 4 the summation can be placed into matrix form D6.1.2 Rectangular Matrix.	$(|A|\delta_j^{\ i}) = (\sum_{k=1}^{n} a_k^{\ i} A_k^{\ j})$
6	From 5 by factoring out |A| T6.3.3 Uniqueness of Scalar Multiplication, from the Identity Matrix left side and factoring matrix A and cofactor right side of equality by A6.2.10 Matrix Multiplication.	$|A|*{}^{t}I = {}^{t}A * (A_i^j)$
7	From 6 by T6.4.1 Uniqueness of Transposition of Matrices and T6.4.5 Distribution of the transpose operation for products	$|A|*{}^{t}({}^{t}I) = {}^{t}({}^{t}A) * {}^{t}(A_i^j)$
8	From 7 by T6.4.3 of the transpose operation	$|A|*I = A * {}^{t}(A_i^j)$
9	From 8 by D6.9.5 Adjoint of a square matrix A	$|A|*I = A * (adj\ A)$
10	From A4.2.2 Equality	$1/|A| = 1/|A|$
11	From 9, 10 and T4.4.2 Equalities: Uniqueness of Multiplication	$|A|(1/|A|)*I = A * (1/|A|) *(adj\ A)$
12	From 11, T4.4.16 Equalities: Product by Division, Common Factor and A4.2.13 Inverse Multp	$1 * I = A * (1/|A|) *(adj\ A)$
∴	From 12, A6.2.9 Matrix Multiplication by a Scalar and A4.2.12 Identity Multp	$I = A * ((adj\ A) / |A|)$ for $| A | \neq 0$

Theorem 6.9.17 Square and Inverse Matrices are Commutative

1g	Given	$A \equiv (a_i^j)$ any square matrix.
2g	and for	$|A| \neq 0$ independent
3g		$I \equiv A * ((adj\ A) / |A|)$
4g		$I \equiv ((adj\ A) / |A|) * A$
Steps	Hypothesis	$I \equiv A * ((adj\ A) / |A|) \equiv ((adj\ A) / |A|) * A$ for $|A| \neq 0$
1	From A4.2.2 Equality	$|A| = |A|$
2	From 3g and 4g	$I = A * ((adj\ A) / |A|)$ $I = ((adj\ A) / |A|) * A$
3	From 1, 2 and T4.4.2 Equalities: Uniqueness of Multiplication	$|A|*I = A * |A|*((adj\ A) / |A|)$ $|A|*I = |A|*((adj\ A) / |A|) * A$
4	From 3, A6.2.9 Matrix Multiplication by a Scalar and D6.9.5 Adjoint of a square matrix A	$(|A|\delta_i^j) = A * ({}^t(A_i^j)\ |A| / |A|)$ $|A|*I = ({}^t(A_i^j)\ |A| / |A|) * A$
5	From 4 and A4.2.12 Identity Multp	$(|A|\delta_i^j) = A * ({}^t(A_i^j)\ 1)$ $|A|*I = ({}^t(A_i^j)\ 1) * A$
6	From 5 and A4.2.12 Identity Multp	$(|A|\delta_i^j) = A * ({}^t(A_i^j))$ $|A|*I = ({}^t(A_i^j)) * A$
7	From 6 and A6.2.9 Matrix Multiplication by a Scalar	$(|A|\delta_i^j) = (a_i^j) * ((A_j^i))$ $|A|*I = ((A_j^i)) * (a_i^j)$
8	From 7 and A6.2.10 Matrix Multiplication	$(|A|\delta_i^j) = (\sum_{k=1}^{n} a_i^k\ {}^tA_k^j)$ $|A|*I = (\sum_{k=1}^{n} {}^tA_i^k\ a_k^j)$
9	From 8 and A6.2.9 Matrix Multiplication by a Scalar	$(|A|\delta_i^j) = (\sum_{k=1}^{n} a_i^k\ A_j^k)$ $|A|*I = (\sum_{k=1}^{n} A_k^i\ a_k^j)$
10	From 9 and A4.2.10 Commutative Multp	$(|A|\delta_i^j) = (\sum_{k=1}^{n} a_i^k\ A_j^k)$ $|A|*I = (\sum_{k=1}^{n} a_k^j\ A_k^i)$
11	From 10, D6.1.6 Identity Matrix and A6.2.9 Matrix Multiplication by a Scalar	$(|A|\delta_i^j) = (\sum_{k=1}^{n} a_i^k\ A_j^k)$ $(|A|\delta_j^i) = (\sum_{k=1}^{n} a_k^j\ A_k^i)$
12	From 11 since off diagonal elements are zero for $i \neq j$ than the equating diagonal elements for $i = j$ are:	$|A| = \sum_{k=1}^{n} a_i^k\ A_i^k$ for $i = j$ $|A| = \sum_{k=1}^{n} a_k^i\ A_k^i$ for $i = j$
13	From 12 and T6.9.5 Expanding Determinant [A] of order [n] about a Row or Column, Adjoint of a square matrix A	$|A| = \sum_{k=1}^{n} a_i^k\ A_i^k$ for $i = j$ $|A| = \sum_{k=1}^{n} a_i^k\ A_i^k$ for $i = j$
$\therefore$	From 13 and A6.2.7 Equality of Matrices	$I \equiv A * ((adj\ A) / |A|) \equiv ((adj\ A) / |A|) * A$ Every commutative element is equal to every other.

Tensor Calculus & Physics: A General Treatise

Theorem 6.9.18 Inverse Square Matrix for Multiplication

Steps	Hypothesis	$A^{-1} \equiv \text{adj } A \,/\, |A| \text{ for } |A| \neq 0$
1	From T6.9.17 Square and Inverse Matrices are Commutative	$I = A * ((\text{adj } A) / |A|) = ((\text{adj } A) / |A|) * A$
$\therefore$	From 1 it must be concluded that the quantity $((\text{adj } A) / |A|)$ is the inverse matrix to A, because it obtains the unity matrix, D6.9.6.	$A^{-1} = \text{adj } A / |A| \text{ for } |A| \neq 0$

Theorem 6.9.19 Closure with respect to the Identity Matrix for Multiplication

Steps	Hypothesis	$I = A * A^{-1} = A^{-1} * A \text{ for } |A| \neq 0$
$\therefore$	From T6.9.17 Square and Inverse Matrices are Commutative, T6.9.18 Inverse Square Matrix for Multiplication and A4.2.3 Substitution	$I = A * A^{-1} = A^{-1} * A \text{ for } |A| \neq 0$

Corollary 6.9.19.1 Zero Exponential for a Square Matrix

Steps	Hypothesis	$I = A^{0}$ for an invertible square matrix
1	From T6.9.19 Closure with respect to the Identity Matrix for Multiplication and T6.3.21 Matrix Exponential Notation	$I = A^{1} * A^{-1} = A^{-1} * A^{1}$ for $|A| \neq 0$
2	From 1 and D6.1.13 Matrix Subtraction Expediential Notation	$I \equiv A^{(1-1)} = A^{(-1+1)}$
$\therefore$	From 2 and D4.1.19 Primitive Definition for Rational Arithmetic	$I = A^{0}$ for an invertible square matrix

Theorem 6.9.20 Determinant of a Triangular Matrix

1g 2g	Given	$T \equiv (a_i{}^j)$ $t_i{}^j \equiv 0$ for upper $i < j$ or $i > j$ lower.
Steps	Hypothesis	$|T| \equiv \prod\limits_{i=1}^{n} t_i{}^i$
1	From D6.9.1D	$|T| = \sum\limits_{j=1}^{n!} \text{sgn}(\text{parityn}(\sigma(j))) \prod\limits_{i=1}^{n} t_i{}^{j_i}$
2	From 1 by expanding the summation to the first term for $\sigma(1) = \varepsilon$ and C6.7.3.2 Sign of the Identity Permutation	$|T| = \prod\limits_{i=1}^{n} t_i{}^i + \sum\limits_{j=2}^{n!} \text{sgn}(\text{parityn}(\sigma(j))) \prod\limits_{i=1}^{n} t_i{}^{j_i}$
$\therefore$	From 2 2g all off terms will have at least one $a_j^i = 0$ factor for $i \neq j$	$|T| = \prod\limits_{i=1}^{n} t_i{}^i$

Corollary 6.9.20.1 Determinate of a Diagonal Matrix

Steps	Hypothesis	$\mid D \mid = \Pi^n d_i^i$
$\therefore$	From T6.9.20 Determinant of a Triangular Matrix and D6.1.5A Diagonal Matrix	$\mid D \mid = \Pi^n d_i^i$

Corollary 6.9.20.2 Determinant of a Identity Matrix

Steps	Hypothesis	$\mid I \mid = 1$
1	From D6.1.6 Identity Matrix and C6.9.20.1 Determinate of a Diagonal Matrix	$\mid I \mid = \Pi^n \delta_i^i$
$\therefore$	From 1 for all diagonal elements 1	$\mid I \mid = 1$

Theorem 6.9.21 Lower Triangular Inverse Matrix

1g	Given	$T_L = (t_i^j)$ for $T_L \in {}_\varepsilon E$
2g		$t_i^j = 0$ for $i > j$
3g		$t_i^j \neq 0$ for $i < j$

Steps	Hypothesis	$T_L^{-1} \equiv (sgn(i + j)M_i^j / \mid T_L \mid \setminus (1/t_i^i) \setminus 0)$ for $M_i^j \neq 0$ and $i < j$
1	From 1g and T6.9.18 Inverse Square Matrix for Multiplication	$T_L^{-1} = adj\, T_L / \mid T_L \mid$
2	From 1 and D6.9.5 Adjoint of a square matrix A	$adj\, T_L = {}^t(T_{Li}^j)$
3	From 2 and D6.9.4 Cofactor of a square matrix A	$adj\, T_L = {}^t(\, sgn(i + j)\, \mid M_i^j \mid)$
4	From 3 following through with transposition	$adj\, T_L = (\, sgn(j + i)\, \mid M_j^i \mid)$
5	From 4 and expanding the cofactor about a row or column such that 2g applies.	$\mid M_j^i \mid = \sum_{k=1}^{n-1} t_j^k Q_k^i \qquad$ for $j > i$
6	From 5	$\mid M_j^i \mid = \sum_{k=1}^{n-1} 0 * Q_k^i = 0 \qquad$ for $j > i$
7	From 4 and 2g yield diagonal cofactors that are determinates of triangular matrices. Using T6.9.20 Determinant of a Triangular Matrix, yields	$\mid M_i^i \mid = t_i^1 * t_i^2 * \ldots * t_i^{(i-1)} * t_i^{(i+1)} * \ldots * t_i^n$ for $j = i$
8	From 1 and T6.9.20 Determinant of a Triangular Matrix	$\mid T_L \mid = \Pi^n t_i^i$
9	From 1, 7 and 8	$\mid M_i^i \mid / \mid T_L \mid = 1 / t_i^i \qquad$ for $j = i$
10	From 4 and 3g	$\mid M_i^i \mid \neq 0 \qquad$ for $i < j$
11	From 8 and 10	$\mid M_j^i \mid / \mid T_L \mid \neq 0 \qquad$ for $i < j$
$\therefore$	From 1, 6, 9 and 11	$T_L^{-1} = (sgn(i + j)M_i^j / \mid T_L \mid \setminus (1/t_i^i) \setminus 0)$ for $M_i^j \neq 0$ and $i < j$

Theorem 6.9.22 Upper Triangular Inverse Matrix

Steps	Hypothesis	
1g	Given	T_L and $T_U \in {}_\varepsilon E$
2g		$T_L{}^\tau = T_U$ transpose of lower triangular matrix is equivalent to an upper triangular matrix.
Steps	Hypothesis	$T_U{}^{-1} \equiv (T_L{}^{-1})^\tau$
1	From 1g and T6.9.19 Closure with respect to the Identity Matrix for Multiplication	$I = T_L * T_L{}^{-1}$
2	From 1 and T6.4.4 Transpose of a constant	${}^\tau I = {}^\tau T_L * {}^\tau (T_L{}^{-1})$
3	From 2, 2g, T6.4.5 Distribution of the transpose operation for products and T6.4.6 Transpose of the Identity Matrix	$I = T_U * {}^\tau (T_L{}^{-1})$
$\therefore$	From 3 it is concluded	$T_U{}^{-1} = {}^\tau (T_L{}^{-1})$

Corollary 6.9.22.1 Diagonal Inverse Matrix

Steps	Hypothesis	
1g	Given	$T_L = (t_i{}^j)$ for $T_L \in {}_\varepsilon E$
2g		$t_i{}^j = 0$ for $i > j$ and $i < j$ for a diagonal matrix
3g		$T_L = D$ special case
4g		$t_i{}^j = d_i{}^j$ special case
Steps	Hypothesis	$D^{-1} = (0 \setminus (1/d_i{}^i) \setminus 0)$
1	From T6.9.21 Lower Triangular Inverse Matrix	$T_L{}^{-1} = (\text{sgn}(i + j) M_i{}^j / |T_L| \setminus (1/t_i{}^i) \setminus 0)$ for $M_i{}^j \neq 0$ and $i < j$
2	From 1 and expanding the cofactor about a row or column	$|M_j{}^i| = \sum\limits_{k=1}^{n-1} t_j{}^k Q_k{}^i \qquad$ for $i > j$ and $i < j$
3	From 2 and 2g	$|M_j{}^i| = \sum\limits_{k=1}^{n-1} 0 * Q_k{}^i = 0 \qquad$ for $i > j$ and $i < j$
$\therefore$	From 1, 3, 3g, 4g and A4.2.3 Substitution	$D^{-1} = (0 \setminus (1/d_i{}^i) \setminus 0)$

Tensor Calculus & Physics: A General Treatise

Theorem 6.9.23 Determinate of Upper and Lower Inverse Triangular Matrix

1g	Given	T_L and $T_U \in {}_\varepsilon E$
Steps	Hypothesis	$\mid T_U^{-1} \mid \equiv \mid T_L^{-1} \mid \equiv 1 / \mid T \mid$
1	From 1g, T6.9.22 Upper Triangular Inverse Matrix and T6.9.1 Uniqueness of a Determinate	$\mid T_U^{-1} \mid = \mid {}^{\tau}(T_L^{-1}) \mid$
2	From 1 and T6.9.22 Upper Triangular Inverse Matrix	$\mid T_U^{-1} \mid = \mid {}^{\tau}(\text{sgn}(i+j)M_i^j / |T_L| \setminus (1/t_i^i) \setminus 0) \mid$
3	From 2 and A6.2.9 Matrix Multiplication by a Scalar	$\mid T_U^{-1} \mid = \mid (0 \setminus (1/t_i^i) \setminus \text{sgn}(j+i)M_j^i / |T_L|) \mid$
4	From 3 and T6.9.20 Determinant of a Triangular Matrix	$\mid T_U^{-1} \mid = 1 / \prod t_i^i$
5	From 4 and T6.9.20 Determinant of a Triangular Matrix	$\mid T_U^{-1} \mid = 1 / \mid T \mid$
6	From 1g and T6.9.21 Lower Triangular Inverse Matrix	$\mid T_L^{-1} \mid = \mid (\text{sgn}(i+j)M_i^j / |T_L| \setminus (1/t_i^i) \setminus 0) \mid$
7	From 3 and T6.9.20 Determinant of a Triangular Matrix	$\mid T_L^{-1} \mid = 1 / \prod t_i^i$
8	From 4 and T6.9.20 Determinant of a Triangular Matrix	$\mid T_L^{-1} \mid = 1 / \mid T \mid$
$\therefore$	From 5 and 8	$\mid T_U^{-1} \mid \equiv \mid T_L^{-1} \mid \equiv \mid T \mid^{-1} \equiv 1 / \mid T \mid$

Theorem 6.9.24 Determinate of a constant times a row or column for a Triangular Matrix

1g 2g	Given	T_L and $T_U \in {}_\varepsilon E$ k any non-zero number
Steps	Hypothesis	$\mid kT \mid \equiv k \prod^n t_i^i$ for kT x (R_i or C^j)
1	From 1g, D6.1.11 Elementary Matrices	$\mid kT \mid = \mid kT \mid$ for kT x (R_i or C^j)
2	From 1 and T6.9.11 Multiply a row or column of a determinant [A] by a scalar k	$\mid kT \mid = k \mid T \mid$
$\therefore$	From 2 and T6.9.20 Determinant of a Triangular Matrix	$\mid kT \mid = k \prod^n t_i^i$ for kT x (R_i or C^j)

Theorem 6.9.25 Determinate of a constant times a Triangular Matrix

1g 2g	Given	T_L and $T_U \in {}_\varepsilon E$ k any non-zero number
Steps	Hypothesis	$\mid kT \mid \equiv k^n \mid T \mid \equiv k^n \prod^n t_i^i$
1	From 1g, D6.1.11 Elementary Matrices	$\mid kT \mid = \mid kT \mid$
2	From 1 and T6.9.20 Determinant of a Triangular Matrix	$\mid kT \mid = \prod^n k \, t_i^i$
3	From 2 factoring out all k's in the product	$\mid kT \mid = \prod^n k \prod^n t_i^i$
4	From 3 and D4.1.17 Exponential Notation	$\mid kT \mid = k^n \prod^n t_i^i$
5	From 4 and T6.9.20 Determinant of a Triangular Matrix	$\mid kT \mid = k^n \mid T \mid$
$\therefore$	From 4 and 5	$\mid kT \mid \equiv k^n \mid T \mid \equiv k^n \prod^n t_i^i$

Lema 6.9.2 Determinant of an elementary matrix

1g	Given	$E \equiv (e_i^j) \in {}_\varepsilon E$
Steps	Hypothesis	$|E| = \Pi^n e_i^i$
∴	From 1g, D6.1.11 Elementary Matrices and T6.9.20 Determinant of a Triangular Matrix on all types of elementary matrices.	$|E| = \Pi^n e_i^i$

Theorem 6.9.26 Elementary and matrix products can be factored as individual determinants

1g	Given	$A \equiv (a_i^j)$
2g		$E \equiv (e_i^j)$ for $E \in {}_\varepsilon E$
3g		$p_{\sigma n}(j) \equiv \text{parityn}(\sigma(j))$
4g	Product of A * E elements	$c_i^j \equiv \sum\limits_{k=1}^{n} a_k^j e_i^k = a_i^j e_i^i$
5g		$|A * E|$
Steps	Hypothesis	$|A * E| \equiv |A| * |E|$
1	From 3g and 5g by D6.9.1D	$|A * E| = \sum\limits_{j=1}^{n!} \text{sgn}(p_{\sigma n}(j)) \prod\limits_{i=1}^{n} c_i^{j_i}$
2	From 1 by substitution of 4g	$|A * E| = \sum\limits_{j=1}^{n!} \text{sgn}(p_{\sigma n}(j)) \prod\limits_{i=1}^{n} (\sum\limits_{k=1}^{n} a_k^{j_i} e_i^k)$
3	From 2 by 4g	$|A * E| = \sum\limits_{j=1}^{n!} \text{sgn}(p_{\sigma n}(j)) \prod\limits_{i=1}^{n} e_i^i a_i^{j_i}$
4	From 3 by factoring e_i^i and A4.2.14 Distribution	$|A * E| = (\sum\limits_{j=1}^{n!} \text{sgn}(p_{\sigma n}(j)) \prod\limits_{i=1}^{n} a_i^{j_i}) * (\prod\limits_{i=1}^{n} e_i^i)$
∴	From 4 by D6.9.1D and T6.9.20 Determinant of a Triangular Matrix	$|A * E| = |A| * |E|$

Corollary 6.9.26.1 Determinant of a Matrix made from Elementary Matrices

1g	Given	$A \equiv \Pi^n E^i$ partitioned into Elementary Matrices
2g		$|E^i| \equiv \Pi^n E^i$ elements $\leq i$ are missing or $(n-i \times n-i)$ elementary matrix of A
Steps	Hypothesis	$|A| \equiv \Pi^n |E^i|$
1	From 1g and T6.9.1 Uniqueness of a Determinate	$|A| = |\prod_n E^i|$
2	From 1 expanding one matrix product	$|A| = |E^1 * (\prod^{2,n} E^i)|$
3	From 2, 2g and A4.2.3 Substitution	$|A| = |E^1 * |E^1||$
4	From 2 by T6.9.11 Multiply a row or column of a determinant [A] by a scalar k	$|A| = |E^1| * |E^1|$
∴	From 4 by repeat steps 2 and 4 for all i:	$|A| = \Pi^n |E^i|$

Theorem 6.9.27 The product of two matrices can be factored as individual determinants

1g	Given	$|A * B| \equiv |A * B|$
Steps	Hypothesis	$|A * B| \equiv |A| * |B|$
1	From 1g and A6.2.12 Matrix Decomposition into Elementary Matrices	$|A * B| = |\Pi^n E^i * B|$
2	From 1 and C6.9.26.1 Determinant of a Matrix made from Elementary Matrices	$|A * B| = ||A| B|$
∴	From 2 and T6.9.11 Multiply a row or column of a determinant [A] by a scalar k	$|A * B| = |A| * |B|$

Corollary 6.9.27.1 Reciprocal of the Inverse Matrix

Steps	Hypothesis	$|A^{-1}| = |A|^{-1}$
1	From T6.9.19 Closure with respect to the Identity Matrix for Multiplication	$I = A * A^{-1}$
2	From 1 and T6.9.1 Uniqueness of a Determinate	$|I| = |A * A^{-1}|$
3	From 2, C6.9.20.2 Determinant of a Identity Matrix and T6.9.26 Elementary and matrix products can be factored as individual determinants	$1 = |A||A^{-1}|$
4	From 3, A4.2.10 Commutative Multp and T4.4.4A Equalities: Right Cancellation by Multiplication	$1/|A| = |A^{-1}|$
∴	From 4, A4.2.2B Equality and D4.1.18 Negative Exponential	$|A^{-1}| = |A|^{-1}$

Theorem 6.9.28 The inverse determinate of an elementary matrix

1g	Given	$E = (e_i^j)$ for $E \in {}_\varepsilon E$
Steps	Hypothesis	$|E^{-1}| \equiv |E|^{-1} \equiv 1/\Pi^n e_i^i$
∴	From D6.1.11 Elementary Matrices, T6.9.23 Determinate of Upper and Lower Inverse Triangular Matrix and T6.9.20 Determinant of a Triangular Matrix	$|E^{-1}| = |E|^{-1} = 1/\Pi^n e_i^i$

Tensor Calculus & Physics: A General Treatise

Theorem 6.9.29 Matrix Exponentials and Determinate Relationship

Steps	Hypothesis	$|A^n| = |A|^n$
1	From 1g and T6.3.21 Matrix Exponential Notation	$A^n = \Pi^n *A$
2	From 1 and T6.9.1 Uniqueness of a Determinate	$|A^n| = |\Pi^n *A|$
3	From 2 and T6.9.27 The product of two matrices can be factored as individual determinants	$|A^n| = \Pi^n |A|$
$\therefore$	From 3 and D4.1.17 Exponential Notation	$|A^n| = |A|^n$

Theorem 6.9.30 Triangular Matrix Exponentials and Determinate Relationship

1g	Given	T Triangler Matrix
Steps	Hypothesis	$|T^n| = \Pi^n (t_i{}^i)^n$
1	From T6.9.29 Matrix Exponentials and Determinate Relationship	$|T^n| = |T|^n$
2	From 1 and T4.8.6 Integer Exponents: Uniqueness of Exponents	$|T^n| = (|T|)^n$
3	From 2 and T6.9.20 Determinant of a Triangular Matrix	$|T^n| = (\Pi^n t_i{}^i)^n$
$\therefore$	From 3 and A4.2.21 Distribution Exp	$|T^n| = \Pi^n (t_i{}^i)^n$

Theorem 6.9.31 Form for the Product of n Lower Triangular Matrices

1g	Given	T_L Lower Triangular Matrix
Steps	Hypothesis	$T_L{}^n = (. p_i{}^j . \backslash . (t_i{}^i)^n . \backslash . 0 .)$
1	From T6.9.30 Triangular Matrix Exponentials and Determinate Relationship	$|T_L{}^n| = \Pi^n (t_i{}^i)^n$
2	From 1, T6.9.20 Determinant of a Triangular Matrix and A6.2.10 Matrix Multiplication where	$|T_L{}^n| = |. p_i{}^j . \backslash . (t_i{}^i)^n . \backslash . 0 .|$ $p_i{}^j$ is the non-zero product terms for $i < j$
$\therefore$	From 2 and T6.9.1 Uniqueness of a Determinate	$T_L{}^n = (. p_i{}^j . \backslash . (t_i{}^i)^n . \backslash . 0 .)$

Theorem 6.9.32 Form for the Product of n Upper Triangular Matrices

1g	Given	T_U Lower Triangular Matrix
Steps	Hypothesis	$T_U{}^n = (. 0 . \backslash . (t_i{}^i)^n . \backslash . p_i{}^j .)$
1	From T6.9.30 Triangular Matrix Exponentials and Determinate Relationship	$|T_U{}^n| = \Pi^n (t_i{}^i)^n$
2	From 1, T6.9.20 Determinant of a Triangular Matrix and A6.2.10 Matrix Multiplication where	$|T_U{}^n| = |. 0 . \backslash . (t_i{}^i)^n . \backslash . p_i{}^j .|$ $p_i{}^j$ is the non-zero product terms for $i > j$
$\therefore$	From 2 and T6.9.1 Uniqueness of a Determinate	$T_U{}^n = (. 0 . \backslash . (t_i{}^i)^n . \backslash . p_i{}^j .)$

Theorem 6.9.33 Form for the Product of n Diagonal Matrices

1g	Given	D Diagonal Matrix is a special case of a Triangular
Steps	Hypothesis	$D^n = (. \, 0 \, . \, \backslash \, . \, (d_i^i \,)^n \, . \, \backslash \, . \, 0 \, .)$
$\therefore$	From 1g, D6.1.5 Diagonal Matrix and T6.9.32 Form for the Product of n Upper Triangular Matrices	$D^n = (. \, 0 \, . \, \backslash \, . \, (d_i^i \,)^n \, . \, \backslash \, . \, 0 \, .)$

Tensor Calculus & Physics: A General Treatise

Section 6.10 Short Cuts and other Missilanous Theorems

Theorem 6.10.1 Cancellation Law for Square Matrix Multiplication

1g	Given	$A\,C = B\,C$ for A, B and C square matrices
Steps	Hypothesis	$A = B$
1	From A6.2.7 Equality of Matrices	$C^{-1} = C^{-1}$
2	From 1, 1g and T6.3.4 Uniqueness of Matrix Multiplication	$A\,C\,C^{-1} = B\,C\,C^{-1}$
3	From 2 and T6.3.11 Matrix Association for Multiplication	$A\,(C\,C^{-1}) = B\,(C\,C^{-1})$
4	From 3 and T6.9.19 Closure with respect to the Identity Matrix for Multiplication	$A\,I = B\,I$
∴	From 4 and T6.3.12 Matrix Identity right of Multiplication	$A = B$

Theorem 6.10.2 Determinate of a 2x2 Matrix

Steps	Hypothesis	$|A_{2x2}| = ad - cb$ EQ A	$|A_{2x2}| = \begin{vmatrix} a & b \\ c & d \end{vmatrix}$ EQ B
1	From D6.9.7 Matrix 2 x 2	$A_{2x2} = \begin{pmatrix} a & b \\ c & d \end{pmatrix}$	
2	From 1 and T6.9.1 Uniqueness of a Determinate	$|A_{2x2}| = \begin{vmatrix} a & b \\ c & d \end{vmatrix}$	
3	From 2 and D6.91C Determinant of A	$|A_{2x2}| = \text{sgn}(\sigma(1))\,ab + \text{sgn}(\sigma(2))\,cd$	
∴	From 3 and D6.6.3B Association of Sign to Parity	$|A_{2x2}| = ad - cb$ EQ A	$|A_{2x2}| = \begin{vmatrix} a & b \\ c & d \end{vmatrix}$ EQ B

Theorem 6.10.3 Inverse of a 2x2 Matrix

Steps	Hypothesis	$A_{2x2}^{-1} = \dfrac{1}{(ad - bc)} \begin{pmatrix} +d & -b \\ -c & +a \end{pmatrix}$ for $(ad - bc) \neq 0$
1	From D6.9.7 Matrix 2 x 2	$A_{2x2} = \begin{pmatrix} a & b \\ c & d \end{pmatrix}$
2	From 1 and T6.9.18 Inverse Square Matrix for Multiplication	$A_{2x2}^{-1} = \text{adj } A_{2x2} / |A_{2x2}|$ for $|A_{2x2}| \neq 0$
3	From 2 and D6.9.5 Adjoint of a Square Matrix A	$A_{2x2}^{-1} = {}^t(A_i^j) / |A_{2x2}|$
4	From 3 and D6.9.4 Cofactor of a Square Matrix A	$A_{2x2}^{-1} = {}^t(\text{sgn}(i + j) \, |M_i^j|) / |A_{2x2}|$
5	From 4 and D6.9.3 Minor of a Square Matrix A evaluating the Minor	$A_{2x2}^{-1} = {}^t\begin{pmatrix} \text{sgn}(2)\,d & \text{sgn}(3)\,c \\ \text{sgn}(3)\,b & \text{sgn}(4)\,a \end{pmatrix} / |A_{2x2}|$
6	From 5 and D6.6.3A Association of Sign to Parity	$A_{2x2}^{-1} = {}^t\begin{pmatrix} +d & -c \\ -b & +a \end{pmatrix} / |A_{2x2}|$
∴	From 6, A6.2.11 Matrix Transposition, T6.10.2 Determinate of a 2x2 Matrix and A4.2.3 Substitution	$A_{2x2}^{-1} = \dfrac{1}{(ad - bc)} \begin{pmatrix} +d & -b \\ -c & +a \end{pmatrix}$ for $(ad - bc) \neq 0$

Section 6.11 Column Vector Operations

Table 6.11.1 Legal Column Vector Multiplications

	E (n x 1)	F (1 x n)
A (1 x n)	AE (1 x n)	AF (1 x n)
B (n x 1)	BE (n x 1)	BF (n x n)

Note: White is legal based on axioms A6.2.6 "Column Vector Multiplication", gray not.

Theorem 6.11.1 Commute Column Vectors under Conservation of Order

1g	Given	A an n x n square matrix
2g		B_v (n x 1) and C_v (n x 1) column vectors
3g		$A = B_v^T C_v \qquad$ (n x n) $\neq$ (1 x n)(n x 1) = (1 x n)
Step	Hypothesis	$A = C_v B_v^T$ for matrix form (n x n) = (n x 1)(1 x n)
1	From 3g	$A = B_v^T C_v \qquad$ (n x n) $\neq$ (1 x n)(n x 1) = (1 x n)
∴	From 1, In order to maintain the order on the left hand side the column vectors must be commuted	$A = C_v B_v^T$ for matrix form (n x n) = (n x 1)(1 x n)

Theorem 6.11.2 Commute Column Vectors under Conservation of Transposition

1g	Given	A an n x n square matrix
2g		B_v (n x 1) and C_v (n x 1) column vectors
3g		$A = B_v C_v^T \qquad$ (n x n) = (n x 1) (1 x n)
Step	Hypothesis	$A = C_v B_v^T$ for matrix form (n x n) = (n x 1)(1 x n)
1	From 3g	$A = B_v C_v^T \qquad$ (n x n) = (n x 1) (1 x n)
2	From 1 and T6.4.1 Uniqueness of Transposition of Matrices	$A^T = (B_v C_v^T)^T$
3	From 2, T6.4.5 Distribution of the transpose operation for products and T6.4.3 Invariance of the transpose operation	$A = B_v^T C_v^{T\,T}$ for matrix form (n x n) $\neq$ (1 x n)(n x 1) = (1 x n)
∴	From 3, In order to maintain the order on the left hand side the column vectors must be commuted	$A = C_v B_v^T$ for matrix form (n x n) = (n x 1)(1 x n)

Section 6.12 Permutation System: Fundamental Properties

Permuatation sets have a varity of fundamental properties, which classifies them as axioms. From appendix-F those properties are restated as formal axioms.

First proposition the cyclic property of permutation of a set. From observation OF.1.3 "Cyclic construction of permutation numbers, C3 to C[n-2]:", hence sets of permutations are cyclic and have closure.

Axiom 6.12.1 Closure of Permutation of Sets Under Any Operation

Any set of permutations S_n is cyclic, hence has closer under any operation $[◊]$, $(\sigma ◊ \tau) \in S_n$.

Axiom 6.12.2 Transposition of Any Two Adjacent Indices Changes Parity by One

For any parityn-a and count-a then any transposition of any two adjacent indices gives,

$$\text{parityn-b} \equiv \begin{cases} (\text{parityn - a}) - 1 & \text{for (count index - a)} > \text{(count index - b)} \\ (\text{parityn - a}) + 1 & \text{for (count index - a)} < \text{(count index - b)} \end{cases}$$

Axiom 6.12.3 Transposition of Any Two Indices Gives an Opposite Odd or Even Parity

For any parity-a (even/odd) then any transposition of any two indices gives an opposite parity-b (odd/even).

Tensor Calculus & Physics: A General Treatise

Table of Contents

Tensor Calculus & Physics: A General Treatise

Tensor Calculus & Physics: A General Treatise

List of Tables

List of Figures

List of Givens

List of Observations

Chapter 7 Principles of Modern Vectors

Section 7.1 Algebra of Complex Numbers

With Newton's formal reaffirmation of the parallelogram law, vector addition had been anchored down. Everybody was pretty, much happy, except: [BRI60, Vol 6, pg 181]

Since certain algebraic equations such as $x^2 + 1 = 0$ have no solutions in real numbers, early mathematicians were led to consider purely formal solutions involving square roots of negative numbers. Thus Heron of Alexandria (c. ADE 100) obtained the solution $\sqrt{-63}$, and Girolamo Cardan (1545) wrote

$$(5 + \sqrt{-15}) * (5 - \sqrt{-15}) = 25 + (\sqrt{-15}) - (\sqrt{-15}) - (-15) = 40.$$

These numbers were considered to be, quite meaningless, hence, the term *imaginary* was applied to them.

Geometric Representation:

If [**i**] is as Euler (1707-1783) defined it a solution of the equation $x^2 + 1 = 0$, numbers of the form $a + ib$ where [a] and [b] are real then the compound number is called ***complex*** (Karl Friedrich Gauss, 1831). The modern development of complex numbers begins with the discovery of a geometric interpretation for them. This was indistinctly set forth by John Wallis (1685), and in completely satisfactory form by Caspar Wessel (1799)

Wessel's work received no attention and was rediscovered by Jean Robert Argand (1806), and again by Gauss (1831). It is now frequently called the Argand diagram.

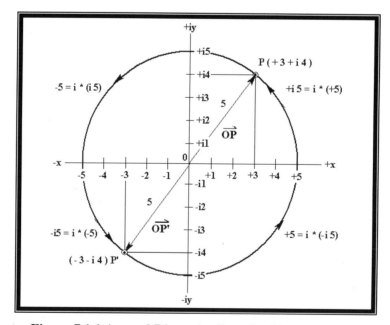

Figure 7.1.1 Argand Diagram, Complex Number Plane

As can be seen from the above figure complex numbers have some interesting properties. Multiplying by [i] rotates a complex number $90°$ and by [-1] $180°$.

Section 7.2 Complex Numbers: Definitions

Definition 7.2.1 Complex Field

Let $\exists$ an ordered product set C^2 called a ***complex field*** being the set of all complex numbers, than

$$C^2 \equiv R \times IM \equiv \{ (a, b) \mid (a, b) \in R, ib \in IM, a + ib\} \text{ where } [a + ib] \text{ is called a } \textbf{\textit{complex number}}.$$

By its very definition as being an ordered set:

$$R \times IM \neq IM \times R, \text{ hence non-commutative.}$$

Any point P in the complex plane can be described as a set of ordered pairs or if thought of as a vector then the compound Gaussian number can be used.

Definition 7.2.2 Complex Vector and Order Points

$$P(a, b) \equiv P(a + ib) \equiv P(\overrightarrow{OP}) \text{ or as a true vector}$$
$$\overrightarrow{OP} = (a, b) = a + ib$$

Now the complex number has been called a vector and not just, because like its real number counter part it can be multiplied by a [-1], thereby changing direction in the opposite way, but for the following reason as show in the diagram below.

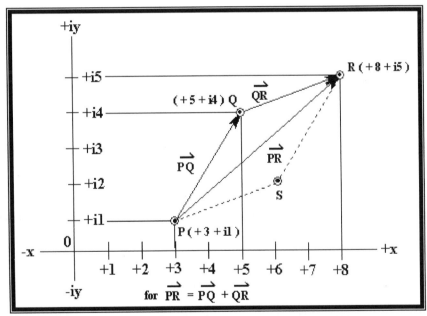

Figure 7.2.1 Parallelogram in the Complex Plane

As can be seen from the above figure complex numbers add and subtract with the Newton-Euclidian parallelograms, SPQR, hence can be considered a parallelogram of vectors. It also reveals something else all complex vectors can be broken down into simple constituent components that are unique to this vector system. They have properties that are **orthogonal** (ninety degrees to each other), and **unitary** (having a magnitude of one), (1, i).

Complex Vector Geometry:

Using Newton-Euclidian vectors was difficult, because of the geometric constructions required, now that same geometry is referenced to a standardized set of base unit vectors [ê] for a coordinate system rather than non-standard un-referenced set of component vectors. The vectors can now be converted algebraically making them easier to manipulate, which intern allows magnitude and angle to be quickly computed, and still be visual.

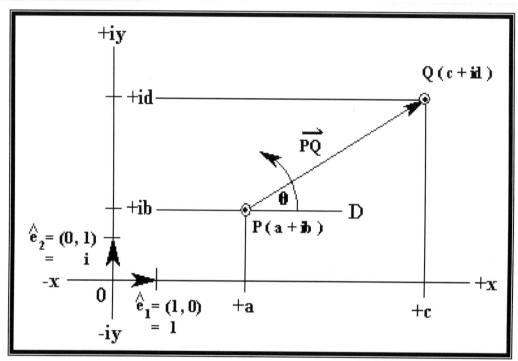

Figure 7.2.2 Unit Base Vectors and Elementary Geometry of the complex plane

From the above geometry the magnitude of the vector is

$$|\vec{PQ}| = \sqrt{(c - a)^2 + (d - b)^2} \text{ by the Pythagorean theorem.}$$

$$\angle QPD = \theta = \arctan[(d - b) / (c - a)]$$

One of the nice things about complex numbers is the ability to multiply them. As a natural extension of real numbers they carry with them all the properties of multiplication including some additional unique properties, see T7.4.6 Conjugation: Complex Vector Magnitude.

So, the magnitude of complex number or vector can be found by simply multiplying itself with a sign change to the imaginary term.

Tensor Calculus & Physics: A General Treatise

In order to start with complex numbers some additional definitions have to be implemented:

Given 7.2.1 The Following Complex Parameters

1g	Given	$\exists$ a set of quantities $(z, w) \in C^2$
2g	such that	$z \equiv a + ib$
3g	and	$w \equiv c + id,$
4g	were	$(a, b, c, d) \in R$

Definition 7.2.3 **Real and Imaginary Component Operators**

$a = Re\{z\}$ the real operator retrieves the real value of a complex number.
Equation A

and

$b = Im\{z\}$ the imaginary operator retrieves the pure imaginary value of a complex number.
Equation B

Definition 7.2.4 **Conjugant of a Complex Number**

if $z = a + ib$ than $z^* \equiv a - ib$ with superscripted asterisk denotes the conjugated operation on a complex number.

Definition 7.2.5 **Magnitude of a Complex Number**

$| z | \equiv | a + ib | = \sqrt{a^2 + b^2}$

where |z| is called the ***modulus*** or ***absolute magnitude*** of the vector.

Definition 7.2.6 **Phase Angle of a Complex Number**

$\theta = \arctan(Im\{z\} / Re\{z\})$

Definition 7.2.7 **Division of Complex Numbers**

$z \div w \equiv z / w$

In order to be forwards compatible with true vectors two new multiplication operators have been added.

Definition 7.2.8 **Dot or Inner Product**

$z \bullet w \equiv |z| \, |w| \cos(\theta)$ for $\theta \equiv \arctan(Im\{zw\} / Re\{zw\})$

Definition 7.2.9 **Cross product**

$z \times w \equiv |z| \, |w| \sin(\theta)$ for $\theta \equiv \arctan(Im\{zw\} / Re\{zw\})$

Definition 7.2.10 **Phasor or Polar Form of a Complex Number**

$z \equiv |z| \angle \varphi$

where $\angle \varphi$ is a short hand notation called a ***phasor*** and replaces the Euler's complex phase expression [$e^{i\varphi}$]. The word phasor coming from the angular phase difference as the polar form spins about the point of its tail in the complex plane with an angle $\angle \varphi$.

Tensor Calculus & Physics: A General Treatise

Definition 7.2.11 **Primitive Definition for Imaginary Arithmetic**

+	+i	+1	+0	-1	-i
+i	0 + 2i	1 + i	0 + i	-1 + i	0 + i0
+1	1 + i	2 + i0	1 + i0	0 + i0	1 − i
+0	0 + i	1 + i0	0 + i0	-1 + i0	0 − i
-1	-1+ i	0 + i0	-1 + i0	-2 + i0	-1 − i
-i	0 − i	1 − i	0 − i	-1 − i	0 − 2i

-	+i	+1	+0	-1	-i
+i	0 + i0	-1 + i	0 + i	1 + i	0 + 2i
+1	1 − i	0 + i0	+1 + i0	+2 + i0	+1 + i
+0	0 − i	-1 + i0	+0 + i0	+1 + i0	0 + i
-1	-1 − i	-2 + i0	-1 + i0	+0 + i0	-1 + i
-i	0 − 2i	-1 − i	0 − i	1 − i	0 + i0

×	+i	+1	+0	-1	-i
+i	-1	+i	+0i	-i	+1
+1	+i	+1	+0	-1	-i
+0	+0	+0	+0	+0	+0
-1	-i	-1	+0	+1	+i
-i	+1	-i	+0i	+i	-1

÷	+i	+1	+0	-1	-i
+i	+1	+i	U	-i	-1
+1	-i	+1	U	-1	+i
+0	+0	+0	U	-0	-0
-1	+i	-1	U	+1	-i
-i	-1	-i	U	+i	+1

U is undefined for 0/0, ±1/0 and ±i /0

Notice that the Primitive definition for real arithmetic had to be expanded to account for the new quantity the pure imaginary number also observes ordination of zero is implied as 0 + i0.

Section 7.3 Complex Numbers: Axioms

Complex numbers are a natural extension to real numbers so it is not surprising that every axiom from real numbers should apply to the complex domain:

Axiom 7.3.1 Continuity between Real and Complex Numbers

All axioms A4.2.1 to A4.2.27 apply to complex numbers. In order to keep the argument proof list a little simpler, A7.3.1 "Continuity between Real and Complex Numbers" will be implied and understood that it is there when ever a proof requires an axiom or theorem from the algebraic system 4.1.x.

Axiom 7.3.2 Counter Clockwise Cyclic Permutation for 2-Space Unit Base Vectors

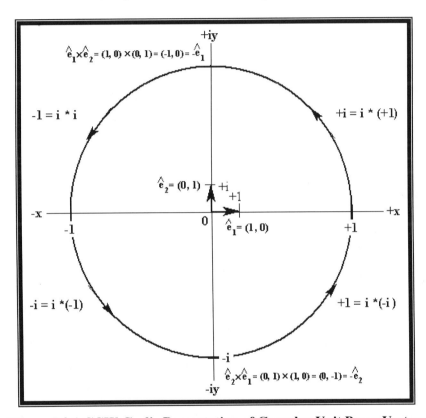

Figure 7.3.1 CCW Cyclic Permutation of Complex Unit Bases Vectors

Clearly, the ordered combination of the bases vectors forms 2-dimensional type of cyclic 2-by-2 permutation 180° apart.

$$\hat{e}_1 \times \hat{e}_2 = -\hat{e}_1$$
$$\hat{e}_2 \times \hat{e}_1 = -\hat{e}_2$$

While this does not play an important role in complex numbers, however when William Rowan Hamilton (1805-1865) [EVE76, pg 395] tried to expand these ideas into three dimensions it was a major factor in the development of his Quaternions.

Hamilton was an excellent mathematician having developed algebra for complex numbers and presented a paper in 1833 to the Irish Academy in which the algebra of complex numbers appears as algebra of ordered pairs of real numbers. In 1835, he was knighted.

Section 7.4 Complex Numbers: Minor Theorems for Arithmetic and Algebra

Theorem 7.4.1 Imaginary [i] Raised to 2m for m even

1g	Given	$0 < m$ and even for $m \in IG$
Steps	Hypothesis	$i^{2m} = +1$ for m even
1	From D2.2.28A Imaginary Field and T4.8.6 Integer Exponents: Uniqueness of Exponents	$i^{2m} = i^{2m}$
2	From 1 and A4.2.21 Distribution Exp	$i^{2m} = (i^2)^m$
3	From 2 and D2.2.28B Imaginary Field	$i^{2m} = (-1)^m$
∴	From 3 and T4.8.3 Integer Exponents: Negative One Raised to an Even Number	$i^{2m} = +1$

Theorem 7.4.2 Imaginary [i] Raised to 2m for m odd

1g	Given	$0 < m$ and odd for $m \in IG$
Steps	Hypothesis	$i^{2m} = -1$ for m odd
1	From D2.2.28A Imaginary Field and T4.8.6 Integer Exponents: Uniqueness of Exponents	$i^{2m} = i^{2m}$
2	From 1 and A4.2.21 Distribution Exp	$i^{2m} = (i^2)^m$
3	From 2 and D2.2.28B Imaginary Field	$i^{2m} = (-1)^m$
∴	From 3 and T4.8.4 Integer Exponents: Negative One Raised to an Odd Number	$i^{2m} = -1$

Theorem 7.4.3 Imaginary [i] Raised to 2m+1 for m even

1g	Given	$0 < m$ and even for $m \in IG$
Steps	Hypothesis	$i^{2m+1} = +i$ for m even
1	From D2.2.28A Imaginary Field and T4.8.6 Integer Exponents: Uniqueness of Exponents	$i^{2m+1} = i^{2m+1}$
2	From 1 and A4.2.18 Summation Exp	$i^{2m+1} = i^{2m} i^1$
3	From 2 and A4.2.23 Identity Exp	$i^{2m+1} = (i^{2m}) i$
∴	From 3 and T7.3.1 Imaginary [i] Raised to 2m for m even	$i^{2m+1} = +i$

Theorem 7.4.4 Imaginary [i] Raised to 2m+1 for m odd

1g	Given	$0 < m$ and odd for $m \in IG$
Steps	Hypothesis	$i^{2m+1} = -i$ for m odd
1	From D2.2.28A Imaginary Field and T4.8.6 Integer Exponents: Uniqueness of Exponents	$i^{2m+1} = i^{2m+1}$
2	From 1 and A4.2.18 Summation Exp	$i^{2m+1} = i^{2m} \, i^1$
3	From 2 and A4.2.23 Identity Exp	$i^{2m+1} = (i^{2m}) \, i$
∴	From 3 and T7.4.2 Imaginary [i] Raised to 2m for m odd	$i^{2m+1} = -i$

Theorem 7.4.5 Imaginary [i] raised to the zero[th] power

1g	Given	$0 < m$ and odd for $m \in IG$
Steps	Hypothesis	$i^0 = 1$
1	From D2.2.28A Imaginary Field and T4.8.6 Integer Exponents: Uniqueness of Exponents	$i^0 = i^0$
2	From 1 and D2.2.28A Imaginary Field	$i^0 = (\sqrt{-1})^0$
3	From 2 and T4.10.2 Radicals: Distribution Across a Product	$i^0 = \sqrt{(-1)^0}$
4	From 3 and A4.2.24 Inverse Exp	$i^0 = \sqrt{+1}$
∴	From 4, A4.2.25 Reciprocal Exp and T4.9.1	$i^0 = 1$

Theorem 7.4.6 Euler's Complex Function for e^{ix}

1g	Given	$y = ix$
Steps	Hypothesis	$e^{ix} = \cos(x) + i\sin(x)$
1	From LxE.3.11	$e^y = \sum\limits_{n=0}^{\infty} \dfrac{y^n}{n!}$
2	From 1, 1g and A4.2.3 Substitution	$e^{ix} = \sum\limits_{n=0}^{\infty} \dfrac{(ix)^n}{n!}$
3	From 2 separating summation into odd and even terms	$e^{ix} = \sum\limits_{n=0}^{\infty} \dfrac{(ix)^{2n}}{2n!} + \dfrac{(ix)^{2n+1}}{(2n+1)!}$
4	From 3 and A4.2.6 Associative Add term by term over an infinite series.	$e^{ix} = \sum\limits_{m=0}^{\infty} \dfrac{(ix)^{2m}}{2m!} + \sum\limits_{m=0}^{\infty} \dfrac{(ix)^{2m+1}}{(2m+1)!}$
5	From 4 and A4.2.21 Distribution Exp	$e^{ix} = \sum\limits_{m=0}^{\infty} \dfrac{(i)^{2m} x^{2m}}{2m!} + \sum\limits_{m=0}^{\infty} \dfrac{(i)^{2m+1} x^{2m+1}}{(2m+1)!}$
6	From 5 and A4.2.18 Summation Exp	$e^{ix} = \sum\limits_{m=0}^{\infty} \dfrac{(i)^{2m} x^{2m}}{2m!} + \sum\limits_{m=0}^{\infty} \dfrac{i\,(i)^{2m} x^{2m+1}}{(2m+1)!}$
7	From 5 and A4.2.14 Distribution	$e^{ix} = \sum\limits_{m=0}^{\infty} \dfrac{(i)^{2m} x^{2m}}{2m!} + i\sum\limits_{m=0}^{\infty} \dfrac{(i)^{2m} x^{2m+1}}{(2m+1)!}$
8	From 7 and A4.2.22 Product Exp	$e^{ix} = \sum\limits_{m=0}^{\infty} \dfrac{(i^2)^m x^{2m}}{2m!} + i\sum\limits_{m=0}^{\infty} \dfrac{(i^2)^m x^{2m+1}}{(2m+1)!}$
9	From 8 and D2.2.28B Imaginary Field	$e^{ix} = \sum\limits_{m=0}^{\infty} \dfrac{(-1)^m x^{2m}}{2m!} + i\sum\limits_{m=0}^{\infty} \dfrac{(-1)^m x^{2m+1}}{(2m+1)!}$
∴	From 8, LxE.3.22, LxE.3.23 and A4.2.3 Substitution	$e^{ix} = \cos(x) + i\sin(x)$

Theorem 7.4.7 Product of Two Complex Numbers [z w]

Steps	Hypothesis	$z\,w = (ac - bd) + i(ad + bc)$
1	From A7.3.1 Continuity between Real and Complex Numbers and A4.2.2A Equality	$z\,w = z\,w$
2	From 2, 2g, 3g, A7.3.1 Continuity between Real and Complex Numbers and A4.2.3 Substitution	$z\,w = (a + ib)\,(c + id)$
3	From 2, T4.12.4	$z\,w = ac + iad + ibc + iibd$
4	From 3, D4.1.17 Exponential Notation, A4.2.5 Commutative Add, A4.2.6 Associative Add and A4.2.14 Distribution	$z\,w = ac + i^2bd + i(ad + bc)$
5	From 4 and D7.2.11 Primitive Definition for Imaginary Arithmetic	$z\,w = ac - bd + i(ad + bc)$
∴	From 5 and A4.2.6 Associative Add	$z\,w = (ac - bd) + i(ad + bc)$

Theorem 7.4.8 Product of Two Complex Numbers [z w*]

Steps	Hypothesis	$z\,w^{*} = (ac + bd) + i(bc - ad)$
1	From A7.3.1 and A4.2.2A Equality	$z\,w^{*} = z\,w^{*}$
2	From 2, 2g, 3g, A7.3.1 and A4.2.3 Substitution	$z\,w^{*} = (a + ib)\,(c - id)$
3	From 2, T4.12.4 Polynomial Quadratic: Product of N-Quantities and D4.1.19 Primitive Definition for Rational Arithmetic	$z\,w^{*} = ac - iad + ibc - iibd$
4	From 3, D4.1.17 Exponential Notation, A4.2.5 Commutative Add, A4.2.6 Associative Add and A4.2.14 Distribution	$z\,w^{*} = ac - i^2bd + i(bc - ad)$
5	From 4 and D7.2.11 Primitive Definition for Imaginary Arithmetic	$z\,w^{*} = ac + bd + i(bc - ad)$
∴	From 5 and A4.2.6 Associative Add	$z\,w^{*} = (ac + bd) + i(bc - ad)$

Theorem 7.4.9 Product of Two Complex Numbers [z* w]

Steps	Hypothesis	$z^* w = (ac + bd) + i(ad - bc)$
1	From A7.3.1 Continuity between Real and Complex Numbers and A4.2.2A Equality	$z^* w = z^* w$
2	From 2, 2g, 3g, A7.3.1 Continuity between Real and Complex Numbers and A4.2.3 Substitution	$z^* w = (a - ib)(c + id)$
3	From 2, T4.12.4 Polynomial Quadratic: Product of N-Quantities and D4.1.19 Primitive Definition for Rational Arithmetic	$z^* w = ac + iad - ibc - iibd$
4	From 3, D4.1.17 Exponential Notation, A4.2.5 Commutative Add, A4.2.6 Associative Add and A4.2.14 Distribution	$z^* w = ac - i^2bd + i(ad - bc)$
5	From 4 and D7.2.11 Primitive Definition for Imaginary Arithmetic	$z^* w = ac + bd + i(ad - bc)$
∴	From 5 and A4.2.6 Associative Add	$z^* w = (ac + bd) + i(ad - bc)$

Theorem 7.4.10 Product of Two Complex Numbers [$z^* w^*$]

Steps	Hypothesis	$z^* w^* = (ac - bd) - i(ad + bc)$
1	From A7.3.1 Continuity between Real and Complex Numbers and A4.2.2A Equality	$z^* w^* = z^* w^*$
2	From 2, 2g, 3g, A7.3.1 Continuity between Real and Complex Numbers and A4.2.3 Substitution	$z^* w^* = (a - ib)(c - id)$
3	From 2, T4.12.4 Polynomial Quadratic: Product of N-Quantities and D4.1.19 Primitive Definition for Rational Arithmetic	$z^* w^* = ac - iad - ibc + iibd$
4	From 3, D4.1.17 Exponential Notation, A4.2.5 Commutative Add, A4.2.6 Associative Add and A4.2.14	$z^* w^* = ac + i^2bd - i(ad + bc)$
5	From 4 and D7.2.11 Primitive Definition for Imaginary Arithmetic	$z^* w^* = ac - bd - i(ad + bc)$
$\therefore$	From 5 and A4.2.6 Associative Add	$z^* w^* = (ac - bd) - i(ad + bc)$

Theorem 7.4.11 Complex Difference of Two Squares

1g	Given	$x = i = \sqrt{-1}$
Steps	Hypothesis	$(a + ib)(a - ib) = a^2 + b^2$
1	From T4.12.3 Polynomial Quadratic: Difference of Two Squares	$(a + xb)(a - xb) = a^2 - (xb)^2$
2	From 1, 1g and A4.2.3 Substitution	$(a + ib)(a - ib) = a^2 - (ib)^2$
3	From 2 and A4.2.21 Distribution Exp	$(a + ib)(a - ib) = a^2 - i^2b^2$
$\therefore$	From 3 and D2.2.28A Imaginary Field	$(a + ib)(a - ib) = a^2 + b^2$

Theorem 7.4.12 Magnitude: Complex Difference of Two Squares

Steps	Hypothesis	$|z| = \sqrt{(a + ib)(a - ib)}$
1	From D7.3.5 Magnitude of a Complex Number	$|z| = \sqrt{a^2 + b^2}$
$\therefore$	From 1, T7.4.11 Complex Difference of Two Squares and A4.2.3 Substitution	$|z| = \sqrt{(a + ib)(a - ib)}$

Theorem 7.4.13 Magnitude: Conjugated

Steps	Hypothesis	$|z| = \sqrt{z\, z^*}$
1	From T7.4.12 Magnitude: Complex Difference of Two Squares	$|z| = \sqrt{(a + ib)\,(a - ib)}$
$\therefore$	From 2, 2g, D7.3.4 Conjugant of a Complex Number and A4.2.3 Substitution	$|z| = \sqrt{z\, z^*}$

Corllary 7.4.13.1 Magnitude: Squared

Steps	Hypothesis	$|z|^2 = z\, z^*$
1	From T7.4.13 Magnitude: Conjugated	$|z| = \sqrt{z\, z^*}$
2	From 1 and T4.8.6 Integer Exponents: Uniqueness of Exponents	$|z|^2 = (\sqrt{z\, z^*})^2$
$\therefore$	From 2, T4.10.2 Radicals: Distribution Across a Product and T4.10.3 Radicals: Identity	$|z|^2 = z\, z^*$

Theorem 7.4.14 Conjugation: Invariance

Steps	Hypothesis	$z = z^{**}$
1	From A7.3.1 Continuity between Real and Complex Numbers and A4.2.2A Equality	$a + ib = a + ib$
2	From 1 and D7.3.4 Conjugant of a Complex Number	$a + ib = (a - ib)^*$
3	From 2 and D7.3.4 Conjugant of a Complex Number	$a + ib = (a + ib)^{**}$
$\therefore$	From 3, 2g, A7.3.1 Continuity between Real and Complex Numbers and A4.2.2A Equality	$z = z^{**}$

Theorem 7.4.15 Conjugation: Distribution Across a Complex Summation

1g	Given	$(z + w)^* = z^* + w^*$ assume
Steps	Hypothesis	$(z + w)^* = z^* + w^*$
1	From 1g	$(z + w)^* = z^* + w^*$
2	From 2, 2g, 3g, A7.3.1 Continuity between Real and Complex Numbers and A4.2.3 Substitution	$((a + ib) + (c + id))^* = (a + ib)^* + (c + id)^*$
3	From 2, A4.2.5 Commutative Add, A4.2.6 Associative Add and D7.3.4 Conjugant of a Complex Number	$((a + c) + (ib + id))^* = (a - ib) + (c - id)$
4	From 3, A4.2.5 Commutative Add, A4.2.6 Associative Add and A4.2.14 Distribution	$((a + c) + i(b + d))^* = (a + c) + (-ib - id)$
5	From 4, D7.3.4 Conjugant of a Complex Number and A4.2.14 Distribution	$(a + c) - i(b + d) = (a + c) - i(b + d)$
∴	From 1g, 5 and by identity	$(z + w)^* = z^* + w^*$

Theorem 7.4.16 Conjugation: Distribution Across a Complex Product

Steps	Hypothesis	$(z\,w)^* = z^*\,w^*$
1	From Hypothesis	$(z\,w)^* = z^*\,w^*$
2	From 2, 2g, 3g, A7.3.1 Continuity between Real and Complex Numbers and A4.2.3 Substitution	$((a + ib)(c + id))^* = (a + ib)^*(c + id)^*$
3	From 2, D7.3.4 Conjugant of a Complex Number and T7.4.7 Product of Two Complex Numbers [z w]	$((ac - bd) + i(ad + bc))^* = (a - ib)(c - id)$
∴	From 3, D7.3.4 Conjugant of a Complex Number and T7.4.10 Product of Two Complex Numbers [z* w*]	$(ac - bd) - i(ad + bc) = (ac - bd) - i(ad + bc)$

Theorem 7.4.17 Conjugation: Division

Steps	Hypothesis	$zw^* / |w|^2 = z / w$
1	From Hypothesis	$zw^* / |w|^2 = z / w$
2	From 1, A4.2.11 Associative Multp and A4.2.12 Identity Multp	$zw^* / |w|^2 = (z / w)\ 1$
3	From 2, A7.3.1 Continuity between Real and Complex Numbers and A4.2.12 Identity Multp	$zw^* / |w|^2 = (z / w)\ (w^* / w^*)$
4	From 3, A7.3.1 Continuity between Real and Complex Numbers and T4.5.4 Equalities: Product of Two Rational Fractions	$zw^* / |w|^2 = (z\ w^*) / (w\ w^*)$
∴	From 4 and C7.4.13.1 Magnitude: Squared	$zw^* / |w|^2 = zw^* / |w|^2$

Theorem 7.4.18 Complex Products: Components of $Re\{z\ w^*\} = Re\{z^*\ w\}$

Steps	Hypothesis	$Re\{z\ w^*\} = Re\{z^*\ w\} = \frac{1}{2}(z^*w + z\ w^*) = ac + bd$
1	From Hypothesis, D7.3.3A Real and Imaginary Component Operators and T7.4.8 Product of Two Complex Numbers [z w*]	$ac + bd = Re\{z\ w^*\}$
2	From Hypothesis, D7.3.3A Real and Imaginary Component Operators and T7.4.9 Product of Two Complex Numbers [z* w]	$ac + bd = Re\{z^*\ w\}$
3	From Hypothesis, T7.4.9 Product of Two Complex Numbers [z* w], T7.4.8 Product of Two Complex Numbers [z w*], A7.3.1 Continuity between Real and Complex Numbers and A4.2.3 Substitution	$ac + bd = \frac{1}{2}(\ (ac + bd) + i(ad - bc) + (ac + bd) + i(bc - ad)\ )$
4	From 3, A7.3.1 Continuity between Real and Complex Numbers, A4.2.5 Commutative Add and A4.2.14 Distribution	$ac + bd = \frac{1}{2}(\ (1 + 1)(ac + bd) + i(ad - bc + bc - ad)\ )$
5	From 4, D4.1.19 Primitive Definition for Rational Arithmetic, A4.2.5 Commutative Add and A4.2.8 Inverse Add	$ac + bd = \frac{1}{2}(\ 2(ac + bd) + i\ 0\ )$

6	From 5 and D7.3.11 Primitive Definition for Imaginary Arithmetic	$ac + bd = \tfrac{1}{2}\,2(ac + bd)$
7	From 6 and D4.1.19 Primitive Definition for Rational Arithmetic	$ac + bd = ac + bd$
∴	From 1, 2 and 7	$\mathrm{Re}\{z\,w^{*}\} = \mathrm{Re}\{z^{*}w\} = \tfrac{1}{2}\,(z^{*}w + z\,w^{*}) = ac + bd$

Theorem 7.4.19 Complex Products: Components of $\mathrm{Im}\{z\,w^{*}\} = -\mathrm{Im}\{z^{*}w\}$

Steps	Hypothesis	$\mathrm{Im}\{^{*}z\,w\} = -\mathrm{Im}\{z\,w^{*}\} = \tfrac{1}{2}i\,(z^{*}w - z\,w^{*}) = ad - bc$
1	From Hypothesis, D7.3.3B Real and Imaginary Component Operators and T7.4.8 Product of Two Complex Numbers [z w*]	$ad - bc = +\mathrm{Im}\{z^{*}w\}$
2	From Hypothesis, D7.3.3B Real and Imaginary Component Operators, T7.4.9 Product of Two Complex Numbers [z* w], A7.3.1 Continuity between Real and Complex Numbers and A4.2.14 Distribution	$ad - bc = -\mathrm{Im}\{z\,w^{*}\}$
3	From Hypothesis, T7.4.9 Product of Two Complex Numbers [z* w], T7.4.8 Product of Two Complex Numbers [z w*], A7.3.1 Continuity between Real and Complex Numbers and A4.2.3 Substitution	$ad - bc = \tfrac{1}{2}i\,(\,(ac + bd) + i(ad - bc) - (\,(ac + bd) + i(bc - ad)\,)\,)$
4	From 3, A7.3.1 Continuity between Real and Complex Numbers and A4.2.14 Distribution	$ad - bc = \tfrac{1}{2}i\,(\,(ac + bd) + i(ad - bc) - (ac + bd) - i(bc - ad)\,)$
5	From 4, D4.1.17 Exponential Notation, A4.2.5 Commutative Add and A4.2.8 Inverse Add	$ad - bc = \tfrac{1}{2}i\,(\,0 + i(ad - bc) + i(ad - bc)\,)$
6	From 5, A4.2.7 Identity Add and A4.2.14 Distribution	$ad - bc = \tfrac{1}{2}i\,(\,(1 + 1)\,i(ad - bc)\,)$
7	From 6 and D4.1.19 Primitive Definition for Rational Arithmetic	$ad - bc = \tfrac{1}{2}i\,2i\,(ad - bc)$
8	From 6, A4.2.13 Inverse Multp and A4.2.12 Identity Multp	$ad - bc = ad - bc$
∴	From 1, 2 and 8	$\mathrm{Im}\{^{*}z\,w\} = -\mathrm{Im}\{z\,w^{*}\} = \tfrac{1}{2}i\,(z^{*}w - z\,w^{*}) = ad - bc$

Theorem 7.4.20 Complex Number: Orthogonal Components

Steps	Hypothesis	$a = c$ and $b = d$, hence $w = z$ components are orthogonal.	
1	From A7.3.1 Continuity between Real and Complex Numbers and A4.2.2A Equality	$w = z$	
2	From A7.3.1 Continuity between Real and Complex Numbers and A4.2.2A Equality	$-w = -w$	
3	From 1, 2, A7.3.1 Continuity between Real and Complex Numbers and T4.3.1 Equalities: Uniqueness of Addition	$w - w = z - w$	
4	From 3, A7.3.1 Continuity between Real and Complex Numbers, A4.2.8 Inverse Add and D7.3.11 Primitive Definition for Imaginary Arithmetic	$0 + i0 = z - w$	
5	From 4, 2g, 3g, A7.3.1 Continuity between Real and Complex Numbers and A4.2.3 Substitution	$0 + i0 = (a + ib) - (c + id)$	
6	From 5, A7.3.1 Continuity between Real and Complex Numbers and A4.2.14 Distribution	$0 + i0 = a + ib - c - id$	
7	From 6, A7.3.1 Continuity between Real and Complex Numbers and A4.2.5 Commutative Add	$0 + i0 = a - c + ib - id$	
8	From 6, A7.3.1 Continuity between Real and Complex Numbers, A4.2.6 Associative Add and A4.2.14 Distribution	$0 + i0 = (a - c) + i(b - d)$	
9	From 8 and equating terms	$0 = a - c$ and $0 = b - d$	
∴	From 9 and T4.3.4 Equalities: Reversal of Right Cancellation by Addition	$a = c$ and $b = d$	

Theorem 7.4.21 Complex Number: Uniqueness of Real and Imaginary Operators

Steps	Hypothesis	$a = c$ and $b = d$, hence $w = z$ components are orthogonal.	
∴	From T7.4.20 Complex Number: Orthogonal Components and D7.3.3 Real and Imaginary Component Operators	$\text{Re}\{z\} = \text{Re}\{w\}$ and $\text{Im}\{z\} = \text{Im}\{w\}$	

Theorem 7.4.22 Complex Polar Form: Inverse Transformation

Steps	Hypothesis	$a = Re\{z\} = |z| \cos(\varphi)$ and $b = Im\{z\} = |z| \sin(\varphi)$	
1	From D7.3.10 Phasor Polar Form of a Complex Number	$z = |z| \, e^{i\varphi}$	
2	From 1, T7.4.6 Euler's Complex Function for e^{ix}, A7.3.1 Continuity between Real and Complex Numbers and A4.2.14 Distribution	$z = |z| \cos(\varphi) + i|z| \cos(\varphi)$	
3	From 2 and 2g	$a + ib = z = |z| \cos(\varphi) + i|z| \cos(\varphi)$	
∴	From 3 and T7.4.21 Complex Number: Uniqueness of Real and Imaginary Operators	$a = Re\{z\} = |z| \cos(\varphi)$ and $b = Im\{z\} = |z| \sin(\varphi)$	

Theorem 7.4.23 Complex Polar Form: Distribution of Real and Imaginary Operators

1g 2g	Given	$z = a + ib$ $w = c + id$	
Steps	Hypothesis	$Re\{z + w\} = Re\{z\} + Re\{w\}$	$Im\{z + w\} = Im\{z\} + Im\{w\}$
1	From 1g, 2g, A4.2.3 Substitution, A4.2.5 Commutative Add and A4.2.14 Distribution	$Re\{z + w\} = Re\{a + c + i(b + d)\}$	$Im\{z + w\} = Im\{ a + c + i(b + d)\}$
2	From 1, D7.3.3 Real and Imaginary Component Operators	$Re\{z + w\} = a + c$	$Im\{z + w\} = b + d$
∴	From 1g, 2g, D7.3.3 Real and Imaginary Component Operators and A4.2.3 Substitution	$Re\{z + w\} = Re\{z\} + Re\{w\}$	$Im\{z + w\} = Im\{z\} + Im\{w\}$

Theorem 7.4.24 Complex Polar Form: Real Constant of Real and Imaginary Operators

1g 2g	Given	$z = a + ib$ $k \in R$ constant real number	
Steps	Hypothesis	$Re\{k\,z\} = k\,Re\{z\}$	$Im\{k\,z\} = k\,Im\{z\}$
1	From A4.2.2A Equality and A4.2.3 Substitution	$Re\{ k\,z \} = Re\{k\,(a + i\,b)\}$	$Im\{ k\,z \} = Im\{k\,(a + i\,b)\}$
2	From 1 and A4.2.14 Distribution	$Re\{ k\,z \} = Re\{k\,a + i\,k\,b)\}$	$Im\{ k\,z \} = Im\{k\,a + i\,k\,b)\}$
3	From 2 and D7.3.3 Real and Imaginary Component Operators	$Re\{ k\,z \} = k\,a$	$Im\{ k\,z \} = k\,b$
∴	From 1g and D7.3.3 Real and Imaginary Component Operators	$Re\{ k\,z \} = k\,Re\{z\}$	$Im\{ k\,z \} = k\,Im\{z\}$

Theorem 7.4.25 Complex Polar Form: Phase Equation

Steps	Hypothesis	$\varphi = -i \ln(z / |z|)$
1	From D7.3.10 Phasor Polar Form of a Complex Number	$z = |z| \, e^{i\varphi}$
2	From 1, A7.3.1 Continuity between Real and Complex Numbers and T4.4.4 Equalities: Right Cancellation by Multiplication	$e^{i\varphi} = z / |z|$
3	From 2, A4.2.27 Correspondence Exp and T4.11.1 Real Exponents: Uniqueness of Logarithms, now taking the natural log of both sides.	$i\varphi = \ln(z / |z|)$
4	From A7.3.1 Continuity between Real and Complex Numbers and A4.2.2A Equality	$-i = -i$
5	From 3, 4, T4.3.7 Equalities: Uniqueness of Subtraction	$-i \, i\varphi = -i \ln(z / |z|)$
$\therefore$	From 5 and D7.3.11 Primitive Definition for Imaginary Arithmetic	$\varphi = -i \ln(z / |z|)$

Theorem 7.4.26 Complex Polar Form: COSINE defined as Complex Number

1g	Given	$\cos(\varphi) = \frac{1}{2}(e^{j(\varphi)} + e^{-j(\varphi)})$ assume
Steps	Hypothesis	$\cos(\varphi) = \frac{1}{2}(e^{j(\varphi)} + e^{-j(\varphi)})$
1	From 1g, A4.2.3 Substitution, T7.4.6 Euler's Complex Function for e^{ix}, A4.2.5 Commutative Add and D4.1.20 Negative Coefficient	$\cos(\varphi) = \frac{1}{2}(\cos(\varphi) + i\sin(\varphi) + \cos(\varphi) + i\sin(-\varphi))$
2	From 1 and DxE.1.8.1	$\cos(\varphi) = \frac{1}{2}(\cos(\varphi) + i\sin(\varphi) + \cos(\varphi) - i\sin(\varphi))$
3	From 2, A4.2.5 Commutative Add, A4.2.14 Distribution, A4.2.8 Inverse Add and T4.3.9 Equalities: Summation of Repeated Terms	$\cos(\varphi) = \frac{1}{2}(2\cos(\varphi) + i\,0)$ for n = 2 terms
4	From 3, T4.4.1 Equalities: Any Quantity Multiplied by Zero is Zero and A4.2.7 Identity Add	$\cos(\varphi) = \frac{1}{2}(2\cos(\varphi))$
5	From 4, D4.1.4A Rational Numbers, A4.2.13 Inverse Multp and A4.2.12 Identity Multp	$\cos(\varphi) = \cos(\varphi)$
$\therefore$	From 1g, 5 and by identity	$\cos(\varphi) = \frac{1}{2}(e^{j(\varphi)} + e^{-j(\varphi)})$

Theorem 7.4.27 Complex Polar Form: SINE defined as Complex Number

1g	Given	$\sin(\varphi) = (e^{i(\varphi)} - e^{-i(\varphi)}) / i2$
Steps	Hypothesis	$\sin(\varphi) = (e^{i(\varphi)} - e^{-i(\varphi)}) / i2$
1	From 1g, A4.2.3 Substitution, T7.4.6 Euler's Complex Function for e^{ix}, A4.2.5 Commutative Add and D4.1.20 Negative Coefficient	$\sin(\varphi) = (\cos(\varphi) + i \sin(\varphi) - \cos(\varphi) - i \sin(-\varphi)) / i2$
2	From 1, DxE.1.8.1 and D4.1.19 Primitive Definition for Rational Arithmetic	$\sin(\varphi) = (\cos(\varphi) + i \sin(\varphi) - \cos(\varphi) + i \sin(\varphi)) / i2$
3	From 2, A4.2.5 Commutative Add, A4.2.14 Distribution, A4.2.8 Inverse Add and T4.3.9 Equalities: Summation of Repeated Terms	$\sin(\varphi) = (0 + i2 \sin(\varphi)) / i2$ for n = 2 terms
4	From 3, A4.2.7 Identity Add, A4.2.10 Commutative Multp and D4.1.4C Rational Numbers	$\sin(\varphi) = (i2 / i2) \sin(\varphi)$
5	From 4, A4.2.13 Inverse Multp and A4.2.12 Identity Multp	$\sin(\varphi) = \sin(\varphi)$
∴	From 1g, 5 and by identity	$\sin(\varphi) = (e^{i(\varphi)} - e^{-i(\varphi)}) / i2$

Theorem 7.4.28 Complex Polar Form: Double COSINE defined as Complex Number

Steps	Hypothesis	$2\cos(\varphi) = e^{j(\varphi)} + e^{-j(\varphi)}$
1	From A4.2.2A Equality	$2 = 2$
2	From 1, T7.4.26 Complex Polar Form: COSINE defined as Complex Number and T4.4.2 Equalities: Uniqueness of Multiplication	$2 \cos(\varphi) = 2 (½) (e^{j(\varphi)} + e^{-j(\varphi)})$
∴	From 2, D4.1.4A Rational Numbers, A4.2.13 Inverse Multp and A4.2.12 Identity Multp	$2 \cos(\varphi) = e^{j(\varphi)} + e^{-j(\varphi)}$

Theorem 7.4.29 Complex Polar Form: Double SINE defined as Complex Number

Steps	Hypothesis	$i2 \sin(\varphi) = e^{j(\varphi)} - e^{-j(\varphi)}$
1	From A4.2.2A Equality	$i2 = i2$
2	From 1, T7.4.27 Complex Polar Form: SINE defined as Complex Number and T4.4.2 Equalities: Uniqueness of Multiplication	$i2 \sin(\varphi) = (e^{j(\varphi)} - e^{-j(\varphi)}) (1/i2) (i2)$
∴	From 2, D4.1.4A Rational Numbers, A4.2.13 Inverse Multp and A4.2.12 Identity Multp	$i2 \sin(\varphi) = e^{j(\varphi)} - e^{-j(\varphi)}$

Complex numbers, its algebra and calculus, is a vast topic and goes beyond the scope of this treatise so I refer the reader to Murray R. Spiegel [SPI64] and leave this topic simply as an introduction.

Theorem 7.4.30 Complex Roots: De Moivre's Identity of Periodicity[7.4.1]

Steps	Hypothesis	
1g	Given	$re^{j(\theta)} = re^{j(\theta + 2\pi m)}$
2g		$m \in IG$ an integer
Steps	Hypothesis	$re^{j(\theta)} = re^{j(\theta)}$
1	From 1g	$re^{j(\theta)} = re^{j(\theta + 2\pi m)}$
2	From 1, T7.4.6 Euler's Complex Function for e^{ix} and A4.2.3 Substitution	$re^{j(\theta)} = r [\cos (\theta + 2\pi m) + j \sin (\theta + 2\pi m)]$
3	From 2, LxE.1.24 Sum of Angles for Cosine and LxE.1.22 Sum of Angles for Sine	$re^{j(\theta)} = r [\cos(\theta)\cos(2\pi m) - \sin(\theta)\sin(2\pi m)$ $+ j \sin(\theta)\cos(2\pi m) + \cos(\theta)\sin(2\pi m)]$
4	From 2g, 3, DxE.1.6.15 Periodicity of Special Angles and A4.2.3 Substitution	$re^{j(\theta)} = r [\cos(\theta) 1 - \sin(\theta) 0$ $+ j \sin(\theta) 1 + \cos(\theta) 0]$
5	From 4, A4.2.12 Identity Multp, T4.4.1 Equalities: Any Quantity Multiplied by Zero is Zero and A4.2.7 Identity Add	$re^{j(\theta)} = r [\cos(\theta) + j \sin(\theta)]$
∴	From 5, T7.4.6 Euler's Complex Function for e^{ix} and A4.2.3 Substitution	$re^{j(\theta)} = re^{j(\theta)}$

[7.4.1]Note: Periodicity of 2π leaves complex numbers invariant.

Theorem 7.4.31 Complex Roots: De Moivre's Complex Number Raised to a Power

1g	Given	$re^{j(\theta)} = re^{j(\theta + 2\pi m)}$
2g		$(m, n) \in IG$ an integer
Steps	Hypothesis	$[re^{j(\theta)}]^n = r^n\, e^{j(\theta n)}$
1	From 1g and T4.8.6 Integer Exponents: Uniqueness of Exponents	$[re^{j(\theta)}]^n = [re^{j(\theta + 2\pi m)}\,]^n$
2	From 1, A4.2.22 Product Exp	$[re^{j(\theta)}]^n = r^n\, e^{j(\theta + 2\pi m)\,n}$
3	From 2g, 2 and A4.2.14 Distribution	$[re^{j(\theta)}]^n = r^n\, e^{j(\theta n + 2\pi m\,n)}$
$\therefore$	From 3 and T7.4.30 Complex Planes: De Moivre's Identity of Periodicity	$[re^{j(\theta)}]^n = r^n\, e^{j(\theta n)}$

Theorem 7.4.32 Complex Roots: De Moivre's Root of a Complex Number

1g	Given	$re^{j(\theta)} = re^{j(\theta + 2\pi m)}$
2g		$(m, n) \in IG$ an integer
Steps	Hypothesis	$\sqrt[n]{[re^{j(\theta)}]} = \sqrt[n]{r}\, e^{j(\theta/n + 2\pi p/n)}$ for each value of p yielding a corresponding root as a solution
1	From 1g, 2g and T4.10.1 Radicals: Uniqueness of Radicals	$\sqrt[n]{[re^{j(\theta)}]} = \sqrt[n]{[re^{j(\theta + 2\pi m)}\,]}$
2	From 1 and T4.10.5 Radicals: Distribution Across a Product	$\sqrt[n]{[re^{j(\theta)}]} = \sqrt[n]{r}\ \sqrt[n]{e^{j(\theta + 2\pi m)}}\,]$
3	From 2 and A4.2.25 Reciprocal Exp	$\sqrt[n]{[re^{j(\theta)}]} = \sqrt[n]{r}\, [\, e^{j(\theta + 2\pi m)}\,]^{1/n}$
4	From 3 and A4.2.22 Product Exp	$\sqrt[n]{[re^{j(\theta)}]} = \sqrt[n]{r}\, e^{j(\theta + 2\pi m)\,(1/n)}$
5	From 4, A4.2.14 Distribution and D4.1.4 (D, A)	$\sqrt[n]{[re^{j(\theta)}]} = \sqrt[n]{r}\, e^{j(\theta/n + 2\pi m/n)}$
6	From 5, EQ A.2.2 Quotient, EQ A.2.3 Rational Remainder and EQ A.2.5 Modulus Remainder	$m/n = q + p/n$ for p = 0, 1, 2, …, (n − 1) all possible modulus remainders and q a lesser integer than m and n
7	From 5, 6, A4.2.3 Substitution and A4.2.14 Distribution	$\sqrt[n]{[re^{j(\theta)}]} = \sqrt[n]{r}\, e^{j(\theta/n + 2\pi q + 2\pi p/n)}$
$\therefore$	From 7 and T7.4.30 Complex Planes: De Moivre's Identity of Periodicity	$\sqrt[n]{[re^{j(\theta)}]} = \sqrt[n]{r}\, e^{j(\theta/n + 2\pi p/n)}$ for each value of p yielding a corresponding root as a solution

Theorem 7.4.33 Complex Roots: De Moivre's Root of One

1g	Given	$re^{j(\theta)} = 1$
2g		$r = 1$
3g		$\theta = 0$ radians
Steps	Hypothesis	$\sqrt[n]{1} = e^{j(2\pi p/n)}$ for $p = 0, 1, \ldots, (n-1)$ and n roots
1	From T7.4.32 Complex Roots: De Moivre's Root of a Complex Number	$\sqrt[n]{[re^{j(\theta)}]} = \sqrt[n]{r}\, e^{j(\theta/n + 2\pi p/n)}$ for $p = 0, 1, \ldots, (n-1)$
2	From 1g, 2g, 3g, 1 and A4.2.3 Substitution	$\sqrt[n]{1} = \sqrt[n]{1}\, e^{j(0/n + 2\pi p/n)}$
$\therefore$	From 2, A4.2.25 Reciprocal Exp, A4.2.12 Identity Multp and A4.2.7 Identity Add	$\sqrt[n]{1} = e^{j(2\pi p/n)}$ for $p = 0, 1, \ldots, (n-1)$

Theorem 7.4.34 Complex Roots: De Moivre's Root of a Negative One

1g	Given	$re^{j(\theta)} = -1$
2g		$r = 1$
3g		$\theta = \pi$
Steps	Hypothesis	$\sqrt[n]{-1} = = e^{j(2p+1)\pi/n}$ for $p = 0, 1, \ldots, (n-1)$ and n roots
1	From T7.4.32 Complex Planes: De Moivre's Root of a Complex Number	$\sqrt[n]{[re^{j(\theta)}]} = \sqrt[n]{r}\, e^{j(\theta/n + 2\pi p/n)}$ for $p = 0, 1, \ldots, (n-1)$
2	From 1g, 2g, 3g, 1 and A4.2.3 Substitution	$\sqrt[n]{-1} = \sqrt[n]{1}\, e^{j(\pi/n + 2\pi p/n)}$
$\therefore$	From 2, A4.2.25 Reciprocal Exp, A4.2.12 Identity Multp and A4.2.7 Identity Add	$\sqrt[n]{-1} = e^{j(2p+1)\pi/n}$ for $p = 0, 1, \ldots, (n-1)$

Theorem 7.4.35 Complex Roots: Square Root of One

1g	Given	$n = 2$
Steps	Hypothesis	$\sqrt{1} = +1, -1$ or ± 1
1	From T7.4.34 Complex Roots: De Moivre's Root of a Negative One	$\sqrt[n]{1} = e^{j(2\pi p/n)}$ for $p = 0, 1, \ldots, (n-1)$
2	From 1g and A4.2.3 Substitution	$\sqrt{1} = e^{j(2\pi p/2)}$ for $p = 0, 1$
3	From 2, A4.2.10 Commutative Multp, A4.2.13 Inverse Multp and A4.2.12 Identity Multp	$\sqrt{1} = e^{j(\pi p)}$ for $p = 0, 1$
4	From 3 and A4.2.3 Substitution	$\sqrt{1} = e^{j(0)}, e^{j(\pi)}$
5	From 4, T7.4.6 Euler's Complex Function for e^{ix}, DxE.1.6.1 zero of sine and cosine, DxE.1.6.13 pi of sine and cosine	$\sqrt{1} = 1 + j0, -1 + j0$
∴	From 5 and D7.2.11 Primitive Definition for Imaginary Arithmetic	$\sqrt{1} = +1, -1$ or ± 1

Theorem 7.4.36 Complex Roots: Square Root of Negative One

1g	Given	$n = 2$
Steps	Hypothesis	$\sqrt{-1} = +j, -j$ or $\pm j$
1	From T7.4.33 Complex Roots: De Moivre's Root of One	$\sqrt[n]{-1} = e^{j(2p+1)\pi/n}$ for $p = 0, 1, \ldots, (n-1)$
2	From 1g and A4.2.3 Substitution	$\sqrt{-1} = e^{j(2p+1)\pi/2}$ for $p = 0, 1$
3	From 2, D4.1.19 Primitive Definition for Rational Arithmetic and A4.2.3 Substitution	$\sqrt{-1} = e^{j(1)\pi/2}, e^{j(3)\pi/2}$
4	From 3 and A4.2.12 Identity Multp	$\sqrt{-1} = e^{j\pi/2}, e^{j(3\pi/2)}$
5	From 4, T7.4.6 Euler's Complex Function for e^{ix}, DxE.1.6.7 half-pi of sine and cosine, DxE.1.6.14 three-halves pi of sine and cosine	$\sqrt{-1} = 0 + j1, 0 + j(-1)$
∴	From 5 and D7.2.11 Primitive Definition for Imaginary Arithmetic	$\sqrt{-1} = +j, -j$ or $\pm j$

Theorems 7.4.35 and 7.4.36 are formal proofs of square root of one and negative one and replace theorems T4.9.4 Rational Exponent: Square Root of a Positive One and T4.9.5 Rational Exponent: Square Root of a Negative One.

Section 7.5 The Hamiltonian Quaternion

In Hamilton's effort to expand complex numbers from 2 to 3-dimensions he focused on definition D7.3.11 "Primitive Definition for Imaginary Arithmetic" specifically the multiplication part of the table:

×	+0	+1	−1	+i
+0	+0	+0	+0	+0
+1	+0	+1	−1	+i
−1	+0	−1	+1	−i
+i	+0	+i	−i	−1

What he had asked himself was how would this table look if the idea of complex numbers where to be taken into three dimensions.

×	+0	+1	+i	+j	+k	−1	−j	−k
+0	+0	+0	+0			+0		
+1	+0	+1	+i			−1		
+i	+0	+i	−1			−i		
+j								
+k								
−1	+0	−1	−i			+1		
−j								
−k								

Already having one bases vector [i] he would need two new unit bases vectors that would be pure imaginary to properly span the extra two dimensions in three-space, altogether:

$$i^2 = j^2 = k^2 = -1$$

Than he started to fill in the blank boxes:

×	+0	+1	+i	+j	+k	−1	−j	−k
+0	+0	+0	+0	+0	+0	+0	+0	+0
+1	+0	+1	+i	+j	+k	−1	−j	−k
+i	+0	+i	−1	ji=?	ki=?	−i	−ji=?	−ki
+j	+0	+j	ij=?	−1	kj=?	−j	+1	−kj
+k	+0	+k	ik=?	jk=?	−1	−k	−jk=?	+1
−1	+0	−1	−i	−j	−k	+1	+j	+k
−j	+0	−j	−ij=?	+1	−kj=?	+j	−1	kj=?
−k	+0	−k	−ik=?	−jk	+1	+k	+jk	−1

This is a bit messy so lets eliminate those columns and rows that differ by a negative number and zero reducing the repetitive clutter.

×	+1	+i	+j	+k
+1	+1	+i	+j	+k
+i	+i	−1	ji=?	ki=?
+j	+j	ij=?	−1	kj=?
+k	+k	ik=?	jk=?	−1

What is left is what Hamilton needed to discover in terms of the products of his new quantities.

So, he made a list of his unknowns as follows:

$$ij \propto ji \equiv k$$
$$ik \propto ki \equiv j$$
$$jk \propto kj \equiv i$$

He knew somehow they would have to be cyclic, rotated to the next imaginary number, and be equivalent. With the publication of his 1833 paper, he resolved this problem. He had spent years puzzling on and off over this and could not resolve the relationships. He was always stymied on the matter of how to define multiplication to preserve the familiar laws of that operation. Finally, in a flash of insight while standing on a bridge along the Royal Canal outside Dublin, it occurred to him that he was demanding too much of establishing equivalency for his vector products, trying to make them symmetrically commutative. So he threw out the Commutative Law, followed the natural cyclic permutation with the signs and everything fell into place. He was so struck by this revelation that he carved the gist of the following table into the stones on the bridge. With that insight non-commutative algebras where born.

$$-ij \ = +ji \ = +k$$
$$+ik = -ki \ = +j$$
$$-jk \ = +kj = +i$$

Now he substituted back into the table and it looked like this.

Table 7.5.1 Hamilton's Multiplication Table for Quaternion Arithmetic

×	+1	+i	+j	+k
+1	+1	+i	+j	+k
+i	+i	−1	+k	−j
+j	+j	−k	−1	+i
+k	+k	+j	−i	−1

When Hamilton used cyclic permutation he was thinking of something like what was used in 2-dimensional complex numbers as seen in Figure 7.3.4. So, he drew a simple cycle diagram based on a counter clockwise rotation for his new basis vectors as follows.

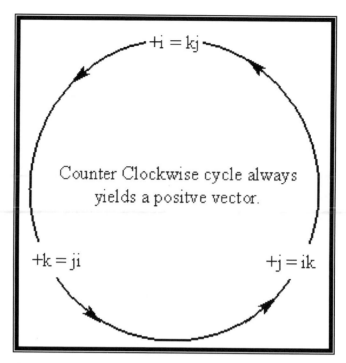

Figure 7.5.1 Counter Clockwise Vector Permutation Cycle Diagram

Of course going counter to the cycle yields a negative number, a simple technique to select the correct sign.

During the remaining 10-some years of his life, Hamilton expended most of his time and energy in developing his Quaternions, which he felt would be of revolutionary significance in mathematical physics, he was right, Maxwell made good use of the calculus portion he developed for analyzing electromagnetism. In his great work, "Treatise on Quaternions", appeared in 1835, after which he devoted himself to preparing an enlarged Elements of Quaternions, but died in Dublin in 1865, essentially from poor health before the work was quite complete.

The number that he created was an order quadruplet, hence the name **Quaternion**, and can be written like this:

$$v = a + ib + jc + kd \text{ or } (a, b, c, d)$$

The multiplication of quaternions is not only limited to the natural multiplication of numbers, but there are other operators on quaternions that do other things, such as finding magnitude for a quaternion in an alternate non-Hamiltonian form.

Definition 7.5.1 Four-Dimensional Quaternion, Imaginary-Vector

Four-dimensional quaternion:

$$\vec{a}_4 \equiv (\vec{a}, \text{id}) \qquad \text{Equation A}$$

$$(\vec{a}, \text{id}) \equiv (a, b, c, \text{id}) \qquad \text{Equation B}$$

for $i = \sqrt{-1}$.

Definition 7.5.2 Inner Product Of Four-Dimensional Quaternions

These four-dimensional quaternion vectors follow the rules of vector inner product.

$$(a, b, c, \text{id})_4 \bullet (e, f, g, \text{ih})_4 \equiv ae + bf + cg - dh \quad \text{Equation A}$$

$$(\vec{a}, \text{id})_4 \bullet (\vec{b}, \text{ih})_4 \equiv \vec{a} \bullet \vec{b} - dh \qquad \text{Equation B}$$

Multiplication of Conjugated Quaternions yields a similar magnitude

$$(\vec{a}, \text{id})_4 \bullet (\vec{b}, \text{ih})_4^* \equiv \vec{a} \bullet \vec{b} + dh \qquad \text{Equation C}$$

Tensor Calculus & Physics: A General Treatise

Section 7.6 Vector Law: Definitions

Now from complex numbers a dual vector is established as an ordered pair, while the quaternion becomes an ordered quadruplet

In 2-dimensions a + ib
In 3-dimensions a + ib + jc + kd

Well all very nice but the quaternion quadruplet prevents a natural extrapolation into a more general form of n-dimensions.

Looking at the two above equations it would be kind of nice to simply to clip off the real part to the number and simply define a set of bases vectors to span the 2 and 3-space and thereby delineate their dimensionality in a general way.

In 2-dimensions ai + bj
In 3-dimensions ai + bj + ck

That was exactly what a man, a continent away in, Germany, was thinking Hermann Günther Grassmann (1809-1877) in his 1844 paper ***Ausdehnungslehre***, in which were developed classes of algebras of much greater generality than Hamilton's quaternion algebra. Instead of considering just ordered sets of quadruples of real numbers, Grassmann considered ordered sets of n-real numbers.

Now couple this idea with Hamilton's product of unit vectors gives rise to the modern vector operation called the cross product operator for a three-dimensional orthogonal vector system.

Table 7.6.1 Grassmann's Multiplication Table for Cross Product Operator

×	$+\hat{i}$	$+\hat{j}$	$+\hat{k}$
$+\hat{i}$	0	$+\hat{k}$	$-\hat{j}$
$+\hat{j}$	$-\hat{k}$	0	$+\hat{i}$
$+\hat{k}$	$+\hat{j}$	$-\hat{i}$	0

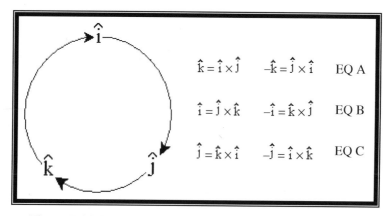

$$\hat{k}=\hat{i}\times\hat{j} \quad -\hat{k}=\hat{j}\times\hat{i} \quad \text{EQ A}$$
$$\hat{i}=\hat{j}\times\hat{k} \quad -\hat{i}=\hat{k}\times\hat{j} \quad \text{EQ B}$$
$$\hat{j}=\hat{k}\times\hat{i} \quad -\hat{j}=\hat{i}\times\hat{k} \quad \text{EQ C}$$

Figure 7.6.1 Grassmann's Product of Permuted Unit Vectors

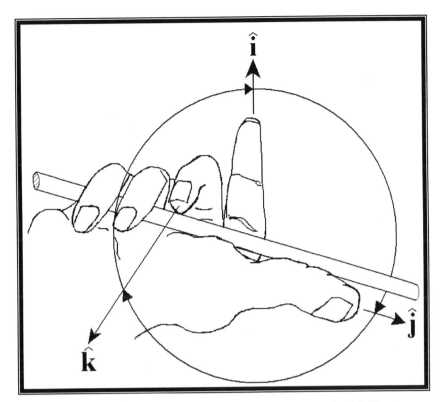

Figure 7.6.2 Right-hand Rule and Directed Permutated Unit Vectors

Definition 7.6.1 **The Right-hand Rule**

Another analogous way of visually seeing the representation of cyclic permutation is to grasp a long straight wire, rod, or pencil, with your thumb lying along side the axis of the object. Curl your three index fingers around the rod with your first index finger pointing up the direction of the unit-**i** vector. Your thumb is in the direction of the unit-**j** vector; the curled second index finger is pointing out. Now if you cross or rotate the first index finger in the direction of the thumb the motion yields the third directed unit-**k** vector. With this method the bearing of a magnetic field can be found knowing the direction of current flow down a conductor (straight wire), physicists have dubbed this the ***right-hand rule***.

Tensor Calculus & Physics: A General Treatise

The downside of the cross product operator is that in this form it cannot be naturally expanded to higher dimensions. In order to expand it to higher dimensions the following definitions of n-spaces are required.

Definition 7.6.2 Magnitude of a Vector

$\lvert \vec{v} \rvert = \lvert \mathbf{V} \rvert$	for vectors in bold face type	Equation A
$\lvert \mathbf{V} \rvert = V$	for $0 \leq V$ such that $V \in R$	Equation B
$\mathbf{V} = V\,\mathbf{v}$	for unit vectors bold face lower case	Equation C

Where [V] is a scalar, absolute, distance that spans from the tail end of the vector to it's head. Now $\lvert \mathbf{V} \rvert$ is called the *modulus* or *absolute magnitude* of the vector.

Definition 7.6.3 Base Vectors Spanning a Space

A collection of unique, non-collinear, vectors spanning or covering a space, such that each one is uniquely associated with each coordinate-axis. These are special vectors because all other vectors are constructed from them; hence, it is important that they do not degenerate to a non-collinear case. Sometimes they are called the *coordinate vectors* for the E^n system otherwise they are known as *basis vectors* ($._\bullet \mathbf{b}_i {}_\bullet$).

$$B^n \equiv \{\ \mathcal{S}^n,\ (._\bullet \mathbf{b}_i {}_\bullet)\ \lvert\ \text{where for all } i,\ \mathbf{b}_i \rightarrow (x_i \in A_i)\ \} \qquad \text{Basis Vector Set}$$

the *basis vector set* spanning $\mathcal{S}^n$ space.

Definition 7.6.4 Unit Vector

A vector having a magnitude of unity, $\lvert \vec{U} \rvert = 1$.

The vector can also represent by a boldfaced capital U, $[\vec{U} = \mathbf{U}]$

Definition 7.6.5 Zero Vector

A vector having zero magnitude, $\lvert \vec{0} \rvert = 0$.	Equation A
The vector can also be represented by a boldfaced zero, $[\vec{0} = \mathbf{0}]$.	Equation B
$\mathbf{0} = \sum_i z_i \mathbf{b}_i$ for $z_i = 0$ see T7.10.4 Grassmann's Zero Vector	Equation C

Definition 7.6.6 Orthogonal Vector Pairs

Orthogonal Vector Pairs have an angle between them of 90°.

Definition 7.6.7 Collinear Vector Pairs

Collinear Vector Pairs have an angle between them of 0°.

Axiom 7.6.1 Euclidian Vector Space E^n of n-Dimensions

Let $\exists$ a set of **basis vectors** ($._{\bullet} b_i ._{\bullet}$) and ($._{\bullet} x_i ._{\bullet}$) called **vector coefficients**, **measured numbers** or **vector components** when associated with the respective axial sets ($._{\bullet} A_i ._{\bullet}$), than the set E^n is called the **Euclidian Vector Space (n-vector space)**, having the following relationships and properties:

$$E^n \equiv \{ \; B^n , D \mid D \rightarrow (B^n \in V) \; \}$$
Equation A

A vector in that space can be represented either as an ordered set of numbers $V(x_1, x_2, ..., x_i, ..., x_n)$, or abbreviated $V_n(._{\bullet} x_i ._{\bullet})$, or as a Grassmann complex number called a **vector**,

$$V = \sum_i x_i b_i$$
Equation B

The point at the tail and the point at the head of the vector define paired points. This forces the question to be asked how far away is one point from the other? This requires a definition to be devised that will allow that distance to be measured. That magnitude is characterized by the Pythagorean formula for a Euclidian n-space.

$$D = \sqrt{\sum_i x_i^2} = |V|$$
Equation C

Where the **bases scaling factor** is

$$| b_i | > 0 \text{ for all } [i]$$
Equation D

and when the tail of the vector is at the origin of the coordinate system the vector points at the coordinate point $P_n(._{\bullet} x_i ._{\bullet})$.

Now the bottom line for Euclidian bases vectors ($._{\bullet} b_i ._{\bullet}$) in Space E^3 is that (i, j, k), which form a bases vector set, can now be associated with each axis

$$(b_x \equiv i) \rightarrow {}_x A, (b_y \equiv j) \rightarrow {}_y A \text{ and } (b_z \equiv k) \rightarrow A_z \text{ axis's.}$$
Equation E

Essentially base vectors are needed to act as frame or structure to shape the space, or another way of looking at them, is without them the E^n space would degenerate into a simple 0-space scalar S^n with no vector properties.

Now taking the squared magnitude of the Euclidian bases vectors

$$| b_x |^2 = i^2 = | b_y |^2 = j^2 = | b_z |^2 = k^2 = 1.$$

or taking the square root for each positive magnitude

$$| b_x | = | b_y | = | b_z | = 1.$$
Equation F

Section 7.7 Observations about Ridged Parallelograms

There are other things to notice in the ridged parallelogram. The lower triangle has the diagonal opposing the upward shift on the right side lath. The opposition of the diagonal slat is said to be under *compression*. Now the upper triangle pulls away in the opposite direction as the downward motion of the left side slat, thereby stretching it. The extend slat is said to be under *tension*.

There is something else the rigid parallelogram also provides in its triangles and that is a sense of orientation in the Euclidian Plane. That is the parallelogram provides us with two adjoining triangles with a common side and having a unique properties. So if they are under compression as shown in Figure 7.6.2 the bottom triangle has a rotational direction that is counter-clockwise and this intern imparts the same counter-clockwise direction in the above triangle. If on the other hand the compression on the slate is down instead of up, it then gives an opposite sense of direction imparted to both triangles in a clockwise direction. With this tool any plane's orient ability can be measured, hence if the parallelogram is moved over the entire surface and than moved back to its origin and the orientation has not changed, such a surface is said to be *oriented*. This is one of the key characteristics of a Euclidian Plane.

Observation 7.7.1 Orientation of a Surface

Orientation of a plane can be looked at in another way that is as an embedded property. If the ridged parallelogram where replicated over the entire surface and the orientation still has not changed this would also be proof of the surface's orientation, hence a natural property of the surface.

Observation 7.7.2 Invariant Quantities

When a quantity is left unchanged by a transformation, as the moving ridged parallelogram, that quantity is said to be *invariant*. So as an example the sense of orientation for a surface is an invariant quantity, since it is left unchanged.

Tensor Calculus & Physics: A General Treatise

Section 7.8 Parallelogram Law: Axioms

With the new symbolism, we can update Newton's Ridged Parallelogram as follows:

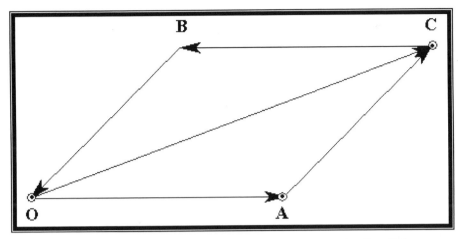

Figure 7.8.1 Modernized vector symbolism for the Newtonian Parallelogram

The diagram can be simplified further by removing the upper triangle since it contributes nothing to the net vectors and leave dotted lines to preserve the parallelogram geometry. It can also be seen that the diagonal is comprised of the other two vectors establishing how their addition, thereby achieve a resultant.

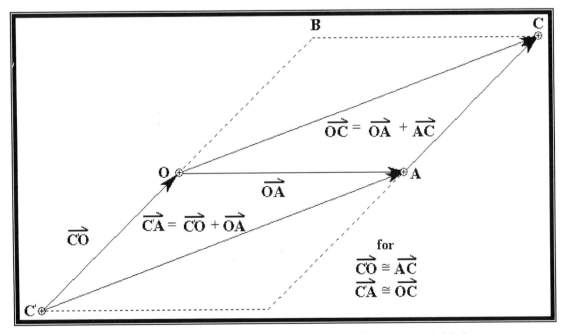

Figure 7.8.2 Parallelogram Showing How Vectors are Added

So far, the resultant diagonal has been the main vector of consideration and provided a geometric interpretation of addition of vectors, but what of the cross diagonal? The vector in opposition is now $\overrightarrow{AC}$, hence the addition of its other components. When solved for the cross diagonal turns out to be the difference of the vector components, thereby demonstrating geometrically the subtraction of vectors.

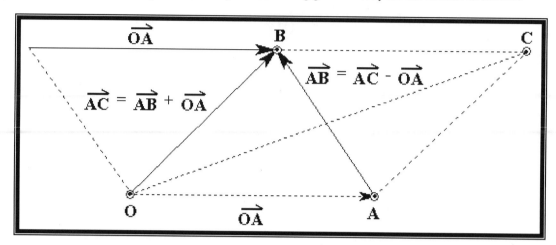

Figure 7.8.3 Parallelogram Showing How Vectors are subtracted

Observe that in Figure 7.8.2 vector components $\overrightarrow{AC}$ and $\overrightarrow{OA}$ add by aiding one another while in Figure 7.8.3 they subtract by being in opposition to one another. The result in either case is a vector, hence the addition of vectors yield a vector and likewise the subtraction of vectors concluded with a vector.

Axiom 7.8.1 Parallelogram Law: Closer of Addition for Vectors

$(A + B \wedge B + A) \in V$. Where V is the set of all vectors within the parallelogram frame.

Axiom 7.8.2 Parallelogram Law: Closer of Subtraction for Vectors

$(A - B \wedge B - A) \in V$. Where V is the set of all vectors within the parallelogram frame.

Another interesting property of a parallelogram is when the resultant vector is scaled by a real number [k].

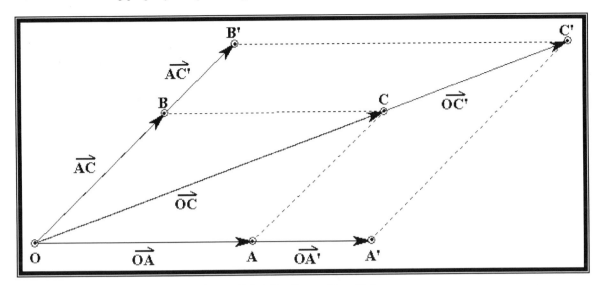

Figure 7.8.4 Scaling a Parallelogram

Notice real number [a] can be found to scale the components comprising vectors in the parallelogram thereby not altering the direction of the resultant vector regardless how it is commutative with the vector. The scaled vectors can be rewritten as follows:

$$\text{If } \overrightarrow{OC'} = a * \overrightarrow{OC} = \overrightarrow{OC} * a \text{ then}$$
$$\overrightarrow{AC'} = a * \overrightarrow{AC} = \overrightarrow{AC} * a \text{ and}$$
$$\overrightarrow{OA'} = a * \overrightarrow{OA} = \overrightarrow{OA} * a$$

Actually this property can be proven from the fundamental principle of similar triangles, hence Δ**COA** is similar to Δ**C'OA'**, because all interior angles are equal, therefore $\overrightarrow{OC}$ is scaled by [a] satisfying the above vector relations. It is from this proof that the notion of scaled ***collinear vectors*** arises.

Axiom 7.8.3 Scaling Collinear Vectors

        If a **A** = a **B** then **A** = **B**   Equation A

        If **A** a = **B** a then **A** = **B**   Equation B

        for [**A**, **B**] collinear vectors and [a] a scalar

Axiom 7.8.4 Converse of Scaling Collinear Vectors

        If **A** = **B** then a **A** = a **B**   Equation A

        If **A** = **B** then **A** a = **B** a   Equation B

Axiom 7.8.5 Parallelogram Law: Commutative and Scalability of Collinear Vectors

        a **A** = **A** a

vectors multiplied by scalar numbers leave the direction of the vector invariant but change the magnitude. This intern satisfies scalability of the parallelograms.

Again, another interesting property of a parallelogram is when the resultant vector is rotated 180° about its origin point O.

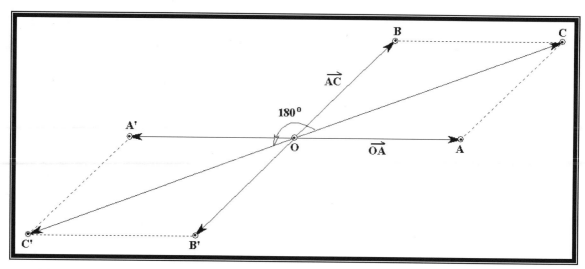

Figure 7.8.5 Mirror reflection of a parallelogram by rotating the resultant vector 180°

Notice that the reconstruction of all component vectors forces them to be opposite and equal to them selves. The reflected components can be rewritten as follows:

If the resultant vector is rotated 180° such that
$$\overrightarrow{OC'} = -\overrightarrow{OC} \text{ than}$$
$$\overrightarrow{OA'} = -\overrightarrow{OA} \text{ and}$$
$$\overrightarrow{OB'} = -\overrightarrow{OB}$$

Thereby establishing mirror symmetry by multiplying a negative number times a vector.

One can find hundreds of such relationships with parallelograms, but these are key to what follows next, which is setting up a system for vector algebra. The above geometric proofs while not totally rigorous do give a degree of confidence for the validity of the following vector axioms.

Axiom 7.8.6 Parallelogram Law: Multiplying a Negative Number Times a Vector
$$-\mathbf{A} = (-1)\mathbf{A}$$

Multiplying a negative number times a vector reflects or rotates the vector in the opposite direction by 180°.

Through the addition, of vectors using the parallelogram law, one other thing that can be gleaned from the diagram is how vectors move. Notice that if $\overrightarrow{AC}$ is moved parallel to the opposite side of the diagram such that $\overrightarrow{OB} = \overrightarrow{AC}$ is left unaltered in magnitude and direction.

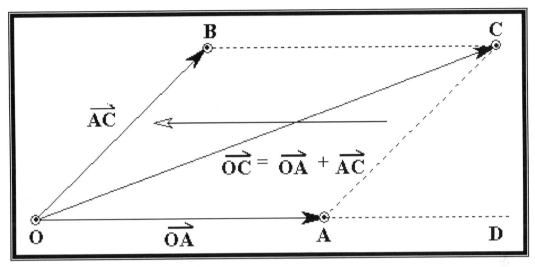

Figure 7.8.6 Lateral Movement of a Vector

This form is more common because it shows the resulting vector $\overrightarrow{OC}$ clearly in its constituent components. The above notation of equivalence of vectors comes from two properties of vectors.

$$\overrightarrow{OB} = \overrightarrow{AC}$$

iff

$$\overline{OB} = \overline{AC} \qquad \text{magnitude and}$$
$$\angle DOB = \angle DAC \qquad \text{interior angles}$$

Hence, the concept of equivalence, or translating vectors, is really constructing vectors in a new position. Also, notice equivalence of vectors allows the construction of one vector on top of another or congruence of geometric objects. This is a given requirement of vectors because without it affine geometry would not exist, which is an important property of tensors.

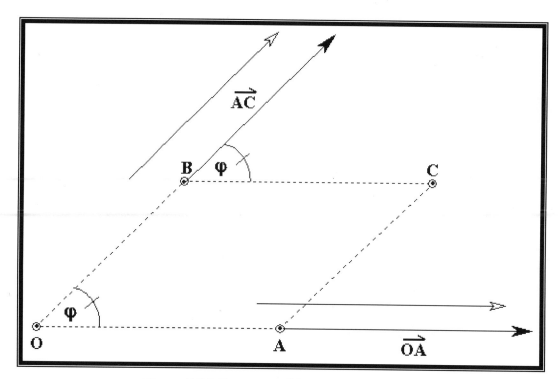

Figure 7.8.7 Transverse Movement of Vectors

This geometric movement of vectors, are always ***parallel*** within the framework of the parallelogram, called an ***Affine Transformation***, thereby leaving the vector unchanged under the transformation.

Axiom 7.8.7 Parallelogram Law: Affine Transformation of Vectors

Affine transformation is the geometric movement of vectors, always parallel within the framework of the parallelogram, this type of transformation leaves the vector unchanged in direction and magnitude. When vectors are left unaltered, they are said to be ***invariant***. See Figure 7.8.6 and 7.

Section 7.9 Vector Law: Axioms as a Group

It turns out that if we look at the geometrical constructions in Figures 7.8.1 to 7 and realize our perspective is within the plane of the parallelogram there really is no way that we could tell if we are in two-dimensions or a plane embedded within n-space. In other words, these diagrammatic drawings are dimensionally independent; hence, the principles they demonstrate are general. This is true of the following diagrammatic development of axioms, so we can now set down the rules of vectors in a comprehensive way. Treating a vector as a number gives rise to the following axioms.

Axiom 7.9.1 Equivalence of Vector

A = B by Observation 7.1.2 "Equivalence of Vectors". That is two vectors are equal if their magnitude and direction are the same. Also vectors remain equivalent under equal and commutative scaling of the vectors, see A7.8.5 "Parallelogram Law: Commutative and Scalability of Collinear Vectors".

Notice that the notion of geometric congruence must be excepted if this is to work. Congruence being that a vector is not a different vector if it is moved. Only that it is different if it is placed on top of another with the same direction and magnitude. It follows that the attributes of direction and magnitude between the two vectors is what is equivalent.

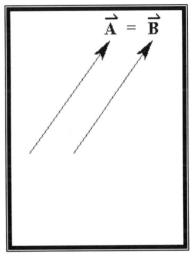

Figure 7.9.1: Vector Equivalence

Axiom 7.9.2 Substitution of Vectors

A = B
B = C
∴A = C by Figure 7.6.7 "Transverse Movement of Vectors"

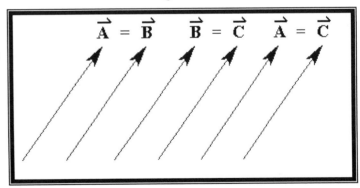

Figure 7.9.2: Substitution of Vectors

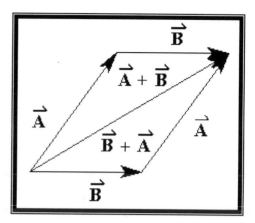

Axiom 7.9.3 Commutative by Vector Addition

A + B = B + A

by Figure 7.6.2 "Parallelogram Showing How Vectors are Added"

Figure 7.9.3: Commutative by Addition

Axiom 7.9.4 Associative by Vector Addition

A + (B + C) = (A + B) + C Figure 7.6.2 "Parallelogram Showing How Vectors are Added" in repeated parts.

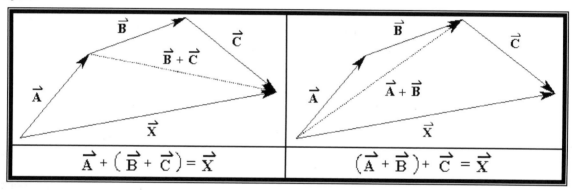

Figure 7.9.4: Associative by Vector Addition

Tensor Calculus & Physics: A General Treatise

Now treating vectors as independent objects the above axioms parallel to the axioms on congruency.

Table 7.9.1: Axioms for Vector Objects

Axiom	Axioms of Quantities	Description
Axiom 7.9.5 Existence of Vector Object	$\exists$ a field $V = \{\ _\bullet x_i \wedge \mathbf{b}_i \ _\bullet \mid [\ P(_\bullet x_{i\bullet}) \in (\ \mathbf{R} \vee \mathbf{C}^2)\ ] \wedge V(_\bullet \mathbf{b}_i \ _\bullet) \wedge (|\mathbf{b}_i| \sim\vee=1)$ for all [i]\}	Existence of vector objects in a field
Axiom 7.9.6 Reflexiveness of Vector Object	$A \cong B$ Equation A $B \cong A$ Equation B	Congruent reflexiveness of vector objects
Axiom 7.9.7 Transitivity of Vector Objects	$B \cong A$ and $C \cong B$ $\therefore C \cong A$	Transitivity of Vector Objects

Table 7.9.2 Axioms for Vector Objects under Addition

Axiom	Axioms by Addition	Description
Axiom 7.9.8 Closure Vector Object Add	$(A + B \wedge B + A) \in V$	Closure with respect to addition
Axiom 7.9.9 Commutative Vector Object Add	$A + B \cong B + C$	Commutative by Addition
Axiom 7.9.10 Associative Vector Object Add	$A + (B + C) \cong (A + B) + C$	Associative by Addition
Axiom 7.9.11 Identity Vector Object Add	$A + 0 \cong A$	Identity by Addition
Axiom 7.9.12 Inverse Vector Object Add	$A - A \cong 0$	Inverse by Addition

Table 7.9.3 Axioms Equivalency of Equality and Congruency

Axiom	Axioms by Addition	Description
Axiom 7.9.13 Vector Numbers and Object Equivalency	Theorems on vector equivalency are the same for congruency, hence interchangeable.	Proofs on vectors as numbers are the same for objects since both identically are based on the parallelogram law.

Section 7.10 Vector Law: Theorems with Scalars

Theorem 7.10.1 Uniqueness of Scalar Multiplication to Vectors

	Given				
1g	Given	$(\mathbf{A}, \mathbf{B}) \in E^n$			
2g		$(a, b) \in R$			
3g		$a = b$			
4g		$\mathbf{A} = \mathbf{B}$			
Steps	Hypothesis	$a\mathbf{A} = b\mathbf{B}$ EQ A		$\mathbf{A}a = \mathbf{B}b$ EQ B	
1	From A7.9.1 Vector Equivalence	$a\mathbf{A} = a\mathbf{A}$		$\mathbf{A}a = \mathbf{A}a$	
2	From 3g, 1 and A4.2.3 Substitution	$a\mathbf{A} = b\mathbf{A}$		$\mathbf{A}a = \mathbf{A}b$	
∴	From 2 and A7.9.2 Substitution of Vectors	$a\mathbf{A} = b\mathbf{B}$ EQ A		$\mathbf{A}a = \mathbf{B}b$ EQ B	

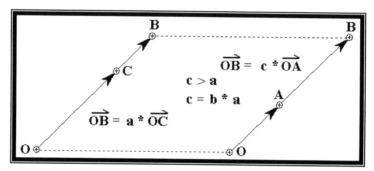

Figure 7.10.1 Associative Scalars by Multiplication

Theorem 7.10.2 Associative Scalar Multiplication to Vectors

1g	Given	the above geometry of Figure F7.10.1
Steps	Hypothesis	$a(b\mathbf{A}) = (ab)\mathbf{A}$
1	From 1g	$\mathbf{B} = a\mathbf{C}$
2	From 1g	$\mathbf{B} = c\mathbf{A}$
3	From 1g	$c > a$ implying $c \longleftrightarrow a$
4	From 3 and D4.1.23C Ratio of Quantities	$c = b\,a$
5	From 2, 4 and A4.2.3 Substitution	$\mathbf{B} = b\,a\,\mathbf{A}$
6	From 1, 5 and A7.9.2 Substitution of Vectors	$a\,\mathbf{C} = b\,a\,\mathbf{A}$
7	From 6 and T4.4.3 Equalities: Cancellation by Multiplication	$\mathbf{C} = b\,\mathbf{A}$
8	From 7 and A4.2.11 Associative Multp	$a(\mathbf{C}) = (b\,a)\mathbf{A}$
9	From 7, 8 and A7.9.2 Substitution of Vectors	$a(b\mathbf{A}) = (b\,a)\mathbf{A}$
∴	From 9 and A4.2.10 Commutative Multp	$a(b\mathbf{A}) = (ab)\mathbf{A}$

Theorem 7.10.3 Distribution of a Scalar over Addition of Vectors

1g 2g	Given	$(\mathbf{A}, \mathbf{B}) \in E^n$ $a \in R$
Steps	Hypothesis	$a\,(\mathbf{A} + \mathbf{B}) = a\mathbf{A} + a\mathbf{B}$
1	From A7.9.1 Vector Equivalence	$a\,(\mathbf{A} + \mathbf{B}) = a\,(\mathbf{A} + \mathbf{B})$
$\therefore$	From 1 and A7.8.5 Parallelogram Law: Commutative and Scalability of Collinear Vectors	$a\,(\mathbf{A} + \mathbf{B}) = a\,\mathbf{A} + a\,\mathbf{B}$ the scalar can be distributed across the vectors because the resulting parallelogram scales proportionally. So, the sum of their direction remains the same.

Theorem 7.10.4 Grassmann's Zero Vector

1g	Given	$\mathbf{0} \in E^n$
Steps	Hypothesis	$\mathbf{0} = \sum_i 0\,\mathbf{b}^i$
1	From D7.6.6 Zero Vector	$0 = \mid \mathbf{0} \mid$
2	From 1g, 1, A7.6.1C Euclidian Vector Space E^n of n-Dimensions and A4.3.3 Substitution	$0 = \sqrt{\sum_i a_i^2}$
3	From 2 the result can only be valid since all the values are absolute positive numbers, iff	$a_i = 0$ for all i.
$\therefore$	From 1g, 3, A7.6.1B Euclidian Vector Space E^n of n-Dimensions and A4.3.3 Substitution	$\mathbf{0} = \sum_i 0\,\mathbf{b}^i$

Theorem 7.10.5 Identity with Scalar Multiplication to Vectors

1g	Given	$A \in E^n$
Steps	Hypothesis	$1\,A = A$
1	From A7.9.1 Vector Equivalence	$1\,A = 1\,A$
2	From 1g, 1 and A7.6.1B Euclidian Vector Space E^n of n-Dimensions	$1\,A = 1 \sum_i a_i \mathbf{b}_i$
3	From 2, T7.10.3 Distribution of a Scalar over Addition of Vectors and A4.2.11 Associative Multp	$1\,A = \sum_i (1\ a_i) \mathbf{b}_i$
4	From 3 and A4.2.12 Identity Multp	$1\,A = \sum_i a_i \mathbf{b}_i$
∴	From 4 and A7.6.1B Euclidian Vector Space E^n of n-Dimensions	$1\,A = A$

Theorem 7.10.6 Magnitude Squared

1g	Given	$A \in E^n$
Steps	Hypothesis	$|A|^2 = \sum_i a_i^2$
1	From 1g and D4.1.4C Rational Numbers and A4.2.10 Commutative Multp	$|A| = \sqrt{\sum_i a_i^2}$
∴	From 1, T4.8.6 Integer Exponents: Uniqueness of Exponents and T4.10.2 Radicals: Commutative Product and T4.10.3 Radicals: Identity	$|A|^2 = \sum_i a_i^2$

Tensor Calculus & Physics: A General Treatise

Theorem 7.10.7 Inverse by Normalized Magnitude of a Vector

1g	Given	$\mathbf{A} \in E^n$	
Steps	Hypothesis	$\mathbf{A} / |\mathbf{A}| = \mathbf{a}$	$\dfrac{\mathbf{A}}{|\mathbf{A}|} = \mathbf{a}$
1	From A7.9.1 Vector Equivalence	$\mathbf{A} / |\mathbf{A}| = \mathbf{A} / |\mathbf{A}|$	
2	From 1, D4.1.4(A, B) Rational Numbers and A4.2.10 Commutative Multp	$\mathbf{A} / |\mathbf{A}| = (1 / |\mathbf{A}|)\, \mathbf{A}$	
3	From 2, A7.6.1B Euclidian Vector Space E^n of n-Dimensions and A4.3.3 Substitution	$\mathbf{A} / |\mathbf{A}| = (1 / |\mathbf{A}|)\, \sum_i a_i \mathbf{b}_i$	
4	From 3, T7.10.3 Distribution of a Scalar over Addition of Vectors and A4.2.11 Associative Multp	$\mathbf{A} / |\mathbf{A}| = \sum_i ((1 / |\mathbf{A}|)\, a_i)\, \mathbf{b}_i$	
5	From 4 and D4.1.4C Rational Numbers and A4.2.10 Commutative Multp	$|(\mathbf{A} / |\mathbf{A}|)| = \sqrt{\sum_i ((1 / |\mathbf{A}|)\, a_i)^2}$	
6	From 5 and A4.2.21 Distribution Exp	$|(\mathbf{A} / |\mathbf{A}|)| = \sqrt{\sum_i (1 / |\mathbf{A}|)^2\, a_i{}^2}$	
7	From 6 and A4.2.14 Distribution	$|(\mathbf{A} / |\mathbf{A}|)| = \sqrt{(1 / |\mathbf{A}|)^2 \sum_i a_i{}^2}$	
8	From 7, T7.10.6 Magnitude Squared and A4.2.3 Substitution	$|(\mathbf{A} / |\mathbf{A}|)| = \sqrt{(1 / |\mathbf{A}|)^2 |\mathbf{A}|^2}$	
9	From 8, A4.2.1 Distribution Exp, T4.10.3 Radicals: Identity	$|(\mathbf{A} / |\mathbf{A}|)| = (1 / |\mathbf{A}|)\, |\mathbf{A}|$	
10	From 9 and T4.4.16 Equalities: Product by Division, Common Factor	$|(\mathbf{A} / |\mathbf{A}|)| = 1$	
∴	From 10, A7.6.1 Unit Vector and D4.1.4(A, C) Rational Numbers	$\mathbf{A} / |\mathbf{A}| = \mathbf{a}$	$\dfrac{\mathbf{A}}{|\mathbf{A}|} = \mathbf{a}$

Theorem 7.10.8 Vector Based on Unit Vector

Steps	Hypothesis	$\mathbf{A} = |\mathbf{A}|\, \mathbf{a}$
∴	From 10, T7.10.7 Inverse by Normalized Magnitude of a Vector and T4.4.7B Equalities: Reversal of Left Cancellation by Multiplication	$\mathbf{A} = |\mathbf{A}|\, \mathbf{a}$

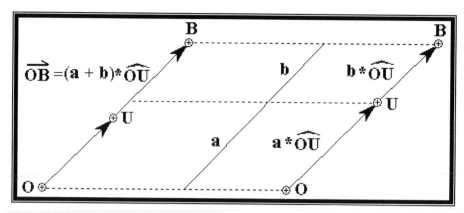

Figure 7.10.2 Distribution of Vector over Addition of Scalars

Theorem 7.10.9 Distribution of Vector over Addition of Scalars

1g 2g	Given	the above geometry of Figure F7.10.2 $(a, b) \in R$
Steps	Hypothesis	$(a + b)\mathbf{A} = a\mathbf{A} + b\mathbf{A}$
1	From 1g and 2g	$\mathbf{B} = (a + b)\,\mathbf{u}$
2	From 1g	$\mathbf{B} = a\,\mathbf{u} + b\,\mathbf{u}$
3	From 1, 2 and A7.9.1 Vector Equivalence	$(a + b)\,\mathbf{u} = a\,\mathbf{u} + b\,\mathbf{u}$
4	From A4.2.2A Equality	$|\mathbf{A}| = |\mathbf{A}|$ by the magnitude of any given vector
5	From 3, 4, T4.4.2 Equalities: Uniqueness of Multiplication	$|\mathbf{A}| (a + b)\,\mathbf{u} = |\mathbf{A}| (a\,\mathbf{u} + b\,\mathbf{u})$
6	From 5, A4.2.10 Commutative Multp and T7.10.3 Distribution of a Scalar over Addition of Vectors	$(a + b) |\mathbf{A}|\mathbf{u} = a |\mathbf{A}|\mathbf{u} + b |\mathbf{A}|\mathbf{u}$
∴	From 6 and T7.10.8 Vector Based on Unit Vector	$(a + b)\mathbf{A} = a\mathbf{A} + b\mathbf{A}$

Section 7.11 Vector Law: Theorem on Vectors

Theorem 7.11.1 Vector Addition: Uniqueness for Addition

1g	Given	$A = C$
2g		$B = D$
Steps	Hypothesis	$A + B = C + D$ for $A = B$ and $C = D$
1	From A7.8.1 Parallelogram Law: Closer of Addition for Vectors	$A + B = F$
2	From 1g, 2g and A7.9.7 Transitivity	$C + D = F$
∴	From 1, 2 and A7.9.1 Equivalence of Vector	$A + B = C + D$ for $A = B$ and $C = D$

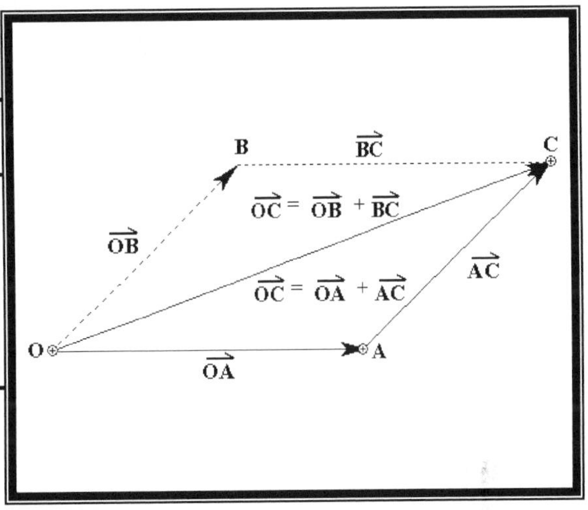

Figure 7.11.1: Uniqueness of Vector Addition

Theorem 7.11.2 Vector Addition: Identity

1g	Given	$A \in E^n$
Steps	Hypothesis	$A + 0 = A$
1	From 1g and A7.6.1B Euclidian Vector Space E^n of n-Dimensions	$A = \sum_i a_i \mathbf{b}_i$
2	From A7.9.1 Vector Equivalence	$A + 0 = A + 0$
3	From 1, 2, T7.10.4 Grassmann's Zero Vector and A7.9.2 Substitution of Vectors	$A + 0 = \sum_i a_i \mathbf{b}_i + \sum_i 0\, \mathbf{b}_i$
4	From 3 and A7.9.4 Associative by Vector Addition	$A + 0 = \sum_i (a_i \mathbf{b}_i + 0\, \mathbf{b}_i)$
5	From 4 and T7.10.9 Distribution of Vector over Addition of Scalars	$A + 0 = \sum_i (a_i + 0\,)\mathbf{b}_i$
6	From 5 and A4.2.7 Identity Add	$A + 0 = \sum_i a_i \mathbf{b}_i$
∴	From 1, 6 and A7.9.1 Substitution of Vectors	$A + 0 = A$

Theorem 7.11.3 Vector Addition: Inverse

1g	Given	$A \in E^n$		
Steps	Hypothesis	$A - A = 0$		
1	From A7.9.1 Vector Equivalence	$A - A = A - A$		
2	From 1g, 1 and A7.6.1B Euclidian Vector Space E^n of n-Dimensions	$A - A = \sum_i a_i \mathbf{b}^i - \sum_i a_i \mathbf{b}_i$		
3	From 2 and A7.9.4 Associative by Vector Addition	$A - A = \sum_i (a_i \mathbf{b}_i - a_i \mathbf{b}_i)$		
4	From 3 and T7.10.9 Distribution of Vector over Addition of Scalars	$A - A = \sum_i (a_i - a_i) \mathbf{b}_i$		
5	From 4 and A4.2.8 Inverse Add	$A - A = \sum_i 0 \, \mathbf{b}_i$		
∴	From 5, T7.10.4 Grassmann's Zero Vector and A7.9.2 Substitution of Vectors	$A - A = 0$		

Theorem 7.11.4 Vector Addition: Right Cancellation by Addition

1g	Given	$A = B + C$			
Steps	Hypothesis	$A - C = B$	Equation A	$B = A - C$	Equation B
1	From A7.9.6A Congruency	$-C = -C$			
2	From 1, 1g and T7.11.1 Vector Addition: Uniqueness for Addition	$A - C = (B + C) - C$			
3	From 2 and A7.9.10 Associative Vector Object Add	$A - C = B + (C - C)$			
4	From 3 and A7.9.12 Inverse Vector Object Add	$A - C = B + 0$			
∴	From 4 and A7.9.11 Inverse Vector Object Add	$A - C = B$	by A7.9.6A	$B = A - C$	by A7.9.6B

Theorem 7.11.5 Vector Addition: Reversal of Right Cancellation by Addition

1g	Given	$B = A - C$			
Steps	Hypothesis	$B + C = A$	Equation A	$A = B + C$	Equation B
∴	From T7.11.4B Vector Addition: Right Cancellation by Addition and T3.7.2 Reversibility of a Logical Argument	$B + C = A$		$A = B + C$	

Theorem 7.11.6 Vector Addition: Left Cancellation by Addition

1g	Given	$B + C = A$			
Steps	Hypothesis	$A - C = B$	Equation A	$B = A - C$	Equation B
1	From 1g and A4.2.2B Equality	$A = B + C$			
∴	From 1 and A4.2.4 Closure Add	$A - C = B$	by A7.9.6A	$B = A - C$	by A7.9.6B

Theorem 7.11.7 Vector Addition: Reversal of Left Cancellation by Addition

1g	Given	$A - C = B$			
Steps	Hypothesis	$A = B + C$	Equation A	$B + C = A$	Equation B
∴	From T7.11.6A Equalities: Left Cancellation by Addition and T3.7.2 Reversibility of a Logical Argument	$A = B + C$	by A7.9.6A	$B + C = A$	by A7.9.6B

Theorem 7.11.8 Vector Addition: Uniqueness of Subtraction

1g	Given	$A = B$	
2g		$C = D$	
Steps	Hypothesis	$A - C = B - D$ for $A = B$ and $C = D$	
1	From A7.8.2 Parallelogram Law: Closer of Subtraction for Vectors	$A - C = F$	
2	From 1g, 2g, 1 and A7.9.7 Transitivity	$B - D = F$	
∴	From 1, 2 and A7.9.1 Equivalence of Vector	$A - C = B - D$ for $A = B$ and $C = D$	

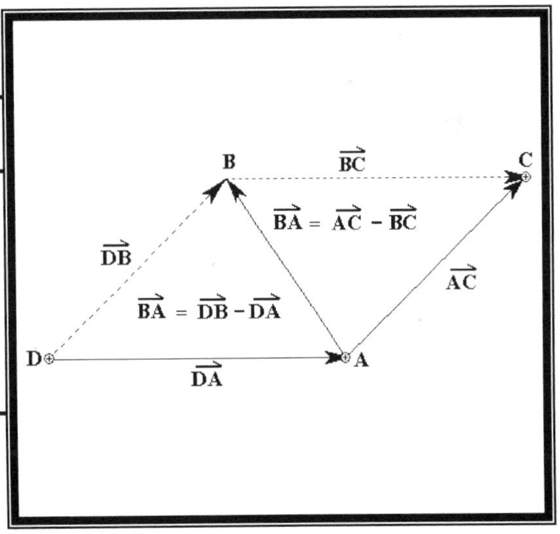

Figure 7.11.2 Uniqueness of Vector Subtraction

Theorem 7.11.9 Vector Addition: Equivalency by Matching Components

1g	Given	$A \in E^n$	
Steps	Hypothesis	$A_i = B_i$	
1	From A7.9.1 Equivalence of Vector	$\mathbf{A} = \mathbf{B}$	
2	From 1 and T7.11.2 Vector Addition: Identity and T7.11.4B Vector Addition: Right Cancellation by Addition	$\mathbf{0} = \mathbf{B} - \mathbf{A}$	
3	From 1g, 2 and A7.6.1B Euclidian Vector Space E^n of n-Dimensions	$\mathbf{0} = \sum_i B_i\mathbf{b}_i - \sum_i A_i\mathbf{b}_i$	
4	From 3 and T7.10.9 Distribution of Vector over Addition of Scalars	$\mathbf{0} = \sum_i (B_i - A_i)\mathbf{b}_i$	
5	From 4 and T7.10.4 Grassmann's Zero Vector	$0 = B_i - A_i$	
$\therefore$	From 5, T4.3.4A Equalities: Reversal of Right Cancellation by Addition and A4.2.7 Identity Add	$A_i = B_i$	

Section 7.12 Vector Law: Axioms on Linear Dependence and Independence

The question now is how does all of these new pieces of the puzzle fit together and allow the construction of an n-dimensional vector. First lets establish the following axioms and theorems:

Axiom 7.12.1 Linear Dependence for a Set of Vectors (. b_i .)

A set of vectors (. b_i .) is called *linearly dependent* if there exist numbers (. m_i .), not all of which are zero, such that

$$\alpha V + \sum^n \alpha_i b_i = 0 \text{ for not all (. } m_i \neq 0 \text{ .)} \qquad \text{Equation A}$$

Another and more intuitive way of proving dependences is for a set of vectors spanning n-space that the determinate of their coefficients is zero, $0 = |a_i^j| = |A|$. Implying repeatable or collinear vectors in the determinate.

$$|A| = 0 \qquad \text{Equation B}$$

Axiom 7.12.2 Linear Independence for a Set of Vectors (. b_i .)

If vector coefficients (. m_i .) do not exist for a set of bases vectors, that is all zeros, the vectors are said to be *linearly independent*. [SOK67, pg 6] Another and more intuitive way of proving independents is for a set of vectors spanning n-space that the determinate of their coefficients is not zero, $0 \neq |a_i^j|$.

$$\alpha V + \sum^n \alpha_i b_i = 0 \text{ for all (. } m_i = 0 \text{ .)} \qquad \text{Equation A}$$

and

$$|A| \neq 0 \qquad \text{Equation B}$$

Section 7.13 Vector Law: Theorems on Linear Dependence and Independence

Theorem 7.13.1 Clearing the Right to Left Hand Side of the Sweep Vector

1g	Given	$(\bullet\, b^i \,\bullet) \in E^n$ i = 1, 2, ..., n
2g		$V = \sum^n m_i \mathbf{b}_i$
3g		$m_i = -\alpha_i/\alpha$
Steps	Hypothesis	$\alpha V + \sum_n \alpha_i \mathbf{b}_i = 0$ for all i
1	From 2g, 3g and A4.2.3 Substitution	$V = \sum^n (-\alpha_i/\alpha) \mathbf{b}^i$
2	From 1 and A4.2.2A Equality	$\alpha = \alpha$
3	From 2 and T4.3.7 Equalities: Uniqueness of Subtraction	$\alpha V = \alpha \sum^n (-\alpha_i/\alpha) \mathbf{b}_i$
4	From 3 and A4.2.14 Distribution	$\alpha V = \sum^n \alpha(-\alpha_i/\alpha) \mathbf{b}_i$
5	From 4 and A4.2.10 Commutative Multp	$\alpha V = \sum^n -\alpha(\alpha_i/\alpha) \mathbf{b}_i$
6	From 5 and A4.2.10 Commutative Multp	$\alpha V = \sum^n -\alpha(1/\alpha)\alpha_i \mathbf{b}_i$
7	From 6, D4.1.20 Negative Coefficient and A4.2.14 Distribution	$\alpha V = -\sum^n \alpha(1/\alpha)\alpha_i\ \mathbf{b}_i$
8	From 7 and A4.2.11 Associative Multp	$\alpha V = -\sum^n (\alpha(1/\alpha))\alpha_i\ \mathbf{b}_i$
9	From 8 and A4.2.13 Inverse Multp	$\alpha V = -\sum^n 1\ \alpha_i\ \mathbf{b}_i$
10	From 9 and A4.2.12 Identity Multp	$\alpha V = -\sum^n \alpha_i\ \mathbf{b}_i$
11	From 10 and A4.2.2A Equality	$\sum^n \alpha_i\ \mathbf{b}_i = \sum^n \alpha_i\ \mathbf{b}_i$
12	From 10, 11 and T4.3.1 Equalities: Uniqueness of Addition	$\alpha V + \sum^n \alpha_i\ \mathbf{b}_i = -\sum^n \alpha_i\ \mathbf{b}_i + \sum^n \alpha_i\ \mathbf{b}_i$
∴	From 12, 1g and A4.2.8 Inverse Add	$\alpha V + \sum^n \alpha_i\mathbf{b}_i = 0$ for all i

Theorem 7.13.2 Construction of N-Dimensional Dependent Vectors

1g	Given	$(\bullet\, b_i\, \bullet) \in E^n\ i = 1, 2, \ldots, n$	
2g		$m_i = -\alpha_i/\alpha$	
3g		$m_{i+1} = -\alpha_{i+1}/\alpha$	
n-space	Hypothesis	$\mathbf{V} = \sum^n m_i \mathbf{a}_i$	$\alpha\mathbf{V} + \sum^n \alpha_i\mathbf{a}_i = 0$
	Real Vector Coefficients	$\mathbf{V}$ Sweep Vector	Non-Collinear Base Vectors
1	$m_1 = -\alpha_1/\alpha$	$\mathbf{V} = m_1\mathbf{b}_1$	$\alpha\mathbf{V} + \alpha_1\mathbf{b}_1 = 0$
2	$m_1 = -\alpha_1/\alpha,\ m_2 = -\alpha_2/\alpha$	$\mathbf{V} = m_1\mathbf{b}_1 + m_2 b_2$	$\alpha\mathbf{V} + \alpha_1\mathbf{b}_1 + \alpha_2\mathbf{b}_2 = 0$
3	$m_1 = -\alpha_1/\alpha,\ m_2 = -\alpha_2/\alpha,\ m_3 = -\alpha_3/\alpha$	$\mathbf{V} = m_1\mathbf{b}_1 + m_2 b_2 + m_3 b_3$	$\alpha\mathbf{V} + \alpha_1\mathbf{b}_1 + \alpha_2\mathbf{b}_2 + \alpha_3\mathbf{b}_3 = 0$
...	...	...	... From T7.7.1
i	$m_i = -\alpha_i/\alpha$ From 2g	$\mathbf{V} = \sum^n m_i\mathbf{b}_i$	$\alpha\mathbf{V} + \sum^n \alpha_i\mathbf{b}_i = 0$
i+1	$m_{i+1} = -\alpha_{i+1}/\alpha$ From 3g	$\mathbf{V} = \sum^n m_{i+1}\mathbf{b}_{i+1}$	$\alpha\mathbf{V} + \sum^n \alpha_{i+1}\mathbf{b}_{i+1} = 0$
...	...	...	...
n	$m_i = -\alpha_i/\alpha$	$\mathbf{V} = \sum^n m_i\mathbf{b}_i$	$\alpha\mathbf{V} + \sum^n \alpha_i\mathbf{b}_i = 0$
∴	From A3.11.1 Induction Property of Integers by ♭₂	$\mathbf{V} = \sum^n m_i\mathbf{b}_i$	$\alpha\mathbf{V} + \sum^n \alpha_i\mathbf{b}_i = 0$

This is truly a milestone this is the first n-dimensional vector proof in this paper, but what does it all mean?

One thing that is clear from T7.13.2 "Construction of N-Dimensional Dependent Vectors" the sweep vector is comprised of a set of uniquely defined sub-vectors $[\mathbf{b}_i]$ and for any value $[m_i]$ associated with them $[\mathbf{V}]$ can uniquely specify a point $(\bullet\, m_i\, \bullet)$ within n-space. Also, the sweep Vector is clearly dependent on the base vectors $(\bullet\, \mathbf{b}_i\, \bullet)$ as long as there at least some non-zero vector coefficients.

What if, two or more of the base vectors in the framework where collapsed so that they had become collinear and dependent on one another?

Theorem 7.13.3 Degenerate ($_\bullet\mathbf{b}^{i+1}{}_\bullet$) System

1g	Given	$(_\bullet, \mathbf{b}_i, _\bullet) \in E^n$ i = 1, 2, ..., n
2g		$_{bi} = \mathbf{b}_{i+1}$
3g		$\beta_i = \alpha_i + \alpha_{i+1}$
Steps	Hypothesis	if any pair of base vectors are collinear than $(_\bullet\mathbf{b}^{i+1}{}_\bullet)$ are independent under n-space, but form a new (n-1)-space.
1	From T7.13.2 Construction of N-Dimensional Dependent Vectors and expand about (i, i+1)	$\alpha\mathbf{V} + \text{LOT} + \alpha_i\mathbf{b}\,i + \alpha_{i+1}\mathbf{b}_{i+1} + \text{HOT} = 0$
2	From 1 and 2g	$\alpha\mathbf{V} + \text{LOT} + \alpha_i\mathbf{b}_i + \alpha_{i+1}\mathbf{b}_i + \text{HOT} = 0$
3	From 2 and T7.10.3 Distribution of Vector over Addition of Scalars	$\alpha\mathbf{V} + \text{LOT} + (\alpha_i + \alpha_{i+1})\mathbf{b}_i + \text{HOT} = 0$
4	From 3, 3g and A4.2.3 Substitution	$\alpha\mathbf{V} + \text{LOT} + \beta_i\mathbf{b}_i + \text{HOT} = 0$
∴	From 4 there is no longer a number that can be found for $\mathbf{b}_{i+1}$, hence a complete set of $(_\bullet, \alpha_i, _\bullet)$ cannot be found and by A7.12.2 Linear Independence for a Set of Vectors $(_\bullet, \mathbf{b}_i, _\bullet)$, however a new dependent set of $(_\bullet, \beta_{i+1}, _\bullet)$ define an (n-1)-space by A7.12.1 Linear Dependence for a Set of Vectors $(_\bullet, \mathbf{b}_i, _\bullet)$	If any pair of base vectors are collinear than $(_\bullet, \mathbf{b}_{i+1}, _\bullet)$ are independent under n-space, but form a new dependent (n − 1)-space.

Table of Contents

Tensor Calculus &Physics: A General Treatise

Tensor Calculus &Physics: A General Treatise

List of Figures

List of Tables

List of Observations

Chapter 8 First and Second Products: N-Space Vector Multiplication

Section 8.1 Distance

I have not even talked about, probably the most important, and complicated property, vector multiplication, but have interjected the idea of multiple systems of vectors with the notation of tensors, why? The reason is quite literally for centuries vectors have always been developed in 2 and 3-space so vector multiplication required two vectors spanning a 3-space. The reality is it takes (n-1)-vectors spanning an n-space to carryout vector multiplication in a general way (see T8.4.6). Since this is a treatises and not an elementary introduction into vectors I have decided not to pull any punches and jump right into the topic with a complete n-dimensional development.

One other point in this chapter all proofs, are based on orthogonal Euclidian coordinates. While sets of vectors spanning a space are considered, where generality does not conflict, they should not be considered base vectors for the coordinate system. Non-orthogonal systems will be considered in chapter 10 in a generalization of tensors.

Now how distance is defined determines the type of geometrical space that is to be studied for instance if distance where measured as a taxicab driver sees his world than distance is measured in a days travel by city blocks:

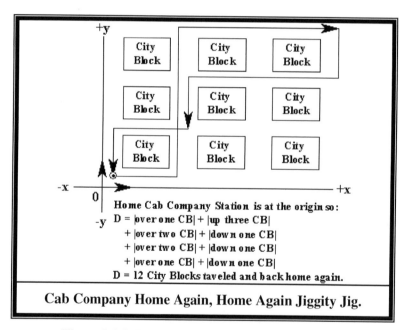

Figure 8.1.1: Distance Measured by Taxicab Driver

Another example of distance defined determines how space is perceived between parity numbers if you remember C6.8.3.1 "The Arithmetic of parity numbers predict composition parity" predicting parity distance

$$\text{parityn}(\eta°\sigma) = |\ \text{parityn}(\eta) \pm \text{parityn}(\sigma)\ | \mp \text{parityn}(\eta x \sigma)$$

has no real geometric imagery, but means something very significant for distance when measuring between parity numbers.

Of course, there is an elastic distance found with relativistic magnitude where it is not a fixed number, but stretchable contracting at a rate relative to the velocity of light in the direction of travel, see TB.2.3 "LTD: Lorentz Length Contraction":

$$\mathbf{D} = \mathbf{D}_0 \sqrt{1 - (v/c)^2}$$

Section 8.2 Magnitude

In the world of tensors distance is measured in a more classical manner with the Pythagoras's Theorem. So, Pythagoras will define vector magnitude and will be extended to read:

Definition 8.2.1 **Pythagoras's Theorem for Magnitude of a Vector**

$$|V| \equiv D = \sqrt{a^2 + b^2} \text{ for } V(a, b)$$
(see T5.8.8 "Right Triangle: The Hypotenuse")

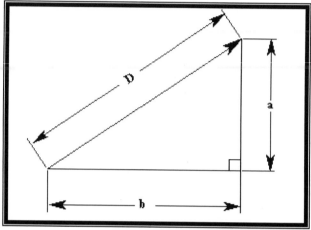

Figure 8.2.1: Pythagorean Magnitude for a Vector

Tensor Calculus &Physics: A General Treatise

The next step is to build a general version of Pythagoras's Theorem in an n-space coordinate system.

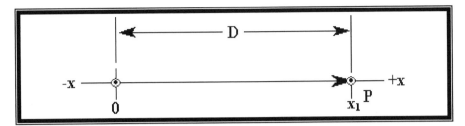

Figure 8.2.2: Vector Magnitude D in 1-Dimension

Theorem 8.2.1 Pythagorean Theorem for Vector Magnitude D in One Dimension

1g 2g	Given	the above geometry of Figure F8.2.2 $OP = x_1$
Steps	Hypothesis	$D = \sqrt{x_1^2} = |x_1|$
1	From 1g and 2g	$D = \sqrt{x_1^2} + 0$
2	From 1 and A4.2.7 Identity Add	$D = \sqrt{x_1^2}$
∴	From 2 and D4.1.11 Absolute Value	$D = \sqrt{x_1^2} = |x_1|$

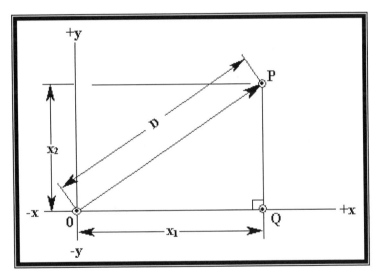

Figure 8.2.3: Vector Magnitude D in 2-Dimensions

Theorem 8.2.2 Pythagorean Theorem for Vector Magnitude D in Two Dimensions

1g 2g 3g	Given	the above geometry of Figure F8.2.3 $OQ = x_1$ $QP = x_2$
Steps	Hypothesis	$D = \sqrt{x_1^2 + x_2^2}$
1	From 1g, and D8.2.1 Pythagoras's Theorem for Magnitude of a Vector; hypotenuse of ΔOPQ	$D = \sqrt{OQ^2 + QP^2}$
∴	From 1, 2g, 3g and A4.2.3 Substitution	$D = \sqrt{x_1^2 + x_2^2}$

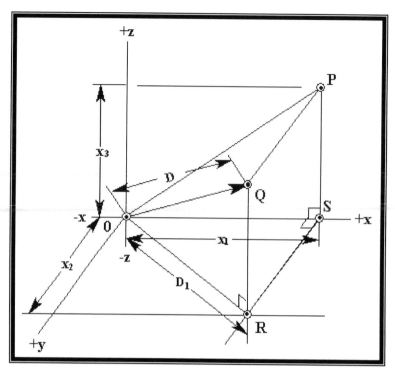

Figure 8.2.4: Vector Magnitude D in 3-Dimensions

Theorem 8.2.3 Pythagorean Theorem for Vector Magnitude D in Three Dimensions

1g	Given	the above geometry of Figure F8.2.4
2g		$OS = x_1$
3g		$SR = x_2$
4g		$OR = D_1$
5g		$RQ = x_3$
Steps	Hypothesis	$D = \sqrt{x_1{}^2 + x_2{}^2 + x_3{}^2}$
1	From 1g, ΔOSR and D8.2.1	$D_1 = \sqrt{OS^2 + SR^2}$
2	From 1, 2g, 3g and A4.2.3 Substitution	$D_1 = \sqrt{x_1{}^2 + x_2{}^2}$
3	From F8.2.4, ΔORQ and D8.2.1	$D = \sqrt{OR^2 + RQ^2}$
4	From 3, 4g, 5g and A4.2.3 Substitution	$D = \sqrt{D_1{}^2 + x_3{}^2}$
$\therefore$	From 2, 4, and A4.2.3 Substitution	$D = \sqrt{x_1{}^2 + x_2{}^2 + x_3{}^2}$

There is a clear pattern of building one dimension upon another from right triangle to right triangle. So a count from 1 to 3-dimensions has been established as correct. Lets assume that this dimensional progress continues for D_i and on to D_{i+1} as follows:

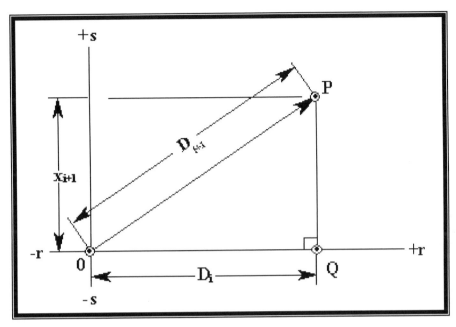

Figure 8.2.5: Vector Magnitude D in (i + 1)-Dimensions

Theorem 8.2.4 Pythagorean Theorem for Vector Magnitude D in n-Dimensions

1g	Given	the above geometry of Figure F8.2.5
2g		$OQ^2 = D_i^2 = \sum_i x_i^2$
3g		$QP = x_{i+1}$
Steps	Hypothesis	$D_n = \sqrt{\sum_n x_i^2}$
1	From 1g, ΔOQP and D8.2.1	$D_{i+1} = \sqrt{OQ^2 + QP^2}$
2	From 1, 2g, 3g and A4.2.3 Substitution	$D_{i+1} = \sqrt{x_{i+1}^2 + D_i^2}$
3	From 2, 2g and A4.2.3 Substitution	$D_{i+1} = \sqrt{x_{i+1}^2 + \sum_i x_j^2}$
4	From 3 and summing over j to i+1	$D_{i+1} = \sqrt{\sum_{i+1} x_j^2}$
$\therefore$	From 4 and A3.11.1 Induction Property of Integers	$D_n = \sqrt{\sum_n x_i^2}$

Section 8.3 The First Product: The Inner or Dot

In this chapter all vector analysis is done in contravariant format, since a covariant study mirrors contravariance.

Definition 8.3.1 The Inner or Dot Product Operator

$$\mathbf{V} \bullet \mathbf{U} \equiv |\mathbf{V}| \, |\mathbf{U}| \cos(\theta_{vu})$$

However, the origin of the inner product is not so mysterious as reconstructed in T8.3.6 "Dot Product: Analytic Relationship". The construction F8.3.1 "Constructing the Angle Between Two Vectors" is laid out on a flat hyper-plane in n-space, so there are no distortions of magnitude and angle, it follows than that all geometrical properties are preserved and the angle between the two vectors is $\angle QOP = \theta_{vu}$. Let O, P, Q be constructed on the hyper-plane where point O is at the origin and associated with each point is a set of ordered n-tuple, then $O_n(\bullet \, 0 \, \bullet)$, $P_n(\bullet \, y_i \, \bullet)$ and $Q_n(\bullet \, x_i \, \bullet)$. The straight-line segment constructed between the points has a magnitude as specified by T8.3.6 "Dot Product: Defined by the Law of Cosine". From the geometry, the angle can be derived.

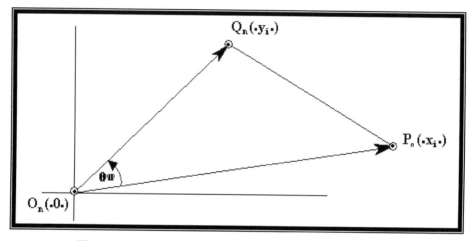

Figure 8.3.1 Constructing the Angle Between Two Vectors

Definition 8.3.2 Fundamental Base Vector Dot Product

Base vectors are said to be fundamental if vectors within their space can only be comprised from them and them a loan, this requirement makes them linearly independent satisfying A7.12.2 "Linear Independence for a Set of Vectors $(\bullet \, \mathbf{b}_i \, \bullet)$". Relative to one another they can have vector coefficients that can be placed into a ***vector coefficient matrix***:

$$\mathbf{b}_i \bullet \mathbf{b}_j = |\mathbf{b}_i| \, |\mathbf{b}_j| \cos(\theta_{ij}) \qquad \text{Equation A}$$
$$\mathbf{b}_i \bullet \mathbf{b}_j = b_i \, b_j \cos(\theta_{ij}) \qquad \text{Equation B}$$

Definition 8.3.3 Orthogonal Bases Vectors

Two vectors are said to be orthogonal, when are parallel [$0°$] for i = j and they have a relative angle of ninety degrees [$90°$] for i ≠ j, between them.

$$\mathbf{b}_i \bullet \mathbf{b}_j = b_i \, b_j \cos(\, 0°) = b_i{}^2 \quad \text{for } \mathbf{b}_i \parallel \mathbf{b}_j \;\; i = j, \;\; \text{Equation A}$$
$$\mathbf{b}_i \bullet \mathbf{b}_j = b_i \, b_j \cos(90°) = 0 \quad \text{for } \mathbf{b}_i \perp \mathbf{b}_j. \; i \neq j \;\; \text{Equation B}$$

or all together under the Kronecker Delta matrix operator by D6.1.6 "Identity Matrix",

$$\mathbf{b}_i \bullet \mathbf{b}_j = b_i \, b_j \, \delta_{ij} \qquad \text{Equation C}$$

Definition 8.3.4 Concurrent Vectors

Two vectors are said to be concurrent iff they are collinear, lying on the same line drawn through them, such that they have a relative angle of zero degrees [0°] between them.

Definition 8.3.5 Orthonormal Vector Set

For a set of n-vectors $\mathbf{v}_1$, $\mathbf{v}_2$, $\mathbf{v}_3$ …, $\mathbf{v}_i$, …,$\mathbf{v}_n$ are orthonormal iff for any i = j than $\mathbf{v}_i$ is concurrent with $\mathbf{v}_j$ and i ≠ j $\mathbf{v}_i$ is orthogonal to $\mathbf{v}_j$.

Definition 8.3.6 Orthonormal Unit Base Vectors

Unit base vectors are an orthonormal vector set, such that their magnitudes are unity, having an identity coefficient matrix of their inner products as follows:

$$U \equiv (\mathbf{e}_i \bullet \mathbf{e}_j) = (\delta_{ij}) = I.$$

Definition 8.3.7 Perpendicular Vectors

Perpendicular vectors are at right angles, 90° to one another and represented by the symbol ⊥.

Theorem 8.3.1 Dot Product: Uniqueness Operation

1g 2g	Given	$\mathbf{A} = \mathbf{B}$ $\mathbf{C} = \mathbf{D}$
Steps	Hypothesis	$\mathbf{A} \bullet \mathbf{C} = \mathbf{B} \bullet \mathbf{D}$
1	From 1g, 2g and D8.3.1 The Inner or Dot Product Operator	$\mathbf{A} \bullet \mathbf{C} = \mid \mathbf{A} \mid \mid \mathbf{C} \mid \cos(\theta_{AC})$
2	From 1g, 2g and D8.3.1 The Inner or Dot Product Operator	$\mathbf{B} \bullet \mathbf{D} = \mid \mathbf{B} \mid \mid \mathbf{D} \mid \cos(\theta_{BD})$
3	From 1g, 1, A7.9.2 Substitution of Vectors and D8.3.1 The Inner or Dot Product Operator	$\mathbf{A} \bullet \mathbf{C} = \mid \mathbf{B} \mid \mid \mathbf{C} \mid \cos(\theta_{BC})$
4	From 2g, 3, A7.9.2 Substitution of Vectors and D8.3.1 The Inner or Dot Product Operator	$\mathbf{A} \bullet \mathbf{C} = \mid \mathbf{B} \mid \mid \mathbf{D} \mid \cos(\theta_{BD})$
∴	From 2, 4 and A4.3.3 Substitution	$\mathbf{A} \bullet \mathbf{C} = \mathbf{B} \bullet \mathbf{D}$

Theorem 8.3.2 Dot Product: Equality

1g 2g	Given	$\mathbf{A} = \mathbf{B}$ $\mathbf{C} = \mathbf{D}$
Steps	Hypothesis	$\mathbf{A} \bullet \mathbf{B} = \mathbf{A} \bullet \mathbf{B}$
1	From D8.3.1 The Inner or Dot Product Operator	$\mathbf{A} \bullet \mathbf{B} = \mid \mathbf{A} \mid \mid \mathbf{B} \mid \cos(\theta_{AB})$
∴	From 1 and A4.3.3 Substitution	$\mathbf{A} \bullet \mathbf{B} = \mathbf{A} \bullet \mathbf{B}$

Theorem 8.3.3 Dot Product: Commutative Operation

1g	Given	$\theta_{AB} \cong \theta_{BA}$ positive quadrant range for the angle of the cosine.
Steps	Hypothesis	$\mathbf{A} \bullet \mathbf{B} = \mathbf{B} \bullet \mathbf{A}$
1	From T8.3.2 Dot Product: Equality	$\mathbf{A} \bullet \mathbf{B} = \mathbf{A} \bullet \mathbf{B}$
2	From 1 and D8.3.1 The Inner or Dot Product Operator	$\mathbf{A} \bullet \mathbf{B} = \mid \mathbf{A} \mid \mid \mathbf{B} \mid \cos(\theta_{AB})$
3	From 1g, 2, A4.2.10 Commutative Multp, and A5.2.18 GIV Transitivity of Angles	$\mathbf{A} \bullet \mathbf{B} = \mid \mathbf{B} \mid \mid \mathbf{A} \mid \cos(\theta_{BA})$
∴	From 3 and D8.3.1 The Inner or Dot Product Operator	$\mathbf{A} \bullet \mathbf{B} = \mathbf{B} \bullet \mathbf{A}$

Theorem 8.3.4 Dot Product: Magnitude

1g	Given	$\mathbf{B} = \mathbf{A}$
2g		$\theta_{AA} \cong 0°$ no angle for a vector onto itself.
Steps	Hypothesis	$\mathbf{A} \bullet \mathbf{A} = \mid \mathbf{A} \mid^2$
1	From D8.3.1 The Inner or Dot Product Operator	$\mathbf{A} \bullet \mathbf{B} = \mid \mathbf{A} \mid \mid \mathbf{B} \mid \cos(\theta_{AB})$
2	From 1g, 2g, 1 and A7.9.2 Substitution of Vectors	$\mathbf{A} \bullet \mathbf{A} = \mid \mathbf{A} \mid \mid \mathbf{A} \mid \cos(\theta_{AA})$
3	From 2 and A5.2.18 GIV Transitivity of Angles	$\mathbf{A} \bullet \mathbf{A} = \mid \mathbf{A} \mid \mid \mathbf{A} \mid \cos(0°)$
4	From 3, D4.1.17 Exponential Notation and DxE.1.6.1	$\mathbf{A} \bullet \mathbf{A} = \mid \mathbf{A} \mid^2 1$
∴	From 4 and A4.2.12 Identity Multp	$\mathbf{A} \bullet \mathbf{A} = \mid \mathbf{A} \mid^2$

Theorem 8.3.5 Dot Product: Law of Cosine

1g	Given	the above geometry of Figure F8.3.1
2g		$\vec{OQ} = \mathbf{A}$
3g		$\vec{OP} = \mathbf{B}$
4g		$\vec{QP} = \mathbf{C}$
Steps	Hypothesis	$\mathbf{C} \bullet \mathbf{C} = \mathbf{A} \bullet \mathbf{A} + \mathbf{B} \bullet \mathbf{B} - 2\,\mathbf{A} \bullet \mathbf{B}$
1	From 1g and LxE.2.2 Law of Cosines	$\mid\vec{QP}\mid^2 = \mid\vec{OQ}\mid^2 + \mid\vec{OP}\mid^2 - 2\mid\vec{OQ}\mid\,\mid\vec{OP}\mid \cos(\theta_{qp})$
2	From 2g, 3g, 4g, 1 and A7.9.2 Substitution of Vectors	$\mid\mathbf{C}\mid^2 = \mid\mathbf{A}\mid^2 + \mid\mathbf{B}\mid^2 - 2\mid\mathbf{A}\mid\,\mid\mathbf{B}\mid \cos(\theta_{qp})$
∴	From 2, T8.3.4 Dot Product: Magnitude and D8.3.1 The Inner or Dot Product Operator	$\mathbf{C} \bullet \mathbf{C} = \mathbf{A} \bullet \mathbf{A} + \mathbf{B} \bullet \mathbf{B} - 2\,\mathbf{A} \bullet \mathbf{B}$

Theorem 8.3.6 Dot Product: Defined by the Law of Cosine

Steps	Hypothesis	$A \bullet B = \frac{1}{2} (A \bullet A + B \bullet B - C \bullet C)$
1	From T8.3.5 Dot Product: Law of Cosine	$C \bullet C = A \bullet A + B \bullet B - 2 A \bullet B$
2	From 1, T4.3.6B Equalities: Reversal of Left Cancellation by Addition and T4.3.6B Equalities: Left Cancellation by Addition	$2 A \bullet B = A \bullet A + B \bullet B - C \bullet C$
$\therefore$	From 2 and T4.4.6 Equalities: Left Cancellation by Multiplication	$A \bullet B = \frac{1}{2} (A \bullet A + B \bullet B - C \bullet C)$

Theorem 8.3.7 Dot Product: Distribution of Dot Product Across Addition of Vectors

1g	Given	$V \bullet \sum^n U_i \equiv \sum_n (V \bullet U_i)$ assume
2g		$R = \sum^{n-1} U_i$ vector comprised of the sum of vectors
Steps	Hypothesis	$V \bullet \sum^n U_i \equiv \sum_n (V \bullet U_i)$ The Dot Product is a Complementary Operator to Addition
1	From 1g	$V \bullet \sum^{n-1} U_i = \sum_{n-1} (V \bullet U_i)$
2	From 1, 2g and A4.2.3 Substitution	$V \bullet R = \sum^{n-1} (V \bullet U_i)$
3	From 2 and D8.3.1 The Inner or Dot Product Operator	$|V| \, |R| \cos(\theta_{vr}) = \sum^{n-1} |V| \, |U_i| \cos(\theta_{vi})$
4	From A4.2.2A Equality	$1/|V| = 1/|V|$
5	From 3, 4 and T4.4.2 Equalities: Uniqueness of Multiplication	$(1/|V|) \, |V| \, |R| \cos(\theta_{vr}) =$ $\qquad (1/|V|) \sum^{n-1} |V| \, |U_i| \cos(\theta_{vi})$
6	From 5 and A4.2.14 Distribution	$(1/|V|) \, |V| \, |R| \cos(\theta^{vr}) =$ $\qquad (1/|V|) \, |V| \sum^{n-1} |U^i| \cos(\theta^{vi})$
7	From 6 and A4.2.13 Inverse Multp	$1 \, |R| \cos(\theta_{vr}) = 1 \sum^{n-1} |U_i| \cos(\theta_{vi})$
8	From 7 and A4.2.12 Identity Multp	$|R| \cos(\theta_{vr}) = \sum^{n-1} |U_i| \cos(\theta_{vi})$
9	From 8 and we are dealing with a polygon and have established a $[\gamma]$ relative to the vector $[V]$ as a construction line of the projection of the sides of the polygon.	$\theta_{vr} = \gamma - \alpha_n$ and $\theta_{vr} = \gamma - \alpha_i$ where $i = 1, 2, \dots n-1$ and α is the base angle.
10	From 9 and A5.2.18 GIV Transitivity of Angles	$|R| \cos(\gamma - \alpha_n) = \sum^{n-1} |U_i| \cos(\gamma - \alpha_i)$
$\therefore$	From 10 an irregular polygon, but the equality holds under the geometry of irregular polygons, T4.24.12 Irregular Polygons: Summation of Projected Sides by Langlands's Hypothesis, hence the Dot Product is a complementary operator to addition and holds in general.	$V \bullet \sum^n U_i = \sum^n (V \bullet U_i)$ distribution holds.

Theorem 8.3.8 Dot Product: Distribution of Dot Product Across Another Vector

1g	Given	$V = \sum_i V_i$
Steps	Hypothesis	$(\sum_i V_i) \bullet (\sum_j U_j) = \sum_j \sum_i V_i \bullet U_j$
1	From T8.3.7 Dot Product: Distribution of Dot Product Across Addition of Vectors	$V \bullet (\sum_j U_j) = \sum_j (V \bullet U_j)$
2	From 1g, 1 and A7.9.2 Substitution of Vectors	$(\sum_i V_i) \bullet (\sum_j U_j) = \sum_j [(\sum_i V_i) \bullet U_j]$
3	From 2 and T8.3.7 Dot Product: Distribution of Dot Product Across Addition of Vectors	$(\sum_i V_i) \bullet (\sum_j U_j) = \sum_j [(\sum_i V_i \bullet U_j)]$
∴	From 3 and D8.3.1 The Inner or Dot Product Operator; is scalar hence has no preferred order of summation can be transposed.	$(\sum_i V_i) \bullet (\sum_j U_j) = \sum_i \sum_j V_i \bullet U_j$

Theorem 8.3.9 Dot Product: Left Scalar-Vector Association

Steps	Hypothesis	$a (A \bullet B) = (a A) \bullet B$ EQ A	$(a A) \bullet B = a (A \bullet B)$ EQ B
1	From T8.3.2 Dot Product: Equality	$a (A \bullet B) = a (A \bullet B)$	
2	From 1, D8.3.2 Fundamental Base Vector Dot Product and A4.2.3 Substitution	$a (A \bullet B) = a [|A| |B| \cos (\theta_{ab})]$	
3	From 2 and A4.2.11 Associative Multp	$a (A \bullet B) = (a |A|) |B| \cos (\theta_{ab})$	
∴	From 3 and D8.3.2 Fundamental Base Vector Dot Product	$a (A \bullet B) = (a A) \bullet B$ EQ A	$(a A) \bullet B = a (A \bullet B)$ EQ B

Theorem 8.3.10 Dot Product: Right Scalar-Vector Association

Steps	Hypothesis	$a (A \bullet B) = A \bullet (aB)$ EQ A	$A \bullet (aB) = a (A \bullet B)$ EQ B
1	From T8.3.2 Dot Product: Equality	$a (A \bullet B) = a (A \bullet B)$	
2	From 1, D8.3.2 Fundamental Base Vector Dot Product and A4.2.3 Substitution	$a (A \bullet B) = a [|A| |B| \cos (\theta_{ab})]$	
3	From 2, A4.2.10 Commutative and A4.2.11 Associative Multp	$a (A \bullet B) = |A| (a |B|) \cos (\theta_{ab})$	
∴	From 3 and D8.3.2 Fundamental Base Vector Dot Product	$a (A \bullet B) = A \bullet (aB)$ EQ A	$A \bullet (aB) = a (A \bullet B)$ EQ B

Theorem 8.3.11 Dot Product: Perpendicular Vectors

Steps	Hypothesis	$\mathbf{A} \bullet \mathbf{B} = 0$ for $\mathbf{A} \perp \mathbf{B}$
1	From D8.3.1 The Inner or Dot Product Operator	$\mathbf{A} \bullet \mathbf{B} = \mid \mathbf{A} \mid \mid \mathbf{B} \mid \cos(\theta_{AB})$
2	From D8.3.7 Perpendicular Vectors	$\theta_{AB} = 90°$ for $\mathbf{A} \perp \mathbf{B}$
3	From 2 and A4.3.3 Substitution	$\mathbf{A} \bullet \mathbf{B} = \mid \mathbf{A} \mid \mid \mathbf{B} \mid \cos(90°)$
∴	From 3, DxE.1.6.7 Cosine of 90° and T4.4.1 Equalities: Any Quantity Multiplied by Zero is Zero	$\mathbf{A} \bullet \mathbf{B} = 0$ for $\mathbf{A} \perp \mathbf{B}$

Section 8.4 The First Product: The Inner or Dot Analytical

Theorem 8.4.1 Dot Product: Analytic Representation for Angle Between Two Vectors

1g	Given	the above geometry of Figure F8.3.1
2g		$\overrightarrow{OP} = Q_n(\bullet\, x_i\, \bullet) = \sum_n x_i \mathbf{b}_i$
3g		$\overrightarrow{OQ} = P_n(\bullet\, y_i\, \bullet) = \sum^n y_i \mathbf{b}_i$
4g	LxE.2.2 and F8.3.1	$\overline{QP}^2 = \overline{OQ}^2 + \overline{OP}^2 - 2\,\overline{OQ}\;\overline{OP}\;\cos(\theta_{qp})$
Steps	Hypothesis	$\overline{OQ}\;\overline{OP}\;\cos(\theta^{qp}) = \sum_n x_i\, y_i$
1	From 2g, 3g and A7.9.1 Vector Equivalence	$\overrightarrow{OQ} - \overrightarrow{OP} = \overrightarrow{OQ} - \overrightarrow{OP}$
2	From 2g, 3g, 1 and A7.9.2 Substitution of Vectors	$\overrightarrow{OQ} - \overrightarrow{OP} = \sum^n y_i \mathbf{b}_i - \sum_n x_i \mathbf{b}_i$
3	From 2, A4.2.6 Associative Add and A4.2.14 Distribution	$\overrightarrow{OQ} - \overrightarrow{OP} = \sum^n (y_i - x_i)\mathbf{b}_i$
4	From 3, D8.2.1 Pythagoras's Theorem for Magnitude of a Vector and T8.2.4 Pythagorean Theorem for Vector Magnitude D in n-Dimensions	$\overline{QP}^2 = \sum^n (y_i - x_i)^2$
5	From 2g, D8.2.1 Pythagoras's Theorem for Magnitude of a Vector and T8.2.4 Pythagorean Theorem for Vector Magnitude D in n-Dimensions	$\overline{OQ}^2 = \sum^n y_i^2$
6	From 3g, D8.2.1 Pythagoras's Theorem for Magnitude of a Vector and T8.2.4	$\overline{OP}^2 = \sum^n x_i^2$
7	From A4.2.2A Equality	$-\overline{QP}^2 + 2\,\overline{OQ}\;\overline{OP}\;\cos(\theta_{qp}) = -\overline{QP}^2 + 2\,\overline{OQ}\;\overline{OP}\;\cos(\theta_{qp})$
8	From 7, 4g and T4.3.1 Equalities: Uniqueness of Addition	$\overline{QP}^2 - \overline{QP}^2 + 2\,\overline{OQ}\;\overline{OP}\;\cos(\theta_{qp}) = \overline{OQ}^2 + \overline{OP}^2 - 2\,\overline{OQ}\;\overline{OP}\;\cos(\theta_{qp}) - \overline{QP}^2 + 2\,\overline{OQ}\;\overline{OP}\;\cos(\theta_{qp})$
9	From 8 and A4.2.8 Inverse Add	$0 + 2\,\overline{OQ}\;\overline{OP}\;\cos(\theta_{qp}) = \overline{OQ}^2 + \overline{OP}^2 - 2\,\overline{OQ}\;\overline{OP}\;\cos(\theta_{qp}) - \overline{QP}^2 + 2\,\overline{OQ}\;\overline{OP}\;\cos(\theta_{qp})$
10	From 9, A4.2.7 Identity Add and A4.2.6 Associative Add	$2\,\overline{OQ}\;\overline{OP}\;\cos(\theta_{qp}) = \overline{OQ}^2 + \overline{OP}^2 - 2\,\overline{OQ}\;\overline{OP}\;\cos(\theta_{qp}) + 2\,\overline{OQ}\;\overline{OP}\;\cos(\theta_{qp}) - \overline{QP}^2$
11	From 10 and A4.2.8 Inverse Add	$2\,\overline{OQ}\;\overline{OP}\;\cos(\theta_{qp}) = \overline{OQ}^2 + \overline{OP}^2 + 0 - \overline{QP}^2$
12	From 11 and A4.2.7 Identity Add	$2\,\overline{OQ}\;\overline{OP}\;\cos(\theta_{qp}) = \overline{OQ}^2 + \overline{OP}^2 - \overline{QP}^2$
13	From 12, 4, 5, 6 and A4.2.3 Substitution	$2\,\overline{OQ}\;\overline{OP}\;\cos(\theta_{qp}) = \sum^n y_i^2 + \sum_n x_i^2 - \sum_n (y_i - x_i)^2$
14	From 13 and A4.2.6 Associative Add	$2\,\overline{OQ}\;\overline{OP}\;\cos(\theta_{qp}) = \sum^n (y_i^2 + x_i^2 - (y_i - x_i)^2)$
15	From 14 and T4.12.2 Polynomial Quadratic: The Perfect Square by Difference	$2\,\overline{OQ}\;\overline{OP}\;\cos(\theta_{qp}) = \sum^n (y_i^2 + x_i^2 - y_i^2 - x_i^2 + 2x_i\, y_i)$
16	From 15 and A4.2.5 Commutative Add	$2\,\overline{OQ}\;\overline{OP}\;\cos(\theta_{qp}) = \sum^n (y_i^2 - y_i^2 + x_i^2 - x_i^2 + 2x_i\, y_i)$
17	From 16 and A4.2.8 Inverse Add	$2\,\overline{OQ}\;\overline{OP}\;\cos(\theta_{qp}) = \sum^n (0 + 0 + 2x_i\, y_i)$
18	From 17, A4.2.7 Identity Add and A4.2.14 Distribution	$2\,\overline{OQ}\;\overline{OP}\;\cos(\theta_{qp}) = 2\sum^n x_i\, y_i$
∴	From 18 and T4.4.3 Equalities: Cancellation by Multiplication	$\overline{OQ}\;\overline{OP}\;\cos(\theta_{qp}) = \sum^n x_i\, y_i$

Theorem 8.4.2 Dot Product: Analytic Relationship

Steps	Hypothesis	$|V| \, |U| \cos(\theta_{vu}) = V \bullet U = \sum^n x_i \, y_i$
1	From D8.3.1 The Inner or Dot Product Operator	$V \bullet U \equiv |V| \, |U| \cos(\theta_{vu})$
2	From T8.4.1 Dot Product: Analytic Representation for Angle Between Two Vectors	$|V| \, |U| \cos(\theta_{vu}) = \sum^n x_i \, y_i$
∴	From 1, 2 and A4.2.3 Substitution	$|V| \, |U| \cos(\theta_{vu}) = V \bullet U = \sum^n x_i \, y_i$

Theorem 8.4.3 Dot Product: Orthogonal Unitary Base Vectors

1g	Given	$V(\bullet \, x_i \, \bullet) = \sum^n x_i \mathbf{e}_i$
2g		$U(\bullet \, y_i \, \bullet) = \sum^n y_i \mathbf{e}_i$
3g		for $\mathbf{e}_i \bullet \mathbf{e}_j = \delta_{ij}$
Steps	Hypothesis	$V \bullet U = \sum^n x_i \, y_i$ for $\mathbf{e}_i \bullet \mathbf{e}_j = \delta_{ij}$
1	From A4.2.2A Equality	$V \bullet U = V \bullet U$
2	From 1g, 2g, 1, T8.4.2 Dot Product: Analytic Relationship, A7.9.2 Substitution of Vectors and A4.2.3 Substitution	$V \bullet U = (\sum^n x_i \mathbf{e}_i) \bullet (\sum^n y_j \mathbf{e}_j)$
3	From 2 and T8.3.8 Dot Product: Distribution of Dot Product Across Another Vector	$V \bullet U = \sum^n \sum^n x_i \mathbf{e}_i \bullet y_j \mathbf{e}_j$
4	From 3 and A7.8.5 Parallelogram Law: Commutative and Scalability of Collinear Vectors	$V \bullet U = \sum^n \sum^n x_i \, y_j \, \mathbf{e}_i \bullet \mathbf{e}_j$
5	From 4, 3g and A4.2.3 Substitution	$V \bullet U = \sum^n \sum^n x_i \, y_j \, \delta_{ij}$
∴	From 5 and evaluating the Kronecker Delta term-by-term.	$V \bullet U = \sum^n x_i \, y_i$

Theorem 8.4.4 Dot Product: Magnitude of a Vector

1g	Given	$U = V$
2g		$y_i = x_i$
3g		$\theta_{uu} = 0°$
Steps	Hypothesis	$|V|^2 = V \bullet V = \sum^n x_i^2$
1	From 1g, 2g, T8.4.2 Dot Product: Analytic Relationship, A7.9.2 Substitution of Vectors and A4.2.3 Substitution	$|V| \, |V| \cos(\theta_{VV}) = V \bullet V = \sum^n x_i \, x_i$
2	From 3g, 1, A5.2.18 GIV Transitivity of Angles, DxE.1.6.1 and D4.1.17 Exponential Notation	$|V|^2 \, 1 = \sum^n x_i^2$
∴	From 2 and A4.2.12 Identity Multp	$|V|^2 = V \bullet V = \sum^n x_i^2$

Theorem 8.4.5 Dot Product: Unit Vector

1g 2g	Given	Any vector A such that U = A / |A|
Steps	Hypothesis	$U \bullet U = 1$ for $U = A / |A|$
1	From A4.2.2A Equality	$U \bullet U = U \bullet U$
2	From 1, 2g and A7.9.2 Substitution of Vectors	$U \bullet U = (A / |A|) \bullet (A / |A|)$
3	From 2 and A4.2.10 Commutative Multp	$U \bullet U = A \bullet A / (|A||A|)$
4	From 3, D4.1.17 Exponential Notation and T8.4.4 Dot Product: Magnitude of a Vector	$U \bullet U = |A|^2 / |A|^2$
∴	From 4 and A4.2.13 Inverse Multp	$U \bullet U = 1$ for $U = A / |A|$

Theorem 8.4.6 Dot Product: Vector and Unit Base Vector

1g 2g 3g	Given	$V = \sum^n x_i \, \mathbf{e}_i$ $U = \mathbf{e}_i$ for $\mathbf{e}_i \bullet e^j = \delta_{ij}$
Steps	Hypothesis	$V \bullet \mathbf{e}_i = |V| \cos(\theta_{vi}) = x^i$
1	From A4.2.2A Equality	$V \bullet U = V \bullet U$
2	From 1g, 2g, 1 and A7.9.2 Substitution of Vectors	$V \bullet \mathbf{e}_j = (\sum^n x_i \, \mathbf{e}_i) \bullet \mathbf{e}_j$
3	From 2 and T8.3.7 Dot Product: Distribution of Dot Product Across Addition of Vectors	$V \bullet \mathbf{e}_j = \sum^n x_i \, \mathbf{e}_i \bullet \mathbf{e}_j$
4	From 3g, 3 and A4.2.3 Substitution	$V \bullet \mathbf{e}_j = \sum^n x_i \, \delta_{ij}$
5	From 4 and evaluating the Kronecker Delta term-by-term.	$V \bullet \mathbf{e}_j = x_j$
6	From 1 and D8.3.1 The Inner or Dot Product Operator	$V \bullet U = |V||U| \cos(\theta_{vu})$
7	From 1g, 2g, 6 and A7.9.2 Substitution of Vectors	$V \bullet \mathbf{e}_j = |V|| e_j | \cos(\theta_{vj})$
8	From 7 and T8.4.5 Dot Product: Unit Vector	$V \bullet \mathbf{e}_j = |V| \, 1 \, \cos(\theta_{vj})$
9	From 8 and A4.2.12 Identity Multp	$V \bullet \mathbf{e}_j = |V| \cos(\theta_{vj})$
∴	From 5, 9 and A4.2.2A Equality	$V \bullet \mathbf{e}_i = |V| \cos(\theta_{vi}) = x_i$

Theorem 8.4.7 Dot Product: Directional Cosine

Steps	Hypothesis	$\cos(\theta_{vi}) = x_i / |V|$
1	From T8.4.6 Dot Product: Vector and Unit Base Vector	$|V| \cos(\theta_{vi}) = x_i$
∴	From 1 and T4.4.6B Equalities: Left Cancellation by Multiplication	$\cos(\theta_{vi}) = x_i / |V|$

Theorem 8.4.8 Dot Product: Normalization

Steps	Hypothesis	$\cos(\theta_{vu}) = \sum^n (x_i / |V|)(y_i / |U|)$
1	From T8.4.2 Dot Product: Analytic Relationship	$|V| |U| \cos(\theta_{vu}) = \sum^n x_i y_i$
2	From 1 and T4.4.6B Equalities: Left Cancellation by Multiplication	$\cos(\theta_{vu}) = (\sum^n x_i y_i) / |V| |U|$
∴	From 2, A4.2.14 Distribution and A4.2.11 Associative Multp	$\cos(\theta_{vu}) = \sum^n (x_i / |V|)(y_i / |U|)$

Theorem 8.4.9 Dot Product: Comprised of Orthogonal Directional Cosines

Steps	Hypothesis	$\cos(\theta_{vu}) = \sum^n \cos(\theta_{iv}) \cos(\theta_{iu})$
1	From T8.4.7 Dot Product: Directional Cosine	$\cos(\theta_{iv}) = x_i / |V|$
2	From T8.4.7 Dot Product: Directional Cosine	$\cos(\theta_{iu}) = y_i / |U|$
∴	From 1, 2, T8.4.8 Dot Product: Normalization and A4.2.3 Substitution	$\cos(\theta_{vu}) = \sum^n \cos(\theta_{iv}) \cos(\theta_{iu})$

Theorem 8.4.10 Dot Product: Trigonometric Identity for a Unit Hyper-Sphere

1g	Given	$V = U$ hence $\theta_{vu} = 0^o$ and $\theta_{iv} = \theta_{iu}$
Steps	Hypothesis	$1 = \sum^n \cos(\theta_{iv})^2$
1	From T8.4.9 Dot Product: Comprised of Orthogonal Directional Cosines	$\cos(\theta_{vu}) = \sum^n \cos(\theta_{iv}) \cos(\theta_{iu})$
2	From 1, 1g, A5.2.18 GIV Transitivity of Angles and A7.9.2 Substitution of Vectors	$\cos(0^o) = \sum^n \cos(\theta_{iv}) \cos(\theta_{iv})$
∴	From 2, DxE.1.6.1 and D4.1.17 Exponential Notation	$1 = \sum^n \cos(\theta_{iv})^2$

What is interesting about this theorem is that like complex variables with its unit circle n-dimensional vectors have a unit hyper-sphere.

Section 8.5 The Second Product: Orthogonal Product Operators

For any two vectors, the angle between them can be found as given by the normalized dot product in T8.3.12 "Dot Product: Normalization". So far so good but if the cosine is used for multiplying vectors what of its orthogonal counter part the sine? If the sine is considered then what is gained in its use? In order to answers, these equations I've defined two functions that will operate on any two given vectors.

Definition 8.5.1 **Orthogonal Cosine Operator**

$$\mathbf{u}_p \odot \mathbf{u}_q \equiv \cos(\theta_{pq}) = \sum_n (x_k / |B_p|)(y_k / |B_q|)$$

where $\mathbf{u}_i$ and $\mathbf{u}_j$ are any two unit vectors not necessarily unit base vectors[8.5.1].

Definition 8.5.2 **Orthogonal Sine Operator**

$$\mathbf{u}_p \otimes \mathbf{u}_q \equiv \sin(\theta_{pq})$$

Graphically they can be represented in the following way:

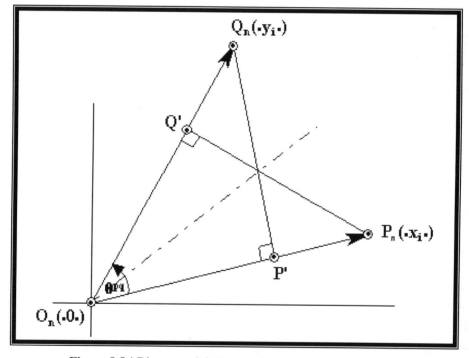

Figure 8.5.1 Diagram of Orthogonal Cosine and Sine Operators

[8.5.1]Note: The "Orthogonal Cosine Operator" is represented by [◉] from 5x15-"Wingdings 2" Font set.

Definition 8.5.3 **Domain for the Relative Angle Between Two Vectors**

With the introduction of the skewed, asymmetry, minus sign the domain is relegated to the upper half of the plane. However there are certain degenerate cases that have to be handled separately such as 0° and 180° so the finial domain is between those limits. Used with manifolds than beyond 180° the shape of a manifold space maybe periodically duplicated and convoluted at a point, however as a tensor field there may arise a situation where that domain may have to be broken. For this treatise to prevent such conflicts the angle θ_{pq} between vectors is placed in the upper half of the plane and the range of the angle defined for:	$$0 \leq \theta_{pq} \leq \pi$$

Definition 8.5.4 **Associated Parity Between any Two Vectors ($\gamma[p], \gamma[q]$)**

$$\theta pr(\gamma[p], \gamma[q]) \equiv \begin{cases} 1 & \text{for } (\gamma[q], \gamma[p]) \text{ turning CW} \\ 0 & \text{for } (\gamma[p], \gamma[q]) \text{ turning CCW} \end{cases}$$

$$\text{such that } (p, q) \leq n \text{ and } p \neq q.$$

Observation 8.5.1 Directed Sign for a Common Angle

Notice that Figure 8.4.1 constructs two nested, similar, right triangles ($\Delta QOP' \sim \Delta POQ'$) sharing a common angle ($\angle QOP' \cong \theta_{pq}$) opposite to their vertical sides. The triangles share the same angle, hence they have a dual role where orientation of the two triangles, $\Delta QOP'$ and $\Delta POQ'$, are relative to one another. These two triangles are constructed on their respective vectors $\gamma(p)$ and $\gamma(q)$ giving a relative orientation to them. Let triangle $\Delta QOP'$ follow quadrature convention for angles, having vector $\gamma(p)$ sweep through a positive arc ($\angle POQ' \cong \theta_{pq}$), CCW to vector $\gamma(q)$, $\gamma(p) \rightarrow \gamma(q)$, since $\Delta POQ'$ is flipped to the left verses the other's right than vector $\gamma(q)$ sweep through a negative arc θ_{qp} (CW) to vector $\gamma(p)$, $\gamma(q) \rightarrow \gamma(p)$, hence

$$\theta_{pq} \cong -\theta_{qp}$$

Axiom 8.5.1 Directed Sign Operator between any Two Vectors

$\exists!$ between any two vectors, $\gamma(p)$ and $\gamma(q)$, a common angle θ_{pq} and θ_{qp} from which can be constructed two similar right triangles ($\Delta QOP' \sim \Delta POQ'$). Angle θ_{pq} has a positive sense (CCW) for $\Delta QOP'$ and angle θ_{qp} has a negative sense (CCW) for $\Delta POQ'$, hence

$$\theta_{pq} \cong -\theta_{qp} \qquad \text{Equation A}$$

$$\angle QOP' \cong \theta_{pq} \qquad \text{Equation B}$$

$$\angle POQ' \cong \theta_{pq} \qquad \text{Equation C}$$

for right triangles

$$\Delta QOP' \sim \Delta POQ' \qquad \text{Equation D}$$

Tensor Calculus &Physics: A General Treatise

Theorem 8.5.1 Orthogonal Cosine Operator: Geometrical Interpretation

1g	Given	the above geometry of Figure F8.4.1	
2g		$\mathbf{b}_p \equiv \overrightarrow{OP}$	
3g		$\mathbf{b}_q \equiv \overrightarrow{OQ}$	
4g		$\overline{OP} = |\overrightarrow{OP}| = |\mathbf{b}_p|$	
5g		$\overline{OQ} = |\overrightarrow{OQ}| = |\mathbf{b}_q|$	
6g		$\widehat{OP} = \overrightarrow{OP} / \overline{OP} = \overrightarrow{OP} / |\mathbf{b}^p| = \mathbf{u}_p$	
7g		$\widehat{OQ} = \overrightarrow{OQ} / \overline{OQ} = \overrightarrow{OQ} / |\mathbf{b}_q| = \mathbf{u}_q$	
Steps	Hypothesis	$\overline{OP}' = \mathbf{u}_p \bullet \mathbf{b}_q = \mathbf{u}_p \odot \mathbf{b}_q$ $= |\mathbf{b}_q| \cos(\theta_{pq})$	$\overline{OQ}' = \mathbf{u}_q \bullet \mathbf{b}_p = \mathbf{u}_q \odot \mathbf{b}_p$ $= |\mathbf{b}_p| \cos(\theta_{qp})$
1	From 1g	$\overline{OP}' = |\mathbf{b}_q| \cos(\theta_{pq})$	$\overline{OQ}' = |\mathbf{b}_p| \cos(\theta_{qp})$
2	From A4.2.2A Equality	$|\mathbf{b}_p| = |\mathbf{b}_p|$	$|\mathbf{b}_q| = |\mathbf{b}_q|$
3	From 2, 3 and T4.4.2 Equalities: Uniqueness of Multiplication	$|\mathbf{b}_p| \, \overline{OP}' = |\mathbf{b}_p| \, |\mathbf{b}_q| \cos(\theta_{pq})$	$|\mathbf{b}_q| \, \overline{OQ}' = |\mathbf{b}_q| \, |\mathbf{b}_p| \cos(\theta_{qp})$
4	From 3 and D8.3.1 The Inner or Dot Product Operator	$|\mathbf{b}_p| \, \overline{OP}' = \overrightarrow{OP} \bullet \mathbf{b}_q$	$|\mathbf{b}_q| \, \overline{OQ}' = \overrightarrow{OQ} \bullet \mathbf{b}_p$
5	From 4 and D8.5.1 Orthogonal Cosine Operator	$|\mathbf{b}_p| \, \overline{OP}' = \overrightarrow{OP} \odot \mathbf{b}_q$	$|\mathbf{b}_q| \, \overline{OQ}' = \overrightarrow{OQ} \odot \mathbf{b}_p$
6	From 5, T4.4.6B Equalities: Left Cancellation by Multiplication and A4.2.11 Associative Multp	$\overline{OP}' = (\overrightarrow{OP} / |\mathbf{b}_p|) \odot \mathbf{b}_q$	$\overline{OQ}' = (\overrightarrow{OQ} / |\mathbf{b}_q|) \odot \mathbf{b}_p$
7	From 6g, 7g, 6 and A4.2.3 Substitution	$\overline{OP}' = \mathbf{u}_p \bullet \mathbf{b}_q = \mathbf{u}_p \odot \mathbf{b}_q$	$\overline{OQ}' = \mathbf{u}_q \bullet \mathbf{b}_p = \mathbf{u}_q \odot \mathbf{b}_p$
∴	From 1, 7 and A4.2.2A Equality	$\overline{OP}' = \mathbf{u}_p \bullet \mathbf{b}_q = \mathbf{u}_p \odot \mathbf{b}_q$ $= |\mathbf{b}_q| \cos(\theta_{pq})$	$\overline{OQ}' = \mathbf{u}_q \bullet \mathbf{b}_p = \mathbf{u}_q \odot \mathbf{b}_p$ $= |\mathbf{b}_p| \cos(\theta_{qp})$

Theorem 8.5.2 Orthogonal Cosine Operator: Unitary Cosine Operator

1g	Given	the above geometry of Figure F8.4.1	
2g		$\mathbf{b}_p \equiv \overrightarrow{OP}$	
3g		$\mathbf{b}_q \equiv \overrightarrow{OQ}$	
4g		$\overline{OP} = |\overrightarrow{OP}| = |\mathbf{b}_p|$	
5g		$\overline{OQ} = |\overrightarrow{OQ}| = |\mathbf{b}_q|$	
6g		$\widehat{OP} = \overrightarrow{OP} / \overline{OP} = \overrightarrow{OP} / |\mathbf{b}_p| = \mathbf{u}_p$	
7g		$\widehat{OQ} = \overrightarrow{OQ} / \overline{OQ} = \overrightarrow{OQ} / |\mathbf{b}_q| = \mathbf{u}_q$	
Steps	Hypothesis	$\mathbf{u}_p \bullet \mathbf{u}_q = \mathbf{u}_p \odot \mathbf{u}_q$ $= \cos(\theta_{pq})$	$\mathbf{u}_q \bullet \mathbf{u}_p = \mathbf{u}_q \odot \mathbf{u}_p$ $= \cos(\theta_{qp})$
1	From T8.5.1 Orthogonal Cosine Operator: Geometrical Interpretation	$\mathbf{u}_p \bullet \mathbf{b}_q = \mathbf{u}_p \odot \mathbf{b}_q$ $= |\mathbf{b}_q| \cos(\theta_{pq})$	$\mathbf{u}_q \bullet \mathbf{b}_p = \mathbf{u}_q \odot \mathbf{b}_p$ $= |\mathbf{b}_p| \cos(\theta_{qp})$
2	From 1, T4.4.6B Equalities: Left Cancellation by Multiplication and A4.2.11 Associative Multp	$\overline{OP}' = \mathbf{u}_p \bullet (\mathbf{b}_q / |\mathbf{b}_q|)$ $= \mathbf{u}_p \odot (\mathbf{b}_q / |\mathbf{b}_q|)$ $= \cos(\theta_{pq})$	$\overline{OQ}' = \mathbf{u}_q \bullet (\mathbf{b}_p / |\mathbf{b}_p|)$ $= \mathbf{u}_q \odot (\mathbf{b}_p / |\mathbf{b}_p|)$ $= \cos(\theta_{qp})$
3	From 6g, 7g, 2 and A4.2.3 Substitution	$\overline{OP}' = \mathbf{u}_p \bullet \mathbf{u}_q$ $= \mathbf{u}_p \odot \mathbf{u}_q$ $= \cos(\theta_{pq})$	$\overline{OQ}' = \mathbf{u}_q \bullet \mathbf{u}_p$ $= \mathbf{u}_q \odot \mathbf{u}_p$ $= \cos(\theta_{qp})$
∴	From 3 and A4.2.2A Equality	$\mathbf{u}_p \bullet \mathbf{u}_q = \mathbf{u}_p \odot \mathbf{u}_q$ $= \cos(\theta_{pq})$	$\mathbf{u}_q \bullet \mathbf{u}_p = \mathbf{u}_q \odot \mathbf{u}_p$ $= \cos(\theta_{qp})$

Theorem 8.5.3 Orthogonal Cosine Operator: Absolute Symmetry of Value

Steps	Hypothesis	$\mathbf{u}_p \bullet \mathbf{u}_q = \mathbf{u}_p \odot \mathbf{u}_q$ $= \cos(\theta_{qp})$ $= \cos(\theta_{pq})$ EQ A	$\mathbf{u}_q \bullet \mathbf{u}_p = \mathbf{u}_q \odot \mathbf{u}_p$ $= \cos(\theta_{pq})$ $= \cos(\theta_{qp})$ EQ B
1	From T8.5.2 Orthogonal Cosine Operator: Unitary Cosine Operator	$\mathbf{u}_p \bullet \mathbf{u}_q = \mathbf{u}_p \odot \mathbf{u}_q$ $= \cos(\theta_{pq})$	$\mathbf{u}_q \bullet \mathbf{u}_p = \mathbf{u}_q \odot \mathbf{u}_p$ $= \cos(\theta_{qp})$
2	From 1, A8.5.1A Directed Sign Operator between any Two Vectors and A5.2.18 GIV Transitivity of Angles	$\mathbf{u}_p \bullet \mathbf{u}_q = \mathbf{u}_p \odot \mathbf{u}_q$ $= \cos(-\theta_{qp})$	$\mathbf{u}_q \bullet \mathbf{u}_p = \mathbf{u}_q \odot \mathbf{u}_p$ $= \cos(-\theta_{pq})$
∴	From 1, 2, DxE.1.7.1 and A4.2.2A Equality	$\mathbf{u}_p \bullet \mathbf{u}_q = \mathbf{u}_p \odot \mathbf{u}_q$ $= \cos(\theta_{qp})$ $= \cos(\theta_{pq})$ EQ A	$\mathbf{u}_q \bullet \mathbf{u}_p = \mathbf{u}_q \odot \mathbf{u}_p$ $= \cos(\theta_{pq})$ $= \cos(\theta_{qp})$ EQ B

Theorem 8.5.4 Orthogonal Cosine Operator: Commutative Absolute Symmetry

Steps	Hypothesis	$\mathbf{u}_p \odot \mathbf{u}_q = \mathbf{u}_q \odot \mathbf{u}_p$ $= \mathbf{u}_p \bullet \mathbf{u}_q = \mathbf{u}_q \bullet \mathbf{u}_p$
1	From A8.5.1A Directed Sign Operator between any Two Vectors	$\theta_{pq} \cong -\theta_{qp}$
2	From 1 and uniqueness of cosine operator	$\cos(\theta_{pq}) = \cos(-\theta_{qp})$
3	From 2 and DxE.1.7.1	$\cos(\theta_{pq}) = \cos(\theta_{qp})$
∴	From 3, T8.5.3 Orthogonal Cosine Operator: Absolute Symmetry of Value and A4.2.3 Substitution	$\mathbf{u}_p \odot \mathbf{u}_q = \mathbf{u}_q \odot \mathbf{u}_p$ $= \mathbf{u}_p \bullet \mathbf{u}_q = \mathbf{u}_q \bullet \mathbf{u}_p$

Theorem 8.5.5 Orthogonal Sine Operator: Geometrical Interpretation

	Given	the above geometry of Figure F8.4.1	
1g	Given	the above geometry of Figure F8.4.1	
2g		$\mathbf{b}_p \equiv \vec{OP}$	
3g		$\mathbf{b}_q \equiv \vec{OQ}$	
4g		$\overline{OP} = |\vec{OP}| = |\mathbf{b}_p|$	
5g		$\overline{OQ} = |\vec{OQ}| = |\mathbf{b}_q|$	
6g		$\hat{OP} = \vec{OP} / \overline{OP} = \vec{OP} / |\mathbf{b}_p| = \mathbf{e}_p$	
7g		$\hat{OQ} = \vec{OQ} / \overline{OQ} = \vec{OQ} / |\mathbf{b}_q| = \mathbf{u}_q$	
Steps	Hypothesis	$\overline{QP}' = \mathbf{u}_p \otimes \mathbf{b}_q = |\mathbf{b}_q| \sin(\theta_{pq})$ EQ A	$\overline{PQ}' = \mathbf{u}_q \otimes \mathbf{b}_p = |\mathbf{b}_p| \sin(\theta_{qp})$ EQ B
1	From 1g	$\overline{QP}' = |\mathbf{b}_q| \sin(\theta_{pq})$	$\overline{PQ}' = |\mathbf{b}_p| \sin(\theta_{qp})$
2	From A4.2.2A Equality	$|\mathbf{b}_p| = |\mathbf{b}_p|$	$|\mathbf{b}_q| = |\mathbf{b}_q|$
3	From 2, 3 and T4.4.2 Equalities: Uniqueness of Multiplication	$|\mathbf{b}_p| \, \overline{QP}' = |\mathbf{b}_p| \, |\mathbf{b}_q| \sin(\theta_{pq})$	$|\mathbf{b}_q| \, \overline{PQ}' = |\mathbf{b}_q| \, |\mathbf{b}_p| \sin(\theta_{qp})$
4	From 3 and D8.5.2 Orthogonal Sine Operator	$|\mathbf{b}_p| \, \overline{QP}' = \vec{OP} \otimes \mathbf{b}_q$	$|\mathbf{b}_q| \, \overline{PQ}' = \vec{OQ} \otimes \mathbf{b}_p$
5	From 4, T4.4.6B Equalities: Left Cancellation by Multiplication and A4.2.11 Associative Multp	$\overline{QP}' = (\overline{QP}' / |\mathbf{b}_p|) \otimes \mathbf{b}_q$	$\overline{PQ}' = (\vec{OQ} / |\mathbf{b}_q|) \otimes \mathbf{b}_p$
∴	From 6g, 7g, 1, 6, A4.2.3 Substitution and A4.2.2A Equality	$\overline{QP}' = \mathbf{u}_p \otimes \mathbf{b}_q = |\mathbf{b}_q| \sin(\theta_{pq})$ EQ A	$\overline{PQ}' = \mathbf{u}_q \otimes \mathbf{b}_p = |\mathbf{b}_p| \sin(\theta_{qp})$ EQ B

Theorem 8.5.6 Orthogonal Sine Operator: Unitary Sine Operator

	Given	the above geometry of Figure F8.4.1	
1g	Given	the above geometry of Figure F8.4.1	
2g		$\mathbf{b}_p \equiv \vec{OP}$	
3g		$\mathbf{b}_q \equiv \vec{OQ}$	
4g		$\overline{OP} = |\vec{OP}| = |\mathbf{b}_p|$	
5g		$\overline{OQ} = |\vec{OQ}| = |\mathbf{b}_q|$	
6g		$\hat{OP} = \vec{OP} / \overline{OP} = \vec{OP} / |\mathbf{b}_p| = \mathbf{u}_p$	
7g		$\hat{OQ} = \vec{OQ} / \overline{OQ} = \vec{OQ} / |\mathbf{b}_q| = \mathbf{u}_q$	
Steps	Hypothesis	$\mathbf{u}_p \otimes \mathbf{u}_q = \sin(\theta_{pq})$ EQ A	$\mathbf{u}_q \otimes \mathbf{u}_p = \sin(\theta_{qp})$ EQ B
1	From T8.5.5 Orthogonal Sine Operator: Geometrical Interpretation	$\overline{QP}' = \mathbf{u}_p \otimes \mathbf{b}_q = |\mathbf{b}^q| \sin(\theta_{pq})$	$\overline{PQ}' = \mathbf{u}_q \otimes \mathbf{b}_p = |\mathbf{b}_p| \sin(\theta_{qp})$
2	From 1, T4.4.6B Equalities: Left Cancellation by Multiplication and A4.2.11 Associative Multp	$\overline{QP}' = \mathbf{u}_p \otimes (\mathbf{b}_q / |\mathbf{b}_q|)$ $= \sin(\theta_{pq})$	$\overline{PQ}' = \mathbf{u}_q \otimes (\mathbf{b}_p / |\mathbf{b}_p|)$ $= \sin(\theta_{pq})$
∴	From 2 and A4.2.2A Equality	$\mathbf{u}_p \otimes \mathbf{u}_q = \sin(\theta_{pq})$ EQ A	$\mathbf{u}_q \otimes \mathbf{u}_p = \sin(\theta_{qp})$ EQ B

Theorem 8.5.7 Orthogonal Sine Operator: Asymmetry of Skewed Value

Steps	Hypothesis	$\mathbf{u}_p \otimes \mathbf{u}_q = -\sin(\theta_{qp})$ $= \sin(\theta_{pq})$ EQ A	$\mathbf{u}_q \otimes \mathbf{u}_p = -\sin(\theta_{pq})$ $= \sin(\theta_{qp})$ EQ B
1	From T8.5.6 Orthogonal Sine Operator: Unitary Sine Operator	$\mathbf{u}_p \otimes \mathbf{u}_q = \sin(\theta_{pq})$	$\mathbf{u}_q \otimes \mathbf{u}_p = \sin(\theta_{qp})$
2	From 1, A8.5.1A Directed Sign Operator between any Two Vectors and A5.2.18 GIV Transitivity of Angles	$\mathbf{u}_p \otimes \mathbf{u}_q = \sin(-\theta_{qp})$	$\mathbf{u}_q \otimes \mathbf{u}_p = \sin(-\theta_{pq})$
∴	From 1, 2, DxE.1.7.1 and A4.2.2A Equality	$\mathbf{u}_p \otimes \mathbf{u}_q = -\sin(\theta_{qp})$ $= \sin(\theta_{pq})$ EQ A	$\mathbf{u}_q \otimes \mathbf{u}_p = -\sin(\theta_{pq})$ $= \sin(\theta_{qp})$ EQ B

Theorem 8.5.8 Orthogonal Sine Operator: Commutative Asymmetrically Skewed

Steps	Hypothesis	$\mathbf{u}_p \otimes \mathbf{u}_q \equiv (-1)^{\theta pr} \mathbf{u}_q \otimes \mathbf{u}_p$
1	From A8.5.1A Directed Sign Operator between any Two Vectors	$\theta_{pq} \cong -\theta_{qp}$
2	From 1 and uniqueness of sine operator	$\sin(\theta_{pq}) = \sin(-\theta_{qp})$
3	From 2 and DxE.1.7.1	$\sin(\theta_{pq}) = -\sin(\theta_{qp})$
4	From 3 and D8.5.4 Associated Parity Between any Two	$\sin(\theta_{pq}) = (-1)^{\theta pr} \mid \sin(\theta_{pq}) \mid$ regardless of transposition of indices
∴	From 2, T8.5.7 Orthogonal Sine Operator: Asymmetry of Skewed Value and A4.2.3 Substitution	$\mathbf{u}_p \otimes \mathbf{u}_q = (-1)^{\theta pr} \mid \mathbf{u}_q \otimes \mathbf{u}_p \mid$

The paired theorem on Orthogonal Cosine and Sine operators above are the core to tensors this is the point at which the vector set of a tensor are glued together and becomes a tensor. Placing these theorems in the context of an ordered system of vectors, such as tensors, forms an ordered index that can determine the asymmetric relationship between a set of relative vectors, D4.7.1 "Inequalities: Transitive Law" can now be invoked see D8.4.4 "Associated Parity Between any Two Vectors".

From D8.4.1 "Orthogonal Cosine Operator" there is an analytic interpretation for the cosine operator, but what of the sine operator?

Theorem 8.5.9 Orthogonal Sine Operator: Analytic Relationship

Steps	Hypothesis	$\mathbf{u}_p \otimes \mathbf{u}_q = (\,(-1)^{\theta pr} / (|\mathbf{B}_p||\mathbf{B}_q|)\,)\sqrt{\sum_{i<j}(\,(x_j\,y_i)^2 - (y_j\,x_i)^2\,)}$
1	From T8.5.8 Orthogonal Sine Operator: Commutative Asymmetrically Skewed	$\mathbf{u}_p \otimes \mathbf{u}_q = (-1)^{\theta pr}\,|\,\mathbf{u}_q \otimes \mathbf{u}_p\,|$
2	From 1 and D8.5.2 Orthogonal Sine Operator	$\mathbf{u}_p \otimes \mathbf{u}_q = (-1)^{\theta pr}\,|\sin(\theta_{qp})\,|$
3	From 2 and LxE.1.19	$\mathbf{u}_p \otimes \mathbf{u}_q = (-1)^{\theta pr}\sqrt{1 - \cos(\theta^{qp})^2}$
4	From 3 and D8.5.1 Orthogonal Cosine Operator	$\mathbf{u}_p \otimes \mathbf{u}_q = (-1)^{\theta pr}\sqrt{1 - (\sum^n (x_k/|\mathbf{B}_p|)\,(y_k/|\mathbf{B}_q|))^2}$
5	From 4 and D4.1.17 Exponential Notation	$\mathbf{u}_p \otimes \mathbf{u}_q = (-1)^{\theta pr}\sqrt{1 - (\sum^n (x_k/|\mathbf{B}_p|)\,(y_k/|\mathbf{B}_q|))\,(\sum^n (x_k/|\mathbf{B}_p|)\,(y_k/|\mathbf{B}_q|))}$
6	From 5, A4.2.13 Inverse Multp and A4.2.14 Distribution	$\mathbf{u}_p \otimes \mathbf{u}_q = (-1)^{\theta pr}\sqrt{(\,(|\mathbf{B}_p||\mathbf{B}_q|)^2 / (|\mathbf{B}_p||\mathbf{B}_q|)^2\,) - (1/(|\mathbf{B}_p||\mathbf{B}_q|\,|\mathbf{B}_p||\mathbf{B}_q|))\,(\sum^n x_k\,y_k)\,(\sum^n x_k\,y_k)}$
7	From 6 and D4.1.17 Exponential Notation	$\mathbf{u}_p \otimes \mathbf{u}_q = (-1)^{\theta pr}\sqrt{(\,(|\mathbf{B}_p||\mathbf{B}_q|)^2 / (|\mathbf{B}_p||\mathbf{B}_q|)^2\,) - (1/(|\mathbf{B}_p||\mathbf{B}_q|)^2)\,(\sum^n x_k\,y_k)^2}$
8	From 6 and A4.2.14 Distribution	$\mathbf{u}_p \otimes \mathbf{u}_q = (-1)^{\theta pr}\sqrt{(\,1/(|\mathbf{B}_p||\mathbf{B}_q|)^2\,)\,((|\mathbf{B}_p||\mathbf{B}_q|)^2 - (\sum^n x_k\,y_k)^2\,)}$
9	From 7 and T4.10.4 Radicals: Distribution Across a Product	$\mathbf{u}_p \otimes \mathbf{u}_q = (-1)^{\theta pr}\sqrt{(1/(|\mathbf{B}_p||\mathbf{B}_q|)^2)}\sqrt{(\,(|\mathbf{B}_p||\mathbf{B}_q|)^2 - (\sum^n x_k\,y_k)^2\,)}$
10	From 8 and T4.8.7 Integer Exponents: Distribution Across a Rational Number	$\mathbf{u}_p \otimes \mathbf{u}_q = (-1)^{\theta pr}\sqrt{(1/(|\mathbf{B}_p||\mathbf{B}_q|))^2}\sqrt{(|\mathbf{B}_p||\mathbf{B}_q|)^2 - (\sum^n x_k\,y_k)^2}$
11	From 9 and T4.10.3 Radicals: Identity	$\mathbf{u}_p \otimes \mathbf{u}_q = (-1)^{\theta pr} / (|\mathbf{B}_p||\mathbf{B}_q|)\sqrt{(|\mathbf{B}_p||\mathbf{B}_q|)^2 - (\sum^n x_k\,y_k)^2}$
13	From 11, and A4.2.21 Distribution Exp	$\mathbf{u}_p \otimes \mathbf{u}_q = (-1)^{\theta pr} / (|\mathbf{B}_p||\mathbf{B}_q|)\sqrt{|\mathbf{B}_p|^2|\mathbf{B}_q|^2 - (\sum^n x_k\,y_k)^2}$
14	From 13, T8.4.4 Dot Product: Magnitude of a Vector and A4.2.3 Substitution	$\mathbf{u}_p \otimes \mathbf{u}_q = (-1)^{\theta pr} / (|\mathbf{B}_p||\mathbf{B}_q|)\sqrt{(\sum^n x_k^2)(\sum^n y_k^2) - (\sum^n x_k\,y_k)(\sum^n x_k\,y_k))}$
15	From 14 and T4.12.4 Polynomial Quadratic: Product of N-Quantities	$\mathbf{u}_p \otimes \mathbf{u}_q = (-1)^{\theta pr} / (|\mathbf{B}_p||\mathbf{B}_q|)\sqrt{(\sum^n x_k^2\,y_k^2) + \sum_{i<j}(x_i^2\,y_j^2 + x_j^2\,y_i^2) - (\sum^n x_k\,y_k\,x_k\,y_k) - \sum_{i<j}(x_i\,y_i\,x_j\,y_j + x_j\,y_j\,x_i y_i)}$
16	From 15, A4.2.10 Commutative Multp and A4.2.5 Commutative Add	$\mathbf{u}_p \otimes \mathbf{u}_q = (-1)^{\theta pr} / (|\mathbf{B}_p||\mathbf{B}_q|)\sqrt{(\sum^n x_k^2\,y_k^2) - (\sum^n x_k\,x_k\,y_k\,y_k) + \sum_{i<j}(x_i^2\,y_j^2 + x_j^2\,y_i^2) - \sum_{i<j}(x_i\,y_i\,x_j\,y_j + x_j\,y_j\,x_i y_i)}$
17	From 16 and D4.1.17 Exponential Notation	$\mathbf{u}_p \otimes \mathbf{u}_q = (-1)^{\theta pr} / (|\mathbf{B}_p||\mathbf{B}_q|)\sqrt{(\sum^n x_k^2\,y_k^2) - (\sum_n x_k^2\,y_k^2) + \sum_{i<j}(x_i^2\,y_j^2 + x_j^2\,y_i^2) - \sum_{i<j}(x_i\,y_i\,x_j\,y_j + x_j\,y_j\,x_i y_i)}$
18	From 17 and A4.2.8 Inverse Add	$u^p \otimes u^q = (-1)^{\theta pr} / (|\mathbf{B}_p||\mathbf{B}_q|)\sqrt{0 + \sum_{i<j}(x_i^2\,y_j^2 + x_j^2\,y_i^2) - \sum_{i<j}(x_i\,y_i\,x_j\,y_j + x_j\,y_j\,x_i y_i)}$
19	From 18, A4.2.7 Identity Add, A4.2.6 Associative Add and A4.2.14 Distribution	$\mathbf{u}_p \otimes \mathbf{u}_q = (-1)^{\theta pr} / (|\mathbf{B}_p||\mathbf{B}_q|)\sqrt{\sum_{i<j}(x_i^2\,y_j^2 + x_j^2\,y_i^2 - x_i\,y_i\,x_j\,y_j - x_j\,y_j\,x_i y_i)}$
20	From 19 and A4.2.5 Commutative Add	$\mathbf{u}_p \otimes \mathbf{u}_q = (-1)^{\theta pr} / (|\mathbf{B}_p||\mathbf{B}_q|)\sqrt{\sum_{i<j}(x_i^2\,y_j^2 - x_i\,y_i\,x_j\,y_j + x_j^2\,y_i^2 - x_i\,y_i\,x_i y_i)}$

21	From 20 and A4.2.14 Distribution	$\mathbf{u}_p \otimes \mathbf{u}_q = (-1)^{\theta pr} / (	\mathbf{B}_p		\mathbf{B}_q	)$ $\sqrt{\sum_{i<j}(y_i x_i (x_i y_i - y_i x_i) + x_i y_i (x_i y_i - y_i x_i))}$
22	From 21 and A4.2.14 Distribution	$\mathbf{u}_p \otimes \mathbf{u}_q = (-1)^{\theta pr} / (	\mathbf{B}_p		\mathbf{B}_q	)$ $\sqrt{\sum_{i<j}(y_i x_i + x_i y_i)(x_i y_i - y_i x_i))}$
23	From 22 and A4.2.5 Commutative Add	$\mathbf{u}_p \otimes \mathbf{u}_q = (-1)^{\theta pr} / (	\mathbf{B}_p		\mathbf{B}_q	)$ $\sqrt{\sum_{i<j}(x_i y_i + y_i x_i)(x_i y_i - y_i x_i))}$
∴	From 23 and T4.12.3 Polynomial Quadratic: Difference of Two Squares	$\mathbf{u}_p \otimes \mathbf{u}_q = ((-1)^{\theta pr} / (	\mathbf{B}_p		\mathbf{B}_q	)) \sqrt{\sum_{i<j}((x_j y_i)^2 - (y_j x_i)^2)}$

We now have the tools to finish the development of the concept of linearly dependent and independent vectors spanning an n-space.

Theorem 8.5.10 Orthogonal Independence: The First Unitary Base Vector [e₁]

1g	Given	A set of vectors $(. x_j .)$ for $1 \le j$ spanning n-space, such that $x_i \bullet x_k \ne 0$ and $j \ne k$.
2g	by inspection	$y_1 = x_1$
3g		$|e_1| = 1$
Steps	Hypothesis	$e_1 = x_1 / |x_1|$
∴	From 1g, 2g, 3g and T8.4.5 Dot Product: Unit Vector	$e_1 = x_1 / |x_1|$

Theorem 8.5.11 Orthogonal Independence: Unitary Base Vector [e_2]

1g	Given		A set of vectors $(\mathbf{e}_1, {}_\bullet \mathbf{x}_j {}_\bullet)$ for $1 < j$ spanning n-space, such that $\mathbf{x}_i \bullet \mathbf{x}_k \neq 0$ and $j \neq k$.	
2g	by inspection		$\mathbf{y}_2 = \mathbf{x}_2 - (\mathbf{x}_2 \bullet \mathbf{e}_1)\,\mathbf{e}_1$	
3g			$|\mathbf{e}_2| = 1$	
Steps	Hypothesis		$\mathbf{e}_2 = (\mathbf{x}_2 - (\mathbf{x}_2 \bullet \mathbf{e}_1)\,\mathbf{e}_1) / \sqrt{|\mathbf{x}_2|^2 - 2(\mathbf{x}_2 \bullet \mathbf{e}_1)^2 + (\mathbf{x}_2 \bullet \mathbf{e}_1)^2}$	
1	From A4.2.2A Equality	From T8.4.4 Dot Product: Magnitude of a Vector, for y^2	$\mathbf{y}_2 \bullet \mathbf{e}_1 = \mathbf{y}_2 \bullet \mathbf{e}_1$	$|\mathbf{y}_2|^2 = \mathbf{y}_2 \bullet \mathbf{y}_2$
2	From 1, 2g and A4.2.3 Substitution	From 1, 2g and A4.2.3 Substitution	$\mathbf{y}_2 \bullet \mathbf{e}_1 =$ $(\mathbf{x}_2 - (\mathbf{x}_2 \bullet \mathbf{e}_1)\,\mathbf{e}_1) \bullet \mathbf{e}_1$	$|\mathbf{y}_2|^2 =(\mathbf{x}_2 - (\mathbf{x}_2 \bullet \mathbf{e}_1)\,\mathbf{e}_1) \bullet (\mathbf{x}_2 - (\mathbf{x}_2 \bullet \mathbf{e}_1)\,\mathbf{e}_1)$
3	From 2 and T8.3.7 Dot Product: Distribution of Dot Product Across Addition of Vectors	From 2 and T8.3.7 Dot Product: Distribution of Dot Product Across Addition of Vectors	$\mathbf{y}_2 \bullet \mathbf{e}_1 =$ $\mathbf{x}_2 \bullet \mathbf{e}_1 - (\mathbf{x}_2 \bullet \mathbf{e}_1)\,\mathbf{e}_1 \bullet \mathbf{e}_1$	$|\mathbf{y}_2|^2 = \mathbf{x}_2 \bullet \mathbf{x}_2 - (\mathbf{x}_2 \bullet \mathbf{e}_1)\,(\mathbf{x}_2 \bullet \mathbf{e}_1) - (\mathbf{x}_2 \bullet \mathbf{e}_1)\,(\mathbf{e}_1 \bullet \mathbf{x}_2) + (\mathbf{x}_2 \bullet \mathbf{e}_1)\,(\mathbf{x}_2 \bullet \mathbf{e}_1)\,(\mathbf{e}_1 \bullet \mathbf{e}_1)$
4	From 3 and T8.4.5 Dot Product: Unit Vector	From 3, D4.1.17 Exponential Notation and T8.4.5 Dot Product: Unit Vector	$\mathbf{y}_2 \bullet \mathbf{e}_1 = \mathbf{x}_2 \bullet \mathbf{e}_1 - (\mathbf{x}_2 \bullet \mathbf{e}_1)\,1$	$|\mathbf{y}_2|^2 = \mathbf{x}_2 \bullet \mathbf{x}_2 - (\mathbf{x}_2 \bullet \mathbf{e}_1)^2 - (\mathbf{x}_2 \bullet \mathbf{e}_1)^2 + (\mathbf{x}_2 \bullet \mathbf{e}_1)^2\,1$
5	From 4 and A4.2.12 Identity Multp	From 4 A4.2.12 Identity Multp, and A4.2.14 Distribution	$\mathbf{y}_2 \bullet \mathbf{e}_1 = \mathbf{x}_2 \bullet \mathbf{e}_1 - \mathbf{x}_2 \bullet \mathbf{e}_1$	$|\mathbf{y}_2|^2 = \mathbf{x}_2 \bullet \mathbf{x}_2 + (-1 -1\ )\,(\mathbf{x}_2 \bullet \mathbf{e}_1)^2 + (\mathbf{x}_2 \bullet \mathbf{e}_1)^2$
6	From 5 and A4.2.8 Inverse Add	From 5 and D4.1.19 Primitive Definition for Rational Arithmetic	$\mathbf{y}_2 \bullet \mathbf{e}_1 = 0$	$|\mathbf{y}_2|^2 = \mathbf{x}_2 \bullet \mathbf{x}_2 - 2(\mathbf{x}_2 \bullet \mathbf{e}_1)^2 + (\mathbf{x}_2 \bullet \mathbf{e}_1)^2$
7		From 6 and T8.4.4 Dot Product: Magnitude of a Vector		$|\mathbf{y}_2|^2 = |\mathbf{x}_2|^2 - 2(\mathbf{x}_2 \bullet \mathbf{e}_1)^2 + (\mathbf{x}_2 \bullet \mathbf{e}_1)^2$
8	From T8.4.5 Dot Product: Unit Vector		$\mathbf{e}_2 = \mathbf{y}_2 / |\mathbf{y}_2|$	
∴	From 2g, 7, 8 and A4.2.3 Substitution		$\mathbf{e}_2 = (\mathbf{x}_2 - (\mathbf{x}_2 \bullet \mathbf{e}_1)\,\mathbf{e}_1) / \sqrt{|\mathbf{x}_2|^2 - 2(\mathbf{x}_2 \bullet \mathbf{e}_1)^2 + (\mathbf{x}_2 \bullet \mathbf{e}_1)^2}$	

Theorem 8.5.12 Orthogonal Independence: Unitary Base Vector [e₃]

1g	Given		A set of vectors $(\mathbf{e}_1, \mathbf{e}_2, \bullet \, \mathbf{x}_j \, \bullet)$ for $2 < j$ spanning n-space, such that $\mathbf{x}_i \bullet \mathbf{x}_k \neq 0$ and $i \neq k$.	
2g	by inspection		$\mathbf{y}_3 = \mathbf{x}_3 - (\mathbf{x}_3 \bullet \mathbf{e}_2) \, \mathbf{e}_2$	
3g			$|\mathbf{e}_3| = 1$	
Steps	Hypothesis		$\mathbf{e}_3 = (\mathbf{x}_3 - (\mathbf{x}_3 \bullet \mathbf{e}_2) \, \mathbf{e}_2) / \sqrt{|\mathbf{x}_3|^2 - 2(\mathbf{x}_3 \bullet \mathbf{e}_2)^2 + (\mathbf{x}_3 \bullet \mathbf{e}_2)^2}$	
1	From A4.2.2A Equality	From T8.4.4 Dot Product: Magnitude of a Vector for y^3	$\mathbf{y}_3 \bullet \mathbf{e}_2 = \mathbf{y}_3 \bullet \mathbf{e}_2$	$|\mathbf{y}_3|^2 = \mathbf{y}_3 \bullet \mathbf{y}_3$
2	From 1, 2g and A4.2.3 Substitution	From 1, 2g and A4.2.3 Substitution	$\mathbf{y}_3 \bullet \mathbf{e}_2 = (\mathbf{x}_3 - (\mathbf{x}_3 \bullet \mathbf{e}_2) \, \mathbf{e}_2) \bullet \mathbf{e}_2$	$|\mathbf{y}_3|^2 = (\mathbf{x}_3 - (\mathbf{x}_3 \bullet \mathbf{e}_2) \, \mathbf{e}_2) \bullet (\mathbf{x}_3 - (\mathbf{x}_3 \bullet \mathbf{e}_2) \, \mathbf{e}_2)$
3	From 2 and T8.3.7 Dot Product: Distribution of Dot Product Across Addition of Vectors	From 2 and T8.3.7 Dot Product: Distribution of Dot Product Across Addition of Vectors	$\mathbf{y}_3 \bullet \mathbf{e}_2 = \mathbf{x}_3 \bullet \mathbf{e}_2 - (\mathbf{x}_3 \bullet \mathbf{e}_2) \, \mathbf{e}_2 \bullet \mathbf{e}_2$	$|\mathbf{y}_3|^2 = \mathbf{x}_3 \bullet \mathbf{x}_3 - (\mathbf{x}_3 \bullet \mathbf{e}_2)(\mathbf{x}_3 \bullet \mathbf{e}_2) - (\mathbf{x}_3 \bullet \mathbf{e}_2)(\mathbf{e}_2 \bullet \mathbf{x}_3) + (\mathbf{x}_3 \bullet \mathbf{e}_2)(\mathbf{x}_3 \bullet \mathbf{e}_2)(\mathbf{e}_2 \bullet \mathbf{e}_2)$
4	From 3 and T8.4.5 Dot Product: Unit Vector	From 3, D4.1.17 Exponential Notation and T8.4.5 Dot Product: Unit Vector	$\mathbf{y}_3 \bullet \mathbf{e}_2 = \mathbf{x}_3 \bullet \mathbf{e}_2 - (\mathbf{x}_3 \bullet \mathbf{e}_2) \, 1$	$|\mathbf{y}_3|^2 = \mathbf{x}_3 \bullet \mathbf{x}_3 - (\mathbf{x}_3 \bullet \mathbf{e}_2)^2 - (\mathbf{x}_3 \bullet \mathbf{e}_2)^2 + (\mathbf{x}_3 \bullet \mathbf{e}_2)^2 \, 1$
5	From 4 and A4.2.12 Identity Multp	From 4 A4.2.12 Identity Multp, and A4.2.14 Distribution	$\mathbf{y}_3 \bullet \mathbf{e}_2 = \mathbf{x}^3 \bullet \mathbf{e}^2 - \mathbf{x}^3 \bullet \mathbf{e}^2$	$|\mathbf{y}_3|^2 = \mathbf{x}_3 \bullet \mathbf{x}_3 + (-1 - 1)(\mathbf{x}_3 \bullet \mathbf{e}_2)^2 + (\mathbf{x}_3 \bullet \mathbf{e}_2)^2$
6	From 5 and A4.2.8 Inverse Add	From 5 and D4.1.19 Primitive Definition for Rational Arithmetic	$\mathbf{y}_3 \bullet \mathbf{e}_2 = 0$	$|\mathbf{y}_3|^2 = \mathbf{x}_3 \bullet \mathbf{x}_3 - 2(\mathbf{x}_3 \bullet \mathbf{e}_2)^2 + (\mathbf{x}_3 \bullet \mathbf{e}_2)^2$
7		From 6 and T8.4.4 Dot Product: Magnitude of a Vector		$|\mathbf{y}_3|^2 = |\mathbf{x}_3|^2 - 2(\mathbf{x}_3 \bullet \mathbf{e}_2)^2 + (\mathbf{x}_3 \bullet \mathbf{e}_2)^2$
8	From T8.4.5 Dot Product: Unit Vector		$\mathbf{e}_3 = \mathbf{y}_3 / |\mathbf{y}_3|$	
∴	From 2g, 7, 8 and A4.2.3 Substitution		$\mathbf{e}_3 = (\mathbf{x}_3 - (\mathbf{x}_3 \bullet \mathbf{e}_2) \, \mathbf{e}_2) / \sqrt{|\mathbf{x}_3|^2 - 2(\mathbf{x}_3 \bullet \mathbf{e}_2)^2 + (\mathbf{x}_3 \bullet \mathbf{e}_2)^2}$	

Theorem 8.5.13 Orthogonal Independence: Unitary Base Vector [e_{p-1}]

1g	Given	A set of vectors ($._\bullet e_i ._\bullet ,_\bullet x_j ._\bullet$) for i = 1, … , p-1 and p-1 < j spanning n-space, such that $x_i \bullet x_k \neq 0$ for and $j \neq k$.
2g	by inspection	$y_{p-1} = x_{p-1} - (x_{p-1} \bullet e_{p-2}) e_{p-2}$
3g		$|e_{p-1}| = 1$
Steps	Hypothesis	$e_{p-1} = y_{p-1} / |y_{p-1}|$
∴	From T8.4.5 Dot Product: Unit Vector	$e_{p-1} = y_{p-1} / |y_{p-1}|$

Theorem 8.5.14 Orthogonal Independence: Unitary Base Vector [e_p]

1g	Given		A set of vectors $(\bullet\, e_i\, \bullet\,,\, \bullet\, x_j\, \bullet)$ for $i = 1, \ldots, p$ and $p < j$ spanning n-space, such that $x_i \bullet x_k \neq 0$ for and $j \neq k$.
2g	by inspection		$y_p = x_p - (x_p \bullet e_{p-1})\, e_{p-1}$
3g			$|e_p| = 1$
Steps	Hypothesis		$e_p = (x_p - (x_p \bullet e_{p-1})\, e_{p-1}) / \sqrt{|x_p|^2 - 2(x_p \bullet e_{p-1})^2 + (x_p \bullet e_{p-1})^2}$
1	From A4.2.2A Equality	From T8.4.4 Dot Product: Magnitude of a Vector, for y^p	$y_p \bullet e_{p-1} = y_p \bullet e_{p-1}$ $|y_p|^2 = y_p \bullet y_p$
2	From 1, 2g and A4.2.3 Substitution	From 1, 2g and A4.2.3 Substitution	$y_p \bullet e_{p-1} = (x_p - (x_p \bullet e_{p-1})\, e_{p-1}) \bullet e_{p-1}$ $|y_p|^2 = (x_p - (x_p \bullet e_{p-1})\, e_{p-1}) \bullet (x_p - (x_p \bullet e_{p-1})\, e_{p-1})$
3	From 2 and T8.3.7 Dot Product: Distribution of Dot Product Across Addition of Vectors	From 2 and T8.3.7 Dot Product: Distribution of Dot Product Across Addition of Vectors	$y_p \bullet e_{p-1} = x_p \bullet e_{p-1} - (x_p \bullet e_{p-1})\, e_{p-1} \bullet e_{p-1}$ $|y_p|^2 = x_p \bullet x_p - (x_p \bullet e_{p-1})(x_p \bullet e_{p-1}) - (x_p \bullet e_{p-1})(e_{p-1} \bullet x_p) + (x_p \bullet e_{p-1})(x_p \bullet e_{p-1})(e_{p-1} \bullet e_{p-1})$
4	From 3 and T8.4.5 Dot Product: Unit Vector	From 3, D4.1.17 Exponential Notation and T8.4.5 Dot Product: Unit Vector	$y_p \bullet e_{p-1} = x_p \bullet e_{p-1} - (x_p \bullet e_{p-1})\, 1$ $|y_p|^2 = x_p \bullet x_p - (x_p \bullet e_{p-1})^2 - (x_p \bullet e_{p-1})^2 + (x_p \bullet e_{p-1})^2\, 1$
5	From 4 and A4.2.12 Identity Multp	From 4 A4.2.12 Identity Multp, and A4.2.14 Distribution	$y_p \bullet e_{p-1} = x_p \bullet e_{p-1} - x_p \bullet e_{p-1}$ $|y_p|^2 = x_p \bullet x_p + (-1 -1)(x_p \bullet e_{p-1})^2 + (x_p \bullet e_{p-1})^2$
6	From 5 and A4.2.8 Inverse Add	From 5 and D4.1.19 Primitive Definition for Rational Arithmetic	$y_p \bullet e_{p-1} = 0$ $|y_p|^2 = x_p \bullet x_p - 2(x_p \bullet e_{p-1})^2 + (x_p \bullet e_{p-1})^2$
7		From 6 and T8.4.4 Dot Product: Magnitude of a Vector	$|y_p|^2 = |x_p|^2 - 2(x_p \bullet e_{p-1})^2 + (x_p \bullet e_{p-1})^2$
8	From T8.4.5 Dot Product: Unit Vector		$e_p = y_p / |y_p|$
∴	From 2g, 7, 8 and A4.2.3 Substitution		$e_p = (x_p - (x_p \bullet e_{p-1})\, e_{p-1}) / \sqrt{|x_p|^2 - 2(x_p \bullet e_{p-1})^2 + (x_p \bullet e_{p-1})^2}$

Theorem 8.5.15 Orthogonal Independence: Unitary Base Vector [e_n]

1g	Given	A set of vectors $(\bullet\, \mathbf{e}_i \,\bullet)$ for $i = 1, \ldots , n$.
2g	by inspection	$\mathbf{y}_n = \mathbf{x}_n - (\mathbf{x}_n \bullet \mathbf{e}_{n-1})\, \mathbf{e}_{n-1}$
3g		$|\mathbf{e}_n| = 1$
Steps	Hypothesis	$\mathbf{e}_n = (\mathbf{x}_n - (\mathbf{x}_n \bullet \mathbf{e}_{n-1})\, \mathbf{e}_{n-1}) / \sqrt{\; |\mathbf{x}_n|^2 - 2(\mathbf{x}_n \bullet \mathbf{e}_{n-1})^2 + (\mathbf{x}_n \bullet \mathbf{e}_{n-1})^2}$
1	From T8.5.10 to 12 is true. Assumed T8.5.13 Orthogonal Independence: Unitary Base Vector [$\mathbf{e}_{p-1}$] and T8.5.14 Orthogonal Independence: Unitary Base Vector [e^p]	$\mathbf{y}_n \bullet \mathbf{e}_{n-1} = 0$ $\qquad$ $|\mathbf{y}_n|^2 = |\mathbf{x}_n|^2 - 2(\mathbf{x}_n \bullet \mathbf{e}_{n-1})^2 + (\mathbf{x}_n \bullet \mathbf{e}_{n-1})^2$
2	From T8.4.5 Dot Product: Unit Vector	$\mathbf{e}_n = \mathbf{y}_n / |\mathbf{y}_n|$
$\therefore$	From 1 and 2 and A3.11.1 Induction Property of Integers	$\mathbf{e}_n = (\mathbf{x}_n - (\mathbf{x}_n \bullet \mathbf{e}_{n-1})\, \mathbf{e}_{n-1}) / \sqrt{\; |\mathbf{x}_n|^2 - 2(\mathbf{x}_n \bullet \mathbf{e}_{n-1})^2 + (\mathbf{x}_n \bullet \mathbf{e}_{n-1})^2}$

Observation 8.5.2 Unit Bases Vectors: Existence of Construction and Their Independence

This last theorem proves a couple of points:

1)   A set of orthogonal unitary base vectors $(\bullet\, \mathbf{e}_i \,\bullet)$ can always be constructed from any set of fundamental (linearly independent) vectors spanning n-space.

2)   Orthogonal unitary base vectors are in themselves independent.

Section 8.6 Angles and Distance from a Point to a Plane

Definition 8.6.1 **Direction of a Line**

If $(\bullet \alpha_i, \bullet)$ are ***direction angles*** of a directed line $\vec{L}$, then $(\bullet \cos \alpha_i \bullet)$ are called ***directed cosines*** of a directed line $\vec{L}$ in n-dimensions.

$$x_{i1} < x_{i0} < x_{i2} \qquad \text{Equation A}$$
$$\cos \alpha_i \equiv x_{i0} / d \qquad \text{Equation B}$$
$$\cos \alpha_i \equiv (x_{i2} - x_{i1}) / d \qquad \text{Equation C}$$
$$\cos \alpha_i \equiv \gamma_i \qquad \text{Equation D}$$
$$\text{for all } i.$$

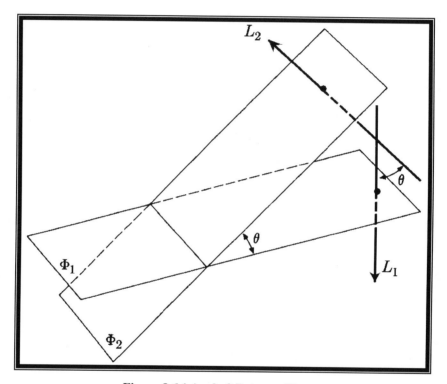

Figure 8.6.1 Angle θ Between Planes

Theorem 8.6.1 Angle between Hyperplanes Φ_{12}

1g	Given	$\chi_i = x_i / |L_1|$ directed cosines per axis for plane-Φ_1
2g		$\zeta_i = y_i / |L_2|$ directed cosines per axis for plane-Φ_2
		for x_i, y_i perpendicular altitude numbers to their planes
Steps	Hypothesis	$\cos(\theta_{12}) = \sum^n \chi_i \zeta_i$
1	From the dot product of the directed lines $\vec{L}_1$ and $\vec{L}_2$ by D8.3.1 The Inner or Dot Product Operator	$L_1 \bullet L_2 \equiv |L_1||L_2| \cos(\theta_{12})$
2	From 1 and T8.4.2 Dot Product: Analytic Relationship	$|L_1||L_2| \cos(\theta_{12}) = \sum^n x_i \, y_i$
3	From 2, T4.4.4B Equalities: Right Cancellation by Multiplication, A4.2.14 Distribution, A4.2.12 Identity Multp and A4.2.11 Associative Multp	$\cos(\theta_{12}) = \sum^n (x_i / |L_1|)(y_i / |L_2|)$
$\therefore$	From 1g, 2g, 2 and A4.2.3 Substitution	$\cos(\theta_{12}) = \sum^n \chi_i \zeta_i$

Tensor Calculus & Physics: A General Treatise

Volume I Appendix I

Table of Contents

List of Tables

Tensor Calculus & Physics: A General Treatise

List of Observations

List of Equations

List of Figures

Appendix A –The Mathematics of Multinomial Summation Notation

Section A.1 *Evaluation of Multinomial Product*

Evaluation of multinomial product $[_nM_m]$ is one of the problems that continually arise in a variety of mathematical disciplines; tensor calculus is one such subject others are in sampling theory in probability and statistics. So here, I set aside a study of this mathematical enigma. Let's consider the following general multinomial equation:

$$_nM_m = (\sum^n x_i)^m$$

EQ A.1.1A

expanded for n-terms and m-products of the multinomial;

$$_nM_m = \sum_{\rho(k)\in Lr} \prod^m x_{\chi[i,\,k]}$$
$$\text{for } r = n^m$$

EQ A.1.1B

Evaluating equation A.1.1 requires an understanding how to generate sets of long indices. In turn, this is dependent on understanding the theory of positional number theory.

Before that is considered, lets define the objective. What is wanted is a way of taking the non-linear multinomial equation $_nM_m$ and find away of expanding it term-by-term into some near linear form similar to EQ A.1.1B and EQ 1.7.3.1for determinates by using set notation. The following analysis leads to that objective now lets consider the following.

Definition A.1.1 Multinomial Components

The components are defined here as the x's, the series of terms within the multinomial product.

Definition A.1.2 Multinomial Terms

The term is the resulting product $\prod^m x_{\chi[i,\,k]}$ from the expansion of the multinomial.

Tensor Calculus & Physics: A General Treatise

Section A.2 *Numerical Principles of Coordinate Indices as Manifolds*

Lets take two manifolds ^{s}L for linear integer sequential set of numbers and $_{d}C^{n}$ for an ordered n-space of integers. From number theory, it has been known that these two manifolds are true. That is there exists a transformation between the two, if one accepts modulus arithmetic for the division operator [÷] D4.1.4E, the resulting quotient is now broken up into two new operators the modulus operator, returning the rational integer numerator of the remainder using the forward slash [/] D4.1.4A. Also the backslash [\ ] returns only the whole integer number of the quotient, that is it clips off any fractional part behind the decimal, and is know as integer division.

$$q = i_q + f_q \text{ for } 0 \leq f < 1 \qquad\qquad \text{EQ A.2.1}$$

that is

$$i_q = a \setminus m \qquad\qquad \text{EQ A.2.2}$$

and

$$f_q = r \,/\, m \qquad\qquad \text{EQ A.2.3}$$

where

$$q = a \,/\, m \qquad\qquad \text{EQ A.2.4}$$

and

$$r = a \div m \text{ for } 0 \leq r < a \qquad\qquad \text{EQ A.2.5}$$

such that

$$a = q\, m + r \qquad\qquad \text{EQ A.2.6}$$

where [q] is the quotient, [r] the modulus remainder, [a] the dividend and [m] the modulus divisor, such that

$$0 \leq (a, m, q, r, i_q) \in \text{IG, hence positive integers.} \qquad\qquad \text{EQ A.2.7}$$

For these fields to be true manifolds there must exist the following topological transformation:

$_{d}C^{n}$ maps onto but ~1-1 ^{s}L and the inverse exists such that ^{s}L maps onto but ~1-1 this is a B.1.1.9 product successor and B.1.1.11 product processor function with inverse.

$$_{d}C^{n}: \rightarrow {}^{s}L \qquad\qquad \text{EQ A.2.8}$$

and

$$^{s}L: \rightarrow {}_{d}C^{n} \qquad\qquad \text{EQ A.2.9}$$

This property of numbers while not well understood in the mathematical community is one of the most important. Finding use in computer science in mapping linear memory space in computers, called logical space or memory, to n-dimensional coordinate systems, developing time lines for calendar systems and many more uses. Weather it be electronic memory or the surface media of a hard disk drive.

Before this discussion can go further a better understanding of the principle behind it needs to be discussed.

Tensor Calculus & Physics: A General Treatise

Section A.3 *The Positional Numeral and the Decimal System*

It was only after the adoption of the Hindu-Arabic, or decimal, notation that the easy positional counting methods in use for adding, subtracting, multiplying and dividing were possible. In this notation, every number is written as a linear combination of powers to ten [BRI60, Vol. 2, pg 357]. Thus

$$32158 = 3 \times 10^4 + 2 \times 10^3 + 1 \times 10^2 + 5 \times 10^1 + 8 \times 10^0$$

where each integer coefficient can generally be represented by [a] where $0 \le a \le 9$.

The use of ten as a base is doubtless due to the fact that early man used his ten digits of fingers as counters. There are vestiges of systems using the base 12 (the dozen and gross), the base 20 (counting on all indexes, fingers and toes) and the Babylonian system with base 60, which still survives in our system of dividing the angular degree into minutes. From time to time there has been some agitation to supplant the decimal system with the duodecimal system having 12 as its base, but the advantages of this system over the decimal system are scarcely material. In modern times, decimal systems have been popularized, since the advent of the electronic computer in the mid part of the last century. Multiple bases of two have become popular, 2 binary, 8 octal and 16 hexadecimal base systems. Octal has sort of fallen by the wayside but hex holds out since it is a nice power of two, 2^4, and acts as an easy shorthand to represent laboriously written binary numbers. Thus

$$111110110011110 = 1 \times 2^{14} + 1 \times 2^{13} + 1 \times 2^{12} + 1 \times 2^{11} + 1 \times 2^{10} + 0 \times 2^9 +$$
$$1 \times 2^8 + 1 \times 2^7 + 0 \times 2^6 + 0 \times 2^5 + 1 \times 2^4 + 1 \times 2^3 +$$
$$1 \times 2^2 + 1 \times 2^1 + 0 \times 2^0$$

here the binary integer coefficient [a] is $0 \le a \le 1$,

$$76636 = 7 \times 8^4 + 6 \times 8^3 + 6 \times 8^2 + 3 \times 8^1 + 6 \times 8^0$$

as above each octal integer coefficient [a] is defined as $0 \le a \le 7$,

$$7D9E = 7 \times 16^3 + D \times 16^2 + 9 \times 16^1 + E \times 16^0$$

Evaluation of multinomial product $[_nM_m]$ is one of the problems that continually arise in a variety of mathematical disciplines; tensor calculus is one such subject others are in sampling theory in probability and statistics. So here, I set aside a study of this mathematical enigma. Let's consider the following general multinomial equation:

$$_nM_m = (\textstyle\sum^n x_i)^m \qquad\qquad \text{EQ A.1.1A}$$

expanded for n-terms and m-products of the multinomial;

$$_nM_m = \sum_{\rho(k)\in Lr} \textstyle\prod^m x_{\chi[i,\,k]} \qquad\qquad \text{EQ A.1.1B}$$
for $r = n^m$

Evaluating equation A.1.1 requires an understanding how to generate sets of long indices. In turn, this is dependent on understanding the theory of positional number theory.

Before that is considered, lets define the objective. What is wanted is a way of taking the non-linear multinomial equation $_nM_m$ and find away of expanding it term-by-term into some near linear form similar to EQ A.1.1B and EQ 1.7.3.1for determinates by using set notation. The following analysis leads to that objective now lets consider the following.

Lets take two manifolds ^sL for linear integer sequential set of numbers and $_dC^n$ for an ordered n-space of integers. From number theory, it has been known that these two manifolds are true. That is there exists a transformation between the two, if one accepts modulus arithmetic for the division operator [÷] D4.1.4E, the resulting quotient is now broken up into two new operators the modulus operator, returning the rational integer numerator of the remainder using the forward slash [/] D4.1.4A. Also the backslash [\ ] returns only the whole integer number of the quotient, that is it clips off any fractional part behind the decimal, and is know as integer division.

where [q] is the quotient, [r] the modulus remainder, [a] the dividend and [m] the modulus divisor, such that

$$0 \le (a, m, q, r, i_q) \in IG, \text{ hence positive integers.}$$ EQ A.2.7

For these fields to be true manifolds there must exist the following topological transformation:

$_dC^n$ maps onto but ~1-1 ^sL and the inverse exists such that ^sL maps onto but ~1-1 this is a B.1.1.9 product successor and B.1.1.11 product processor function with inverse.

$$_dC^n: \rightarrow {}^sL$$ EQ A.2.8

and

$$^sL: \rightarrow {}_dC^n$$ EQ A.2.9

This property of numbers while not well understood in the mathematical community is one of the most important. Finding use in computer science in mapping linear memory space in computers, called logical space or memory, to n-dimensional coordinate systems, developing time lines for calendar systems and many more uses. Weather it be electronic memory or the surface media of a hard disk drive.

Before this discussion can go further a better understanding of the principle behind it needs to be discussed. It was only after the adoption of the Hindu-Arabic, or decimal, notation that the easy positional counting methods in use for adding, subtracting, multiplying and dividing were possible. In this notation, every number is written as a linear combination of powers to ten [BRI60, Vol. 2, pg 357]. Thus

$$32158 = 3 \times 10^4 + 2 \times 10^3 + 1 \times 10^2 + 5 \times 10^1 + 8 \times 10^0$$

where each integer coefficient can generally be represented by [a] where $0 \le a \le 9$.

The use of ten as a base is doubtless due to the fact that early man used his ten digits of fingers as counters. There are vestiges of systems using the base 12 (the dozen and gross), the base 20 (counting on all indexes, fingers and toes) and the Babylonian system with base 60, which still survives in our system of dividing the angular degree into minutes. From time to time there has been some agitation to supplant the decimal system with the duodecimal system having 12 as its base, but the advantages of this system over the decimal system are scarcely material. In modern times, decimal systems have been popularized, since the advent of the electronic computer in the mid part of the last century. Multiple bases of two have become popular, 2 binary, 8 octal and 16 hexadecimal base systems. Octal has sort of fallen by the wayside but hex holds out since it is a nice power of two, 2^4, and acts as an easy shorthand to represent laboriously written binary numbers. Thus

$$111110110011110 = 1 \times 2^{14} + 1 \times 2^{13} + 1 \times 2^{12} + 1 \times 2^{11} + 1 \times 2^{10} + 0 \times 2^9 +$$
$$1 \times 2^8 + 1 \times 2^7 + 0 \times 2^6 + 0 \times 2^5 + 1 \times 2^4 + 1 \times 2^3 +$$
$$1 \times 2^2 + 1 \times 2^1 + 0 \times 2^0$$

here the binary integer coefficient [a] is $0 \le a \le 1$,

$$76636 = 7 \times 8^4 + 6 \times 8^3 + 6 \times 8^2 + 3 \times 8^1 + 6 \times 8^0$$

as above each octal integer coefficient [a] is defined as $0 \le a \le 7$, again [a] takes on sixteen digits 0 through 9 and six alphabetic letters representing the last six digits, A through F, such that each hex coefficient is $0 \leq a \leq F$.

With the introduction of the decimal fraction, the decimal system has been extended so that a decimal number can approximate all positive numbers, integral, fractional and irrational, to any desired degree of accuracy. Introducing negative exponents terms to hold position in an infinite series, generalizing the abstracted coefficients to an unlimited number and likewise the base does this. So, our rational number [32158] can now be written as shown above in series form.

Symbolically a delineator was introduced to separate the integer from the fractional part after an evolutionary process from about 1492 with Pellos, Rudolff's work "Exemplel-Büchlin" of 1530, De Thiende 1585, Stevin of the same year. Finally the first writer to use a decimal point with full understanding of its significance seems to have been Clavius in his work on sine's printed for the astrolabe, also in the year of 1585 [BRI60, Vol. 9, pg 576]. Now the above number could be characteristically represented as,

$$321.58$$

where the decimal point is placed just after the power of zero from where the fraction starts.

Tensor Calculus & Physics: A General Treatise

The above based systems can be generalized into any arbitrary **base**, by letting the base be represented by [b] than each integer coefficient [a] is $0 \leq a \leq b - 1$, such that

$$(\bullet a_i \bullet , b) \in B_{ig} \text{ for } i = 0, \pm1, \pm2, \ldots$$

then

$$\ldots + a_2 \times b^2 + a_1 \times b^1 + a_0 \times b^0 + a_{-1} \times b^{-1} + a_{-2} \times b^{-2} + \ldots$$

A number of a particular base is said to be the **radix of that number**, hence the number $(\bullet a_i \bullet)$ is said to have a radix of b [EVE76, pg 24]:

$$(\bullet a_i \bullet)_b$$

All in all this type of number system is said to be positional and of a homogenous base [b], or radix. It's said to be of homogenous base because every position has the same number for a base, however there is no reason why non-homogenous bases can be considered. This leads to the original proposition of how to transform from coordinate indices between manifolds and can be generally written;

$$\ldots + a_2 \times b_2^2 + a_1 \times b_1^1 + a_0 \times b_0^0 + a_{-1} \times b_{-1}^{-1} + a_{-2} \times b_{-2}^{-2} + \ldots$$

An example of such a system can be found in disk drives for computers. Lets look at the coordinate media space delineated by the physical geometry of the **C**ylinder, **H**ead and **S**ector abbreviated [CHS] for hard disk drives and its transformation to an easer, more logical system for the end user, **L**ogical **B**lock **A**ddressing [LBA] based on linear sectors.

A sector originally came from the number of possible partitions or pie shaped sections about a single revolution of the circular plate of the disk drive. The size of a sector came to be known by an evolutionary process that has become to be defined by the disk drive industry as 512 bytes, where one byte is comprised of 8 bits, the fundamental unit that a magnetic head can resolve representing a one or zero. A disk drive is really comprised of a stack of plates or disks, each plate being assigned a magnetic reading head. These heads are mounted together on a single prong, rack, or armature, that moves in and out from the center of spin. The cylinder came about because as the, heads hold a fixed position, they trace out a stack of rings, which all-together comprising a vertical cylinder. This physical system constitutes a natural coordinate system known, as stated above, the CHS.

The CHS is an awkward system to use, and find your way around on the disk drive, so people wanted a simpler system where the sectors could simply be logically ordered 0 to the maximum number of sectors the disk drive could support. This brought about the **L**inear **B**lock **A**ddress coordinate system the [LBA].

This transformation from CHS to LBA is represented as follows:

$$lba = cyld * (maxsec * maxhds) + head * maxsec + sect - 1 \qquad \text{EQ A.3.1}$$

where

cyld	is the cylinder coordinate number.
head	is the head coordinate number.
sect	is the sector coordinate number.
maxsec	is the maximum number of sectors in a single revolution of the disk
maxhds	is the maximum number of heads on the armature corresponding to the appropriate disk.

The inverse transformation from LBA to CHS is represented as follows:

$$\text{cyld} = (\,\text{lba} \,/\, (\text{maxsec} * \text{maxhds})\,) \qquad \text{EQ A.3.2}$$
$$\text{head} = (\,(\text{lba} \,/\, \text{maxsec}) \div \text{maxhds}\,) \qquad \text{EQ A.3.3}$$
$$\text{sect} = (\,\text{lba} \div \text{maxsec}\,) + 1 \qquad \text{EQ A.3.4}$$

Notice that there is a subtraction and addition of one in EQ A.3.1 and EQ A.3.4 this comes about because the sector count for a disk drive logically always starts at zero. The origins of this are deep in the symbiosis relationship between disk drives and the programming language C, which always starts the array count at zero. As time has gone by the zero sector has taken on the important meaning of containing the essential information to initiate or start the disk drive up when you turn the power on and now goes by the name of the **boot sector**. However, do not confuse the LBA sector with the CHS sector, which does not start at zero. The more natural coordinate number of one is 0 LBA $\equiv$ (1, 1, 1) CHS not (1, 1, 0), which physically makes no sense, since cylinders and heads being physical start their count at one.

Now for computer languages, which generally move from one to n-dimensions, following the above development EQ A.3.1, yields the following equation [SHVβ01, Lesson 9 pg 8]:

$$
\begin{aligned}
\text{idl} = \;& \text{idx}_1 + (\text{idx}_2 * \text{idm}_1) \\
& + (\text{idx}_3 * \text{idm}_1 * \text{idm}_2) \\
& + (\text{idx}_4 * \text{idm}_1 * \text{idm}_2 * \text{idm}_3) + \ldots \\
& + (\text{idx}_k * \text{idm}_1 * \text{idm}_2 * \ldots * \text{idm}_{(k-1)}) + \ldots \\
& + (\text{idx}_n * \text{idm}_1 * \text{idm}_2 * \ldots * \text{idm}_{(n-1)})
\end{aligned}
$$

$$\text{idl} = \sum_{k=1}^{n} \text{idx}_k \prod_{i=1}^{k-1} \text{idm}_i \qquad \text{EQ A.3.5}$$

The inverse operation can be created if we realize the modulus of [idl] yields the first term since all higher indices are multiples of the index dimension, hence do not have a modulus of the first dimension and are filtered out.

$$\text{idx1} = (\text{idl} \div \text{idm1}) \qquad \text{EQ A.3.6}$$

now dividing through by any given dimensional product of a given term creates a fraction for the lower order terms which is zeroed by the integer division. Now the term being sought is cleared and of the correct dimensional modulus range. Since all numbers for the higher order, terms and their products are greater than the modulus number of the dimension sought than taking the modulus of the appropriate dimension filters those terms and leaves the appropriate index number.

$$\text{idx2} = (\text{idl} \setminus \text{idm1}) \div \text{idm2}$$
$$\text{idx3} = (\text{idl} \setminus (\text{idm1} * \text{idm2})) \div \text{idm3}$$
$$\text{idx4} = (\text{idl} \setminus (\text{idm1} * \text{idm2} * \text{idm3})) \div \text{idm4}$$

In general

$$\text{idx}_k = (\text{idl} \setminus (\,\prod_{j=1}^{k-1} \text{idm}_j\,)) \div \text{idm}_k \qquad \text{EQ A.3.7A}$$
$$\text{for } k = 1, 2, \ldots, n \qquad \text{EQ A.3.7B}$$

Now this system starts at $(0, 0, \ldots, 0)$ if one wishes to start at a different index you have to translate down the index of the linear manifold ${}^{s}L$, to start at $(1, 1, \ldots, 1)$. In order to start-off at that point simply put the ordered index into equation A.3.5 and initiate that value for idl_p.

$$idl_p = \sum_{k=1}^{n} (1) \prod_{i=1}^{k-1} idm_i$$

and start here;

$$idl_p = \sum_{k=1}^{n} \prod_{i=1}^{k-1} idm_i \qquad\qquad \text{EQ A.3.8}$$

$$idx_k = ((idl + idl_p - 1) \setminus (\prod_{j=1}^{k-1} idm_j)) \div idm_k \qquad\qquad \text{EQ A.3.9}$$

for

$$idl = 1, \ldots, n^m \qquad\qquad \text{EQ A.3.10}$$

Compensating for the modulus of idm_k when zero the following statements must be added to the above equations:

If $idx_{k-1} = 0$ Then $idx_{k-1} = idm_k$: If $0 < idx_k$ Then $idx_k = idx_k - 1$ $\qquad\qquad$ EQ A.3.11

With this algorithm, all indices are accounted for and calculated.

Section A.4 *Computer Analysis: Building Tables of Long and Combined Indices*

This is a program to help with the analysis of delving into the principles of expanding a multinomial equation $_nM_m$. It is divided into four major parts adjacent to it contracting the long resulting index set, expediential index, count of common terms and finally collecting all counted terms. The program is developed using Visual Basic Microsoft Excel program.

```
    Option Base 1
    Option Explicit
    Const shtNum As String = "2"  '"2" '"3" '"4" '"5" '"6" 'm-products
    Const ntrms  As Integer = 3              'n-terms
'   Const cpars  As Integer = 32
    Const binsz  As Integer = 9              'n-index digits to search
    Const shtrw  As Integer = 1              'Shift construction
    Const HIDIT  As Boolean = False, SEEIT As Boolean = True
    Const MOVDN  As Boolean = False, MOVUP As Boolean = True

    Dim idn, idm           As Integer
    Dim npars              As Integer
'   Dim parseOnes(cpars) As String


    Type dig
        Bin(binsz) As Integer
        Hid(binsz) As Boolean
    End Type


    Type bitrm
        stridx As String
        strpwr As String
        ndigts As Integer
        ctally As Integer
    End Type

'^^^^^^^^^^^^^^^^^^^^^^^^Multinomial to Combinatorial^^^^^^^^^^^^^^^^^^^^^^^^
Private Sub Multinomial2Combinatorial()

    Dim sht1 As Range
    Dim rngRows As Integer
    Dim idu, idv, idw, idx, idy, idz As Integer
    Dim idm1, idm2, idm3, idm4, idm5, idm6, nterms As Integer
    Dim idl, idk, idc As Integer ', idh As Integer
    Dim tmpidx, tmppwr As String
    Dim Bucket As dig
    Dim source As bitrm
    Dim target As bitrm
```

```
    Worksheets("sheet" + shtNum).Activate  'can't select unless
                                           'the sheet is active
    Sheets("Sheet" + shtNum).Select
    Set sht1 = Range("A1:CT15625")
    sht1.Select
    rngRows = sht1.Rows.Count + 1

    idn = ntrms
    idm = Val(shtNum)

'^^^^^^^^^^^^^^^^^^^^^^^Construct Set of Long Indices^^^^^^^^^^^^^^^^^^^^^^^
'Positional Long Indices Algorithm
'Trans = 1 + idm1 + (idm1 * idm2) + (idm1 * idm2 * idm3) +
'                      (idm1 * idm2 * idm3 * idm4)
'idp = Trans
'
'idz2 = idp Mod idm1
'idy2 = (idp \ idm1) Mod idm2
'idx2 = (idp \ (idm1 * idm2)) Mod idm3
'idw2 = (idp \ (idm1 * idm2 * idm3)) Mod idm4
'idv2 = (idp \ (idm1 * idm2 * idm3 * idm4)) Mod idm5

'If idz2 = 0 Then idz2 = idm1: If 0 < idy2 Then idy2 = idy2 - 1
'If idy2 = 0 Then idy2 = idm2: If 0 < idx2 Then idx2 = idx2 - 1
'If idx2 = 0 Then idx2 = idm3: If 0 < idw2 Then idw2 = idw2 - 1
'If idw2 = 0 Then idw2 = idm4: If 0 < idv2 Then idv2 = idv2 - 1
'If idv2 = 0 Then idv2 = idm5
```

```
'^^^^^^^^^^^^^^^^^^^^Generate all possible long indices^^^^^^^^^^^^^^^^^^^
idm1 = idn
idm2 = idn
idm3 = idn
idm4 = idn
idm5 = idn
idm6 = idn
idl = 0

For idu = 1 To idm6
For idv = 1 To idm5
For idw = 1 To idm4
For idx = 1 To idm3
For idy = 1 To idm2
For idz = 1 To idm1

    sht1.Cells(idl + 1, 1).Value = Str$(idl + 1)
    'sht1.Cells(idl + 1, 2).Value = Str$(idy) + Str$(idz)
    'sht1.Cells(idl + 1, 2).Value = Str$(idx) + Str$(idy) + Str$(idz)
    'sht1.Cells(idl + 1, 2).Value = Str$(idw) + Str$(idx) + _
                                    Str$(idy) + Str$(idz)
    'sht1.Cells(idl + 1, 2).Value = Str$(idv) + Str$(idw) + _
    '                               Str$(idx) + Str$(idy) + Str$(idz)
    sht1.Cells(idl + 1, 2).Value = Str$(idu) + Str$(idv) + _
                                   Str$(idw) + Str$(idx) + _
                                   Str$(idy) + Str$(idz)

    idl = idl + 1

Next idz
Next idy
Next idx
Next idw
Next idv
Next idu
```

```
'^^^^^^^^^Contracting Long to Combinatorial Indices by Exponent^^^^^^^^^
npars = 2 ^ (idm - 1)
'If cpars < npars Then
'    MsgBox "Array out of bounds parseOnes(" + TrimUp(Str$(cpars)) + ")
actual value <" + _
'                    TrimUp(Str$(npars)) + ">", vbOKOnly + vbInformation,
"Multinomial"
'    Exit Sub
'End If

nterms = idn ^ idm
For idk = 1 To nterms
    tmpidx = ""
    tmppwr = ""
    Call LoadBin(Bucket, sht1.Cells(idk + shtrw, 2).Value)
    For idc = 1 To binsz
        If Not (0 = Bucket.Bin(idc)) Then
            tmpidx = tmpidx + Str$(idc)
            tmppwr = tmppwr + Str$(Bucket.Bin(idc))
        End If
    Next idc
    sht1.Cells(idk + shtrw, 3).Value = RTrim(tmpidx)
    sht1.Cells(idk + shtrw, 4).Value = RTrim(tmppwr)
Next idk

'^^^^^^^^^^Counting Terms by Common Indices and Exponents^^^^^^^^^^^^^
For idl = 1 To nterms
    If Not ("x" = Mid(sht1.Cells(idl + shtrw, 2).Value, 1, 1)) Then
        source.stridx = sht1.Cells(idl + shtrw, 3).Value
        source.strpwr = sht1.Cells(idl + shtrw, 4).Value
        source.ndigts = Len(TrimUp(sht1.Cells(idl + shtrw, 3).Value))
        source.ctally = 0
        For idk = 1 To nterms
            If Not ("x" = Mid(sht1.Cells(idk + shtrw, 2).Value, _
                    1, 1)) Then
                target.stridx = sht1.Cells(idk + shtrw, 3).Value
                target.strpwr = sht1.Cells(idk + shtrw, 4).Value
                target.ndigts = Len(TrimUp(sht1.Cells(idk + _
                                    shtrw, 3).Value))
                If CompareTrms(source, target) Then
                    source.ctally = source.ctally + 1
                    sht1.Cells(idk + shtrw, 2).Value = "x" + _
                    sht1.Cells(idk + shtrw, 2).Value 'Mark with x do _
                                                    'not check again
                End If
            End If
        Next idk
        sht1.Cells(idl + shtrw, 5).Value = source.ctally
    End If
Next idl
```

```
'^^^^^^^^^^^^^^^^^^^^^^^^Collect All Counted Terms^^^^^^^^^^^^^^^^^^^^^^^^^^^^
idk = 1 'Collect terms
For idl = 1 To nterms
    If Not (0 = sht1.Cells(idl + shtrw, 5).Value) Then
        sht1.Cells(idk + 1, 7).Value = sht1.Cells(idl + shtrw, 1).Value
        sht1.Cells(idk + 1, 8).Value = sht1.Cells(idl + shtrw, 2).Value
        sht1.Cells(idk + 1, 9).Value = sht1.Cells(idl + shtrw, 3).Value
        sht1.Cells(idk + 1, 10).Value = sht1.Cells(idl + shtrw, 4).Value
        sht1.Cells(idk + 1, 11).Value = sht1.Cells(idl + shtrw, 5).Value
        idk = idk + 1
    End If
Next idl

End Sub
```

```
'^^^^^^^^^^^^^^^^^^^^^^^^Calculate Power per Index^^^^^^^^^^^^^^^^^^^^^^^^^
Private Sub LoadBin(Bucket As dig, DigSource As String)
    Dim idk As Integer

    Call ZeroDig(Bucket)

    'Calculate the power per index
    For idk = 1 To Len(DigSource)
        Select Case Val(Mid(DigSource, idk, 1))
                Case 1
                    Bucket.Bin(1) = Bucket.Bin(1) + 1
                Case 2
                    Bucket.Bin(2) = Bucket.Bin(2) + 1
                Case 3
                    Bucket.Bin(3) = Bucket.Bin(3) + 1
                Case 4
                    Bucket.Bin(4) = Bucket.Bin(4) + 1
                Case 5
                    Bucket.Bin(5) = Bucket.Bin(5) + 1
                Case 6
                    Bucket.Bin(6) = Bucket.Bin(6) + 1
                Case 7
                    Bucket.Bin(7) = Bucket.Bin(7) + 1
                Case 8
                    Bucket.Bin(8) = Bucket.Bin(8) + 1
                Case 9
                    Bucket.Bin(9) = Bucket.Bin(9) + 1
        End Select
    Next idk

End Sub
'^^^^^^^^^^^^^^^^^^^^^^^^^Compare Multinomial Terms^^^^^^^^^^^^^^^^^^^^^^^^
Private Function CompareTrms(source As bitrm, target As bitrm) As
Boolean

        CompareTrms = True
        If Not (source.ndigts = target.ndigts) Then
            CompareTrms = False
            Exit Function
        End If
        If Not (0 = StrComp(source.stridx, target.stridx)) Then
            CompareTrms = False
            Exit Function
        End If
        If Not (0 = StrComp(source.strpwr, target.strpwr)) Then
            CompareTrms = False
            Exit Function
        End If

End Function
```

```
'^^^^^^^^^^^^^^^^^^^^^^^^^Zero Digit Data Type^^^^^^^^^^^^^^^^^^^^^^^^^
Private Sub ZeroDig(Flush As dig)
Dim idx As Integer
    For idx = 0 To binsz
        Flush.Bin(idx) = 0
        Flush.Hid(idx) = SEEIT
    Next idx
End Sub
'^^^^^^^^^^^^^^^^^^Delete all White Spaces in a String^^^^^^^^^^^^^^^^^^
Private Function TrimUp(CleanUp As String) As String
Dim idx As Integer
    TrimUp = ""
    For idx = 1 To Len(CleanUp)
        If Not (" " = Mid(CleanUp, idx, 1)) Then
            TrimUp = TrimUp + Mid(CleanUp, idx, 1)
        End If
    Next idx
End Function

'^^^^Alternately Insert White Spaces between Characters in a String^^^^
Private Function SpaceOut(FillUp As String) As String
Dim idx As Integer
    FillUp = TrimUp(FillUp)
    SpaceOut = " "
    For idx = 1 To Len(FillUp)
        SpaceOut = SpaceOut + Mid(FillUp, idx, 1) + " "
    Next idx
End Function
'^^^^^^^^^^^^Reflect Characters in a String about Its Center^^^^^^^^^^^^
Private Function Reflect(mirror As String) As String
Dim idx, lng As Integer
Dim refMirror As String
    lng = Len(mirror)
    Reflect = ""

    If lng < 2 Then
        Exit Function
    End If

    For idx = 1 To lng
        Reflect = Reflect + Mid(mirror, lng - idx + 1, 1)
    Next idx
End Function
```

```
'^^^^^^^^^^^^^^^^^^^^^^^^^^^^^^Factorial of N^^^^^^^^^^^^^^^^^^^^^^^^^^^^^^^
Private Function Factorial(n As Integer) As Integer
Dim idx, lng As Integer

    If n < 2 Then
        Factorial = 1
        Exit Function
    End If

    Factorial = 1
    For idx = 2 To n
        Factorial = Factorial * idx
    Next idx

End Function
'^^^^^^^^^^^^^^^^^^^^^^^^^^^^Combinatorial for Multinomial^^^^^^^^^^^^^^^^^^^^^
Private Function combinatorial(m As Integer, multiways As String)
                              As Integer
Dim idx, idt, OneFact As Integer
Dim mways As String

    mways = TrimUp(multiways) 'Test for NULL or blank string.
    If 0 = Len(mways) Then
        combinitoral = 0
        Exit Function
    End If

    If m < 2 Then 'Test for negative 0 or 1 argument for a factorial
        combinitoral = 1
        Exit Function
    End If

    OneFact = 0 'Test for all ones.
    For idx = 1 To Len(mways)
        OneFact = OneFact + (1 - Val(Mid(mways, idx, 1)))
    Next idx

    If 0 = OneFact Then
        combinitoral = 1
        Exit Function
    End If

    combinitoral = Factorial(m)
    For idx = 1 To Len(mways)
        combinitoral = combinitoral /
                      Factorial(Val(Mid(mways, idx, 1)))
    Next idx

End Function
```

The program is devoted to counting common terms, thereby computing the combinatorial coefficients and organizing them into common permuted terms. Such contracted terms are positional being marked by the component indices. So $[a_1a_2a_1a_3] \rightarrow [1,2,3,4]$ contracts to $[a_1{}^2a_2a_3] \rightarrow [1,2,3]$ for component $[a_1]$ position 1 and $[a_1]$ position 3 are the same, hence can be exponentially combined and $[1,2,3] \rightarrow [2,1,1]$.

The sorting algorithm is singular and as a result limited giving an index sorted set of terms. As such, it gives an uneven distribution of terms, yet accurately the count of all terms adds to n^m. The interesting thing is that it yields clear rules for organizing terms with common coefficients, thereby allowing distribution leaving a permuted spine of terms as can be seen in the following tables.

Section A.5 *Resulting Properties of Contracting Long Sets to Combinatorial*

The program, while it can generate all terms in any expanded combinatorial series, in itself is not a definitive mathematical proof, but explores and revels the fundamental properties of a combinatorial expansion. From the use of this program the following tables where generated for $_nM_m$ where n = 2, 3 and m = 1, 2, 3, 4, 5, 6.

Table A.5.1: $n^m \rightarrow {}_nM_m = {}_2M_2$ Combinatorial Terms

Count	Long Term Indices	Contracted Indices	Exponent Indices	Common Terms
1	1 1	1	2	1
2	1 2	1 2	1 1	2
4	2 2	2	2	1

Table A.5.2: $n^m \rightarrow {}_nM_m = {}_2M_3$ Combinatorial Terms

Count	Long Term Indices	Contracted Indices	Exponent Indices	Common Terms
1	1 1 1	1	3	1
2	1 1 2	1 2	2 1	3
4	1 2 2	1 2	1 2	3
8	2 2 2	2	3	1

Table A.5.3: $n^m \rightarrow {}_nM_m = {}_2M_4$ Combinatorial Terms

Count	Long Term Indices	Contracted Indices	Exponent Indices	Common Terms
1	1 1 1 1	1	4	1
2	1 1 1 2	1 2	3 1	4
4	1 1 2 2	1 2	2 2	6
8	1 2 2 2	1 2	1 3	4
16	2 2 2 2	2	4	1

Table A.5.4: $n^m \rightarrow {}_nM_m = {}_2M_5$ Combinatorial Terms

Count	Long Term Indices	Contracted Indices	Exponent Indices	Common Terms
1	1 1 1 1 1	1	5	1
2	1 1 1 1 2	1 2	4 1	5
4	1 1 1 2 2	1 2	3 2	10
8	1 1 2 2 2	1 2	2 3	10
16	1 2 2 2 2	1 2	1 4	5
32	2 2 2 2 2	2	5	1

Table A.5.5: $n^m \rightarrow {}_nM_m = {}_2M_6$ Combinatorial Terms

Count	Long Term Indices	Contracted Indices	Exponent Indices	Common Terms
1	1 1 1 1 1 1	1	6	1
2	1 1 1 1 1 2	1 2	5 1	6
4	1 1 1 1 2 2	1 2	4 2	15
8	1 1 1 2 2 2	1 2	3 3	20
16	1 1 2 2 2 2	1 2	2 4	15
32	1 2 2 2 2 2	1 2	1 5	6
64	2 2 2 2 2 2	2	6	1

What is of interest here is that the program yields the classical Pascal's triangle of coefficients affirming the finite binomial expansion equation LxE.3.5 and confidence that the computer program reflects the correct principles of how terms are counted, grouped and combined.

Table A.5.6: $n^m \rightarrow {}_nM_m = {}_3M_2$ Combinatorial Terms

Count	Term Indices	Contracted Indices	Exponent Indices	Common Terms
1	1 1	1	2	1
2	1 2	1 2	1 1	2
3	1 3	1 3	1 1	2
5	2 2	2	2	1
6	2 3	2 3	1 1	2
9	3 3	3	2	1

Sorting contracted indices by increasing order within the corresponding rising exponent products, lines up common terms with the same coefficients.

Table A.5.7: $n^m \rightarrow {}_nM_m = {}_3M_2$ Combinatorial Terms

Count	Long Term Indices	Contracted Indices	Exponent Indices	Common Terms
1	1 1	1	2	1
5	2 2	2	2	1
9	3 3	3	2	1
2	1 2	1 2	1 1	2
3	1 3	1 3	1 1	2
6	2 3	2 3	1 1	2

With the common coefficients aligned, they can be distributed out leaving the permuted spine of terms behind.

Table A.5.8: $n^m \rightarrow {}_nM_m = {}_3M_3$ Combinatorial Terms

Count	Long Term Indices	Contracted Indices	Exponent Indices	Common Terms
1	1 1 1	1	3	1
14	2 2 2	2	3	1
27	3 3 3	3	3	1
2	1 1 2	1 2	2 1	3
5	1 2 2	1 2	1 2	3
3	1 1 3	1 3	2 1	3
9	1 3 3	1 3	1 2	3
15	2 2 3	2 3	2 1	3
18	2 3 3	2 3	1 2	3
6	1 2 3	1 2 3	1 1 1	6

Further spine structure can be seen with ${}_3M_3$ where duplicated indices having constant exponents.

Table A.5.9: $n^m \rightarrow {}_nM_m = {}_3M_3$ Combinatorial Terms

Count	Long Term Indices	Contracted Indices	Exponent Indices	Common Terms
1	1 1 1	1	3	1
14	2 2 2	2	3	1
27	3 3 3	3	3	1
5	1 2 2	1 2	1 2	3
9	1 3 3	1 3	1 2	3
18	2 3 3	2 3	1 2	3
2	1 1 2	1 2	2 1	3
3	1 1 3	1 3	2 1	3
15	2 2 3	2 3	2 1	3
6	1 2 3	1 2 3	1 1 1	6

The duplicated indices can be organized into two increasing ordered groups of spines having constant exponents.

All of these rules for ordering sets of combinatorial terms can now be used to build the following tables:

Table A.5.10: $n^m \rightarrow {}_nM_m = {}_3M_4$ Combinatorial Terms

Count	Long Term Indices	Contracted Indices	Exponent Indices	Common Terms
1	1 1 1 1	1	4	1
41	2 2 2 2	2	4	1
81	3 3 3 3	3	4	1
14	1 2 2 2	1 2	1 3	4
27	1 3 3 3	1 3	1 3	4
54	2 3 3 3	2 3	1 3	4
2	1 1 1 2	1 2	3 1	4
3	1 1 1 3	1 3	3 1	4
42	2 2 2 3	2 3	3 1	4
5	1 1 2 2	1 2	2 2	6
9	1 1 3 3	1 3	2 2	6
45	2 2 3 3	2 3	2 2	6
18	1 2 3 3	1 2 3	1 1 2	12
15	1 2 2 3	1 2 3	1 2 1	12
6	1 1 2 3	1 2 3	2 1 1	12

Table A.5.11: $n^m \rightarrow {}_nM_m = {}_3M_5$ Combinatorial Terms

Count	Long Term Indices	Contracted Indices	Exponent Indices	Common Terms
1	1 1 1 1 1	1	5	1
122	2 2 2 2 2	2	5	1
243	3 3 3 3 3	3	5	1
41	1 2 2 2 2	1 2	1 4	5
81	1 3 3 3 3	1 3	1 4	5
162	2 3 3 3 3	2 3	1 4	5
2	1 1 1 1 2	1 2	4 1	5
3	1 1 1 1 3	1 3	4 1	5
123	2 2 2 2 3	2 3	4 1	5
14	1 1 2 2 2	1 2	2 3	10
27	1 1 3 3 3	1 3	2 3	10
135	2 2 3 3 3	2 3	2 3	10
5	1 1 1 2 2	1 2	3 2	10
9	1 1 1 3 3	1 3	3 2	10
126	2 2 2 3 3	2 3	3 2	10
54	1 2 3 3 3	1 2 3	1 1 3	20
42	1 2 2 2 3	1 2 3	1 3 1	20
6	1 1 1 2 3	1 2 3	3 1 1	20
45	1 2 2 3 3	1 2 3	1 2 2	30
18	1 1 2 3 3	1 2 3	2 1 2	30
15	1 1 2 2 3	1 2 3	2 2 1	30

Table A.5.12: $n^m \rightarrow {}_nM_m = {}_3M_6$ Combinatorial Terms

Count	Long Term Indices	Contracted Indices	Exponent Indices	Common Terms
1	1 1 1 1 1 1	1	6	1
365	2 2 2 2 2 2	2	6	1
729	3 3 3 3 3 3	3	6	1
122	1 2 2 2 2 2	1 2	1 5	6
243	1 3 3 3 3 3	1 3	1 5	6
486	2 3 3 3 3 3	2 3	1 5	6
2	1 1 1 1 1 2	1 2	5 1	6
3	1 1 1 1 1 3	1 3	5 1	6
366	2 2 2 2 2 3	2 3	5 1	6
41	1 1 2 2 2 2	1 2	2 4	15
81	1 1 3 3 3 3	1 3	2 4	15
405	2 2 3 3 3 3	2 3	2 4	15
5	1 1 1 1 2 2	1 2	4 2	15
9	1 1 1 1 3 3	1 3	4 2	15
369	2 2 2 2 3 3	2 3	4 2	15
14	1 1 1 2 2 2	1 2	3 3	20
27	1 1 1 3 3 3	1 3	3 3	20
378	2 2 2 3 3 3	2 3	3 3	20
162	1 2 3 3 3 3	1 2 3	1 1 4	30
123	1 2 2 2 2 3	1 2 3	1 4 1	30
6	1 1 1 1 2 3	1 2 3	4 1 1	30
135	1 2 2 3 3 3	1 2 3	1 2 3	60
126	1 2 2 2 3 3	1 2 3	1 3 2	60
54	1 1 2 3 3 3	1 2 3	2 1 3	60
42	1 1 2 2 2 3	1 2 3	2 3 1	60
18	1 1 1 2 3 3	1 2 3	3 1 2	60
15	1 1 1 2 2 3	1 2 3	3 2 1	60
45	1 1 2 2 3 3	1 2 3	2 2 2	90

Table A.5.13: $n^m \rightarrow {}_nM_m = {}_4M_5$ Combinatorial Terms

Count	Long Term Indices	Contracted Indices	Exponent Indices	Common Terms
1	1 1 1 1 1	1	5	1
342	2 2 2 2 2	2	5	1
683	3 3 3 3 3	3	5	1
1024	4 4 4 4 4	4	5	1
86	1 2 2 2 2	1 2	1 4	5
171	1 3 3 3 3	1 3	1 4	5
256	1 4 4 4 4	1 4	1 4	5
427	2 3 3 3 3	2 3	1 4	5
512	2 4 4 4 4	2 4	1 4	5
768	3 4 4 4 4	3 4	1 4	5
2	1 1 1 1 2	1 2	4 1	5
3	1 1 1 1 3	1 3	4 1	5
4	1 1 1 1 4	1 4	4 1	5
343	2 2 2 2 3	2 3	4 1	5
344	2 2 2 2 4	2 4	4 1	5
684	3 3 3 3 4	3 4	4 1	5
22	1 1 2 2 2	1 2	2 3	10
43	1 1 3 3 3	1 3	2 3	10
64	1 1 4 4 4	1 4	2 3	10
363	2 2 3 3 3	2 3	2 3	10
384	2 2 4 4 4	2 4	2 3	10
704	3 3 4 4 4	3 4	2 3	10
6	1 1 1 2 2	1 2	3 2	10
11	1 1 1 3 3	1 3	3 2	10
16	1 1 1 4 4	1 4	3 2	10
347	2 2 2 3 3	2 3	3 2	10
352	2 2 2 4 4	2 4	3 2	10
688	3 3 3 4 4	3 4	3 2	10
107	1 2 3 3 3	1 2 3	1 1 3	20
128	1 2 4 4 4	1 2 4	1 1 3	20
192	1 3 4 4 4	1 3 4	1 1 3	20
448	2 3 4 4 4	2 3 4	1 1 3	20
87	1 2 2 2 3	1 2 3	1 3 1	20
88	1 2 2 2 4	1 2 4	1 3 1	20
172	1 3 3 3 4	1 3 4	1 3 1	20
428	2 3 3 3 4	2 3 4	1 3 1	20
7	1 1 1 2 3	1 2 3	3 1 1	20
8	1 1 1 2 4	1 2 4	3 1 1	20
12	1 1 1 3 4	1 3 4	3 1 1	20
348	2 2 2 3 4	2 3 4	3 1 1	20
91	1 2 2 3 3	1 2 3	1 2 2	30
96	1 2 2 4 4	1 2 4	1 2 2	30
176	1 3 3 4 4	1 3 4	1 2 2	30
432	2 3 3 4 4	2 3 4	1 2 2	30

Count	Long Term Indices	Contracted Indices	Exponent Indices	Common Terms
27	1 1 2 3 3	1 2 3	2 1 2	30
32	1 1 2 4 4	1 2 4	2 1 2	30
48	1 1 3 4 4	1 3 4	2 1 2	30
368	2 2 3 4 4	2 3 4	2 1 2	30
23	1 1 2 2 3	1 2 3	2 2 1	30
24	1 1 2 2 4	1 2 4	2 2 1	30
44	1 1 3 3 4	1 3 4	2 2 1	30
364	2 2 3 3 4	2 3 4	2 2 1	30
112	1 2 3 4 4	1 2 3 4	1 1 1 2	60
108	1 2 3 3 4	1 2 3 4	1 1 2 1	60
92	1 2 2 3 4	1 2 3 4	1 2 1 1	60
28	1 1 2 3 4	1 2 3 4	2 1 1 1	60

Definition A.5.1 **Binomial Coefficient** [PAR60 pg 36]

$$_nC_m \equiv \frac{n!}{m!(n-m)!} \quad \text{for}$$

EQ A.5.1A

$$0! \equiv 1$$

EQ A.5.1B

Definition A.5.2 **Permutation Coefficient** [PAR60 pg 35]

$$_nP_s \equiv \frac{n!}{(n-s)!}$$

EQ A.5.2A

or substituting EQ A.5.1 into EQ A.5.5A

$$_nP_s = s!\ _nC_s$$

EQ A.5.2B

Observation A.5.1: Sorting Coefficients and Spines

Step 1) Order contracted indices in increasing order [$j_i < j_{(i+1)}$] for i = 1, 2, ... s, lines up common terms with the same coefficients. As an example [$(1)_1 < (3)_2]_8$ lines up coefficients of [6] see $_3M_6$.

Step 2) Duplicated indices can be separated into two increasing ordered groups of spines having **constant** exponents [$\prod k_i = p]_k$ and k + 1. As an example [$(1)_1 < (3)_2$ and $(5)_1$ x $(1)_2 = 5]_6 < [(1)_1 < (3)_2$ and $(1)_1$ x $(5)_2 = 5]_7$

Observation A.5.2: Parsing by Ones

From $_3M_6$ the sum of any given exponential indices always add to 6:

6	15	51	24	42	33	114	141	411	123	132	213	231	321	312	222

The process of contracting starts with a fixed number of components-m, hence ideally each component before contraction has an exponent of one. After contraction, the set of exponents will be less by the number of contractions [c], but must always add to the original number of components-m. Essentially what the contraction is doing is taking the set of exponential ones and parsing them in all possible permutations starting with [$(m)_1$] and ending with [$(1)_1,(1)_2,(1)_3, ..., (1)_n$] m-ones. In general given the contracted subset of exponential indices [$k_1,k_2,k_3, ..., k_{(m-c)}$] add to m:

$$m \equiv k_1 + k_2 + k_3 + ... + k_s$$

EQ A.5.2A

$$s \equiv m - c \qquad \text{the number of exponent digits}$$

EQ A.5.2B

Observation A.5.3: Counting Parsed Ones

From $_3M_6$ develops only the first row of exponential indices, but actually it can be extended to its full length completing all possible parses of ones:

6	15	51	24	42	33	114	141	411	123	132	213	231	321	312	222
1	1	2	3	4	5	1	2	3	4	5	6	7	8	9	10

1122	1212	1221	2121	2112	2211	1113	1131	1311	3111
1	2	3	4	5	6	7	8	9	10

11112	11121	11211	12111	21111	111111
1	2	3	4	5	1

The question is now how many ways can the exponential ones be parsed? This gives a new perspective in organizing the expansion. Since $_3M_6$ gives a large number of parsed ones its seen that it starts with an exponential set of 1 digit comprised of 3 sets, 2 digits comprised of 5 sets, 3 digits comprised of 10 sets, 4 digits comprised of 10 sets and finally 5 digits comprised of 5 sets and 6 digits comprised of 1 set. So the total number of parsed ones are:

$$1 + 5 + 10 + 10 + 5 + 1 = 32$$

This looks familiar, but lets first build a table of $n^m \rightarrow {}_nM_m = {}_5M_m$ for m = 1,2,3,4, and 5.

Table A.5.14: Counting Parsed Ones for m = 1,2,3,4,5,6,7,8

m / s	1	2	3	4	5	6	7	8	=	Tally
1	1								=	1
2	1	1							=	2
3	1	2	1						=	4
4	1	3	3	1					=	8
5	1	4	6	4	1				=	16
6	1	5	10	10	5	1			=	32
7	1	6	15	20	15	6	1		=	64
8	1	7	21	35	35	21	7	1	=	128

Clearly, these coefficients are from Pascal's Triangle. So, the above table can be rewritten as:

Table A.5.15: Counting Parsed Ones for m = 1, … , 8 by Number of Digits-s

m \ s	1	2	3	4	5	6	7	8	=	Tally
1	$_0C_0$								=	2^0
2	$_1C_0$	$_1C_1$							=	2^1
3	$_2C_0$	$_2C_1$	$_2C_2$						=	2^2
4	$_3C_0$	$_3C_1$	$_3C_2$	$_3C_3$					=	2^3
5	$_4C_0$	$_4C_1$	$_4C_2$	$_4C_3$	$_4C_4$				=	2^4
6	$_5C_0$	$_5C_1$	$_5C_2$	$_5C_3$	$_5C_4$	$_5C_5$			=	2^5
7	$_6C_0$	$_6C_1$	$_6C_2$	$_6C_3$	$_6C_4$	$_6C_5$	$_6C_6$		=	2^6
8	$_7C_0$	$_7C_1$	$_7C_2$	$_7C_3$	$_7C_4$	$_7C_5$	$_7C_6$	$_7C_7$	=	2^7

From T3.11.6 these coefficients add up to the total number of possible ways to parse ones by the Polynomial Identity: Binomial-2^n and can be generally written:

$$2^{(m-1)} = \sum_{s=1}^{m} {}_{(m-1)}C_{(s-1)}$$

EQ A.5.3A

The number of ways to parse m-ones given k exponents is

$$_mN_s \equiv {}_{(m-1)}C_{(s-1)}$$

EQ A.5.3B

Alternate definition for multiplying factor
$$Q(m, s) \equiv {}_mN_s$$

EQ A.5.3C

Observation A.5.4: Counting Permutated Spine Terms

$_nM_m$\s	1	2	3	4	5	6	7
$_2M_2$	2	1					
$_2M_3$	2	1					
$_2M_4$	2	1					
$_2M_5$	2	1					
$_2M_6$	2	1					
$_3M_2$	3	3					
$_3M_3$	3	3	1				
$_3M_4$	3	3	1				
$_3M_5$	3	3	1				
$_3M_6$	3	3	1				
$_4M_5$	4	6	4	1			
$_5M_?$	5	10	NC	NC	1		
$_6M_?$	6	15	NC	NC	NC	1	
$_7M_?$	7	21	NC	NC	NC	NC	1

$_nM_m$\s	1	2	3	4	5	6	7
$_2M_2$	$_2C_1$	$_2C_2$					
$_2M_3$	$_2C_1$	$_2C_2$					
$_2M_4$	$_2C_1$	$_2C_2$					
$_2M_5$	$_2C_1$	$_2C_2$					
$_2M_6$	$_2C_1$	$_2C_2$					
$_3M_2$	$_3C_1$	$_3C_2$	$n \le s$				
$_3M_3$	$_3C_1$	$_3C_2$	$_3C_3$				
$_3M_4$	$_3C_1$	$_3C_2$	$_3C_3$				
$_3M_5$	$_3C_1$	$_3C_2$	$_3C_3$				
$_3M_6$	$_3C_1$	$_3C_2$	$_3C_3$				
$_4M_5$	$_4C_1$	$_4C_2$	$_4C_3$	$_4C_4$			
$_5M_?$	$_5C_1$	$_5C_2$	NC	NC	$_5C_5$		
$_6M_?$	$_6C_1$	$_6C_2$	NC	NC	NC	$_6C_6$	
$_7M_?$	$_7C_1$	$_7C_2$	NC	NC	NC	NC	$_7C_7$

From combinatorial tables $_nC_s$, [BEY88, pg 62], also note calculation of combinatorials are defined only for $m \ge s$ other wise they are independent of [m]. Where Not Calculated is [NC]. In general, the equation for calculating the permutated spine terms are given by:

$$_nS_s \equiv S(n, s) \begin{cases} _nC_s & \text{for } n \ge s & \text{EQ A.5.4A} \\ \text{undefined} & \text{for } n < s & \text{EQ A.5.4B} \end{cases}$$

Observation A.5.5: Counting All Terms by Number of Digits-s

The total number of terms [$_nT_m$] that can be calculated from EQ A.5.5 is summarized by using Table-A.5.15 in the following table:

Table A.5.16: Counting Terms by Multiplying Factor $_mN_s$

$_nM_m \setminus s$	1	2	3	4
$_3M_1$	1			
$_3M_2$	1	1		
$_3M_3$	1	2	1	
$_3M_4$	1	3	3	
$_3M_5$	1	4	6	
$_3M_6$	1	5	10	
$_4M_5$	1	4	6	4
$_nM_m \setminus s$	1	2	3	4
$_3M_1$	$_0C_0$			
$_3M_2$	$_1C_0$	$_1C_1$		
$_3M_3$	$_2C_0$	$_2C_1$	$_2C_2$	
$_3M_4$	$_3C_0$	$_3C_1$	$_3C_2$	
$_3M_5$	$_4C_0$	$_4C_1$	$_4C_2$	
$_3M_6$	$_5C_0$	$_5C_1$	$_5C_2$	
$_4M_5$	$_4C_0$	$_4C_1$	$_4C_2$	$_4C_3$

So, the above table provides the multiplier factor $_mN_s$ for the number of terms and when multiplied by the number of permutated spine terms gives the total number of terms, $_nT_m$.

total terms can be computed by

$$_nT_m \equiv \sum_{s=1}^{n} {}_{(m-1)}C_{(s-1)}\, {}_nC_s \qquad \text{EQ A.5.5A}$$

or substituting EQ A.5.3B

$$_nT_m \equiv \sum_{s=1}^{n} {}_mN_s\, {}_nC_s \qquad \text{EQ A.5.5B}$$

Observation A.5.6: Calculating Coefficients

From the above table of coefficients for $_3M_{1,2,3,4,5}$ the multinomial numbers can be shown to be equivalent:

Table A.5.17: Comparison of Parsed Exponential Term for $_3M_{1,2,3,4,5}$

Count	1
Exponents	1
Coefficients	1

Count	1	2
Exponents	2	11
Coefficients	1	2

Count	1	2	3	4
Exponents	3	12	21	111
Coefficients	1	3	3	6

Count	1	2	3	4	5	6	7
Exponents	4	13	31	22	112	121	211
Coefficients	1	4	4	6	12	12	12

Count	1	2	3	4	5	6	7	8	9	10	11
Exponents	5	14	41	23	32	113	131	311	122	212	221
Coefficients	1	5	5	10	10	20	20	20	30	30	30

$_1C_{[1]} = 1$

$_2C_{[2]} = 1;\ _2C_{[1,1]} = 2$

$_3C_{[3]} = 1;\ _3C_{[1,2]} = _3C_{[2,1]} = 3;\ _3C_{[1,1,1]} = 6$

$_4C_{[4]} = 1;\ _4C_{[1,3]} = _4C_{[3,1]} = 4;\ _4C_{[2,2]} = 6;\ _4C_{[1,1,2]} = _4C_{[1,2,1]} = _4C_{[2,1,1]} = 12$

$_5C_{[5]} = 1;\ _5C_{[1,4]} = _5C_{[4,1]} = 5;\ _5C_{[2,3]} = _5C_{[3,2]} = 10;\ _5C_{[1,1,3]} = _5C_{[1,3,1]} = _5C_{[3,1,1]} = 20;$
$$_5C_{[1,2,2]} = _5C_{[2,1,2]} = _5C_{[2,2,1]} = 30$$

This bears out for all $_nM_m$ constructed by the computer program. In general, it can be extrapolated that the coefficients of a multinomial expansion is simply the multinomial coefficient itself, $_mC_{K(s,\ t)}$. This is also boron out from Parzen in his development of his formal equation for the expansion of the multinomial equation, [PAR60 pg 40].

Observation A.5.7: Termination of Terms

Table A.5.18: Comparison of Parsed Exponential Term for $n^m \rightarrow {}_nM_m = {}_3M_{1,2,3,4,5}$

${}_3M_1$: Count	1
Exponents	1
Coefficients	1

${}_3M_2$: Count	1	2
Exponents	2	11
Coefficients	1	2

${}_3M_3$: Count	1	2	3	4
Exponents	3	12	21	111
Coefficients	1	3	3	6

${}_3M_4$: Count	1	2	3	4	5	6	7	8
Exponents	4	13	31	22	112	121	211	1111
Coefficients	1	4	4	6	12	12	12	0

${}_3M_5$: Count	1	2	3	4	5	6	7	8	9	10	11	12	13	14	15	16
Exponents	5	14	41	23	32	113	131	311	122	212	221	1112	1121	1211	2111	11111
Coefficients	1	5	5	10	10	20	20	20	30	30	30	0	0	0	0	0

${}_3M_6$: Count	1	2	3	4	5	6	7	8	9	10	11	12	13	14	15	16
Exponents	6	15	51	24	42	33	114	141	411	123	132	213	231	321	312	222
Coefficients	1	6	6	15	15	20	30	30	30	60	60	60	60	60	60	90

Count	17	18	19	20	21	22	23	24	25	26	27	28	29	30
Exponents	1122	1212	1221	2121	2112	2211	1113	1131	1311	3111	11112	11121	11211	12111
Coefficients	0	0	0	0	0	0	0	0	0	0	0	0	0	0

Count	31	32
Exponents	21111	111111
Coefficients	0	0

Clearly, conditions on the Multinomial Coefficient constrain it to being set to zero, but what conditions? Certainly, n^m limits the number of terms for an upper limit, but combined with permuted spine terms comprise a lower limit. From Observation A.5.4, "Counting Permutated Spine Terms" shows that it is zero when $m < s$ and $s = n$ for $m < n$. Do to m being less than n limits the number of possible terms.

From Observation A.5.6 the Multinomial Coefficient is constrain to an upper limit the number of permuted spine terms, which are independent of the multinomial expansion being true for an number of terms and permutated to a specified number of digits. Here $n < m \leq s$ and $n < s$ so no terms can be found that can be categorized with the appropriate ordering of exponents, hence the number of terms are zero. This is not unexpected since when $s = m$ there is exactly as many digits as possible parsed exponents, hence no contraction can occur as seen with $n = m = 3$ in the above table and any $m < s$ no terms exist to be sorted and as such zero:

$$_mC_{\kappa(s,\,t)} \neq 0 \qquad \text{for } s \leq n \qquad\qquad\qquad\qquad \text{EQ A.5.4A}$$
$$_mC_{\kappa(s,\,t)} = 0 \qquad \text{for } (m < s \text{ and } s = n) \text{ or } (n < m \leq s \text{ and } n < s) \qquad \text{EQ A.5.4B}$$

This can be affirmed from EQA.5.4B as can be seen $_nC_s \sim 1 / (n - s)!$ and when $s > n$ the factorial is undefined, hence can not exist, physically a permutation cannot be constructed for digits greater than the number to be permutated. This than is the limitation that forces the multinomial coefficient to be set to zero, since the permutated spine terms form the backbone of the multinomial expansion this than becomes a major limiting factor.

Allowing for zero modifies the multinomial coefficient, which is improper and a new number needs to be defined, hence

$$_mK_{\kappa(s,\,t)} \equiv \begin{cases} _mC_{\kappa(s,t)} & \text{for} \qquad\qquad\qquad\qquad\quad s \leq n \\ 0 & \text{for} \quad (m < s \text{ and } s = n) \text{ or } (n < m \leq s \text{ and } n < s) \end{cases} \qquad \text{EQ A.5.4C}$$

Observation A.5.8: The Permutated Spine and Expediential Sets

Consider $_4M_5$ and the following comparison Table:

$\zeta(1, 1) = \{(1)_1, (2)_2, (3)_3, ([4]_{1=v})_{3=u}\}$ $\longleftrightarrow$ $\kappa(1, 1) = \{[5]_{1=v}\}$

$\zeta(2, 1) = \{(1, 2)_1, (1, 3)_2, (1, 4)_3, (2, 3)_4, (2, 4)_5, ([3]_1, [4]_{2=v})_{6=u}\}$ $\longleftrightarrow$ $\kappa(2, 1) = \{[1]_1, [4]_{2=v}\}$

$\zeta(2, 2) = \{(1, 2)_1, (1, 3)_2, (1, 4)_3, (2, 3)_4, (2, 4)_5, (3, 4)_6\}$ $\longleftrightarrow$ $\kappa(2, 2) = \{4, 1\}$

$\zeta(2, 3) = \{(1, 2)_1, (1, 3)_2, (1, 4)_3, (2, 3)_4, (2, 4)_5, (3, 4)_6\}$ $\longleftrightarrow$ $\kappa(2, 3) = \{2, 3\}$

$\zeta(2, 4) = \{(1, 2)_1, (1, 3)_2, (1, 4)_3, (2, 3)_4, (2, 4)_5, (3, 4)_6\}$ $\longleftrightarrow$ $\kappa(2, 4) = \{3, 2\}$

$\zeta(3, 1) = \{(1, 2, 3)_1, (1, 2, 4)_2, (1, 3, 4)_3, (2, 3, 4)_{4=u}\}$ $\longleftrightarrow$ $\kappa(3, 1) = \{1, 1, 3\}$

$\zeta(3, 2) = \{(1, 2, 3)_1, (1, 2, 4)_2, (1, 3, 4)_3, (2, 3, 4)_4\}$ $\longleftrightarrow$ $\kappa(3, 2) = \{1, 3, 1\}$

$\zeta(3, 3) = \{(1, 2, 3)_1, (1, 2, 4)_2, (1, 3, 4)_3, (2, 3, 4)_4\}$ $\longleftrightarrow$ $\kappa(3, 3) = \{3, 1, 1\}$

$\zeta(3, 4) = \{(1, 2, 3)_1, (1, 2, 4)_2, (1, 3, 4)_3, (2, 3, 4)_4\}$ $\longleftrightarrow$ $\kappa(3, 4) = \{1, 2, 2\}$

$\zeta(3, 5) = \{(1, 2, 3)_1, (1, 2, 4)_2, (1, 3, 4)_3, (2, 3, 4)_4\}$ $\longleftrightarrow$ $\kappa(3, 5) = \{2, 1, 2\}$

$\zeta(3, 6) = \{(1, 2, 3)_1, (1, 2, 4)_2, (1, 3, 4)_3, (2, 3, 4)_4\}$ $\longleftrightarrow$ $\kappa(3, 6) = \{2, 2, 1\}$

$\zeta(4, 1) = \{(1, 2, 3, 4)_{1=u}\}$ $\longleftrightarrow$ $\kappa(4, 1) = \{1, 1, 1, 2\}$

$\zeta(4, 2) = \{(1, 2, 3, 4)_1\}$ $\longleftrightarrow$ $\kappa(4, 2) = \{1, 1, 2, 1\}$

$\zeta(4, 3) = \{(1, 2, 3, 4)_1\}$ $\longleftrightarrow$ $\kappa(4, 3) = \{1, 2, 1, 1\}$

$\zeta(4, 4) = \{(1, 2, 3, 4)_1\}$ $\longleftrightarrow$ $\kappa(4, 4) = \{2, 1, 1, 1\}$

$\zeta(s, t) = \{(\gamma[s, t, u, v],), \}$ $\longleftrightarrow$ $\kappa(s, t) = \{\eta[s, t, v], \}$

$$_4M_5 = \sum_{s=1}^{4} \sum_{t=1}^{Q(5, s)} {}_5C_{\kappa(s, t)} \sum_{u=1}^{S(4, s)} \prod_{v=1}^{s} a_{\gamma[s, t, u, v]}{}^{\eta[s, t, v]}$$ EQ A.5.8

where
$s = 1, \ldots, n$	Digit Size	EQ A. 5.8C
$t = 1, \ldots, Q(m, s)$	Multiple Spine	EQ A. 5.8D
$u = 1, \ldots, S(n, s)$	Spine Element	EQ A. 5.8E
$v = 1, \ldots, s$	Exponent	EQ A. 5.8F

Section A.6 : *Properties and Axioms Associated with Multinomial Expansion*

Definition A.6.1 **Multinomial Coefficient** [PAR60 pg 40]

The number of ways in which one can partition a set of size [m] into [s] ordered subsets so that the first subset has size k_1, the second subset has size k_2, an so on, where $m = \sum_{i=1}^{s} k_i$ yields the following quantity known as a multinomial product. Such quantities arise frequently, and a special notation is introduced to denote them:

$$_mC_{[\bullet ki\bullet]} \equiv \begin{pmatrix} m \\ \bullet\, k_i\, \bullet \end{pmatrix} \equiv \frac{m!}{\prod_{i=1}^{s} (k_i)!} \text{ for} \qquad\qquad \text{EQ A.6.1A}$$

$$s = 1, 2, ..., n \qquad\qquad\qquad \text{EQ A.6.1B}$$

From Observations: OA.5.1 to 8 the base principles of count of digits can be extrapolated to construct the following axiom:

Axiom A.6.1 Digit Count Set

$\exists$ an ordered set of number of digits counted per product factors. Being an ordered set these; counts have a one-to-one correspondence to the number of ways of parsing ones:

$$\zeta(s, t) \equiv \{ s, t \,|\, \ni,$$

$q_u = \gamma[s, t, u, v]$ Permutated Index EQ A.6.1A

such that

$1 \leq q_{u < q_{u+1}} \leq m$ EQ A.6.1B

for $q_u \in$ IG and indices are

$s = 1, ..., n$ Digit Size EQ A.6.1C

$t = 1, ..., Q(m, s)$ Multiple Spine EQ A.6.1D

$u = 1, ..., S(n, s)$ Spine Element EQ A.6.1E

$v = 1, ..., s$ Exponent EQ A.6.1F

$\}$

Axiom A.6.2 Parsed Exponential Set

$\exists$ a set of parsed exponentials created by parsing ones in an ordered combinatorial count:

$$\kappa(s, t) \equiv \{ s, t \,|\, \bullet\, k_v\, \bullet, \ni, m = \sum_v^s k_v,$$ EQ A.6.2A

where

$k_v \equiv \eta[s, t, v]$ Exponent EQ A.6.2B

$1 \leq k_v \leq m$ EQ A.6.2C

for $k_v \in$ IG and indices are

$s = 1, ..., n$ Digit Size EQ A.6.1D

$t = 1, ..., Q(m, s)$ Multiple Spine EQ A.6.1E

$v = 1, ..., s$ Exponent EQ A.6.1F

$\}$

From Observation: OA.5.8 and EQ A.5.8

Axiom A.6.3 Summation Notation over Combinatorial Indices

$$_nM_m \equiv \sum_{s=1}^{n} \sum_{t=1}^{Q(m, s)} {}_mK_{\kappa(s, t)} \sum_{u=1}^{S(n, s)} \prod_{v=1}^{s} a_{\gamma[s, t, u, v]}{}^{\eta[s, t, v]}$$

AA.6.3 summation is actually the summation over the combinatorial set C_r and its properties specified in the following axiom.

Axiom A.6.4 Combinatorial Set C_r

$$C_r \equiv \{ \; r \mid \text{summed over the sets } \zeta(s, t) \wedge \kappa(s, t) \text{ of } r = {}_nT_m, \text{ terms where}$$

$s = 1, \ldots, n$	Digit Size	EQ A.6.4A
$t = 1, \ldots, Q(m, s)$	Multiple Spine	EQ A.6.4B

$$\}$$

Theorem A.6.1 Partition Contraction into the Multinomial Coefficient

1g	Given	${}_nC_{k1\ (n-k1)}C_{k2\ (n-k1-k2)}C_{k3} \cdots {}_{(n-k1-k2-\ldots-k(r-1))}C_{kr}$
Steps	Hypothesis	${}_nC_{k1\ (n-k1)}C_{k2\ (n-k1-k2)}C_{k3} \cdots {}_{(n-k1-k2-\ldots-k(r-1))}C_{kr} \equiv {}_mC_{[\bullet ki\bullet]}$
1	From 1g	${}_mC_{k1} \qquad {}_{(m-k1)}C_{k2} \qquad {}_{(m-k1-\ldots-k(r-1))}C_{kr}$
2	From 1 and DA.5.1 Binomial Coefficient	$\dfrac{m!}{(m-k_1)!\,k_1!} \qquad \dfrac{(m-k_1)!}{(m-k_1-k_2)!\,k_2!} \qquad \dfrac{(m-k_1-\ldots-k_{(r-1)})!}{(m-k_1-k_2-\ldots-k_r)!\,k_r!}$
3	From 2, A4.2.13 Inverse Multp and T4.4.9 Equalities: Any Quantity Divided by One is that Quantity	$\dfrac{m!}{k_1!} \qquad \dfrac{1}{k_2!} \qquad \dfrac{1}{(m-k_1-k_2-\ldots-k_r)!\,k_r!}$
4	From 3 and AA.6.2 Parsed Exponential Set	$\dfrac{m!}{k_1!} \qquad \dfrac{1}{k_2!} \qquad \dfrac{1}{(m-m)!\,k_r!}$
5	From 4 and A4.2.8 Inverse Add	$\dfrac{m!}{k_1!} \qquad \dfrac{1}{k_2!} \qquad \dfrac{1}{(0)!\,k_r!}$
6	From 5 and DA.5.1B Binomial Coefficient	$\dfrac{m!}{k_1!} \qquad \dfrac{1}{k_2!} \qquad \dfrac{1}{1\,k_r!}$
7	From 6 and T4.5.4 Equalities: Product of Two Rational Fractions	$\dfrac{m!}{k_1!\ k_2!\ \ldots\ k_r!}$
$\therefore$	From 1, 7 and DA.6.1 Multinomial Coefficient	${}_nC_{k1\ (n-k1)}C_{k2\ (n-k1-k2)}C_{k3} \cdots {}_{(n-k1-k2-\ldots-k(r-1))}C_{kr} \equiv {}_mC_{[\bullet ki\bullet]}$

Observation A.6.1: Origins of Counting Terms for a Multinomial Expansion

From the multinomial expansion, EQA.1.1B the number of total terms is n^m, which must hold even after the foreshorten contraction of the new count.

$_nM_m \backslash s$											
$_3M_1$	1										
	3										
$3^1 = 1$	1										
$_3M_2$	2	1 1									
	3	3									
$3^2 = 9$	1	2									
$_3M_3$	3	1 2	2 1	1 1 1							
	3	3	3	1							
$3^3 = 27$	1	3	3	6							
$_3M_4$	4	1 3	3 1	2 2	1 1 2	1 2 1	2 1 1				
	3	3	3	3	3	0	0				
$3^4 = 81$	1	4	4	6	12	12	12				
$_3M_5$	5	1 4	4 1	2 3	3 2	1 1 3	1 3 1	3 1 1	1 2 2	2 1 2	2 2 1
	3	3	3	3	3	2	2	2	1	0	0
$3^5 = 243$	1	5	5	10	10	20	20	20	30	30	30

$_nM_m \backslash s$											
$_4M_5$	5	1 4	4 1	2 3	3 2	1 1 3	1 3 1	3 1 1	1 2 2	2 1 2	2 2 1
	4	4	4	4	4	4	4	4	4	4	4
	1	5	5	10	10	20	20	20	30	30	30
	1 1 1 2	1 1 2 1	1 2 1 1	2 1 1 1							
	1	0	0	0							
$4^5 = 1024$	60	60	60	60							

As seen from $_4M_5$ an expanded representation leads to a generalization for finding the total of all terms:

$1 \times 4 + 5 \times 12 + 10 \times 12 + 20 \times 12 + 30 \times 12 + 60 \times 4$

$4 + 12 \times (5 + 10 + 20 + 30) + 240$

$244 + 12 \times 65$

$1024 = 4^5$

$4^5 = {}_5C_{[5]} \times {}_4P_1 + {}_5C_{[1,\,4]} \times {}_4P_2 + {}_5C_{[2,\,3]} \times {}_4P_2 + {}_5C_{[1,\,1,\,3]} \times {}_4P_2 + {}_5C_{[1,\,2,\,2]} \times {}_4P_2 + {}_5C_{[1,\,1,\,1,\,2]} \times {}_4P_1$

$n^m = {}_mC_{[m]} \times {}_nP_1 + \sum {}_mC_{[\bullet ki\bullet]} \times {}_nP_2 + {}_mC_{[\bullet kn\bullet]} \times {}_nP_1$ EQ A.5.7A

$n^m = {}_mC_{[m]} \times {}_nP_1 + {}_nP_2 \times \sum {}_mC_{[\bullet ki\bullet]} + {}_mC_{[\bullet kn\bullet]} \times {}_nP_1$ EQ A.5.7B

Tensor Calculus & Physics: A General Treatise

Table of Contents

List of Tables

List of Figures

List of Observations

Section B.1 *Principles of Special Sets Called Groups*

Groups are special sets of quantities that have the following properties under a binary operator [o].

Definition B.1.1 **Propositions of Quantity: Expanded Galois Group**

Axioms by Group Quantities	Description
$\exists$ a field G[o] $\ni$, quantities $(e, a, a^{-1}, b, c) \in$ G[o]	Existence of a Group Field where [e] is called the identity element and has the unique identity property of leaving quantities being left unaltered under the operation of the group system.
$a = b$ Equation A $b = a$ Equation B	Equality of quantities
$b = a$ and $c = b$ $\therefore c = a$	Substitution of quantities

Definition B.1.2 **Propositions of a Non-Abelian Group**

Axioms by Group Operator	Description
$(a \circ b \wedge b \circ a) \in$ G[o]	Closure with respect to Group Operator
$a \circ b \neq b \circ c$	Commutative by Group Operator
$a \circ (b \circ c) = (a \circ b) \circ c$	Associative by Group Operator
$a \circ e = a$	Identity by Group Operator
$a \circ a^{-1} = e$	Inverse Group Operator by reciprocal Quantity

Definition B.1.3 **Propositions of an Abelian Group**

Axioms by Group Operator	Description
$(a \circ b \wedge b \circ a) \in$ G[o]	Closure with respect to Group Operator
$a \circ b = b \circ c$	Commutative by Group Operator
$a \circ (b \circ c) = (a \circ b) \circ c$	Associative by Group Operator
$a \circ e = a$	Identity by Group Operator
$a \circ a^{-1} = e$	Inverse Group Operator by reciprocal Quantity

Definition B.1.4 **De Morgan Distribution Over Intersection of Binary Group Operators**

Axioms by Complement	Description
$a \lozenge (b \circ c) = (a \lozenge b) \circ (a \lozenge c)$ $a \circ (b \lozenge c) = (a \circ b) \lozenge (a \circ c)$	Distribution by complementary operator across group operator

By the method of exhaustion with definitions D2.2.17, D2.2.18 and D2.2.19, these are all of the possibilities exploring the existence of functions and their inverse.

Table compiled based on the work of Symour Lipschutz. [LIP64, pg 45], given

  Predecessor Function $\equiv p$
  Successor   Function $\equiv q$

Table B.1.1 Product--Inverse Functions using Successor and Processor Operations

Successor: Given p then q requires a 1-1 condition for an inverse to exist.											
Map Types	1-1	onto B	A	to by p	B	to by q	A	1-1	onto A	Inverse Exists?	Inverse for
			a	→	x	→	a				Given p
B.1.1.1	1-1	onto B	b	→	y	→	b	1-1	onto A	YES	$p \circ q = e$
			c	→	z	→	c				
			a	→	x	→	a				Given p
B.1.1.2	1-1	~onto B	b	→	y	→	b	~1-1	onto A	YES	$p \circ q = e$
					z	↗					
			a	→	x	→	a				Given p
B.1.1.3	~1-1	onto B	b	→	y	→	b	1-1	~onto A	NO	$p \circ q \neq e$
			c	↗			c				
			a	→	x	→	a				Given p
B.1.1.4	~1-1	~onto B	b	→	y	→	b	1-1	onto A	NO	$p \circ q \neq e$
			c	↗	z	→	c				
Predecessor: Given the q then p requires an, onto condition for an inverse to exist.											
			a	→	x	→	a				Given q
B.1.1.5	1-1	onto B	b	→	y	→	b	1-1	onto A	YES	$q \circ p = e$
			c	→	z	→	c				
			a	→	x	→	a				Given q
B.1.1.6	1-1	~onto B	b	→	y	→	b	~1-1	onto A	YES	$q \circ p = e$
			c		z	↗	c				
			a	→	x	→	a				Given q
B.1.1.7	~1-1	onto B	b	→	y	→	b	1-1	~onto A	NO	$q \circ p \neq e$
			c	↗	z		c				
			a	→	x	→	a				Given q
B.1.1.8	~1-1	~onto B	b	↗→	y	→	b	~1-1	~onto A	NO	$q \circ p \neq e$
	1-1	onto B	c	↗	z	↗	c				

(Continued) Existence of Product Function with Inverse ($p \circ q$) : A → A for Unequal Sets

Successor: Given p then q requires an, onto condition for an inverse to exist.

Map Types	1-1	AontoB	A	to by p	B	to by q	A	1-1	BontoA	Inverse Exists?	Inverse for
B.1.1.9	1-1	onto	a	→	x	→	a	1-1	onto	YES	Given p
			b	→	y	→	b				$p \circ q = e$
			c				c				
B.1.1.10	1-1	~onto	a	→	x	→	a	1-1	onto	NO	Given p
			b	→	y	→	b				$p \circ q \neq e$
			c				c				

Predecessor: Given the q then p requires an, onto condition for an inverse to exist.

			A	to by p	B	to by q	A	1-1	BontoA	Inverse Exists?	Inverse for
B.1.11	1-1	onto	a	→	x	→	a	1-1	onto	YES	Given q
			b	→	y	→	b				$q \circ p = e$
			c				c				
B.1.1.12	1-1	onto	a	→	x	→	a	~1-1	~onto	NO	Given q
			b	→	y	→	b				$q \circ p \neq e$
			c				c				

Tensor Calculus & Physics: A General Treatise

Table of Contents

Tensor Calculus & Physics: A General Treatise

List of Tables

List of Figures

List of Observations

Appendix D –Arithmetic Algorithms

An algorithm is a procedure for solving a mathematical problem in a finite number of steps that frequently involves repetition of an operation, broadly: a step-by-step procedure for solving a problem or accomplishing some desired end-result. Here I present five of some of the most famous algorithms that form the cornerstone to arithmetic and mathematics, as we know it. These algorithms come from the Encyclopedia Britannica on arithmetic [BRI60 Vol. 2, pg 357] and allow implementation of these fundamental arithmetic operators.

Since decimal system numbers deal with an infinite field above the decimal point to an infinitesimally small number below the decimal, algorithms need to handle this range. The algorithms on addition, subtraction and multiplication create numbers that are sequential that when calculated will lie either above or below the original number being operated on, yielding some value on the number line. However algorithms on division and roots can calculate a number between rational and irrational numbers and since in between numbers are most likely irrational an algorithm is needed that can approximate irrational numbers. If they are irrational than the algorithm must be iterative so that on the finial, iteration the number will be exact, or another way of phrasing this is that it converges onto that number. This is said because just as infinite series and products must converge onto a number to obtain a valid result here to there must be convergences to obtain the targeted number. After all the decimal number in reality is an infinite series in itself, so any operator, operating on this series must provide a convergent result. This is why the algorithms of addition, subtraction and multiplication start with a set of exact numbers that also results in an exact value, while division and roots starting with an approximate number to be iterative onto the limit, but still only approximates the targeted as a final number.

Section D.1 *Algorithm for Decimal System Addition*

Theorem D.1.1 Iterative Algorithm: Addition by Carrying

1g 2g 3g 4g 5g 6g	Given	$(s_n, ta_n, tb_n) \in R$ Summation $\equiv$ TermA + TermB TermA $\equiv \sum_{n=-\infty}^{m} ta_n \, 10^n$ TermB $\equiv \sum_{n=-\infty}^{m} tb_n \, 10^n$ Sum $\equiv \sum_{n=-\infty}^{m} s_n \, 10^n$ $0 \le (s_n, ta_n, tb_n) \le 9$ decimal numbers	
Steps	Hypothesis	Sum $= \sum_{n=-\infty}^{m} s_n \, 10^n$ for $0 \le s_n \le 9$ do nothing	Sum $= \sum_{n=-\infty}^{m} (1*10^{n+1} + sd * 10^n)$ for $10 \le s_n \le 18$ carry the one to the next decimal place.
1	From 2g, 3g, 4g and A4.2.3 Substitution	$\sum_{n=-\infty}^{m} s_n \, 10^n = \sum_{n=-\infty}^{m} ta_n \, 10^n + \sum_{n=-\infty}^{m} tb_n \, 10^n$	
2	From 1 and A4.2.6 Associative Add term-by-term	$\sum_{n=-\infty}^{m} s_n \, 10^n = \sum_{n=-\infty}^{m} (ta_n \, 10^n + tb_n \, 10^n)$	
3	From 2 and A4.2.14 Distribution	$\sum_{n=-\infty}^{m} s_n \, 10^n = \sum_{n=-\infty}^{m} (ta_n + tb_n) \, 10^n$	
4	From 3, equating term-by-term and T4.4.3 Equalities: Cancellation by Multiplication	$s_n = ta_n + tb_n$	
5	From 6g and 4	$0 \le s_n \le 9$	
6	From 4, 5 and A4.2.3 Substitution	$0 \le ta_n + tb_n \le 9$ do nothing	
7	From 4, 5 and A4.2.3 Substitution	$10 \le ta_n + tb_n \le 18$ for $\quad ta_n = 1$ and $tb_n = 9$ lower bound or $\quad\quad ta_n = 9$ and $tb_n = 9$ upper bound	
8	From 4, 7 and A4.2.3 Substitution	$10 \le s_n \le 18$	
9	From 8 and expanding	$s_n = 1*10^1 + sd \, 10^0$ for $0 \le sd \le 8$	
10	From 5g, 9 and A4.2.3 Substitution	Sum $= \sum_{n=-\infty}^{m} (1*10^1 + sd \, 10^0) \, 10^n$	
11	From 10 and A4.2.14 Distribution	Sum $= \sum_{n=-\infty}^{m} (1*10^1 10^n + sd \, 10^0 10^n)$	
12	From 11 and A4.2.18 Summation Exp	Sum $= \sum_{n=-\infty}^{m} (1*10^{n+1} + sd * 10^n)$ carry the one to the next decimal place.	
$\therefore$	From 2g, 5, 6, 8 and 12	Sum $= \sum_{n=-\infty}^{m} s_n \, 10^n$ for $0 \le s_n \le 9$ do nothing	Sum $= \sum_{n=-\infty}^{m} (1*10^{n+1} + sd * 10^n)$ for $10 \le s_n \le 18$ carry the one to the next decimal place.

Section D.2 *Algorithm for Decimal System Subtraction*

Theorem D.2.1 Iterative Algorithm: Subtraction by Borrowing

1g 2g 3g 4g 5g 6g 7g	Given	$(d_n, ta_n, tb_n) \in R$ Difference $\equiv \text{TermB} - \text{TermA}$ TermA $\equiv \sum_{n=-\infty}^{m} ta_n 10^n$ TermB $\equiv \sum_{n=-\infty}^{m} tb_n 10^n$ Diff $\equiv \sum_{n=-\infty}^{m} d_n 10^n$ $0 \le (d_n, ta_n, tb_n) \le 9$ decimal numbers $0 \le \text{Difference}$
Steps	Hypothesis	$((10 + tb_k) - ta_k) + ((tb_{k+1} - 1) - ta_{k+1})\,10$ iff $tb_k < ta_k$ for the kth term but iff $1 \le tb_{k+1}$ borrowing one from the upper decimal place
1	From 2g, 7g, A4.2.3 Substitution and D4.17 Greater Than Inequality [>]	$\text{TermA} \le \text{TermB}$ where the ordering of terms are freely selected so this is always true.
2	From 2g, 3g, 4g and A4.2.3 Substitution	$\sum_{n=-\infty}^{m} d_n 10^n = \sum_{n=-\infty}^{m} tb_n 10^n - \sum_{n=-\infty}^{m} ta_n 10^n$
3	From 2 and A4.2.6 Associative Add term-by-term	$\sum_{n=-\infty}^{m} d_n 10^n = \sum_{n=-\infty}^{m} (tb_n 10^n - ta_n 10^n)$
4	From 3 and A4.2.14 Distribution	$\sum_{n=-\infty}^{m} d_n 10^n = \sum_{n=-\infty}^{m} (tb_n - ta_n)\,10^n$
5	From 4 by comparing term-by-term	$d_n = tb_n - ta_n$
6	From 5g and 7g	$0 \le \sum_{n=-\infty}^{m} d_n 10^n$
7	From 6 and expand about the $(k+1)^{th}$ and k^{th} terms	$0 \le \sum_{n=-\infty}^{k-1} d_n 10^n$ $\qquad + d_k 10^k + d_{k+1} 10^{k+1}$ iff $tb_k < ta_k$ for the kth term $\qquad + \sum_{n=k+2}^{m} d_n 10^n$
8	From 5, 7, A4.2.3 Substitution, A4.2.18 Summation Exp and A4.2.14 Distribution	$+\, ((tb_k - ta_k) + (tb_{k+1} - ta_{k+1})\,10)10^k$ iff $tb_k < ta_k$ for the kth term but iff $1 \le tb_{k+1}$
9	From 8, A4.2.7 Identity Add and A4.2.8 Inverse Add	$(tb_k - ta_k) + (tb_{k+1} - 1 + 1 - ta_{k+1})\,10$ iff $tb_k < ta_k$ for the kth term but iff $1 \le tb_{k+1}$
10	From 9, A4.2.5 Commutative Add and A4.2.14 Distribution	$(tb_k - ta_k) + (tb_{k+1} - 1 - ta_{k+1})\,10 + 1*10$ iff $tb_k < ta_k$ for the kth term but iff $1 \le tb_{k+1}$
11	From 10, A4.2.5 Commutative Add and A4.2.6 Associative Add	$(10 + tb_k - ta_k) + (tb_{k+1} - 1 - ta_{k+1})\,10$ iff $tb_k < ta_k$ for the kth term but iff $1 \le tb_{k+1}$
∴	From 11 and A4.2.6 Associative Add	$((10 + tb_k) - ta_k) + ((tb_{k+1} - 1) - ta_{k+1})\,10$ iff $tb_k < ta_k$ for the kth term but iff $1 \le tb_{k+1}$ borrowing one from the upper decimal place

Section D.3 *Algorithm for Decimal System Multiplication*

Theorem D.3.1 Iterative Algorithm: Long Multiplication

1g	Given	$(p_w, c_n, m_k) \in R$
2g		product $\equiv$ multiplicand * multiplier
3g		multiplicand $\equiv \sum_{n=-r}^{u} c_n 10^n$
4g		multiplier $\equiv \sum_{k=-s}^{v} m_k 10^k$
5g		product $\equiv \sum_{w=-r-s}^{(u+v+1)} p_w 10^w$
6g		$0 \leq (p_w, c_n, m_k) \leq 9$ decimal numbers
7g		$g = n + r$
8g		$h = k + s$
Steps	Hypothesis	product $= 10^{-(r+s)} \sum_{h=0}^{v+s} (\sum_{g=0}^{u+r} c_{g-r} * m_{h-s} 10^g) 10^h$
1	From A4.2.2A Equality	$10^r = 10^r$
2	From 3g, 1 and T4.4.2 Equalities: Uniqueness of Multiplication	10^r multiplicand $= 10^r \sum_{n=-r}^{u} c_n 10^n$
3	From A4.2.2A Equality	$10^s = 10^s$
4	From 4g, 3 and T4.4.2 Equalities: Uniqueness of Multiplication	10^s multiplier $= 10^s \sum_{k=-s}^{v} m_k 10^k$
5	From 2, A4.2.14 Distribution and A4.2.18 Summation Exp	10^r multiplicand $= \sum_{n=-r}^{u} c_n 10^{n+r}$
6	From 4, A4.2.14 Distribution and A4.2.18 Summation Exp	10^s multiplier $= \sum_{k=-s}^{v} m_k 10^{k+s}$
7	From 7g and T4.3.3B Equalities: Right Cancellation by Addition	$n = g - r$
8	From 8g and T4.3.3B Equalities: Right Cancellation by Addition	$k = h - s$
9	From 5, 7, A4.2.3 Substitution and reinitializing index	10^r multiplicand $= \sum_{g=0}^{u+r} c_{g-r} 10^g$
10	From 6, 8, A4.2.3 Substitution and reinitializing index	10^s multiplier $= \sum_{h=0}^{v+s} m_{h-s} 10^h$
11	From A4.2.2A Equality	$10^{r+s} = 10^{r+s}$
12	From 2g, 11, T4.4.2 Equalities: Uniqueness of Multiplication, A4.2.18 Summation Exp and A4.210 Commutative Multp	10^{r+s} product $= 10^r$ multiplicand $* 10^s$ multiplier
13	From 9, 10, 12 and A4.2.3 Substitution	10^{r+s} product $= (\sum_{g=0}^{u+r} c_{g-r} 10^g) * (\sum_{h=0}^{v+s} m_{h-s} 10^h)$

14	From 13, A4.2.14 Distribution and summing over h	10^{r+s} product $= \sum_{h=0}^{v+s} (\sum_{g=0}^{u+r} c_{g-r} * m_{h-s} \, 10^g) \, 10^h$
15	From A4.2.2A Equality	$10^{-(r+s)} = 10^{-(r+s)}$
$\therefore$	From 14, 15, T4.4.2 Equalities: Uniqueness of Multiplication, A4.2.18 Summation Exp, A4.2.8 Inverse Add, A4.2.24 Inverse Exp and A4.2.12 Identity Multp	product $= 10^{-(r+s)} \sum_{h=0}^{v+s} (\sum_{g=0}^{u+r} c_{g-r} * m_{h-s} \, 10^g) \, 10^h$

This is the algorithm for long multiplication:

Step 1) Multiply the multiplicand and multiplier by the power of ten raised to the number of digits behind the decimal converting them to integer numbers.

Step 2) For every digit of the multiplier, multiply the resulting product line by ten, raised to the power of its digital position.

Step 3) Add the resulting product lines.

Step 4) Divide the summation by ten, raised to the power of the sum of the number of digits to the right of the decimal point for the multiplicand and multiplier

Section D.4 *Algorithm for Decimal System Division*

The formula for division is given as a multiplication product:

Theorem D.4.1 Iterative Algorithm: Division by Digit

1g	Given	$(d_n, v_k, q_s, p_k) \in R$
2g		Dividend $\equiv$ Quotient * Divisor + Remainder
3g		Dividend $\equiv \sum_{n=-\infty}^{m} d_n \, 10^n$
4g		Quotient * Divisor $\equiv \sum_{k=u}^{m} p_k \, 10^k$ for $m \geq u$
5g		Divisor $\equiv \sum_{k=u}^{m} v_k \, 10^k$
6g		Quotient $\equiv \sum_{s=-\infty}^{r} q_s \, 10^s$
7g		$0 \leq (d_n, v_k, q_s) \leq 9$ decimal numbers
Steps	Hypothesis	$\sum_{k=u}^{m} (d_k - q_s * v_k) \, 10^k = \rho_s \geq 0$
1	From 2g, T4.3.3 Equalities: Right Cancellation by Addition	Dividend – Quotient * Divisor = Remainder
2	From 3g, 4g, 1 and A4.2.3 Substitution	$(\sum_{n=-\infty}^{m} d_n \, 10^n) - (\sum_{k=u}^{m} p_k \, 10^k) =$ Remainder
3	From 2 and summation by parts	$(\sum_{k=u}^{m} d_k \, 10^k + \sum_{t=-\infty}^{u-1} d_t \, 10^t) - (\sum_{k=u}^{m} p_k \, 10^k) =$ Remainder
4	From 3, A4.2.5 Commutative Add and A4.2.6 Associative Add by summing term-by-term	$(\sum_{k=u}^{m} (d_k \, 10^k - p_k \, 10^k)) + (\sum_{t=-\infty}^{u-1} d_t \, 10^t)(\sum_{k=u}^{m}) =$ Remainder
5	From 4 and A4.2.14 Distribution	$(\sum_{k=u}^{m} (d_k - p_k) \, 10^k) + (\sum_{t=-\infty}^{u-1} d_t \, 10^t) =$ Remainder
6	From 4g, 5g, 6g, 5 and A4.2.3 Substitution	$q_s * (\sum_{k=u}^{m} v_k \, 10^k) = \sum_{k=u}^{m} p_k \, 10^k$ for $s \longleftrightarrow$ Divisor number
7	From 6 and A4.2.14 Distribution	$\sum_{k=u}^{m} q_s * v_k \, 10^k = \sum_{k=u}^{m} p_k \, 10^k$
8	From 7, A4.2.2 Equality term-by-term and T4.4.3 Equalities: Cancellation by Multiplication	$p_k = q_s * v_k$
$\therefore$	From 5, 8 and A4.2.3 Substitution	$\sum_{k=u}^{m} (d_k - q_s * v_k) \, 10^k = \rho_s \geq 0 \qquad$ Remainder

Corllary D.4.1.1 Size of Quotient Coefficients

1g	Given	$(a, b, c, d, f) \in R$
2g		Dividend $\equiv$ Quotient * Divisor + Remainder
Steps	Hypothesis	$d_k \geq q_s * v_k$
1	From TD.4.1 Algorithm: Division	$\sum_{k=u}^{m} (d_k - q_s * v_k) \, 10^k = \rho_s \geq 0$ Remainder
2	From 1 and remainder must be positive	$0 \leq d_k - q_s * v_k$
$\therefore$	From 2 and D4.1.7 Greater Than Inequality [>]	$d_k \geq q_s * v_k$

Since the set of decimal d_k's are known, but the quotient coefficient q_s is not, than the algorithm part requires the q_s to be found by inspection such that the product is chosen as close to the magnitude of the set of decimal d_k's as possible.

Theorem D.4.2 Iterative Algorithm: Division

1g	Given	$(d_n, v_k, q_s, p_k) \in R$
2g		Dividend $\equiv$ Quotient * Divisor + Remainder
3g		Dividend $\equiv \sum_{n=-\infty}^{m} d_n \, 10^n$
4g		Divisor $\equiv \sum_{k=u}^{m} v_k \, 10^k$
5g		Quotient $\equiv \sum_{s=-\infty}^{r} q_s \, 10^s$
6g		$0 \leq (d_n, v_k, q_s) \leq 9$ decimal numbers
Steps	Hypothesis	$\text{Remainder}_{k+1} = (\text{Remainder}_k + d_{m-(k+1)}10^0) - q_{r-(k+1)} * \text{Divisor}$ for k = 0, 1, 2 …
1	From 2g and T4.3.3B Equalities: Right Cancellation by Addition	$\text{Remainder} = \text{Dividend} - \text{Quotient} * \text{Divisor}$
2	From 5g and 1	$\text{Remainder}_0 = \text{Dividend} - q_r * \text{Divisor}$
3	From 1, 2 for iterated dividend	$\text{Remainder}_1 = (\text{Remainder}_0 + d_{m-1}10^0) - q_{r-1} * \text{Divisor}$
4	From 1, 3 for iterated dividend	$\text{Remainder}_2 = (\text{Remainder}_1 + d_{m-2}10^0) - q_{r-2} * \text{Divisor}$
5	From 1, 4 for iterated dividend	$\text{Remainder}_3 = (\text{Remainder}_2 + d_{m-3}10^0) - q_{r-3} * \text{Divisor}$
$\therefore$	From 2, 3, 4, 5 and extrapolating index terms	$\text{Remainder}_{k+1} = (\text{Remainder}_k + d_{m-(k+1)}10^0) - q_{r-(k+1)} * \text{Divisor}$ for k = 0, 1, 2 …

Section D.5 *Multiple Cultural Styles of Division found Worldwide*

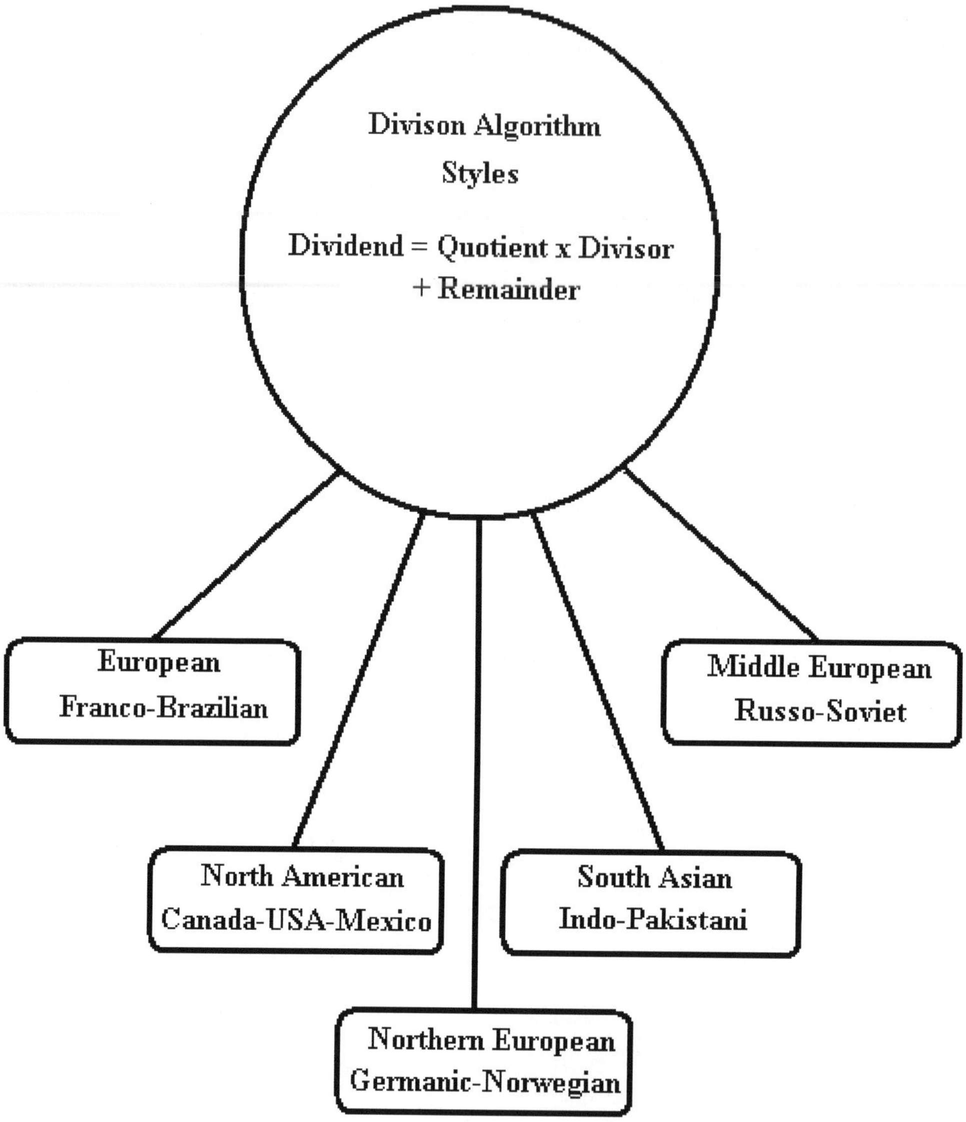

Divison Algorithm Styles

Dividend = Quotient x Divisor + Remainder

European Franco-Brazilian

Middle European Russo-Soviet

North American Canada-USA-Mexico

South Asian Indo-Pakistani

Northern European Germanic-Norwegian

Figure D.5.1 Sources of Multicultural Styles of Division

While the iterative algorithm for division is universal and does not change how various cultures around the world interrupt the style is unconstrained, imaginative and varied.

They're five styles that have been identified by Orey: Canada-USA-Mexico; Franco-Brazilian; Indo-Pakistani; and Russo-Soviet. A fifth, introduced to Orey by a colleague in Norway, Germanic-Norwegian, has not been found in the samples gathered from this part of the world as of this writing. [OREβ03]

Tensor Calculus & Physics: A General Treatise

Examples of Typical Cultural Styles:

Canada-USA-Mexico Style

```
                                              1  2  3 . 4
5  )  6  1  7 . 0  ( 1  2  3 . 4       5 ⌐  6  1  7 . 0
   -  5                                    -  5
         1  1                                    1  1
      -  1  0                                  -  1  0
            1  7                                     1  7
         -  1  5                                  -  1  5
               2     0                                  2     0
            -  2     0                               -  2     0
                     0                                        0
```

Notice that the American style of division to the left, though rear, has placed the quotient to the right for brevity and the use of common type rather than the icon symbol of long division.

Russo-Soviet Style

```
      6   1   7   | 5
   -  5           | 1   2   3 , 4
         1   1
      -  1   0
            1   7
         -  1   5
               2   0
               2   0
                   0
```

Franco-Brazilian Style

```
   6   1   7      | 5
   1   1            1   2   3 , 4
      1   7
         2   0
            0
```

Notice that the subtraction factors for the quotient have been skipped.

Germanic-Norwegian Style

```
6  1  7      :  5  =  1  2  3  ,  4
5
‾‾‾
1  1
1  0
   ‾‾‾‾‾‾
      1  7
      1  5
      ‾‾‾‾‾‾
         2  0
         2  0
         ‾‾‾‾‾‾
            0
```

Notice the use if the colon [:] as a delineator between the dividend and divisor likewise the equal sign [=] between the divisor and quotient. Also, the subtraction sign has been dropped for brevity.

Netherlands' Style

```
5    |  1  2  3  ,  4
‾‾‾‾‾‾‾‾‾‾‾‾‾‾‾‾‾‾‾‾‾‾
6  1  7  ,  0
5  ^
‾‾‾‾
1  1
1  0
‾‾‾‾‾‾
   1  7
   1  5
   ‾‾‾‾‾‾
      2        0
      2        0
      ‾‾‾‾‾‾‾‾‾‾
               0
```

Notice that the subtraction symbol is not used and the carat [^] or a raised dash [ˉ] symbol is used as a placeholder reminder. Here too the subtraction sign has been dropped for brevity.

Section D.6 *Cultural Mathematical Icons*

Orey has also identified that icons or what we take for math symbols can vary from country-to-country and culture-to-culture for example:

the long division sign used by the English vs. Ukraine use root sign

or the decimal point the English vs. European use the comma

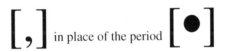

Section D.7 *Algorithm for Decimal System Square Root*

Theorem D.7.1 Iterative Algorithm: Square Root

1g	Given	$(a, b, d_n, r_k) \in R$
2g		Dividend $\equiv \text{Root}^2 \equiv (a + b)^2$
3g		Dividend $\equiv \sum_{n=-\infty}^{m} d_n 10^n$
4g		Root $\equiv \sum_{k=-\infty}^{m-2} r_k 10^k$
5g		$0 \le (d_n, r_k) \le 9$ decimal numbers
Steps	Hypothesis	$\text{Remainder}_{k+1} = \text{Remainder}_k - (2 \sum_{s=0}^{k} r_{m-(k+s)} 10^{k-s})$ $+ r_{m-(2k+1)}) r_{m-(2k+1)}$
1	From 2g, A4.2.25 Reciprocal Exp and T4.10.3 Radicals: Identity	root $= \sqrt{\text{Dividend}} = a + b$
2	From 2g and T4.12.1 Polynomial Quadratic: The Perfect Square	$(a + b)^2 = a^2 + 2ab + b^2$
3	From 2g, 2, T4.3.3B Equalities: Right Cancellation by Addition and A4.2.14 Distribution	Remainder $= \text{Dividend} - a^2 - (2a + b) b$
4	From 2g, 4g, 3 for first term	$\text{Remainder}_0 = \text{Dividend} - r_{m-2}^2$
5	From 3, 4 for second term	$\text{Remainder}_1 = \text{Remainder}_0 - (2 (r_{m-2} 10^0) + r_{m-3}) r_{m-3}$
6	From 3, 5 for second term	$\text{Remainder}_2 = \text{Remainder}_1 - (2 (r_{m-2}10^1 + r_{m-3} 10^0) + r_{m-4}) r_{m-4}$
7	From 3, 6 for second term	$\text{Remainder}_3 = \text{Remainder}_2 - (2 (r_{m-2} 10^2 + r_{m-3} 10^1 + r_{m-4} 10^0) + r_{m-5}) r_{m-5}$
∴	From 4, 5, 6 and extrapolating index terms	$\text{Remainder}_{k+1} = \text{Remainder}_k - (2 \sum_{s=0}^{k} r_{m-(k+s)} 10^{k-s})$ $+ r_{m-(2k+1)}) r_{m-(2k+1)}$

Since the set of decimal d_k's are known, but the root coefficient r_k are not, than the algorithm part requires the r_k to be found by inspection such that the iterated roots are chosen as close to the magnitude of the set of decimals of the iterated remainders as possible.

Example:

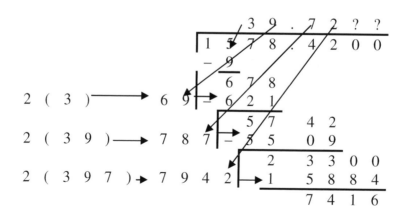

Section D.8 *Algorithm for Decimal System Cube Root*

Theorem D.8.1 Iterative Algorithm: Cube Root

1g	Given	$(a, b, d_n, r_k) \in R$
2g		Dividend $\equiv \text{Root}^3 \equiv (a + b)^3$
3g		Dividend $\equiv \sum_{n = -\infty}^{m} d_n\, 10^n$
4g		Root $\equiv \sum_{k = -\infty}^{m-2} r_k\, 10^k$
5g		$0 \leq (d_n, r_k) \leq 9$ decimal numbers
Steps	Hypothesis	$\text{Remainder}_{k+1} = \text{Remainder}_k - (3(\sum_{s=0}^{k} r_{m-(k+s)}\, 10^{k-s})^2)$ $\quad + 3(\sum_{s=0}^{k} r_{m-(k+s)}\, 10^{k-s})\, r_{m-(2k+1)} + r_{m-(2k+1)}^2)\, r_{m-(2k+1)}$
1	From 2g, A4.2.25 Reciprocal Exp and T4.10.3 Radicals: Identity	$\text{root} = \sqrt[3]{\text{Dividend}} = a + b$
2	From 2g and T4.12.1 Polynomial Quadratic: The Perfect Square	$(a + b)^3 = a^3 + 3a^2 b + 3ab^2 + b^3$
3	From 2g, 2, T4.3.3B Equalities: Right Cancellation by Addition and A4.2.14 Distribution	$\text{Remainder} = \text{Dividend} - a^3 - (3a^2 + 3ab + b^2)\, b$
4	From 2g, 4g, 3 for first term	$\text{Remainder}_0 = \text{Dividend} - r_{m-2}^3$
5	From 3, 4 for second term	$\text{Remainder}_1 = \text{Remainder}_0 - (3\,(r_{m-2}\, 10^0)^2 + 3r_{m-2}\, r_{m-3} + r_{m-3}^2)\, r_{m-3}$
6	From 3, 5 for second term	$\text{Remainder}_2 = \text{Remainder}_1 - (3\,(r_{m-2} 10^1 + r_{m-3}\, 10^0)$ $\quad + (r_{m-2} 10^1 + r_{m-3}\, 10^0)\, r_{m-4} + r_{m-4}^2)\, r_{m-4}$
7	From 3, 6 for second term	$\text{Remainder}_3 = \text{Remainder}_2 - (3\,(r_{m-2}\, 10^2 + r_{m-3}\, 10^1 + r_{m-4}\, 10^0)$ $\quad + 3(r_{m-2}\, 10^2 + r_{m-3}\, 10^1 + r_{m-4}\, 10^0)\, r_{m-5} + r_{m-5}^2)\, r_{m-5}$
∴	From 4, 5, 6 and extrapolating index terms	$\text{Remainder}_{k+1} = \text{Remainder}_k - (3(\sum_{s=0}^{k} r_{m-(k+s)}\, 10^{k-s})^2)$ $\quad + 3(\sum_{s=0}^{k} r_{m-(k+s)}\, 10^{k-s})\, r_{m-(2k+1)} + r_{m-(2k+1)}^2)\, r_{m-(2k+1)}$

Since the set of decimal d_k's are known, but the cube root coefficient r_k are not, than the algorithm part requires the r_k to be found by inspection such that the iterated roots are chosen as close to the magnitude of the set of decimal of the iterated remainders as possible.

Example:

$3\ (\ 6\ )^2 \longrightarrow 1\ 0\ 8\ 0\ 0$

$3\ (\ 6\ )\ 5 \longrightarrow 9\ 0\ 0$

$5^2 \longrightarrow +\ 2\ 5$

$1\ 1\ 7\ 2\ 5$

$3\ (\ 6\ 5\ )^2 \longrightarrow 1\ 2\ 6\ 7\ 5\ 0\ 0$

$3\ (6\ 5\ )\ 3 \longrightarrow 5\ 8\ 5\ 0$

$3^2 \longrightarrow +\ 9$

$1\ 2\ 7\ 3\ 3\ 5\ 9$

$3\ (\ 6\ 5\ 3\ )^2 \longrightarrow 1\ 2\ 7\ 9\ 2\ 2\ 7\ 0\ 0$

$3\ (\ 6\ 5\ 3\ )\ 8 \longrightarrow 1\ 5\ 6\ 7\ 2\ 0$

$8^2 + \longrightarrow 6\ 4$

$1\ 2\ 8\ 0\ 7\ 9\ 4\ 8\ 4$

```
                                    6  5  .  3  8
                       2  7  9  4  6  3  .  0  0  0  0  0  0
                     - 2  1  6
                       6  3  4  6  3
                                   -  5  8  6  2  5
                                      4  8  3  8     0  0  0
                                   -  3  8  2  0        0  7  7
                                      1  0  1  7     9  2  3  0  0  0
                                   -  1  0  2  4     6  3  5  8  7  2
```

Clearly by using the binomial expansion for n-multiples any n^{th} –root $(a + b)^n$ algorithm can be derived.

Tensor Calculus & Physics: A General Treatise

Table of Contents

Tensor Calculus & Physics: A General Treatise

List of Tables

List of Figures

List of Observations

Tensor Calculus & Physics: A General Treatise

Appendix E –Trigonometric, Algebraic and Mensuration Tables

Section E.1 *Plane Trigonometry: Right Triangles*

Excerpt from [BEY88, pg 134] CRC Standard Mathematical Tables

Definition E.1.1 **One Degree**

An angle of one degree is one part out of 360 equal parts of a circle.

Definition E.1.2 **Straight Angle**

A straight angle is an angle of 180° (180 degrees) or half the circumference around a circle.

Definition E.1.3 **Right Angle**

A right angle is an angle of 90° or one quarter the circumference around a circle.

Definition E.1.4 **Acute Angle**

An acute angle is an angle between 0° and 90°.

Definition E.1.5 **Obtuse Angle**

An obtuse angle is an angle between $90°^+$ and 180°.

Definition E.1.6 **Radian**

A radian is an angle subtended at the center of a circle by an arc whose length is equal to that of the radius.

180°	$= \pi$ radians:
1°	$= (\pi / 180°)$ radians:
1 radian	$= (180° / \pi)$ degrees.

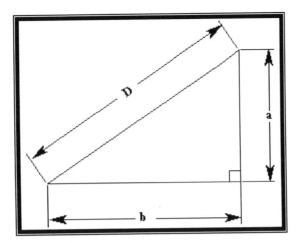

Figure E.1.1 Pythagorean Theorem

Definition E.1.7 **Pythagorean Theorem: Algebraic Version**

$$D = \sqrt{a^2 + b^2} \qquad \text{Equation A}$$
$$a = \sqrt{D^2 - b^2} \qquad \text{Equation B}$$
$$b = \sqrt{D^2 - a^2} \qquad \text{Equation C}$$

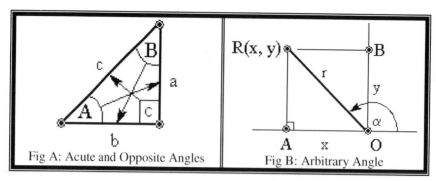

Figure E.1.2 Trigonometric Functions of Acute and Arbitrary Angles

Table E.1.1 Trigonometric Functions of Acute Angles

DxE.1.1.1	sine	of A	=	sin(A)	=	a/c
DxE.1.1.2	cosine	of A	=	cos(A)	=	b/c
DxE.1.1.3	tangent	of A	=	tan(A)	=	a/b
DxE.1.1.4	cosecant	of A	=	csc(A)	=	c/a
DxE.1.1.5	secant	of A	=	sec(A)	=	c/b
DxE.1.1.6	cotangent	of A	=	cot(A)	=	b/a
DxE.1.1.7	exsecant	of A	=	exsec(A)	=	sec(A) − 1
DxE.1.1.8	versine	of A	=	vers(A)	=	1 − cos(A)
DxE.1.1.9	coversine	of A	=	covers(A)	=	1 − sin(A)
DxE.1.1.10	haversine	of A	=	hav(A)	=	½ vers(A)

Table E.1.2 Trigonometric Functions of Arbitrary Angles

DxE.1.2.1	sine	of α	=	sin(α)	=	y/r
DxE.1.2.2	cosine	of α	=	cos(α)	=	x/r
DxE.1.2.3	tangent	of α	=	tan(α)	=	y/x
DxE.1.2.4	cosecant	of α	=	csc(α)	=	r/y
DxE.1.2.5	secant	of α	=	sec(α)	=	r/x
DxE.1.2.6	cotangent	of α	=	cot(α)	=	x/y
DxE.1.2.7	exsecant	of α	=	exsec(α)	=	sec(α) − 1
DxE.1.2.8	versine	of α	=	vers(α)	=	1 − cos(α)
DxE.1.2.9	coversine	of α	=	covers(α)	=	1 − sin(α)
DxE.1.2.10[E.1.1]	haversine	of α	=	hav(α)	=	½ vers(α)

for A = 180° − α (an acute angle)

[E.1.1]Note [DxE.] is read as <u>D</u>efinition of Appendi<u>x</u>-<u>E</u>.

Table E.1.3 Relationship between Trigonometric and Algebraic Functions

then	given	sin(x) = a	cos(x) = a	tan(x) = a	csc(x) = a	sec(x) = a	cot(x) = a
sin(x)	=	$\pm a$	$\pm\sqrt{1-a^2}$	$\pm a/\sqrt{1+a^2}$	$\pm 1/a$	$(\pm\sqrt{a^2-1})/a$	$\pm 1/\sqrt{1+a^2}$
cos(x)	=	$\pm\sqrt{1-a^2}$	$\pm a$	$\pm 1/\sqrt{1+a^2}$	$(\pm\sqrt{a^2-1})/a$	$\pm 1/a$	$\pm a/\sqrt{1+a^2}$
tan(x)	=	$\pm a/\sqrt{1-a^2}$	$(\pm\sqrt{1-a^2})/a$	$\pm a$	$\pm 1/\sqrt{a^2-1}$	$\pm\sqrt{a^2-1}$	$\pm 1/a$
csc(x)	=	$\pm 1/a$	$\pm 1/\sqrt{1-a^2}$	$(\pm\sqrt{1+a^2})/a$	$\pm a$	$\pm a/\sqrt{a^2-1}$	$\pm\sqrt{1+a^2}$
sec(x)	=	$\pm 1/\sqrt{1-a^2}$	$\pm 1/a$	$\pm\sqrt{1+a^2}$	$\pm a/\sqrt{a^2-1}$	$\pm a$	$(\pm\sqrt{1+a^2})/a$
cot(x)	=	$(\pm\sqrt{1-a^2})/a$	$\pm a/\sqrt{1-a^2}$	$\pm 1/a$	$\pm\sqrt{a^2-1}$	$\pm 1/\sqrt{a^2-1}$	$\pm a$

for $|x| \le \pi$, [±] determined by quadrant and [a] is a normalized ratio see FE.1.A and B

Table E.1.4 Quick Lookup: Signs of the Trigonometric Functions

Quadrant	sin	cos	tan	cot	sec	csc
I	+	+	+	+	+	+
II	+	−	−	−	−	+
III	−	−	+	+	−	−
IV	−	+	−	−	+	−

Table E.1.5 Quick Lookup: Variations of the Trigonometric Functions

Quadrant	sin	cos	tan	cot	sec	csc
I	$0 \to +1$	$+1 \to 0$	$0 \to +\infty$	$+\infty \to 0$	$+1 \to +\infty$	$+\infty \to +1$
II	$+1 \to 0$	$0 \to -1$	$-\infty \to 0$	$0 \to -\infty$	$-\infty \to -1$	$+1 \to +\infty$
III	$0 \to -1$	$-1 \to 0$	$0 \to +\infty$	$+\infty \to 0$	$-1 \to -\infty$	$-\infty \to -1$
IV	$-1 \to 0$	$0 \to +1$	$-\infty \to 0$	$0 \to -\infty$	$+\infty \to +1$	$-1 \to -\infty$

Table E.1.6 Quick Lookup: Trigonometric Functions of Some Special Angles

	Angle	sin	cos	tan	cot	sec	csc
DxE.1.6.1	$0° = 0$	0	1	0	$+\infty$	1	$+\infty$
DxE.1.6.2	$15° = \pi/12$	$\sqrt{2}(\sqrt{3}-1)/4$	$\sqrt{2}(\sqrt{3}+1)/4$	$2-\sqrt{3}$	$2+\sqrt{3}$	$\sqrt{2}(\sqrt{3}-1)$	$\sqrt{2}(\sqrt{3}+1)$
DxE.1.6.3	$30° = \pi/6$	$1/2$	$\sqrt{3}/2$	$\sqrt{3}/3$	$\sqrt{3}$	$2\sqrt{3}/3$	2
DxE.1.6.4	$45° = \pi/4$	$\sqrt{2}/2$	$\sqrt{2}/2$	1	1	$\sqrt{2}$	$\sqrt{2}$
DxE.1.6.5	$60° = \pi/3$	$\sqrt{3}/2$	$1/2$	$\sqrt{3}$	$\sqrt{3}/3$	2	$2\sqrt{3}/3$
DxE.1.6.6	$75° = 5\pi/12$	$\sqrt{2}(\sqrt{3}+1)/4$	$\sqrt{2}(\sqrt{3}-1)/4$	$2+\sqrt{3}$	$2-\sqrt{3}$	$\sqrt{2}(\sqrt{3}+1)$	$\sqrt{2}(\sqrt{3}-1)$
DxE.1.6.7	$90° = \pi/2$	1	0	$+\infty$	0	$+\infty$	1
DxE.1.6.8	$105° = 7\pi/12$	$\sqrt{2}(\sqrt{3}+1)/4$	$-\sqrt{2}(\sqrt{3}-1)/4$	$-(2+\sqrt{3})$	$-(2-\sqrt{3})$	$-\sqrt{2}(\sqrt{3}+1)$	$\sqrt{2}(\sqrt{3}-1)$
DxE.1.6.9	$120° = 2\pi/3$	$\sqrt{3}/2$	$-1/2$	$-\sqrt{3}$	$-\sqrt{3}/3$	-2	$2\sqrt{3}/3$
DxE.1.6.10	$135° = 3\pi/4$	$\sqrt{2}/2$	$-\sqrt{2}/2$	-1	-1	$-\sqrt{2}$	$\sqrt{2}$
DxE.1.6.11	$150° = 5\pi/6$	$1/2$	$-\sqrt{3}/2$	$-\sqrt{3}/3$	$-\sqrt{3}$	$-2\sqrt{3}/3$	2
DxE.1.6.12	$165° = 11\pi/12$	$\sqrt{2}(\sqrt{3}-1)/4$	$-\sqrt{2}(\sqrt{3}+1)/4$	$-(2-\sqrt{3})$	$-(2+\sqrt{3})$	$-\sqrt{2}(\sqrt{3}-1)$	$\sqrt{2}(\sqrt{3}+1)$
DxE.1.6.13	$180° = \pi$	0	-1	0	$-\infty$	-1	$+\infty$
DxE.1.6.14	$270° = 3\pi/2$	-1	0	$+\infty$	0	$-\infty$	-1
DxE.1.6.15	$360° = 2\pi n$[E.1.2]	0	1	0	$+\infty$	1	$+\infty$

[E.1.2]Note [n] is any integer and holds true do to periodicity of trigonometric functions.

Table E.1.7 Quick Lookup: Periodic Reduction Formulas for Sign

	Angle	sin	cos	tan	cot	sec	csc
DxE.1.7.1	$0° - \alpha$	$-\sin(\alpha)$	$+\cos(\alpha)$	$-\tan(\alpha)$	$-\cot(\alpha)$	$+\sec(\alpha)$	$-\csc(\alpha)$
DxE.1.7.2	$90° + \alpha$	$+\cos(\alpha)$	$-\sin(\alpha)$	$-\cot(\alpha)$	$-\tan(\alpha)$	$-\csc(\alpha)$	$+\sec(\alpha)$
DxE.1.7.3	$90° - \alpha$	$+\cos(\alpha)$	$+\sin(\alpha)$	$+\cot(\alpha)$	$+\tan(\alpha)$	$+\csc(\alpha)$	$+\sec(\alpha)$
DxE.1.7.4	$180° + \alpha$	$-\sin(\alpha)$	$-\cos(\alpha)$	$+\tan(\alpha)$	$+\cot(\alpha)$	$-\sec(\alpha)$	$-\csc(\alpha)$
DxE.1.7.5	$180° - \alpha$	$+\sin(\alpha)$	$-\cos(\alpha)$	$-\tan(\alpha)$	$-\cot(\alpha)$	$-\sec(\alpha)$	$+\csc(\alpha)$
DxE.1.7.6	$270° + \alpha$	$-\cos(\alpha)$	$+\sin(\alpha)$	$-\cot(\alpha)$	$-\tan(\alpha)$	$+\csc(\alpha)$	$-\sec(\alpha)$
DxE.1.7.7	$270° - \alpha$	$-\cos(\alpha)$	$-\sin(\alpha)$	$+\cot(\alpha)$	$+\tan(\alpha)$	$-\csc(\alpha)$	$-\sec(\alpha)$
DxE.1.7.8	$360° + \alpha$	$+\sin(\alpha)$	$+\cos(\alpha)$	$+\tan(\alpha)$	$+\cot(\alpha)$	$+\sec(\alpha)$	$+\csc(\alpha)$
DxE.1.7.9	$360° - \alpha$	$-\sin(\alpha)$	$+\cos(\alpha)$	$-\tan(\alpha)$	$-\cot(\alpha)$	$+\sec(\alpha)$	$-\csc(\alpha)$

Table E.1.8 Quick Lookup: Periodic Reduction Formulas for Multiples of Lagging 90°

DxE.1.8.1	$\sin(\alpha)$	$=$	$+\cos(\alpha - 90°)$	$=$	$-\sin(\alpha - 180°)$	$=$	$-\cos(\alpha - 270°)$
DxE.1.8.2	$\cos(\alpha)$	$=$	$-\sin(\alpha - 90°)$	$=$	$-\cos(\alpha - 180°)$	$=$	$+\sin(\alpha - 270°)$
DxE.1.8.3	$\tan(\alpha)$	$=$	$-\cot(\alpha - 90°)$	$=$	$+\tan(\alpha - 180°)$	$=$	$-\cot(\alpha - 270°)$
DxE.1.8.4	$\csc(\alpha)$	$=$	$+\sec(\alpha - 90°)$	$=$	$-\csc(\alpha - 180°)$	$=$	$-\sec(\alpha - 270°)$
DxE.1.8.5	$\sec(\alpha)$	$=$	$-\csc(\alpha - 90°)$	$=$	$-\sec(\alpha - 180°)$	$=$	$+\csc(\alpha - 270°)$
DxE.1.8.6	$\cot(\alpha)$	$=$	$-\tan(\alpha - 90°)$	$=$	$+\cot(\alpha - 180°)$	$=$	$-\tan(\alpha - 270°)$

Table E.1.9 Quick Lookup: Periodic Reduction Formulas for Multiples of Leading 90°

DxE.1.9.1	$\sin(\alpha)$	$=$	$-\sin(-\alpha)$	$=$	$+\cos(90° - \alpha)$	$=$	$+\sin(180° - \alpha)$	$=$	$-\cos(270° - \alpha)$
DxE.1.9.2	$\cos(\alpha)$	$=$	$+\cos(-\alpha)$	$=$	$+\sin(90° - \alpha)$	$=$	$-\cos(180° - \alpha)$	$=$	$-\sin(270° - \alpha)$
DxE.1.9.3	$\tan(\alpha)$	$=$	$-\tan(-\alpha)$	$=$	$+\cot(90° - \alpha)$	$=$	$-\tan(180° - \alpha)$	$=$	$+\cot(270° - \alpha)$
DxE.1.9.4	$\csc(\alpha)$	$=$	$-\csc(-\alpha)$	$=$	$+\sec(90° - \alpha)$	$=$	$+\csc(180° - \alpha)$	$=$	$-\sec(270° - \alpha)$
DxE.1.9.5	$\sec(\alpha)$	$=$	$+\sec(-\alpha)$	$=$	$+\csc(90° - \alpha)$	$=$	$-\sec(180° - \alpha)$	$=$	$-\csc(270° - \alpha)$
DxE.1.9.6	$\cot(\alpha)$	$=$	$-\cot(-\alpha)$	$=$	$+\tan(90° - \alpha)$	$=$	$-\cot(180° - \alpha)$	$=$	$+\tan(270° - \alpha)$

Axiom E.1.1 General Reduction Formula

$$\text{function}(\pm \alpha + n\,90°) = \begin{cases} (-1)^m & \text{function} \quad (\alpha) \quad n = 2m \quad \text{even} \\ (-1)^m & \text{cofunction}\,(\alpha) \quad n = 2m+1 \text{ odd} \end{cases} \quad \text{for } m = 0, \pm1, \pm2, \ldots$$

Axiom E.1.2 Uniqueness of all Trigonometric Functions Operating on Equal Angles

$\pm \alpha + n\,90° = \pm \alpha + n\,90°$ Angle   Equation A

$\text{function}(\pm \alpha + n\,90°) = \text{function}(\pm \alpha + n\,90°)$ Trig. Operation   Equation B

Axiom E.1.3 Change in Direction of Angular Measurement Changes Sign

It is seen from Figure E.1.2B "Trigonometric Functions of Acute and Arbitrary Angles" the sign changes when measurement is made in the opposite direction [$-\alpha$].

Table E.1.10 Quick Lookup: Trigonometric Function verses coFunction

function	sin	cos	tan	csc	sec	cot
cofunction	cos	sin	cot	sec	csc	tan

Tensor Calculus & Physics: A General Treatise

Axiom E.1.1 indicates that the trigonometric functions are periodic such segments of repeated functions are considered unique functions in the trigonometric space. So, the inverse functions and complex numbers are treated independent of the continuous function.

Definition E.1.8 Branch Point

Wherever a trigonometric function starts to repeat itself that segment is considered a new and a complete function for that period this unique position on the function is called the ***branch point***.

Definition E.1.9 Branch Segment

Wherever a trigonometric function starts to repeat itself that segment is considered a new and a complete function for that period this unique segment of the function is called the ***branch segment***.

Table E.1.11 Quick Lookup: Periodicity of Trigonometric Functions

function	sin	cos	tan
n of 90° is	odd to the right of 0°	even to the right of 0°	odd about 0°

Table E.1.12 Quick Lookup: Periodicity of Reciprocal Trigonometric Functions

function	csc	sec	cot
n of 90° is	every other even to the right 0°	every other odd about 0°	even to the right of 0°

Table E.1.13 Quick Lookup: Inverse Trigonometric Functions

Inverse function	$\alpha = \sin^{-1}(a)$	$\alpha = \cos^{-1}(a)$	$\alpha = \tan^{-1}(a)$
Domain	$-1 \le a \le +1$	$-1 \le a \le +1$	$-\infty \le a \le +\infty$
Range	$-90° \le \alpha \le +90°$	$0° \le \alpha \le +180°$	$-90° \le \alpha \le +90°$

Table E.1.14 Quick Lookup: Inverse Reciprocal Trigonometric Functions

Inverse function	$\alpha = \csc^{-1}(a)$	$\alpha = \sec^{-1}(a)$	$\alpha = \cot^{-1}(a)$
Domain	$-\infty \le a \le -1$ or $+1 \le a \le +\infty$	$-\infty \le a \le -1$ or $+1 \le a \le +\infty$	$-\infty \le a \le +\infty$
Range	$-90° \le \alpha \le +90°$	$0° \le \alpha \le +180°$	$0° \le \alpha \le +180°$

Tensor Calculus & Physics: A General Treatise

From these elementary propositions and definitions, the following Lemas can be proven by identity.

Table E.1.15 Quick Lookup Trigonometric Formulas: Right Triangles by Identity

	TE.15: Reciprocal Relations		
LxE.3.1.1	$\sin(\alpha)$	=	$1 / \csc(\alpha)$
LxE.3.1.2	$\cos(\alpha)$	=	$1 / \sec(\alpha)$
LxE.3.1.3	$\tan(\alpha)$	=	$1 / \cot(\alpha)$
LxE.3.1.4	$\csc(\alpha)$	=	$1 / \sin(\alpha)$
LxE.3.1.5	$\sec(\alpha)$	=	$1 / \cos(\alpha)$
LxE.3.1.6	$\cot(\alpha)$	=	$1 / \tan(\alpha)$
	TE.15: Product Relations		
LxE.3.1.7	$\sin(\alpha)$	=	$\tan(\alpha)\cos(\alpha)$
LxE.3.1.8	$\cos(\alpha)$	=	$\cot(\alpha)\sin(\alpha)$
LxE.3.1.9	$\tan(\alpha)$	=	$\sin(\alpha)\sec(\alpha)$
LxE.3.1.10	$\csc(\alpha)$	=	$\sec(\alpha)\cot(\alpha)$
LxE.3.1.11	$\sec(\alpha)$	=	$\csc(\alpha)\tan(\alpha)$
LxE.3.1.12	$\cot(\alpha)$	=	$\cos(\alpha)\csc(\alpha)$
	TE.15: Quotient Relations		
LxE.3.1.13	$\sin(\alpha)$	=	$\tan(\alpha) / \sec(\alpha)$
LxE.3.1.14	$\cos(\alpha)$	=	$\cot(\alpha) / \csc(\alpha)$
LxE.3.1.15	$\tan(\alpha)$	=	$\sin(\alpha) / \cos(\alpha)$
LxE.3.1.16	$\csc(\alpha)$	=	$\sec(\alpha) / \tan(\alpha)$
LxE.3.1.17	$\sec(\alpha)$	=	$\csc(\alpha) / \cot(\alpha)$
LxE.3.1.18	$\cot(\alpha)$	=	$\cos(\alpha) / \sin(\alpha)$
	TE.15: Pythagorean Relations		
LxE.3.1.19A	1	=	$\sin^2(\alpha) + \cos^2(\alpha)$
LxE.3.1.19B	$\sin^2(\alpha)$	=	$1 - \cos^2(\alpha)$
LxE.3.1.19C	$\cos^2(\alpha)$	=	$1 - \sin^2(\alpha)$
LxE.3.1.20	$\sec^2(\alpha)$	=	$1 + \tan^2(\alpha)$
LxE.3.1.21	$\csc^2(\alpha)$	=	$1 + \cot^2(\alpha)$
	TE.15: Angle-Sum and Angle-Difference Relations		
LxE.3.1.22	$\sin(\alpha + \beta)$	=	$\sin(\alpha)\cos(\beta) + \cos(\alpha)\sin(\beta)$
LxE.3.1.23	$\sin(\alpha - \beta)$	=	$\sin(\alpha)\cos(\beta) - \cos(\alpha)\sin(\beta)$
LxE.3.1.24	$\cos(\alpha + \beta)$	=	$\cos(\alpha)\cos(\beta) - \sin(\alpha)\sin(\beta)$
LxE.3.1.25	$\cos(\alpha - \beta)$	=	$\cos(\alpha)\cos(\beta) + \sin(\alpha)\sin(\beta)$
LxE.3.1.26	$\tan(\alpha + \beta)$	=	$(\tan(\alpha) + \tan(\beta)) / (1 - \tan(\alpha)\tan(\beta))$
LxE.3.1.27	$\tan(\alpha - \beta)$	=	$(\tan(\alpha) - \tan(\beta)) / (1 + \tan(\alpha)\tan(\beta))$
LxE.3.1.28	$\cot(\alpha + \beta)$	=	$(\cot(\alpha)\cot(\beta) - 1)/(\cot(\beta) + \cot(\alpha))$
LxE.3.1.29	$\cot(\alpha - \beta)$	=	$(\cot(\alpha)\cot(\beta) + 1)/(\cot(\beta) - \cot(\alpha))$
LxE.3.1.30	$\sin(\alpha + \beta)\sin(\alpha - \beta)$	=	$\sin^2(\alpha) - \sin^2(\beta)$
LxE.3.1.31	$\sin(\alpha + \beta)\sin(\alpha - \beta)$	=	$\cos^2(\beta) - \cos^2(\alpha)$
LxE.3.1.32	$\cos(\alpha + \beta)\cos(\alpha - \beta)$	=	$\cos^2(\alpha) - \sin^2(\beta)$
LxE.3.1.33	$\cos(\alpha + \beta)\cos(\alpha - \beta)$	=	$\cos^2(\beta) - \sin^2(\alpha)$

Tensor Calculus & Physics: A General Treatise

TE.15: Multiple-Angle Relations			
LxE.3.1.34	$\sin(2\alpha)$	=	$2\sin(\alpha)\cos(\alpha)$
LxE.3.1.35	$\sin(2\alpha)$	=	$2\tan(\alpha) / (1 + \tan^2(\alpha))$
LxE.3.1.36	$\sin(3\alpha)$	=	$3\sin(\alpha) - 4\sin^3(\alpha)$
LxE.3.1.37	$\sin(4\alpha)$	=	$4\sin(\alpha)\cos(\alpha) - 8\sin^3(\alpha)\cos(\alpha)$
LxE.3.1.38	$\sin(5\alpha)$	=	$5\sin(\alpha) - 20\sin^3(\alpha) + 16\sin^5(\alpha)$
LxE.3.1.39	$\sin(6\alpha)$	=	$32\sin(\alpha)\cos^5(\alpha) - 32\sin(\alpha)\cos^3(\alpha) + 6\sin(\alpha)\cos(\alpha)$
LxE.3.1.40	$\sin(n\alpha)$	=	$2\sin((n-1)\alpha)\cos(\alpha) - \sin((n-2)\alpha)$
LxE.3.1.41	$\cos(2\alpha)$	=	$\cos^2(\alpha) - \sin^2(\alpha)$
LxE.3.1.42	$\cos(2\alpha)$	=	$2\cos^2(\alpha) - 1$
LxE.3.1.43	$\cos(2\alpha)$	=	$1 - 2\sin^2(\alpha)$
LxE.3.1.44	$\cos(2\alpha)$	=	$(1 - \tan^2(\alpha)) / (1 + \tan^2(\alpha))$
LxE.3.1.45	$\cos(3\alpha)$	=	$4\cos^3(\alpha) - 3\cos(\alpha)$
LxE.3.1.46	$\cos(4\alpha)$	=	$8\cos^4(\alpha) - 8\cos^2(\alpha) + 1$
LxE.3.1.47	$\cos(5\alpha)$	=	$16\cos^5(\alpha) - 20\cos^3(\alpha) + 5\cos(\alpha)$
LxE.3.1.48	$\cos(6\alpha)$	=	$32\cos^6(\alpha) - 48\cos^4(\alpha) + 18\cos^2(\alpha) - 1$
LxE.3.1.49	$\cos(n\alpha)$	=	$2\cos((n-1)\alpha)\cos(\alpha) - \cos((n-2)\alpha)$
LxE.3.1.50	$\tan(2\alpha)$	=	$2\tan(\alpha) / (1 - \tan^2(\alpha))$
LxE.3.1.51	$\tan(3\alpha)$	=	$(3\tan(\alpha) - \tan^3(\alpha)) / (1 - 3\tan^2(\alpha))$
LxE.3.1.52	$\tan(4\alpha)$	=	$(4\tan(\alpha) - 4\tan^3(\alpha)) / (1 - 6\tan^2(\alpha) + \tan^4(\alpha))$
LxE.3.1.53	$\tan(n\alpha)$	=	$(\tan((n-1)\alpha) + \tan(\alpha)) / (1 - \tan((n-1)\alpha)\tan(\alpha))$
LxE.3.1.54	$\cot(2\alpha)$	=	$(\cot^2(\beta) - 1) / 2\cot(\alpha)$
TE.15: Product of Function Relations			
LxE.3.1.55	$\sin(\alpha)\sin(\beta)$	=	$\tfrac{1}{2}\cos(\alpha - \beta) - \tfrac{1}{2}\cos(\alpha + \beta)$
LxE.3.1.56	$\cos(\alpha)\cos(\beta)$	=	$\tfrac{1}{2}\cos(\alpha - \beta) + \tfrac{1}{2}\cos(\alpha + \beta)$
LxE.3.1.57	$\sin(\alpha)\cos(\beta)$	=	$\tfrac{1}{2}\sin(\alpha + \beta) + \tfrac{1}{2}\sin(\alpha - \beta)$
LxE.3.1.59	$\cos(\alpha)\sin(\beta)$	=	$\tfrac{1}{2}\sin(\alpha + \beta) - \tfrac{1}{2}\sin(\alpha - \beta)$
TE.15: Sum and Difference Function Relations			
LxE.3.1.60	$\cos(\alpha) + \cos(\beta)$	=	$+2\cos\tfrac{1}{2}(\alpha + \beta)\cos\tfrac{1}{2}(\alpha - \beta)$
LxE.3.1.61	$\cos(\alpha) - \cos(\beta)$	=	$-2\sin\tfrac{1}{2}(\alpha + \beta)\sin\tfrac{1}{2}(\alpha - \beta)$
LxE.3.1.62	$\sin(\alpha) + \sin(\beta)$	=	$+2\sin\tfrac{1}{2}(\alpha + \beta)\cos\tfrac{1}{2}(\alpha - \beta)$
LxE.3.1.63	$\sin(\alpha) - \sin(\beta)$	=	$+2\cos\tfrac{1}{2}(\alpha + \beta)\sin\tfrac{1}{2}(\alpha - \beta)$
LxE.3.1.64	$\tan(\alpha) + \tan(\beta)$	=	$+\sin(\alpha + \beta) / (\cos(\alpha)\cos(\beta))$
LxE.3.1.65	$\tan(\alpha) - \tan(\beta)$	=	$+\sin(\alpha - \beta) / (\cos(\alpha)\cos(\beta))$
LxE.3.1.66	$\cot(\alpha) + \cot(\beta)$	=	$+\sin(\alpha + \beta) / (\sin(\alpha)\sin(\beta))$
LxE.3.1.67	$\cot(\alpha) - \cot(\beta)$	=	$-\sin(\alpha - \beta) / (\sin(\alpha)\sin(\beta))$
TE.15: Half-Angle Relations			
LxE.3.1.68	$\sin(\tfrac{1}{2}\alpha)$	=	$\pm\sqrt{\tfrac{1}{2}(1 - \cos(\alpha))}$
LxE.3.1.69	$\cos(\tfrac{1}{2}\alpha)$	=	$\pm\sqrt{\tfrac{1}{2}(1 + \cos(\alpha))}$
LxE.3.1.70	$\tan(\tfrac{1}{2}\alpha)$	=	$\pm\sqrt{\tfrac{1}{2}(1 - \cos(\alpha)) / (1 + \cos(\alpha))}$
LxE.3.1.71	$\tan(\tfrac{1}{2}\alpha)$	=	$(1 - \cos(\alpha)) / \sin(\alpha)$
LxE.3.1.72	$\tan(\tfrac{1}{2}\alpha)$	=	$\sin(\alpha) / (1 + \cos(\alpha))$
LxE.3.1.73	$\cot(\tfrac{1}{2}\alpha)$	=	$\pm\sqrt{\tfrac{1}{2}(1 + \cos(\alpha)) / (1 - \cos(\alpha))}$
LxE.3.1.74	$\cot(\tfrac{1}{2}\alpha)$	=	$\sin(\alpha) / (1 - \cos(\alpha))$
LxE.3.1.75	$\cot(\tfrac{1}{2}\alpha)$	=	$(1 + \cos(\alpha)) / \sin(\alpha)$

TE.15: Power Relations		
LxE.3.1.76	$\sin^2(\alpha)$ =	$\frac{1}{2}(1 - \cos(2\alpha))$
LxE.3.1.77	$\sin^3(\alpha)$ =	$\frac{1}{4}(3\sin(\alpha) - \sin(3\alpha))$
LxE.3.1.78	$\sin^4(\alpha)$ =	$\frac{1}{8}(3 - 4\cos(2\alpha) - \cos(4\alpha))$
LxE.3.1.79	$\cos^2(\alpha)$ =	$\frac{1}{2}(1 + \cos(2\alpha))$
LxE.3.1.80	$\cos^3(\alpha)$ =	$\frac{1}{4}(3\cos(\alpha) + \cos(3\alpha))$
LxE.3.1.81	$\cos^4(\alpha)$ =	$\frac{1}{8}(3 + 4\cos(2\alpha) + \cos(4\alpha))$
LxE.3.1.82	$\tan^2(\alpha)$ =	$(1 - \cos(2\alpha)) / (1 + \cos(2\alpha))$
LxE.3.1.83	$\cot^2(\alpha)$ =	$(1 + \cos(2\alpha)) / (1 - \cos(2\alpha))$
TE.15: Area of a Right Triangle		
LxE.3.1.84	Area =	$\frac{1}{2}$ ab

Section E.2 *Plane Trigonometry: Oblique Triangles*

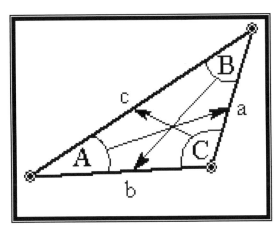

Figure E.2.1 Trigonometric of Oblique Triangle

Definition E.2.1 **Isosceles Triangle**

Any triangle having two equal sides and corresponding angles. If $\angle A = \theta$ than $\angle B = \angle C = \frac{1}{2}\theta$ and $a = b$.

Definition E.2.2 **Equilateral Triangle**

Any triangle with three sides and angles that is equal. If $\angle A = \theta$ than $\angle B = \angle C = \theta$ and $a = b = c$.

Table E.2.1 Quick Lookup Trigonometric Formulas: Oblique Triangles by Identity

TE.16: Law of Sines					
LxE.2.1.1	$\sin(A)/a$	$=$	$\sin(B)/b$	$=$	$\sin(C)/c$
LxE.2.1.2	$a/\sin(A)$	$=$	$b/\sin(B)$	$=$	$c/\sin(C)$
TE.16: Law of Cosines					
LxE.2.1.3	a^2	$=$	$b^2 + c^2 - 2bc\cos(A)$		
LxE.2.1.4	b^2	$=$	$a^2 + c^2 - 2ac\cos(B)$		
LxE.2.1.5	c^2	$=$	$a^2 + b^2 - 2ab\cos(C)$		
LxE.2.1.6	$\cos(A)$	$=$	$(b^2 + c^2 - a^2)/2bc$		
LxE.2.1.7	$\cos(B)$	$=$	$(a^2 + c^2 - b^2)/2ac$		
LxE.2.1.8	$\cos(C)$	$=$	$(a^2 + b^2 - c^2)/2ab$		
TE.16: Law of Tangents					
LxE.2.1.9	$(b-c)/(b+c)$	$=$	$\tan\frac{1}{2}(B-C)/\tan\frac{1}{2}(B+C)$		
LxE.2.1.10	$(c-a)/(c+a)$	$=$	$\tan\frac{1}{2}(C-A)/\tan\frac{1}{2}(C+A)$		
LxE.2.1.11	$(a-b)/(a+b)$	$=$	$\tan\frac{1}{2}(A-B)/\tan\frac{1}{2}(A+B)$		
TE.16: Area of a Triangle					
LxE.2.1.12	$\frac{1}{2}bc\sin(A)$	$=$	$\frac{1}{2}ac\sin(B)$	$=$	$\frac{1}{2}ab\sin(C)$
LxE.2.1.13	$\dfrac{a^2\sin(B)\sin(C)}{2\sin(A)}$	$=$	$\dfrac{b^2\sin(A)\sin(C)}{2\sin(B)}$	$=$	$\dfrac{c^2\sin(A)\sin(B)}{2\sin(C)}$
LxE.2.1.14	$\sqrt{s(s-a)(s-b)(s-c)}$	for	$s = \frac{1}{2}(a+b+c)$ Heron's formula		

Section E.3 *Algebraic Series*

Definition E.3.1 **LOT**

LOT is Lower Order Terms and used to collect the lower series terms in an expansion providing a short hand notation:

$$\text{LOT} = \sum_{j=1}^{i} x_j \text{ for } j = 1, 2, 3, \ldots, i$$

Definition E.3.2 **HOT**

HOT is Higher Order Terms and used to collect the higher series terms in an expansion providing a short hand notation:

$$\text{HOT} = \sum_{j=i}^{n} x_j \text{ for } j = i, i+1, \ldots, n$$

Table E.3.1 Quick Lookup Series

TE.17: Maclaurin Series [BEY88 pg 298]			
LxE.3.1.1	$f(x)$	$=$	$\sum_{k=0}^{n} f^{k}(0)\, x^{k} / k! + R_{n+1}$
LxE.3.1.2	R_{n+1}	$=$	$f^{n+1}(\theta_x)\, x^{n+1} / (n+1)!$ for $0 < \theta_x < 1$
TE.17: Taylor Series [BEY88 pg 298]			
LxE.3.1.3	$f(x)$	$=$	$\sum_{k=0}^{n} f^{k}(a)\, (x-a)^{k} / k! + R_{n+1}$
LxE.3.1.4	R_{n+1}	$=$	$f^{n+1}(a)\, (\theta_x - a)^{n+1} / (n+1)!$ for $0 < \theta_x - a < 1$
TE.17: Binomial Series [BEY88 pg 297]			
LxE.3.1.5	$(y + x)^n$	$=$	$\sum_{k=0}^{n} \binom{n}{k} y^{n-k} x^{k}$ for $x^2 < y^2$ $0 < n \in IG$
LxE.3.1.6	$(1 \pm x)^n$	$=$	$1 + \sum_{k=1}^{\infty} (\pm 1)^{k} \prod_{i=1}^{k} (n - k + 1)\, x^{k} / k!$ for $x^2 < 1$ $0 < n \in IG$
LxE.3.1.7	$(1 \pm x)^{-n}$	$=$	$1 + \sum_{k=1}^{\infty} (\mp 1)^{k} \prod_{i=1}^{k} (n + k - 1)\, x^{k} / k!$ for $x^2 < 1$ $0 < n \in IG$
LxE.3.1.8	$(1 \pm x)^m$	$=$	$1 + \sum_{k=1}^{\infty} \prod_{i=1}^{k} (m - i + 1)\, (\pm x)^{k} / k!$ for $x^2 < 1$ $m \in R$
LxE.3.1.9	$(1 \pm x^n)^m$	$=$	$1 + \sum_{k=1}^{\infty} \prod_{i=1}^{k} (m - i + 1)\, (\pm x^n)^{k} / k!$ for $x^2 < 1$ $m \in R$ and $0 < n \in IG$
LxE.3.1.10	$(1 \pm x^2)^{\frac{1}{2}}$	$=$	$1 + \sum_{k=1}^{\infty} (\pm 1)^{k} \prod_{i=2}^{k} (2k - 3)\, (\tfrac{1}{2} x^2)^{k} / k!$ for $x^2 < 1$
LxE.3.1.11	$(1 \pm x^2)^{-\frac{1}{2}}$	$=$	$1 + \sum_{k=1}^{\infty} (\mp 1)^{k} \prod_{i=1}^{k} (2k - 1)\, (\tfrac{1}{2} x^2)^{k} / k!$ for $x^2 < 1$
TE.17: Exponential [BEY88 pg 299]			
LxE.3.1.12	e	$=$	$\sum_{k=0}^{\infty} 1/k!$ for all x
LxE.3.1.13	e^x	$=$	$\sum_{k=0}^{\infty} x^{k}/k!$ for all x
LxE.3.1.14	a^x	$=$	$\sum_{k=0}^{\infty} (x \log_e a)^{k} / k!$ for all x
LxE.3.1.15	e^x	$=$	$e^{a}\sum_{k=0}^{\infty} (x - a)^{k}/k!$ for all x

		TE.17: Logarithmic [BEY88 pg 299]		
LxE.3.1.16	$\log_e x$	$=$	$\sum_{k=1}^{\infty} ((x-1)/x)^k / k$	for $x > \frac{1}{2}$
LxE.3.1.17	$\log_e x$	$=$	$\sum_{k=1}^{\infty} (-1)^{k+1} (x-1)^k / k$	for $2 \geq x > 0$
LxE.3.1.18	$\log_e x$	$=$	$2 \sum_{k=1}^{\infty} ((x-1)/(x+1))^{2k-1} / (2k-1)$ for $x > 0$	
LxE.3.1.19	$\log_e (1+x)$	$=$	$\sum_{k=1}^{\infty} (-1)^{k+1} x^k / k$	for $-1 < x \leq +1$
LxE.3.1.20	$\log_e (n+1)/(n-1)$	$=$	$2\sum_{k=1}^{\infty} 1 / (2k-1)n^{2k-1}$	
LxE.3.1.21	$\log_e (a+x)$	$=$	$\log_e a + 2\sum_{k=1}^{\infty} (x/(2a+x))^{2k-1} / (2k-1)$	
		for $a > 0$, $-a < x < +\infty$		
LxE.3.1.22	$\log_e (x+1)/(x-1)$	$=$	$\sum_{k=1}^{\infty} x^{2k-1} / (2k-1)$	for $-1 < x < +1$
LxE.3.1.23	$\log_e x$	$=$	$\log_e a + \sum_{k=1}^{\infty} (-1)^{k+1} (x-a)^k / (2a)$ for $0 < x \leq 2a$	

		TE.17: Trigonometric [BEY88 pg 299, 300]		
LxE.3.1.24	$\sin(x)$	$=$	$\sum_{k=1}^{\infty} (-1)^{k+1} x^{2k-1} / (2k-1)!$	for all x
LxE.3.1.25	$\cos(x)$	$=$	$\sum_{k=0}^{\infty} (-1)^k x^{2k} / (2k)!$	for all x
LxE.3.1.26	$\tan(x)$	$=$	$\sum_{k=1}^{\infty} (-1)^{k-1} 2^{2k} (2^{2k} - 1) B_{2k} x^{2k-1} / (2k)!$ $x^2 < \pi^2 / 4$, B_k represents the kth Bernoulli number.	
LxE.3.1.27	$\cot(x)$	$=$	$\sum_{k=1}^{\infty} (-1)^{k-1} 2^{2k} B_{2k} x^{2k-1} / (2k)!$ $x^2 < \pi^2$, B_k represents the kth Bernoulli number.	
LxE.3.1.28	$\sec(x)$	$=$	$\sum_{k=1}^{\infty} (-1)^k E_{2k} x^{2k} / (2k)!$ $x^2 < \pi^2/4$, E_k represents the kth Euler number.	
LxE.3.1.29	$\csc(x)$	$=$	$\sum_{k=0}^{\infty} (-1)^{k-1} 2 (2^{2k-1} - 1) B_{2k} x^{2k-1} / (2k)!$ $x^2 < \pi^2$, B_k represents the kth Bernoulli number.	

		TE.17: Hyperbolic Functions [BEY88 pg 173, 174]		
LxE.3.1.30	$\sinh(x)$	$=$	$\sum_{k=1}^{\infty} x^{2k-1} / (2k-1)!$	for all x
LxE.3.1.31	$\cosh(x)$	$=$	$\sum_{k=0}^{\infty} x^{2k} / (2k)!$	for all x
LxE.3.1.32	$\tanh(x)$	$=$	$\sum_{k=1}^{\infty} 2^{2k} (2^{2k} - 1) B_{2k} x^{2k-1} / (2k)!$ $x^2 < \pi^2 / 4$, B_k represents the kth Bernoulli number.	
LxE.3.1.33	$\coth(x)$	$=$	$\sum_{k=1}^{\infty} 2^{2k} B_{2k} x^{2k-1} / (2k)!$ $x^2 < \pi^2$, B_k represents the kth Bernoulli number.	
LxE.3.1.34	$\text{sech}(x)$	$=$	$\sum_{k=1}^{\infty} E_{2k} x^{2k} / (2k)!$ $x^2 < \pi^2/4$, E_k represents the kth Euler number.	
LxE.3.1.35	$\text{csch}(x)$	$=$	$\sum_{k=0}^{\infty} 2 (2^{2k-1} - 1) B_{2k} x^{2k-1} / (2k)!$ $x^2 < \pi^2$, B_k represents the kth Bernoulli number.	

TE.17: Sums of Powers of N for the First n-Integers [BEY88 pg 54, 55]			
LxE.3.1.36	$\sum^n 0$	=	0 $\qquad$ sum of zeros
LxE.3.1.37	$\sum^n 1$	=	n
LxE.3.1.38	$\sum^n k$	=	$\frac{1}{2} n(n + 1)$
LxE.3.1.39	$\sum^n k^2$	=	$n(n+1)(2n+1) / 6$
LxE.3.1.40	$\sum^n k^3$	=	$n^2(n+1)^2 / 4$
LxE.3.1.41	$\sum^n k^4$	=	$n(n+1)(2n+1)(3n^2 + 3n - 1) / 30$
LxE.3.1.42	$\sum^n k^5$	=	$n^2(n+1)^2(2n^2 + 2n - 1) / 12$
LxE.3.1.43	$\sum^n k^6$	=	$n(n+1)(2n+1)(3n^4 + 6n^3 - 3n + 1) / 42$
LxE.3.1.44	$\sum^n k^7$	=	$n^2(n+1)^2(3n^4 + 6n^3 - n^2 - 4n + 2) / 24$
LxE.3.1.45	$\sum^n k^8$	=	$n(n+1)(2n+1)$ $(5n^6 + 15n^4 + 5n^2 - 15n^3 - n^2 + 9n - 3) / 90$
LxE.3.1.46	$\sum^n k^9$	=	$n^2(n+1)^2(2n^6 + 6n^5 + n^4 - 8n^3 + n^2 + 6n - 3) / 20$
LxE.3.1.47	$\sum^n k^{10}$	=	$n(n+1)(2n+1)$ $(3n^8 + 12n^7 + 8n^6 - 18n^5 - 10n^4 + 24n^3 +$ $2n^2 - 15n + 5) / 66$
LxE.3.1.48	$\sum^n k^p$	=	$(B_{p+1}(n+1) - B_{p+1}(0)) / (p+1)$ B_{p+1} represents the $(p+1)^{th}$ Bernoulli number.
LxE.3.1.49	$\sum^n (-1)^{n-k}k^p$	=	$\frac{1}{2} (E_p(n+1) + (-1)^n E_p(0))$ E_{p+1} represents the $(p+1)^{th}$ Bernoulli number.

TE.17: Bernoulli and Euler Numbers [BEY88 pg 392–398]		
LxE.3.1.50	$B_p(x)$	$= -2\,(p!\,/\,(2\pi)^p)\sum_{k=1}^{\infty}\cos(\,2\pi kx - \tfrac{1}{2}\pi p\,)\,/\,k^p$ $n > 1,\ 1 \ge x \ge 0$ or $n = 1,\ 1 > x > 0$
LxE.3.1.51	$E_p(x)$	$= 4\,(p!\,/\,\pi^{p+1})\sum_{k=0}^{\infty}\sin(\,(2k+1)\pi x - \tfrac{1}{2}\pi p\,)\,/\,(2k+1)^{p+1}$ $n > 0,\ 1 \ge x \ge 0$ or $n = 0,\ 1 > x > 0$
LxE.3.1.52	$B_p(x)$	$= \sum_{k=0}^{p} b_k x^k$ Bernoulli's Polynomial[E.3.1]
LxE.3.1.53	$E_p(x)$	$= \sum_{k=0}^{p} e_k x^k$ Euler's Polynomial[E.3.1]
LxE.3.1.54	$x\,/\,e^x - 1$	$= \sum_{k=0}^{\infty} b_k x^k\,/\,k!$ Generates Bernoulli Coefficients
LxE.3.1.55	x^p	$= \tfrac{1}{2}\,(E_p(x + 1) + E_p(x))$ For $E_0(x) = 1$, Euler's Polynomial recursion relation generates Euler Coefficients

TE.17: Riemann Zeta Function [ABR72 pg 807,808]			
LxE.3.1.56	$\zeta(p)$		$= \sum_{k=1}^{\infty} k^{-p}$ for $p > 1$
LxE.3.1.57	$\zeta(-2n)$		$= 0$
LxE.3.1.58	$\zeta(1 - 2n)$		$= -b_{2n}\,/\,2n$
LxE.3.1.59	$\zeta(2n)$		$= (\,(2\pi)^{2n}\,|b_{2n}|\,)\,/\,(\,2(2n)!\,)$
LxE.3.1.60	$\zeta(2n + 1)$		$= [((-1)^{n+1}\,(2\pi)^{2n+1}\,)\,/\,(\,2(2n + 1)!\,)]$ $\displaystyle\int_0^1 B_{2n+1}(x)\cot(\pi x)dx$
LxE.3.1.61	$\eta(p)$	$= (1 - 2^{1-p})\,\zeta(p)$	$= \sum_{k=0}^{\infty} (-1)^k k^{-p}$
LxE.3.1.62	$\lambda(p)$	$= (1 - 2^{-p})\,\zeta(p)$	$= \sum_{k=1}^{\infty} (2k - 1)^{-p}$
LxE.3.1.63	$\beta(p)$		$= \sum_{k=1}^{\infty} (-1)^k (2k + 1)^{-p}$
LxE.3.1.64	$\zeta(0)$		$= -\tfrac{1}{2}$
LxE.3.1.65	$\zeta(1)$		$= \infty$
LxE.3.1.66	$\zeta(2)$		$= \pi^2\,/\,6$
LxE.3.1.67	$\zeta(4)$		$= \pi^4\,/\,90$
LxE.3.1.68	$\eta(2)$		$= \pi^2\,/\,12$
LxE.3.1.69	$\eta(4)$		$= \pi^4\,/\,720$
LxE.3.1.70	$\lambda(2)$		$= \pi^2\,/\,8$
LxE.3.1.71	$\lambda(4)$		$= \pi^4\,/\,96$
LxE.3.1.72	$\beta(1)$		$= \pi\,/\,4$
LxE.3.1.73	$\beta(3)$		$= \pi^3\,/\,32$

	TE.17: Geometric Progression [BEY88 pg 8]		
LxE.3.1.74	$(1 - r^{n+1}) / (1 - r)$	$=$	$\sum_{k=0}^{n} r^{-k}$
LxE.3.1.75	$(1 + (-1)^n r^{n+1}) / (1 + r)$	$=$	$\sum_{k=0}^{n} (-1)^k r^{-k}$
LxE.3.1.76	$r (1 + r^n) / (1 - r)$	$=$	$\sum_{k=1}^{n} r^{-k}$
LxE.3.1.77	$r (-1 + (-1)^n r^n)/(1 + r)$	$=$	$\sum_{k=1}^{n} (-1)^k r^{-k}$
LxE.3.1.78	$1 / (1 - r)$	$=$	$\sum_{k=0}^{\infty} r^{-k}$ $\quad$ for $|r| < 1$
LxE.3.1.79	$1 / (1 + r)$	$=$	$\sum_{k=0}^{\infty} (-1)^k r^{-k}$ $\quad$ for $|r| < 1$
	TE.17: Harmonic Geometric Progression (see TE.3.1)		
LxE.3.1.80	$-\tfrac{1}{2} + (-1)^M \cos((2M + 1)\kappa) / 2\cos(\kappa)$	$=$	$\sum_{m=1}^{M} (-1)^m \cos(m\, 2\kappa)$
LxE.3.1.81	$-\tfrac{1}{2}\tan(\kappa) + (-1)^M \sin((2M + 1)\kappa) / 2\cos(\kappa)$	$=$	$\sum_{m=1}^{M} (-1)^m \sin(m\, 2\kappa)$

[E.3.1]Note: tables for Bernoulli and Euler coefficient can be found in [BEY88, pg 394].

[E.3.2]Note: tables for "Certain sums of trigonometric and hyperbolic functions" see "Special cases" can be found in [GRA65, pg30].

Theorem E.3.1 Series: Harmonic Geometric Progression

1g	Given	$Z_0 \equiv X_0 + iY_0 = e^{i\theta_0}$	
2g		$X_0 \equiv \sum_{m=1}^{M} (-1)^m \cos(m\,2\kappa)$	
3g		$Y_0 \equiv \sum_{m=1}^{M} (-1)^m \sin(m\,2\kappa)$	
4g		$r \equiv e^{i\,2\kappa}$	
Steps	Hypothesis	$X_0 \equiv -\frac{1}{2} +$ $(-1)^M \cos((2M+1)\kappa)\,/\,2\cos(\kappa)$	$Y_0 \equiv -\frac{1}{2}\tan(\kappa) +$ $(-1)^M \sin((2M+1)\kappa)\,/\,2\cos(\kappa)$
1	From 1g, 2g, 3g, A3.2.3 Substitution and A3.2.14 Distribution	$Z_0 \equiv \sum_{m=1}^{M} (-1)^m (\cos(m\,2\kappa) + i\sin(m\,2\kappa))$	
2	From 1 and T8.3.6 Conjugation: Euler's Complex Function for e^{ix}	$Z_0 \equiv \sum_{m=1}^{M} (-1)^m e^{i\,m\,2\kappa}$	
3	From 4g, LxE.3.1.77D Geometric Progression and A3.2.3 Substitution	$Z_0 \equiv e^{i(2\kappa)} (-1 + (-1)^M e^{i(M\,2\kappa)})\,/\,(1 + e^{i(2\kappa)})$	
4	From 3 and A3.2.24	$Z_0 \equiv e^{i(\kappa+\kappa)} (-1 + (-1)^M e^{i(M\,2\kappa)})\,/\,(e^{i(0)} + e^{i(\kappa+\kappa)})$	
5	From 4 and A3.2.8 Inverse Add	$Z_0 \equiv e^{i(\kappa+\kappa)} (-1 + (-1)^M e^{i(M\,2\kappa)})\,/\,(e^{i(\kappa-\kappa)} + e^{i(\kappa+\kappa)})$	
6	From 5, A3.2.14 Distribution and A3.2.18 Summation Exp	$Z_0 \equiv e^{i(\kappa)} e^{i(\kappa)} (-1 + (-1)^M e^{i(M\,2\kappa)})\,/\,(e^{i(\kappa)} e^{-i(\kappa)} + e^{i(\kappa)} e^{i(\kappa)})$	
7	From 6, A3.2.14 Distribution and A3.2.5 Commutative Add	$Z_0 \equiv e^{i(\kappa)} e^{i(\kappa)} (-1 + (-1)^M e^{i(M\,2\kappa)})\,/\,e^{i(\kappa)} (e^{i(\kappa)} + e^{-i(\kappa)})$	
8	From 7, A3.2.13 Inverse Multp and A3.2.12 Identity Multp	$Z_0 \equiv e^{i(\kappa)} (-1 + (-1)^M e^{i(M\,2\kappa)})\,/\,(e^{i(\kappa)} + e^{-i(\kappa)})$	
9	From 8, D3.1.19-A Negative Coefficient and A3.2.14 Distribution	$Z_0 \equiv (-e^{i(\kappa)} + (-1)^M e^{i(M\,2\kappa)} e^{i(\kappa)})\,/\,(e^{i(\kappa)} + e^{-i(\kappa)})$	
10	From 9, A3.2.18 Summation Exp, A3.2.12 Identity Multp and A3.2.14 Distribution	$Z_0 \equiv (-e^{i(\kappa)} + (-1)^M e^{i(M\,2+1)\kappa})\,/\,(e^{i(\kappa)} + e^{-i(\kappa)})$	
11	From 10, T8.3.28 and A3.2.3 Substitution	$Z_0 \equiv (-e^{i(\kappa)} + (-1)^M e^{i(M\,2+1)\kappa})\,/\,2\cos(\kappa)$	

12	From 11, T8.3.21 Complex Number: Uniqueness of Real and Imaginary Operators and T8.3.24 Complex Polar Form: Real Constant of Real and Imaginary Operators	$\mathrm{Re}\{Z_0\} \equiv \mathrm{Re}\{(-e^{i(\kappa)} + (-1)^M e^{i(M2+1)\kappa)}\} / 2\cos(\kappa)$	$\mathrm{Im}\{Z_0\} \equiv \mathrm{Im}\{(-e^{i(\kappa)} + (-1)^M e^{i(M2+1)\kappa)}\} / 2\cos(\kappa)$
13	From 12, T8.3.23 Complex Polar Form: Distribution of Real and Imaginary Operators and D8.3.3 Real, Imaginary Component Operators and A3.2.10 Commutative Multp	$\mathrm{Re}\{Z_0\} \equiv (-\cos(\kappa) + (-1)^M \cos((2M+1)\kappa)) / 2\cos(\kappa)$	$\mathrm{Im}\{Z_0\} \equiv (-\sin(\kappa) + (-1)^M \sin((2M+1)\kappa)) / 2\cos(\kappa)$
14	From 13, A3.2.14 Distribution, A3.2.13, A3.2.12	$\mathrm{Re}\{Z_0\} \equiv -\tfrac{1}{2} + (-1)^M \cos((2M+1)\kappa) / 2\cos(\kappa)$	$\mathrm{Im}\{Z_0\} \equiv -\tfrac{1}{2}\tan(\kappa) + (-1)^M \sin((2M+1)\kappa) / 2\cos(\kappa)$
∴	From 14, 1g, 2g, 3g and D8.3.3 Real and Imaginary Component Operators; derivation see LxE.3.1.80 Harmonic Geometric Progression Real Part and LxE.3.1.81 Geometric Progression Imaginary Part as finial solution	$X_0 \equiv -\tfrac{1}{2} + (-1)^M \cos((2M+1)\kappa) / 2\cos(\kappa)$	$Y_0 \equiv -\tfrac{1}{2}\tan(\kappa) + (-1)^M \sin((2M+1)\kappa) / 2\cos(\kappa)$

Theorem E.3.2 Series: Summing Number of Triangular Elements

1g	Given	[a] a constant	
Steps	Hypothesis	$\sum_{i<j} a = \tfrac{1}{2} n (n-1) a$ EQ A	$\sum_{i<j} a = a \tfrac{1}{2} n (n-1)$ EQ B
1	From A4.2.2A Equality	$\sum_{i<j} a = \sum_{i<j} a$	
2	From 1g, 1, A4.2.12 Identity Multp and A4.2.14 Distribution	$\sum_{i<j} a = a \sum_{i<j} 1$	
3	From 2 and triangular summation (inequality) is based on a square matrix summation form	$\sum_{i<j} a = a \tfrac{1}{2} [\, \sum_i \sum_j 1 \text{ (summing all square elements)} - \sum_j 1 \text{ (diagonal)}]$	
4	From 3, and taking only the upper half of a square matrix (nxn) having n^2 elements and subtracting off the [n] diagonal elements, yields the following results	$\sum_{i<j} a = a \tfrac{1}{2} (n^2 - n)$	

5	From 4, D4.1.17 Exponential Notation, A4.2.12 Identity Multp and A4.2.14 Distribution	$\sum_{i<j} a = a\,\frac{1}{2}\,n\,(n-1)$	
∴	From 5 and A4.2.10 Commutative Multp	$\sum_{i<j} a = \frac{1}{2}\,n\,(n-1)\,a$ EQ A	$\sum_{i<j} a = a\,\frac{1}{2}\,n\,(n-1)$ EQ B

Section E.4 *Mensuration Formulas*

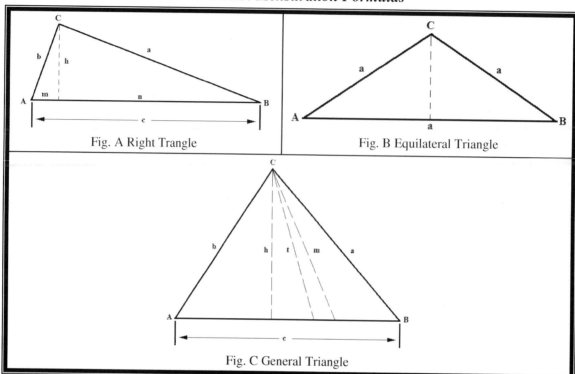

Fig. A Right Trangle

Fig. B Equilateral Triangle

Fig. C General Triangle

Figure E.4.1 Mensuration Triangles

Table E.4.1 Quick Lookup Mensuration Formulas: Triangles

TE.18: Right [BEY88 pg 121] Fig. E.4.1A				
LxE.4.1.1	$90° = C$	$=$	$A + B$	Sum of angles for a triangle
LxE.4.1.2	c^2	$=$	$a^2 + b^2$	Pythagorean relation
LxE.4.1.3	a	$=$	$\sqrt{(c + b)(c - b)}$	Leg
LxE.4.1.4	K	$=$	½ ab	Area
LxE.4.1.5	r	$=$	$\dfrac{ab}{a + b + c}$	Radius of inscribed circle
LxE.4.1.6	R	$=$	½ c	Radius of circumscribed circle
LxE.4.1.7	h	$=$	ab / c	Altitude
LxE.4.1.8	m	$=$	b^2 / c	Left base split of hypotenuse
LxE.4.1.9	n	$=$	a^2 / c	Right base split of hypotenuse
TE.18: Equilateral [BEY88 pg 121] Fig. E.4.1B				
LxE.4.1.10	$60° = C$	$=$	$A = B$	Angles for a triangle
LxE.4.1.11	K	$=$	¼ $a^2 \sqrt{3}$	Area
LxE.4.1.12	r	$=$	1/6 a $\sqrt{3}$	Radius of inscribed circle
LxE.4.1.13	R	$=$	1/3 a $\sqrt{3}$	Radius of circumscribed circle
LxE.4.1.14	h	$=$	½ c $\sqrt{3}$	Altitude
LxE.4.1.15	m	$=$	b^2 / c	Left base split of hypotenuse

TE.18: General [BEY88 pg 121] Fig. E.4.1C				
LxE.4.1.10	$180°$	$=$	$A + B + C$	Sum of angles for a triangle
LxE.4.1.11	c^2	$=$	$a^2 + b^2 - 2ab \cos C$	Law of cosines
LxE.4.1.12	K	$=$	$\frac{1}{2} ab \sin C$	Area
LxE.4.1.13	K	$=$	$\sqrt{s(s-a)(s-b)(s-c)}$	Heron's formula
LxE.4.1.14	r	$=$	K / s	Radius of inscribed circle
LxE.4.1.15	R	$=$	$abc / 4K$	Radius of circumscribed circle
LxE.4.1.16	h	$=$	$a \sin B = b \sin A$	Altitude
LxE.4.1.17	m	$=$	$\sqrt{\frac{1}{2}(a^2 + b^2 - \frac{1}{2}c^2)}$	Right bisect of interior angle C
LxE.4.1.18	t	$=$	$\dfrac{2ab}{a+b} \cos \frac{1}{2} C$	Right split of interior angle C

TE.18: General [BEY88 pg 121] Fig. E.4.1C				
LxE.4.1.10	$180°$	$=$	$A + B + C$	Sum of angles for a triangle
LxE.4.1.11	c^2	$=$	$a^2 + b^2 - 2ab \cos C$	Law of cosines
LxE.4.1.12	K	$=$	$\frac{1}{2} ab \sin C$	Area
LxE.4.1.13	K	$=$	$\sqrt{s(s-a)(s-b)(s-c)}$	Heron's formula
LxE.4.1.14	r	$=$	K / s	Radius of inscribed circle
LxE.4.1.15	R	$=$	$abc / 4K$	Radius of circumscribed circle
LxE.4.1.16	h	$=$	$a \sin B = b \sin A$	Altitude
LxE.4.1.17	m	$=$	$\sqrt{\frac{1}{2}(a^2 + b^2 - \frac{1}{2}c^2)}$	Right bisect of interior angle C
LxE.4.1.18	t	$=$	$\dfrac{2ab}{a+b} \cos \frac{1}{2} C$	Right split of interior angle C

TE.18: General [BEY88 pg 121] Fig. E.4.1C				
LxE.4.1.10	$180°$	$=$	$A + B + C$	Sum of angles for a triangle
LxE.4.1.11	c^2	$=$	$a^2 + b^2 - 2ab \cos C$	Law of cosines
LxE.4.1.12	K	$=$	$\frac{1}{2} ab \sin C$	Area
LxE.4.1.13	K	$=$	$\sqrt{s(s-a)(s-b)(s-c)}$	Heron's formula
LxE.4.1.14	r	$=$	K / s	Radius of inscribed circle
LxE.4.1.15	R	$=$	$abc / 4K$	Radius of circumscribed circle
LxE.4.1.16	h	$=$	$a \sin B = b \sin A$	Altitude
LxE.4.1.17	m	$=$	$\sqrt{\frac{1}{2}(a^2 + b^2 - \frac{1}{2}c^2)}$	Right bisect of interior angle C
LxE.4.1.18	t	$=$	$\dfrac{2ab}{a+b} \cos \frac{1}{2} C$	Right split of interior angle C

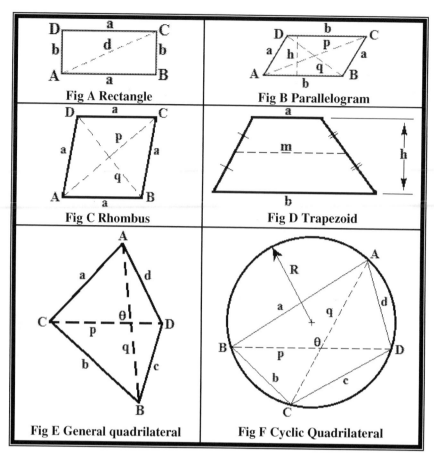

Figure E.4.2 Quadrilaterals

Table E.4.2 Quick Lookup Mensuration Formulas: Quadrilaterals

TE.19: Rectangle [BEY88 pg 122] Fig. E.4.2A			
LxE.4.2.1	$90°$ =	$A = B = C = D$	Angle
LxE.4.2.2	K =	ab	Area
	p =	$a^2 + b^2$	Diagonal
TE.19: Parallelogram [BEY88 pg 123] Fig. E.4.2B			
LxE.4.2.3	A =	C	Angle
LxE.4.2.4	B =	D	Angle
LxE.4.2.5	$180°$ =	$A + B$	Sum of Angles
LxE.4.2.6	p =	$\sqrt{a^2 + b^2 - 2ab \cos A}$	Positive Slope Diagonal
LxE.4.2.7	q =	$\sqrt{a^2 + b^2 - 2ab \cos B}$	Negative Slope Diagonal
LxE.4.2.8	h =	$a \sin A$	Altitude
LxE.4.2.9	h =	$a \sin B$	Altitude
LxE.4.2.10	K =	bh	Area
TE.19: Rhombus [BEY88 pg 123] Fig. E.4.2C			
LxE.4.2.5	$4a^2$ =	$p^2 + q^2$	Side Diagonal Relation
LxE.4.2.6	K =	$½pq$	Area

TE.19: Trapezoid [BEY88 pg 123] Fig. E.4.2D				
LxE.4.2.8	m	=	$\frac{1}{2}(a + b)$	Median Length
LxE.4.2.9	K	=	$\frac{1}{2}(a + b)h$	Area
TE.19: General quadrilateral [BEY88 pg 123] Fig. E.4.2E				
LxE.4.2.10	K	=	$\frac{1}{2}pq \sin \theta$	Area
LxE.4.2.11	s	=	$\frac{1}{2}(a + b + c + d)$	Parametric Parameter
LxE.4.2.12	K	=	$\sqrt{(s - a)(s - b)(s - c)}$	Parametric Area
			$(s - d) - abcd \cos^2 \frac{1}{2}(A + B)$	Bretschneider's Formula
TE.19: Cyclic Quadrilateral [BEY88 pg 123] Fig. E.4.2F				
LxE.4.2.13	180°	=	$A + C$	Vertical Supplemental Angles
LxE.4.2.14	$A + C$	=	$B + D$	Horizontal Supplemental Angles
LxE.4.2.15	K	=	$\sqrt{(s - a)(s - b)(s - c)(s - d)}$	Brahmagupta's Formula of Area
LxE.4.2.16	K	=	$\dfrac{\sqrt{(ac + bd)(ad + bc)(ab + cd)}}{4R}$	Area
LxE.4.2.17	p	=	$\sqrt{\dfrac{(ac + bd)(ab + cd)}{ad + bc}}$	Positive Slope Diagonal
LxE.4.2.18	q	=	$\sqrt{\dfrac{(ac + bd)(ad + bc)}{ab + cd}}$	Negative Slope Diagonal
LxE.4.2.19	$\sin \theta$	=	$\dfrac{2K}{ac + bd}$	Quadrilateral Skew Angle
TE.19: Rectangle [BEY88 pg 122] Fig. E.4.2A				
LxE.4.2.1	90°	=	$A = B = C = D$	Angle
LxE.4.2.2	K	=	ab	Area
	p	=	$a^2 + b^2$	Diagonal
TE.19: Parallelogram [BEY88 pg 123] Fig. E.4.2B				
LxE.4.2.3	A	=	C	Angle
LxE.4.2.4	B	=	D	Angle
LxE.4.2.5	180°	=	$A + B$	Sum of Angles
LxE.4.2.6	p	=	$\sqrt{a^2 + b^2 - 2ab \cos A}$	Positive Slope Diagonal
LxE.4.2.7	q	=	$\sqrt{a^2 + b^2 - 2ab \cos B}$	Negative Slope Diagonal
LxE.4.2.8	h	=	$a \sin A$	Altitude
LxE.4.2.9	h	=	$a \sin B$	Altitude
LxE.4.2.10	K	=	bh	Area
TE.19: Rhombus [BEY88 pg 123] Fig. E.4.2C				
LxE.4.2.5	$4a^2$	=	$p^2 + q^2$	Side Diagonal Relation
LxE.4.2.6	K	=	$\frac{1}{2}pq$	Area
TE.19: Trapezoid [BEY88 pg 123] Fig. E.4.2D				
LxE.4.2.8	m	=	$\frac{1}{2}(a + b)$	Median Length
LxE.4.2.9	K	=	$\frac{1}{2}(a + b)h$	Area

TE.19: General quadrilateral [BEY88 pg 123] Fig. E.4.2E		
LxE.4.2.10	$K = \frac{1}{2}pq \sin\theta$	Area
LxE.4.2.11	$s = \frac{1}{2}(a + b + c + d)$	Parametric Parameter
LxE.4.2.12	$K = \sqrt{(s-a)(s-b)(s-c)}$	Parametric Area
	$(s-d) - abcd \cos^2 \frac{1}{2}(A+B)$	Bretschneider's
		Formula
TE.19: Cyclic Quadrilateral [BEY88 pg 123] Fig. E.4.2F		
LxE.4.2.13	$180° = A + C$	Vertical Supplemental Angles
LxE.4.2.14	$A + C = B + D$	Horizontal Supplemental Angles
LxE.4.2.15	$K = \sqrt{(s-a)(s-b)(s-c)(s-d)}$	Brahmagupta's
		Formula of Area
LxE.4.2.16	$K = \dfrac{\sqrt{(ac+bd)(ad+bc)(ab+cd)}}{4R}$	Area
LxE.4.2.17	$p = \sqrt{\dfrac{(ac+bd)(ab+cd)}{ad+bc}}$	Positive Slope Diagonal
LxE.4.2.18	$q = \sqrt{\dfrac{(ac+bd)(ad+bc)}{ab+cd}}$	Negative Slope Diagonal
LxE.4.2.19	$\sin\theta = \dfrac{2K}{ac+bd}$	Quadrilateral Skew Angle

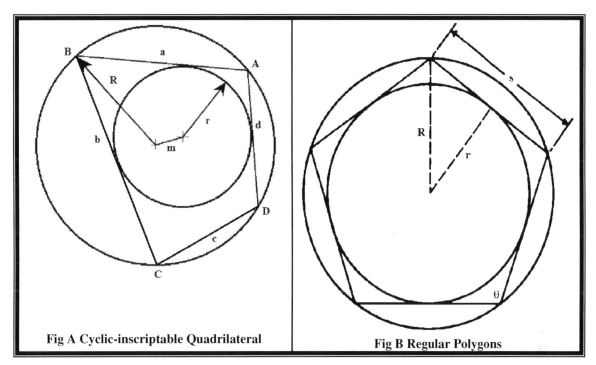

Fig A Cyclic-inscriptable Quadrilateral   Fig B Regular Polygons

Figure E.4.3 Cyclic Quadrilateral

TE.19: Cyclic-inscriptable Quadrilateral [BEY88 pg 124] Fig. E.4.3A				
LxE.4.2.20	$180°$	$=$	$A + C$	Vertical Supplemental Angles
LxE.4.2.21	$A + C$	$=$	$B + D$	Horizontal Supplemental Angles
LxE.4.2.22	$a + c$	$=$	$b + d$	Opposite Sides
LxE.4.2.23	K	$=$	$\sqrt{abcd}$	Area
LxE.4.2.24	r	$=$	$\dfrac{K}{s}$	Inscribed Radius
LxE.4.2.25	R	$=$	$\frac{1}{4}\sqrt{\dfrac{(ac + bd)(ad + bc)(ab + cd)}{abcd}}$	Circumscribed Radius

TE.19: Regular Polygons [BEY88 pg 124] Fig. E.4.3B				
LxE.4.2.26	θ	$=$	$180°(\dfrac{n-2}{n})$	One Angular Vertex
LxE.4.2.27	s	$=$	$2R \sin 180° / n$	Length of each side
LxE.4.2.28	p	$=$	ns	Perimeter
LxE.4.2.29	K	$=$	$\frac{1}{2}nR^2 \sin 360° / n$	Area
LxE.4.2.30	r	$=$	$\frac{1}{2} s \cot \dfrac{180°}{n}$	Inscribed Circle Radius
LxE.4.2.31	R	$=$	$\frac{1}{2} s \csc \dfrac{180°}{n}$	Circumscribed Circle Radius

TE.19: Cyclic-inscriptable Quadrilateral [BEY88 pg 124] Fig. E.4.3A				
LxE.4.2.20	$180°$	$=$	$A + C$	Vertical Supplemental Angles
LxE.4.2.21	$A + C$	$=$	$B + D$	Horizontal Supplemental Angles
LxE.4.2.22	$a + c$	$=$	$b + d$	Opposite Sides
LxE.4.2.23	K	$=$	$\sqrt{abcd}$	Area
LxE.4.2.24	r	$=$	$\dfrac{K}{s}$	Inscribed Radius
LxE.4.2.25	R	$=$	$\frac{1}{4}\sqrt{\dfrac{(ac + bd)(ad + bc)(ab + cd)}{abcd}}$	Circumscribed Radius

TE.19: Regular Polygons [BEY88 pg 124] Fig. E.4.3B				
LxE.4.2.26	θ	$=$	$180°\left(\dfrac{n - 2}{n}\right)$	One Angular Vertex
LxE.4.2.27	s	$=$	$2R \sin(180° / n)$	Length of each side
LxE.4.2.28	p	$=$	ns	Perimeter
LxE.4.2.29	K	$=$	$\frac{1}{2}nR^2 \sin(360° / n)$	Area
LxE.4.2.30	r	$=$	$\frac{1}{2} s \cot \dfrac{180°}{n}$	Inscribed Circle Radius
LxE.4.2.31	R	$=$	$\frac{1}{2} s \csc \dfrac{180°}{n}$	Circumscribed Circle Radius

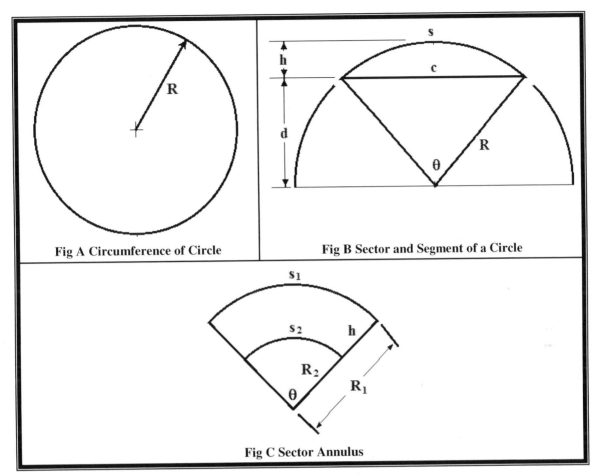

Fig A Circumference of Circle   **Fig B Sector and Segment of a Circle**

Fig C Sector Annulus

Figure E.4.4 Mensuration Formulas for a Circle

Table E.4.3 Quick Lookup Mensuration Formulas: Circles

TE.20: Circumference and Area of a Circle [BEY88 pg 125] Fig E.4.4A		
LxE.4.3.1	$C = \pi D$	Circumference
LxE.4.2.2	$D = 2R$	Radius as Diameter
LxE.4.2.3	$K = \frac{1}{4}\pi D^2$	Area as Diameter
LxE.4.2.4	$K = \pi R^2$	Area as Radius
TE.20: Sector and Segment of a Circle [BEY88 pg 125] Fig E.4.4B		
LxE.4.3.5	$s = \theta R$ (in radian angle)	Arclength of a Sector
LxE.4.3.6	$K = \frac{1}{2}\theta R^2$ (in radian angle)	Area of a Sector
TE.20: Sector of an Annulus [BEY88 pg 126] Fig E.4.4C		
LxE.4.3.5	$h = R_1 - R_2$	Annulus Gap
LxE.4.3.6	$K = \frac{1}{2}\theta(R_1 + R_2)(R_1 - R_2)$	Area Radial Annulus
LxE.4.3.7	$K = \frac{1}{2}\theta h(s_1 + s_2)$	Area Arclength Annulus

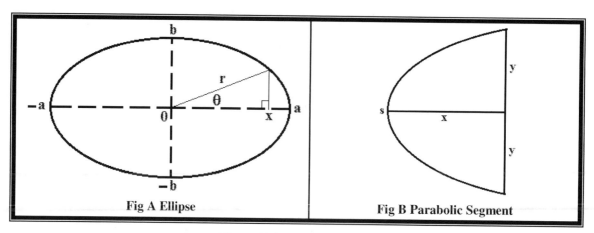

Figure E.4.5 Conic Sections

Table E.4.4 Quick Lookup Mensuration Formulas: Conic Sections

	TE.21: Ellipse and Definite Integrals [BEY88 pg 126] Fig E.4.5A			
LxE.4.4.1	k	$=$	$\sqrt{1 - \tau^2}$	Elliptic Modulus
LxE.4.4.2	τ	$=$	b/a for	Elliptic Coefficient
LxE.4.4.3	c_e	$=$	$4\, E_s(k, \tfrac{1}{2}\pi)$	Circumference[E.4.1]
LxE.4.4.4	K_e	$=$	πab	Area
LxE.4.4.5	$F_s(k, \phi)$	$=$	$(\tau/k)F_c(\tau, \varphi)$	Elliptic Ratio $\tau 2k$
LxE.4.4.6	$E_s(k, \phi)$	$=$	$(k/\tau)E_c(\tau, \varphi)$	Elliptic Ratio $k2\tau$
LxE.4.4.7	$F_s(k, \phi + m\pi)$	$=$	$m\, F_s(k, \pi) + F_s(k, \phi)$ m=0,1…	Elliptic Periodicity
LxE.4.4.8	$E_s(k, \phi + m\pi)$	$=$	$m\, E_s(k, \pi) + E_s(k, \phi)$ m=0,1…	Elliptic Periodicity
	TE.21: Parabolic Segment [BEY88 pg 126] Fig E.4.5B			
LxE.4.4.9	s	$=$	$\sqrt{(4x^2 + y^2)} + (y^2 / 2x)$	
		$+$	$\ln\left(\dfrac{2x + \sqrt{(4x^2 + y^2)}}{y} \right)$	Vertex Distance

Table E.4.5 Quick Lookup Mensuration Formulas: Solids Bounded by Planes

	TE.22: Cube [BEY88 pg 127]			
LxE.4.5.1	T	$=$	$6a^2$	Total Surface
LxE.4.5.2	Diag. Face	$=$	$a\sqrt{2}$	Diagonal edge face
LxE.4.5.3	V		a^3	Volume
LxE.4.5.4	Diag. Cube	$=$	$a\sqrt{3}$	Diagonal edge cube
	TE.22: Rectangular Parallelepiped (or box) [BEY88 pg 127]			
LxE.4.5.5	T	$=$	$2(ab + bc + ca)$	Total Surface
LxE.4.5.6	V	$=$	abc	Volume
LxE.4.5.7	D	$=$	$\sqrt{a^2 + b^2 + c^2}$	Diagonal

[E.4.1]Note: Elliptic Integral of the second kind.

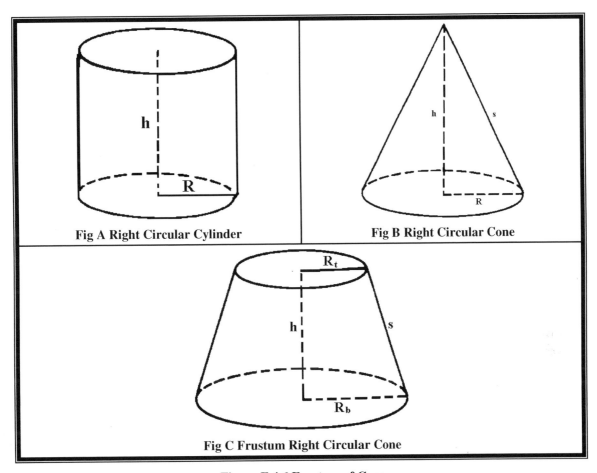

Figure E.4.6 Frustum of Cone

Table E.4.6 Quick Lookup Mensuration Formulas: Cylinders and Cones

TE.23: Right Circular Cylinder [BEY88 pg 129] **Fig E.4.6A**			
LxE.4.6.1	$S = 2\pi Rh$		Surface
LxE.4.6.2	$T = 2\pi R(R + h)$		Total Surface
LxE.4.6.3	$V = \pi R^2 h$		Volume
TE.23: Cone [BEY88 pg 129] **Fig E.4.6B**			
LxE.4.6.4	$V = $ 1/3 (Area of Lower Base) (Altitude or Height)		
LxE.4.6.5	General Formula		
TE.23: Frustum Right Circular Cone [BEY88 pg 129] **Fig E.4.6C**			
LxE.4.6.6	$s = \sqrt{R^2 + h^2})$		Slant Height
LxE.4.6.7	$S = \pi Rs$		Surface
LxE.4.6.8	$T = \pi R(R + s)$		Total Surface
LxE.4.6.9	$V = $ 1/3$\pi R^2 h$		Volume

TE.23: Frustum of Cone [BEY88 pg 129]		
LxE.4.6.10	$V \quad = \quad$ 1/3h(Base Area$_b$ + Base Area$_t$)	
	$\sqrt{}$(Base Area$_b$)(Base Area$_t$)	
TE.23: Frustum of Right Circular Cone [BEY88 pg 129]		
LxE.4.6.10	$s \quad = \quad \sqrt{(R_b - R_t)^2 + h^2}$	Slant Height
LxE.4.6.11	$S \quad = \quad \pi(R_b + R_t)s$	Surface
LxE.4.6.12	$T \quad = \quad \pi[R_b^2 + R_t^2 + (R_b + R_t)s]$	Total Surface
LxE.4.6.13	$V \quad = \quad 1/3\pi(R_b^2 + R_t^2 + R_b R_t)h$	Volume

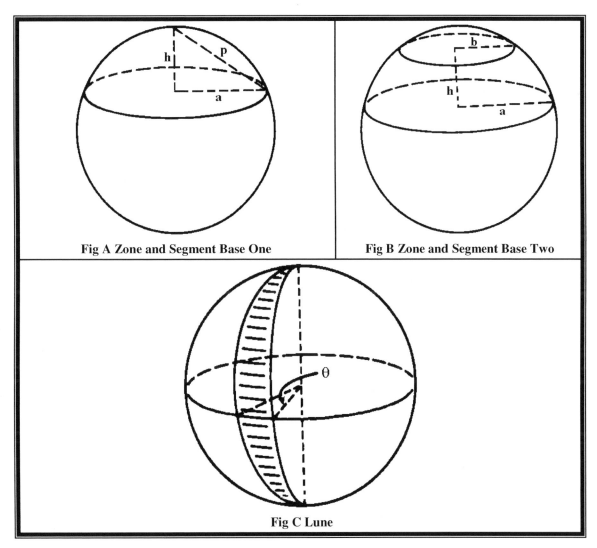

Figure E.4.7 Spherical Shapes

Table E.4.7 Quick Lookup Mensuration Formulas: Spherical Figures

TE.24: Sphere [BEY88 pg 130]		
LxE.4.7.1	$S = 4\pi R^2$	Surface
LxE.4.7.2	$V = 4/3\,\pi R^3$	Volume
TE.24: Zone and Segment of One Base [BEY88 pg 130] Fig E.4.7A		
LxE.4.7.3	$S = 2\pi Rh$	Surface
LxE.4.7.4	$V = 1/3\,\pi h^2 (3R - h)$	Volume
TE.24: Zone and Segment of Two Bases [BEY88 pg 130] Fig E.4.7B		
LxE.4.7.5	$S = 2\pi Rh$	Surface
LxE.4.7.6	$V = 1/6\,\pi h^2 (3R - h)$	Volume
TE.24: Lune [BEY88 pg 130] Fig E.4.7C		
LxE.4.7.7	$S = 2R^2\theta$ (Radian Angular Slice)	Surface

Section E.5 Analytic Geometry of (x, y) Conic Sections

TE.1: Lines [PRO70 pg 283]		
LxE.5.1.1	$0 \quad = \quad Ax + By + C$	First degree equation
LxE.5.1.2	$y \quad = \quad mx + d$	Line Intercept
TE.2: Parabola [PRO70 pg 299] Fig E.4.5B		
LxE.5.1.3	$0 \quad = \quad Ay^2 + Bxy + Cx + Dy + E$	Second degree equation
LxE.5.1.4	$(y - b)^2 \quad = \quad 2p(x - a)$	Parabola
TE.3: Ellipse [PRO70 pg 303] Fig E.4.5A		
LxE.5.1.5	$0 \quad = \quad Ay^2 + Bxy + Cx^2 + Dx + Ey + F$	Second degree equation
LxE.5.1.6	$1 \quad = \quad \dfrac{x^2}{a^2} + \dfrac{y^2}{b^2}$	Ellipse
LxE.5.1.7	$c^2 \quad = \quad a^2 - b^2$	Intercept on Axis
LxE.5.1.8	$a \quad = \quad b = r \text{ (radius) } e = 0$	Circle
LxE.5.1.9	$e \quad = \quad c/a < 1$	Eccentricity
TE.24: Hyperbola [PRO70 pg 303]		
LxE.5.1.10	$0 \quad = \quad -Ay^2 + Bxy + Cx^2 + Dx + Ey + F$	Second degree equation
LxE.5.1.11	$1 \quad = \quad \dfrac{x^2}{a^2} - \dfrac{y^2}{b^2}$	Hyperbola
LxE.5.1.12	$c^2 \quad = \quad a^2 + b^2$	Intercept on Axis
LxE.5.1.13	$e \quad = \quad c/a > 1$	Eccentricity
TE.25: Line Segment		
LxE.5.1.14	$L \quad = \quad \{(x, y): Ax + By + C = 0\}$	Line Segment
LxE.5.1.15	$e \quad = \quad c/a = 1, b = 0$	Eccentricity

Section E.6 Elliptic Integrals

Theorem E.6.1 Algebraic 2nd Kind Elliptic Integral

1g	Given	$f' = dy / dx$	
2g		$k^2 = 1 - (b^2/a^2)$	Elliptic Modulus
Steps	Hypothesis	$s = \int_0^a \sqrt{\left(\dfrac{1 - k^2 x^2/a^2}{1 - x^2/a^2} \right)} \, dx$	
1	From LxE.5.1.6 Ellipse	$1 = \dfrac{x^2}{a^2} + \dfrac{y^2}{b^2}$	
2	From 1, T9.3.9 Exact Differential Parametric Chain Rule, Lx9.3.3.1B Differential of a Constant, Lx9.3.2. 3B Distribute Constant in/out of Product and Lx9.3.3.3B Differential of a power	$0 = (2x/a^2)\,(dx/dx) + (2y/b^2)\,(dy/dx)$	
3	From 2, 2g, Lx9.3.3.2A Differential identity and A4.2.3 Substitution	$0 = 2x/a^2\,1 + 2y/b^2\,f'$	
4	From 3, A4.2.12 Identity Multp, A4.2.14 Distribution and T4.4.1 Equalities: Any Quantity Multiplied by Zero is Zero	$0 = x/a^2 + y/b^2\,f'$	
5	From 4, T4.3.3A Equalities: Right Cancellation by Addition and A4.2.7 Identity Add	$y/b^2\,f' = -x/a^2$	
6	From 5, T4.4.4B Equalities: Right Cancellation by Multiplication and T4.4.5B Equalities: Reversal of Right Cancellation by Multiplication	$f' = -(b^2/a^2)\,x/y$	
7	From T12.13.1A Curvature On Curve-C: Calculating Curvature in a Flat Euclidian Plane	$s = \int_0^a \sqrt{1 + f'^2}\, dx$	
8	From 6, 7, T4.8.6 Integer Exponents: Uniqueness of Exponents, D4.1.20 Negative Coefficient, T4.8.2 Integer Exponents: Negative One Squared, T4.8.7 Integer Exponents: Distribution Across a Rational Number and A4.2.3 Substitution	$s = \int_0^a \left(\sqrt{1 + (b^2/a^2)^2\,\dfrac{x^2}{y^2}} \right) dx$	

9	From 8, A4.2.13 Inverse Multp and A4.2.14 Distribution	$s = \int_0^a \sqrt{\left(\dfrac{y^2 + (b^2/a^2)^2 x^2}{y^2} \right)}\, dx$
10	From 1 and T4.3.3B Equalities: Right Cancellation by Addition	$y^2/b^2 = 1 - x^2/a^2$
11	From 10, T4.4.5B Equalities: Reversal of Right Cancellation by Multiplication	$y^2 = b^2(1 - x^2/a^2)$
12	From 9, 11 and A4.2.3 Substitution	$s = \int_0^a \sqrt{\left(\dfrac{b^2(1 - x/a^2) + (b^2/a^2)^2 x^2}{b^2(1 - x^2/a^2)} \right)}\, dx$
13	From 12, A4.2.14 Distribution, A4.2.14 Inverse Multp, A4.2.12 Identity Multp, A4.2.21 Distribution Exp and A4.2.11 Associative Multp	$s = \int_0^a \sqrt{\left(\dfrac{(1 - x/a^2) + (b^2/a^2) x^2/a^2}{(1 - x^2/a^2)} \right)}\, dx$
14	From 13, D4.1.20 Negative Coefficient, A4.2.11 Associative Multp and A4.2.14 Distribution	$s = \int_0^a \sqrt{\left(\dfrac{1 - [1 - (b^2/a^2)] x^2/a^2}{(1 - x^2/a^2)} \right)}\, dx$
∴	From 2g, 14 and A4.2.3 Substitution	$s = \int_0^a \sqrt{\left(\dfrac{1 - k^2 x^2/a^2}{1 - x^2/a^2} \right)}\, dx$

Observation E.6.1 Ratio of X-Axis to Elliptic Intersection Axis

The ratio x::a forms a number directly proportional to a trigonometric angle

$x = 0$ then $\eta = 0$	Equation A
$x = a$ then $\eta = \frac{1}{2}\pi$	Equation B
$0 \le \xi \le \frac{1}{2}\pi$	Equation C
$x/a \equiv \xi = \sin \eta.$	Equation D

Theorem E.6.2 Trigonometric 2nd Kind Elliptic Integral

1g	Given	$x/a \equiv \xi$
2g		$\xi = \sin \eta$
Steps	Hypothesis	$s = \int_o^\phi \sqrt{1 - k^2 \sin \eta^2}\, d\eta$
1	From TE.6.1 Algebraic 2nd Kind Elliptic Integral	$s = \int_0^a \sqrt{\left(\dfrac{1 - k^2 x^2/a^2}{1 - x^2/a^2} \right)}\, dx$
2	From 1, OE.6.1A,B,D Ratio of X-Axis to Elliptic Intersection Axis and A4.2.3 Substitution	$s = \int_o^\phi \sqrt{\left(\dfrac{1 - k^2 \sin \eta^2}{1 - \sin \eta^2} \right)} \cos \eta\, d\eta$

3	From 2, LxE.3.1.19C TE.15: Pythagorean Relations and A4.2.3 Substitution	$s = \int_0^\phi \sqrt{\left(\dfrac{1 - k^2 \sin \eta^2}{\cos^2 \eta} \right)} \cos \eta \, d\eta$
4	From 3, T4.10.3 Radicals: Identity Power Raised to a Radical and A4.2.11 Associative Multp	$s = \int_0^\phi \sqrt{\left(1 - k^2 \sin \eta^2 \right)} \dfrac{\cos \eta \, d\eta}{\cos \eta}$
∴	From 4, A4.2.13 Inverse Multp and A4.2.12 Identity Multp	$s = \int_0^\phi \sqrt{1 - k^2 \sin \eta^2} \, d\eta$

Observation E.6.2 Defining an Elliptic Integral for the Second Kind

Theorems E.6.1 "Algebraic 2nd Kind Elliptic Integral" and E.6.2 "Trigonometric 2nd Kind Elliptic Integral" have characteristic parameters that the integral can be represented by and is defined as follows with the elliptic modulus and the angle of arc ϕ.

$$E(k, a) = \int_0^a \sqrt{\left(\frac{1 - k^2 x^2/a^2}{1 - x^2/a^2} \right)} \, dx \qquad \qquad \text{Equation A}$$

$$E(k, \phi) = \int_0^\phi \sqrt{1 - k^2 \sin \eta^2} \, d\eta \qquad \qquad \text{Equation B}$$